INTERMEDIATE ALGEBRA

a straightforward approach for college students

symbols

		See Page
$\begin{vmatrix} a_1 & b_1 & c_1 \\ a_2 & b_2 & c_2 \\ a_3 & b_3 & c_3 \end{vmatrix}$	3-by-3 determinant	319
a^{-m}	$\frac{1}{a^m}$	330
a^0	1	330
$a^{1/2}$	the square root of a	344
$a^{1/n}$	the nth root of a	346
$a^{m/n}$	$(a^{1/n})^m$	346
$\sqrt[n]{a}$	$a^{1/n}$	352
i	$\sqrt{-1}$	364
$\log_b n$	the logarithm of n to the base b	375
$\log n$	$\log_{10} n$	387
antilog m		394
$\approx$	is approximately	416
$\pm a$	plus and minus a	430
$a < x < b$, $a \leq x \leq b$	$a < x$ and $x < b$, $a \leq x$ and $x \leq b$	457
(a, b), $[a, b]$	open interval, closed interval	457
(a, ∞), $[a, \infty)$	open right ray, closed right ray	459
$(-\infty, b)$, $(-\infty, b]$	open left ray, closed left ray	459
$A \cup B$	A union B	478
dist (P_1, P_2)	the distance between P_1 and P_2	503
a_i	$a(i)$, the ith term of a sequence	560
$\sum_{i=1}^{n} a_i$	$a_1 + a_2 + a_3 + \cdots + a_n$	566
$n!$	$n(n-1)(n-2)\ldots3\cdot2\cdot1$	585
$\binom{n}{i}$	$\frac{n!}{i!(n-i)!}$	587

INTERMEDIATE ALGEBRA

a straightforward approach for college students

MARTIN M. ZUCKERMAN

City College of the City University of New York

W · W · NORTON & COMPANY

New York · London

First Edition

Library of Congress Cataloging in Publication Data
Zuckerman, Martin M.
Intermediate algebra.
Includes index.
1. Algebra. I. Title.
QA152.2.Z83 1976 512.9′042 75–38989
ISBN 0–393–09207–0

Published simultaneously in Canada
by George J. McLeod Limited, Toronto

Printed in the United States of America

5 6 7 8 9

In memory of my mother

contents

**Indicates optional topic.*

preface

While teaching Intermediate Algebra for the past five years, I have become increasingly dissatisfied with the textbooks available. It seems to me that they emphasize the wrong things—axiomatic foundations and set-theoretic notation—while they fail to provide enough basic drill. *Intermediate Algebra: A Straightforward Approach for College Students* attempts to overcome these objections. Its approach is essentially intuitive, with brief explanations given in simple, yet precise, language. Common learning difficulties are recognized and dealt with as they occur. Set-theoretic notation is used sparingly.

The textual material is interspersed with an abundance of illustrative examples—in fact, over 600, many with multiple parts. There are extensive drill exercises at the end of each section and also a number of more challenging problems. A *Workbook* is available for the student who may need additional practice. At the end of each chapter there are review exercises and a sample test. Altogether there are almost 5,000 student exercises (in the text) from which to choose.

The first chapter includes a thorough review of arithmetic. These sections can either be presented in class or assigned for self-study. The next few chapters comprise a comprehensive review of elementary algebra.

Linear equations are treated in Chapter 6 so that material on fractions and rational expressions (Chapters 4 and 5) can be utilized. If preferred, equations can be introduced earlier. In fact, Sections 6.1 and 6.2 can be covered any time after taking Section 2.5. See page 6 of the accompanying Instructor's Manual for the minor modifications that should be made for an earlier introduction to equations.

Throughout the book there is an emphasis on "word problems". Chapter 7, in particular, is devoted to practical applications of linear equations. The student is first shown how to translate a problem into algebraic symbolism, and then how to solve the problem by solving the resulting equation. Section 13.5 presents applications of quadratic equations.

Graphical techniques are stressed wherever they are useful. There are about 300 figures accompanying the textual material, illustrative examples, exercises, and solutions.

The accompanying flow chart indicates which chapters depend on material in prior chapters (or sections). For instance, before covering Chapter 13, the student should have a firm grasp of Sections 1.1–1.2, 1.3–1.7 (if a review of arithmetic is needed), and Chapters 2–7 and 11.

Several sections treat more advanced topics. These sections are designated by an asterisk and may be omitted. No later topics depend on them.

For a detailed analysis of the contents of each chapter and for suggestions on presenting the material, see the "Chapter by Chapter Comments" in the *Instructor's Manual.*

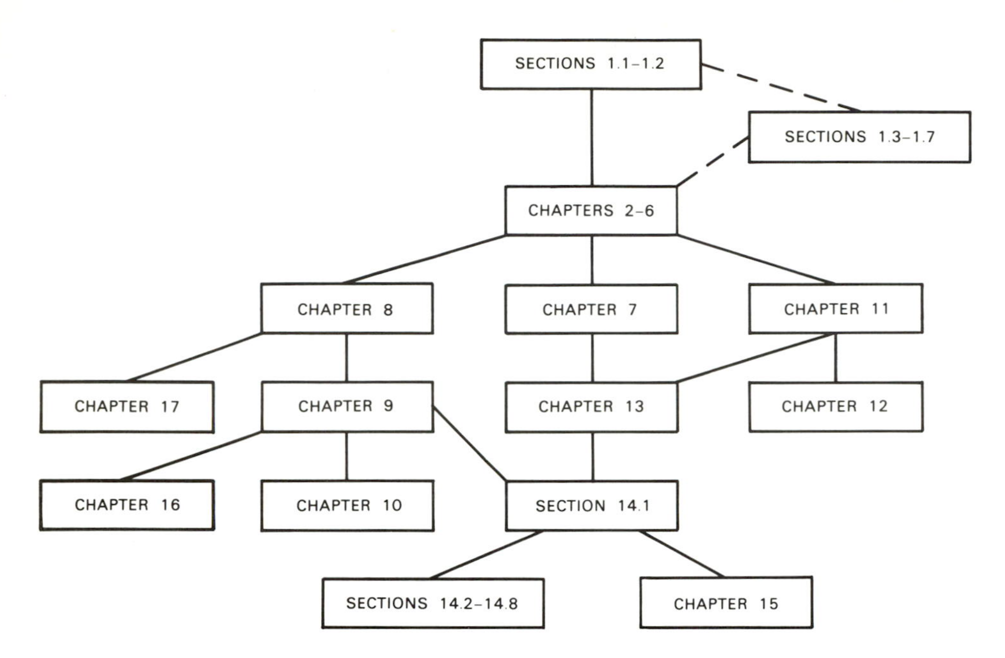
SECTIONS 1.1–1.2
SECTIONS 1.3–1.7
CHAPTERS 2–6
CHAPTER 8
CHAPTER 7
CHAPTER 11
CHAPTER 17
CHAPTER 9
CHAPTER 13
CHAPTER 12
CHAPTER 16
CHAPTER 10
SECTION 14.1
SECTIONS 14.2–14.8
CHAPTER 15

acknowledgments

The author gratefully acknowledges the many helpful comments of Ward Bouwsma of the University of Southern Illinois who read through, and improved upon, the final version of the manuscript; as well as those of Howard Banilower of Baruch College of the City University of New York; the late Leonard Cohen of City College of the City University of New York; Leonard Gillman of the University of Texas at Austin; M. Kenneth Hathaway, Jr., of the Stratford (Conn.) High School and of Housatonic Community College; and Donald R. Ostberg of Northern Illinois University. I also wish to express my appreciation to Joseph B. Janson II, my editor, as well as to Katherine Hyde, Mary Shuford, and Roy Tedoff, all of Norton, and to Elsa Ann Danenberg, who designed the book.

INTERMEDIATE ALGEBRA

a straightforward approach for college students

1

arithmetic review

1.1 THE REAL LINE

Algebra is primarily concerned with properties of **real numbers.** It will be convenient to regard real numbers from a geometric point of view.

A horizontal line, L, extends indefinitely in both directions. Real numbers will be identified with points of L as follows. Choose an arbitrary point O on L and call this the **origin.** This point O represents the number 0. Choose another point P on L to the *right* of O; the point P represents the number 1. That part of the line L between O and P is called a **line segment,** and is denoted by $\overline{OP}$. The length of the line segment $\overline{OP}$ determines the basic **unit of distance.** [See Figure 1.1.] The number 2 is located 1 (distance) unit to the right of P; the number 3 is 1 unit farther to the right. This process can be continued indefinitely.

The point corresponding to the number -1 is located on L 1 unit to the *left* of 0; the point corresponding to -2 is located 1 unit farther to the left, and so on.

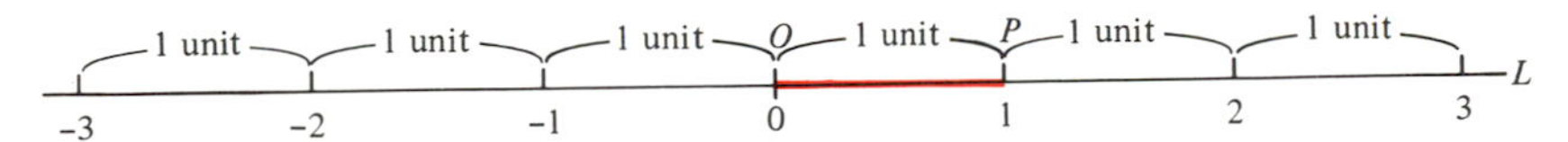

FIGURE 1.1 The line segment $\overline{OP}$ is in color. The length of $\overline{OP}$ is the unit of distance on L.

DEFINITION. Numbers that can be represented in this manner are called **integers.**

EXAMPLE 1. On the line L:

(a) The integer 0 corresponds to the point O.
(b) The integer 63 corresponds to the point 63 units to the right of O.

(c) The integer -15 corresponds to the point 15 units to the left of O.
(d) Which integer corresponds to the point 31 units to the right of O?
(e) Which integer corresponds to the point 78 units to the left of O? ■

The symbol

$$=$$

is read "is equal to"; the symbol

$$\neq$$

is read "is not equal to".

Midway between O and P on L is the point corresponding to the *rational number* $\frac{1}{2}$. [See Figure 1.2.]

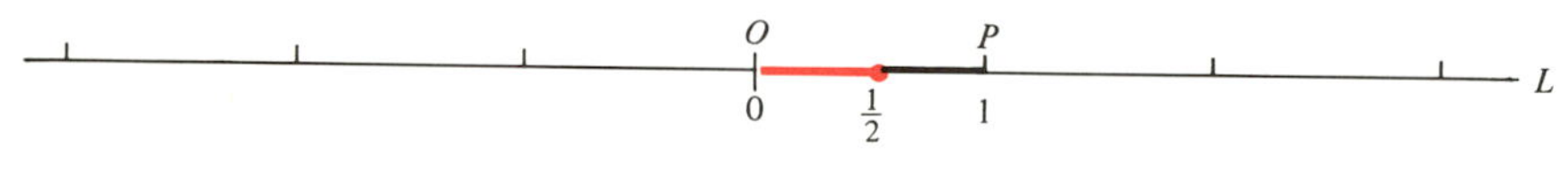

FIGURE 1.2

DEFINITION. **Rational numbers** are real numbers that can be expressed in the form $\frac{M}{N}$, where M and N are integers and $N \neq 0$. In this expression, M is called the **numerator** and N the **denominator.**

Thus rational numbers are real numbers that can be expressed as the **ratio** (or quotient) of two integers. For instance,

$$\frac{1}{5}, \quad \frac{3}{4}, \quad \frac{-2}{3}, \quad \frac{3}{2}$$

are rational numbers.

An integer, such as 4, is a rational number

$$4 = \frac{4}{1}.$$

Thus 4 can be expressed in the form $\frac{M}{N}$ with $M = 4$ and $N = 1$.

Decimals, such as

$$.5, \qquad -.25, \qquad 1.42,$$

are rational numbers because

$$.5 = \frac{5}{10}, \qquad -.25 = \frac{-25}{100}, \qquad 1.42 = \frac{142}{100}.$$

Every **infinite repeating decimal**, such as

$$.6666\ldots, \qquad .09\,09\,09\ldots$$

The 3 dots are read, "and so on".

in which one or more of the **digits**

$$0, 1, 2, 3, 4, 5, 6, 7, 8, 9,$$

repeat, is a rational number. For, with more advanced techniques, it can be shown that every repeating decimal can be expressed in the form $\frac{M}{N}$, where M and $N(\neq 0)$ are integers. Observe that

$$3\overline{)2.0000\ldots}^{\;.6666\ldots}$$

and

$$11\overline{)1.00\;00\;00\ldots}^{\;.09\;09\;09\ldots}$$

Thus

$$.6666\ldots = \frac{2}{3}$$

and

$$.09\,09\,09\ldots = \frac{1}{11}$$

Every rational number can be expressed as an infinite repeating decimal, possibly with 0's repeating, as in .5 000 000 . . . .

EXAMPLE 2.

(a) Observe that $1 = \frac{1}{1}$, $0 = \frac{0}{1}$, $7 = \frac{7}{1}$, and $-8 = \frac{-8}{1}$. Thus 1, 0, 7, and -8 are rational numbers.

(b) $\frac{2}{5}$, $\frac{-17}{4}$, $\frac{9}{5}$ are rational numbers.

(c) $3\frac{1}{7}$ is a rational number because you can also write it as $\frac{22}{7}$. ■

Hereafter, identify each real number a with the point on L that represents a, and speak of the "point a". Thus the "point 0" really means the point O, the "point 1" means the point P, and "the line segment from 0 to 1" is $\overline{OP}$.

In order to represent rational numbers on L, you will have to divide the line segments between integer points, as in Example 3.

EXAMPLE 3.

(a) To represent the rational number $\frac{1}{5}$, divide the line segment from 0 to 1 into 5 equal parts. The first point of division to the *right* of 0 represents $\frac{1}{5}$. [See Figure 1.3(a).]

(b) To represent the rational number $\frac{3}{4}$, divide the line segment from 0 to 3 into 4 equal parts. The first point of division to the *right* of 0 represents $\frac{3}{4}$. [See Figure 1.3(b).]

(c) To represent the rational number $\frac{5}{4}$, or $1\frac{1}{4}$, divide the line segment from 0 to 5 into 4 equal parts. The *first* point of division to the *right* of 0 represents $\frac{5}{4}$. [See Figure 1.3(c).]

(d) To represent $\frac{-2}{3}$, divide the line segment from -2 to 0 into 3 equal parts. The first point of division to the *left* of 0 represents $\frac{-2}{3}$. [See Figure 1.3(d).] ∎

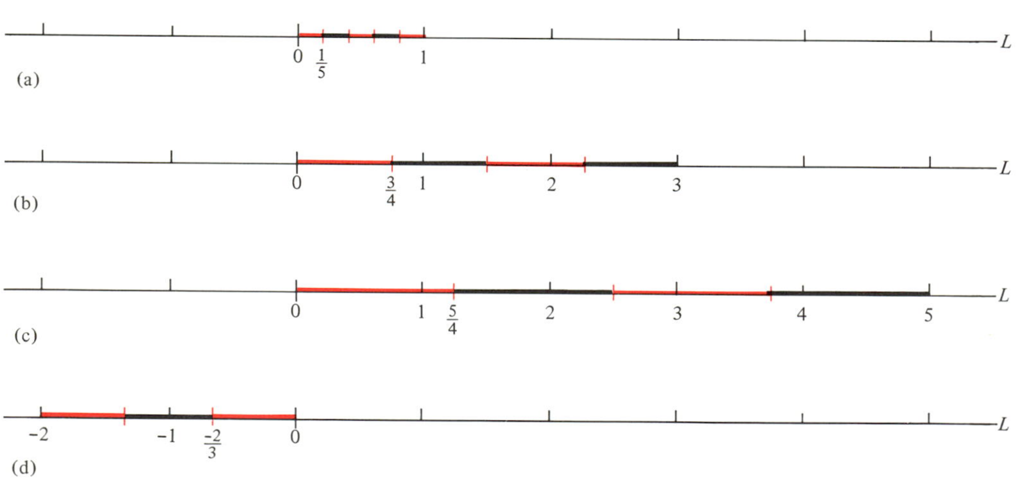

FIGURE 1.3

It may surprise you to learn that there are points on the line L that do not represent rational numbers. These points correspond to **irrational numbers.** An example of an irrational number is $\sqrt{2}$, the **square root of** 2; that is, the number that when multiplied by itself yields 2.

$$\sqrt{2} \cdot \sqrt{2} = 2$$

Irrational numbers can be expressed as **infinite nonrepeating decimals**, in which the digits do *not* repeat. For instance, the decimal representation of $\sqrt{2}$ begins 1.414 21 . . . Thus $\sqrt{2}$ lies between the rational numbers 1.41 and 1.42 . Another irrational number is π, whose decimal representation begins 3.141 59 . . . . See Figure 1.4. (Recall that the circumference of a circle is $2\pi r$, where r is the length of the radius.)

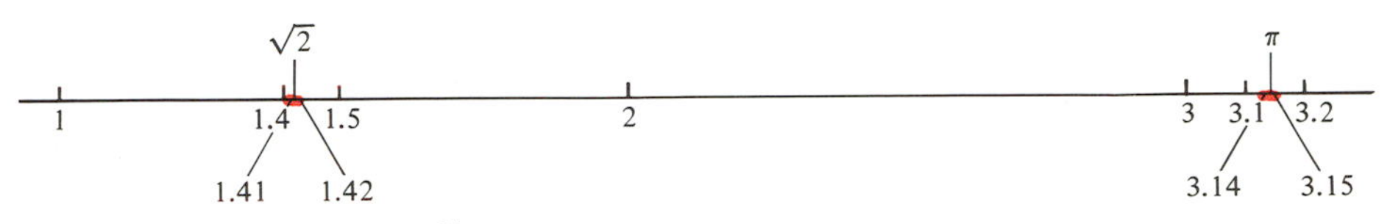

FIGURE 1.4 The irrational $\sqrt{2}$ lies between the rational numbers 1.41 and 1.42. The irrational number π lies between 3.14 and 3.15.

Every point on L represents exactly one real number, and every real number corresponds to exactly one point on L. It is this correspondence between real numbers and points on the line L that enables you to picture numbers geometrically.

You have considered three types of real numbers:

(a) **Integers**—such as 0, 1, 17, -1, -208

(b) **Rational numbers**—such as $\frac{2}{3}$, $\frac{-8}{5}$, 3. Every integer n is a rational number because $n = \frac{n}{1}$. Rational numbers can be expressed as infinite repeating decimals.

(c) **Irrational numbers**—such as $\sqrt{2}$, π. Irrational numbers can be expressed as infinite nonrepeating decimals.

Sometimes "number" will be used to mean "real number".

DEFINITION. A real number is said to be **positive** if it lies to the right of 0 on L, and **negative** if it lies to the left of 0. (The number 0 is neither positive nor negative.) The **nonnegative numbers** consist of 0 together with the positive numbers.

Thus the numbers 5, $\frac{3}{4}$, $\sqrt{2}$ are positive, whereas -2, $-\frac{14}{9}$, $-\pi$ are negative.

INEQUALITIES

Everyone knows that

3 *is less than* 7.

But do you also know that

-7 *is less than* -3?

DEFINITION. Let a and b be real numbers. Say that "**a is less than b**" and write

$$a < b$$

if a lies to the *left* of b *(on L)*. In this case you also say that "**b is greater than a**", and write

$$b > a.$$

(Observe that in this notation, the symbols

$$<, >$$

point to the *smaller* number.)

Let a, b, and c be real numbers. If

$$a < b \quad \text{and} \quad b < c,$$

then

$$a < c$$

For, a lies to the left of b and b lies to the left of c; hence a lies to the left of c. [See Figure 1.5.]

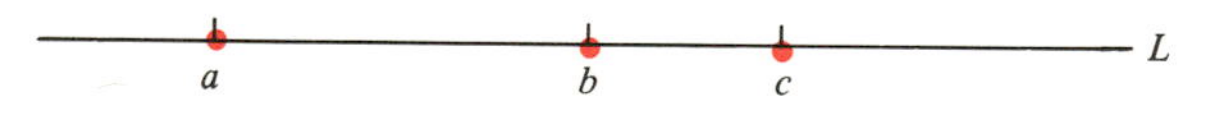

FIGURE 1.5 $a < b$ and $b < c$. a lies to the left of c. Therefore $a < c$

EXAMPLE 4. Refer to Figure 1.6:

(a) $2 < 6$ because 2 lies to the left of 6 on L. You can also write $6 > 2$.
(b) $-3 < 3$ because -3 lies to the left of 3.
(c) $-4 < -2$ because -4 lies to the left of -2.
(d) if $a < -4$, then because $-4 < -2$, it follows that

$$a < -2.$$

■

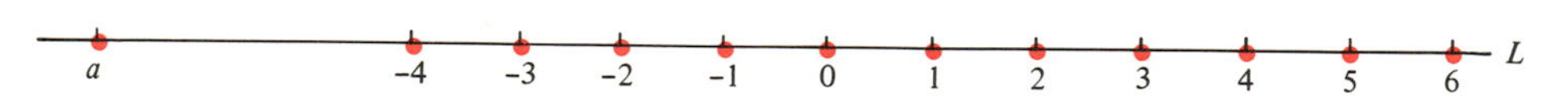

FIGURE 1.6 (a) $2 < 6$ (2 lies to the left of 6.) (b) $-3 < 3$ (-3 lies to the left of 3.) (c) $-4 < -2$ (-4 lies to the left of -2.) (d) If $a < -4$, then $a < -2$

A negative number a is less than any positive number b. In fact,

$$a < 0 \quad \text{and} \quad 0 < b$$

Hence

$$a < b$$

For every two real numbers a and b exactly one of the following holds:

$$a = b; \qquad a < b; \qquad a > b$$

This is because on the real line L,

a and b are the same point

or

a lies to the left of b

or

a lies to the right of b.

DEFINITION. Let a and b be real numbers. Write

$$a \leq b$$

if either $a < b$ or $a = b$. Thus $a \leq b$ indicates that a **is less than or equal to** b. Similarly, write

$$a \geq b$$

if either $a > b$ or $a = b$. Then $a \geq b$ indicates that a **is greater than or equal to** b.

Observe that for every number a,

$$a \leq a \qquad \text{and} \qquad a \geq a$$

are both true.

EXAMPLE 5.

(a) $5 \leq 6$ because $5 < 6$
(b) $9 \geq 3$ because $9 > 3$
(c) $-7 \leq -4$ because $-7 < -4$ ■

EXAMPLE 6.

(a) $2 \leq 2$ because $2 = 2$; similarly, $2 \geq 2$ because $2 = 2$
(b) $-5 \leq -5$ and $-5 \geq -5$, simultaneously, because $-5 = -5$ ■

The real line L extends indefinitely both to the right and to the left. Correspondingly, there are larger and larger positive numbers, as well as smaller and smaller negative numbers.

EXERCISES

1. On a large sheet of paper draw a horizontal line L to represent the real numbers. Plot the following numbers on L:

(a) 0 (b) 1 (c) 2 (d) 3

(e) -1 (f) -2 (g) $\frac{1}{4}$ (h) $\frac{3}{4}$

(i) $1\frac{1}{2}$ (j) $\frac{7}{5}$ (k) π (l) 1.6

(m) -1.2 (n) 2.9

2. Each of the points P, Q, R, S, T, U in Figure 1.7 represents one of the following numbers. Indicate which number is represented by each point.

$$.9, \quad -.9, \quad -1.05, \quad \frac{2}{5}, \quad \frac{1}{2}, \quad \frac{-1}{3}$$

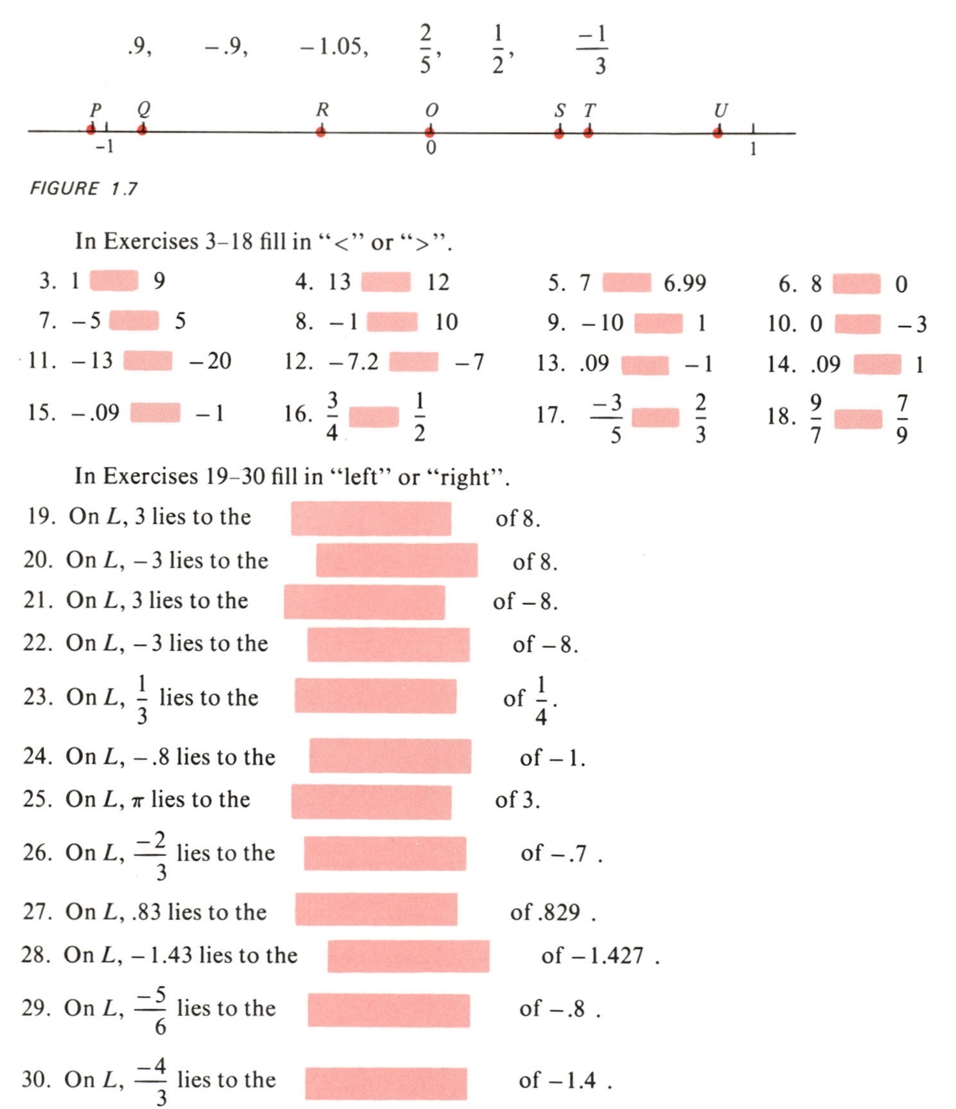

FIGURE 1.7

In Exercises 3–18 fill in “<” or “>”.

3. 1 ____ 9
4. 13 ____ 12
5. 7 ____ 6.99
6. 8 ____ 0
7. -5 ____ 5
8. -1 ____ 10
9. -10 ____ 1
10. 0 ____ -3
11. -13 ____ -20
12. -7.2 ____ -7
13. .09 ____ -1
14. .09 ____ 1
15. $-.09$ ____ -1
16. $\frac{3}{4}$ ____ $\frac{1}{2}$
17. $\frac{-3}{5}$ ____ $\frac{2}{3}$
18. $\frac{9}{7}$ ____ $\frac{7}{9}$

In Exercises 19–30 fill in “left” or “right”.

19. On L, 3 lies to the ____ of 8.
20. On L, -3 lies to the ____ of 8.
21. On L, 3 lies to the ____ of -8.
22. On L, -3 lies to the ____ of -8.
23. On L, $\frac{1}{3}$ lies to the ____ of $\frac{1}{4}$.
24. On L, $-.8$ lies to the ____ of -1.
25. On L, π lies to the ____ of 3.
26. On L, $\frac{-2}{3}$ lies to the ____ of $-.7$.
27. On L, .83 lies to the ____ of .829 .
28. On L, -1.43 lies to the ____ of -1.427 .
29. On L, $\frac{-5}{6}$ lies to the ____ of $-.8$.
30. On L, $\frac{-4}{3}$ lies to the ____ of -1.4 .

In Exercises 31–34 rearrange the numbers so that you can write "<" between any two numbers.

31. $2,\ 8,\ -8,\ \frac{1}{2},\ 0,\ -6$

32. 1.5, 1.439, 1.498, 1.501, 1.44, 1.51

33. −2.7, −2.689, −2.69, −2.693, −2.701, −2.71

34. $\frac{3}{8},\ \frac{3}{4},\ \frac{1}{4},\ \frac{5}{8},\ \frac{7}{8},\ 2,\ \frac{1}{2}$

In Exercises 35–38 rearrange the numbers so that you can write ">" between any two numbers.

35. 21, 23, −23, 27, −22, −24

36. −2.8, −2.85, −2.79, −2.789, −2.9, −2.89

37. $\frac{-5}{3},\ \frac{-7}{3},\ \frac{-4}{3},\ \frac{-1}{3},\ -2,\ \frac{-2}{3}$

38. $\frac{4}{3},\ .7,\ .77,\ .81,\ .79,\ \frac{4}{5}$

In exercises 39–50 fill in "≤" or "≥". If both symbols apply, write "both".

39. 1 ______ 17 40. −2 ______ 5 41. 6 ______ 3

42. −6 ______ 3 43. −6 ______ −3 44. −3 ______ −6

45. 0 ______ −8 46. .1 ______ .01 47. −12 ______ 12

48. 9 ______ 9 49. .099 ______ 1 50. −3 ______ −3

1.2 ABSOLUTE VALUE

The numbers 2 and −2 lie on opposite sides of the origin, 0, but are the same distance from 0. The positive number 2 is 2 units to the right of 0, whereas the negative number −2 is 2 units to the left of 0. Similarly, the numbers $\frac{3}{4}$ and $-\frac{3}{4}$ lie on opposite sides of, but are the same distance from, the origin. [See Figure 1.8.]

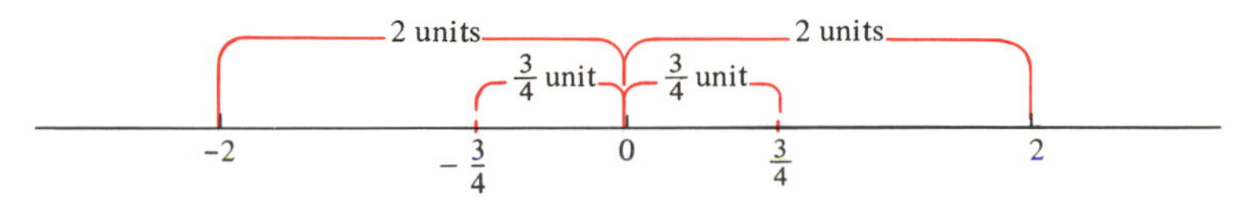

FIGURE 1.8

DEFINITION. Let n be a nonzero number. The **inverse of** n is the number that lies on the opposite side of the origin and is the same distance from the origin as is n. The **inverse of** 0 is 0 itself.

Extend the previous notation by letting

$$-n$$

denote the inverse of n *for every number n, positive, negative, or zero.*

The inverse of a positive number is negative; the inverse of a negative number is positive.

EXAMPLE 1.

(a) The inverse of the positive number 3 is the negative number -3.
(b) The inverse of the negative number -3 is the positive number 3.

$$-(-3) = 3$$

■

DEFINITION. The **absolute value of a number** n is its distance from the origin.

Let

$$|n|$$

denote the absolute value of n.

For every number n,

$$|n| = |-n|$$

because n and its inverse $-n$ are the same distance from the origin. Recall that (if $n \neq 0$) n and $-n$ are on opposite sides of the origin. Thus absolute value measures distance from the origin, but neglects direction.

EXAMPLE 2.

(a) $|6| = 6$ because 6 is 6 units to the right of the origin.
(b) $|-6| = 6$ because -6 is 6 units to the left of the origin.
(See Figure 1.9.) ■

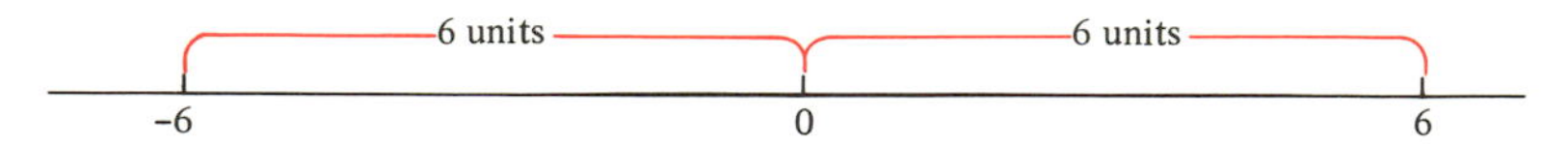

FIGURE 1.9 6 is 6 units to the right of the origin. −6 is 6 units to the left of the origin. Therefore

$$|6| = |-6| = 6$$

EXAMPLE 3. A Volkswagen gets 20 miles to the gallon. It uses the same 3 gallons of gas to travel 60 miles west as it would to travel 60 miles east. [See Figure 1.10.] Thus gas consumption is measured in terms of absolute value; you consider distance, but neglect direction. ■

FIGURE 1.10

The absolute value of a number n can be described in the following way:

$$|n| = n, \qquad \text{if } n \text{ is positive or } 0$$

$$|n| = -n, \qquad \text{if } n \text{ is negative}$$

For, if n is negative, then $-n$ is positive; $-n$ is the distance from n to the origin. [See Figure 1.11 (b).]

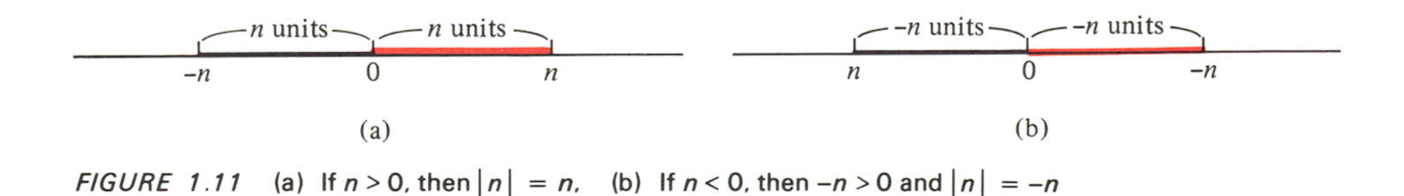

FIGURE 1.11 (a) If $n > 0$, then $|n| = n$. (b) If $n < 0$, then $-n > 0$ and $|n| = -n$

EXAMPLE 4.

(a) $|8| = 8$ because 8 is positive.
(b) $|-8| = -(-8) = 8$ because -8 is negative.
(c) $|0| = 0$
(d) $|-1.1| = 1.1$
(e) $-|-10| = -10$ because $|-10| = 10$ ■

Absolute values will be discussed further in Chapter 14.

EXERCISES

In Exercises 1–10 determine the inverse of each number.

1. 4
2. -3
3. $\frac{2}{3}$
4. $-.12$
5. $|-7|$
6. 0
7. -0
8. $-|-2|$
9. $\left|\frac{-3}{5}\right|$
10. π

In Exercises 11–19 determine each absolute value.

11. $|19|$
12. $|-19|$
13. $|0|$
14. $\left|\frac{1}{4}\right|$
15. $\left|\frac{-2}{5}\right|$
16. $|7.2|$
17. $\left|\frac{-1}{10}\right|$
18. $|-.001|$
19. $|-2\,000\,000|$

In Exercises 20–29 fill in “<”, “>”, or “=”.

20. $|-2| \underline{\qquad} -2$
21. $|-1| \underline{\qquad} |1|$
22. $\left|\frac{2}{3}\right| \underline{\qquad} \frac{2}{3}$
23. $-|7| \underline{\qquad} |-7|$
24. $|-5| \underline{\qquad} -(-5)$
25. $-(-5) \underline{\qquad} -|5|$
26. $|-10| \underline{\qquad} 1$
27. $|-.01| \underline{\qquad} |-.1|$
28. $|-1.2| \underline{\qquad} \frac{6}{5}$
29. $|-\pi| \underline{\qquad} |-3|$
30. Which positive number is 5 units from the origin?
31. Which negative number is 7 units from the origin?
32. Which numbers are $\frac{2}{3}$ of a unit from the origin?

1.3 ADDITION AND SUBTRACTION OF TWO NUMBERS

In the remainder of the chapter the arithmetic of real numbers will be sketched. Addition of numbers will be introduced geometrically. Subtraction and multiplication will be defined in terms of addition, division and exponentiation, in terms of multiplication. Examples and exercises will review the basic computational techniques for *integers.*[1] In Chapter 5 the arithmetic of fractions and decimals will be discussed.

[1]For more examples and exercises on arithmetic, see M. Zuckerman, *Workbook to Accompany Intermediate Algebra: A Straightforward Approach,* Sections 1.3–1.7.

When two numbers a and b are added, the resulting **sum** is written

$$a + b.$$

When a *positive* number p is added to a, the sum, $a + p$, lies p units to the *right* of a. [See Figure 1.12 (a).] When a *negative* number n is added to a, the sum, $a + n$, lies $|n|$ units to the *left* of a. [See Figure 1.12 (b).] Thus $3 + 4$, or 7, lies 4 units to the right of 3, whereas $3 + (-2)$, or 1, lies 2 units to the left of 3. [See Figure 1.12 (c).] Also, $2 + \frac{1}{2}$ or $2\frac{1}{2}$ or $\frac{5}{2}$ lies $\frac{1}{2}$ unit to the right of 2. [See Figure 1.12 (d).] And $\pi + 1$ lies 1 unit to the right of π. [See Figure 1.12(e).]

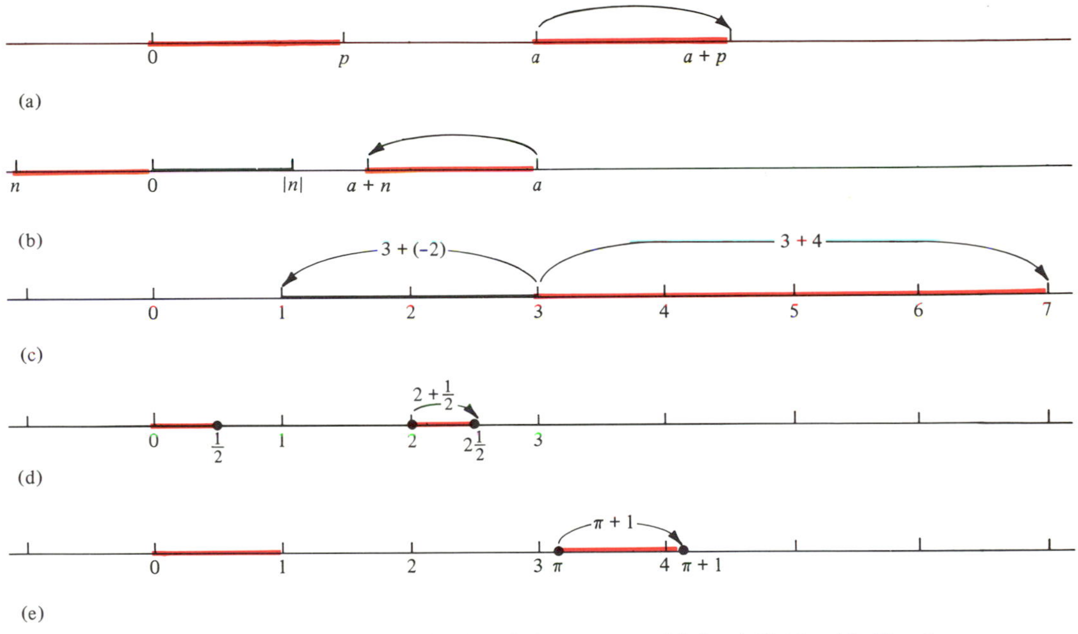

FIGURE 1.12 (a) If $p > 0$, then $a < a + p$. (b) If $n < 0$, then $a + n < a$. (c) $3 + (-2) < 3$ and $3 < 3 + 4$.

The **difference between two numbers** a and b (in this order) is written

$$a - b.$$

Note that this is a different use of the symbol

$$-$$

than in the expression

$$-n.$$

the inverse of n

Subtraction can be defined in terms of addition.

DEFINITION. Let a and b be any real numbers. Define

$$a - b = a + (-b).$$

Thus *subtracting a number is the same as adding its inverse.* Consequently, when a *positive* number p is subtracted from a, the negative number $-p$ is, in effect, added to a. The difference, $a - p$, lies to the *left* of a. [See Figure 1.13 (a).] When a *negative* number n is subtracted from a, the positive number $-n$ is added to a. The difference, $a - n$, lies to the *right* of a. [See Figure 1.13 (b).] For example,

$$5 - 3 = 5 + (-3) = 2$$

and $5 - 3$ lies to the *left* of 5;

$$5 - (-3) = 5 + 3 = 8$$

and $5 - (-3)$ lies to the *right* of 5. [See Figure 1.13 (c).]

Finally, subtraction is the "inverse operation" of addition in that it "undoes" addition. Thus

$$2 + 3 = 5$$

whereas

$$5 - 3 = 2$$

[See Figure 1.13 (d).]

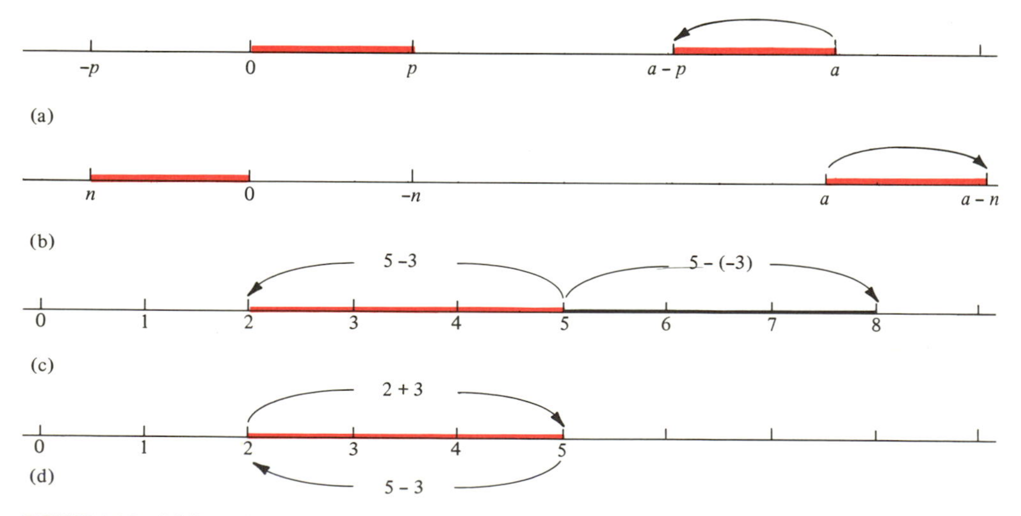

FIGURE 1.13 (a) If $p > 0$, then $a - p < a$, (b) If $n < 0$, then $a < a - n$, (c) $5 - 3 < 5$ and $5 < 5 - (-3)$, (d) Subtraction is the "inverse operation" of addition.

You can add real numbers in either order. Thus

$$a + b = b + a$$

This is known as the **Commutative Law of addition.** However, in subtraction, the order is essential.

$$a - b \neq b - a$$

unless $a = b$. In fact, for all a and b,

$$a - b = -(b - a)$$

Therefore

$$|a - b| = |b - a|$$

Thus $a - b$ and $b - a$ differ in sign but have the same absolute value.

EXAMPLE 1.

(a) $6 + 4 = 4 + 6 = 10$
(b) $6 - 4 = 6 + (-4) = 2; \quad 4 - 6 = 4 + (-6) = -2$
Therefore, $4 - 6 = -(6 - 4)$ ■

Clearly, for every number a,

$$a + 0 = 0 + a = a; \qquad a - 0 = a$$

whereas

$$0 - a = 0 + (-a) = -a$$

When two *positive* numbers p and q are added, the sum, $p + q$, lies to the *right* of each number. [See Figure 1.14 (a).] Thus

$$p < p + q, \qquad q < p + q \qquad (p \text{ and } q \text{ positive})$$

When two negative numbers m and n are added, the sum, $m + n$, lies to the *left* of each number. [See Figure 1.14 (b).] Thus

$$m + n < m, \qquad m + n < n \qquad (m \text{ and } n \text{ negative})$$

FIGURE 1.14 (a) If $p > 0$ and $q > 0$, then $p < p + q$ and $q < p + q$. (b) If $m < 0$ and $n < 0$, then $m + n < m$ and $m + n < n$

EXAMPLE 2. If Pete loses 10 dollars at the trotters and then loses another 5 dollars, he is behind 15 dollars. Pete's *winnings* are given by:

$$-\$15 = (-\$10) + (-\$5)$$

Note that

$$-(10 + 5) = (-10) + (-5).$$

■

EXAMPLE 3. In Figure 1.15, observe that

$$(-5) + (-3) = -8$$

■

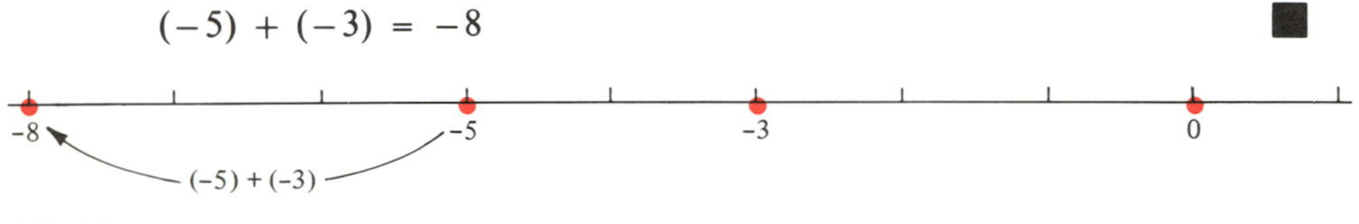

FIGURE 1.15

In general, to add two negative numbers, add their absolute values and prefix a minus sign. In the case of $(-5) + (-3)$, you could say:

The absolute value of -5 is 5; the absolute value of -3 is 3.

$$5 + 3 = 8$$

Thus

$$(-5) + (-3) = -8$$

Normally, you would not write down everything, but would proceed as in Example 4.

EXAMPLE 4.

$$(-9) + (-6) = -(9 + 6) = -15$$

■

A different situation occurs when you *add numbers with different signs.*

Recall that for $a \neq 0$, a and $-a$ are each the same distance from 0, but are on opposite sides of 0. Thus

$$a + (-a) = (-a) + a = 0$$

[See Figure 1.16.] In other words, *the inverse of a is the number added to a to ob-*

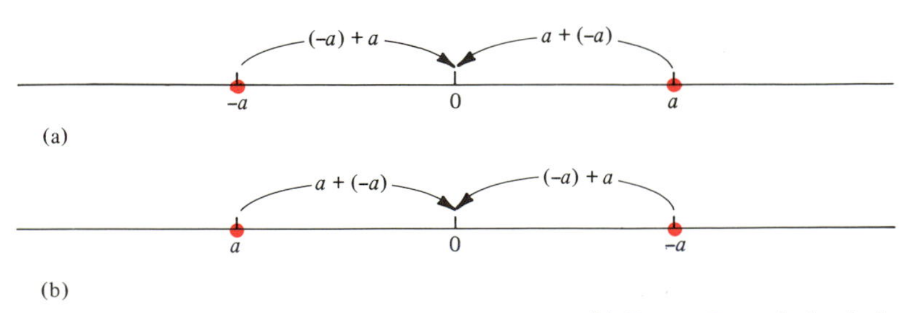

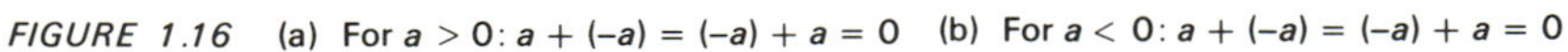

FIGURE 1.16 (a) For $a > 0$: $a + (-a) = (-a) + a = 0$ (b) For $a < 0$: $a + (-a) = (-a) + a = 0$

tain 0. Note that

$$|a| = |-a|.$$

Thus *when you add two numbers with different signs, but with the same absolute value, the sum is* 0.

Next, add numbers with *different signs and different absolute values.*

EXAMPLE 5. One day Pete wins 20 dollars on one race and loses 15 dollars on another. The next day Pete loses 18 dollars on one race and wins 12 dollars on another. Can you determine his winnings for each day?

SOLUTION.

(a) Winnings for first day: \$20 − \$15 = \$5. [See Figure 1.17 (a).]
(b) The second day Pete is a loser. *Subtract* his winnings from his losings:

$$\begin{array}{r} \$18 \\ \underline{\$12} \\ \$\ 6 \end{array}$$

His *losings* are \$6. Consequently, his winnings are

$$-\$6.$$

[See Figure 1.17 (b).] ■

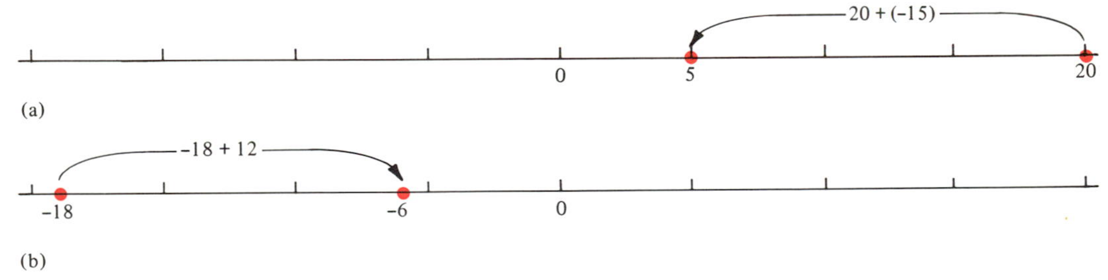

FIGURE 1.17 (a) $20 + (-15) = 20 - 15 = 5$ (b) $-18 + 12 = -6$

Observe that in both parts of Example 5, *the absolute value of the sum is the difference of the absolute values. Thus*

(a) $|20 + (-15)| = |20| - |-15| = 20 - 15 = 5$
(b) $|-18 + 12| = |-18| - |12| = 18 - 12 = 6$

Prefix the sign of the number whose absolute value is larger. Thus

(a) $20 + (-15) = 5$ because $|20| > |-15|$
(b) $(-18) + 12 = -6$ because $|-18| > |12|$

EXAMPLE 6. Add: $(-23) + 36$

SOLUTION. $|36| = 36$, $|-23| = 23$, $36 > 23$

Subtract:

$$\begin{array}{r} 36 \\ \underline{23} \\ 13 \end{array}$$

The *positive* sign of 36, the larger absolute value, prevails. Therefore,

$$(-23) + 36 = 13$$ ■

EXAMPLE 7. Add: $381 + (-414)$

SOLUTION. $|-414| = 414$, $|381| = 381$, $414 > 381$

Subtract:

$$\begin{array}{r} 414 \\ \underline{381} \\ 33 \end{array}$$

The larger absolute value, 414, corresponds to the *negative* number, -414. Thus

$$381 + (-414) = -33$$ ■

For any real numbers a and b,

$$-(a + b) = (-a) + (-b)$$

and

$$-(a - b) = -a + b = b - a$$

These equalities hold regardless of the signs of a and b.

EXAMPLE 8.

(a) $-(6 + 4) = -10 = (-6) + (-4)$
(b) $-[(-6) + (-4)] = -[-10] = 10 = [-(-6)] + [-(-4)]$
(c) $-[(-6) + 4] = -[-2] = 2 = [-(-6)] + [-4]$
(d) $-(6 - 4) = -2 = 4 - 6$
(e) $-[6 - (-4)] = -[6 + 4] = -10 = -4 - 6$ ■

EXERCISES

In Exercises 1–16 add, as indicated:

1. $197 + 298$
2. $2703 + 1940$
3. $1819 + 9999$
4. $29 + (-92)$
5. $(-202) + 107$
6. $172 + (-284)$
7. $(-284) + 19$
8. $(-608) + (-17)$
9. $\begin{array}{r} 1084 \\ \underline{394} \end{array}$
10. $\begin{array}{r} 6803 \\ \underline{9174} \end{array}$
11. $\begin{array}{r} 621 \\ \underline{-109} \end{array}$
12. $\begin{array}{r} -83 \\ \underline{-49} \end{array}$
13. $\begin{array}{r} -8374 \\ \underline{6218} \end{array}$
14. $\begin{array}{r} -5083 \\ \underline{6921} \end{array}$
15. $\begin{array}{r} 815 \\ \underline{-962} \end{array}$
16. $\begin{array}{r} -8148 \\ \underline{-2095} \end{array}$

In Exercises 17–26 subtract, as indicated:

17. $170 - 34$
18. $19 - 48$
19. $208 - 194$
20. $172 - 1296$
21. $2749 - (-2821)$
22. $-483 - 282$
23. $-1143 - 2891$
24. $-8121 - (-7163)$
25. $-207 - 0$
26. $318 - 460$

In Exercises 27–38 *subtract* the bottom number from the top one:

27.	28.	29.	30.
5	5	−5	−5
2	−2	2	−2

31.	32.	33.	34.
181	317	−2084	29 193
72	509	1897	−9 209

35.	36.	37.	38.
−7146	61 263	−631	−699
−4183	−5 384	−837	700

39. Add -27 and -381.
40. Subtract 29 from 92.
41. Subtract 207 from 193.
42. Subtract -137 from 360.
43. A car travels 40 miles east of Denver. It then reverses its direction and travels 60 miles west. Where does the trip finish?
44. A car travels 124 miles north of Omaha. The driver finds he has gone 36 miles beyond his destination. Determine his destination in relation to Omaha.

1.4 ADDITION AND SUBTRACTION OF THREE OR MORE NUMBERS

Let a, b, and c be any real numbers. When you first add a and b, and then add c, you obtain the same sum as when you first add b and c, and then add this to a:

$$(a + b) + c = a + (b + c)$$

This is known as the **Associative Law of addition.** Because of this equality, you may omit parentheses and write

$$a + b + c$$

for either sum.

EXAMPLE 1.

$$(3 + 2) + 7 = 3 + (2 + 7) = 12$$ ■

EXAMPLE 2.

$$[(-5) + (-9)] + (-2) = (-5) + [-9 + (-2)] = -16$$ ■

The Associative and Commutative Laws of addition are utilized in the following.

To add 3 or more numbers, not all of the same sign:

1. Rearrange and add the positive numbers and the negative numbers separately.
2. Let $a(>0)$ be the sum of the positive numbers and let $-b(b > 0)$ be the sum of the negative numbers. Then the total sum is $a + (-b)$, which equals

$$a - b.$$

EXAMPLE 3.

$$\begin{aligned} 4 + (-3) + 7 &= 4 + [(-3) + 7] \\ &= 4 + [7 + (-3)] && \text{by the Commutative Law} \\ &= [4 + 7] + (-3) && \text{by the Associative Law} \\ &= 11 + (-3) \\ &= 8 \end{aligned}$$

■

EXAMPLE 4.

$$\begin{aligned} 41 + (-21) + 33 + (-8) + (-15) &= (41 + 33) \\ &\quad + [(-21) + (-8) + (-15)] \\ &= 74 + (-44) \\ &= 30 \end{aligned}$$

■

EXAMPLE 5. Add:

$$\begin{array}{r} -13 \\ 16 \\ -23 \\ -39 \\ 8 \\ \underline{51} \end{array}$$

SOLUTION. Rearrange, and add the positive and negative numbers separately:

$$\begin{array}{rcr} 16 & \qquad & -13 \\ 8 & & -23 \\ \underline{51} & & \underline{-39} \\ 75 & & -75 \end{array}$$

The resulting sum, $75 + (-75)$, equals 0. ■

Subtraction, unlike addition, is not associative. In general,

$$(a - b) - c \neq a - (b - c)$$

Can you see that parentheses are necessary to indicate which subtraction precedes the other? *If no parentheses are written, subtract in the order written, from left to right.* Thus

$$a - b - c = (a - b) - c$$

EXAMPLE 6.

(a) $\underbrace{(10 - 3)}_{7} - 2 = 7 - 2 = 5$

(b) $10 - \underbrace{(3 - 2)}_{1} = 10 - 1 = 9$

(c) Therefore $(10 - 3) - 2 \neq 10 - (3 - 2)$

(d) $10 - 3 - 2 = (10 - 3) - 2 = 5$ ■

Addition and subtraction can be combined in the same example. Here parentheses or brackets are sometimes necessary to indicate the intended meaning.

EXAMPLE 7. Find the value of $(4 - 8) - (3 + 7 - 2)$.

SOLUTION. First perform the operations within each pair of parentheses.

$$(4 - 8) - (3 + 7 - 2) = -4 - 8 = -12$$ ■

When pairs of parentheses or brackets are enclosed within other pairs, work from the innermost pairs outward.

EXAMPLE 8. Find the value of $-[13 - (3 - 8 + 10)]$.

SOLUTION. Here the value of $3 - 8 + 10$ is subtracted from 13. You then determine the inverse of this difference.

$$-[13 - \underbrace{(3 - 8 + 10)}_{5}] = -[13 - 5] = -8$$ ■

EXAMPLE 9. Find the value of $20 - [6 - (3 - 2) - (8 - 12)]$.

SOLUTION. Here, two pairs of parentheses are enclosed within an outer pair of brackets. Remove the two inner pairs simultaneously.

$$\begin{aligned} 20 - [6 - (3 - 2) - (8 - 12)] &= 20 - [6 - 1 - (-4)] \\ &= 20 - (6 - 1 + 4) \\ &= 20 - 9 \\ &= 11 \end{aligned}$$ ■

EXERCISES

In Examples 1–15 find the value of each sum.

1. $51 + 32 + 81 + 7$

2. $9 + 8 + 7 + 15 + 14 + 6$

3. $$\begin{array}{r} 57 \\ 82 \\ 193 \\ 212 \\ \hline \end{array}$$

4. $$\begin{array}{r} 20\,143 \\ 9\,620 \\ 50\,300 \\ 1\,749 \\ \hline \end{array}$$

5. $(-55) + (-9) + (-33)$

6. $(-463) + (-18) + (-36) + (-15)$

7. $(-192) + (-74) + (-1) + (-330)$

8. $$\begin{array}{r} -586 \\ -207 \\ -141 \\ -803 \\ \hline \end{array}$$

9. $$\begin{array}{r} -\ 763 \\ 892 \\ -1704 \\ \hline \end{array}$$

10. $$\begin{array}{r} -9638 \\ 432 \\ -\ 128 \\ 7001 \\ \hline \end{array}$$

11. $$\begin{array}{r} 94\,328 \\ -\quad 541 \\ 82\,416 \\ -\quad 436 \\ \hline \end{array}$$

12. $$\begin{array}{r} 14\,284 \\ -\ 9\,526 \\ 724\,083 \\ 401\,001 \\ 360\,036 \\ \hline \end{array}$$

13. $$\begin{array}{r} 814\,930 \\ -\ 178\,200 \\ -\ 358\,916 \\ 1\,130\,081 \\ 728\,993 \\ -\quad 70\,820 \\ \hline \end{array}$$

14. $52 + (-17) + (-19) + (-18)$

15. $373 + (-378) + 606 + 204 + (-294)$

In Exercises 16–24 subtract, as indicated.

16. $8 - 3 - 4$
17. $18 - 9 - 4 - 1$
18. $41 - 21 - 7 - 3 - 19$
19. $6 - 8 - (4 - 3)$
20. $(18 - 9) - (4 - 13)$
21. $(12 - 5) - 7 - 2$
22. $(13 - 10 - 7) - (8 - 5)$
23. $12 - (11 - 4 - 3) - (7 - 2)$
24. $42 - 21 - (7 - 9) - 2 - 8 - (1 - 4)$

In Exercises 25–36 perform the indicated addition and subtraction.

25. $8 - 5 + 9 - 3$
26. $18 + 15 - 3 - 7 + 1$
27. $5 + (8 - 3) + (6 - 7)$
28. $15 - (80 - 3) - (6 + 2)$
29. $(81 - 9 + 7) - (18 - 15) + 6$

30. $7 - 10 - (4 - 10) + (8 - 5) - 3 - (1 - 2)$
31. $7 - (8 - 5 + 2) - (6 + 5 + 17 - 2)$
32. $18 - [15 - (4 - 7)]$
33. $21 - (8 - 3) - [40 - (3 + 6)]$
34. $17 + [8 - 5 - (2 - 3)] - 17$
35. $43 - (24 - [47 - (31 - 19)])$
36. $38 - (21 - [17 - 14] - [7 - (8 - 19)])$
37. Subtract 7 from the sum of 13 and 9.
38. Subtract the sum of -18 and -85 from 800.
39. Subtract the sum of -272 and 183 from the sum of 19 and -161.
40. Subtract 15 minus -8 from the sum of 27 and -34.
41. A man wins 10 dollars one day, loses 5 dollars the next day, and loses 4 dollars the third day. Determine his total winnings.
42. You owe Jim 10 dollars and you owe Juan 8 dollars. On the other hand, Carlos owes you 6 dollars and Bill owes you 20 dollars. If everybody pays his debts how much money do you pay or get paid?

1.5 MULTIPLICATION AND DIVISION

The **product** of a and b is written in each of the following ways:

$$a \cdot b \qquad ab \qquad (a)(b) \qquad a(b) \qquad (a)b \qquad a \times b$$

When two or more numbers are multiplied, they are called **factors** of the product. Thus a and b are each factors of $a \cdot b$.

When the factor b is a *positive integer,* multiplication by b amounts to repeated addition.

DEFINITION. Let a be any real number and let b be a positive integer. Then

$$a \cdot b$$

means

$$\underbrace{a + a + a + \cdots + a}_{b \text{ times}}.$$

EXAMPLE 1.

(a) $2 \cdot 3 = 2 + 2 + 2 = 6$ [See Figure 1.18(a).]
(b) $(-4)2 = (-4) + (-4) = -8$ [See Figure 1.18(a).]
(c) $\frac{1}{3} \cdot 2 = \frac{1}{3} + \frac{1}{3}$

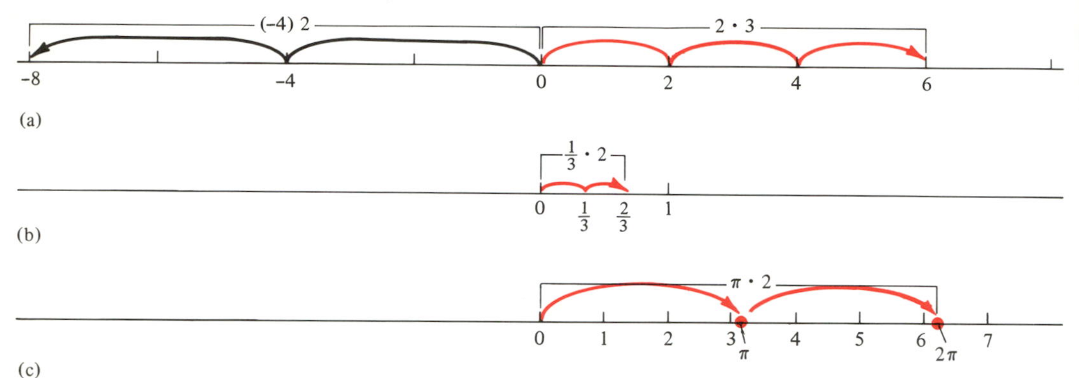

FIGURE 1.18

As you will see in Chapter 5,

$$\frac{1}{3} \cdot 2 = \frac{1}{3} + \frac{1}{3} = \frac{2}{3}$$ [See Figure 1.18(b).]

(d) $\pi \cdot 2 = \pi + \pi$ [See Figure 1.18(c).]

This number is denoted by 2π. ■

DEFINITION. Let a be any real number. Define

(1) $a(-1)$ to be $-a$.

Thus *multiplication of a by −1 yields the inverse of a.*

It can be shown that multiplication is both **commutative** and **associative.** Thus let a, b, c be real numbers. Then

$a \cdot b = b \cdot a$ (**Commutative Law**)

$(ab)c = a(bc)$ (**Associative Law**)

In order to discover the rule for multiplying a positive number by a negative number, fill in the blanks in Table 1.1 by following the pattern in the right-hand column. Refer to Figure 1.19.

$5 \times 3 =$	15
$5 \times 2 =$	10
$5 \times 1 =$	5
$5 \times 0 =$	0
$5(-1) =$	
$5(-2) =$	
$5(-3) =$	

TABLE 1.1

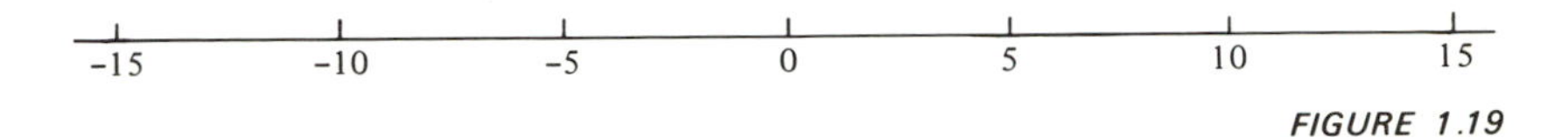

FIGURE 1.19

Let p and q be positive. Then $-q$ is negative, and

$$-q = q(-1). \qquad \text{by (1)}$$

Consequently,

$$\begin{aligned} p(-q) &= p[q(-1)] \\ &= (pq)(-1) && \text{by the Associative Law} \\ &= -(pq) && \text{by (1)} \end{aligned}$$

Thus $p(-q)$ is the inverse of the positive number pq, and is therefore negative. Similarly,

$$(-p)q = -(pq)$$

Hence $(-p)q$ is negative.

For example,

$$3(-2) = (-3)2 = -(3 \cdot 2) = -6 \qquad \text{[See Figure 1.20.]}$$

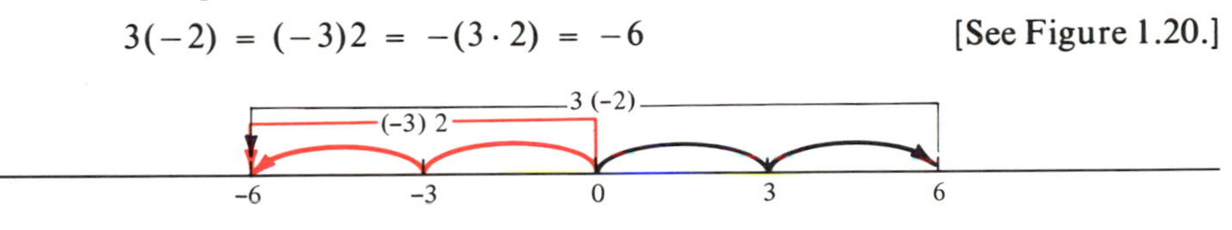

FIGURE 1.20 $3(-2) = (-3)2 = -6$

In order to discover the rule for multiplying two negative numbers, fill in the blanks in Table 1.2 by considering the pattern in the right-hand column. Again refer to Figure 1.19.

$(-5) \times 3 =$	-15
$(-5) \times 2 =$	-10
$(-5) \times 1 =$	-5
$(-5) \times 0 =$	0
$(-5) \times (-1) =$	
$(-5) \times (-2) =$	
$(-5) \times (-3) =$	

TABLE *1.2*

Let p and q be positive. Then $-p$ and $-q$ are both negative, and

$$\begin{aligned} (-p)(-q) &= [(-1)p][(-1)q] && \text{by (1)} \\ &= [\underbrace{(-1)(-1)}_{1}][pq] && \text{by the Associative and Commutative Laws} \\ &= pq \end{aligned}$$

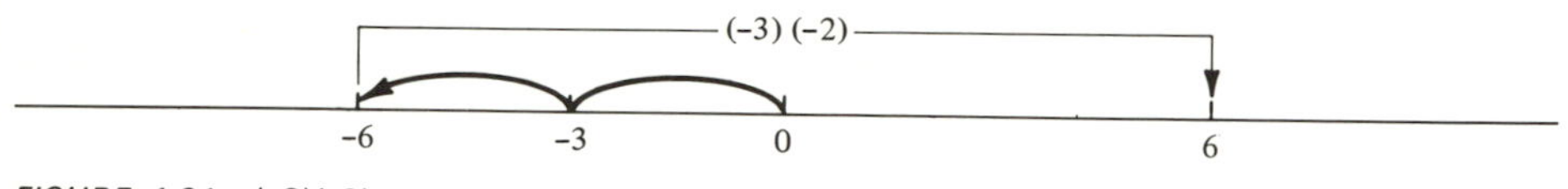

FIGURE 1.21 $(-3)(-2) = 6$

For example,

$$(-3)(-2) = -[-(3 \cdot 2)] = 6 \qquad \text{[See Figure 1.21.]}$$

Table 1.3 indicates the possibilities for the sign of the product.

a	b	ab
positive	positive	positive
positive	negative	negative
negative	positive	negative
negative	negative	positive

TABLE 1.3

It is worth mentioning that

$$a(-b) = -ab, \qquad (-a)b = -ab, \qquad \text{and} \qquad (-a)(-b) = ab$$

for all integers a and b, regardless of their signs.

EXAMPLE 2.

(a) $5 \cdot 12 = 60$

(b) $5(-12) = -60$

(c) $(-5)12 = -60$

(d) $(-5)(-12) = 60$ ■

DEFINITION. Let a be any real number. Define:

$$a \cdot 0 = 0$$

EXAMPLE 3.

(a) $8 \cdot 0 = 0$

(b) $0 \cdot 8 = 0$

(c) $0(-8) = 0$

(d) $0 \cdot 0 = 0$ ■

EXAMPLE 4.

(a) $(7)(0)(-4)(-8) = 0$

$$\underbrace{\underbrace{\underbrace{(7)(0)}_{0}(-4)}_{0}(-8)}_{0}$$

(b) $(-2)(-3)(4)(10) = 240$

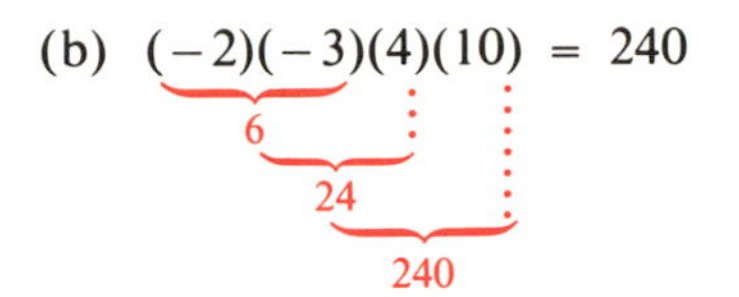

There are two negative factors. The product is positive.

(c) $5(-2)(-1)(2)(-3) = -60$

-10
10
20
-60

There are three negative factors. The product is negative.

(d) $(-1)(-2)(-1)(3)(-5) = 30$

2
-2
-6
30

■

When two or more numbers are multiplied, the product is zero if at least one of the factors is 0; otherwise, the product is nonzero. Suppose all of the factors are nonzero. You can multiply in any order. The product of two negative numbers is positive. A third negative factor yields a negative product; a fourth negative factor yields a positive product, and so forth.

1. The product is positive if there is an even number (possibly 0) of negative factors.
2. The product is negative if there is an odd number of negative factors.

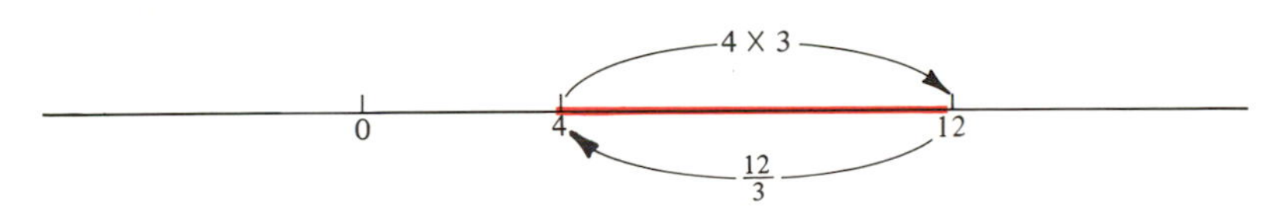

FIGURE 1.22

Division is the inverse operation of multiplication. As is suggested in Figure 1.22, division is defined in terms of multiplication, just as subtraction was defined in terms of addition.

DEFINITION. Let a, b, c be real numbers, and suppose $b \neq 0$. Then

$$\frac{a}{b} = c, \qquad \text{if} \qquad a = bc$$

Here a is called the **dividend** and b the **divisor**; c is called the **quotient.**

You also write

$$a \div b = c \qquad \text{or} \qquad b\overline{)a}^{\,c}.$$

EXAMPLE 5.

$$\frac{6}{2} = 3 \qquad \text{because} \qquad 6 = 2 \cdot 3$$

Here, $a = 6, b = 2, c = 3$ ■

Geometrically, if $\frac{a}{b} = c$, then both $\frac{a}{b}$ and c correspond to the same point on the number line L. Thus in Example 5,

$$\frac{6}{2} = 3$$

Divide the line segment between 0 and 6 into 2 equal parts. [See Figure 1.23.] The point of division, which represents $\frac{6}{2}$, is 3.

0 1 2 3 4 5 6 L

FIGURE 1.23 $\frac{6}{2} = 3$

Division is defined for *all* real numbers a and $b(\neq 0)$. The quotient $\frac{a}{b}$ is always a *real number*. However, *even when both a and b are integers,* the quotient $\frac{a}{b}$ need not be an integer $\left(\text{as in the case of } \frac{5}{4}\right)$.

EXAMPLE 6.

(a) $\frac{6}{3} = 2$ because $6 = 3 \cdot 2$

(b) $\frac{6}{-3} = -2$ because $6 = (-3)(-2)$

(c) $\frac{-6}{3} = -2$ because $-6 = 3(-2)$

(d) $\frac{-6}{-3} = 2$ because $-6 = (-3)2$ ■

Let a and b be real numbers and let

$$\frac{a}{b} = c.$$

Because of the definition of division in terms of multiplication, the quotient c is positive if a and b have the same sign; c is negative if a and b have different signs.

Division by 0 *is undefined.* For suppose $a \neq 0$.

$$\frac{a}{0} = c \qquad \text{would mean} \qquad a = 0 \cdot c = 0,$$

contradicting the assumption that $a \neq 0$. Moreover,

$$\frac{0}{0} \text{ is undefined.}$$

In fact,

$$\frac{0}{0} = c \qquad \text{would mean} \qquad 0 = 0 \cdot c.$$

But *every* real number c is such that

$$0 = 0 \cdot c.$$

For example,

$$0 = 0 \cdot 3, \qquad 0 = 0(-3), \qquad 0 = 0 \cdot 0$$

There is no way to choose which number $\frac{0}{0}$ should be. Therefore, $\frac{0}{0}$ is undefined.

However, $\frac{0}{a}$ is defined for $a \neq 0$. In this case,

$$\frac{0}{a} = 0 \qquad \text{because} \qquad 0 = a \cdot 0$$

Thus

$$\frac{0}{3} = 0 \qquad \text{and} \qquad \frac{0}{-3} = 0$$

You can divide products of numbers.

EXAMPLE 7.

(a) $$\frac{\overbrace{(-4)(-1)(-2)}^{\text{dividend}}}{\underbrace{(-3)5}_{\text{divisor}}} = \frac{-8}{-15} = \frac{8}{15}$$

Note that there are 4 negative factors altogether: 3 in the dividend and 1 in the divisor. The quotient is positive.

(b) $\dfrac{(-4)(-1)(-2)}{(-3)(-5)} = \dfrac{-8}{15} = -\dfrac{8}{15}$ ■

Here there are a total of 5 negative factors: 3 in the dividend and 2 in the divisor. The quotient is negative.

When you divide products of numbers, count up the total number of negative factors in both dividend and divisor. *If each factor is nonzero, then the quotient is*

1. *positive, if there is an even number of negative factors, possibly none at all, in dividend and divisor;*
2. *negative, if there is an odd number of negative factors in dividend and divisor.*

EXAMPLE 8.

(a) $\dfrac{(-7)(5)(-2)}{-3} = -\dfrac{70}{3}$

There are altogether 3 negative factors: 2 in the dividend and 1 in the divisor. Because 3 is odd, the quotient is negative.

(b) $\dfrac{7 \cdot 5 \cdot 2}{3} = \dfrac{70}{3}$

Here there are no negative factors. Thus the quotient is positive. ■

EXERCISES

In Exercises 1–22 multiply:

1. $13 \cdot 14$
2. $(-13)14$
3. $13(-14)$
4. $(-13)(-14)$
5. $19(-23)$
6. $(-37)(-21)$
7. $(18)(-4)(0)$
8. $(21)(35)(-7)$
9. $(4)(-3)(-5)$
10. $(-2)(-9)(-8)$
11. $(4)(-1)(-2)$
12. $(-7)(19)(-8)$
13. $(6)(-1)(-4)(2)$
14. $(-8)(-3)(-1)(-7)$
15. $(-2)(-4)(-2)(-3)(-1)$
16. $(7)(-1)(3)(-2)(-1)$
17. $\begin{array}{r} -164 \\ \times\ -\ 17 \\ \hline \end{array}$
18. $\begin{array}{r} -155 \\ \times\ -101 \\ \hline \end{array}$
19. $\begin{array}{r} -891 \\ \times\ \ 12 \\ \hline \end{array}$
20. $\begin{array}{r} 97 \\ \times\ -11 \\ \hline \end{array}$
21. $\begin{array}{r} -6693 \\ \times\ -\ 749 \\ \hline \end{array}$
22. $\begin{array}{r} -8083 \\ \times\ \ 7405 \\ \hline \end{array}$

In Exercises 23–38 divide, or indicate that division is undefined.

23. $\frac{-18}{9}$ 24. $\frac{18}{-9}$ 25. $\frac{-18}{-9}$ 26. $\frac{0}{-9}$

27. $\frac{-16}{-8}$ 28. $\frac{-63}{9}$ 29. $\frac{0}{8}$ 30. $\frac{8}{0}$

31. $\frac{-8}{0}$ 32. $\frac{0}{0}$ 33. $37\overline{)111}$ 34. $-24\overline{)96}$

35. $-43\overline{)-645}$ 36. $-213\overline{)23\,856}$ 37. $607\overline{)-23\,673}$ 38. $1001\overline{)18\,018}$

In Exercises 39–44 multiply and divide, as indicated.

39. $\frac{3 \cdot 4}{6}$ 40. $\frac{(-2)(-1)}{2}$ 41. $\frac{(4)(9)(-3)}{-12}$

42. $\frac{-8 \cdot 0}{-6}$ 43. $\frac{28}{(-4)(-7)}$ 44. $\frac{(-7)(-2)}{2 \cdot 7}$

45. A daily double ticket nets \$815 to a group of 5 betting partners. How much is each partner's share?

46. Fourteen students each contribute \$12 to aid victims of a tornado. How much money is contributed by the students?

47. The product of 8 and 14 is divided by 56. Determine the quotient.

48. The quotient of 100 divided by 25 is multiplied by the product of 9 and 16. Determine the result.

1.6 COMBINING OPERATIONS

Several arithmetic operations—addition, subtraction, multiplication, and division—can be combined in the same example. The order in which these operations are applied is frequently crucial.

Let a, b, c be real numbers. Then

$$a + b \cdot c$$

is understood to mean:

1. First multiply b by c.
2. Add this product to a.

In other words, here *multiplication precedes addition.* Similarly,

$$a - b \cdot c$$

is understood to mean:

1. First multiply b by c.
2. Subtract this product from a.

EXAMPLE 1.

(a) $6 + 4 \cdot 5 = 6 + 20 = 26$
(b) $14 - 2 \cdot 3 = 14 - 6 = 8$
(c) $7 \cdot 2 + 1 = 14 + 1 = 15$ ■

Similarly, *if not otherwise indicated, divide before you add or subtract.*

EXAMPLE 2.

(a) $10 + \frac{8}{4} = 10 + 2 = 12$ (b) $\frac{9}{-3} - 7 = -3 - 7 = -10$ ■

If addition or subtraction is to precede multiplication (or division), you must indicate this by means of parentheses. Thus

$$a(b + c)$$

indicates the product of a and the sum of b and c. The **Distributive Laws** allow you to rewrite such products as follows:

$$a(b + c) = a \cdot b + a \cdot c$$

$$(b + c)a = b \cdot a + c \cdot a$$

Similarly,

$$a(b - c) = a \cdot b - a \cdot c$$

$$(b - c)a = b \cdot a - c \cdot a$$

EXAMPLE 3. By the Distributive Laws,

$$\begin{aligned} 31 \cdot 14 &= 31(4 + 10) \\ &= 31 \cdot 4 + 31 \cdot 10 \\ &= 124 + 310 \\ &= 434 \end{aligned}$$

Note that this is actually used in the multiplication process:

$$\begin{array}{r} 31 \\ \underline{14} \\ 124 \\ \underline{310} \\ 434 \end{array}$$

■

The Distributive Laws can be extended to such cases as

$$a(b + c + d) = ab + ac + ad,$$

$$a(b - c + d - e) = ab - ac + ad - ae,$$

and so on.

In the remaining examples, observe the use of parentheses to clarify the intended meaning.

EXAMPLE 4.

$$\frac{(-8)(5 - 7) + 9}{5} = \frac{(-8)(-2) + 9}{5}$$

$$= \frac{16 + 9}{5} = \frac{25}{5} = 5 \quad \blacksquare$$

EXAMPLE 5.

$$(8 + 2 - 7)(3 - 9) + 10 = 3(-6) + 10$$

$$= -18 + 10 = -8 \quad \blacksquare$$

EXAMPLE 6.

$$\frac{8 + (-4)(-5) - 7 \cdot 4}{6 \cdot 8 + 5(-7)} = \frac{8 + 20 - 28}{48 - 35} = \frac{0}{13} = 0 \quad \blacksquare$$

EXERCISES

In Exercises 1–32 find the value of the given expression.

1. $8 + 5 \cdot 2$
2. $8 - 5 \cdot 2$
3. $(8 + 5)2$
4. $8(5 + 2)$
5. $(5 - 8)(6 - 3)$
6. $6[3 + (5 - 7)]$
7. $\dfrac{6 + 3}{-1}$
8. $\dfrac{9 + 6}{-5}$
9. $(8 - 3 \cdot 2) - 5 + 7$
10. $5 - (8 + 3)(-1)$
11. $5 - 2(4 - 7)$
12. $(-3 + 8)(6 - 12)$
13. $\dfrac{5 + 25}{-4 - 2}$
14. $\dfrac{-8 + 3(5 - 1)}{(-2)(-2)}$
15. $9(5 - 4) + 8(-7 + 3)$
16. $\dfrac{-4(6 - 3) + 8}{-4}$
17. $\dfrac{9 + 8 + 3}{6 + 5 - 1}$
18. $\dfrac{9 + 9 - 2}{5 + (18 - 15)}$
19. $\dfrac{9 \cdot 0 + 14}{(-2)(2) + 3 \cdot 2}$
20. $\dfrac{-25}{5 + 8 - (2 \cdot 4)}$
21. $6 - 3 \cdot 4 + 9(-7 + 8)$
22. $14 - (3 + 8[4(-7 - 3)])$

23. $4(8 - 5) + 6(9 - 5) + 2(6 - 3)$
24. $0(8 - 3) + 9(6 + 3 - 9) + 8(-1 + 3)$
25. $6(4 + 8 + 3) + (-7)(-3 + 10) - 5 \cdot 2 \cdot 4$
26. $9 \cdot 2(6 - 5) + 8[-5(3 - 4)] + 7$
27. $12([4 - 8] - [4 - (3 - 2)])$
28. $-9[-2 + 5(8 - 3)2]$
29. $\dfrac{-9 + 3}{2} + 8(4 - 7)$
30. $\dfrac{10 + 5}{3} - \dfrac{8 - 9(2 + 2)}{-7}$
31. $\dfrac{24 + 16 - 5}{-4 - 3} \cdot \dfrac{8 + 2}{2 + 3}$
32. $3\left(\dfrac{6 + 5}{-3 - 8}\right) + 5\left(\dfrac{9 - 6}{-1 - 2}\right)$
33. Multiply the sum of 8, 4, and 2 by 5.
34. Subtract 5 from the product of 3 and 10.
35. Add the sum of 6 and 4 to the product of -1 and 4.
36. Divide the product of 6 and -4 by the sum of 3 and 5.

1.7 POSITIVE INTEGRAL EXPONENTS

Let a be an arbitrary real number. Define

$$a^2 \quad \text{to mean} \quad a \cdot a.$$

Here

$$a^2$$

is read

a* squared** or **the square of *a or **the second power of *a***

or *a* to the second.

EXAMPLE 1.

(a) $5^2 = 5 \cdot 5 = 25$
(b) $40^2 = (40)(40) = 1600$
(c) $0^2 = 0$
(d) $(-3)^2 = (-3)(-3) = 9$ ■

In Example 1(d) you found the square of -3. On the other hand, by convention

$$-3^2 = -(3 \cdot 3) = -9$$

Here you determine the additive inverse of the square of 3.

DEFINITION. In general, let a be a real number. Then a^1 is defined to be a. If n

is an *integer* greater than 1,

$$a^n \quad \text{means} \quad \underbrace{a \cdot a \cdot a \ldots a}_{n \text{ times}}.$$

There are n factors of a. Call a^n

the nth **power of** a or a **to the** nth.

In particular, the third power of a, a^3, is also called

a **cubed** or **the cube of** a.

EXAMPLE 2.

(a) $(4+2)^2 = 6^2 = 36$
(b) $4 + 2^2 = 4 + 4 = 8$
(c) $4^2 + 2^2 = 16 + 4 = 20$
(d) $(4 \cdot 2)^2 = 8^2 = 64$
(e) $4 \cdot 2^2 = 4 \cdot 4 = 16$ ■

EXAMPLE 3.

(a) $2^1 = 2$
(b) $2^3 = 2 \cdot 2 \cdot 2 = 8$
(c) $2^5 = 2 \cdot 2 \cdot 2 \cdot 2 \cdot 2 = 32$ ■

EXAMPLE 4.

(a) $3^3 = 3 \cdot 3 \cdot 3 = 27$ (Thus 27 is the cube of 3.)
(b) $3^4 = 3 \cdot 3 \cdot 3 \cdot 3 = 81$
(c) $3^5 = 3 \cdot 3 \cdot 3 \cdot 3 \cdot 3 = 243$ ■

Observe that

$$3^3 \cdot 3^2 = (3 \cdot 3 \cdot 3)(3 \cdot 3) = 3^5.$$

Thus

$$3^3 3^2 = 3^5 = 3^{3+2}$$

DEFINITION. In the expression

$$a^n,$$

a is known as the **base** and n as the **exponent** (whereas a^n *is the* nth *power of* a).

In Example 4 (b), you found that

$$3^4 = 81.$$

Thus 3 is the *base*, 4 is the *exponent* (of 3), and 81 is the 4th *power of* 3. **Exponentiation** is the operation of determining a power of a number.

In the remaining examples exponentiation is combined with other arithmetic operations.

EXAMPLE 5.

$$\begin{aligned} 3^4 - 2^3 &= (3 \cdot 3 \cdot 3 \cdot 3) - (2 \cdot 2 \cdot 2) \\ &= 81 - 8 \\ &= 73 \end{aligned}$$

■

EXAMPLE 6. Find the value of $3 \cdot 4^3 + (2 \cdot 3)^2$.

SOLUTION. Only the 4 is cubed; the product of 2 and 3 is squared. Thus

$$\begin{aligned} 3 \cdot 4^3 + (2 \cdot 3)^2 &= 3(4 \cdot 4 \cdot 4) + 6^2 \\ &= 3 \cdot 64 + 36 \\ &= 192 + 36 \\ &= 228 \end{aligned}$$

■

Frequently, the central role in an algebraic expression is played by *powers of one number,* as in the final two examples.

EXAMPLE 7. Find the value of $2^5 - 3 \cdot 2^4 + 5 \cdot 2^3 - 2$.

SOLUTION.

$$\begin{aligned} 2^5 - 3 \cdot 2^4 + 5 \cdot 2^3 - 2 &= 32 - 3 \cdot 16 + 5 \cdot 8 - 2 \\ &= 32 - 48 + 40 - 2 \\ &= 22 \end{aligned}$$

■

EXAMPLE 8. Find the value of

$$\frac{3^3 - 2 \cdot 3^2 - 1}{3^2 + 2 \cdot 3^1}.$$

SOLUTION.

$$\begin{aligned} \frac{3^3 - 2 \cdot 3^2 - 1}{3^2 + 2 \cdot 3^1} &= \frac{27 - 2 \cdot 9 - 1}{9 + 2 \cdot 3} \\ &= \frac{27 - 18 - 1}{9 + 6} \\ &= \frac{8}{15} \end{aligned}$$

■

EXERCISES

In Exercises 1–44 find each value.

1. 4^3
2. 2^6
3. 8^1
4. $(-7)^2$
5. 0^{10}
6. 9^3
7. 10^4
8. $(-4)^3$
9. $(-2)^5$
10. 3^7
11. 2^8
12. 12^2
13. $(-6)^2$
14. -6^2
15. $(-6)^3$
16. -6^3
17. $(-30)^2$
18. 100^3
19. $(5 + 2)^2$
20. $5^2 + 2^2$
21. $5 + 2^2$
22. $(5 - 2)^2$
23. $5 - 2^2$
24. $5^2 - 2^2$
25. $(5 \cdot 2)^2$
26. $5 \cdot 2^2$
27. $4^2 + 3^3$
28. $(-2)^5 + 5^2$
29. $6^2 + 2(-3)^2$
30. $0^4 + 8^1 - 2^4$
31. $9^2 + 6 \cdot 3^3$
32. $4^4 - 3^3 + 2^2$
33. $4(3^2) - 3(2^4)$
34. $3(5^2) + 5(3^2)$
35. $2^5 - 4 \cdot 2^3 + 1$
36. $3^4 + (2 \cdot 3)^3 - 3^2 + 3^1$
37. $(-5)^3 - 4(-5)^2 + (-5)^1 - 2$
38. $7^2 + 3 \cdot 7^1 - 5$
39. $2^6 - 5 \cdot 2^5 + 3 \cdot 2^4 - 2^3 + 2^2 - 7 \cdot 2^1 + 1$
40. $2^7 - (3 \cdot 2)^4 + 7 \cdot 2^3 - 5 \cdot 2^2 + 3 \cdot 2^1 + 3$
41. $\dfrac{2^4 + 3 \cdot 2^2 + 2^1 - 1}{2^3 - 2^2 + 2^1 - 3}$
42. $\dfrac{5^3 + 3 \cdot 5^2 - 5^1 + 1}{5^2 - 5^1 + 3}$
43. $\dfrac{(-2)^4 + (-2)^3 - 3(-2)^2 + 1}{(-2)^3 - 5(-2)}$
44. $\dfrac{(-3)^4 + 7(-3)^2 - (-3)^1 + 3}{(-3)^3 - (-3)^2 + 1}$
45. Add the square of 3 and the cube of 2.
46. Subtract the cube of -2 from the product of 5 and the square of -1.
47. Divide the fourth power of 4 by the cube of 2.
48. Divide the square of 12 by three times the cube of 5.

REVIEW EXERCISES FOR CHAPTER 1

1.1 THE REAL LINE

1. Fill in "$<$" or "$>$".

   (a) -3 ▭ 0 (b) -6 ▭ 3

   (c) -9 ▭ -4 (d) $\dfrac{-3}{4}$ ▭ $\dfrac{-4}{3}$

2. Fill in "left" or "right".
   (a) On L, -8 lies to the ______ of -5.
   (b) On L, .01 lies to the ______ of .1 .
   (c) On L, $\frac{1}{4}$ lies to the ______ of $\frac{1}{2}$.
   (d) On L, $\frac{-1}{4}$ lies to the ______ of $\frac{-1}{2}$.
3. Rearrange the following numbers so that you can write "<" between any two numbers.

   $$\frac{1}{6}, \quad \frac{1}{3}, \quad \frac{2}{5}, \quad \frac{1}{2}, \quad \frac{1}{5}, \quad \frac{2}{3}$$

4. Rearrange the following numbers so that you can write ">" between any two numbers.

   -2.9, -3, -2.99, -2.91, -2.09, -2.009

1.2 ABSOLUTE VALUE

5. Determine the additive inverse of each number.
   (a) 5 (b) -5 (c) 0 (d) $|-5|$
6. Determine each absolute value.
   (a) $|12|$ (b) $|-4|$ (c) $|0|$ (d) $\left|\frac{-3}{4}\right|$
7. Fill in "<", ">", or "=".
   (a) $|-5|$ ______ 5 (b) $|2|$ ______ 2
   (c) $|.02|$ ______ $|-.2|$ (d) $|-8|$ ______ $|-3|$
8. (a) Which positive number is 9 units from the origin?
   (b) Which negative number is 6 units from the origin?

1.3 ADDITION AND SUBTRACTION OF TWO NUMBERS

9. Add: $-281 + 400$
10. Add: $\begin{array}{r} 5017 \\ \underline{9393} \end{array}$
11. *Subtract* the bottom number from the top one: $\begin{array}{r} 5024 \\ \underline{438} \end{array}$
12. *Subtract* the bottom number from the top one: $\begin{array}{r} -\ 6\,749 \\ \underline{23\,089} \end{array}$
13. A car travels 50 miles north of Tulsa. It then reverses its direction and travels 85 miles south. Where does the trip finish in relation to Tulsa?

1.4 ADDITIONAL SUBTRACTION OF THREE OR MORE NUMBERS

14. Determine the sum: $\begin{array}{r} -968 \\ -309 \\ -1292 \\ \hline \end{array}$

15. Determine the sum: $\begin{array}{r} 8083 \\ -974 \\ -939 \\ 594 \\ \hline \end{array}$

In Exercises 16–18 find the indicated values.

16. $43 + (-19) + (-34) + (-24)$

17. $39 - (24 - 15) - (7 - 12)$

18. $(84 - 19 + 7) - (4 - 29) + 6 - 9 - 13$

19. Subtract the sum of -27 and 43 from the sum of -19 and -39.

1.5 MULTIPLICATION AND DIVISION

In Exercises 20–24 multiply or divide, as indicated.

20. $7(-4)(-5)(-3)(-1)$

21. *Multiply:* $\begin{array}{r} -2043 \\ 405 \\ \hline \end{array}$

22. $\dfrac{-36}{-18}$

23. $-305\overline{)27\,450}$

24. $\dfrac{(-3)(-5)}{2 \cdot 4}$

25. The product of 25 and 36 is divided by -30. Determine the resulting quotient.

1.6 COMBINING OPERATIONS

In Exercises 26–28 perform the indicated operations.

26. $\dfrac{6 - 5 \cdot 2}{4 + 7 \cdot 3}$

27. $(3 - 5)(15 - 3)$

28. $-4(5 - 3 - 6) + 12[-2 - 3(1 - 7)] - 3(1 - 4)$

29. Multiply the sum of -2 and -5 by the sum of 8 and -3.

1.7 POSITIVE INTEGRAL EXPONENTS

In Exercises 30–33 evaluate each expression.

30. $(-7)^2$

31. -7^2

32. $2^8 - 2^6 + 3^2 - 3^4$

33. $\dfrac{4^2 - 3^2}{5(10^2) + (5 \cdot 10)^2}$

34. Add the square of -2 and the cube of -3.

TEST YOURSELF ON CHAPTER 1.

1. Rearrange the following numbers so that you can write “<” between any two numbers.

   $$-3.7, \qquad -3.8, \qquad -3.78, \qquad -3.77, \qquad -3.87, \qquad -3.88$$

2. Determine each absolute value.

   (a) $|-1.4|$ (b) $|0|$ (c) $\left|\frac{1}{2}\right|$

3. Determine all real numbers that are 8 units from the origin.

4. Add:
   $$\begin{array}{r} 444 \\ -609 \\ 3217 \\ \underline{-599} \end{array}$$

In Exercises 5–11 evaluate each expression.

5. $6 - [-1 - (3 + 8) + 2] - (4 - 3)$

6. *Multiply:* $\begin{array}{r} -414 \\ \underline{-873} \end{array}$

7. $-914\overline{)3656}$

8. $\dfrac{12 - 4 \cdot 2}{(3 - 1)2}$

9. $(10 - 4)(5 - 8)$

10. $(-2)(-3)(-2)(-1)(-4)$

11. $\dfrac{3^2 - 5^2}{(3 - 5)^2}$

12. Multiply the sum of 6 and -2 by the square of 10.

2

polynomials

2.1 TERMS AND POLYNOMIALS

In algebra, symbols are used to represent numbers.

DEFINITION. A **constant** is a symbol or a combination of symbols that designates a *specific* number.

The familiar symbols

$$0, \quad 1, \quad 2, \quad -1, \quad \frac{1}{2}, \quad \sqrt{2}, \quad \pi$$

are constants. Each of these symbols designates a *specific* number. Sometimes a letter, such as c or K, will also be used as a constant to designate a *specific* number.

DEFINITION. A **variable** is a symbol that represents *any* one of a given collection of numbers.

The letters x, y, z, and t will be used as variables; other letters will also be used.

Sometimes a variable, such as x, will represent *any* real number. However, a variable can also be used to stand for only *some* of the real numbers. For example, a variable can stand only for nonzero numbers or only for positive integers.

EXAMPLE 1.

(a) When you say,

"The sum is 9"

the symbol

9

is used to designate a *specific* number. Thus

9

is a constant.[1]

(b) When you say,

"The product is -6"

the combination of symbols

-6

designates a *specific* number, the inverse of 6. Therefore

-6

is a constant.

(c) When you say,

"Let x be a real number"

the letter "x" is used as a variable. It is understood that x can be *any real number*.

(d) When you say,

"Let y be a positive integer"

the letter "y" is used as a variable. It represents *any one of the integers* 1, 2, 3, 4, *and so on.* ■

EXAMPLE 2. Each of the expressions

$$4 + 4, \qquad 4 \cdot 2, \qquad \frac{16}{2}$$

designates the number 8; each of these expressions is a constant. ■

[1]In ordinary usage, you do not distinguish between a number and the constant that designates (or names) it. Thus you say

the number 6

when, in fact, you mean

the number designated by 6.

Sometimes it is convenient to attach **subscripts** to letters. For example, in a problem involving several positive and negative numbers you might distinguish between them by letting

$p_1, p_2, p_3,$ and so on read: p-sub-1, p-sub-2, p-sub-3, and so on

represent the *positive* numbers and

$n_1, n_2, n_3,$ and so on

represent the *negative* numbers. Here, $p_1, p_2,$ and p_3 are regarded as being *different symbols,* just as x, y, and z are different symbols. You may also write

a' read: a-prime

and consider this different than

$a.$

Because constants and variables represent numbers, you can add, subtract, multiply, and divide them. The quantities to be added, multiplied, and so on, can be either constants or variables. You can also consider exponentiation involving variables. You will consider such "algebraic expressions" as

$$x + 4, \qquad x + y, \qquad \frac{x-1}{x+2}, \qquad x^3, \qquad x^4 - 3x^2 + 4, \qquad 2xy.$$

The arithmetic rules discussed in Chapter 1 apply to these more general expressions. Thus you can combine expressions, as for example,

$$(x + 2) + 3 = x + (2 + 3) = x + 5$$

$$2(x + y) = 2x + 2y, \text{ and so on.}$$

The building blocks of algebraic expressions are *terms.*

DEFINITION. A **term** is either

a constant,

a variable,

or

a *product* of constants and variables.

EXAMPLE 3. Each of the following is a term:

(a) -2 (This is a constant.)

(b) $\frac{1}{3}$ (This is a constant.)

(c) 0 (This is a constant.)

(d) π (This is a constant.)

(e) x (This is a variable.)
(f) $5x$ (This is the product of the constant 5 and the variable x.)
(g) $-3y$ (This is the product of the constant -3 and the variable y.)
(h) xyz (This is the product of the variables x, y, and z.) ■

EXAMPLE 4. Each of the following is a term:

(a) x^2 $x^2 = x \cdot x$
(b) y^6 $y^6 = y \cdot y \cdot y \cdot y \cdot y \cdot y$
(c) x^2y^3 $x^2y^3 = x \cdot x \cdot y \cdot y \cdot y$
(d) $(3x)(2x)$ ■

You can simplify (d) by means of the Associative and Commutative Laws of multiplication. Thus

$$(3x)(2x) = (3 \cdot 2)(x \cdot x) = 6x^2$$

EXAMPLE 5.

(a) The expression

$$\frac{x}{4}$$

is a term because

$$\frac{x}{4} = \frac{1}{4}x.$$

Thus $\frac{x}{4}$ is the product of the constant $\frac{1}{4}$ and the variable x.

(b) Neither of the expressions

$$x + y \qquad \text{nor} \qquad \frac{x}{y}$$

is a term. The definition of a term permits neither addition nor division of *variables*. Note that $\frac{x}{4}$, with the *constant* 4 as divisor, is a term, whereas $\frac{x}{y}$, with the *variable* y as divisor, is not a term. ■

A term can be expressed in various ways. Thus, as you will see in Section 2.4, each of the expressions

$$(4x)(3x), \qquad 6(2x^2), \qquad 2(2x)(3x) \qquad \text{can be } \textit{simplified} \text{ to} \qquad 12x^2.$$

Note that in each of these expressions, the product of all occurrences of constants is 12.

DEFINITION. The **(numerical) coefficient of a term** is the single occurrence of a constant when the term has been simplified. If only variables occur, the coefficient of the term is understood to be 1 or -1.

EXAMPLE 6.

(a) The coefficient of $4x^2$ is 4.
(b) The coefficient of $-3xy$ is -3.
(c) The coefficient of 10 is 10.
(d) The coefficient of $(3x)(2y)$ is 6.
(e) The coefficient of $(3 + 2)x$ is 5.
(f) The coefficient of x is understood to be 1.
(g) The coefficient of $-x^2$ is understood to be -1.
(h) The coefficient of $\frac{x}{2}$ is $\frac{1}{2}$.
(i) If a term has 0 as its coefficient, the term reduces to 0. For, when you multiply by 0, the product is always 0. Thus, $3 \cdot 0x^2 = 0$ and $0bc^3 = 0$ ■

DEFINITION. A **polynomial** is defined to be either

a term

or

a sum of terms.

EXAMPLE 7.

(a) x^2, $3x$, xy, and π are each nonzero terms; each of these is a polynomial.
(b) $x^2 + 3x$ is a polynomial. It is the sum of the terms x^2 and $3x$.
(c) $x^2 + xy + \pi$ is a polynomial in 2 variables, x and y. It is the sum of the terms x^2, xy, and π. ■

EXAMPLE 8. Each of the following is a polynomial:

(a) $5xy + 2xz$
(b) $y^8 - 7y^6 + 3y^2 - y + \frac{1}{2}$
(c) $x_1{}^2 + x_2$ read: x-sub-1 squared plus x-sub-2
(d) $x^4 + y^3 - z^2$
(e) $\frac{3x^2 + 1}{3}$, with the *constant* 3 as divisor, is a polynomial.

For, as you will see,

$$\frac{3x^2 + 1}{3} = \frac{3}{3}x^2 + \frac{1}{3} = x^2 + \frac{1}{3}$$ ■

EXAMPLE 9. $\frac{3x^2 + 1}{x + 3}$, with the *nonconstant polynomial* $x + 3$ as divisor, is not a polynomial. It is, rather, the quotient of polynomials, and is known as a *rational expression.* Rational expressions will be discussed in Chapters 4 and 5. ■

EXERCISES

Which expressions in Exercises 1–12 are terms?

1. $9x$
2. $-3x_1x_2y_1y_2$
3. -1
4. $x + 1$
5. a^2
6. $2x^2y^3$
7. $(x - 1)^2$
8. $\frac{m}{x}$
9. $\frac{x + 1}{2}$
10. $\frac{x^2}{2}$
11. 0
12. $200rst^2x^2yz$

In Exercises 13–24 determine the coefficient of each term.

13. $15x$
14. $-19xy^2$
15. $\frac{z}{3}$
16. 7
17. -7
18. $\frac{-x}{7}$
19. 0
20. πxy
21. $(5x)(8y)$
22. $\frac{2x_1x_2x_3}{3}$
23. $x^2 \cdot \frac{y^3}{3}$
24. $x^2(0 \cdot y)$

In Exercises 25–28 rewrite each term so that its coefficient appears first. Also, simplify this coefficient.

25. $(5x)(2y)$
26. $\frac{x^2}{2}$
27. $x(-y)$
28. $3x^2 \cdot 5y \cdot 6z^2$

Which expressions in Exercises 29–38 are polynomials?

29. $2xy + x^2$
30. $\frac{-1}{x^2}$
31. $\frac{x^2}{5}$
32. $\frac{x^2}{5y}$
33. $\frac{x + 3}{7}$
34. $x^3 + yz + z^3$
35. $x^4 + 3x^2 + 5x - 7 + \frac{1}{x}$
36. π
37. $\frac{x^2 + x}{3}$
38. $\frac{3}{x^2 + x}$

2.2 NUMERICAL EVALUATION

Frequently, you evaluate a (nonconstant) polynomial for specific values of the variable (or variables). To do this, replace the variables by these numbers. If a variable occurs more than once, each time it occurs, replace it by the same number.

EXAMPLE 1. Evaluate $x^2 - 3x + 2$ when $x = 2$

SOLUTION. Replace each occurrence of x by 2, and obtain

$$2^2 - 3 \cdot 2 + 2 = 4 - 6 + 2 = 0.$$ ■

EXAMPLE 2. Evaluate

$$x^3 - 5x^2 + 3x - 2$$

for the following values of x:

(a) 1 (b) -1

SOLUTION.

(a) When $x = 1$, the polynomial

$$x^3 - 5x^2 + 3x - 2$$

becomes

$$1^3 - 5(1)^2 + 3 \cdot 1 - 2 = 1 - 5 + 3 - 2 = -3.$$

(b) $x = -1$:

$$(-1)^3 - 5(-1)^2 + 3(-1) - 2 = -1 - 5 - 3 - 2 = -11$$ ■

EXAMPLE 3. Evaluate $x^2 - 5xy + 8y^2$ when $x = 3$ and $y = 4$.

SOLUTION. Replace each occurrence of x by 3 and each occurrence of **y** by **4:**

$$3^2 - 5(3)(\mathbf{4}) + 8(\mathbf{4})^2 = 9 - 60 + 128 = 77$$ ■

EXAMPLE 4. Evaluate $x^3yz + 4(y^2 - z^2)$ when $x = -1,\ y = 2,\ z = 10$.

SOLUTION. The polynomial becomes

$$\begin{aligned}(-1)^3(2)(10) + 4[(2)^2 - (10)^2] &= -20 + 4[4 - 100]\\ &= -20 - 384\\ &= -404\end{aligned}$$ ■

EXAMPLE 5. Evaluate $t_1^2 - 4t_2$ when $t_1 = 3$ and $t_2 = -2$.

SOLUTION. Replace t_1 by 3 and $\mathbf{t_2}$ by **−2:**

$$3^2 - 4(\mathbf{-2}) = 9 + 8 = 17$$ ■

EXAMPLE 6. The area of a circle is given by

$$\pi r^2$$

where r is the length of the radius. [See Figure 2.1.] Determine the area when

(a) r is 4 inches, (b) r is 9 inches.

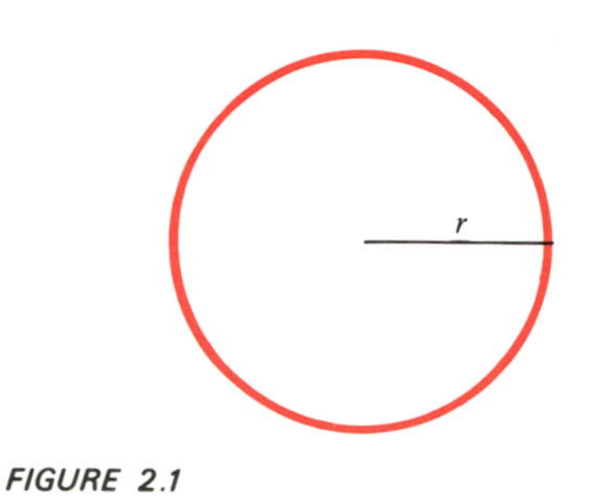

FIGURE 2.1

SOLUTION.

(a) The area is $\pi \cdot 4^2$, or 16π, square inches.
(b) The area is $\pi \cdot 9^2$, or 81π, square inches. ■

EXERCISES

In Exercises 1–12 evaluate each polynomial for the specified value of the variable.

1. $x + 10 \quad [x = 7]$
2. $3x - 5 \quad [x = -1]$
3. $5y^2 - 7y + 1 \quad [y = 2]$
4. $3z^2 + z - 11 \quad [z = 0]$
5. $5x^2 - x + 9 \quad [x = 10]$
6. $x^3 - x^2 + 3x + 2 \quad [x = 1]$
7. $x^3 + x + 1 \quad [x = -1]$
8. $\frac{1}{3}(t^3 + 3) \quad [t = 3]$
9. $x^4 - x^2 + x + 5 \quad [x = -2]$
10. $(a^2 + 5a + 3) \quad [a = 10]$
11. $x^8 - x^6 + 5x^3 - 1 \quad [x = -1]$
12. $3x^5 + x^3 - x^2 + 2x - 2 \quad [x = 2]$
13. Evaluate $x + 8$ when
    (a) $x = 2$, (b) $x = -2$, (c) $x = 92$.
14. Evaluate $x^2 + 4$ when
    (a) $x = 0$, (b) $x = -1$, (c) $x = 4$.
15. Evaluate $y^2 - 2y + 5$ when
    (a) $y = 3$, (b) $y = -5$, (c) $y = 100$.
16. Evaluate $z^4 - 3z + 1$ when
    (a) $z = -2$, (b) $z = 3$, (c) $z = 10$.
17. Evaluate $t^3 + 10t^2 - t + 1$ when
    (a) $t = 10$, (b) $t = -10$, (c) $t = 100$.
18. Evaluate $x^7 - x^6 + x^2$ when
    (a) $x = 1$, (b) $x = -1$, (c) $x = 10$.

In Exercises 19–32 evaluate each polynomial for the specified values of the variables.

19. $x^2 - 2y \quad [x = 1, y = 2]$

20. $2x + 3xy - 2y$ $[x = 2, y = 1]$
21. $(2x)^2 + 3y^2$ $[x = -2, y = 4]$
22. $3(x - y) + 2$ $[x = 3, y = 4]$
23. $5(x - y) - 4(y - x)$ $[x = 6, y = 0]$
24. $1 - x_1x_2$ $[x_1 = 2, x_2 = -2]$
25. $-y_1^2 + (-y_2)^2 + 1$ $[y_1 = 3, y_2 = -3]$
26. $4t^2 + 2st - 3s + s^2$ $[s = 1, t = 2]$
27. $10a^3 + 3a^2b - 3ab^2 + b^3$ $[a = 2, b = 3]$
28. $u^5 - 4uv + v^3$ $[u = 2, v = -1]$
29. $(m - n) - (m + n)$ $[m = 1, n = 4]$
30. $a^2bc + 4ab^2c^3$ $[a = 5, b = 2, c = 3]$
31. $3x^2y + 2xz - yz^2$ $[x = 4, y = -3, z = 1]$
32. $2xyz - 3xy + xz - 10yz$ $[x = -2, y = -1, z = 4]$
33. Evaluate $2xy + 3y^2$ when
    (a) $x = 2, y = -1$; (b) $x = 3, y = -2$.
34. Evaluate $1 - 2x^2y + y^2$ when
    (a) $x = -1, y = -2$; (b) $x = 2, y = 3$.
35. Evaluate $3abc + 2a^2 - b^2$ when
    (a) $a = 4, b = 2, c = -2$; (b) $a = 1, b = 2, c = -1$.
36. Evaluate $m^2 - 2mn + n^2 - t^2$ when
    (a) $m = 8, n = 6, t = 10$; (b) $m = -3, n = 0, t = 9$.
37. The area of a square is given by s^2, where s is the length of a side. Determine the area when s is 11 feet. [See Figure 2.2.]

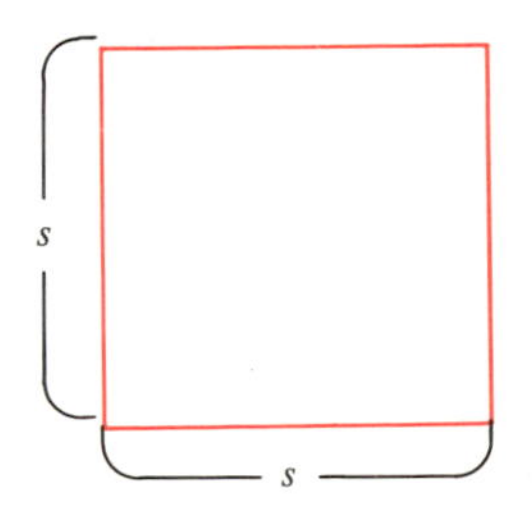

FIGURE 2.2

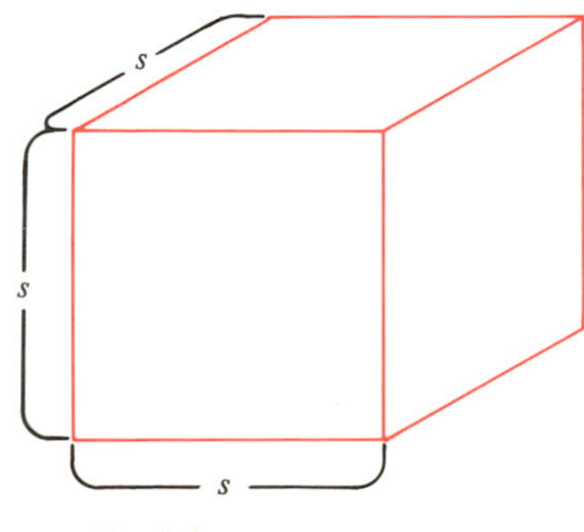

FIGURE 2.3

38. The volume of a cube is given by s^3, where s is the length of a side. Determine the volume when s is 10 feet. [See Figure 2.3.]
39. The surface area of a box without a top is given by

$$lw + 2lh + 2wh,$$

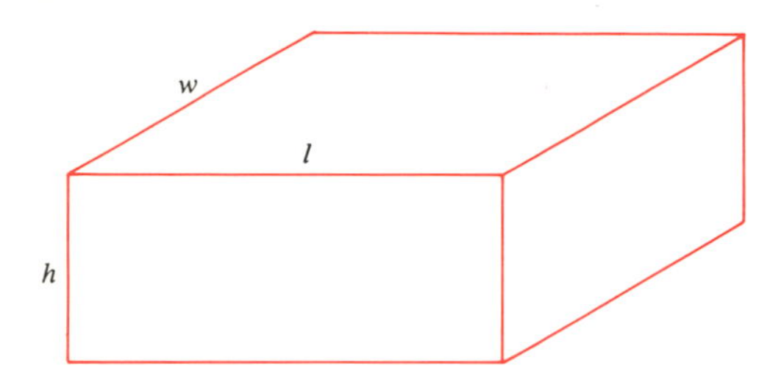

FIGURE 2.4

where l is the length of the base, w is the width of the base, and h is the height of the box. Determine the surface area when l is 25 inches, w is 10 inches, and h is 6 inches. [See Figure 2.4 .]

40. The distance (in miles) traveled by a car going at a constant rate is given by rt, where r is the number of miles per hour and t is the number of hours traveled. Determine the distance traveled if $r = 60$ and $t = 4$.

41. Suppose that an object in motion travels a distance given by

$$v_0(A + Bt),$$

where v_0 is its initial velocity and t is its time in motion. Determine the distance traveled when v_0 is 200 feet per second, A is 9 seconds, B is 13 , and t is 7 seconds.

42. The circumference of a circle is given by $2\pi r$, where r is the length of the radius. Determine the circumference when r is 5 inches.

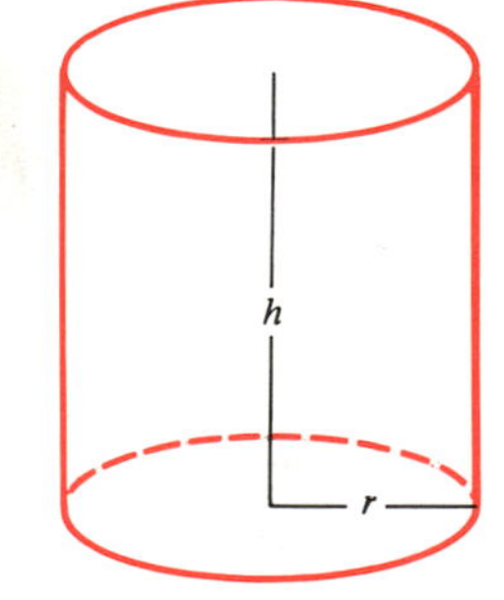

FIGURE 2.5

43. The volume of a right circular cylinder is given by $\pi r^2 h$, where r is the length of the base radius and h is the length of the altitude. Determine the volume when r is 10 inches and h is 12 inches. [See Figure 2.5.]

44. The lateral surface area of a right circular cylinder is given by $2\pi rh$, where r and h are as in Exercise 43. Determine the lateral surface area when r is 10 inches and h is 12 inches. [See Figure 2.5.]

2.3 ADDITION AND SUBTRACTION

The term

$$x^3y^4x^2$$

can be simplified to

$$x^5y^4,$$

because, as you will see in Section 2.4,

$$\begin{aligned} x^3y^4x^2 &= x^3x^2y^4 \\ &= xxx \cdot xx \cdot y^4 \\ &= x^5y^4. \end{aligned}$$

DEFINITION. Let x be a variable that occurs in a nonzero term. Then n is called the **exponent of** x if x^n is the highest power of x that occurs when the term has been simplified.

EXAMPLE 1. Consider the term $3x^2zyz^3$.

(a) The exponent of x is 2.
(b) The exponent of y is 1.
(c) The exponent of z is 4.

Note that $3x^2zyz^3 = 3x^2yz^4$.

■

DEFINITION. Two or more terms are said to be **similar** if

1. they contain precisely the same variables

and

2. each variable has the same exponent in each term.

Thus *similar terms can differ only in their numerical coefficients or in the order of their variables.*

EXAMPLE 2.

(a) $2xy$ and $-3xy$ are similar.
(b) $5x^2z$ and zx^2 are similar.
(c) $2zx^2$, $2x^2z$, and $2xzx$ are similar. (In fact, they are equal.)
(d) x^2y and xy are dissimilar (not similar) because the two terms differ in the exponent of x.
(e) Any two constant terms are similar because neither contains any variables.

■

Similar terms can be added and subtracted by extending the Distributive Laws. For example, let a and b be constants and let x be a variable. Then

$$ax + bx = (a + b)x$$ **Addition of similar terms**

$$-(bx) = (-b)x$$ **the inverse of the term bx**

so that

$$\begin{aligned} ax - bx &= ax + [-(bx)] \\ &= ax + [(-b)x] \\ &= [a + (-b)]x \\ &= (a - b)x \end{aligned}$$ **Subtraction of similar terms**

EXAMPLE 3.

$$10x + 3x = (10 + 3)x = 13x$$ ■

EXAMPLE 4.

$$7x^2y - 5x^2y = (7 - 5)x^2y = 2x^2y$$ ■

Terms that are not similar cannot be combined in the above manner.

EXAMPLE 5. $5x + 3y$ cannot be further simplified. For, $5x$ and $3y$ are not similar. The Distributive Laws cannot be used, as in Examples 3 and 4. ■

The Commutative and Associative Laws apply to the addition and subtraction of terms. Thus when several terms are to be combined, arrange these so that similar terms are grouped together.

EXAMPLE 6.

$$\begin{aligned} 3x + 7y + 5x &= 3x + (7y + 5x) \\ &= 3x + (5x + 7y) && \text{by the Commutative Law} \\ &= (3x + 5x) + 7y && \text{by the Associative Law} \\ &= 8x + 7y \end{aligned}$$ ■

When there are more terms involved, you may prefer to list similar terms in the same column, as in Example 7.

EXAMPLE 7. Simplify:

$$5x + 3y - 2z + 5x - 2y + z$$

SOLUTION. Arrange the terms in 3 columns:

$$\begin{array}{r} 5x + 3y - 2z \\ 5x - 2y + z \\ \hline 10x + y - z \end{array}$$ ■

From now on, you will speak of a polynomial as a sum of terms, even when it contains just one term.

Polynomials are added in the same way as terms are. Thus to **add polynomials,** arrange their terms so that similar terms are grouped together. Write

$$P + Q$$

for the sum of the polynomials P and Q.

EXAMPLE 8. Let

$$\begin{aligned} P &= 4x + 3y - 5z, \\ Q &= 2x - 4y + 7z, \\ R &= -4x - 3y + 5z. \end{aligned}$$

Determine:

(a) $P + Q$ (b) $P + R$

SOLUTION.

(a)
$$\begin{array}{rr} P: & 4x + 3y - 5z \\ Q: & \underline{2x - 4y + 7z} \\ P + Q: & 6x - y + 2z \end{array}$$

(b)
$$\begin{array}{rr} P: & 4x + 3y - 5z \\ R: & \underline{-4x - 3y + 5z} \\ P + R: & 0 \end{array}$$

■

Example 8(b) suggests the following definition.

DEFINITION. Let P be a polynomial. Then the **inverse of** P is the polynomial added to P to obtain 0.

Write

$$-P$$

for the inverse of P. Thus

$$P + (-P) = 0$$

As you see from Example 8(b), *to obtain* $-P$, *replace the coefficient of each term of* P *by its inverse.*

EXAMPLE 9.

$$-(-3 + x^2 + y - 2z) = 3 - x^2 - y + 2z$$ ■

To **subtract polynomials,** define

$$P - Q \quad \text{to be} \quad P + (-Q).$$

EXAMPLE 10. Subtract $x^2 + 3xy - y^2$ from $2x^2 + 6xy + 5y^2$.

SOLUTION. You want

$$2x^2 + 6xy + 5y^2 - (x^2 + 3xy - y^2)$$

or

$$2x^2 + 6xy + 5y^2 + [-(x^2 + 3xy - y^2)].$$

The inverse of the polynomial $x^2 + 3xy - y^2$ is obtained by replacing each numerical coefficient by its inverse:

$$\begin{array}{r} 2x^2 + 6xy + 5y^2 \\ \overset{-}{+}\ x^2 \overset{-}{+} 3xy \overset{+}{-}\ \ y^2 \\ \hline x^2 + 3xy + 6y^2 \end{array}$$ ■

The Commutative and Associative Laws apply to the addition of polynomials: Let P, Q, R be polynomials.

$$P + Q = Q + P \qquad \textbf{(Commutative Law)}$$

$$(P + Q) + R = P + (Q + R) \qquad \textbf{(Associative Law)}$$

Because of the Associative Law, you may omit parentheses, and write

$$P + Q + R.$$

EXAMPLE 11. Find the sum $P + Q + R$, where

$$P = 5x^2y^2 + 8x^2y + 3xy, \qquad Q = x^2y^2 - x^2y + 2xy^2,$$

$$R = 4x^2y + 6xy^2.$$

SOLUTION. Rearrange the terms so that similar terms are in the same column.

$$\begin{array}{rllll} P: & 5x^2y^2 & + 8x^2y & & + 3xy \\ Q: & x^2y^2 & - x^2y & + 2xy^2 & \\ R: & & 4x^2y & + 6xy^2 & \\ \hline P + Q + R: & 6x^2y^2 & + 11x^2y & + 8xy^2 & + 3xy \end{array}$$

■

EXAMPLE 12.

Simplify: $5x - [3 - (4x - 5)]$

SOLUTION. As in Section 1.4, *begin with the innermost parentheses and work outward.*

$$\begin{aligned} 5x - [3 - (4x - 5)] &= 5x - (3 - 4x + 5) \\ &= 5x - 3 + 4x - 5 \\ &= (5x + 4x) + (-3 - 5) \\ &= 9x - 8 \end{aligned}$$

■

As you develop facility in handling algebraic expressions, you may wish to skip elementary steps. For instance, in Example 12, you can combine steps, and write

$$5x - [3 - (4x - 5)] = 5x - 3 + 4x - 5 = 9x - 8.$$

Finally, a *nonzero* term is also called a **monomial.** The sum of two *dissimilar* monomials is called a **binomial.** The sum of three (mutually) *dissimilar* terms is called a **trinomial.** Thus $4x$ is a monomial, $2x + 5y$ is a binomial, and $x + y - z$ is a trinomial. Note that $x + 2x$ is a monomial because it is the sum of *similar* monomials.

EXERCISES

In Exercises 1–6 indicate which pairs of terms are similar.

1. $10z$ and $-5z$
2. $3xy$ and $4xz$
3. x^2y^3 and $-x^2y^3$
4. xzy and $4yxz$
5. x^3y^2 and x^2y^3
6. 19 and 35

In Exercises 7–12 group similar terms together.

7. $-xy,\ x,\ -x,\ 2xy,\ 3x$ *Answer:* $-xy,\ 2xy \quad | \quad x,\ -x,\ 3x$
8. $x^2y^2,\ -2x^2y,\ 3xy,\ -9x^2y^2,\ 5x^2y$
9. $14,\ x^2,\ x,\ -x^2,\ \frac{1}{2}x,\ \frac{1}{2}$
10. $2xy,\ -4xz,\ 3x^2z,\ 5zx,\ 6xz$
11. $-r^2st,\ rs^2t,\ 2r^2ts,\ 4rst^2,\ 3s^2rt$
12. $5xyz,\ 2xz,\ 3xzy,\ x^2zy,\ 5zy,\ -xyz,\ -2yz$

In Exercises 13–28 simplify wherever possible.

13. $5a + 4a$
14. $x - 6x$
15. $-7x - 2x$
16. $-y^2 + y^2$
17. $9b + 3b + 2b$
18. $y - 2y + 3y$
19. $3xy - 4xy - 6xy$
20. $ab^2 - ab^2 + ab^2$
21. $a + b - a + b + 1$
22. $3x^3 - 2x^2 + x^3 + x^2$
23. $4c - 4d + 4c - d$
24. $s^2 + t^2 - 2s^2 + t^2 + 2t^2$
25. $5r - (2s + 3s)$
26. $3m + 2m + (4m - n)$
27. $2xy - 3xy - (3xy + 2xy)$
28. $x^2 + 2y^2 - (3z^2 + 4x^2 - 5z^2)$

In Exercises 29–34 add the polynomials.

29. $$\begin{array}{l} 2a + 3b \\ \underline{4a + b} \end{array}$$

30. $$\begin{array}{l} a - b + c \\ \underline{2a - c} \end{array}$$

31. $$\begin{array}{l} 2xy - 3xz + yz \\ \underline{-2xy + 3xz + 4yz} \end{array}$$

32. $$\begin{array}{l} w - 3y \\ x + y + z \\ \underline{2w + 2y + z} \end{array}$$

33. $$\begin{array}{l} x^2 - 2xy + y^2 + z \\ - 3xy - y^2 \\ 3x^2 - xy - 2y^2 - 3z \\ \underline{2x^2 + y^2 + z} \end{array}$$

34. $$\begin{array}{l} 5r + s - t \\ s + t - u + v \\ 15r + 3t - v \\ \underline{-3r - t + u + v} \end{array}$$

In Exercises 35–40 *subtract* the bottom polynomial from the top one.

35. $$\begin{array}{l} 3x + 4y \\ \underline{2x + 2y} \end{array}$$

36. $$\begin{array}{l} 16a + 4b \\ \underline{12a + 4b} \end{array}$$

37. $$\begin{array}{l} 3x - 2y + z \\ \underline{2x - 2y - z} \end{array}$$

38. $$\begin{array}{l} a + b - c + 1 \\ \underline{-a - 2b + c - 3} \end{array}$$

39. $$\begin{array}{l} m - n + r - 2s \\ \underline{ 3n - s + t} \end{array}$$

40. $$\begin{array}{l} x^2 - y^2 + z^2 - xyz \\ \underline{ - 2y^2 + xyz} \end{array}$$

In Exercises 41–46 determine the inverse of each polynomial.

41. $2a$
42. $-5b$
43. $2a + 5b$

44. $3a^4 - 2a^2 + 4a$ 45. $-abc + ab - ac + 4$ 46. $x^3y - x^2y + xy - y + 3$

In Exercises 47–54 simplify each polynomial.

47. $a + 2b - [3a - (2a + b)]$
48. $x - (y + z) + (y - z) - (y - z)$
49. $m + (2n - r) - [(2n - 4) - m]$
50. $x^2 - (y^2 - z^2) + [x^2 - (y^2 - z^2)]$
51. $3a - [b - (2a + 3c) - (2a + b)]$
52. $5 + (2s - [3t + (2t - s)])$
53. $2a + 1 - [4b - (3 + 2a - 1)]$
54. $(2a - 3b) - (4a - 3b) - [(4a - 3b) - 2b]$
55. Add $2x + 3y - z$ to the sum of $3x - z$ and $2y + z$.
56. Subtract $2x - 3y$ from $3x - 2y$.
57. Subtract 0 from x^3.
58. Subtract x^2 from 0.
59. Subtract the sum of $3x$ and $5z$ from $2z$.
60. Evaluate $(x^2 + 5x + 1) - (x^2 + 2x - 2)$ when $x = 4$.
61. Evaluate $2x - (3x + 2y) + 8xy$ when $x = 2$ and $y = 3$.
62. Evaluate $x^3 - [x^2y + 2x - (3x + 5y)]$ when $x = -1$ and $y = -2$.

2.4 MULTIPLICATION

First note that

$$2^2 = 4 \quad \text{and} \quad 2^3 = 8.$$

Thus

$$2^2 \cdot 2^3 = 4 \cdot 8 = 32$$

Now observe that

$$32 = 2^5.$$

In other words,

$$2^2 \cdot 2^3 = 2^5 = 2^{2+3}$$

Here is an example of multiplying powers of the same *variable*.

EXAMPLE 1. Multiply a^4 by a^2.

SOLUTION. Observe that if you write out the factors, you obtain

$$a^4 \cdot a^2 = (aaaa)(aa) = aaaaaa = a^6.$$

Thus

$$a^4 \cdot a^2 = a^{4+2} = a^6$$

■

Let m and n be positive integers. *To multiply two powers of the same variable,*

$$a^m \cdot a^n,$$

write down the base a and add the exponents, m + n:

$$a^m \cdot a^n = a^{m+n}$$

This is because

$$a^m = \underbrace{a \cdot a \cdot a \ldots a}_{m \text{ times}} \qquad \text{and} \qquad a^n = \underbrace{a \cdot a \cdot a \ldots a}_{n \text{ times}}$$

Thus

$$a^m \cdot a^n = (\underbrace{a \cdot a \cdot a \ldots a}_{m \text{ times}})(\underbrace{a \cdot a \cdot a \ldots a}_{n \text{ times}}) = \underbrace{a \cdot a \cdot a \ldots a}_{m+n \text{ times}}$$

$$= a^{m+n}$$

EXAMPLE 2.

$$x^{10} \cdot x^7 = x^{17}$$

■

The above rule can be extended to three or more factors. For example, if r is a positive integer,

$$a^m \cdot a^n \cdot a^r = a^{m+n+r}$$

EXAMPLE 3.

$$y^5 \cdot y^3 \cdot y = y^{5+3+1} = y^9$$

■

To multiply monomials, use the Associative and Commutative Laws.

1. Group all numerical coefficients at the beginning.
2. Group powers of the same variable together.

Multiply the coefficients. Also multiply powers of the same variable according to the above rules.

EXAMPLE 4.

$$\begin{aligned}(2xy^3)(5x^4y^2) &= (2 \cdot 5)(x \cdot x^4)(y^3 \cdot y^2)\\ &= 10x^{1+4}y^{3+2}\\ &= 10x^5y^5\end{aligned}$$

■

To multiply two polynomials (not both monomials), first utilize the Disbributive Laws.

EXAMPLE 5.

$$\begin{aligned} 5a(a + b) &= (5a)a + (5a)b \\ &= 5(aa) + 5(ab) \\ &= 5a^2 + 5ab \end{aligned}$$ ■

EXAMPLE 6. Multiply $2x + y$ by $3x + y^2$.

SOLUTION. According to the Distributive Laws,

$$\begin{aligned} (2x + y)(3x + y^2) &= (2x + y)(3x) + (2x + y)(y^2) \\ &= (2x)(3x) + y(3x) + (2x)(y^2) + y(y^2) \\ &= 6x^2 + 3xy + 2xy^2 + y^3 \end{aligned}$$ ■

Thus multiply each term of the first polynomial by each term of the second polynomial. The resulting polynomial is the sum of these products.

In practice, the following method is frequently preferred. Similar terms are placed in the same column, as in Example 7, which follows. These similar terms are then added in the last step.

EXAMPLE 7. Multiply $a^2 + 3ab - b^2$ by $a + 2b$.

SOLUTION.

$$\begin{array}{llll} a^2 + 3ab - b^2 & & & \\ \underline{a \;\; + 2b} & & & \\ a^3 + 3a^2b & - \;\; ab^2 & & \\ & \underline{2a^2b + 6ab^2 - 2b^3} & & \\ a^3 + 5a^2b & + 5ab^2 - 2b^3 & & \end{array}$$ ■

A product that you will frequently encounter is

$$(x + a)(x + b).$$

Of course, the letters may change. For example, you may see

$$(y + m)(y + n).$$

You will often see numbers substituted for a and b, as in

$$(x + 5)(x + 3),$$

where $a = 5$ and $b = 3$.

$$(x + a)(x + b) = xx + ax + xb + ab$$
$$= x^2 + ax + bx + ab = x^2 + (a + b)x + ab$$

Therefore

$$(x + a)(x + b) = x^2 + (a + b)x + ab$$

EXAMPLE 8. Multiply $x + 5$ by $x + 3$.

SOLUTION 1.

$$(x + 5)(x + 3) = x^2 + (5 + 3)x + 5 \cdot 3$$
$$= x^2 + 8x + 15$$

SOLUTION 2.

$$\begin{array}{r} x + 5 \\ x + 3 \\ \hline x^2 + 5x \\ 3x + 15 \\ \hline x^2 + 8x + 15 \end{array}$$ ■

Let $b = a$. Then

$$(x + a)(x + b) = (x + a)(x + a) = (x + a)^2$$

and

$$(x + a)^2 = x^2 + (a + a)x + aa = x^2 + 2ax + a^2$$

Let $b = -a$. Then

$$(x + a)(x + b) = (x + a)(x - a)$$

and

$$(x + a)(x - a) = x^2 + \underbrace{(a - a)}_{0}x + a(-a) = x^2 - a^2$$

$$\begin{array}{r} x + a \\ x - a \\ \hline x^2 + ax \\ - ax - a^2 \\ \hline x^2 - a^2 \end{array}$$

Once again,

$$(x + a)^2 = x^2 + 2ax + a^2$$

$$(x + a)(x - a) = x^2 - a^2$$

EXAMPLE 9. Determine $(x + 4)^2$.

SOLUTION 1.

$$(x + 4)^2 = x^2 + 2 \cdot 4x + 4^2$$
$$= x^2 + 8x + 16$$

SOLUTION 2.

$$\begin{array}{r} x + 4 \\ x + 4 \\ \hline x^2 + 4x \\ 4x + 16 \\ \hline x^2 + 8x + 16 \end{array}$$ ■

EXAMPLE 10.

$$(x + 5)(x - 5) = x^2 - 25$$ ■

EXERCISES

In Exercises 1–54 multiply, as indicated.

1. $x^2 \cdot x$
2. $y^4 \cdot y$
3. $z \cdot z^{10}$
4. $a^2 \cdot a^3 \cdot 6$
5. $b^8 \cdot b^4$
6. $m^{10} \cdot 2 \cdot m^{17} \cdot 7$
7. $z \cdot z^2 \cdot z \cdot z^3 \cdot z^2$
8. $r \cdot r^3 \cdot r^{10} \cdot r^2 \cdot r^2$
9. $(x^2y)y$
10. $a(ab)$
11. $(y^2z)(y^2z^2)$
12. $(2uv)(-5u^2v)$
13. $(-x^2y^3)(-2xy^5)$
14. $(a^3x^2)(-5x^4)$
15. $(7ab)(2a^2b)(3b^2)$
16. $(-2x^2y)(-yz)(xyz)$
17. $(-2txy)(xyz)(3xyz^2)(4t^2xz)$
18. $(rst)(-rst)(r^2st^3)(-2r^3s^4t)$
19. $5(x + y)$
20. $2(x - y)$
21. $3(ax + 5y)$
22. $-4(2x + 3y^2)$
23. $x(x + 3)$
24. $a(a^2b + a^3)$
25. $2x(x^2 + x^3)$
26. $-5x^2(x + xy)$
27. $3abc(a^2 + b^2)$
28. $2abc^2(a^2c - 4ab + 2ac^2)$
29. $-3r^2st^2(-rs + 6st - 2r^3st^{10})$
30. $7m^2n^3(m^8n^2p - 2p + mnp^4 - 3mp)$
31. $(m + 8)(m + 1)$
32. $(m - 2)(m + 7)$
33. $-2(a - 9)(a + 5)$
34. $-(c - 3)(c + 6)$
35. $(x + 6)^2$
36. $(a - 5)^2$
37. $(x + 9)(x - 9)$
38. $(v + 1)(v - 1)$
39. $(a - 11)(a + 11)$
40. $(x^2 + 3)(x^2 - 3)$
41. $(u^4 + 1)(u^4 - 1)$
42. $(c^3 - 1)(c^3 + 6)$
43. $(x^2 + x - 1)(x + 5)$
44. $(y^2 + 4y - 3)(y - 1)$
45. $(x - 2y)(x + y)$
46. $(2x + 2y)(x - 4y)$
47. $(5a - 3b)(-2a + 6b)$
48. $(20m - 3n)(m - 2n)$
49. $(x + y)(x + y + z)$
50. $(x - y)(x - y + z)$
51. $(a + 3b)(2a + b + 5c)$
52. $(2x + y - 3)(x - y)$
53. $(x^2 + 4xy + 3y^2)(x^2 + y^2)$
54. $(a^3 - 3a^2b - 5ab^2 - 3b^3)(a - 2b)$

In Exercises 55–62 simplify each polynomial.

55. $2a - 3(b + 2c)$
56. $5(6 - 2a) + 2(a + 1)$
57. $3x[x - 2(y - 2)]$
58. $x(x - y) + y(y - x)$
59. $(x - 2)(x + 1) + (y - 1)(y + 2)$
60. $x(x - [3 + (2 - a)])$
61. $(2a^2 + b^2) - 2a[3a - (2b + a)]$
62. $2y^2(y - 3) + 5[y^2 - y(2y + 1)]$
63. Multiply $2x + 5$ by the sum of $3x + 1$ and $-2x$.
64. The square of $x - 1$ is multiplied by $2x - 3$. Determine the resulting polynomial.
65. Evaluate $(x + 8)(x - 3)$ when $x = 6$.

66. Evaluate $(a^2 + 2a - 3)(a + 5)$ when $a = -1$.
67. Evaluate $(2x + 3y)(x - 5y)$ when $x = 4$ and $y = 6$.
68. Evaluate $(2x^2 - y)(x^2 + xy - y^2)$ when $x = 1$ and $y = -1$.

2.5 DIVISION

Division of polynomials can be defined in terms of multiplication (of polynomials) just as division of real numbers was defined in terms of multiplication (of real numbers).

DEFINITION. Let P_1, P_2, and Q be polynomials and assume $P_2 \neq 0$. Then

$$\frac{P_1}{P_2} = Q, \qquad \text{if} \qquad P_1 = P_2 \cdot Q$$

In this case, P_1 is called the **dividend,** P_2 the **divisor,** and Q the **quotient.**

You *check* a division example by multiplying P_2, the divisor, by Q, the quotient. The product should be P_1, the dividend.

EXAMPLE 1.

$$\frac{6x^2}{2x} = 3x \qquad \text{because} \qquad \underbrace{6x^2}_{P_1} = \underbrace{2x}_{P_2} \cdot \underbrace{3x}_{Q}$$

■

Only a few simple examples of division will be attempted here. The subject will be further developed later.

First observe that

$$\frac{a^m}{a^m} = \frac{\overbrace{\not{a} \cdot \not{a} \cdot \not{a} \ldots \not{a}}^{m \text{ factors}}}{\underbrace{\not{a} \cdot \not{a} \cdot \not{a} \ldots \not{a}}_{m \text{ factors}}} = \frac{1}{1} = 1.$$

Thus

$$\frac{a^m}{a^m} = 1$$

Next,

$$\frac{a^5}{a^3} = \frac{\not{a} \cdot \not{a} \cdot \not{a} \cdot a \cdot a}{\not{a} \cdot \not{a} \cdot \not{a}} = \frac{a^2}{1} = a^2$$

In general, if $m > n$, then $m = n + (m - n)$, and

$$\frac{a^m}{a^n} = \frac{\overbrace{a \cdot a \cdot a \ldots a}^{m \text{ factors}}}{\underbrace{a \cdot a \cdot a \ldots a}_{n \text{ factors}}} = \frac{\overbrace{\overbrace{\not{a} \cdot \not{a} \cdot \not{a} \ldots \not{a}}^{n \text{ factors}} \cdot \overbrace{a \cdot a \cdot a \ldots a}^{m-n \text{ factors}}}^{m \text{ factors}}}{\underbrace{\not{a} \cdot \not{a} \cdot \not{a} \ldots \not{a}}_{n \text{ factors}}}$$

$$= \frac{a^{m-n}}{1}$$

$$= a^{m-n}$$

Thus

$$\frac{a^m}{a^n} = a^{m-n}, \qquad \textbf{if } m > n$$

EXAMPLE 2.

(a) $\dfrac{x^7}{x^4} = x^{7-4} = x^3$ (b) $\dfrac{y^{20}}{y^{12}} = y^{20-12} = y^8$ ■

EXAMPLE 3. Determine: $\dfrac{20x^9}{5x^3}$

SOLUTION. Observe that

$$\frac{20}{5} = 4 \quad \text{and} \quad \frac{x^9}{x^3} = x^6.$$

Clearly,

$$\frac{20x^9}{5x^3} = \frac{20}{5} \cdot \frac{x^9}{x^3} = 4x^6$$

To check this, note that

$$\underbrace{20x^9}_{P_1} = \underbrace{5x^3}_{P_2} \cdot \underbrace{4x^6}_{Q}.$$

■

EXAMPLE 4.

$$\frac{12x^3}{4x^3} = \frac{12}{4} \cdot \frac{x^3}{x^3} = 3 \cdot 1 = 3$$ ■

The last example will be used in Section 2.7. It involves multiplication of fractions, which will be discussed in Chapter 5.

EXAMPLE 5.

$$\frac{x}{2x} = \frac{1 \cdot x}{2 \cdot x} = \frac{1}{2} \cdot \frac{x}{x} = \frac{1}{2} \cdot 1 = \frac{1}{2}$$ ■

EXERCISES

Simplify:

1. $\frac{x^2}{x}$
2. $\frac{y^4}{y^3}$
3. $\frac{z^5}{z^2}$
4. $\frac{a^8}{a^4}$
5. $\frac{x^6}{x^2}$
6. $\frac{x^9}{x^7}$
7. $\frac{y^8}{y^3}$
8. $\frac{z^9}{z}$
9. $\frac{a^{12}}{a^7}$
10. $\frac{x^{20}}{x^{13}}$
11. $\frac{x^{26}}{x^{15}}$
12. $\frac{b^{100}}{b^{50}}$
13. $\frac{4x^4}{2x^2}$
14. $\frac{6a^4}{3a^3}$
15. $\frac{-4x^3}{2x}$
16. $\frac{8x^2}{2}$
17. $\frac{-9a^5}{-3a^2}$
18. $\frac{7x^9}{x^7}$
19. $\frac{20y^4}{10y}$
20. $\frac{30x^6}{-3x^5}$
21. $\frac{a}{a}$
22. $\frac{-b}{b}$
23. $\frac{c^2}{c^2}$
24. $\frac{x^5}{-x^5}$
25. $\frac{12x^6}{4x^6}$
26. $\frac{-7x^5}{7x^5}$
27. $\frac{3a^4}{-3a^4}$
28. $\frac{a^2}{2a^2}$
29. $\frac{a^3}{3a^2}$
30. $\frac{x^7}{5x^5}$

2.6 DEGREE

DEFINITION. The **degree** of a *nonzero* constant is defined to be 0. If a *nonzero* term contains a single variable, its degree is the exponent of that variable. If a *nonzero* term contains more than one variable, its degree is the *sum* of the exponents of these variables.

For technical reasons, the term 0 has no degree. Every nonconstant term has a positive degree. You speak of **first-degree terms, second-degree terms,** and so on.

EXAMPLE 1.

(a) -7 has degree 0 (or is a 0th-degree term).
(b) $3x^4$ is a fourth-degree term.
(c) $2x^2z$ is a third-degree term. The exponent of z is understood to be 1. Thus the sum of the exponents is $2 + 1$, or 3.
(d) $-5rs^3t^2$ is a sixth-degree term. The sum of the exponents is $1 + 3 + 2$, or 6.
(e) $0 \cdot a^4bc^3d$ equals 0, and thus *has no degree.* ■

DEFINITION. A polynomial in a *single variable* is said to be in **standard form** if terms of the same degree are combined and the resulting terms are arranged in order of decreasing degrees. The term of highest degree is called the **leading term;** the coefficient of this leading term is called the **leading coefficient.**

A monomial in a single variable is automatically in standard form. Its leading term is itself; its leading coefficient is its only coefficient.

EXAMPLE 2.

(a) The standard form of

$$x^5 + x^5 - 8x^7 + 4x^2 - 9 + 10x^2$$

is

$$-8x^7 + 2x^5 + 14x^2 - 9.$$

Its leading term is $-8x^7$, and its leading coefficient is -8.

(b) The monomial

$$-9a^5$$

is already in standard form. Its leading term is $-9a^5$, and its leading coefficient is -9. ■

The following definition applies to polynomials in any number of variables.

DEFINITION. Consider a *nonzero* polynomial in which similar terms have been combined. The **degree of this polynomial** is defined as the highest degree of any of its terms.

Thus *a polynomial in a single variable has the same degree as its leading term.*

EXAMPLE 3.

(a) The monomial $5x^2y^2$ consists of a single term, of fourth degree. Consequently,

$5x^2y^2$ is a *fourth-degree polynomial.*

(b) $10x^2 + 3y - 2$ is a trinomial. Of its three terms,

$10x^2$ is a second-degree term, $3y$ is a first-degree term,

-2 is of degree 0 (or is a 0th-degree term)

The highest degree of any of its terms is 2. Thus, $10x^2 + 3x - 2$ is a *second-degree polynomial* (or a **quadratic polynomial**).

(c) $x^3 + x^2y^2 + x - 1$ is a *fourth-degree polynomial.* The highest-degree term is the second term, x^2y^2.

(d) $x^5 + x^4 + x^3y - 3x^2y^3$ is a *fifth-degree polynomial.* Its first and fourth terms, x^5 and $-3x^2y^3$, are each of fifth degree.

(e) 14 is a polynomial of degree 0. ■

EXERCISES

In Exercises 1–12 determine the degree of each term.

1. x^5
2. xy
3. $5x^2yz$
4. $-2xy^2z^4$
5. 17
6. $\frac{1}{2}x^2$
7. $4rs^2t$
8. $rstxyz$
9. $x^{14}y^{12}$
10. $-x^5y^2z^4$
11. $x^7y^3z^{12}$
12. $r^2s^3t^4uv^5w^5$

In Exercises 13–24 write each polynomial in standard form.

13. $1 - 5x$
14. $3x + x^4 + 1$
15. $18 - x - 2x$
16. $\frac{x}{2} + x^2 - 7x^3$
17. $y^3 - y^2 + 3 - y$
18. $2a^5 - a^{10} + a - a^5$
19. $2m^2 - 4m^4 + 6m^6$
20. $z^3 + z^4 - z + 3$
21. $t^5 - t^8 + t^3 - t^2 + 14 - t^6$
22. $7t^{17} - 4t^{13} + t^{20} + t^2 + 4 - 3t^3 + t^{10}$
23. $6 - \frac{x^2}{2} + 3x - 5x^4 + x^8$
24. $s^{100} - s^{10} + s^2 - s + 1$

In Exercises 25–36 determine the degree of each polynomial.

25. $x^4 - 3x^2 + 5$
26. $xy - x$
27. $x^7 - 3xy^6 + y^5$
28. -19
29. $xyz + 5x^2$
30. $7x^4y^7z$
31. $xyz^4 + x^5 - x^3y^4$
32. $yx^2y^9 - y^{10}$
33. $s^2t^2 + st^3 + t^4$
34. $x^5y^2z + z^7$
35. $20m^2n^8 + 10mn^{10} - 3m^6n^6 + 193$
36. $a^7b^4c - 5b^2c + 17c^{15} - b^3$

2.7 LONG DIVISION

There are procedures for dividing polynomials that resemble those for dividing integers.

If the divisor is a monomial, proceed as in short division.

EXAMPLE 1. Divide $6x^3 + 4x^2 - 2x$ by $2x$.

SOLUTION. Here you divide each term of the dividend by $2x$.

$$\frac{6x^3}{2x} = 3x^2 \qquad \frac{4x^2}{2x} = 2x \qquad \frac{-2x}{2x} = -1$$

Thus

$$2x \overline{\big) 6x^3 + 4x^2 - 2x} \quad \text{quotient: } 3x^2 + 2x - 1$$

You can check the result by multiplying the quotient by the divisor (in either order). You should obtain the dividend.

CHECK.

$$\begin{array}{ll} 3x^2 + 2x - 1 & \longleftarrow \text{quotient} \\ \underline{2x \qquad\qquad\quad} & \longleftarrow \text{divisor} \\ 6x^3 + 4x^2 - 2x & \longleftarrow \text{dividend} \end{array}$$

■

If the divisor has more than one term, proceed as in long division.

EXAMPLE 2.

$$x + 4 \overline{\big) 2x^2 + 11x + 12}$$

SOLUTION. Divide $2x^2$, the first term of the dividend, by x, the first term of the divisor.

$$\frac{2x^2}{x} = 2x$$

Thus the first term of the quotient is $2x$. Multiply this by the divisor, $x + 4$, and subtract the product, $2x^2 + 8x$, from $2x^2 + 11x$. The difference is $3x$. Bring down the next term, 12, and obtain the **first difference polynomial,** $3x + 12$.

$$\begin{array}{rl} 2x \qquad\qquad\quad & \longleftarrow \text{quotient} \\ \text{divisor} \longrightarrow \; x + 4 \overline{\big) 2x^2 + 11x + 12} & \longleftarrow \text{dividend} \\ \underline{\overset{-}{+}2x^2 \overset{-}{+} 8x \qquad} & \\ 3x + 12 & \longleftarrow \text{first difference polynomial} \end{array}$$

Divide $3x$, the first term of the difference polynomial, by x, the first term of the divisor.

$$\frac{3x}{x} = 3$$

Thus the second term of the quotient is 3. Multiply this by the divisor, $x + 4$. Subtract the product, $3x + 12$, from $3x + 12$ to obtain the **second difference polynomial.** Because this is 0, $x + 4$ divides $2x^2 + 11x + 12$ (evenly). The quotient is $2x + 3$.

$$
\begin{array}{r}
2x \quad + \quad 3 \\
x+4 \overline{\big) \; 2x^2 + 11x + 12} \\
\underline{\overset{-}{+}2x^2 \overset{-}{+} 8x} \\
3x + 12 \\
\underline{\overset{-}{+}3x \overset{-}{+} 12} \\
0
\end{array}
\quad \longleftarrow \text{second difference polynomial}
$$

CHECK.

$$
\begin{array}{rl}
2x + 3 & \longleftarrow \text{quotient} \\
\underline{x + 4} & \longleftarrow \text{divisor} \\
2x^2 + 3x & \\
\underline{8x + 12} & \\
2x^2 + 11x + 12 & \longleftarrow \text{dividend}
\end{array}
$$

■

Only division of polynomials in a single variable will be considered. *Such polynomials should always be put into standard form.* Thus the terms are arranged in order of decreasing degrees. *Add* 0's *for the missing terms,* as in the next example.

EXAMPLE 3. Divide $3 - 8y + y^3$ by $3 + y$.

SOLUTION. Put these polynomials into standard form. Write the dividend as

$$y^3 + 0y^2 - 8y + 3$$

and the divisor as

$$y + 3.$$

$$
\begin{array}{rl}
y^2 - 3y + 1 & \longleftarrow \text{quotient} \\
\text{divisor} \longrightarrow \quad y+3 \overline{\big) \; y^3 + 0y^2 - 8y + 3} & \longleftarrow \text{dividend} \\
\underline{\overset{-}{+}y^3 \overset{-}{+} 3y^2} & \\
-3y^2 - 8y & \longleftarrow \text{first difference polynomial} \\
\underline{\overset{+}{-}3y^2 \overset{+}{-} 9y} & \\
y + 3 & \longleftarrow \text{second difference polynomial} \\
\underline{y + 3} & \\
0 & \longleftarrow \text{third difference polynomial}
\end{array}
$$

The quotient is $y^2 - 3y + 1$. ■

In the first three examples, the divisor P_2 divides the dividend P_1 evenly. Thus

$$\frac{P_1}{P_2} = Q,$$

where the quotient Q is a *polynomial.* In general,

$$\frac{P_1}{P_2} = Q + \frac{R}{P_2}$$

Here the polynomial Q is again called the **quotient;** the polynomial R is called the **remainder.** *Either the remainder is* 0 (as in Examples 1, 2, and 3) *or else the degree of the remainder is less than the degree of the divisor.*

In Example 4, which follows, you will see that

$$\frac{2a^3 + 11a^2 - 25a + 10}{2a - 3} = a^2 + 7a - 2 + \frac{4}{2a - 3}.$$

Here $a^2 + 7a - 2$ is the quotient and 4 is the remainder. The degree of the remainder is 0, which is less than 1, the degree of the divisor, $2a - 3$.

Continue the division procedure until either you obtain a 0 *difference polynomial or one whose degree is less than that of the divisor. This difference polynomial will be the remainder. In case the degree of the dividend is less than the degree of the divisor, the quotient is* 0 *and the dividend is the remainder.* (See Example 8.)

In the following examples there is a nonzero remainder.

EXAMPLE 4.

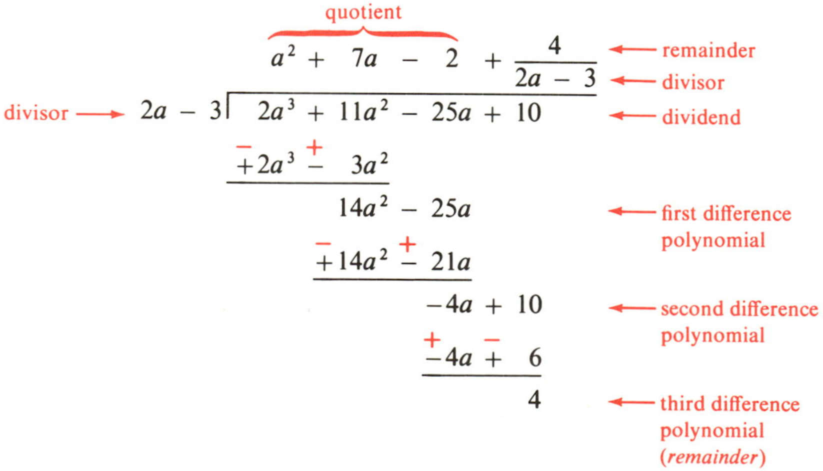

$$
\begin{array}{rrl}
 & \overbrace{a^2 + 7a - 2}^{\text{quotient}} + \dfrac{4}{2a - 3} & \begin{array}{l}\leftarrow \text{remainder}\\ \leftarrow \text{divisor}\end{array} \\
\text{divisor} \longrightarrow 2a - 3 & \overline{\big)\, 2a^3 + 11a^2 - 25a + 10} & \leftarrow \text{dividend} \\
 & \underline{\overset{-}{+}2a^3 \overset{+}{-} 3a^2 \qquad\qquad\quad} & \\
 & 14a^2 - 25a \qquad\quad & \leftarrow \text{first difference polynomial} \\
 & \underline{\overset{-}{+}14a^2 \overset{+}{-} 21a \qquad\quad} & \\
 & -4a + 10 & \leftarrow \text{second difference polynomial} \\
 & \underline{\overset{+}{-}4a \overset{-}{+} \;\, 6} & \\
 & 4 & \leftarrow \text{third difference polynomial } (\textit{remainder})
\end{array}
$$

4 is "left over". This process cannot be continued because the degree, 0, of the third difference polynomial, 4, is less than the degree, 1, of the divisor, $2a - 3$.

Thus 4 is the *remainder* and

$$\frac{2a^3 + 11a^2 - 25a + 10}{2a - 3} = a^2 + 7a - 2 + \frac{4}{2a - 3}.$$

To *check* the result *when there is a remainder,* multiply the quotient by the divisor; then add the remainder. The sum should be the dividend.

CHECK.

$$\begin{array}{rrrrl}
 & a^2 + & 7a & - \ \ 2 & \leftarrow \text{quotient} \\
 & 2a & - \ \ 3 & & \leftarrow \text{divisor} \\
\hline
2a^3 + & 14a^2 - & 4a & & \\
 & -3a^2 - & 21a + & 6 & \\
\hline
2a^3 + & 11a^2 - & 25a + & 6 & \\
+ & & & 4 & \leftarrow \text{remainder} \\
\hline
2a^3 + & 11a^2 - & 25a + & 10 & \leftarrow \text{dividend}
\end{array}$$

■

EXAMPLE 5.

$$x^2 + x + 2 \overline{\big)\, x^3 + 6x^2 + 10x}$$

SOLUTION.

$$\begin{array}{rrrrrl}
 & & & x + 5 & + \dfrac{3x - 10}{x^2 + x + 2} & \\
\text{divisor} \longrightarrow & x^2 + x + 2 \big) & x^3 + 6x^2 & + 10x & + \ 0 & \leftarrow \text{dividend} \\
 & & \underline{+x^3 + x^2} & \underline{+ \ 2x} & & \\
 & & 5x^2 & + \ 8x & + \ 0 & \leftarrow \text{first difference polynomial} \\
 & & \underline{+5x^2} & \underline{+ \ 5x} & \underline{+ 10} & \\
 & & & 3x & - 10 & \leftarrow \text{second difference polynomial } (\textit{remainder})
\end{array}$$

(In the subtraction rows, the signs of $+x^3$, $+x^2$, $+2x$ and of $+5x^2$, $+5x$, $+10$ are shown changed to $-$.)

Note that on the first step, after subtracting $x^3 + x^2 + 2x$ from $x^3 + 6x^2 + 10x$, there are no more terms of the dividend to bring down. The first difference polynomial is therefore $5x^2 + 8x$. *Because its degree,* 2, *equals that of the divisor, the division process continues.*

The degree, 1, of the second difference polynomial, $3x - 10$, is less than the degree, 2, of the divisor, $x^2 + x + 2$. Thus $3x - 10$ is the remainder. The result is

$$x + 5 + \frac{3x - 10}{x^2 + x + 2}.$$

■

EXAMPLE 6. Determine the quotient and remainder:

$$(x^6 - x^4 + 2x^3 - x^2) \div (x^2 + 1)$$

SOLUTION. Write:

$$x^2 + 0x + 1 \overline{\big)\, x^6 + 0x^5 - x^4 + 2x^3 - x^2 + 0x + 0}$$

$$\begin{array}{rl}
 & x^4 \qquad - 2x^2 + 2x + 1 + \dfrac{-2x - 1}{x^2 + 1} \\
x^2 + 0x + 1 & \overline{\big)\, x^6 + 0x^5 - x^4 + 2x^3 - x^2 + 0x + 0} \\
 & \underline{\mp x^6 \qquad \mp x^4} \\
 & \qquad - 2x^4 + 2x^3 - x^2 \quad \leftarrow \text{first difference polynomial} \\
 & \qquad \underline{\pm 2x^4 \qquad \pm 2x^2} \\
 & \qquad\qquad 2x^3 + x^2 + 0x \quad \leftarrow \text{second difference polynomial} \\
 & \qquad\qquad \underline{\mp 2x^3 \qquad \mp 2x} \\
 & \qquad\qquad\qquad x^2 - 2x + 0 \quad \leftarrow \text{third difference polynomial} \\
 & \qquad\qquad\qquad \underline{\mp x^2 \qquad \mp 1} \\
 & \qquad\qquad\qquad\qquad -2x - 1 \quad \leftarrow \text{fourth difference polynomial (remainder)}
\end{array}$$

Because the divisor has three "terms" (including the "zero term", $0x$), the first difference polynomial must also have three terms. Thus bring down $2x^3 - x^2$, so that the first difference polynomial is $-2x^4 + 2x^3 - x^2$. Even though you run out of terms of the dividend after the second step, the degree, 3, of the second difference polynomial is greater than 2, the degree of the divisor. Therefore continue the process. Similarly, the degree, 2, of the third difference polynomial *equals* the degree of the divisor. Continue the process until you obtain a difference polynomial whose degree is *less than* that of the divisor. This occurs on the fourth step. The fourth difference polynomial, $-2x - 1$, is the remainder. Thus

$$\frac{x^6 - x^4 + 2x^3 - x^2}{x^2 + 1} = x^4 - 2x^2 + 2x + 1 + \frac{-2x - 1}{x^2 + 1} \qquad \blacksquare$$

The next example illustrates a very simple, but often misunderstood, point.

EXAMPLE 7.

$$2x + 2 \overline{\big)\, x + 2}$$

SOLUTION. The degree of the dividend *equals* the degree of the divisor. Even though the *leading coefficient,* 1, of the dividend is less than the *leading coefficient,* 2, of the divisor, you must still divide. Obtain $\frac{1}{2}$ as your quotient; your remainder turns out to be 1.

$$\begin{array}{r|l} & \frac{1}{2} + \frac{1}{2x+2} \\ \hline 2x+2 & x+2 \\ & \underline{-x-1} \\ & \phantom{+x+{}}1 \end{array}$$

■

EXAMPLE 8.

$$x^4 + 1 \overline{\big)\, x^3 + x^2 + x + 1}$$

SOLUTION. Here the degree of the dividend is less than that of the divisor. The quotient is therefore 0 and the dividend, $x^3 + x^2 + x + 1$, is the remainder.

$$\begin{array}{r|l} & 0 + \frac{x^3 + x^2 + x + 1}{x^4 + 1} \\ \hline x^4 + 1 & x^3 + x^2 + x + 1 \end{array}$$

■

A remainder can occur even for monomial divisors.

EXAMPLE 9.

$$\frac{2t^4 - 3t^3 + t^2 + 3}{t^2}$$

SOLUTION. Use short division:

$$\begin{array}{r|l} & 2t^2 - 3t + 1 + \frac{3}{t^2} \\ \hline t^2 & 2t^4 - 3t^3 + t^2 + 3 \end{array}$$

The remainder is 3. ■

EXERCISES

In Exercises 1–26 divide. Check the ones so indicated.

1. $x \overline{\big)\, x^2 + x}$ (Check.)
2. $x \overline{\big)\, 5x^3 + 2x^2 - x}$
3. $2x \overline{\big)\, 4x^2 - 6x}$ (Check.)
4. $3y \overline{\big)\, 9y^3 + 18y^2 - 3y}$
5. $-2a \overline{\big)\, 10a^3 + 2a^2 - 2a}$
6. $7b^2 \overline{\big)\, 49b^4 - 7b^2}$
7. $(2x^2 + 6x) \div 2x$
8. $(4x^3 + 20x^2) \div 4x^2$

9. $(a^2 - 9) \div (a + 3)$
10. $(b^2 + 16b + 64) \div (b + 8)$
11. $(z^2 + 13z + 30) \div (z + 3)$
12. $(y^2 + 13y + 40) \div (y + 8)$
13. $a + 4\overline{)a^3 + 5a^2 + 6a + 8}$ (Check.)
14. $m - 3\overline{)m^3 + 2m^2 - 21m + 18}$
15. $\dfrac{35 - 16x + 4x^2 + x^3}{7 + x}$
16. $2x + 1\overline{)2x^2 + 9x + 4}$
17. $3x - 5\overline{)6x^2 + 2x - 20}$
18. $4y + 3\overline{)12y^3 + 17y^2 + 14y + 6}$
19. $\dfrac{3(t^3 + 21) - 13t(t + 1)}{3t - 7}$
20. $t^2 + t + 1\overline{)4t^3 - t^2 - t - 5}$
21. $a^2 - 3a + 2\overline{)a^3 + 2a^2 - 13a + 10}$
22. $m^2 + 2m - 5\overline{)4m^3 + 5m^2 - 26m + 15}$
23. $2m^2 - 5\overline{)2m^3 + 2m^2 - 5m - 5}$ (Check.)
24. $3s^2 - 2s + 1\overline{)9s^3 - s + 2}$
25. $x^2 + 1\overline{)x^4 - x^3 + 3x^2 - x + 2}$
26. $x^2 - x - 2\overline{)x^4 + x^3 - 3x^2 - 5x - 2}$

In Exercises 27–48 divide as indicated. In some cases there will be a remainder. Check the exercises so indicated.

27. $x + 2\overline{)x^2 + 4x + 5}$ (Check.)
28. $y - 3\overline{)y^2 + y + 2}$
29. $\dfrac{x^2 - 9x + 7}{x}$
30. $\dfrac{z^2 - 3z + 8}{z - 4}$
31. $x + 4\overline{)x^2 + 9x + 16}$
32. $a^2\overline{)a^3 - a^2 + a + 1}$
33. $c^2 + 2\overline{)c^3 + 2c^2 + 2c + 4}$
34. $y^2 + 3\overline{)y^4 - 3y^2 + 3y + 9}$
35. $x^2 + x + 1\overline{)2x^4 + 3x^2 - x + 2}$ (Check.)
36. $z^2 - z + 1\overline{)2z^4 + z^2 - 1}$
37. $y^2 - 2\overline{)y^4 - 4y^2 + 2y - 4}$ (Check.)
38. $a^4 - 1\overline{)a^8 - 3a^4 + a^2 - 1}$
39. $\dfrac{z + 2}{z + 1}$
40. $\dfrac{2 - z}{z}$
41. $\dfrac{10t}{10t + 1}$
42. $\dfrac{2t^2}{t^2 + 1}$
43. $t^2 + 2\overline{)t^3 - 3t^2 + 2t - 6}$
44. $z^2 - 5\overline{)z^3 + 6z - 2}$
45. $2z^2 - z + 4\overline{)2z^4 + 3z^3 - z + 8}$
46. $m^2 - 2m\overline{)m^4 - 3m^2 + m - 1}$
47. $a^3 - a^2 + 1\overline{)a^6 + 4a^4 - 2a^2 + 2}$
48. $c^2 + 3\overline{)4c^4 + 6c^2 - 2c + 9}$
49. Suppose the divisor is $x + 2$ and the quotient is $x + 3$ (with no remainder). Determine the dividend.
50. Suppose the divisor is $x + 5$, the quotient is $x - 2$, and the remainder is 1. Determine the dividend.

*2.8 SYNTHETIC DIVISION

There are many steps in the division process that are repetitious. When the divisor is of the form $x + A$ for an integer A, a simplified process, known as **synthetic division,** can be employed. First consider an example worked out by the usual division process.

EXAMPLE 1.

$$
\begin{array}{r|rrrr}
 & x^2 & -\,2x & -\,3 & \\
\hline
x-1 & x^3 & -\,3x^2 & -\,x & +\,3 \\
 & - & + & & \\
 & +x^3 & -\,x^2 & & \\
\cline{2-3}
 & & -\,2x^2 & -\,x & \\
 & & + & - & \\
 & & -\,2x^2 & +\,2x & \\
\cline{3-4}
 & & & -\,3x & +\,3 \\
 & & & + & - \\
 & & & -\,3x & +\,3 \\
\cline{4-5}
\end{array}
$$

1. Arrange the terms of the dividend according to descending degrees. Place 0's for the missing terms, including those "at the end". If the lowest-degree term is of degree 1, place 1 zero after the last nonzero term; if the lowest-degree term is of degree 2, place 2 zeros, and so on. You need only list the coefficients. The last coefficient is the constant term, the next to last is that of the first-degree term, and so on. Thus if x is the variable, then

$$1 \qquad -3 \qquad -1 \qquad 3$$

indicates the above dividend,

$$x^3 \qquad -3x^2 \qquad -\,x \qquad +3.$$

Similarly,

$$2 \qquad 0 \qquad -8 \qquad 1 \qquad 0$$

indicates

$$2x^4 \qquad\qquad -8x^2 \qquad +\,x \qquad (\text{or } 2x^4 + 0x^3 - 8x^2 + x + 0)$$

and

$$1 \qquad 1 \qquad 3 \qquad -2 \quad 0 \quad 0 \quad 0$$

indicates

$$x^6 \qquad +\,x^5 \qquad +3x^4 \qquad -2x^3.$$

*Optional topic.

2. Below the division line, terms of the same degree are in the same column. Thus you need only list the coefficients. Also, the divisor here is always of the form $x + A$. Just write A for the divisor. The division process for the given example can be written as follows:

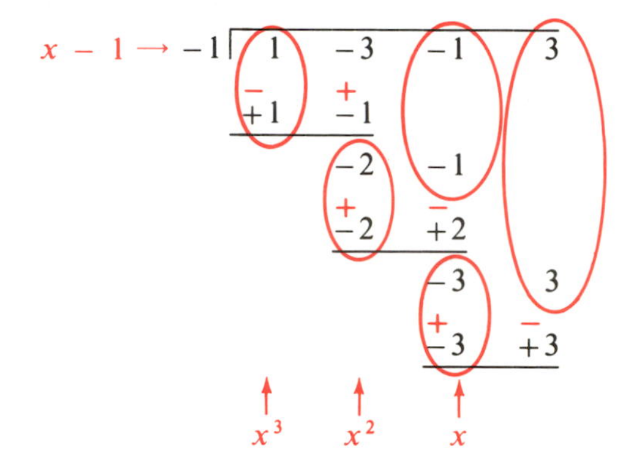

3. Note that in the balloons, the lower numbers repeat the upper numbers. Omit the repetitions:

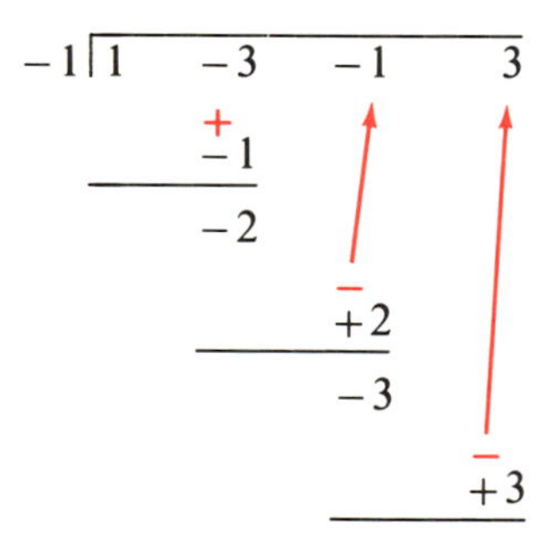

4. Instead of changing signs every time you subtract, write $-A$ when the divisor is $x + A$. List the numbers on the first available space, as indicated by the arrows above. Bring down the leading coefficient of the dividend, as indicated: *Add* the first two lines, column by column. The quotient is now indicated on the third line.

$$\begin{array}{r|rrrr} 1 & 1 & -3 & -1 & 3 \\ & \downarrow & 1 & -2 & -3 \\ \hline & 1 & -2 & -3 & \end{array}$$

Read

$$1 \quad -2 \quad -3$$

as

$$x^2 - 2x - 3. \qquad \blacksquare$$

EXAMPLE 2. Determine $\dfrac{x^3 + x^2 + x - 3}{x - 1}$ by synthetic division.

SOLUTION. Write

$$1 \quad 1 \quad 1 \quad -3$$

for the dividend

$$x^3 + x^2 + x - 3.$$

The divisor is $x - 1$; here $A = -1$ and $-A = 1$. Begin by writing:

$$\underline{1|} \quad 1 \quad 1 \quad 1 \quad -3$$

(The $\underline{1|}$ at the left indicates the divisor, $x - 1$.) After you divide the first term, x^3, with leading coefficient 1, by $x - 1$, the resulting quotient must have leading coefficient 1. In synthetic division, the first entry underneath the horizontal line will always be the same as the number lying above it.

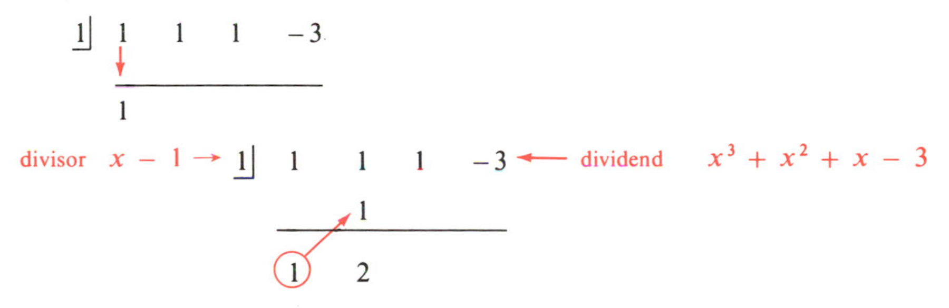

The encircled 1 has been multiplied by the 1 from the divisor. The product is 1, as indicated by the arrow. Adding in the second column, you obtain $1 + 1 = 2$.

Now multiply the encircled 2 by the 1 from the divisor (as shown below) to obtain the product 2 as indicated by the arrow. Adding the numbers in the third column, you obtain $1 + 2 = 3$.

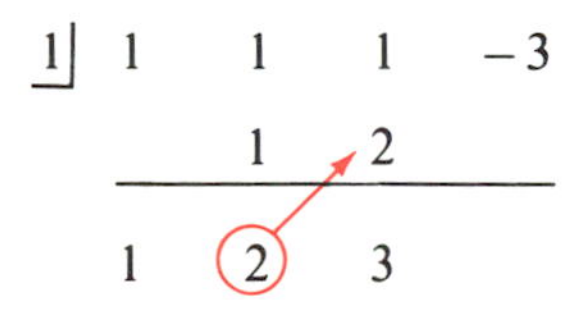

Repeating the process once more, you obtain the final form:

divisor $x - 1 \rightarrow \underline{1|}$ 1 1 1 −3 ← dividend $x^3 + x^2 + x - 3$

1 2 3

quotient $x^2 + 2x + 3 \rightarrow$ 1 2 (3) $\underline{|0}$ ← remainder 0

The numbers below the horizontal line indicate the quotient. Bear in mind that when a nonconstant polynomial is divided by $x - A$, its degree is lowered by 1. Because the first line represents a third-degree polynomial, the bottom line represents a second-degree polynomial. In reading this, note that the last digit represents the remainder. Thus

$$1 \quad 2 \quad 3$$

represents the quotient

$$x^2 + 2x + 3;$$

the $\lfloor 0$ at the end indicates that the remainder is 0. Therefore

$$\frac{x^3 + x^2 + x - 3}{x - 1} = x^2 + 2x + 3$$

■

EXAMPLE 3. Determine $\dfrac{2x^3 - 3x + 10}{x + 2}$ by synthetic division.

SOLUTION. The divisor is $x + 2$. Here $A = 2$ and $-A = -2$. The dividend, which can be written as $2x^3 + 0x^2 - 3x + 10$, is represented by 2 0 −3 10.

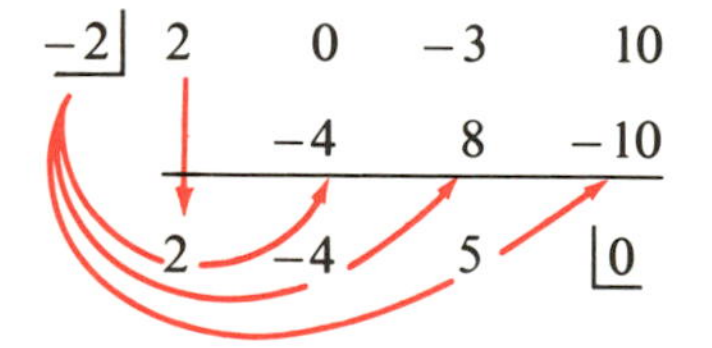

Thus, the quotient is $2x^2 - 4x + 5$; the remainder is 0. ■

EXAMPLE 4. Determine $\dfrac{x^4 + 3x^2 - x}{x + 1}$ by synthetic division.

SOLUTION. There are no third degree or degree 0 (constant) terms in the dividend.

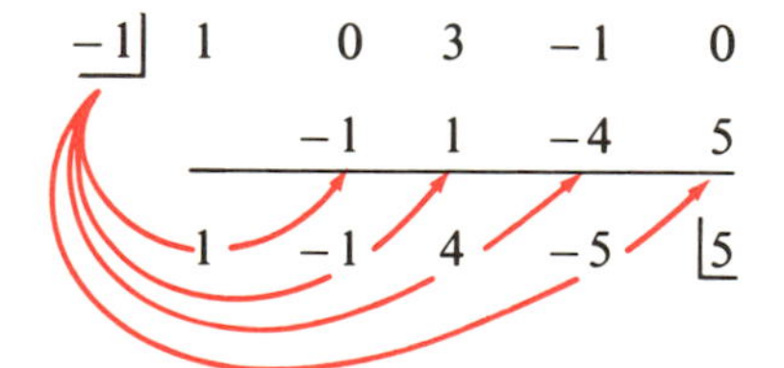

The quotient is $x^3 - x^2 + 4x - 5$, and the remainder is 5. Thus,

$$\frac{x^4 + 3x^2 - x}{x + 1} = x^3 - x^2 + 4x - 5 + \frac{5}{x + 1}$$

■

EXERCISES

In Exercises 1–8 write the first line of synthetic division if the divisor is given by (a) and the dividend by (b).

1. (a) $x + 1$ (b) $x^2 + 2x + 4$
2. (a) $x + 3$ (b) $x^2 - 2x + 1$
3. (a) $x - 2$ (b) $x^3 - 2x^2 + 2x + 4$
4. (a) $x - 5$ (b) $x^3 - 5x^2 + 10x - 5$
5. (a) x (b) $x^4 + 5x^3 + x^2 + x - 1$
6. (a) $x - 10$ (b) $x^4 + 10x^2 - 10$
7. (a) $x + 4$ (b) $x^3 - 3x^2 + 4$
8. (a) $x + 2$ (b) $x^6 + 4x^4 - 2x + 12$

In Exercises 9–16 suppose that the bottom line of a synthetic division example is as indicated. Determine the quotient and if there is one, the remainder.

9. $1 \quad 2 \quad \lfloor 0$
10. $2 \quad 1 \quad \lfloor 0$
11. $2 \quad 2 \quad \lfloor 1$
12. $3 \quad 1 \quad 2 \quad \lfloor 0$
13. $1 \quad 0 \quad 1 \quad 0 \quad \lfloor 0$
14. $2 \quad 0 \quad 1 \quad 0 \quad 0 \quad \lfloor 0$
15. $1 \quad 0 \quad 0 \quad 1 \quad 0 \quad 2 \quad \lfloor 3$
16. $2 \quad 0 \quad 1 \quad 0 \quad 0 \quad 1 \quad 0 \quad 1 \quad \lfloor -1$

In Exercises 17–29 use synthetic division to determine the quotient.

17. $\dfrac{x^2 + 6x + 9}{x + 3}$
18. $\dfrac{x^2 - 12x + 36}{x - 6}$
19. $\dfrac{y^2 - 5y + 6}{y - 2}$
20. $\dfrac{z^2 + 3z - 18}{z + 6}$
21. $x + 2\overline{)x^3 + 3x^2 + 3x + 2}$
22. $x + 1\overline{)x^3 - x}$
23. $t - 1\overline{)t^3 - t^2 + t - 1}$
24. $x - 1\overline{)x^4 - 1}$
25. $\dfrac{x^4 + x^2 - x - 3}{x + 1}$
26. $\dfrac{t^4 - 1}{t + 1}$
27. $\dfrac{a^3 + a^2 + a - 3}{a - 1}$
28. $\dfrac{y^6 - y^5 + y^3 - y^2 + y - 1}{y - 1}$
29. $\dfrac{y^8 - 1}{y - 1}$

In Exercises 30–36 use synthetic division to determine the quotient. In some cases there is a remainder.

30. $\dfrac{z^2 + 3z + 4}{z + 2}$
31. $\dfrac{m^4 + m^2 + 1}{m + 1}$
32. $\dfrac{5u^3 - 25u^2 + u - 5}{u - 5}$
33. $\dfrac{a^3 + 4a^2 + 5a + 2}{a + 2}$
34. $\dfrac{x^6 + x^4 - x^2 + 2}{x - 1}$
35. $\dfrac{z^4 - 9z^2 + 3z}{z - 3}$
36. $\dfrac{6z^3 - 6z^2 - 6}{z - 2}$

REVIEW EXERCISES FOR CHAPTER 2

2.1 TERMS AND POLYNOMIALS

1. Which of the following expressions are terms?

   (a) $3x$ (b) $x + 1$ (c) 4 (d) $\frac{x}{2}$

2. Determine the coefficient of each term.

   (a) $\frac{-x}{4}$ (b) $2x(3x)$ (c) $(4x)(0y)$ (d) z

3. Rewrite each term so that its coefficient appears first. (If more than one constant appears, multiply these.)

   (a) $(2x^2)(3y)$ (b) $4x \cdot 3y \cdot 2z$

   (c) $x \cdot \frac{y}{4}$ (d) $(-2x)(-2y)(-z)$

4. Which of the following expressions are polynomials?

   (a) xy^2 (b) $x + y^2$ (c) $\frac{x}{y}$ (d) $\frac{x}{2} + \frac{y}{4}$

2.2 NUMERICAL EVALUATION

5. Evaluate $x + 3$ when $x = 2$.
6. Evaluate each polynomial for the specified value of the variable.

   (a) $2y + 5 \quad [y = 5]$ (b) $x^2 - 3x + 1 \quad [x = 2]$

   (c) $m^3 + m^2 - m - 1 \quad [m = -1]$

7. Evaluate $x^3 + 2x + 1$ when

   (a) $x = 1$, (b) $x = -1$, (c) $x = 2$, (d) $x = -2$.

8. The area of a rectangle is given by lw, which l is the length and w the width. Determine the area when l is 200 feet and w is 50 feet.
9. Evaluate $x^2y - 2xy$ when $x = 10$ and $y = 5$.

2.3 ADDITION AND SUBTRACTION

10. Group similar terms together: x^2y, xy^2, $-2xy$, $3x^2y$, xy, $4xy$, $10xy^2$
11. Simplify:

    (a) $5m - 3m$ (b) $2x - x + x$

    (c) $5m^2 + 3m - 2m^2 + m$ (d) $5x - 3y + 2z - (2x - 3y + z)$

12. Add:

$$\begin{array}{rrrr} 2t + & x - & y + & z \\ & 3x + & y - & 2z \\ 4t & + & y - & z \\ & & 2y + & z \\ \hline \end{array}$$

13. Determine the inverse of $3x^2y - 2x + 4y$.

14. *Subtract* the bottom polynomial from the top one:

$$\begin{array}{r} 3x - 2y + 4z \\ - 4y + 4z \\ \hline \end{array}$$

15. Simplify: $2x - [4 - (3x + 2y) - (4x + 2y - 1)]$

16. Evaluate $4xy - (5x - 3xy) + 2x$ when $x = 2$ and $y = 3$.

2.4 MULTIPLICATION

In Exercises 17–21 multiply as indicated.

17. $y^4 \cdot y^2 \cdot y$

18. $(-4xy^2z)(-2x^2y^3z)(xz^4)$

19. $x^2(xy^2 + xy - 3x^2)$

20. $2(x + 2)^2$

21. $(2m - n)(m + n)$

22. Simplify: $(x - 2)(x + 1) - 3x(x^2 + x - 1)$

23. Determine the product of x^2 and the square of $x - 1$.

2.5 DIVISION

In Exercises 24–27 divide as indicated.

24. $\dfrac{a^7}{a^5}$ 25. $\dfrac{8x^{11}}{2x^6}$ 26. $\dfrac{-15y^8}{-5y^3}$ 27. $\dfrac{2a^3}{a^3}$

2.6 DEGREE

28. Determine the degree of each term:

(a) y^4 (b) x^2y^3 (c) 13 (d) $-5x^2y^3z^5$

29. Write each polynomial in standard form:

(a) $1 - t^2$ (b) $3x^3 + 4x^4 - 1 + 2x^2 - 2x$

(c) $x^6 - \dfrac{x^9}{3} + x^4 - 8$ (d) $m^4 + m^4 - m^5$

30. Determine the degree of each polynomial:

(a) $x^8 - 3x^6 + 4x^2 - 1$ (b) 10

(c) $xy^2 + xy$ (d) $m^2n + n^2m + 1$

2.7 LONG DIVISION

In Exercises 31–34 divide as indicated. In some cases there will be a remainder.

31. $x + 5\overline{)x^2 + 8x + 15}$ 32. $y - 2\overline{)y^3 - 2y^2 + 3y - 6}$

33. $m^2 + 2\overline{)m^3 + 5m^2 + 2m}$ 34. $c^3 + 4\overline{)c^5 + c^3 + 4c^2 + c + 5}$

2.8 SYNTHETIC DIVISION

In Exercises 35–38 use synthetic division to determine the quotient. In some cases there will be a remainder.

35. $\dfrac{x^2 + 10x + 21}{x + 3}$

36. $t - 9\overline{)t^2 - 5t - 30}$

37. $x - 2\overline{)x^3 - 2x^2 + 4x - 8}$

38. $z - 1\overline{)z^3 + 8z^2 - 4z}$

TEST YOURSELF ON CHAPTER 2.

1. Determine the coefficient of each term.

   (a) $-2xy$ (b) $(2x)(3x)(-x^2)$ (c) $\dfrac{3x^2}{4}$

2. Evaluate each polynomial for the specified value of the variable.

   (a) $4x^2 - 2x + 1$ $[x = 3]$ (b) $y^3 - 2y + 1$ $[y = 1]$

   (c) $t^3 + 8t + 3$ $[t = 10]$

3. Evaluate $x^2y + 5x - 2y$

   (a) when $x = 2$ and $y = -1$,

   (b) when $x = 3$ and $y = 5$.

4. Simplify:

   (a) $3x - 4x$ (b) $x^2 + 2x - x^2 + 4x$

   (c) $4x - (3y + z) - (2x + z)$

5. Add:

$$\begin{array}{rrr} x & +\ 4y & -\ \ z \\ & -\ 2y & +\ 4z \\ 3x & & -\ \ z \\ & 2y & +\ \ z \\ \hline \end{array}$$

6. *Subtract* the bottom polynomial from the top one:

$$\begin{array}{r} 4x^2 + 2x - \ \ 5 \\ 3x^2 - 5x + 12 \\ \hline \end{array}$$

7. Multiply as indicated: $-5x^2 \cdot x \cdot x^3$

8. Multiply as indicated: $(4x + y)(x - y)$

9. Evaluate $(x^2 - 2x + 1)(x - 4)$ when $x = 2$.

10. Determine the degree of each polynomial.

    (a) $x^4 - 3x^2 + 2x - 1$ (b) $xy^2 - yx^2$ (c) $4 - 3x^2 + x^4 - 2x^4$

11. Divide and express the remainder.

$$x - 4\overline{)x^3 - x^2 - 10x - 6}$$

3

factoring

3.1 PRIME FACTORS

In this chapter you will learn how to simplify integers and polynomials. You begin by writing integers as products of integers. For example, 24 can be written as follows:

$$24 = 2 \cdot 2 \cdot 2 \cdot 3$$

DEFINITION. Let a and b ($\neq 0$) be integers. Then b is said to be a **factor of a** (or a **divisor of a**) if

$$a = bc$$

for some integer c. In this case, a is called a **multiple of b** and of c.

For the present, only integer factors will be considered.

EXAMPLE 1.

(a) 7 is a factor of 21 because $21 = 7 \cdot 3$. Clearly, 3 is also a factor of 21.
(b) 5 is a factor of 10 because $10 = 5 \cdot 2$.
(c) -5 is a factor of 10 because $10 = (-5)(-2)$.
(d) -5 is a factor of -10 because $-10 = (-5)2$. ■

If b is a factor of a, then $-b$ is also a factor of a. For suppose b is a factor of a. Then $a = bc$, where c is an integer. It is also true that $a = (-b)(-c)$, so that $-b$ is a factor of a.

"Prime" numbers serve as "atoms" among the integers. By multiplying primes, you can obtain all other positive integers except 1.

DEFINITION. An integer p (≥ 2) is said to be a **prime** if the only positive factors of p are 1 and p. An integer n (≥ 2) that is not a prime is called a **composite.**

Note that (by convention) 1 is neither a prime nor a composite.

EXAMPLE 2.

(a) 2 is a prime because the only positive factors of 2 are 1 and 2.
(b) 3 is a prime because the only positive factors of 3 are 1 and 3.
(c) 4 is a composite because $4 = 2 \cdot 2$. ■

EXAMPLE 3. The first eight primes are 2, 3, 5, 7, 11, 13, 17, and 19. ■

Every composite is expressible as the product of primes.

EXAMPLE 4. Express (a) 10, (b) 28, (c) 144 as the product of primes.

SOLUTION.

(a) $10 = 2 \cdot 5$ (b) $28 = 2 \cdot 2 \cdot 7$
(c) $144 = 12 \cdot 12$

Thus, 144 is not a prime. Moreover, 12 is not a prime because $12 = 2 \cdot 2 \cdot 3$. You can express 144 as the product of primes:

$$144 = (2 \cdot 2 \cdot 3) \cdot (2 \cdot 2 \cdot 3)$$

or

$$144 = 2 \cdot 2 \cdot 2 \cdot 2 \cdot 3 \cdot 3$$

or

$$144 = 3 \cdot 3 \cdot 2 \cdot 2 \cdot 2 \cdot 2$$

and so on. No matter how you express 144 *as the product of primes,* four of the factors are 2, two of the factors are 3, and there are no other factors. ■

Except for the order of the factors, *there is exactly one prime factorization of a composite.* The (only) prime factorization of a prime p is p itself.

It is convenient to use exponents in expressing prime factorization. Thus, in Example 3 (c)

$$144 = 2^4 \cdot 3^2$$

You can express a *negative* integer n (≤ -2) in terms of primes by considering the prime factorization of $|n|$. Thus

$$-144 = -(2^4 \cdot 3^2)$$

In the preceding examples, the prime factorizations were determined almost immediately. There is also a systematic way to determine prime factors.

EXAMPLE 5. Determine the prime factorizations of each:

(a) 60 (b) 54 (c) 72 (d) 23

SOLUTION. Test powers of the primes 2, 3, 5, 7, and so on, in order of magnitude, to determine whether they are factors of the integer in question.

(a) 2, the smallest prime, is a factor of 60. Observe that 4 ($= 2^2$) is the highest power of 2 that is a factor. Next, divide $\frac{60}{4}$, or 15, by 3, and obtain

$$15 = 3 \cdot 5.$$

As you know, 3 and 5 are both primes. Therefore the prime factorization is given by:

$$60 = 2^2 \cdot 3 \cdot 5$$

(b) 2 is a factor of 54, but 4 is not. Divide $\frac{54}{2}$, or 27, by 3. In fact, $27 = 3^3$. Thus

$$54 = 2 \cdot 3^3$$

(c) 2^3 is a factor of 72, but 2^4 is not.

$$\frac{72}{2^3} = \frac{72}{8} = 9 = 3^2$$

Thus

$$72 = 2^3 \cdot 3^2$$

(d) Neither 2 nor 3 is a factor of 23. The next prime is 5. If 23 were not a prime, it would have to have at least two (possibly equal) prime factors. Each of these would be at least as big as 5. Thus their product would be at least 25. Because this is not so, 23 must be a prime. (See Table 3.1 for the prime numbers less than 50.) ■

	2	3	~~4~~	5	~~6~~	7	~~8~~	~~9~~	~~10~~
11	~~12~~	13	~~14~~	~~15~~	~~16~~	17	~~18~~	19	~~20~~
~~21~~	~~22~~	23	~~24~~	~~25~~	~~26~~	~~27~~	~~28~~	29	~~30~~
31	~~32~~	~~33~~	~~34~~	~~35~~	~~36~~	37	~~38~~	~~39~~	~~40~~
41	~~42~~	43	~~44~~	~~45~~	~~46~~	47	~~48~~	~~49~~	

TABLE 3.1 *THE PRIMES LESS THAN 50.*

To obtain these primes, first cross out from the list the *composites* that are multiples of 2, then the composites that are multiples of 3, of 5, and of 7. Each of the remaining integers is a *prime,* because otherwise it would be divisible by $p \cdot q$, where p and q are each primes >7. The integer in question would then be greater than 7^2, or 49.

In order to simplify a fraction, which you will do in Chapter 4, you must determine all factors common to the numerator and denominator.

DEFINITION. Let m and n be integers. Then the integer b is called a **common divisor of m and n** if b is a factor (divisor) of m as well as of n. The largest of the common divisors of m and n is called the **greatest common divisor (GCD) of m and n.**

Write

$$\text{GCD}(m, n)$$

for this greatest common divisor, which is a *positive* integer.

EXAMPLE 6.

(a) The common divisors of 6 and 9 are 1 and 3, and their inverses, -1 and -3. The greatest common divisor of 6 and 9 is 3. Thus

$$\text{GCD}(6, 9) = 3$$

(b) The common divisors of -12 and 24 are 1, 2, 3, 4, 6, 12, and their inverses. Therefore

$$\text{GCD}(-12, 24) = 12$$

(c) The only common divisors of 3 and 4 are 1 and -1. Thus

$$\text{GCD}(3, 4) = 1$$ ■

Example 6(b) illustrates the fact that

if m divides n, then $\text{GCD}(m, n) = |m|$.

DEFINITION. Let $m_1, m_2, \ldots, m_k$ be integers. Then the integer b is called a **common divisor of $m_1, m_2, \ldots, m_k$** if b is a factor of *every one* of the integers $m_1, m_2, \ldots, m_k$. (If b is *not* a factor of *one or more* of these integers, it is not a common divisor of $m_1, m_2, \ldots, m_k$.) The largest of the common divisors of $m_1, m_2, \ldots, m_k$ is called the **greatest common divisor (GCD)** of $m_1, m_2, \ldots, m_k$, and is written

$$\text{GCD}(m_1, m_2, \ldots, m_k).$$

EXAMPLE 7.

(a) The common divisors of 6, 12, and 15 are 1 and 3 and their inverses. (Note that 2 divides 6 and 12, but not 15.) Thus

$$\text{GCD}(6, 12, 15) = 3$$

(b) $\text{GCD}(4, 8, 20, 24) = 4$ ■

If the integers are comparatively simple, their greatest common divisor can be determined at sight, as in the preceding examples. Otherwise, use prime factorization, as in the next example.

EXAMPLE 8. Determine GCD(28, 98, 140).

SOLUTION.

$$28 = 2^2 \cdot 7, \qquad 98 = 2 \cdot 7^2, \qquad 140 = 2^2 \cdot 5 \cdot 7$$

Determine the *smallest* power of each *common prime factor*. Then obtain the *product* of these smallest powers. Thus, the common prime factors are 2 and 7. The smallest power of 2 is 2^1 (in the prime factorization of 98). Were you to consider a higher power of 2, you would not obtain a factor of 98. Similarly, the smallest power of 7 is 7^1. The greatest common divisor is

$$2^1 \cdot 7^1, \quad \text{or} \quad 14. \qquad \blacksquare$$

In the remainder of the chapter you will learn how to factor polynomials. The **coefficients of a polynomial,** that is, the coefficients of its terms, often have factors in common. To factor a polynomial, you must first determine the GCD of its coefficients.

EXAMPLE 9. Determine the greatest common divisor of the coefficients of

$$90x^2 - 54x + 144.$$

SOLUTION. Factor the three coefficients:

$$90 = 2 \cdot 3^2 \cdot 5 \qquad -54 = -(2 \cdot 3^3) \qquad 144 = 2^4 \cdot 3^2$$

Here 2 and 3 are the primes *common to all three* factorizations. The smallest power of 2 that appears is 2^1. The smallest power of 3 that appears is 3^2. Thus the greatest common divisor of the coefficients is

$$2^1 \cdot 3^2, \text{ or } 18. \qquad \blacksquare$$

EXERCISES

In Exercises 1–30 determine the prime factorization of each integer.

1. 8
2. 14
3. 15
4. −16
5. 26
6. −27
7. −30
8. 33
9. −35
10. −36
11. 40
12. −42
13. 60
14. 66
15. −72
16. 84
17. 96
18. 100
19. 108
20. 110
21. 120
22. 121
23. 125
24. 128
25. 132
26. −144
27. 160
28. 225
29. −400
30. 144 000 000

In Exercises 31–48 determine the greatest common divisor of the indicated integers.

31. 8 and 24
32. 10 and 15
33. 16 and 22
34. 9 and 21
35. 36 and 42
36. -30 and -40
37. 28 and 48
38. 77 and -84
39. 132 and 144
40. 112 and 392
41. 2, 8, 15
42. 4, 6, 10
43. 15, 20, 25
44. -8, 20, -36
45. 100, 150, 175
46. 64, 72, 78
47. 12, 18, 28, 36
48. 40, 60, 72, 120

In Exercises 49–56 determine the greatest common divisor of the coefficients of each polynomial.

49. $2x - 6$
50. $3x^2 + 15$
51. $5x^2 - 25$
52. $20x^4 + 44$
53. $-x^3 + 9$
54. $27x^3 + 18x + 81$
55. $4x^3 + 8x^2 - 12x + 2$
56. $7x^2 + 5x - 35$
57. Determine the smallest positive integer that is a product of three distinct primes.
58. Determine the smallest positive integer that is a product of four distinct primes.
59. Determine all primes less than 50.
60. Determine all primes between 50 and 100.
61. Determine all factors of (a) 0, (b) 1, (c) -1.

3.2 COMMON FACTORS

Throughout the chapter all polynomials will have integral coefficients.

DEFINITION. Let P and $Q(\neq 0)$ be polynomials. Then Q is called a **factor of** P if

$$P = Q \cdot R$$

for some polynomial R. In this case P is a **multiple of** Q and of R.

EXAMPLE 1. The (constant) polynomial 3 is a factor of $3x^2 - 6x + 9$ because

$$\underbrace{3x^2 - 6x + 9}_{P} = \underbrace{3}_{Q}\ \underbrace{(x^2 - 2x + 3)}_{R}.$$

Note that 3 is a factor of *every term* of the given polynomial $3x^2 - 6x + 9$. ■

EXAMPLE 2. The polynomial x^2 is a factor of $x^5 + 4x^2$ because

$$\underbrace{x^5 + 4x^2}_{P} = \underbrace{x^2}_{Q}\underbrace{(x^3 + 4)}_{R}.$$

Note that x^2 is a factor of *both terms* of the given polynomial $x^5 + 4x^2$. ■

In Section 3.1 you determined the greatest common divisor of the *coefficients* of a polynomial. In Example 1, 3, the GCD of the *coefficients* of $3x^2 - 6x + 9$, is a *factor of this polynomial.* In Example 2, x^2 is a *factor of the polynomial* $x^5 + 4x^2$. Thus *a power of a variable can also be a factor of a polynomial.* In fact, several different variables, or powers of variables, as well as the GCD of the coefficients can be factors of a polynomial P. *The product of all of these is then a factor of* P, as the next example illustrates.

EXAMPLE 3. Let $P = 4x^4yz^3 + 8x^3y^2z^2$. Each of the following is a factor of *both terms of the polynomial* P; hence, each is a factor of P:

$$4, \quad x^3, \quad y, \quad z^2$$

Moreover, *the product of all of these factors is also a factor of* P because

$$4x^4yz^3 + 8x^3y^2z^2 = 4x^3yz^2(xz + 2y).$$ ■

EXAMPLE 4. Let $Q = 10x^3y + 15x^2y^2 + 5x^2yz$. Then $5 = \text{GCD}(10, 15, 5)$, and is thus a factor of Q. Similarly, x^2 and y are each factors of *all three terms of* Q, and are thus factors of Q. Finally, the *product of these,* $5x^2y$, is a factor of Q. ■

DEFINITION. Let P be an arbitrary *nonzero* polynomial. The **greatest common factor of the terms of** P, or for short, **the common factor of** P, is defined to be the product of

1. the greatest common divisor of the coefficients of P

and

2. the *smallest* power of every variable that occurs in *every* term of P.

Recall that clause 2 is similar to a rule for determining the GCD of integers, in Section 3.1.

In Example 4, the greatest common factor of Q is $5x^2y$.

DEFINITION. To **isolate the common factor** Q **of a polynomial** P, write

$$P = Q \cdot R,$$

where Q is the common factor of P and R is another polynomial. (Of course, R is also a factor of P.)

You isolate the common factor of a polynomial by applying the Distributive Laws.

EXAMPLE 5. Let $P = 7a^2bc^3 - 14ac^4$. Then the common factor Q of P is

$$7ac^3.$$

(Observe that b is not a factor of the second term of P.) Isolate the common factor Q by writing

$$\underbrace{7a^2bc^3 - 14ac^4}_{P} = \underbrace{7ac^3}_{Q}\underbrace{(ab - 2c)}_{R}.$$

■

EXAMPLE 6. Let $P = 3r^2st^3 + 6rt^2u + 7st^2u$. Then GCD(3, 6, 7) = 1. The only variable that occurs in *all three terms of P* is t; the *smallest power of t* that occurs is t^2. Thus Q, the common factor of P, is t^2, and you isolate the common factor by writing

$$\underbrace{3r^2st^3 + 6rt^2u + 7st^2u}_{P} = \underbrace{t^2}_{Q}\underbrace{(3r^2st + 6ru + 7su)}_{R}.$$

■

EXAMPLE 7. Let $P = -4x^4 - 6x^3 - 8x^2$. Here

$$Q = 2x^2$$

It is probably better to consider

$$-Q = -2x^2,$$

and to write

$$\underbrace{-4x^4 - 6x^3 - 8x^2}_{P} = \underbrace{-2x^2}_{-Q}\underbrace{(2x^2 + 3x + 4)}_{-R}$$

so that the coefficients of $-R$ are all *positive*. ■

EXERCISES

Isolate the common factor of each polynomial.

1. $4x + 2$
2. $5a - 15$
3. $20b^2 + 30$
4. $18m^3 + 24$
5. $3x^2 + 5x$
6. $12y^3 - 13y$
7. $a^3 - a^2$
8. $r^8 + r^4$
9. $2x^2 + 8x$
10. $3c^3 - 6c^2$
11. $10a^3 - 10a^2$
12. $12z^4 + 30z^2$
13. $64c^5 - 72c^3$
14. $108z^{10} - 96z^7$
15. $77t^7 + 121t^6$
16. $100x^{47} - 75x^{34}$
17. $x^2y + xy^2$
18. $abc^2 + ac^2$
19. $r^2s^3t - st^2$
20. $l^4m^3n^5 + l^4m^2n^4$
21. $6x^2y - 9xy$
22. $8a^2b^3 + 12a^3b^2$
23. $50m^5n^4 - 30mn^5$
24. $9a^5c^3 + 15a^3c^4$
25. $a^2b^4 + a^3b^3 - 3ab^4$
26. $m^7n^3 + m^6n^4 + m^5n^5$
27. $x^{13}y^4 + x^{10}y^7 + x^7y^{10}$
28. $a^5b^2c - a^3b^3c^3 + b^4c^4$
29. $r^4s^2t^2 + rs^2t^2 - rs^4t^4$
30. $a^4b^3c^2d + a^2b^2c^2d^2 + ab^2c^3d$
31. $8x^2 + 4x + 12$
32. $-5y^3 - 15y - 25$
33. $14z^3 - 21z^2 + 35$
34. $36x^4 - 96x + 16$
35. $75a^2 + 15a - 90$
36. $110c^3 + 55c + 33$
37. $4x^3 + 10x^2 + 12x$
38. $5z^4 + 10z^2 - 5z$
39. $3a^3 - 6a^2 + 3a$

40. $7b^5 + 14b^4 - 7b^2$
41. $5x^2y + 10xy - 15xy^2$
42. $3ab^3 + 18a^2b^2 - 15a^3b$
43. $16m^3n^3 + 4m^2n^2 + 12mn$
44. $25s^2 + 10st + 15st^2$
45. $5xy^4 - 10xy^2z + 20x^2y^2z^2$
46. $7x^2y^3z + 10xyz^4 - 5xz^3$
47. $a^3bc^3d^2 + 4acd - 3abd^2$
48. $15x^4y^6z^4 - 20x^3y^9z^7 + 10xz^6$
49. $16a^3c^2b^4 + 20a^5b - 44a^3c^4 + 40$
50. $9x^2y^2 - 15xy^3 + 18y^4 - 3$

3.3 FACTORING: AN OVERALL VIEW

Polynomials, like integers, can be decomposed into "prime factors".

DEFINITION. Let P be an arbitrary polynomial with integral coefficients. To **factor** P, write

$$P = Q \cdot R_1 \cdot R_2 \ldots R_n,$$

where Q is the common factor of P and where $R_1, R_2, \ldots, R_n$ are polynomials (with integral coefficients) that cannot be further decomposed. (If $Q = 1$, do not write it. See Example 2.)

The preceding definition is illustrated by two examples of factoring. *You are not expected to be able to work out these examples by yourself until after studying later sections.* The techniques of factoring will be the subject of the remainder of this chapter.

EXAMPLE 1. Factor: $4x^4 + 16x^3 + 12x^2$

SOLUTION. First isolate the common factor

$$4x^2.$$

Thus

$$\underbrace{4x^4 + 16x^3 + 12x^2}_{P} = \underbrace{4x^2}_{Q}\underbrace{(x^2 + 4x + 3)}_{R}$$

However, *the factor R can be further decomposed:*

$$x^2 + 4x + 3 = \underbrace{(x + 1)}_{R_1}\underbrace{(x + 3)}_{R_2}$$

The polynomials R_1 and R_2 cannot be further broken down. Thus the factorization is

$$\underbrace{4x^4 + 16x^3 + 12x^2}_{P} = \underbrace{4x^2}_{Q}\underbrace{(x + 1)}_{R_1}\underbrace{(x + 3)}_{R_2}.$$

■

The order in which you write the factors is not important, although you generally write the common factor first. Thus in Example 1, you could have written

$$4x^4 + 16x^3 + 12x^2 = 4x^2(x + 3)(x + 1).$$

Notice also that the common factor can sometimes be further decomposed. In Example 1, you could have written

$$4x^2 = 2x \cdot 2x.$$

By convention, do not further decompose the common factor.

Finally, note that the common factor may be 1, yet the polynomial P can be further decomposed.

EXAMPLE 2. Factor: $a^2 - 25$

SOLUTION. Here the common factor is 1. Observe that

$$a^2 - 25 = \underbrace{(a + 5)}_{R_1} \cdot \underbrace{(a - 5)}_{R_2}.$$ ■

3.4 DIFFERENCE OF SQUARES

One of the easiest types of factoring to recognize is known as the **difference of squares.**

Consider the product:

$$\begin{array}{l} a \;+ b \\ \underline{a \;- b} \\ a^2 + ab \\ \underline{\quad\; - ab - b^2} \\ a^2 \qquad\quad - b^2 \end{array}$$

The cross-terms, ab and $-ab$, are inverses; their sum is 0. Thus

$$a^2 - b^2 = (a + b)(a - b)$$

This method is used when you are given a binomial in which the two terms are squares that are separated by a minus sign.

EXAMPLE 1. Factor: $x^2 - 36$

SOLUTION. $36 = 6^2$. Thus $x^2 - 36$ is the difference of squares:

$$x^2 - 36 = (x + 6)(x - 6)$$ ■

Often you will consider polynomials of the form

$$x^2 - (ab)^2 \qquad \text{or} \qquad (xy)^2 - (ab)^2.$$

By the Associative and Commutative Laws of Multiplication,

$$\begin{aligned}(ab)^2 &= (ab)(ab)\\ &= (a\cdot a)(b\cdot b)\\ &= a^2b^2\end{aligned}$$

The square of a product is the product of the squares.

EXAMPLE 2. Factor: $9x^2 - 16y^2$

SOLUTION.

$$9x^2 = (3x)^2; \qquad 16y^2 = (4y)^2$$

Thus

$$9x^2 - 16y^2 = (3x + 4y)(3x - 4y)$$ ■

The present method applies to the *difference* of squares, but *not to the sum of squares.*

EXAMPLE 3. Factor: $x^2 + y^2$

SOLUTION. Try various combinations of factors to convince yourself that this binomial cannot be factored (as a product of polynomials with integral coefficients). Notice, in particular, that the cross-terms do not cancel in the following product:

$$\begin{array}{lll} x + y & & \\ \underline{x + y} & & \\ x^2 + xy & & \\ \underline{\quad\quad\; xy + y^2} & & \\ x^2 + 2xy + y^2 & & \end{array}$$ ■

Even powers of a number are squares. Observe that

$$(a^2)^2 = a^2\cdot a^2 = a^{2+2} = a^4.$$

Similarly,

$$(a^3)^2 = a^3\cdot a^3 = a^{3+3} = a^6$$

Therefore

$$(a^2)^2 = a^{2\cdot 2}, \qquad (a^3)^2 = a^{2\cdot 3}$$

Similarly,

$$(a^n)^2 = a^{2n}$$

for every (positive) integer n. Thus

$$a^{10} = (a^5)^2, \qquad a^{34} = (a^{17})^2, \qquad 4a^8 = 2^2(a^4)^2 = (2a^4)^2$$

The difference of squares method can be extended to the case of the *difference of even powers.*

EXAMPLE 4. Factor: $a^6 - b^4$

SOLUTION.

$$a^6 = (a^3)^2; \qquad b^4 = (b^2)^2$$

Thus

$$a^6 - b^4 = (a^3 + b^2)(a^3 - b^2)$$

■

EXAMPLE 5. Factor: $25x^{10} - y^2$

SOLUTION.

$$25x^{10} = (5x^5)^2$$

Thus

$$25x^{10} - y^2 = (5x^5 + y)(5x^5 - y)$$

■

EXAMPLE 6. Factor: $x^4 - y^3$

SOLUTION. This is *not* the difference of squares. The present method does not apply. In fact, $x^4 - y^3$ cannot be further decomposed (as the product of polynomials). ■

In some examples first isolate the common factor before applying the present method. This greatly simplifies the remaining factoring.

EXAMPLE 7. Factor: $5x^2 - 20y^2$

SOLUTION.

$$\begin{aligned} 5x^2 - 20y^2 &= 5(x^2 - 4y^2) \\ &= 5(x + 2y)(x - 2y) \end{aligned}$$

■

EXAMPLE 8. Factor: $72s^{12} - 98t^6$

SOLUTION.

$$\begin{aligned} 72s^{12} - 98t^6 &= 2(36s^{12} - 49t^6) \\ &= 2(6s^6 + 7t^3)(6s^6 - 7t^3) \end{aligned}$$

■

The difference of squares method can apply twice in the same example.

EXAMPLE 9.

$$x^4 - y^4 = (x^2 + y^2)\overbrace{(x^2 - y^2)}^{\text{difference of squares}}$$
$$= (x^2 + y^2)(x + y)(x - y)$$ ■

Finally, you may be given the difference between squares of polynomials.

EXAMPLE 10. Factor: $(3a + b)^2 - (x + y)^2$

SOLUTION. Each of the polynomials $(3a + b)^2$ and $(x + y)^2$ is a square. Thus

$$(3a + b)^2 - (x + y)^2 = [(3a + b) + (x + y)][(3a + b) - (x + y)]$$

Remove the inner parentheses and write

$$(3a + b)^2 - (x + y)^2 = (3a + b + x + y)(3a + b - x - y)$$ ■

EXERCISES

Factor each polynomial.

1. $x^2 - 1$
2. $a^2 - 4$
3. $y^2 - 49$
4. $1 - z^2$
5. $x^2 - y^2$
6. $m^2 - n^2$
7. $s^2 - 36$
8. $t^2 - 100$
9. $4x^2 - y^2$
10. $x^2 - 9z^2$
11. $25x^2 - 4a^2$
12. $4m^2 - 81n^2$
13. $100a^2 - 64b^2$
14. $144r^2 - 121t^2$
15. $a^4 - b^2$
16. $x^4 - 25$
17. $9 - y^6$
18. $z^{10} - 36$
19. $a^4 - b^8$
20. $c^6 - d^6$
21. $s^{24} - t^{22}$
22. $x^{100} - y^2$
23. $x^4 - 4y^6$
24. $a^{12} - 9b^{18}$
25. $100 - 49a^6$
26. $36 - 81x^{12}$
27. $9x^2 - 64y^2$
28. $x^6 - 4y^6$
29. $121a^8 - 100b^6$
30. $144x^6 - 169y^2$
31. $2x^2 - 2y^2$
32. $5a^2 - 5b^2$
33. $3a^2 - 12x^2$
34. $7a^4 - 28b^2$
35. $44x^4 - 99y^6$
36. $90m^8 - 1210$
37. $x^3 - x$
38. $3a^3 - 3a$
39. $3a^3 - 12a$
40. $x^5 - x^3$
41. $3ax^2 - 3ay^2$
42. $5by^4 - 20b$
43. $a^4 - b^4$
44. $x^8 - y^4$
45. $m^4 - 16$
46. $a^8 - 81$
47. $16y^{12} - 1$
48. $81z^4 - 16a^8$
49. $(2x + 3)^2 - y^2$
50. $z^2 - (a - 1)^2$
51. $(a + b)^2 - (a - b)^2$
52. $(9x^2 - 1)^3$

3.5 FACTORING QUADRATIC TRINOMIALS, I

DEFINITION. A **quadratic polynomial** is a second-degree polynomial.

EXAMPLE 1.

(a) $x^2 + 3x - 5$, (b) $3y^2 + 10$, (c) $5z^2$

are each quadratic polynomials. In each case, 2 is the highest degree of a term. In particular, polynomial (a), which has three terms, is a **quadratic trinomial.**

■

You will first factor quadratic trinomials in a single variable with leading coefficient 1, such as

$$x^2 + 2x + 1, \qquad y^2 - 5y + 4, \qquad z^2 - z - 6.$$

These are of the form

$$x^2 + Mx + N.$$

Of course, the variable can be y, z, and so on. Also, the coefficients, M and N, will be integers.

To determine the factorization of $x^2 + Mx + N$, consider the product:

$$\begin{array}{lll} x + a & & \\ \underline{x + b} & & \\ x^2 + ax & & \\ \underline{\phantom{x^2 +{}} bx} & \underline{ + ab} & \\ x^2 + (a + b)x + ab & & \end{array}$$

Thus

$$(x + a)(x + b) = x^2 + (a + b)x + ab$$

You must find a and b so that $(x + a)(x + b) = x^2 + \qquad Mx + N$.
If

$$a + b = M \qquad \text{and} \qquad ab = N,$$

then

$$(x + a)(x + b) = x^2 + (a + b)x + ab = x^2 + Mx + N$$

In other words, you want factors, a and b, of N whose sum, $a + b$, is M.

The product of two positive factors, or of two negative factors, is positive. The product of a positive and negative factor (in either order) is negative.

The sum of positive numbers is positive; the sum of negative numbers is negative. The sum of numbers unlike in sign can be positive, negative, or 0.

Table 3.2 is useful in determining the signs of a and b when factoring

$$x^2 + Mx + N, \qquad \text{that is,} \qquad x^2 + (a + b)x + ab.$$

$M(= a + b)$	$N(= ab)$	a and b
positive	positive	both positive
negative	positive	both negative
positive	negative	one positive, one negative
negative	negative	one positive, one negative

TABLE 3.2

EXAMPLE 2. Factor: $x^2 + 2x + 1$

SOLUTION. Here $M = 2$, $N = 1$. Because both M and N are positive, determine *positive* factors a and b of 1. Clearly,

$$a = 1, \qquad b = 1$$

For then

$$ab = 1 \cdot 1 = 1 = N \qquad \text{and} \qquad a + b = 1 + 1 = 2 = M$$

Thus

$$x^2 + 2x + 1 = (x + 1)(x + 1) = (x + 1)^2$$

■

EXAMPLE 3. Factor: $y^2 - 5y + 4$

SOLUTION.

$$y^2 \underbrace{- 5y}_{M < 0} \underbrace{+ 4}_{N > 0}$$

Determine *negative* factors of 4. The possibilities are

$$(-4)(-1) \qquad \text{and} \qquad (-2)(-2).$$

If you try these, you will see that $-4, -1$ is the correct pair:

$$\begin{array}{rr} y - 4 & \\ \underline{y - 1} & \\ y^2 - 4y & \\ \underline{\quad - y + 4} & \\ y^2 - 5y + 4 & \checkmark \end{array} \qquad \begin{array}{rr} y - 2 & \\ \underline{y - 2} & \\ y^2 - 2y & \\ \underline{\quad - 2y + 4} & \\ y^2 - 4y + 4 & \times \end{array}$$

$$y^2 - 5y + 4 = (y - 4)(y - 1)$$

■

EXAMPLE 4. Factor: $z^2 - z - 6$

SOLUTION.

$$z^2 \underbrace{- z}_{M<0} \underbrace{- 6}_{N<0}$$

One factor of -6 must be *positive;* the other factor must be *negative.* The possibilities are

$$6(-1), \qquad (-6)\cdot 1, \qquad 3(-2), \qquad (-3)2.$$

$$\begin{array}{llll}
z + 6 & z - 6 & z + 3 & z - 3 \\
\underline{z - 1} & \underline{z + 1} & \underline{z - 2} & \underline{z + 2} \\
z^2 + 6z & z^2 - 6z & z^2 + 3z & z^2 - 3z \\
\underline{\quad - z - 6} & \underline{\quad z - 6} & \underline{\quad - 2z - 6} & \underline{\quad 2z - 6} \\
z^2 + 5z - 6 & z^2 - 5z - 6 & z^2 + z - 6 & z^2 - z - 6 \\
\times & \times & \times & \checkmark
\end{array}$$

Thus the last pair, $-3, 2$, works.

$$z^2 - z - 6 = (z - 3)(z + 2)$$ ■

EXAMPLE 5. $x^2 + 3x - 5$ cannot be factored (into polynomials with integral coefficients) because neither of the possible pairs of factors of -5,

$$5(-1), \qquad (-5)\cdot 1,$$

"add up to" 3. ■

EXAMPLE 6. Factor: $5x^2 + 30x + 40$

SOLUTION. First isolate the common factor.

$$5x^2 + 30x + 40 = 5(x^2 + 6x + 8)$$

Now factor $x^2 + 6x + 8$.

$$\begin{array}{l}
x + 4 \\
\underline{x + 2} \\
x^2 + 4x \\
\underline{\qquad 2x + 8} \\
x^2 + 6x + 8
\end{array}$$

Thus

$$5x^2 + 30x + 40 = 5(x + 4)(x + 2)$$ ■

EXERCISES

In Exercises 1–32 factor each polynomial.

1. $x^2 + 3x + 2$
2. $y^2 + 6y + 5$
3. $a^2 + 7a + 6$
4. $b^2 + 9b + 8$
5. $x^2 + 4x + 4$
6. $x^2 + 6x + 9$
7. $m^2 + m - 2$
8. $n^2 - n - 2$
9. $s^2 + s - 12$
10. $t^2 - 8t + 12$
11. $a^2 - 10a + 25$
12. $y^2 - 9y + 14$
13. $z^2 + 5z - 14$
14. $b^2 - 9b + 20$
15. $2x^2 + 8x + 6$
16. $5a^2 + 20a + 20$
17. $2t^2 - 12t + 18$
18. $4y^2 - 4y - 8$
19. $x^3 + 6x^2 + 9x$
20. $a^4 + 3a^3 - 10a^2$
21. $2z^5 - 6z^4 - 56z^3$
22. $3x^2y + 36xy + 105y$
23. $5x^2y^2z^2 - 40xy^2z^2 + 60y^2z^2$
24. $56 + x - x^2$
25. $21 + 10a + a^2$
26. $4a^2 - 4a - 48$
27. $-2y^2 - 24y - 70$
28. $-z^2 + 8z - 12$
29. $3t^2 + 33t + 54$
30. $5u^2 + 25u - 30$
31. $9x^4 + 27x^3 + 18x^2$
32. $100x^5 - 700x^4 + 600x^3$

In Exercises 33–42 some of these polynomials cannot be factored by the present method. Factor the polynomial, if possible, or write "DOES NOT FACTOR".

33. $x^2 + x + 1$
34. $y^2 + 2y - 3$
35. $z^2 - 2z - 3$
36. $z^2 - 2z + 3$
37. $m^2 + 4m + 4$
38. $m^2 + 5m + 4$
39. $m^2 - 4m - 4$
40. $m^2 - 3m - 4$
41. $t^2 - 7t - 6$
42. $t^2 + 9t + 10$

3.6 FACTORING QUADRATIC TRINOMIALS, II

In Section 3.5 you factored quadratic trinomials with leading coefficient 1. To factor

$$2x^2 + 5x + 2,$$

first consider the factors of $2x^2$. Clearly

$$2x \cdot x = 2x^2$$

Thus consider:

$$\begin{array}{r} 2x + \square \\ x + \square \\ \hline 2x^2 + \end{array}$$

Try to fill in the boxes with the factors 1 and 2 of the numerical term, 2. Observe that there are two possibilities:

$$\begin{array}{r} 2x \;+\; 2 \\ x \;+\; 1 \\ \hline 2x^2 + 2x \qquad \\ 2x + 2 \\ \hline 2x^2 + 4x + 2 \\ \times \end{array} \qquad \begin{array}{r} 2x \;+\; 1 \\ x \;+\; 2 \\ \hline 2x^2 + \;\; x \qquad \\ 4x + 2 \\ \hline 2x^2 + 5x + 2 \\ \checkmark \end{array}$$

Thus

$$2x^2 + 5x + 2 = (2x + 1)(x + 2)$$

In general, a quadratic trinomial (in the single variable x) is of the form

$$Lx^2 + Mx + N,$$

where $L \neq 0$. Assume $L > 0$. [If the leading coefficient is negative, first factor out -1. For example,

$$-3x^2 + 7x - 9 = -(3x^2 - 7x + 9)]$$

To determine the factors of a quadratic trinomial, consider the product:

$$\begin{array}{l} \quad ax + b \\ \quad cx + d \\ \hline acx^2 + bcx \\ \qquad\qquad adx + bd \\ \hline acx^2 + (bc + ad)x + bd \end{array}$$

Thus

$$(ax + b)(cx + d) = acx^2 + (bc + ad)x + bd$$

If

$$ac = L, \qquad bc + ad = M, \qquad \text{and} \qquad bd = N,$$

then

$$(ax + b)(cx + d) = acx^2 + (bc + ad)x + bd = Lx^2 + Mx + N$$

Determine integers a, b, c, d, such that

$$ac = L \qquad \text{and} \qquad bd = N.$$

Then test these to see whether

$$bc + ad = M.$$

You may have to make several attempts to obtain the right combination.

EXAMPLE 1. Factor: $2x^2 + 7x + 3$

SOLUTION. Here $L = 2, M = 7, N = 3$

Consider only *positive* factors of 2 and of 3. The possible factorizations are

$$2 \cdot 1 = 2 \qquad \text{and} \qquad 3 \cdot 1 = 3.$$

The possible combinations are

$$a = 2, \quad b = 3; \qquad c = 1, \quad d = 1;$$

that is,

$$(2x + 3)(x + 1);$$

or

$$a = 2, \quad b = 1; \qquad c = 1, \quad d = 3;$$

that is,

$$(2x + 1)(x + 3).$$

First consider:

$$\begin{array}{r} 2x + 3 \\ x + 1 \\ \hline 2x^2 + 3x \qquad \\ 2x + 3 \\ \hline 2x^2 + 5x + 3 \\ \times \quad \end{array}$$

The cross-term is wrong!

Next consider:

$$\begin{array}{r} 2x + 1 \\ x + 3 \\ \hline 2x^2 + x \qquad \\ + 6x + 3 \\ \hline 2x^2 + 7x + 3 \\ \checkmark \quad \end{array}$$

Thus

$$2x^2 + 7x + 3 = (2x + 1)(x + 3)$$ ■

EXAMPLE 2. Factor: $5y^2 - 8y - 4$

SOLUTION. Begin by considering:

$$\begin{array}{l} 5y \;+ b \\ \underline{\;y \;\;+ d\;} \\ 5y^2 + \;\; by \\ \underline{\qquad + 5dy \qquad\qquad + bd} \\ 5y^2 + \underbrace{(b + 5d)}_{-8}y + \underbrace{bd}_{-4} \end{array}$$

Because $bd = -4$, you could have:

$$4(-1) \quad \left[\begin{array}{c} 5y + 4 \\ \underline{y - 1} \end{array} \quad \text{or} \quad \begin{array}{c} 5y - 1 \\ \underline{y + 4} \end{array}\right]$$

$$(-4)\cdot 1 \quad \left[\begin{array}{c} 5y - 4 \\ \underline{y + 1} \end{array} \quad \text{or} \quad \begin{array}{c} 5y + 1 \\ \underline{y - 4} \end{array}\right]$$

$$2(-2) \quad \left[\begin{array}{c} 5y + 2 \\ \underline{y - 2} \end{array} \quad \text{or} \quad \begin{array}{c} 5y - 2 \\ \underline{y + 2} \end{array}\right]$$

By trial and error, you will see that

$$b = 2, \qquad d = -2$$

works:

$$\begin{array}{l} 5y \;+\; 2 \\ \underline{\;y \;-\; 2\;} \\ 5y^2 + \;2y \\ \underline{\qquad - 10y - 4} \\ 5y^2 - \;\;8y - 4 \end{array}$$

Thus

$$5y^2 - 8y - 4 = (5y + 2)(y - 2) \qquad \blacksquare$$

The method of this section applies to a polynomial after its common factor has been isolated.

EXAMPLE 3. Factor: $6y^3 - 33y^2 + 36y$

SOLUTION. First isolate the common factor, $3y$. The other factor is a quadratic trinomial that can be further decomposed.

$$6y^3 - 33y^2 + 36y = 3y(2y^2 - 11y + 12)$$

Now factor $2y^2 - 11y + 12$:

$$\begin{array}{l} 2y + b \\ \underline{\ y + d\ } \\ 2y^2 + by \\ \underline{\quad + 2dy \qquad + bd} \\ 2y^2 + \underbrace{(b + 2d)}_{-11}y + \underbrace{bd}_{12} \end{array}$$

Clearly, *both* b and d must be *negative* in order that

$$bd = 12 \qquad \text{and} \qquad b + 2d = -11.$$

The possibilities are:

$$(-12)(-1) \quad \left[\begin{array}{l} 2y - 12 \\ \underline{\ y - 1\ } \end{array} \quad \text{or} \quad \begin{array}{l} 2y - 1 \\ \underline{\ y - 12} \end{array}\right]$$

$$(-6)(-2) \quad \left[\begin{array}{l} 2y - 6 \\ \underline{\ y - 2\ } \end{array} \quad \text{or} \quad \begin{array}{l} 2y - 2 \\ \underline{\ y - 6\ } \end{array}\right]$$

$$(-4)(-3) \quad \left[\begin{array}{l} 2y - 4 \\ \underline{\ y - 3\ } \end{array} \quad \text{or} \quad \begin{array}{l} 2y - 3 \\ \underline{\ y - 4\ } \end{array}\right]$$

The correct combination is $b = -3, d = -4$.

$$\begin{array}{l} 2y - 3 \\ \underline{\ y - 4\ } \\ 2y^2 - 3y \\ \underline{\quad - 8y + 12} \\ 2y^2 - 11y + 12 \end{array}$$

Thus

$$6y^3 - 33y^2 + 36y = 3y(2y - 3)(y - 4)$$ ■

This method applies with very little modification to quadratic trinomials in two variables.

EXAMPLE 4. Factor: $3x^2 + 2xy - y^2$

SOLUTION. Determine a factorization of the form

$$(ax + by)(cx + dy).$$

As above, consider:

$$\begin{array}{l} \quad ax + by \\ \quad cx + dy \\ \hline acx^2 + bcxy \\ \qquad\quad adxy \qquad\qquad + bdy^2 \\ \hline \underbrace{acx^2}_{3} + \underbrace{(bc + ad)}_{2}xy + \underbrace{bdy^2}_{-1} \end{array}$$

Clearly,

$$ac = 3, \qquad bd = -1, \qquad bc + ad = 2$$

By trial and error, you will see that the correct combination is

$$a = 3, c = 1; b = -1, d = 1:$$

$$\begin{array}{l} 3x \;-\; y \\ \;\; x \;+\; y \\ \hline 3x^2 - xy \\ \qquad + 3xy - y^2 \\ \hline 3x^2 + 2xy - y^2 \end{array}$$

Thus

$$3x^2 + 2xy - y^2 = (3x - y)(x + y)$$ ■

EXERCISES

In Exercises 1–44 factor each polynomial.

1. $2x^2 + 3x + 1$
2. $3y^2 + 4y + 1$
3. $2a^2 + 9a + 9$
4. $4b^2 + 6b + 2$
5. $2m^2 + 5m - 3$
6. $2n^2 - 5n - 3$
7. $2a^2 + 7a - 4$
8. $4a^2 + 20a + 25$
9. $9x^2 - 6x + 1$
10. $4z^2 + 4z + 1$
11. $3m^2 - 4m - 7$
12. $4m^2 - 8m + 3$
13. $6a^2 + 5a - 6$
14. $7b^2 + 9b + 2$
15. $10x^2 + 13x + 4$
16. $9y^2 - 36y - 13$
17. $5a^2 + 16a + 12$
18. $2b^2 + 5b - 12$
19. $6b^2 + 10b + 4$
20. $6x^2 + 33x - 18$
21. $45x^2 + 30x + 5$
22. $8z^3 + 4z^2 - 12z$
23. $8y^4 + 20y^3 - 100y^2$
24. $10a^2b + 17ab + 3b$
25. $12a^6 - 25a^5 + 12a^4$
26. $21 - 36z - 12z^2$
27. $x^2 + 2xy + y^2$
28. $a^2 - 2ab + b^2$
29. $4u^2 + 4uv + v^2$
30. $s^2 - 4st + 4t^2$
31. $y^2 + 3yz + 2z^2$
32. $x^2 + 5ax + 6a^2$
33. $a^2 + 3ab - 4b^2$
34. $a^2 + ab - 2b^2$
35. $r^2 + 3rs - 10s^2$
36. $u^2 + 5uv - 14v^2$
37. $9a^2 - 24ab - 20b^2$
38. $16x^2 + 2xy - 3y^2$
39. $a^2b^2 + 2abcd + c^2d^2$
40. $x^2y^2 - 4xyuv + 4u^2v^2$
41. $5x^2 + 30xy + 45y^2$
42. $4x^2 - 12xy + 8y^2$
43. $3ax^2 + 12axy + 12ay^2$
44. $-2a^2x^2 - 14abx^2 - 24b^2x^2$

In Exercises 45–52 some of these polynomials cannot be factored by the present method. Factor the polynomial, if possible, or write "DOES NOT FACTOR".

45. $2x^2 + 2x + 1$
46. $5y^2 + 26y + 5$
47. $3a^2 - 4a + 1$
48. $2b^2 + 3b + 2$
49. $9x^2 - 12x + 4$
50. $6t^2 + t - 5$
51. $a^2 + 2ab + 2b^2$
52. $c^2 + 7cd + 12d^2$

3.7 FACTORING BY GROUPING

The Distributive Laws often enable you to factor polynomials. Rearrange the terms so as to group together terms with a common factor.

EXAMPLE 1. Factor: $ax + ay + bx + by$

SOLUTION. The first two terms have the common factor a; the last two terms have the common factor b.

$$\begin{aligned} ax + ay + bx + by &= (ax + ay) + (bx + by) \\ &= a(x + y) + b(x + y) \end{aligned}$$

Note that $a(x + y)$ and $b(x + y)$ also have a common factor—namely, $x + y$.

$$a(x + y) + b(x + y) = (a + b)(x + y)$$

Thus

$$ax + ay + bx + by = (a + b)(x + y)$$ ■

EXAMPLE 2. Factor: $3a + ab + b^2 + 3b$

SOLUTION.

$$\begin{aligned} 3a + ab + b^2 + 3b &= (3a + ba) + (b^2 + 3b) \\ &= (3 + b)a + (b + 3)b \\ &= (b + 3)a + (b + 3)b \\ &= (b + 3)(a + b) \end{aligned}$$ ■

Sometimes, this method enables you to utilize a previous method of factoring. In the next example you will be able to use the difference of squares.

EXAMPLE 3. Factor: $ax^2 - ay^2 - bx^2 + by^2$

SOLUTION.

$$\begin{aligned} ax^2 - ay^2 - bx^2 + by^2 &= (ax^2 - ay^2) - (bx^2 - by^2) \\ &= a(x^2 - y^2) - b(x^2 - y^2) \\ &= (a - b)(x^2 - y^2) \\ &= (a - b)(x + y)(x - y) \end{aligned}$$ ■

EXAMPLE 4. Factor: $36 + a^2x^2 - 9a^2 - 4x^2$

SOLUTION. First reorder the terms.

$$\begin{aligned} 36 + a^2x^2 - 9a^2 - 4x^2 &= (36 - 9a^2) + (-4x^2 + a^2x^2) \\ &= 9(4 - a^2) - x^2(4 - a^2) \\ &= (9 - x^2)(4 - a^2) \\ &= (3 + x)(3 - x)(2 + a)(2 - a) \end{aligned}$$ ■

EXAMPLE 5. Factor: $sx^2 + 2sxy + sy^2 + tx^2 + 2txy + ty^2$

SOLUTION.

$$\begin{aligned} sx^2 + 2sxy + sy^2 + tx^2 + 2txy + ty^2 &= (sx^2 + 2sxy + sy^2) \\ &\quad + (tx^2 + 2txy + ty^2) \\ &= s(x^2 + 2xy + y^2) \\ &\quad + t(x^2 + 2xy + y^2) \\ &= (s + t)(x^2 + 2xy + y^2) \\ &= (s + t)(x + y)^2 \end{aligned}$$ ■

EXERCISES

Factor each polynomial.

1. $cu + cv + du + dv$
2. $ax + ay - bx - by$
3. $ax - ay - bx + by$
4. $5u + 5v - tu - tv$
5. $y^2 + by - ay - ab$
6. $2ax + 2bx + ay + by$
7. $cx - dx + 3cy - 3dy$
8. $mr + 2ms + 2nr + 4ns$
9. $4ax + 4ay - 6bx - 6by$
10. $5ay + 5az - 2by - 2bz$
11. $18ax - 54ay - 4bx + 12by$
12. $25su + 20sv - 15tu - 12tv$
13. $2ax + 3by + 3ay + 2bx$
14. $5mx - ny + nx - 5my$
15. $xu - 6yv + 2yu - 3xv$
16. $2bx - 14ay + 7ax - 4by$

17. $a^2 + 5a + ab + 5b$

18. $x^2 - x + 6xy - 6y$

19. $uv + u + v + v^2$

20. $r^2 - 3s + rs - 3r$

21. $a^3 + a^2 + ab + b$

22. $ax^2 + by^2 + ay^2 + bx^2$

23. $au^2 + bu^2 - av^2 - bv^2$

24. $a^2x + a^2y - 4x - 4y$

25. $ax^2 - 4ay^2 + x^2 - 4y^2$

26. $cz^2 - c + dz^2 - d$

27. $8ax^2 - 18ay^2 + 4bx^2 - 9by^2$

28. $3ab^2 + c^2x - 3ac^2 - b^2x$

29. $s^2x^2 - s^2y^2 - t^2x^2 + t^2y^2$

30. $4a^2x^2 - a^2y^2 - 4b^2x^2 + b^2y^2$

31. $x^2r^2 + y^2r^2 - x^2s^2 - y^2s^2$

32. $144a^2y^2 - 36a^2z^2 - 64b^2y^2 + 16b^2z^2$

33. $ax^2 + axy + ay^2 + bx^2 + bxy + by^2$

34. $ab^2 + 3abc + 3c^2 + ac^2 + 9bc + 3b^2$

35. $ax^2 + 3ax + 2a + bx^2 + 3bx + 2b$

36. $ay^2 - 5ay + 4a - by^2 + 5by - 4b$

37. $ms^2 + 6mst + 9mt^2 + 2ns^2 + 12nst + 18nt^2$

38. $ay^2 + 8ayz + 15az^2 - by^2 - 8byz - 15bz^2$

39. $a^2x^2 + 2a^2x + a^2 - b^2x^2 - 2b^2x - b^2$

40. $x^2z^2 - 5x^2z + 6x^2 - y^2z^2 + 5y^2z - 6y^2$

41. $x^2y^2 + 7x^2y + 10x^2 - 4y^2 - 28y - 40$

42. $a^2u^2 - 4b^2u^2 - 4a^2v^2 + 16b^2v^2$

43. $axz^2 + ayz^2 + bxz^2 + byz^2$

44. $6ax + 15ay - 18ax - 45ay$

45. $a^2u^3x^2 - b^2u^3x^2 + a^2u^3y^2 - b^2u^3y^2$

46. $3a^2y^3 + 30a^2y^2 + 75a^2y - 3b^2y^3 - 30b^2y^2 - 75b^2y$

3.8 SUMS AND DIFFERENCES OF CUBES

Recall that

$$a^2 - b^2 = (a + b)(a - b).$$

This factoring method was known as the difference of squares. However, *the sum of squares,*

$$a^2 + b^2,$$

cannot be factored (as a product of polynomials with integral coefficients).

In the case of cubes *you can factor both the sum and difference of cubes.* Observe that in the following products, the middle terms cancel:

$$\begin{array}{r}
a^2 - ab + b^2 \\
a + b \\
\hline
a^3 - a^2b + ab^2 \qquad \\
+ a^2b - ab^2 + b^3 \\
\hline
a^3 \qquad\qquad\quad + b^3
\end{array}
\qquad
\begin{array}{r}
a^2 + ab + b^2 \\
a - b \\
\hline
a^3 + a^2b + ab^2 \qquad \\
- a^2b - ab^2 - b^3 \\
\hline
a^3 \qquad\qquad\quad - b^3
\end{array}$$

Thus

$$a^3 + b^3 = (a + b)(a^2 - ab + b^2)$$

$$a^3 - b^3 = (a - b)(a^2 + ab + b^2)$$

EXAMPLE 1. Factor: $x^3 + 8$

SOLUTION.

$$8 = 2^3$$

Here $a = x$ and $b = 2$. Thus

$$x^3 + 8 = (x + 2)(x^2 - 2x + 4)$$ ■

You will consider polynomials of the form

$$x^3 + (ab)^3, \qquad (xy)^3 - (ab)^3, \qquad \text{and so on.}$$

As was the case for squares, *the cube of a product equals the product of the cubes.* For,

$$\begin{aligned} (ab)^3 &= (ab)(ab)(ab) \\ &= (a \cdot a \cdot a)(b \cdot b \cdot b) \\ &= a^3b^3 \end{aligned}$$

EXAMPLE 2. Factor: $x^3 - 27a^3$

SOLUTION.

$$\begin{aligned} x^3 - 27a^3 &= x^3 - (3a)^3 \\ &= (x - 3a)[x^2 + 3ax + (3a)^2] \\ &= (x - 3a)(x^2 + 3ax + 9a^2) \end{aligned}$$ ■

Note that

$$\begin{aligned} (a^2)^3 &= a^2 \cdot a^2 \cdot a^2 \\ &= a^{2+2+2} \\ &= a^6. \end{aligned}$$

Also,

$$\begin{aligned}(a^3)^3 &= a^3 \cdot a^3 \cdot a^3\\ &= a^{3+3+3}\\ &= a^{3\cdot 3}\end{aligned}$$

Similarly,

$$(a^n)^3 = a^{3n}$$

for every (positive) integer n. Thus

$$a^{3n}$$

is a cube. For example,

$$a^{15} = (a^5)^3, \qquad a^{27} = (a^9)^3, \qquad 8a^{12} = 2^3(a^4)^3 = (2a^4)^3$$

Thus the present methods apply to binomials of the forms

$$a^{3m} + b^{3n} \text{ [or } (a^m)^3 + (b^n)^3\text{]}$$

and

$$a^{3m} - b^{3n} \text{ [or } (a^m)^3 - (b^n)^3\text{]}.$$

EXAMPLE 3. Factor: $x^6 - y^9$

SOLUTION.

$$\begin{aligned}x^6 - y^9 &= (x^2)^3 - (y^3)^3\\ &= (x^2 - y^3)[(x^2)^2 + x^2y^3 + (y^3)^2]\\ &= (x^2 - y^3)[x^4 + x^2y^3 + y^6]\end{aligned}$$

■

EXAMPLE 4. Factor: $8a^3 + b^{24}$

SOLUTION.

$$\begin{aligned}8a^3 + b^{24} &= (2a)^3 + (b^8)^3\\ &= (2a + b^8)[(2a)^2 - 2ab^8 + (b^8)^2]\\ &= (2a + b^8)(4a^2 - 2ab^8 + b^{16})\end{aligned}$$

■

The remaining examples combine factoring sums and differences of cubes with other factoring methods.

EXAMPLE 5. Factor: $x^2y^3 - 64x^2$

SOLUTION.

$$\begin{aligned} x^2y^3 - 64x^2 &= x^2(y^3 - 64) \\ &= x^2(y^3 - 4^3) \\ &= x^2(y - 4)(y^2 + 4y + 16) \end{aligned}$$

■

EXAMPLE 6. Factor: $x^3y^2 - 4x^3 + a^3y^2 - 4a^3$

SOLUTION.

$$\begin{aligned} x^3y^2 - 4x^3 + a^3y^2 - 4a^3 &= (x^3y^2 - 4x^3) + (a^3y^2 - 4a^3) \\ &= x^3(y^2 - 4) + a^3(y^2 - 4) \\ &= (x^3 + a^3)(y^2 - 4) \\ &= (x + a)(x^2 - ax + a^2)(y + 2)(y - 2) \end{aligned}$$

■

EXERCISES

Factor each polynomial.

1. $a^3 - 1$
2. $b^3 + 1$
3. $y^3 + 8$
4. $z^3 - 64$
5. $c^3 + 27$
6. $8 - v^3$
7. $x^3 + 1000$
8. $a^3 - 125$
9. $x^3 - 8z^3$
10. $u^3 - (7v)^3$
11. $27a^3 - 64b^3$
12. $1000x^3 + 27y^3$
13. $y^6 - z^6$
14. $m^6 + n^6$
15. $c^9 + 1$
16. $d^{12} - 1$
17. $x^6 - 64$
18. $x^9 - (2y)^9$
19. $(3u)^{15} - 1$
20. $1\,000\,000 - z^9$
21. $(xy)^{18} - 1$
22. $x^{21}y^{24} + 1$
23. $7z^3 - 56y^3$
24. $4a^3 - 108b^3$
25. $a^3x^2 - 1000x^2$
26. $3bc^3 + 81b$
27. $20x^6y^6 + 160$
28. $a^9b^9c^9d^2 - d^2$
29. $ax^3 + 5x^3 - ay^3 - 5y^3$
30. $xy^3 + y^3 + xz^3 + z^3$
31. $am^3 + an^3 + 4m^3 + 4n^3$
32. $ax^3 + by^3 - ay^3 - bx^3$
33. $16ax^3 + 2ay^3 + 8bx^3 + by^3$
34. $16am^3 - 54an^3 - 24bm^3 + 81bn^3$
35. $1000a^3y - 4000a^3z - 27y + 108z$
36. $125c + 250d - 8cx^2 - 16dx^2$
37. $a^2x^3 - b^2x^3 + a^2y^3 - b^2y^3$
38. $x^2y^3 - 9y^3 - x^2z^3 + 9z^3$
39. $5am^3x - 5an^3x + 10am^3y - 10an^3y$
40. $8ay^2 - ax^3y^2 - 16by^2 + 2bx^3y^2$
41. $a^3 - (b + 1)^3$
42. $(x + y)^3 + (a - 2)^3$

REVIEW EXERCISES FOR CHAPTER 3

In Set I the various factoring techniques are separated according to section. It is also important to recognize which technique applies to a given polynomial. Thus in Set II you are asked to select the appropriate factoring technique.

SET I

3.1 PRIME FACTORS

1. Which of the following are primes?

   (a) 11 (b) 12 (c) 13 (d) 23

2. Determine the prime factorization of each integer.

   (a) 30 (b) 40 (c) -144 (d) 900

3. Determine the greatest common divisor of the indicated integers.

   (a) 5 and 20 (b) -16 and 24
   (c) 49 and 14 (d) 50, 75, 125

4. Determine the greatest common divisor of the coefficients of the indicated polynomials.

   (a) $4x^2 - 8$ (b) $12x + 20$
   (c) $7x^2 + 49x - 7$ (d) $3x^2 - 5x + 15$

3.2 COMMON FACTORS

In Exercises 5–8 isolate the common factor of each polynomial.

5. $9x^3 - 9x^2 + 6x$
6. $r^2st^2 - rst^2 + 2r^2st$
7. $15x^2y^2 - 10x^2y + 20x^2$
8. $42a^3b^3c^4 - 24a^2b^3c^3 + 48a^2b^2c^3$

In Exercises 9–32 factor each polynomial.

3.4 DIFFERENCE OF SQUARES

9. $a^2 - 9$
10. $s^2 - t^2$
11. $25m^2 - 16n^2$
12. $a^4 - b^4$
13. $10x^6 - 10y^2$
14. $81x^4 - 1$

3.5 FACTORING QUADRATIC TRINOMIALS, I

15. $x^2 + 4x + 3$
16. $y^2 - 5y + 4$
17. $4m^2 - 28m + 48$

3.6 FACTORING QUADRATIC TRINOMIALS, II

18. $4x^2 - 8x + 3$
19. $6x^2 + 14x + 4$
20. $a^2 + 4ab + 4b^2$

3.7 FACTORING BY GROUPING

21. $ax + ay + bx + by$
22. $3ax + 3ay - 6bx - 6by$
23. $xy + x + y + y^2$
24. $a^2x^2 - a^2y^2 + b^2y^2 - b^2x^2$
25. $4am^2 + 4an^2 + 4bm^2 + 4bn^2$
26. $ax^2 - ay^2 + bx^2 - by^2$

3.8 SUMS AND DIFFERENCES OF CUBES

27. $x^3 + y^3$
28. $x^3 - 8$
29. $a^3 + 1000b^3$
30. $y^3 - 27z^3$
31. $a^6 - b^6$
32. $a^2x^3 + a^2y^3 - 25x^3 - 25y^3$

SET II

Factor each polynomial, if possible. Choose the appropriate technique in each case. Warning: Some of these polynomials cannot be factored by the present methods. Write "DOES NOT FACTOR" in this case.

1. $5x^2 - 25x$
2. $x^2 - 25$
3. $x^2 + 25$
4. $x^2 + 10x + 16$
5. $y^2 - y - 20$
6. $z^2 + 6z + 8$
7. $s^2t^2 - 16s^2$
8. $a^4 - 16$
9. $a^4 + 16$
10. $a^3 + 8$
11. $x^2y^3 + x^2y^2 + x^3y$
12. $5x^2 - 45$
13. $4x^2 + 8x + 3$
14. $10y^2 - 3y - 1$
15. $xy + x + y + 1$
16. $x^2 + 3ax + 2a^2$
17. $2m^2 + 13m + 21$
18. $x^2 - 7x + 7$
19. $y^2 - 7y + 6a^2$
20. $ax + by - ay - bx$

TEST YOURSELF ON CHAPTER 3.

1. Determine the prime factorization of each.

   (a) 12 (b) 80 (c) -132

2. Determine the greatest common divisor of the indicated integers.

   (a) 15 and 25 (b) $-3, 6, -27$

3. Isolate the common factor of each polynomial.

   (a) $18x^2 - 27$ (b) $25x^3 + 40x^2 + 100x$

In Exercises 4–12, factor each polynomial.

4. $9x^2 - 25y^2$
5. $x^2 + x - 6$
6. $5ax + 5ay + 10bx + 10by$
7. $a^3 - 8b^3$
8. $a^4 - 16b^4$
9. $4m^2 - 28mn + 40n^2$
10. $19ax^2 - 19ay^2$
11. $xy - bx - ay + ab$
12. $15a^2 - 7a - 2$

4

rational expressions

4.1 SIMPLIFYING FRACTIONS

DEFINITION. A **fraction** is an expression of the form $\frac{a}{b}$, where a and b ($\neq 0$) are real numbers. a is called the **numerator** and b the **denominator** of the fraction $\frac{a}{b}$.

A fraction expresses division of real numbers.

$$\frac{3}{4}, \quad \frac{-12}{20}, \quad \frac{9}{5}, \quad \frac{6}{6}, \quad \frac{0}{3}$$

are each fractions. *The same notation, $\frac{a}{b}$, is used for the quotient of a divided by b and for the corresponding fraction that expresses this.*

Quotients of real numbers (with nonzero divisors) are real numbers. Thus fractions represent real numbers.

As you know, a rational number is a real number that can be expressed in the form $\frac{M}{N}$, where M and N ($\neq 0$) are integers. Thus rational numbers can be expressed as fractions. In addition, fractions such as

$$\frac{\sqrt{2}}{3}, \quad \frac{1}{\pi}, \quad \frac{\pi}{\sqrt{3}}$$

express irrational numbers.

Recall that when you divide, if $\frac{a}{b} = c$, then $\frac{a}{b}$ and c correspond to the same point on the number line.

DEFINITION. **Two fractions are** said to be **equivalent** if they represent the same real number.

Geometrically, equivalent fractions correspond to the same point on the number line. Write

$$\frac{a}{b} = \frac{c}{d}$$

when these fractions are equivalent.

EXAMPLE 1.

$$\frac{1}{2} = \frac{2}{4}$$

To see this, first divide the line segment between 0 and 1 into 2 equal parts. The point of division represents $\frac{1}{2}$. [Figure 4.1(a)]

Next, divide the line segment between 0 and 2 into 4 equal parts, as in Figure 4.1(b). The first point of division to the right of 0 represents $\frac{2}{4}$. But this is precisely the point that corresponds to $\frac{1}{2}$.

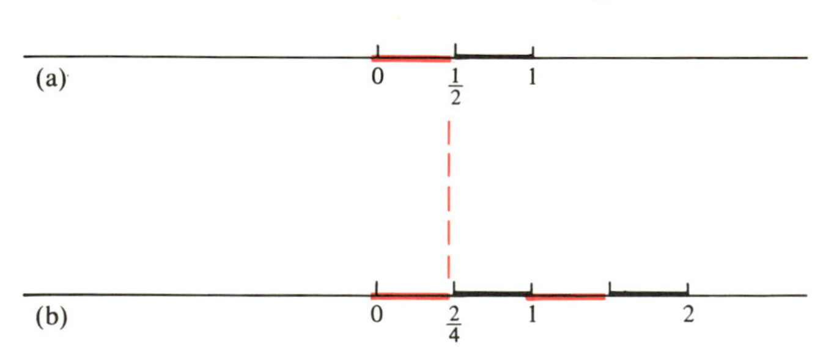

FIGURE 4.1

Observe, finally, that the **cross-products of** these **fractions,**

$$\frac{1}{2} \bowtie \frac{2}{4}$$

are equal, that is,

$$1 \cdot 4 = 2 \cdot 2.$$

In general,

$$\frac{a}{b} = \frac{c}{d} \qquad \textit{if} \qquad ad = bc,$$

and also,

$$ad = bc \quad \text{if} \quad \frac{a}{b} = \frac{c}{d}$$

Thus the equality of the cross-products ad and bc indicates the equivalence of the fractions $\frac{a}{b}$ *and* $\frac{c}{d}$, *and vice versa.*

EXAMPLE 2.

(a) $\frac{6}{9} = \frac{2}{3}$

because the cross-products are equal:

$$\underbrace{6 \cdot 3}_{18} = \underbrace{9 \cdot 2}_{18}$$

(b) $\frac{\sqrt{2}}{1} = \frac{2}{\sqrt{2}}$

because

$$\underbrace{\sqrt{2}\sqrt{2}}_{2} = 1 \cdot 2$$

■

In Example 1, $\frac{2}{4}$ can be obtained from $\frac{1}{2}$ by *multiplying* the numerator and denominator by 2. Thus

$$\frac{1}{2} = \frac{1 \cdot 2}{2 \cdot 2} = \frac{2}{4}$$

In Example 2(a), $\frac{6}{9}$ can be obtained from $\frac{2}{3}$ by *multiplying* the numerator and denominator by 3. Thus

$$\frac{2}{3} = \frac{2 \cdot 3}{3 \cdot 3} = \frac{6}{9}$$

Note that $\frac{2}{3}$ can be obtained from $\frac{6}{9}$ by *dividing* the numerator and denominator by 3. Thus

$$\frac{6}{9} = \frac{2 \cdot \cancel{3}}{3 \cdot \cancel{3}} = \frac{2}{3}$$

FUNDAMENTAL PRINCIPLE OF FRACTIONS

Let $\frac{a}{b}$ be a fraction. Then an equivalent fraction is obtained if the numerator and denominator are each multiplied or divided by the same nonzero real number k. Thus

$$\frac{a}{b} = \frac{ak}{bk}$$

and

$$\frac{ak}{bk} = \frac{a\cancel{k}}{b\cancel{k}} = \frac{a}{b}$$

To show the first equality, observe that by the Commutative and Associative Laws of Multiplication, the cross-products are equal:

$$a(bk) = b(ak)$$

Thus

$$\frac{a}{b} = \frac{ak}{bk}$$

EXAMPLE 3. By the Fundamental Principle,

$$\frac{2}{5} = \frac{6}{15}$$

Here let $a = 2, b = 5, k = 3$. Then

$$\frac{2}{5} = \frac{2 \cdot 3}{5 \cdot 3} = \frac{6}{15}$$ ■

Let a and b be *positive real numbers*. Consider the following *equivalent* ways of expressing the same negative number:

$\frac{-a}{b}$ ("$-$" in the numerator)

$\frac{a}{-b}$ ("$-$" in the denominator)

$-\frac{a}{b}$ ("$-$" before the entire fraction)

Observe that

$$\frac{-a}{b} = \frac{(-a)(-1)}{b(-1)} = \frac{a}{-b}.$$

In fact,

$$\frac{-a}{b} = \frac{a}{-b} = -\frac{a}{b}$$

[See Figure 4.2.] The first form,

$$\frac{-a}{b},$$

is called the **standard form.**

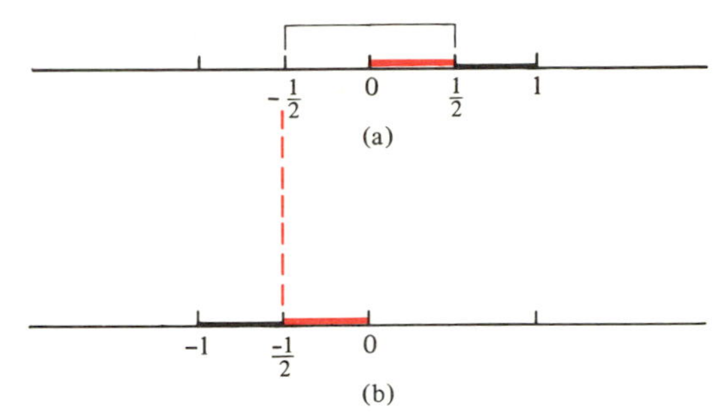

FIGURE 4.2 (a) $\frac{1}{2}$ is obtained by dividing the line segment from 0 to 1 into 2 equal parts. The point of division represents $\frac{1}{2}$. Then $-\frac{1}{2}$ and $\frac{1}{2}$ lie on opposite sides of, but are the same distance from the origin. (b) $\frac{-1}{2}$ is obtained by dividing the line segment from -1 to 0 into 2 equal parts. The point of division represents $\frac{-1}{2}$. Note that $-\frac{1}{2} = \frac{-1}{2}$.

EXAMPLE 4. Express (a) $\frac{7}{-5}$ and (b) $-\frac{4}{9}$ in standard form.

SOLUTION. (a) $\frac{7}{-5} = \frac{-7}{5}$ (b) $-\frac{4}{9} = \frac{-4}{9}$ ■

Observe that

$$\frac{-a}{-b} = \frac{a(-1)}{b(-1)} = \frac{a}{b}.$$

For example,

$$\frac{-3}{-4} = \frac{3}{4}$$

Furthermore,

$$-\frac{-8}{5} = -\left(-\frac{8}{5}\right) = \frac{8}{5} \quad \text{and} \quad -\frac{-3}{-2} = -\frac{3}{2} = \frac{-3}{2}$$

A rational number can be expressed as a fraction in different ways. For example,

$$\frac{6}{12} = \frac{2}{4} = \frac{1}{2} = \frac{-1}{-2}$$

DEFINITION. Let M and N be integers and let $N \neq 0$. The rational number given by the fraction $\frac{M}{N}$ is expressed in **lowest terms** if M and N have no factors in common other than 1 and -1, and if N is positive. To **simplify a fraction of the form** $\frac{M}{N}$ means to express the corresponding rational number in lowest terms.

The above rational number can be expressed in lowest terms as $\frac{1}{2}$. You can use the Fundamental Principle to express a rational number in lowest terms. In any fraction that represents the rational number, divide out all factors that appear in *both* the numerator and denominator. Write a negative rational number in standard form.

EXAMPLE 5. Express in lowest terms the rational number given by $\frac{35}{-55}$.

SOLUTION.

$$\frac{35}{-55} = \frac{7 \cdot \not{5}}{-11 \cdot \not{5}} = \frac{-7}{11}$$ ■

From now on, you need not distinguish between a fraction and the real number (rational or irrational) it represents. You will speak of simplifying fractions instead of reducing rational numbers to lowest terms.

Before considering more difficult examples, here are some basic rules for dividing powers of a number a. The first two of these rules were given in Section 2.5.

Assume a is a *nonzero* number. Let m and n be positive integers. When you divide a^m by a^n, that is,

$$\frac{a^m}{a^n},$$

divide by as many a's as possible common to both numerator and denominator. Thus consider:

$$\frac{\overbrace{a \cdot a \cdot a \ldots a}^{m \text{ times}}}{\underbrace{a \cdot a \cdot a \ldots a}_{n \text{ times}}}$$

1. If $m > n$, all of the factors a will divide out from the denominator. Each time you divide by a, replace it by 1:

$$\frac{\overbrace{\overbrace{\overset{1 \cdot 1 \cdot 1 \ldots 1}{\not{a} \cdot \not{a} \cdot \not{a} \ldots \not{a}}}^{n \text{ times}} \cdot \overbrace{a \cdot a \cdot a \ldots a}^{m - n \text{ times}}}^{m \text{ times}}}{\underbrace{\underset{1 \cdot 1 \cdot 1 \ldots 1}{\not{a} \cdot \not{a} \cdot \not{a} \ldots \not{a}}}_{n \text{ times}}}$$

The denominator is the product of 1's—hence it is 1. The numerator is 1 times the product of $(m - n)$ a's. Hence the quotient is

$$\frac{a^{m-n}}{1}, \quad \text{or} \quad a^{m-n}.$$

Thus **if $m > n$, then $\frac{a^m}{a^n} = a^{m-n}$**

2. If $m = n$, all factors a divide out from both numerator and denominator. Numerator and denominator each reduce to 1. The resulting quotient is therefore 1:

$$\frac{\overbrace{\cancel{a} \cdot \cancel{a} \cdot \cancel{a} \ldots \cancel{a}}^{m \text{ times}}}{\underbrace{\cancel{a} \cdot \cancel{a} \cdot \cancel{a} \ldots \cancel{a}}_{m(=n) \text{ times}}}$$

(each cancelled a replaced by $1 \cdot 1 \cdot 1 \ldots 1$)

Thus, as you would expect,

$$\frac{a^m}{a^m} = 1$$

3. If $m < n$, then all factors a divide out from the numerator. But $n - m$ factors remain in the denominator:

$$\frac{\overbrace{\cancel{a} \cdot \cancel{a} \cdot \cancel{a} \ldots \cancel{a}}^{m \text{ times}}}{\underbrace{\underbrace{\cancel{a} \cdot \cancel{a} \cdot \cancel{a} \ldots \cancel{a}}_{m \text{ times}} \cdot \underbrace{a \cdot a \cdot a \ldots a}_{n - m \text{ times}}}_{n \text{ times}}}$$

(each cancelled a replaced by $1 \cdot 1 \cdot 1 \ldots 1$)

Thus **if $m < n$, then $\frac{a^m}{a^n} = \frac{1}{a^{n-m}}$**

EXAMPLE 6. Simplify:

(a) $\frac{5^7}{5^4}$ (b) $\frac{3^6}{3^6}$ (c) $\frac{11^5}{11^{10}}$

SOLUTION.

(a) $\frac{5^7}{5^4} = 5^{7-4} = 5^3$ (b) $\frac{3^6}{3^6} = 1$ (c) $\frac{11^5}{11^{10}} = \frac{1}{11^{10-5}} = \frac{1}{11^5}$ ■

In Example 6 the fractions involve powers of a single prime. When the numerator or denominator involves more than one prime, it is often useful to determine the prime factors of both numerator and denominator. Divide powers of the same prime by means of the above rules.

EXAMPLE 7. Simplify: $\frac{72}{156}$

SOLUTION.

$$72 = 2^3 \cdot 3^2, \qquad 156 = 2^2 \cdot 3 \cdot 13$$

$$\frac{72}{156} = \frac{2^3 \cdot 3^2}{2^2 \cdot 3^1 \cdot 13} = \frac{2^1 \cdot 3^1}{13^1} = \frac{6}{13}$$ ■

The numerator has been divided by 12 (or $2^2 \cdot 3$) to give 6; the denominator has been divided by 12 to give 13.

EXAMPLE 8. Simplify: $\frac{3^8 \cdot 5^2 \cdot 11^4}{2 \cdot 3^7 \cdot 5^4 \cdot 11^4}$

SOLUTION. Here the given fraction is in factored form.

$$\frac{3^8 \cdot 5^2 \cdot 11^4}{2 \cdot 3^7 \cdot 5^4 \cdot 11^4} = \frac{3^1}{2 \cdot 5^2} = \frac{3}{50}$$ ■

EXERCISES

In Exercises 1–10 express positive fractions without minus signs; express negative fractions in standard form.

1. $\frac{1}{-2}$
2. $\frac{-2}{-3}$
3. $\frac{-\sqrt{3}}{-5}$
4. $\frac{7}{-12}$
5. $-\frac{5}{8}$
6. $-\frac{-15}{-13}$
7. $\frac{-(-5)}{8}$
8. $-\frac{9}{1}$
9. $\frac{2}{-3}$
10. $-\frac{3}{-2}$

In Exercises 11–30 simplify each fraction, that is, express the corresponding rational number in lowest terms.

11. $\frac{2}{4}$
12. $\frac{-3}{-9}$
13. $\frac{4}{8}$
14. $\frac{3}{-15}$
15. $\frac{12}{24}$
16. $\frac{20}{30}$
17. $\frac{-9}{-27}$
18. $\frac{-16}{64}$
19. $\frac{10}{8}$
20. $\frac{64}{32}$
21. $\frac{8}{-32}$
22. $-\frac{14}{20}$
23. $\frac{15}{-24}$
24. $\frac{45}{60}$
25. $\frac{42}{96}$
26. $\frac{124}{142}$
27. $\frac{36}{96}$
28. $\frac{100}{135}$
29. $\frac{80}{125}$
30. $\frac{400}{650}$

In Exercises 31–46 simplify each fraction. (Multiply out the resulting numerator and denominator.)

31. $\frac{2^6}{2^3}$ 32. $\frac{3^4}{3^4}$ 33. $\frac{5^5}{5^7}$ 34. $\frac{2^8}{2^3}$

35. $\frac{3^2}{3^3}$ 36. $\frac{7^{19}}{7^{17}}$ 37. $\frac{3^7}{3^{10}}$ 38. $\frac{5^3}{3^4}$

39. $\frac{2^8 \cdot 3^2}{2^6}$ 40. $\frac{5^2 \cdot 7}{5^3}$ 41. $\frac{2^5}{2^3 \cdot 3}$ 42. $\frac{3^2 \cdot 5 \cdot 7}{3 \cdot 5^2 \cdot 7}$

43. $\frac{2^5 \cdot 3^2 \cdot 5^9}{2 \cdot 3^3 \cdot 5^{10}}$ 44. $\frac{11^3 \cdot 13^2}{11^4 \cdot 13}$ 45. $\frac{2^8 \cdot 5^2 \cdot 7^4}{2^7 \cdot 3 \cdot 5 \cdot 7^5}$ 46. $\frac{2^{13} \cdot 5^{11} \cdot 7^{10}}{2^{14} \cdot 5^9 \cdot 7^8}$

4.2 EVALUATING RATIONAL EXPRESSIONS

A *rational number* was defined to be a real number that can be written in the form $\frac{M}{N}$, where M and $N (\neq 0)$ are integers.

DEFINITION. Let P and $Q (\neq 0)$ be polynomials. Then $\frac{P}{Q}$, which expresses division of polynomials, is called a **rational expression.** P is called the **numerator** and Q the **denominator** of the rational expression $\frac{P}{Q}$.

EXAMPLE 1. Each of the expressions (a)–(e) is a rational expression.

(a) $\frac{x + 7}{x - 2}$ (b) $\frac{x^2 + 5x - 3}{x^3}$

(c) $\frac{7}{x^4 + 3x^2 - 1}$. Here the numerator is the constant polynomial 7.

(d) $\frac{x^4 + 3x^2 - 1}{2}$. Here the denominator is the nonzero constant polynomial 2. ■

Every integer M can be regarded as the rational number $\frac{M}{1}$. Similarly, *every polynomial P can be thought of as the rational expression* $\frac{P}{1}$. Thus the polynomial $x^2 - 5x + 2$ is a rational expression (with denominator 1). Note that *every rational number is a rational expression.* For example, the rational number $\frac{3}{5}$ has the (constant) polynomials 3 and 5 as numerator and denominator, respectively. For that matter, every real number *a can be regarded as the rational expression* $\frac{a}{1}$.

To evaluate a rational expression for given values of the variables, replace the variables by the given numbers. *Each time* a variable occurs, replace it by the *same number.*

EXAMPLE 2. Evaluate $\frac{x^2 - 3x + 10}{x^2 + x + 5}$ when x is (a) 1, (b) 4, (c) -2.

SOLUTION.

(a) When x is 1, the numerator is

$$1^2 - 3 \cdot 1 + 10, \qquad \text{or} \qquad 8,$$

and the denominator is

$$1^2 + 1 + 5, \qquad \text{or} \qquad 7.$$

The value of the rational expression is $\frac{8}{7}$ when $x = 1$.

(b) $x = 4$:

$$\frac{4^2 - 3 \cdot 4 + 10}{4^2 + 4 + 5} = \frac{14}{25}$$

(c) $x = -2$:

$$\frac{(-2)^2 - 3(-2) + 10}{(-2)^2 + (-2) + 5} = \frac{20}{7}$$

■

EXAMPLE 3. Evaluate $\frac{x^2y^2 - 2x + y}{5xy + 5}$ when $x = 2$ and $y = 3$.

SOLUTION. Substitute 2 for each occurrence of x and **3** for each occurrence of **y**:

$$\frac{2^2 \cdot \mathbf{3}^2 - 2 \cdot 2 + \mathbf{3}}{5 \cdot 2 \cdot \mathbf{3} + 5} = \frac{35}{35} = 1$$

For these values of the variables, the value of the rational expression is 1. ■

EXAMPLE 4. Consider the rational expression

$$\frac{0}{x^2 + 5}.$$

When you replace x by any real number, the value of the expression is 0 because the numerator is 0 whereas the value of the denominator is nonzero. Thus $\frac{0}{x^2 + 5}$ reduces to 0. ■

Because division by 0 is undefined, a rational expression $\frac{P}{Q}$ is undefined for values of the variable(s) that make $Q = 0$. For example,

$$\frac{x - 1}{x}$$

is not defined when $x = 0$.

EXERCISES

In Exercises 1–10 evaluate each rational expression for the specified value of the variable.

1. $\frac{3}{x - 1}$ when $x = 2$

2. $\frac{10}{x + 10}$ when $x = 10$

3. $\frac{x + 5}{3}$ when $x = 1$

4. $\frac{y - 9}{7}$ when $y = 2$

5. $\frac{x + 2}{x - 3}$ when $x = 1$

6. $\frac{x + 7}{x + 4}$ when $x = 1$

7. $\frac{z^2 + z - 2}{z + 5}$ when $z = 5$

8. $\frac{3z - 1}{z^2 + 2}$ when $z = 2$

9. $\frac{t^2 + 4t - 3}{t^2 + 6t - 7}$ when $t = 0$

10. $\frac{u^4 - 3u^3 + 5u^2 - 2}{u^3}$ when $u = -1$

11. Evaluate $\frac{x^2 - 2}{x^2 + 2}$ when (a) $x = 0$, (b) $x = 1$, (c) $x = -1$.

12. Evaluate $\frac{y + 3}{y^2 - 2y + 5}$ when (a) $y = 1$, (b) $y = 3$, (c) $y = 10$.

13. Evaluate $\frac{t + 1}{t^4 - 1}$ when (a) $t = 2$, (b) $t = -2$, (c) $t = 3$.

14. Evaluate $\frac{7}{t^3 + t^2 + 1}$ when (a) $t = 0$, (b) $t = -1$, (c) $t = -2$.

15. Evaluate $\frac{y^3 + 2}{5}$ when (a) $y = -1$, (b) $y = 2$, (c) $y = -2$.

16. Evaluate $\frac{t^2 + 3t + 1}{t}$ when (a) $t = 1$, (b) $t = -1$, (c) $t = 2$.

17. Evaluate $\frac{x^{16} + 3x^{13}}{x^{11} - x + 1}$ when (a) $x = 0$, (b) $x = 1$, (c) $x = -1$.

18. Evaluate $\frac{s^4 - 3s^2 + 2s - 1}{s^3 + 2}$ when (a) $s = 2$, (b) $s = 3$, (c) $s = 5$.

19. Evaluate $\frac{z^4 + z^2 - 1}{2z}$ when (a) $z = 10$, (b) $z = -10$, (c) $z = 2$.

20. Evaluate $\dfrac{x}{1 - x^4}$ when (a) $x = 2$, (b) $x = -2$, (c) $x = 3$.

In Exercises 21–30 evaluate each rational expression for the specified values of the variables.

21. $\dfrac{1}{xy}$ when $x = 1, y = 2$

22. $\dfrac{3}{s + t}$ when $s = 1, t = -2$

23. $\dfrac{x}{y + 1}$ when $x = 5, y = 7$

24. $\dfrac{x^2 + z}{1 - z^2}$ when $x = 0, z = 2$

25. $\dfrac{x^2 + xy + y}{3}$ when $x = 1, y = -1$

26. $\dfrac{x^3 + 3y^2 + y^3}{x^2 + xy + y^2}$ when $x = 1, y = 5$

27. $\dfrac{4x^2 + 3x - 5}{y^2 + 2y + 1}$ when $x = 0, y = 10$

28. $\dfrac{5xy}{x^2y^4 - xy^2}$ when $x = 2, y = 2$

29. $\dfrac{1 + xyz}{1 - xyz}$ when $x = 1, y = 2, z = 3$

30. $\dfrac{t^2 - 3rst}{x + 2y}$ when $r = 0, s = t = 1, x = 2, y = 3$

31. Evaluate $\dfrac{1}{x^2 + 3y}$ when (a) $x = 2$, $y = -2$; (b) $x = -2$, $y = 2$.

32. Evaluate $\dfrac{7x - 2y}{5}$ when (a) $x = 0$, $y = 5$; (b) $x = 5$, $y = 0$.

33. Evaluate $\dfrac{x + 1}{y - 3}$ when (a) $x = -1$, $y = 4$; (b) $x = 1$, $y = -4$.

34. Evaluate $\dfrac{x^2 - 5x + 2}{2xy}$ when (a) $x = 3$, $y = 2$; (b) $x = y = -2$.

35. Evaluate $\dfrac{x^2 - 3xy + y^2}{x^2 + y^2}$ when (a) $x = 0$, $y = -1$; (b) $x = y = 10$.

36. Evaluate $\dfrac{xyz}{x^2 + y^2 - 1}$ when (a) $x = 6$, $y = -1$, $z = 1$; (b) $x = 3$, $y = 2$, $z = 10$.

37. Evaluate $\dfrac{r^2 - 2s^2 + 1}{st}$ when (a) $r = 1$, $s = -1$, $t = 2$; (b) $r = 3$, $s = -3$, $t = 1$.

38. Evaluate $\dfrac{u^2 - 3v + w}{uvw + u^2x}$ when (a) $u = 1$, $v = -1$, $w = 2$, $x = 1$; (b) $u = 1$, $v = -1$, $w = 0$, $x = 2$.

39. The acceleration (or change in velocity) of a body is given by $\dfrac{v - u}{t}$, where u is the initial velocity (in feet per second), v is its final velocity (in feet per second), and t is the time in motion (in seconds). Determine the acceleration if $v = 88$, $u = 52$, $t = 4$.

40. An object is attracted to the earth by a force of magnitude $\dfrac{500\,000\,000}{h^2}$ where h is the distance (in miles) of the object from the center of the earth. Determine the magnitude of the force when (a) $h = 5000$, (b) $h = 10\,000$.

41. The number of ties a salesman can sell is given by $\dfrac{4000}{x - a}$, where a is the cost and x the sales price (in dollars). How many ties can be sold if (a) $x = 6$, $a = 4$; (b) $x = 8$, $a = 4$?

42. An ice cream plant is able to produce x gallons of chocolate ice cream and y gallons of vanilla ice cream per day, where $y = \dfrac{800 - 2x}{10 + x}$.

    Determine the number of gallons of vanilla ice cream produced on a day when 40 gallons of chocolate ice cream are made.

4.3 DIVIDING MONOMIALS

A rational expression is of the form $\dfrac{P}{Q}$, where P and Q are polynomials and $Q \neq 0$.

DEFINITION. **Two rational expressions** $\dfrac{P}{Q}$ and $\dfrac{P'}{Q'}$ **are** said to be **equivalent** if their cross-products $\dfrac{P}{Q} \times \dfrac{P'}{Q'}$ are equal:

$$PQ' = P'Q$$

Write

$$\frac{P}{Q} = \frac{P'}{Q'}$$

when these rational expressions are equivalent. To **simplify a rational expression** or to **reduce a rational expression to lowest terms,** write it as an equivalent expression in which numerator and denominator have no common factors other than 1 or -1.

EXAMPLE 1.

$$\frac{2x}{3} = \frac{4x}{6} \qquad \text{because} \qquad 2x \cdot 6 = 4x \cdot 3$$

It can be shown that $\frac{2x}{3}$ and $\frac{4x}{6}$ agree for all values of x. For instance, when $x = 3$, both expressions equal 2; when $x = 5$, both expressions equal $\frac{10}{3}$. Note that

$$\frac{4x}{6} = \frac{2x \cdot \cancel{2}}{3 \cdot \cancel{2}} = \frac{2x}{3}$$

Thus $\frac{4x}{6}$ can be reduced to $\frac{2x}{3}$. ■

Recall that the Fundamental Principle of Fractions asserts that an equivalent fraction is obtained if the numerator a and denominator b of a fraction $\frac{a}{b}$ are each multiplied or divided by the same nonzero number k. Thus

$$\frac{ak}{bk} = \frac{a}{b}$$

where k, a, and b are real numbers and $k \neq 0, b \neq 0$.

The Fundamental Principle of Rational Expressions asserts that an equivalent expression is obtained if the numerator P and denominator Q of the rational expression $\frac{P}{Q}$ are each multiplied or divided by the same nonzero rational expression, R. Thus

$$\frac{P}{Q} = \frac{P \cdot R}{Q \cdot R}$$

and

$$\frac{P \cdot R}{Q \cdot R} = \frac{P \cdot \cancel{R}}{Q \cdot \cancel{R}} = \frac{P}{Q}$$

EXAMPLE 2. Show that $\frac{6x^2y}{8y^2} = \frac{3x^2}{4y}$.

SOLUTION 1. Show the equality of the cross-products, $6x^2 \cdot 4y$ and $3x^2 \cdot 8y^2$. Clearly,

$$6x^2y \cdot 4y = 24x^2y^2 \qquad \text{and} \qquad 3x^2 \cdot 8y^2 = 24x^2y^2$$

Thus the cross-products are equal, and, by definition, the given rational expressions are equal:

$$\frac{6x^2y}{8y^2} = \frac{3x^2}{4y}$$

SOLUTION 2. Use the Fundamental Principle of Rational Expressions. Let

$$\frac{P}{Q} = \frac{3x^2}{4y} \qquad \text{and} \qquad R = 2y.$$

Then

$$\frac{6x^2y}{8y^2} = \frac{(3x^2)(2y)}{(4y)(2y)} = \frac{3x^2}{4y}$$

Thus

$$\frac{6x^2y}{8y^2} = \frac{3x^2}{4y}$$ ■

EXAMPLE 3. Show that $\frac{6x^2}{2x} = 3x$.

SOLUTION. The cross-products

$$\frac{6x^2}{2x} \times \frac{3x}{1}$$

are equal:

$$6x^2 \cdot 1 = 3x \cdot 2x$$ ■

There is a subtle point concerning Example 3. The rational expression $\frac{6x^2}{2x}$ is not defined for $x = 0$, for then the denominator $2x$ also equals 0 (because $2 \cdot 0 = 0$). However, the polynomial $3x \left(\text{or the rational expression } \frac{3x}{1}\right)$ can be defined for $x = 0$. For this reason, a more careful definition of equivalence would have indicated that

$$\frac{6x^2}{2x} = 3x, \qquad \text{for } x \neq 0.$$

However, you need not worry about such subtleties.

The rules, given in Section 4.1, for dividing powers of a number apply also to powers of a variable.

Recall that for $a \neq 0$ and for positive integers m and n:

1. if $m > n$, then $\dfrac{a^m}{a^n} = a^{m-n}$

2. $\dfrac{a^m}{a^m} = 1$

3. if $m < n$, then $\dfrac{a^m}{a^n} = \dfrac{1}{a^{m-n}}$

When both numerator and denominator are monomials, apply these rules in order to simplify this type of rational expression.

EXAMPLE 4.

$$\frac{\overset{3}{\not{9}}x^4y^2}{\underset{2}{\not{6}}x^2y^3} = \frac{3x^{4-2}}{2y^{3-2}} = \frac{3x^2}{2y}$$ ■

Note that you used the Fundamental Principle of Rational Expressions. For, you divided numerator and denominator of $\dfrac{9x^4y^2}{6x^2y^3}$ by $3x^2y^2$.

Factors may appear in a different order in the numerator and denominator.

EXAMPLE 5.

$$\frac{10x^4yz^3}{-2xz^3y^2} = \frac{-5x^{4-1}}{y^{2-1}} = \frac{-5x^3}{y}$$ ■

Certain variables may appear only in the numerator or only in the denominator. Divide only powers of variables that appear in *both* numerator and denominator.

EXAMPLE 6.

$$\frac{3a^2b}{6ac^3} = \frac{ab}{2c^3}$$ ■

The rules for exponents can be applied to polynomial factors common to numerator and denominator.

EXAMPLE 7.

$$\frac{(x+y)^3}{(x+y)^2} = (x+y)^{3-2} = x+y$$ ■

EXAMPLE 8. Simplify: $\dfrac{8(a+b)^2(x-1)^5}{2(a+b)^4(x-1)^3}$

SOLUTION.

$$\frac{8(a+b)^2(x-1)^5}{2(a+b)^4(x-1)^3} = \frac{4(x-1)^{5-3}}{(a+b)^{4-2}} = \frac{4(x-1)^2}{(a+b)^2}$$ ■

EXERCISES

Simplify each rational expression.

1. $\frac{10x^2}{5}$
2. $\frac{9y}{3}$
3. $\frac{25z^3}{5}$
4. $\frac{20c^5}{15}$
5. $\frac{30a^3b}{-15}$
6. $\frac{24z^2}{-16}$
7. $\frac{30ad}{-18}$
8. $\frac{-96xyz}{36}$
9. $\frac{xy}{x}$
10. $\frac{abc}{b}$
11. $\frac{abcd}{-d}$
12. $\frac{-xyz^2}{y}$
13. $\frac{a^3bc^2}{bc}$
14. $\frac{xy^2z}{-xy}$
15. $\frac{rst}{s^2}$
16. $\frac{4a^2b^3c}{2a^2}$
17. $\frac{9m^2n^3}{6mn}$
18. $\frac{18c^6d}{-6c}$
19. $\frac{3abc^3}{6a^2bc}$
20. $\frac{-7a^3bc}{28abc}$
21. $\frac{-25p^2q^2}{5p^3q}$
22. $\frac{-24a^2b^3c}{12a^2b^2c^2}$
23. $\frac{9xz^3}{-18yz^2}$
24. $\frac{9x^6y^6z^2}{-15xz}$
25. $\frac{-4a^2b^2c}{6a^2b^3c^2}$
26. $\frac{abc^4d}{5acd}$
27. $\frac{-x^4y^2z^8}{-x^3y^4z^7}$
28. $\frac{24r^9s^7t}{18r^8s^8t^8}$
29. $\frac{9a^4b^2c^4}{36a^3c^5}$
30. $\frac{3x^2y^2}{10yx}$
31. $\frac{25x^4a^2z}{45ax^3y}$
32. $\frac{132xyz}{-44x^4}$
33. $\frac{-9a^4x^2}{-18a^5x^2}$
34. $\frac{12a^2bc^3}{-12a^2bc^3}$
35. $\frac{(a+b)^2}{a+b}$
36. $\frac{x+y}{(x+y)^3}$
37. $\frac{(a-b)^2}{(a-b)^5}$
38. $\frac{(u+v)^{10}}{(u+v)^5}$
39. $\frac{12(a+c)^3}{3(a+c)^2}$
40. $\frac{x^4(y+z)}{x^2(y+z)^2}$
41. $\frac{(a-b)^3(x+y)^2}{(a-b)^2(x+y)^4}$
42. $\frac{(m+n)(m-n)^3}{(m+n)(m-n)}$
43. $\frac{5xy(x-y)^2}{10x^2(x-y)}$
44. $\frac{(x+y)(m+n)}{(x-y)(m+n)^2}$
45. $\frac{(a+b)(a-b)(p+q)^2}{(a+b)^2(a-b)(p+q)}$
46. $\frac{(a+b)^4(a-b)(p+q)}{(a+b)^3(a-b)^7(p-q)}$
47. $\frac{(a+b)^4(m-n)^3(x+y)^7}{(a+b)^2(m-n)(x+y)^6}$
48. $\frac{16(m-n)^2(p+q)^4(r-s)^3(u+v)^2}{20(m-n)(p+q)(r-s)^{10}}$

4.4 DIVIDING BY MONOMIALS

In Section 4.3, you simplified rational expressions in which both numerator and denominator were monomials. Now suppose the numerator P contains more than one term, but the denominator S is again a monomial. If a constant or a power of a variable is to divide P, it must divide every term of P. Hence it must divide Q, the common factor of P.

EXAMPLE 1. Simplify: $\dfrac{5x^3 + 10x^2}{x^2}$

SOLUTION. First isolate the common factor in the numerator.

$$5x^3 + 10x^2 = 5x^2(x + 2)$$

Thus

$$\frac{5x^3 + 10x^2}{x^2} = \frac{5\cancel{x^2}(x + 2)}{\cancel{x^2}}$$
$$= 5(x + 2)$$ ■

To simplify a rational expression $\dfrac{P}{S}$ in which S is a monomial but P contains more than one term:

1. First isolate the common factor Q in the numerator, and write

$$P = Q \cdot R.$$

2. Reduce $\dfrac{Q}{S}$ to $\dfrac{Q'}{S'}$.

3. Then $\dfrac{P}{S}$ is reduced to $\dfrac{Q'R}{S'}$.

EXAMPLE 2.

$$\frac{\overbrace{4a^3 + 6a^2}^{P}}{\underbrace{8a}_{S}} = \frac{\overbrace{2a^2}^{Q}\overbrace{(2a + 3)}^{R}}{\underbrace{8a}_{S}} = \frac{\overbrace{a}^{Q'}\overbrace{(2a + 3)}^{R}}{\underbrace{4}_{S'}}.$$ ■

Note that you do not actually write out Step 2.

EXAMPLE 3.

$$\frac{3b^2c + 6b}{3bc} = \frac{\overset{1 \cdot 1}{\cancel{3}\cancel{b}}(bc + 2)}{\underset{1 \cdot 1}{\cancel{3}\cancel{b}}c} = \frac{bc + 2}{c}$$ ■

EXAMPLE 4. Simplify: $\dfrac{5x^2y^2 - 15xy^2 + 20y^4}{5x^2y^2}$

SOLUTION.

$$\frac{5x^2y^2 - 15xy^2 + 20y^4}{5x^2y^2} = \frac{\overset{1\cdot 1}{\cancel{5}\cancel{y^2}}(x^2 - 3x + 4y^2)}{\underset{1}{\cancel{5}}x^2\underset{1}{\cancel{y^2}}}$$

$$= \frac{x^2 - 3x + 4y^2}{x^2}$$ ■

EXAMPLE 5. Simplify: $\dfrac{84a^2b^3c^3 + 24ab^4c^2}{72a^2b^2c^2}$

SOLUTION. Consider here the prime factorizations of the coefficients.

$$\frac{84a^2b^3c^3 + 24ab^4c^2}{72a^2b^2c^2} = \frac{2^2 \cdot 3 \cdot 7a^2b^3c^3 + 2^3 \cdot 3ab^4c^2}{2^3 \cdot 3^2a^2b^2c^2}$$

$$= \frac{\overset{1\cdot 1\cdot 1\cdot b\cdot 1}{\cancel{2^2}\,\cancel{3}\,\cancel{a}\,\cancel{b^3}\,\cancel{c^2}}\,(7ac + 2b)}{\underset{2\cdot 3\cdot a\cdot 1\cdot 1}{\cancel{2^3}\,\cancel{3^2}\,\cancel{a^2}\,\cancel{b^2}\,\cancel{c^2}}}$$

Multiply numerical constants that remain in either numerator or denominator.

$$= \frac{b(7ac + 2b)}{6a}$$ ■

It is just as easy to simplify a rational expression in which the numerator is a monomial but the denominator contains more than one term.

EXAMPLE 6. Simplify: $\dfrac{2xy}{4x^2 + 2xy}$

SOLUTION.

$$\frac{2xy}{4x^2 + 2xy} = \frac{\cancel{2}\cancel{x}y}{\cancel{2}\cancel{x}(2x + y)} = \frac{y}{2x + y}$$ ■

EXERCISES

Simplify each rational expression.

1. $\dfrac{a^2b^2 + a^2b}{a}$
2. $\dfrac{m^3n + m^2n^2}{m}$
3. $\dfrac{a^2b^2 + a^2b}{ab}$
4. $\dfrac{cd^4 + c^2d}{cd}$
5. $\dfrac{3x^2 + 5xy}{x}$
6. $\dfrac{z}{y^3z^5 + yz}$

7. $\dfrac{8x^2y + 3y^2}{y^2}$

8. $\dfrac{4x^4y + 8xy}{4}$

9. $\dfrac{9a^2b + 3a^2b^3}{a^2}$

10. $\dfrac{9a^2b + 3a^2b^3}{6}$

11. $\dfrac{x^3y^2 + x^2y^3}{x^3}$

12. $\dfrac{2xy^2 + 4xy}{4}$

13. $\dfrac{8x^2 + 4x}{2x}$

14. $\dfrac{8x^2 + 4x}{14}$

15. $\dfrac{9a^3b^2 + 6b^3}{-3b^2}$

16. $\dfrac{10xy - 15y}{5y}$

17. $\dfrac{a^2b^2 - a^2b}{ab}$

18. $\dfrac{3a^2b^2 + 6ab^4}{9ab}$

19. $\dfrac{a^2b^2 - a^2b}{-ab}$

20. $\dfrac{a^4b^3 - a^2}{a^2b}$

21. $\dfrac{2xy}{2x^3y^2 + 4xy^3}$

22. $\dfrac{9xy^2}{-6x^4y^3 + 15x^2y}$

23. $\dfrac{9x^3y^3 + 15x^2y^2}{3x^2y^2}$

24. $\dfrac{-7ab^5 - 14ab}{21ab}$

25. $\dfrac{5a^3b^3 - 10a^2b}{5ab}$

26. $\dfrac{5ab + 20b^3}{10b^2}$

27. $\dfrac{4x^4y^4 - 8x^2y^2}{8x^2y^2}$

28. $\dfrac{16x^4z^3 - 28x^3z^4}{20x^3z^3}$

29. $\dfrac{10x^3 - 8x^2 + 5x}{x}$

30. $\dfrac{12y^3 - 8y^2 + 4y}{4y}$

31. $\dfrac{3r^4 + 5r^2 - r}{r^2}$

32. $\dfrac{x^2y^3 + x^2y^2 - xy}{xy}$

33. $\dfrac{m^2}{5m^2n + m^2p - m^2}$

34. $\dfrac{3a^2b^2}{12a^4b^2 - 15a^3b^3 + 9a^2b^4}$

35. $\dfrac{10s^4t^2 + 15st^5 + 20t^6}{st^2}$

36. $\dfrac{x^3y^3z^4 + x^2y^3z^5 - x^2y^3z^4}{-x^3y^2z^2}$

37. $\dfrac{a^{14}b^{12}c^{10} - 2a^{10}b^9c^8 + a^6b^6c^6}{2a^5b^5c^5}$

38. $\dfrac{14a^4b^3c^3 + 16a^2b^3c^2 - 28ab^2c^3}{4a^2b^2c^3}$

39. $\dfrac{-24x^3y^2z - 12x^2y^2z^2 - 18xyz^3}{-6xyz}$

40. $\dfrac{42a^3b^2 - 56a^2b^3 + 84a^3b}{14a^2b}$

41. $\dfrac{99x^3yz - 77x^2yz + 121xyz}{-22xyz}$

42. $\dfrac{5x^3y^2 + 9xy^2z - 3xz}{3x^2yz}$

43. $\dfrac{20x^4 - 24x^3 + 16x^2 - 40x}{16x^2}$

44. $\dfrac{3x^4y^3 + 6x^3y^4 - 9x^2y^5 + 3xy^6}{9xy^3}$

45. $\dfrac{-10s^2tu}{25s^2t^3u^3 - 15s^2t^2u^2 + 20s^2tu^2 - 25s^2tu}$

46. $\dfrac{36abc}{24a^4b - 42b^3c}$

47. $\dfrac{108x^3y^2 + 72x^2y^5}{96xy}$

48. $\dfrac{900r^5s^3tu^4 + 360r^4s^4tu^3 - 135r^3su^3 + 45r^2s^2t^2u^3}{90r^2s^2t^2u^3}$

4.5 SIMPLIFYING BY FACTORING

The factoring techniques of Chapter 3 can now be employed to simplify rational expressions in which neither numerator nor denominator is a monomial.

EXAMPLE 1. Simplify: $\dfrac{x^2 - y^2}{(x + y)^2}$

SOLUTION.

$$\frac{x^2 - y^2}{(x + y)^2} = \frac{(x + y)(x - y)}{(x + y)^2} = \frac{x - y}{x + y}$$

Here you divided the numerator and denominator of the middle expression by $x + y$. ■

EXAMPLE 2. Simplify: $\dfrac{c^2x^4 + c^2}{c(x^4 + 1)^3}$

SOLUTION.

$$\frac{c^2x^4 + c^2}{c(x^4 + 1)^3} = \frac{c^2(x^4 + 1)}{c(x^4 + 1)^3} = \frac{c}{(x^4 + 1)^2}$$ ■

EXAMPLE 3. Simplify: $\dfrac{x^2 - 5x + 6}{x^2 - 4}$

SOLUTION.

$$\frac{x^2 - 5x + 6}{x^2 - 4} = \frac{(x - 3)(x - 2)}{(x + 2)(x - 2)} = \frac{x - 3}{x + 2}$$ ■

EXAMPLE 4. Simplify: $\dfrac{a^4 - b^4}{a^2 + 3ab + 2b^2}$

SOLUTION.

$$\begin{aligned}\frac{a^4 - b^4}{a^2 + 3ab + 2b^2} &= \frac{(a^2 + b^2)(a^2 - b^2)}{(a + 2b)(a + b)}\\ &= \frac{(a^2 + b^2)(a + b)(a - b)}{(a + 2b)(a + b)}\\ &= \frac{(a^2 + b^2)(a - b)}{a + 2b}\end{aligned}$$

Leave the answer in factored form. ■

EXAMPLE 5. Simplify: $\dfrac{-25(x^2 - y^2)}{30y^2 - 30x^2}$

SOLUTION.

$$\begin{aligned}\frac{-25(x^2 - y^2)}{30y^2 - 30x^2} &= \frac{-5^2(x + y)(x - y)}{30(y^2 - x^2)}\\ &= \frac{-5^2(x + y)(x - y)}{5 \cdot 6(y + x)(y - x)}\\ &= \frac{-5(x + y)(x - y)}{6(x + y)(-1)(x - y)}\\ &= \frac{5}{6}\end{aligned}$$

Note that

$$y - x = (-1)(x - y).$$

For that matter, you could have observed that

$$y^2 - x^2 = (-1)(x^2 - y^2),$$

and saved some work. ■

EXAMPLE 6. Simplify: $\dfrac{m^3 - n^3}{am - an - m + n}$

SOLUTION. The numerator is the difference of cubes. The denominator can be factored by grouping.

$$\frac{m^3 - n^3}{am - an - m + n} = \frac{(m - n)(m^2 + mn + n^2)}{(m - n)(a - 1)} = \frac{m^2 + mn + n^2}{a - 1}$$ ■

EXERCISES

Simplify, if possible. Leave the answer in factored form.

1. $\dfrac{6a - 6b}{6a + 6b}$
2. $\dfrac{9x^2 + 9y^2}{27x + 27y}$
3. $\dfrac{a^5bc^2 - a^3bc}{a^3c - a}$
4. $\dfrac{abx + ab^2y}{a^2b^2(x + y)}$
5. $\dfrac{a^2 - 4}{a - 2}$
6. $\dfrac{x^2 - a^2}{x - a}$
7. $\dfrac{4m^2 - 4n^2}{2(m + n)^2}$
8. $\dfrac{s^2 + 9s + 8}{(s + 8)(s - 1)}$
9. $\dfrac{x^2 - y^2}{(x + y)^3}$
10. $\dfrac{5a^2c - 5b^2c}{ac + bc}$
11. $\dfrac{a^2 + 7a + 10}{a^2 + 4a + 4}$
12. $\dfrac{b^2 + 8b + 15}{b^2 + 7b + 12}$

13. $\dfrac{x - a}{(a - x)^2}$

14. $\dfrac{x + y - 1}{1 - y - x}$

15. $\dfrac{c^2 - 8c + 12}{c^2 - 5c - 6}$

16. $\dfrac{u^2 - 3u - 10}{u^2 + 3u - 10}$

17. $\dfrac{20 + 16a - 4a^2}{8a^2 + 16a + 8}$

18. $\dfrac{9 - 6x - 3x^2}{3x^2 - 6x + 3}$

19. $\dfrac{ab - ab^2}{(1 - b)^2}$

20. $\dfrac{5u^2 - 125v^2}{(6u - 30v)^2}$

21. $\dfrac{a^3 - z^3}{z^3 - a^3}$

22. $\dfrac{y^4 - 1}{y^6 - 1}$

23. $\dfrac{4 - 9c^2}{(2 - 3c)^3}$

24. $\dfrac{y^2 + z^2}{y^4 - z^4}$

25. $\dfrac{a^2x^3 - 27a^2}{4x^2 - 36}$

26. $\dfrac{b^2 + bc - 2c^2}{b^3 - c^3}$

27. $\dfrac{x^4 + 4x^2y^2 + 4y^4}{(x^2 + 2y^2)^3}$

28. $\dfrac{25 - 5y}{y^2 - 25}$

29. $\dfrac{3z^3 + 3a^3}{6a^2 - 6z^2}$

30. $\dfrac{r^2 - 12r + 35}{r^4 - 7r^3}$

31. $\dfrac{x^2 - y^2}{y^4 - x^4}$

32. $\dfrac{a^4 - 16}{3a^2 + 12}$

33. $\dfrac{ab + ac + b + c}{b^2 - c^2}$

34. $\dfrac{xy + xz - 3y - 3z}{x^2 - 6x + 9}$

35. $\dfrac{r^2 - s^2}{sx - rx - sy + ry}$

36. $\dfrac{ac + bd + ad + bc}{(a^2 + b^2)(c^2 - d^2)}$

37. $\dfrac{x^8 - y^8}{(x^4 + y^4)(x - y)^3}$

38. $\dfrac{(z^2 + 1)(z - 3)}{z^4 - 8z^2 - 9}$

39. $\dfrac{axy - axz + ay - az}{a^2x + a^2}$

40. $\dfrac{x^6 - 64}{8a - ax^3}$

REVIEW EXERCISES FOR CHAPTER 4

4.1 SIMPLIFYING FRACTIONS

1. Which of the following fractions are equivalent to $\frac{1}{4}$?

   (a) $\frac{2}{8}$ (b) $\frac{4}{1}$ (c) $\frac{-1}{-4}$ (d) $\frac{3}{-12}$

2. Express positive fractions without minus signs; express negative fractions in standard form:

   (a) $\frac{-6}{-7}$ (b) $\frac{3}{-4}$ (c) $-\frac{2}{-5}$ (d) $-\frac{-4}{-5}$

3. Simplify each fraction.

   (a) $\frac{3}{6}$ (b) $\frac{-18}{24}$ (c) $\frac{12}{-30}$ (d) $\frac{108}{144}$

4. Simplify each fraction. (Multiply out the resulting numerator and denominator.)

   (a) $\frac{2^8}{2^4}$ (b) $\frac{3^7 \cdot 5^2}{3^5 \cdot 5^3}$

(c) $\dfrac{2 \cdot 3^4 \cdot 7^{10}}{3^2 \cdot 5^2 \cdot 7^{11}}$

(d) $\dfrac{2^8 \cdot 3^5 \cdot 7^2 \cdot 11^3}{2^6 \cdot 3^5 \cdot 7^3 \cdot 11^4}$

4.2 EVALUATING RATIONAL EXPRESSIONS

5. Evaluate each rational expression for the specified value of the variable.

   (a) $\dfrac{4}{x + 2}$ when $x = 2$

   (b) $\dfrac{x - 3}{x + 2}$ when $x = -3$

   (c) $\dfrac{t^2 + 2t + 1}{t^2 - 5t}$ when $t = 10$

6. Evaluate $\dfrac{y^2 + 5}{y - 1}$ when

   (a) $y = 2$, (b) $y = -1$, (c) $y = -5$.

7. Evaluate $\dfrac{2xy}{x + y}$ when $x = 3$ and $y = -2$.

8. Evaluate $\dfrac{u^2 + 2u - w}{v - 2w}$ when $u = 2, v = 1, w = -1$

In Exercises 9–22 simplify each rational expression.

4.3 DIVIDING MONOMIALS

9. $\dfrac{20x^3}{5}$

10. $\dfrac{-12a^2}{6a}$

11. $\dfrac{36x^2yz^2}{9xyz}$

12. $\dfrac{(x + y)^2}{x + y}$

13. $\dfrac{(a + b)^4(a - b)^2}{(a + b)^3(a - b)^3}$

14. $\dfrac{20x^2y(x + y)(a - b)}{15xy^2(x - y)(a - b)}$

4.4 DIVIDING BY MONOMIALS

15. $\dfrac{5x^2 + 10x}{5x}$

16. $\dfrac{6x^2y - 9xy^2}{3xy}$

17. $\dfrac{a^2b^2c^2 + ab^2c^2 - a^2bc}{abc^2}$

18. $\dfrac{12x^2y^3 - 36x^2y^2 + 27xy^3}{9x^2y^2}$

4.5 SIMPLIFYING BY FACTORING

19. $\dfrac{8a^2 - 8b^2}{4a + 4b}$

20. $\dfrac{x^2 + 7x + 6}{x^2 - 36}$

21. $\dfrac{x^2 - a^2}{a^3 - x^3}$

22. $\dfrac{ab + ac + db + dc}{a^2 - d^2}$

TEST YOURSELF ON CHAPTER 4.

1. Simplify each fraction. [In part (c), multiply out the resulting numerator and denominator.]

(a) $\frac{-15}{-25}$ (b) $\frac{2^3 \cdot 3^4}{2^2 \cdot 3^5}$ (c) $\frac{3^4 \cdot 5^2 \cdot 11}{3^3 \cdot 7 \cdot 11^2}$

2. Evaluate each rational expression for the specified value of the variable:

(a) $\frac{5}{x - 2}$ when $x = -3$ (b) $\frac{y + 4}{y^2 - 1}$ when $y = 3$

3. Evaluate $\frac{x^2 - 3xy + 5}{2xy + x - 2}$ when $x = 3$ and $y = -1$.

In Exercises 4–8 simplify each rational expression.

4. $\frac{40x^2y^3}{8x^2y}$ 5. $\frac{(x - y)^3}{(x - y)^2}$ 6. $\frac{18a^2bc^4(a + b)^2(a - c)^3}{12ab^2c^4(a + b)(a + c)}$

7. $\frac{24x^2y^2 + 36xy^2 - 18x^2y}{12x^2y^2}$ 8. $\frac{x^2 - 16}{x^2 - x - 12}$

5

arithmetic of rational expressions

5.1 MULTIPLICATION AND DIVISION

Arithmetic operations can be defined for rational expressions, just as they were for polynomials. You first learn how to multiply and divide fractions. These methods then carry over to more general rational expressions.

As you may know, the area of a rectangle equals its length times its width. Let

$$A = \text{area}, \qquad l = \text{length},$$

$$w = \text{width}.$$

Then

$$A = lw$$

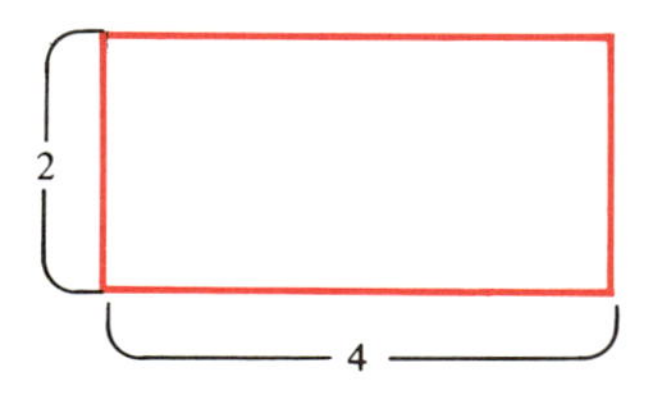

FIGURE 5.1 The area of the rectangle is 4 · 2, or 8, square units.

[See Figure 5.1 .]

Now consider a square each of whose sides is of length 1. Subdivide the square into 15 smaller rectangles, as in Figure 5.2. The area of each small rectangle is $\frac{1}{15}$. The red rectangle with dimensions $\frac{2}{3} \times \frac{2}{5}$ includes 4 small rectangles. It covers $\frac{4}{15}$ of the area of the square. Thus its area should be $\frac{4}{15}$. Because

$$lw = A,$$

it seems reasonable that

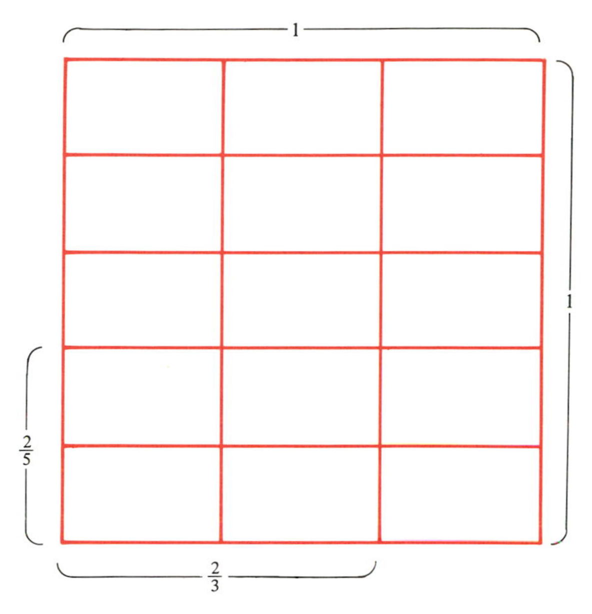

FIGURE 5.2

$$\frac{2}{3} \cdot \frac{2}{5} = \frac{4}{15}.$$

In general, you can define **multiplication of fractions** by

$$\frac{a}{b} \cdot \frac{c}{d} = \frac{ac}{bd}$$

Let us see how this definition ties in with other arithmetic notions. Fractions express division of real numbers. Thus recall how division was defined.

Let a, $b(\neq 0)$, and c be real numbers. According to the definition of division,

$$\text{if } a = bc, \qquad \text{then } \frac{a}{b} = c,$$

and, it is understood,

$$\text{if } \frac{a}{b} = c, \qquad \text{then } a = bc$$

Suppose a', $b'(\neq 0)$, and c' are also real numbers. By using this definition of division, you will see that when you multiply

$$\frac{a}{b} \text{ by } \frac{a'}{b'}, \qquad \text{the product is } \frac{aa'}{bb'}.$$

Note that $bb' \neq 0$.

For, suppose these quotients are given by

(1) $$\frac{a}{b} = c \qquad \text{and} \qquad \frac{a'}{b'} = c'.$$

Then

$$a = bc \qquad \text{and} \qquad a' = b'c'$$

Therefore

$$aa' = (bc)(b'c')$$

$$= (bb')(cc') \qquad \text{by the Commutative and Associative Laws}$$

Consequently,

$$\frac{aa'}{bb'} = cc' = \frac{a}{b} \cdot \frac{a'}{b'} \qquad \text{by the definition of division and by (1)}$$

Thus it is reasonable to define multiplication of fractions by

$$\frac{a}{b} \cdot \frac{a'}{b'} = \frac{aa'}{bb'}.$$

EXAMPLE 1.

(a) $\dfrac{5}{7} \cdot \dfrac{2}{3} = \dfrac{5 \cdot 2}{7 \cdot 3} = \dfrac{10}{21}$ (b) $\dfrac{1}{3} \cdot \dfrac{\pi}{3} = \dfrac{1 \cdot \pi}{3 \cdot 3} = \dfrac{\pi}{9}$ ■

Sometimes the resulting fraction can be simplified by dividing by factors common to both numerator and denominator.

EXAMPLE 2.

$$\frac{4}{9} \cdot \frac{3}{2} = \frac{4 \cdot 3}{9 \cdot 2} = \frac{12}{18} = \frac{2}{3}$$

In practice, *it is generally best to divide by common factors before multiplying:*

$$\frac{\overset{2}{\cancel{4}}}{\underset{3}{\cancel{9}}} \cdot \frac{\overset{1}{\cancel{3}}}{\underset{1}{\cancel{2}}} = \frac{2}{3}$$

■

EXAMPLE 3. Determine: $\dfrac{-384}{25} \cdot \dfrac{125}{-54}$

SOLUTION.

$$\frac{-384}{25} \cdot \frac{125}{-54} = \frac{-2^7 \cdot 3}{5^2} \cdot \frac{5^3}{-2 \cdot 3^3}$$

$$= \frac{2^6 \cdot 5}{3^2}$$

$$= \frac{320}{9}$$

■

DEFINITION. The **reciprocal of a nonzero number** a is defined to be

$$\frac{1}{a}.$$

EXAMPLE 4.

(a) The reciprocal of 5 is $\frac{1}{5}$.

(b) The reciprocal of -5 is $\frac{1}{-5}$. But $\frac{1}{-5} = \frac{-1}{5}$

(c) The reciprocal of π is $\frac{1}{\pi}$. ■

Because $a = \frac{a}{1}$, it follows that

$$a \cdot \frac{1}{a} = \frac{a}{1} \cdot \frac{1}{a} = \frac{a \cdot 1}{1 \cdot a} = \frac{a}{a} = 1.$$

By the Commutative Law of Multiplication,

$$\frac{1}{a} \cdot a = 1$$

Thus the reciprocal of a is the number multiplied by a to obtain 1. (For this reason, the reciprocal of a is often called the "multiplicative inverse of a". Recall that $-a$, the inverse of a, is the number *added to a* to obtain 0.) For this reason, some authors call $-a$ the "additive inverse of a".

Observe that for any a and for $b \neq 0$,

$$\frac{a}{b} = a \cdot \frac{1}{b}$$

[See Figure 5.3.]

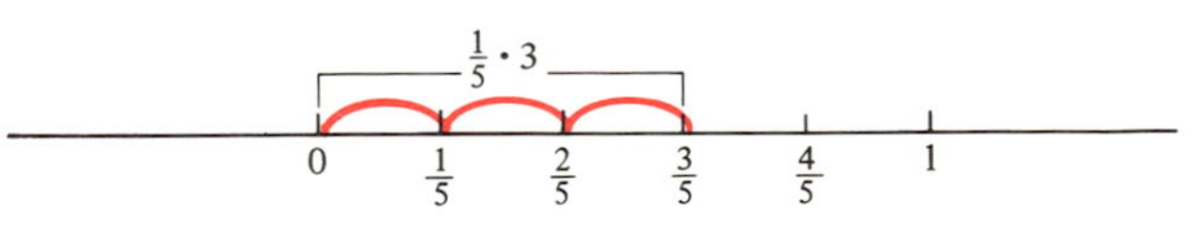

FIGURE 5.3 $\frac{3}{5} = \frac{1}{5} \cdot 3 = 3 \cdot \frac{1}{5}$

For, by definition, $\frac{a}{b}$ is the number c such that

$$a = bc.$$

But

$$b\left(a \cdot \frac{1}{b}\right) = \left(b \cdot \frac{1}{b}\right)a = a$$

In other words, $a \cdot \frac{1}{b}$ is this number c.

Thus *division by b is the same as multiplication by* $\frac{1}{b}$.

Next, observe that

$$\frac{c}{d}\cdot\frac{d}{c} = \frac{cd}{dc} = 1.$$

Thus by the definition of division of real numbers,

$$1 \div \frac{c}{d} = \frac{d}{c}$$

The reciprocal of $\frac{c}{d}$ can be written as $\frac{1}{\frac{c}{d}}$, which means $1 \div \frac{c}{d}$. Thus *the reciprocal of a fraction is obtained by interchanging numerator and denominator.* This is what is meant by "inverting a fraction".

In order to discover the general rule for dividing by a fraction, fill in the blanks in Tables 5.1 and 5.2 by considering the pattern in the right-hand column of each table.

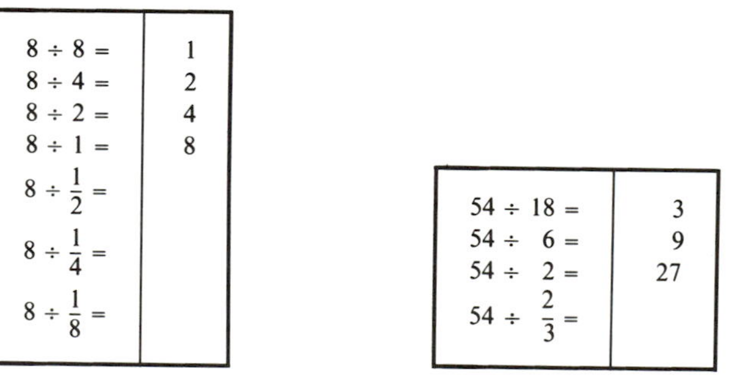

$8 \div 8 =$	1
$8 \div 4 =$	2
$8 \div 2 =$	4
$8 \div 1 =$	8
$8 \div \frac{1}{2} =$	
$8 \div \frac{1}{4} =$	
$8 \div \frac{1}{8} =$	

TABLE 5.1

$54 \div 18 =$	3
$54 \div 6 =$	9
$54 \div 2 =$	27
$54 \div \frac{2}{3} =$	

TABLE 5.2

Observe:

$$8 \div \frac{1}{2} = 8\cdot\frac{2}{1}, \qquad 8 \div \frac{1}{4} = 8\cdot\frac{4}{1}, \qquad 8 \div \frac{1}{8} = 8\cdot\frac{8}{1},$$

$$54 \div \frac{2}{3} = 54\cdot\frac{3}{2}$$

Define **division of fractions** as follows:

$$\frac{a}{b} \div \frac{c}{d} = \frac{ad}{bc}$$

Observe that

$$\frac{c}{d}\cdot\frac{ad}{bc} = \frac{\not{c}a\not{d}}{\not{d}b\not{c}} = \frac{a}{b}.$$

Thus *to divide fractions, invert the divisor and multiply.*

EXAMPLE 5.

$$\frac{12}{-25} \div \frac{42}{-5} = \frac{12}{-25} \cdot \frac{-5}{42}$$
$$= \frac{2^2 \cdot 3}{-5^2} \cdot \frac{-5}{2 \cdot 3 \cdot 7}$$
$$= \frac{2}{5 \cdot 7}$$
$$= \frac{2}{35}$$ ■

The same concepts apply to rational expressions.

DEFINITION. Let P, Q, R, S be polynomials with $Q \neq 0, S \neq 0$. Define

$$\frac{P}{Q} \cdot \frac{R}{S} = \frac{P \cdot R}{Q \cdot S}.$$

Leave the resulting rational expression in factored form.

EXAMPLE 6.

$$\frac{a^4}{2b^2} \cdot \frac{x + y}{x - y} = \frac{a^4(x + y)}{2b^2(x - y)}$$ ■

In the next example simplify wherever possible. Divide by factors common to the numerator and denominator.

EXAMPLE 7. Determine: $\dfrac{x^2 - y^2}{4ax} \cdot \dfrac{2x^3}{x^2 + 2xy + y^2}$

SOLUTION.

$$\frac{x^2 - y^2}{4ax} \cdot \frac{2x^3}{x^2 + 2xy + y^2} = \frac{(x + y)(x - y)}{4ax} \cdot \frac{2x^3}{(x + y)^2}$$
$$= \frac{(x - y) \cdot x^2}{2a(x + y)}$$ ■

Note that for nonzero polynomials P and Q,

$$\frac{P}{Q} \cdot \frac{Q}{P} = 1.$$

Therefore the **reciprocal of** $\dfrac{P}{Q}$, or $\dfrac{1}{\frac{P}{Q}}$, is $\dfrac{Q}{P}$. You obtain the reciprocal of $\dfrac{P}{Q}$ by inverting this rational expression. In particular, a nonzero polynomial P can be considered as the rational expression $\frac{P}{1}$. Thus the reciprocal of P is $\frac{1}{P}$. For example, the reciprocal of $\frac{5x+1}{2x-3}$ is $\frac{2x-3}{5x+1}$ and of $3x^2+5x+9$ is $\frac{1}{3x^2+5x+9}$.

The rule for **division of rational expressions** is the same as for division of rational numbers: *Invert the divisor and multiply.*

$$\frac{P}{Q} \div \frac{R}{S} = \frac{P}{Q} \cdot \frac{S}{R} = \frac{P \cdot S}{Q \cdot R}$$

EXAMPLE 8.

$$\frac{x^2y^3z}{abc} \div \frac{xyz^2}{a^2b} = \frac{x^2y^3z}{abc} \cdot \frac{a^2b}{xyz^2} = \frac{xy^2a}{cz} \qquad \blacksquare$$

EXAMPLE 9. Determine: $\frac{27x^2-27a^2}{x^2+a^2} \div \frac{3a-3x}{x+a}$

SOLUTION.

$$\begin{aligned}\frac{27x^2-27a^2}{x^2+a^2} \div \frac{3a-3x}{x+a} &= \frac{27x^2-27a^2}{x^2+a^2} \cdot \frac{x+a}{3a-3x} \\ &= \frac{27(x+a)(x-a)}{x^2+a^2} \cdot \frac{x+a}{-3(x-a)} \\ &= \frac{-9(x+a)^2}{x^2+a^2} \qquad \blacksquare\end{aligned}$$

EXAMPLE 10. Divide $\frac{x^2+10x+21}{5a^3-5b^3}$ by $\frac{x^2+6x+9}{a^2-b^2}$.

SOLUTION.

$$\begin{aligned}&\frac{x^2+10x+21}{5a^3-5b^3} \div \frac{x^2+6x+9}{a^2-b^2} \\ &= \frac{x^2+10x+21}{5a^3-5b^3} \cdot \frac{a^2-b^2}{x^2+6x+9} \\ &= \frac{(x+7)(x+3)}{5(a-b)(a^2+ab+b^2)} \cdot \frac{(a+b)(a-b)}{(x+3)^2} \\ &= \frac{(x+7)(a+b)}{5(a^2+ab+b^2)(x+3)} \qquad \blacksquare\end{aligned}$$

EXERCISES

In Exercises 1–10 multiply as indicated. Simplify the resulting fractions.

1. $\frac{3}{5} \cdot \frac{1}{2}$
2. $\frac{1}{8} \cdot \frac{5}{2}$
3. $\frac{-2}{3} \cdot \frac{1}{7}$
4. $\frac{-2}{3} \cdot \frac{-4}{9}$

5. $\frac{12}{25} \cdot \frac{-7}{24}$ 6. $\frac{35}{36} \cdot \frac{9}{-14}$ 7. $\frac{48}{-7} \cdot \frac{-49}{18}$ 8. $\frac{200}{3} \cdot \frac{63}{44}$

9. $\frac{144}{169} \cdot \frac{130}{81}$ 10. $\frac{2^5 \cdot 7^2}{3} \cdot \frac{33}{16}$

In Exercises 11–30, multiply as indicated. Simplify and leave the answers in factored form.

11. $\frac{x}{y} \cdot \frac{a}{b}$ 12. $\frac{-x}{y^2} \cdot \frac{1}{a^2}$ 13. $\frac{-2}{x} \cdot \frac{1}{-y}$

14. $\frac{x^2a^2}{y} \cdot \frac{3}{z^2}$ 15. $\frac{x^2a^2}{4} \cdot \frac{2}{x}$ 16. $\frac{a^2b^3c}{3d} \cdot \frac{bcd^2}{a^4}$

17. $\frac{25a^2x^3}{-7y^2z} \cdot \frac{49yz}{5ax}$ 18. $\frac{x+a}{x-a} \cdot \frac{4}{c}$ 19. $\frac{a^2b}{x-1} \cdot \frac{(x-1)^2}{3b}$

20. $\frac{2(x+a)^2}{x-a} \cdot \frac{x+a}{6}$ 21. $\frac{a^2(x-a)^3}{4b} \cdot \frac{12b^3}{a(x-a)^2}$ 22. $\frac{(x+a)y^3z^2}{3-x} \cdot \frac{(x-3)^2}{(x+a)^2yz}$

23. $\frac{a^2 \cdot (x^2-a^2)}{1-x} \cdot \frac{x^2-1}{x+a}$ 24. $\frac{x^2+3x+2}{x^2+1} \cdot \frac{x+1}{(x+2)^2}$

25. $\frac{x^2-5x+4}{5-5a^2} \cdot \frac{a^2+2a+1}{x^2-1}$ 26. $\frac{x^3a^3-ax^3}{a^2+4a+4} \cdot \frac{a^2+5a+6}{a^2-8a+7}$

27. $\frac{4x^2-12x+9}{a^2x+a^2} \cdot \frac{x^2-1}{14x^2-21x}$ 28. $\frac{5x^3+40}{2abc^3} \cdot \frac{ab^2-a^2b}{(b^2-a^2)(x+2)^2}$

29. $\frac{4b^2}{b^2-a^2} \cdot \frac{a^2+2ab+b^2}{12ab^3} \cdot \frac{(a-b)^3}{(a+b)^2}$ 30. $\frac{x^2+8x+15}{x^2-2x+1} \cdot \frac{1-x^2}{x^2+10x+25} \cdot \frac{x^4-1}{1+x^2}$

In Exercises 31–40, divide, as indicated. Simplify the resulting fractions.

31. $\frac{3}{2} \div \frac{1}{5}$ 32. $\frac{7}{3} \div \frac{-5}{2}$ 33. $\frac{-1}{3} \div \frac{1}{2}$ 34. $\frac{-20}{3} \div \frac{7}{12}$

35. $\frac{4}{9} \div \frac{3}{8}$ 36. $\frac{4}{9} \div \frac{8}{3}$ 37. $\frac{-6}{25} \div \frac{36}{125}$ 38. $\frac{24}{49} \div \frac{-12}{35}$

39. $\frac{98}{375} \div \frac{7}{75}$ 40. $\frac{11 \cdot 13^4}{7^3} \div \frac{7^2 \cdot 11^2 \cdot 13}{2}$

In Exercises 41–55, divide, as indicated. Simplify and leave the answers in factored form.

41. $\frac{5a^2}{b} \div \frac{3}{c}$ 42. $\frac{xy}{3} \div \frac{z}{4}$ 43. $\frac{a^2b}{3c} \div \frac{ac}{30b}$

44. $\frac{x^2y^2}{a} \div \frac{xa}{y}$ 45. $\frac{4a^2bx}{9cy} \div \frac{12b^2x^2}{c^2y^2}$ 46. $\frac{36m^2n^4}{25ax^3} \div \frac{-18m^3n^3}{75a^3x}$

47. $\frac{x-a}{x+a} \div \frac{2}{x+a}$ 48. $\frac{3x+y}{1-x} \div \frac{3x+y}{x-1}$ 49. $\frac{(x+1)^2}{x-3} \div \frac{x-3}{x+1}$

50. $\frac{x^2+1}{x+1} \div \frac{x^2+1}{(x+1)^2}$ 51. $\frac{a^2-4}{a^3} \div \frac{a^2+4a+4}{ab}$ 52. $\frac{2ax-2a}{x^2+1} \div \frac{4a-4x}{(x^2+1)^2}$

53. $\dfrac{3y^4 - 27y^2}{4 + 4x^2} \div \dfrac{y^3 + 3y^2}{3x^2 + 3}$

54. $\dfrac{x^4 - a^4}{a^2x^2 + a^2} \div \dfrac{6x^2 - 6a^2}{3x^2 + 3}$

55. $\dfrac{1 - y^2}{a^3 - 1} \div \dfrac{ay + a}{4a^2 - 8a + 4}$

56. Divide $\dfrac{6x^3 - 18x^2}{6a^2}$ by $\dfrac{x^2 - 9}{4ax}$.

57. Divide the product of $\dfrac{(x + 1)^2}{(x - 1)^2}$ and $\dfrac{x^2 - 1}{x + 1}$ by $\dfrac{x^2}{x - 1}$.

58. The quotient of $\dfrac{3x + 1}{x + 3}$ divided by $\dfrac{9x^2 + 3x}{x^2 - 9}$ is multiplied by $\dfrac{x^2}{x - 3}$. Determine the resulting product.

59. Determine the reciprocal of

$$\frac{x^2 - 2x}{(x + 1)^3} \cdot \frac{x^2 - 1}{x^2 + x - 6}.$$

60. What polynomial must be divided by $x^2 + 4x + 3$ to obtain $\dfrac{1}{x + 3}$?

61. What polynomial must be divided by $x^2 - 25$ to obtain $x + 5$?

62. What rational expression must be divided by $\dfrac{x + 3}{x + 4}$ to obtain $\dfrac{(x + 3)^2}{x + 4}$?

In Exercises 63–66, you are given polynomials with *rational* coefficients. Factor each polynomial. (The factors will have rational coefficients.) For example,

$$x^2 - \frac{1}{4} = \left(x + \frac{1}{2}\right)\left(x - \frac{1}{2}\right)$$

63. $a^2 - \dfrac{1}{16}$ 64. $\dfrac{1}{4}x^2 - \dfrac{1}{9}$ 65. $\dfrac{4y^2}{9} - \dfrac{1}{25}$ 66. $\dfrac{u^4}{16} - \dfrac{1}{81}$

5.2 LEAST COMMON MULTIPLES

In order to add and subtract rational expressions, find equivalent expressions with a common denominator. The key to doing this is the notion of "least common multiple".

Let a and b be integers. Recall that a is called a multiple of b ($\neq 0$) if

$$a = bc$$

for some integer c. Thus

a is a multiple of b

means the same as

b is a factor of a.

DEFINITION. **The least common multiple** (**LCM**) of several integers is the smallest positive integer that is a multiple of each of them.

Write

$$\text{LCM}(m_1, m_2, \ldots, m_n)$$

for the least common multiple of $m_1, m_2, \ldots, m_n$.

EXAMPLE 1.

(a) The least common multiple of 2 and 3 is 6.

$$\text{LCM}(2, 3) = 6$$

For, $6 = 2 \cdot 3 = 3 \cdot 2$, and is thus a multiple of 2 as well as of 3. No smaller positive number is a multiple of *both* 2 and 3.

(b) $$\text{LCM}(2, 4) = 4$$

(c) $$\text{LCM}(-4, 6) = 12$$

In fact, 12 is the smallest positive integer that is a multiple of both -4 and 6. ■

To find the LCM of several integers:

1. Determine their prime factorizations.
2. The LCM is the product of the *highest powers* of all primes that occur in these factorizations.

In Example 1(c), $4 = 2^2$, $6 = 2 \cdot 3$. The highest power of 2 that occurs is 2^2. The only other prime that occurs is 3; only 3^1 occurs. Thus the LCM is

$$2^2 \cdot 3, \qquad \text{or} \qquad 12.$$

EXAMPLE 2. Determine: LCM(18, 45, 54)

SOLUTION.

$$18 = 2 \cdot 3^2, \qquad 45 = 3^2 \cdot 5, \qquad 54 = 2 \cdot 3^3$$

$$\text{LCM}(18, 45, 54) = 2 \cdot 3^3 \cdot 5 = 270$$ ■

The same concepts apply to polynomials. Only polynomials with integral coefficients will be considered. *In the remainder of this section, "polynomial" will mean "polynomial with integral coefficients".*

Recall that when P and Q ($\neq 0$) are polynomials, P is said to be a multiple of Q when

$$P = Q \cdot R$$

for some polynomial R. Thus

$$P \text{ is a multiple of } Q$$

means the same as

$$Q \text{ is a factor of } P.$$

For example,

$$3x + 3 \text{ is a multiple of } x + 1$$

because

$$\underbrace{3x + 3}_{P} = \underbrace{(x + 1)}_{Q}\underbrace{3}_{R}.$$

Also,

$$8x^2yz$$

is a multiple of each of the polynomials

$$4, \ 8, \ 8x, \ xy, \ x^2z.$$

The "least common multiple" of several polynomials,

$$\text{LCM}(P_1, P_2, \ldots, P_n)$$

should be a polynomial of lowest degree that is a multiple of each of the polynomials $P_1, P_2, \ldots, P_n$. The absolute values of the coefficients of the (terms of the) LCM should be "as small as possible". Before the precise definition is given, consider an example.

EXAMPLE 3.

(a) $\text{LCM}(x^2, y) = x^2y$. Observe that x^2y is a multiple of x^2 and of y. Any polynomial that is a multiple of both x^2 and y must have degree at least 3; hence x^2y is of lowest possible degree. Also, 1, the coefficient of x^2y, is the smallest positive integer.
(b) $\text{LCM}(xy^2, 3x^2y) = 3x^2y^2$
(c) $\text{LCM}[(x + a), (x - a)] = (x + a)(x - a) = x^2 - a^2$
Note that $a^2 - x^2$, or $-(x^2 - a^2)$, could also be considered as the LCM of $x + a$ and $x - a$.
(d) $\text{LCM}[4(x + 1), -6(x + 1)] = 12(x + 1)$, or $12x + 12$ ■

Note that the coefficients of the LCM are positive and are as small as possible.

DEFINITION. Let $P_1, P_2, \ldots, P_n$ be polynomials. Express each polynomial in factored form.

Then the product of the highest powers of all factors that occur in these factorizations is called **the least common multiple of** $P_1, P_2, \ldots, P_n$, and is written

$$\text{LCM}(P_1, P_2, \ldots, P_n).$$

As in Example 3(c), the inverse of this polynomial can sometimes serve as the LCM.

EXAMPLE 4. Determine: $\text{LCM}(x^2 - 4, x^2 + 4x + 4)$

SOLUTION.

$$x^2 - 4 = (x + 2)(x - 2), \qquad x^2 + 4x + 4 = (x + 2)^2$$

$$\text{LCM}(x^2 - 4, x^2 + 4x + 4) = (x + 2)^2(x - 2)$$ ■

EXAMPLE 5. Determine:

$$\text{LCM}(x^2z^2 - y^2z^2, xz^4 + yz^4, x^2 - 3xy + 2y^2)$$

SOLUTION.

$$x^2z^2 - y^2z^2 = z^2(x^2 - y^2) = z^2(x + y)(x - y)$$

$$xz^4 + yz^4 = z^4(x + y)$$

$$x^2 - 3xy + 2y^2 = (x - 2y)(x - y)$$

$$\text{LCM}(x^2z^2 - y^2z^2, xz^4 + yz^4, x^2 - 3xy + 2y^2)$$

$$= z^4(x + y)(x - y)(x - 2y)$$ ■

DEFINITION. The **least common denominator** (LCD) of several fractions or rational expressions is the LCM of the individual denominators.

EXAMPLE 6. Determine: $\text{LCD}\left(\frac{1}{4}, \frac{2}{3}, \frac{5}{18}\right)$

SOLUTION. Determine the LCM of the *denominators:*

$$4 = 2^2, \qquad 3 = 3, \qquad 18 = 2 \cdot 3^2$$

Thus $\text{LCM}(4, 3, 18) = 2^2 \cdot 3^2 = 36$ and $\text{LCD}\left(\frac{1}{4}, \frac{2}{3}, \frac{5}{18}\right) = 36$

EXAMPLE 7. Determine: $\text{LCD}\left(\frac{1}{(x + 1)^2}, \frac{-1}{x - 3}, \frac{x}{x^2 - 1}\right)$

SOLUTION. Again consider only the denominators:

$$(x + 1)^2, \qquad x - 3, \qquad x^2 - 1 = (x + 1)(x - 1)$$

Thus

$$\text{LCM}[(x+1)^2, x-3, x^2-1] = (x+1)^2(x-3)(x-1)$$

and

$$\text{LCD}\left(\frac{1}{(x+1)^2}, \frac{-1}{x-3}, \frac{x}{x^2-1}\right) = (x+1)^2(x-3)(x-1) \qquad \blacksquare$$

In order to add and subtract rational expressions (or fractions) you first rewrite them as equivalent expressions with the LCD as the denominator. For now, you will just practice writing these equivalent expressions. In Section 5.3 you will perform the actual addition and subtraction.

EXAMPLE 8.

(a) Determine: $\text{LCD}\left(\frac{3}{10}, \frac{1}{4}, \frac{-2}{5}\right)$

(b) Determine equivalent fractions with this LCD as the denominator.

SOLUTION.

(a) First determine LCM(10, 4, 5).

$$10 = 2 \cdot 5, \qquad 4 = 2^2, \qquad 5 = 5$$

$$\text{LCM}(10, 4, 5) = 2^2 \cdot 5 = 20$$

$$\text{LCD}\left(\frac{3}{10}, \frac{1}{4}, \frac{-2}{5}\right) = 20$$

(b) To obtain each equivalent fraction, multiply the numerator and denominator by the appropriate number.

$$\frac{3 \cdot 2}{10 \cdot 2} = \frac{6}{20}, \qquad \frac{1 \cdot 5}{4 \cdot 5} = \frac{5}{20}, \qquad \frac{-2 \cdot 4}{5 \cdot 4} = \frac{-8}{20} \qquad \blacksquare$$

EXAMPLE 9.

(a) Determine: $\text{LCD}\left(\frac{4}{xy^2}, \frac{x-2}{x^2y}\right)$

(b) Determine equivalent rational expressions with this LCD as the denominator.

SOLUTION.

(a) $\text{LCM}(xy^2, x^2y) = x^2y^2$
Thus

$$\text{LCD}\left(\frac{4}{xy^2}, \frac{x-2}{x^2y}\right) = x^2y^2$$

(b) To obtain each equivalent rational expression, multiply the numerator and denominator by the appropriate polynomial.

$$\frac{4}{xy^2} = \frac{4 \cdot x}{xy^2 \cdot x} = \frac{4x}{x^2y^2}$$

$$\frac{x-2}{x^2y} = \frac{(x-2) \cdot y}{x^2y \cdot y} = \frac{xy - 2y}{x^2y^2}$$

In general, multiply out the numerator; this will often be necessary when you add and subtract expressions.

EXAMPLE 10.

(a) Determine: $\text{LCD}\left(\frac{x}{x^2 - a^2}, \frac{a^2}{(x+a)^2}\right)$

(b) Determine equivalent rational expressions with this LCD as the denominator.

SOLUTION.

(a)
$$x^2 - a^2 = (x+a)(x-a), \qquad (x+a)^2 = (x+a)^2$$

$$\text{LCM}[x^2 - a^2, (x+a)^2] = (x+a)^2(x-a)$$

$$\text{LCD}\left(\frac{x}{x^2 - a^2}, \frac{a^2}{(x+a)^2}\right) = (x+a)^2(x-a)$$

(b)
$$\frac{x}{x^2 - a^2} = \frac{x}{(x+a)(x-a)}$$

$$= \frac{x \cdot (x+a)}{(x+a)(x-a) \cdot (x+a)} = \frac{x(x+a)}{(x+a)^2(x-a)}$$

$$= \frac{x^2 + ax}{(x+a)^2(x-a)}.$$

Also,

$$\frac{a^2}{(x+a)^2} = \frac{a^2 \cdot (x-a)}{(x+a)^2 \cdot (x-a)} = \frac{a^2x - a^3}{(x+a)^2(x-a)}$$

■

EXERCISES

In Exercises 1–16 determine the LCM of the indicated integers.

1. 4, 5
2. 3, 8
3. 5, 10
4. 12, 18
5. 15, 20
6. 14, 35
7. 72, 96
8. 75, 90
9. 2, 4, 5
10. 3, 6, 8
11. 10, 12, 15
12. 21, 49, 99
13. 50, 80, 125
14. 132, 144, 168
15. 48, 72, 120
16. 112, 150, 196

In Exercises 17–32 determine an LCM of the indicated polynomials.

17. x, y^3
18. $5, x^2$
19. x, x^2
20. ax, x^2y
21. x^2yz^2, y^2z
22. $8x^2yz^3, 12x^2y^2z$
23. $x - a, (x - a)^2$
24. $x(x + a)^2, (x - a)x^2$
25. $x^2 - 9, ax + 3a$
26. $a^2x^2 - a^2y^2, ax + ay$
27. $x^2 + 5x + 4, x^2 + 2x + 1$
28. $a + b, -b - a$
29. $25a^2x^2 - 100a^2, 10a^4, x^2 - 5x + 6$
30. $u^2 - a^2, (u + a)^3, a^2 - u^2$
31. $y^2 + 4y + 4, y^2 - 4, xy + y - 2x - 2$
32. $z^3 - a^3, z^3 + a^3, z^2 - a^2$

In Exercises 33–46: (a) Determine the LCD of the indicated fractions. (b) Then determine equivalent fractions with this LCD as denominator.

33. $\frac{1}{2}, \frac{1}{3}$
34. $\frac{2}{5}, \frac{1}{10}$
35. $\frac{3}{8}, \frac{-3}{4}$
36. $\frac{5}{8}, \frac{-1}{12}$
37. $\frac{9}{28}, \frac{3}{98}$
38. $\frac{-5}{12}, \frac{-9}{32}$
39. $\frac{1}{2}, \frac{2}{3}, \frac{-3}{4}$
40. $\frac{1}{4}, \frac{5}{6}, \frac{7}{12}$
41. $\frac{5}{16}, \frac{-1}{24}, \frac{11}{42}$
42. $\frac{7}{20}, \frac{-1}{30}, \frac{3}{125}$
43. $\frac{5}{18}, \frac{17}{45}, \frac{7}{60}$
44. $\frac{-2}{3^4 \cdot 7^2}, \frac{1}{3^3 \cdot 7^3}, \frac{-1}{3 \cdot 7^3}$
45. $\frac{8}{3 \cdot 5^2}, \frac{-7}{2 \cdot 3 \cdot 5}, \frac{-3}{2^2 \cdot 5^2}$
46. $\frac{5}{11 \cdot 13^2}, \frac{-3}{11^2 \cdot 13}, \frac{1}{11 \cdot 13}$

In Exercises 47–62: (a) Determine an LCD of the indicated rational expressions. (b) Then determine equivalent rational expressions with this LCD as the denominator.

47. $\frac{1}{a}, \frac{1}{x}$
48. $\frac{-1}{y}, \frac{2}{z}$
49. $\frac{a}{x^2}, \frac{b}{x}$
50. $\frac{1}{xy}, \frac{a}{x^2}$
51. $\frac{a}{x^2y}, \frac{b}{xy^2}$
52. $\frac{-1}{x^2yz}, \frac{2}{xy^2z}$
53. $\frac{1}{x^2(x - a)}, \frac{-1}{x(x - a)^2}$
54. $\frac{x + a}{x^3(x - a)}, \frac{-a}{ax - x^2}$
55. $\frac{x + a}{x - a}, \frac{x - a}{x + a}$
56. $\frac{1}{x^2 + 2ax + a^2}, \frac{-1}{x^2 - a^2}$
57. $\frac{4}{x^2 - 5x + 6}, \frac{-1}{x^2 - 6x + 9}$
58. $\frac{1}{a^2 + b^2}, \frac{2}{-b^2 - a^2}$
59. $\frac{2u}{u^4 - u^2}, \frac{3}{1 - u^2}, \frac{-2}{u^2}$
60. $\frac{1}{9 - x^2}, \frac{x}{x + 3}, \frac{x^2}{x^2 + 6x + 9}$
61. $\frac{y + 2}{y^2 - 7y}, \frac{-1}{y - 7}, \frac{2}{7 - y}$
62. $\frac{u^3}{u^4 - 1}, \frac{-2u}{1 - u^2}, \frac{u - 1}{u^2 + 1}$

5.3 ADDITION AND SUBTRACTION

To add or subtract fractions with *the same denominator D:*

1. Add or subtract the numerators to obtain the numerator N.
2. Simplify the resulting fraction $\frac{N}{D}$, if necessary.

[See Figure 5.4.]

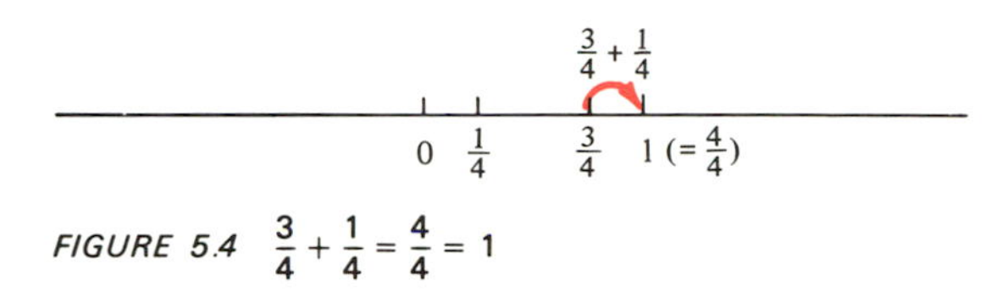

FIGURE 5.4 $\frac{3}{4} + \frac{1}{4} = \frac{4}{4} = 1$

EXAMPLE 1.

$$\frac{3}{10} + \frac{1}{10} - \frac{9}{10} = \frac{3 + 1 - 9}{10} = \frac{-5}{10} = \frac{-1}{2}$$

■

To add or subtract fractions with *different denominators:*

1. Determine their LCD.
2. Determine equivalent fractions with this LCD as denominator.
3. Add or subtract the numerators.
4. Simplify the resulting fraction, if necessary.

[See Figure 5.5.]

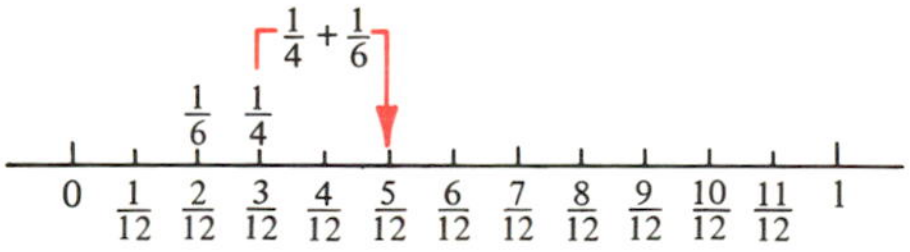

FIGURE 5.5 $\frac{1}{4} + \frac{1}{6} = \frac{5}{12}$. Note that $\frac{1}{4} = \frac{3}{12}$ and $\frac{1}{6} = \frac{2}{12}$. Here $12 = \text{LCD}\left(\frac{1}{4}, \frac{1}{6}\right)$

EXAMPLE 2. Determine: $\frac{1}{2} + \frac{2}{3} - 1$

SOLUTION.

$$\text{LCD}\left(\frac{1}{2}, \frac{2}{3}, 1\right) = \text{LCM}(2, 3, 1) = 6$$

$$\frac{1}{2} = \frac{1 \cdot 3}{2 \cdot 3} = \frac{3}{6}, \qquad \frac{2}{3} = \frac{2 \cdot 2}{3 \cdot 2} = \frac{4}{6}, \qquad 1 = \frac{1}{1} = \frac{1 \cdot 6}{1 \cdot 6} = \frac{6}{6}$$

$$\frac{1}{2} + \frac{2}{3} - 1 = \frac{3}{6} + \frac{4}{6} - \frac{6}{6} = \frac{3 + 4 - 6}{6} = \frac{1}{6} \qquad \blacksquare$$

EXAMPLE 3. Determine: $\dfrac{5}{12} - \dfrac{11}{30}$

SOLUTION.

$$12 = 2^2 \cdot 3, \qquad 30 = 2 \cdot 3 \cdot 5$$

$$\text{LCD}\left(\frac{5}{12}, \frac{11}{30}\right) = \text{LCM}(12, 30) = 2^2 \cdot 3 \cdot 5 = 60$$

$$\frac{5}{12} = \frac{5 \cdot 5}{12 \cdot 5} = \frac{25}{60}, \qquad \frac{11}{30} = \frac{11 \cdot 2}{30 \cdot 2} = \frac{22}{60}$$

$$\frac{5}{12} - \frac{11}{30} = \frac{25 - 22}{60} = \frac{3}{60} = \frac{1}{20} \qquad \blacksquare$$

EXAMPLE 4. Determine: $\dfrac{41}{96} - \left(\dfrac{1}{60} - \dfrac{5}{64}\right)$

SOLUTION.

$$96 = 2^5 \cdot 3, \qquad 60 = 2^2 \cdot 3 \cdot 5, \qquad 64 = 2^6$$

$$\text{LCD}\left(\frac{41}{96}, \frac{1}{60}, \frac{5}{64}\right) = 2^6 \cdot 3 \cdot 5 = 960$$

$$\frac{41}{96} = \frac{41 \cdot 10}{96 \cdot 10} = \frac{410}{960},$$

$$\frac{1}{60} = \frac{1 \cdot 16}{60 \cdot 16} = \frac{16}{960},$$

$$\frac{5}{64} = \frac{5 \cdot 15}{64 \cdot 15} = \frac{75}{960}$$

$$\frac{41}{96} - \left(\frac{1}{60} - \frac{5}{64}\right) = \frac{410 - (16 - 75)}{960}$$

$$= \frac{410 - 16 + 75}{960}$$

$$= \frac{469}{960}$$

(This resulting fraction cannot be simplified because

$$\frac{469}{960} = \frac{7 \cdot 67}{2^6 \cdot 3 \cdot 5}.)$$

■

The same rules, essentially, apply to combining rational expressions.

To add or subtract rational expressions with *the same denominator D:*

1. Add or subtract the numerators to obtain the numerator N.
2. Simplify the resulting fraction $\frac{N}{D}$, if necessary.

EXAMPLE 5.

$$\frac{1}{x^2} - \frac{2}{x^2} + \frac{a}{x^2} = \frac{1 - 2 + a}{x^2} = \frac{a - 1}{x^2}$$

■

To add or subtract rational expressions with *different denominators:*

1. Determine their LCD.
2. Determine equivalent expressions with this LCD as the denominator.
3. Add or subtract the numerators.
4. Simplify the resulting expression, if necessary.

EXAMPLE 6. Subtract: $\frac{a}{x^2yz^2} - \frac{b}{x^3y^2}$

SOLUTION. Here the LCD is $x^3y^2z^2$.

$$\frac{a}{x^2yz^2} - \frac{b}{x^3y^2} = \frac{a \cdot xy}{x^2yz^2 \cdot xy} - \frac{b \cdot z^2}{x^3y^2 \cdot z^2}$$

$$= \frac{axy - bz^2}{x^3y^2z^2}$$

■

EXAMPLE 7. Add: $\frac{a}{x + y} + \frac{a}{x - y}$

SOLUTION. Here the LCD $= (x + y)(x - y) = x^2 - y^2$

$$\frac{a}{x + y} + \frac{a}{x - y} = \frac{a(x - y)}{(x + y)(x - y)} + \frac{a(x + y)}{(x - y)(x + y)}$$

$$= \frac{a(x - y) + a(x + y)}{x^2 - y^2}$$

$$= \frac{ax - \cancel{ay} + ax + \cancel{ay}}{x^2 - y^2}$$

$$= \frac{2ax}{x^2 - y^2}$$

■

EXAMPLE 8. Subtract: $\dfrac{2}{x^2 - 6x + 5} - \dfrac{1}{x^2 - 1}$

SOLUTION.

$$x^2 - 6x + 5 = (x - 5)(x - 1), \qquad x^2 - 1 = (x + 1)(x - 1)$$

The LCD is $(x - 5)(x + 1)(x - 1)$.

$$\begin{aligned}\frac{2}{x^2 - 6x + 5} - \frac{1}{x^2 - 1} &= \frac{2(x + 1)}{(x - 5)(x - 1)(x + 1)} \\ &\quad - \frac{1(x - 5)}{(x + 1)(x - 1)(x - 5)} \\ &= \frac{2(x + 1) - (x - 5)}{(x - 5)(x - 1)(x + 1)} \\ &= \frac{2x + 2 - x + 5}{(x - 5)(x - 1)(x + 1)} \\ &= \frac{x + 7}{(x - 5)(x - 1)(x + 1)}\end{aligned}$$

The denominator, which involves three factors, is left in factored form. ■

EXAMPLE 9. Add: $\dfrac{1}{a^2 - ax} + \dfrac{1}{x^2 - ax}$

SOLUTION.

$$a^2 - ax = a(a - x) = -a(x - a), \qquad x^2 - ax = x(x - a)$$

The LCD is $ax(x - a)$.

$$\begin{aligned}\frac{1}{a^2 - ax} + \frac{1}{x^2 - ax} &= \frac{1 \cdot (-x)}{-a(x - a) \cdot (-x)} + \frac{1 \cdot a}{x(x - a) \cdot a} \\ &= \frac{-x + a}{ax(x - a)} \\ &= \frac{-(x - a)}{ax(x - a)} \\ &= \frac{-1}{ax}\end{aligned}$$

■

EXERCISES

In Exercises 1–60 combine as indicated.

1. $\dfrac{1}{3} + \dfrac{2}{3}$
2. $\dfrac{5}{8} - \dfrac{1}{8}$
3. $\dfrac{7}{10} + \dfrac{1}{10} + \dfrac{3}{10}$

4. $\frac{5}{12} - \frac{1}{12} + \frac{7}{12}$

5. $\frac{19}{24} - \left(\frac{1}{24} + \frac{5}{24}\right)$

6. $\frac{53}{100} - \left(\frac{7}{100} - \frac{9}{100}\right)$

7. $\frac{1}{3} + \frac{1}{4}$

8. $\frac{3}{8} - \frac{1}{4}$

9. $\frac{7}{12} + \frac{3}{4}$

10. $\frac{9}{20} - \frac{3}{10}$

11. $\frac{5}{16} + \frac{1}{24}$

12. $\frac{3}{32} - \frac{5}{48}$

13. $\frac{9}{40} + \frac{7}{30}$

14. $\frac{7}{100} - \frac{3}{80}$

15. $\frac{11}{12} - \frac{1}{4} + \frac{1}{3}$

16. $\frac{1}{6} + \frac{5}{18} - \frac{5}{12}$

17. $\frac{9}{10} + \frac{7}{20} - \frac{1}{40}$

18. $\frac{29}{56} - \left(\frac{5}{28} + \frac{1}{14}\right)$

19. $\frac{13}{14} + \frac{3}{7} - \frac{1}{28}$

20. $\frac{6}{25} + \frac{19}{75} + \frac{12}{125}$

21. $\frac{1}{40} + \frac{1}{25} - \frac{1}{50}$

22. $\frac{7}{144} + \frac{5}{96} + \frac{9}{100}$

23. $\frac{1}{2^3 \cdot 3^2 \cdot 5^4} + \frac{1}{2^5 \cdot 3 \cdot 5^3} - \frac{1}{2^4 \cdot 3^2 \cdot 5^2}$

24. $\frac{2}{3^2 \cdot 5^4 \cdot 7^2} - \frac{1}{2 \cdot 3 \cdot 5^4 \cdot 7^2} + \frac{3}{2 \cdot 5^4 \cdot 7}$

25. $\frac{1}{m} + \frac{2}{m}$

26. $\frac{5}{a} - \frac{3}{a}$

27. $\frac{1}{m} + \frac{1}{n}$

28. $\frac{2}{y} + \frac{3}{x}$

29. $\frac{5}{x^2} + \frac{1}{x}$

30. $\frac{a}{y^2} - \frac{b}{y}$

31. $\frac{2}{xy^2} + \frac{1}{x^2y}$

32. $\frac{5a}{b^2} - \frac{2b}{a^2}$

33. $\frac{1}{a} + \frac{1}{b} - \frac{1}{c}$

34. $\frac{2}{x^2y} + \frac{1}{x^3} + \frac{5}{y^2}$

35. $\frac{b}{tu} + \frac{u}{tv} - \frac{t}{uv}$

36. $\frac{a^2}{x^2y^2} + \frac{b^2}{x^2z^2} + \frac{c^2}{y^2z^2}$

37. $\frac{1}{x - a} + \frac{2}{x - a}$

38. $\frac{5}{x^2 + 1} - \frac{3}{x^2 + 1} + \frac{1}{x^2 + 1}$

39. $\frac{1}{x - y} + \frac{1}{y - x}$

40. $\frac{1}{x^2 - 1} + \frac{1}{x - 1}$

41. $\frac{1}{x^2 - 4} + \frac{2}{x - 2}$

42. $\frac{a}{x^2 - a^2} + \frac{1}{x - a}$

43. $\frac{2}{x^2 + 5x + 4} + \frac{1}{x + 4}$

44. $\frac{x}{x^2 + 7x + 10} + \frac{x^2}{x + 5}$

45. $\frac{1}{x + 3} - \frac{2}{x + 2}$

46. $\frac{3}{x^2} + \frac{1}{x - a}$

47. $\frac{u}{x^2 - u^2} + \frac{x}{x - u} - \frac{1}{x + u}$

48. $\frac{2}{x^2 - 25} + \frac{1}{x + 5} - \frac{1}{x - 5}$

49. $\frac{1}{x + y} - \left(\frac{1}{x - y} - \frac{1}{x^2 - y^2}\right)$

50. $\frac{1}{a^2 - 16} + \frac{1}{a^2 + 5a + 4} + \frac{1}{a^2 - 3a - 4}$

51. $\frac{2}{a^3} + \frac{1}{a^2 - 9a} + \frac{1}{9a^2 - a^3}$

52. $x^2 - \frac{1}{x} + \frac{1}{x^2}$

53. $\dfrac{1}{(x-a)(x-b)} + \dfrac{1}{(x-b)(x-c)} + \dfrac{1}{(x-a)(x-c)}$

54. $\dfrac{1}{u^2 - a^2} - \dfrac{1}{u^2 - b^2}$

55. $\dfrac{x}{x^2 + 6x + 9} + \dfrac{x+1}{x^2 + 3x}$

56. $\dfrac{a}{x^2 y}\left(\dfrac{1}{a^2} + \dfrac{1}{a}\right)$

57. $\dfrac{x}{x+1}\left(\dfrac{x}{x^2 - 1} + \dfrac{1}{x-1}\right)$

58. $\dfrac{x-4}{x^2}\left(\dfrac{1}{x^2 - 16} - \dfrac{1}{x+4}\right)$

59. $\dfrac{t^4}{t^2 - 1} \div \left(\dfrac{1}{t+1} - \dfrac{1}{t-1}\right)$

60. $\dfrac{u^6}{u^2 + 1} \div \left(\dfrac{u}{u^2 + 1} - \dfrac{1}{u+1}\right)$

In Exercises 61–64 factor each polynomial. (The factors will have rational coefficients.)

61. $x^2 + x + \dfrac{1}{4}$

62. $x^2 + \dfrac{5}{6}x + \dfrac{1}{6}$

63. $x^2 - \dfrac{2}{3}x + \dfrac{1}{9}$

64. $x^2 + \dfrac{3}{10}x - \dfrac{1}{10}$

5.4 COMPLEX EXPRESSIONS

DEFINITION. **Complex expressions** are rational expressions that contain other rational expressions either in their numerator or denominator (possibly in both).

Your task is to simplify these complex expressions. You can rewrite such expressions in terms of *division.* The first two examples concern **complex fractions.**

EXAMPLE 1. Simplify: $\dfrac{\frac{1}{2}}{\frac{3}{4}}$

SOLUTION 1. $\dfrac{\frac{1}{2}}{\frac{3}{4}}$ means $\dfrac{1}{2}$ *divided by* $\dfrac{3}{4}$. Thus

$$\frac{\frac{1}{2}}{\frac{3}{4}} = \frac{1}{2} \div \frac{3}{4}$$

$$= \frac{1}{\not{2}} \cdot \frac{\overset{2}{\not{4}}}{3}$$

$$= \frac{2}{3}$$

SOLUTION 2. The LCD of $\frac{1}{2}$ and $\frac{3}{4}$ is 4. Multiply the numerator and denominator of the complex fraction by 4.

$$\frac{\frac{1}{2}}{\frac{3}{4}} = \frac{\frac{1}{\not{2}} \cdot \overset{2}{\not{4}}}{\frac{3}{\not{4}} \cdot \not{4}} = \frac{2}{3}$$

■

EXAMPLE 2. Simplify: $\dfrac{\frac{1}{3} + \frac{1}{6}}{9}$

SOLUTION.

$$\frac{\frac{1}{3} + \frac{1}{6}}{9} = \left(\frac{1}{3} + \frac{1}{6}\right) \div 9$$

$$= \frac{2 + 1}{6} \cdot \frac{1}{9}$$

$$= \frac{\overset{1}{\not{3}}}{\underset{2}{\not{6}}} \cdot \frac{1}{9}$$

$$= \frac{1}{18}$$

■

EXAMPLE 3. Simplify: $\dfrac{1 + \frac{1}{x}}{x^2}$

SOLUTION.

$$\frac{1 + \frac{1}{x}}{x^2} = \left(1 + \frac{1}{x}\right) \div x^2$$

$$= \left(\frac{x}{x} + \frac{1}{x}\right) \div x^2$$

$$= \frac{x + 1}{x} \cdot \frac{1}{x^2}$$

$$= \frac{x + 1}{x^3}$$

■

EXAMPLE 4. Simplify: $\dfrac{\dfrac{x}{x - 1} - \dfrac{1}{x + 1}}{\dfrac{x}{x^2 - 1}}$

SOLUTION.

$$\frac{\dfrac{x}{x - 1} - \dfrac{1}{x + 1}}{\dfrac{x}{x^2 - 1}} = \left(\frac{x}{x - 1} - \frac{1}{x + 1}\right) \div \frac{x}{x^2 - 1}$$

$$= \frac{x(x + 1) - 1 \cdot (x - 1)}{(x - 1)(x + 1)} \div \frac{x}{(x + 1)(x - 1)}$$

$$= \frac{x^2 + x - x + 1}{\cancel{(x - 1)}\cancel{(x + 1)}} \cdot \frac{\cancel{(x + 1)}\cancel{(x - 1)}}{x}$$

$$= \frac{x^2 + 1}{x}$$

■

EXAMPLE 5. Simplify: $\dfrac{\dfrac{y}{y + 1} + \dfrac{y - 1}{y}}{\dfrac{y}{y + 1} - \dfrac{y - 1}{y}}$

SOLUTION.

$$\frac{\dfrac{y}{y + 1} + \dfrac{y - 1}{y}}{\dfrac{y}{y + 1} - \dfrac{y - 1}{y}} = \left(\frac{y}{y + 1} + \frac{y - 1}{y}\right) \div \left(\frac{y}{y + 1} - \frac{y - 1}{y}\right)$$

$$= \frac{y \cdot y + (y-1)(y+1)}{y(y+1)} \div \frac{y \cdot y - (y-1)(y+1)}{y(y+1)}$$

$$= \frac{y^2 + y^2 - 1}{y(y+1)} \cdot \frac{y(y+1)}{y^2 - (y^2 - 1)}$$

$$= \frac{2y^2 - 1}{\cancel{y(y+1)}} \cdot \frac{\cancel{y(y+1)}}{1}$$

$$= 2y^2 - 1$$

■

EXERCISES

Simplify.

1. $\dfrac{\frac{5}{8}}{2}$

2. $\dfrac{\frac{1}{3}}{12}$

3. $\dfrac{4}{\frac{2}{3}}$

4. $\dfrac{-8}{\frac{6}{7}}$

5. $\dfrac{\frac{1}{3}}{\frac{1}{6}}$

6. $\dfrac{\frac{3}{5}}{\frac{-9}{25}}$

7. $\dfrac{\frac{3}{4}}{\frac{15}{8}}$

8. $\dfrac{\frac{7}{10}}{\frac{21}{40}}$

9. $\dfrac{\frac{5}{42}}{\frac{15}{32}}$

10. $\dfrac{\frac{-27}{56}}{\frac{-9}{28}}$

11. $\dfrac{\frac{1}{2} + \frac{1}{4}}{\frac{1}{6} + \frac{5}{6}}$

12. $\dfrac{\frac{2}{3} - \frac{1}{3}}{\frac{3}{4} - \frac{1}{4}}$

13. $\dfrac{\frac{1}{2} + \frac{1}{8}}{\frac{5}{12} - \frac{1}{12}}$

14. $\dfrac{\frac{5}{9} + \frac{2}{9}}{\frac{11}{12} + \frac{1}{4}}$

15. $\dfrac{\frac{1}{16} + \frac{3}{4}}{2 + \frac{1}{8}}$

16. $\dfrac{1 - \frac{1}{4}}{1 + \frac{1}{4}}$

17. $\dfrac{\frac{5}{27} - \frac{2}{9}}{\frac{1}{36} + \frac{5}{18}}$

18. $\dfrac{\frac{3}{100} + \frac{1}{20}}{\frac{7}{75} - \frac{1}{25}}$

19. $\dfrac{a}{\frac{b}{c}}$

20. $\dfrac{\frac{a}{b}}{c}$

21. $\dfrac{\frac{a}{b}}{\frac{c}{d}}$

22. $\dfrac{\frac{ax}{y}}{x}$

23. $\dfrac{\frac{ax^2}{yz^2}}{\frac{a^2x}{yz}}$

24. $\dfrac{\frac{m^2n}{xy^2}}{\frac{mn}{x^2y^2}}$

25. $\dfrac{\frac{abc^3}{xyz^3}}{\frac{c^2}{x^2y^2z^2}}$

26. $\dfrac{\frac{4ax}{yz}}{\frac{12a^2}{xyz}}$

27. $\dfrac{\frac{2abc}{7xy}}{\frac{4a^2b}{21}}$

28. $\dfrac{\frac{49a^2bc^2}{24xy^2}}{\frac{14abc}{9y}}$

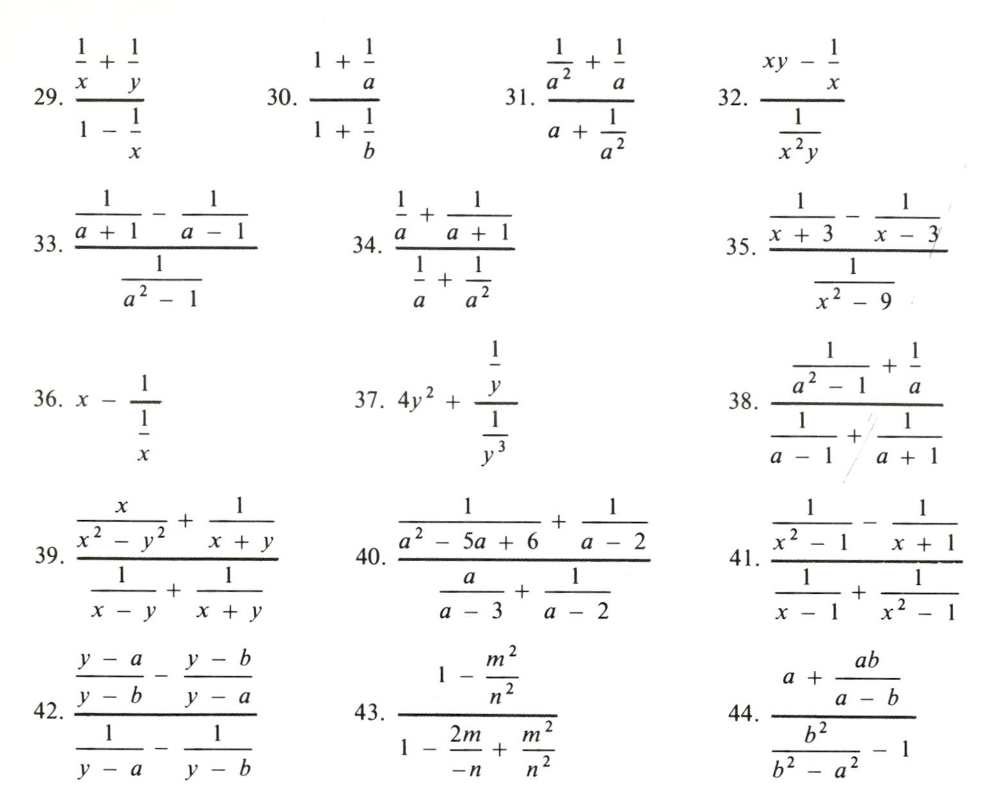

29. $\dfrac{\frac{1}{x} + \frac{1}{y}}{1 - \frac{1}{x}}$

30. $\dfrac{1 + \frac{1}{a}}{1 + \frac{1}{b}}$

31. $\dfrac{\frac{1}{a^2} + \frac{1}{a}}{a + \frac{1}{a^2}}$

32. $\dfrac{xy - \frac{1}{x}}{\frac{1}{x^2y}}$

33. $\dfrac{\frac{1}{a+1} - \frac{1}{a-1}}{\frac{1}{a^2-1}}$

34. $\dfrac{\frac{1}{a} + \frac{1}{a+1}}{\frac{1}{a} + \frac{1}{a^2}}$

35. $\dfrac{\frac{1}{x+3} - \frac{1}{x-3}}{\frac{1}{x^2-9}}$

36. $x - \dfrac{1}{\frac{1}{x}}$

37. $4y^2 + \dfrac{\frac{1}{y}}{\frac{1}{y^3}}$

38. $\dfrac{\frac{1}{a^2-1} + \frac{1}{a}}{\frac{1}{a-1} + \frac{1}{a+1}}$

39. $\dfrac{\frac{x}{x^2-y^2} + \frac{1}{x+y}}{\frac{1}{x-y} + \frac{1}{x+y}}$

40. $\dfrac{\frac{1}{a^2-5a+6} + \frac{1}{a-2}}{\frac{a}{a-3} + \frac{1}{a-2}}$

41. $\dfrac{\frac{1}{x^2-1} - \frac{1}{x+1}}{\frac{1}{x-1} + \frac{1}{x^2-1}}$

42. $\dfrac{\frac{y-a}{y-b} - \frac{y-b}{y-a}}{\frac{1}{y-a} - \frac{1}{y-b}}$

43. $\dfrac{1 - \frac{m^2}{n^2}}{1 - \frac{2m}{-n} + \frac{m^2}{n^2}}$

44. $\dfrac{a + \frac{ab}{a-b}}{\frac{b^2}{b^2-a^2} - 1}$

5.5 CALCULATING WITH DECIMALS

An n-**placed decimal** is a fraction of the form

$$\frac{M}{10^n}.$$

There are n places to the *right* of the decimal point. Thus

$.3 \left(\text{or } \frac{3}{10}\right)$ is a 1-placed decimal;

$1.3 \left(\text{or } \frac{13}{10}\right)$ is a 1-placed decimal;

$-.29 \left(\text{or } \frac{-29}{100}\right)$ is a 2-placed decimal;

and

$$.007 \left(\text{or } \frac{7}{1000}\right) \text{ is a 3-placed decimal.}$$

Each decimal place is occupied by one of the digits

0, 1, 2, 3, 4, 5, 6, 7, 8, 9

If you insert 0's *to the right of the last nonzero digit,* you obtain an equivalent decimal. Thus

$$.3 = .30 = .300$$

or

$$\frac{3}{10} = \frac{3 \times 10}{10 \times 10} = \frac{30}{100} = \frac{30 \times 10}{100 \times 10} = \frac{300}{1000}$$

As you know,

$$\frac{11}{100} + \frac{12}{100} = \frac{11 + 12}{100} = \frac{23}{100}$$

In terms of decimals, express this as follows:

$$\begin{array}{r} .11 \\ +\ .12 \\ \hline .23 \end{array}$$

Add or subtract decimals in columns. If necessary, add 0's to the right of the last nonzero digit, as in Example 1.

EXAMPLE 1. Add: .72 + .436 + .5

SOLUTION. .436 is a 3-placed decimal. Express the other decimals to 3 places:

$$.72 = .720$$

$$.5 = .500$$

Add:

$$\begin{array}{r} .720 \\ .436 \\ .500 \\ \hline 1.656 \end{array}$$

■

Algebraic expressions frequently involve decimals.

EXAMPLE 2. *Subtract:*

$$\begin{array}{r} .26\ \ xyz \\ .178\, xyz \\ \hline \end{array}$$

SOLUTION.

$$.26 = .260$$

Thus

$$\begin{array}{r} .260\,xyz \\ -\ .178\,xyz \\ \hline .082\,xyz \end{array}$$

Next, recall that

$$\frac{21}{100} \times \frac{3}{10} = \frac{21 \times 3}{100 \times 10} = \frac{63}{1000}.$$

Note that

$$\frac{21}{100} = .21 \text{ (a 2-placed decimal)} \quad \text{and} \quad \frac{3}{10} = .3 \text{(a 1-placed decimal.)}$$

The product is $\frac{63}{1000}$ or .063 (a 3-placed decimal). ■

EXAMPLE 3. Multiply .1 by .3 .

SOLUTION.

$$\frac{1}{10} \times \frac{3}{10} = \frac{1 \times 3}{10 \times 10} = \frac{3}{100}$$

In terms of decimals,

$$\underbrace{.1}_{\text{1 place}} \times \underbrace{.3}_{\text{1 place}} = \underbrace{.03}_{\text{2 places}}$$

■

When you multiply an m-placed decimal by an n-placed decimal, the product is an $(m + n)$*-placed decimal. If necessary, move the decimal point to the left and insert* 0's, such as when you multiply .1 by .3 to obtain .03.

EXAMPLE 4. Multiply 8.031 by 9.11.

SOLUTION.

$$\begin{array}{r} 8.031 \\ 9.11 \\ \hline 8031 \\ 8031 \\ 72279 \\ \hline 73.16241 \end{array}$$

There are 3 decimal places in one factor—2 in the other. The product has 5 decimal places. ■

EXAMPLE 5. Determine: $(.04) \times (.2) \times (.001)$

SOLUTION. *Add* the total number of decimal places in the factors.

$$\underbrace{(.04)}_{2 \text{ places}} \times \underbrace{(.2)}_{+\ 1 \text{ place}} \times \underbrace{(.001)}_{+\ 3 \text{ places}}$$

The product is a 6-placed decimal:

$$.04 \times .2 \times .001 = .000\,008$$ ■

EXAMPLE 6. Determine: $(.02)^4$

SOLUTION. Exponentiation involves repeated multiplication. Thus multiply the number of decimal places in the base by the exponent. Because .02 is a 2-placed decimal and the exponent is 4, the product is an 8-placed decimal. Moreover,

$$2^4 = 16$$

Thus

$$(.02)^4 = .000\,000\,16$$ ■

EXAMPLE 7. Evaluate $5x^3 + 3x^2 + 1$ when $x = .1$.

SOLUTION. Replace x by $.1$ to obtain:

$$5(.1)^3 + 3(.1)^2 + 1 = 5(.001) + 3(.01) + 1 = .005 + .03 + 1$$

$$\begin{array}{r} .005 \\ .030 \\ 1.000 \\ \hline 1.035 \end{array}$$ ■

To obtain the rule for division of decimals, consider:

$$.12\overline{)4.8}$$

The quotient can be indicated by the fraction:

$$\frac{4.8}{.12}$$

Multiplying numerator and denominator of this fraction by 100 (to eliminate decimals), you obtain:

$$\frac{4.8}{.12} = \frac{4.8 \times 100}{.12 \times 100} = \frac{480}{12} = 40 \quad \text{or} \quad .12\overline{)4.80}\ \text{quotient } 40$$

Multiplication of the numerator and denominator can be accomplished by moving each decimal point 2 places to the right. Thus you obtain the following rule for division of decimals.

Move the decimal place of the divisor all the way over to the right. Move the decimal place of the dividend the same number of places to the right. This determines the position of the decimal point in the quotient.

EXAMPLE 8.

$$.016\overline{)1.44}$$

SOLUTION. The decimal place is moved 3 places to the right in both the divisor and dividend. This necessitates adding a zero after the second 4 in the dividend.

```
        90.
.016.)1.440.
      1 44
      ----
         0
```

The quotient is 90. ■

EXAMPLE 9. Determine the quotient to 3 decimal places:

$$7.1\overline{)1.815}$$

SOLUTION.

```
        .2556
7.1.)1.8 1500
     1 42
     ----
       395
       355
       ---
        400
        355
        ---
         450
         426
```

You want a 3-placed decimal. By convention, if the digit in the fourth decimal place is 5 or more, increase the digit in the third decimal place by 1. Because 6 is in the fourth decimal place, the quotient to 3 places is

$$.256 .$$

[See Figure 5.6.] ■

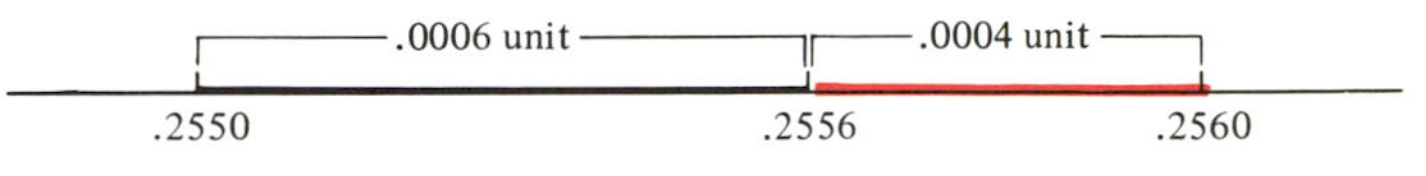

FIGURE 5.6 .2556 is closer to .2560 than to .2550. To 3 decimal places, .2556 is written as .256.

EXERCISES

In Exercises 1–10, add as indicated:

1. $\begin{array}{r} .4 \\ \underline{.2} \end{array}$
2. $\begin{array}{r} .8 \\ \underline{.5} \end{array}$
3. $\begin{array}{r} .27 \\ \underline{.53} \end{array}$
4. $\begin{array}{r} .624 \\ \underline{.585} \end{array}$
5. $\begin{array}{r} 12.85 \\ \underline{3.72} \end{array}$
6. $\begin{array}{r} .83 \\ .34 \\ .62 \\ \underline{.57} \end{array}$
7. $\begin{array}{l} .62 \\ .7 \\ .83 \\ \underline{.4} \end{array}$
8. $\begin{array}{r} .03y^2 \\ .04y^2 \\ \underline{.37y^2} \end{array}$
9. $\begin{array}{l} .43\ abc \\ .008abc \\ \underline{.72\ abc} \end{array}$
10. $\begin{array}{r} 1.53\ x^2y \\ .009x^2y \\ .728x^2y \\ \underline{1.5\ \ x^2y} \end{array}$

In Exercises 11–16 subtract the bottom expression from the top one:

11. $\begin{array}{r} .48 \\ \underline{.12} \end{array}$
12. $\begin{array}{r} .18 \\ \underline{.12} \end{array}$
13. $\begin{array}{r} 1.93 \\ \underline{-\ .827} \end{array}$
14. $\begin{array}{l} 4.435x \\ \underline{3.08\ x} \end{array}$
15. $\begin{array}{l} .47\ \ ab \\ \underline{.3872ab} \end{array}$
16. $\begin{array}{l} 1.5x \\ \underline{1.8x} \end{array}$

In Exercises 17–26, multiply as indicated

17. $\begin{array}{r} 7.3 \\ \underline{\times\ 1.2} \end{array}$
18. $\begin{array}{r} 420 \\ \underline{\times\ .05} \end{array}$
19. $\begin{array}{r} 6000 \\ \underline{\times\ .075} \end{array}$
20. $\begin{array}{r} 1.008x \\ \underline{\times\ \ \ .25} \end{array}$
21. $\begin{array}{r} 212x \\ \underline{\times\ .002x} \end{array}$
22. $\begin{array}{r} .52x \\ \underline{\times\ .21y} \end{array}$
23. $(.2) \times (.5) \times (.3)$
24. $(.04) \times (.05) \times (.2)$
25. $(.3x) \times (.7y) \times (.001z)$
26. $(.001a^2) \times (.001a) \times (.002a)$

In Exercises 27–32, determine each number:

27. $(.1)^5$
28. $(.01)^3$
29. $(.2)^3$

30. $(.3)^2$ 31. $(.4)^2 + (.2)^4$ 32. $(.1)^2 - (.1)^4$

In Exercises 33–38 evaluate each polynomial for the specified values of the variables:

33. $x^2 + 2x \quad [x = .1]$
34. $x^3 + 5x - 2 \quad [x = .2]$
35. $y^4 - y^2 + 3y \quad [y = .1]$
36. $t^2 + 5t - .2 \quad [t = .3]$
37. $3x^2 + 2xy - y^2 \quad [x = .1, y = .2]$
38. $s^3 + s^2t + 4st^2 + t^3 \quad [s = .1, t = .01]$

In Exercises 39–46 divide as indicated:

39. $3.7\overline{)111}$ 40. $-24\overline{)9.6}$ 41. $.17\overline{)-28.9}$

42. $2.5\overline{)-1300}$ 43. $-.0043\overline{)-6.45}$ 44. $-2.13\overline{)23.856}$

45. $6.07\overline{)-236.73}$ 46. $1.001\overline{)1.8018}$

In Exercises 47–50 express the quotient to 3 decimal places.

47. $193\overline{)-29.925}$ 48. $.7043\overline{)971.934}$

49. $6.45\overline{)130.544}$ 50. $80.04\overline{)-80.760\,36}$

REVIEW EXERCISES FOR CHAPTER 5

5.1 MULTIPLICATION AND DIVISION

In Exercises 1–6 multiply or divide, as indicated. Simplify the result. Rational expressions, other than fractions, should be left in factored form.

1. $\frac{12}{35} \cdot \frac{49}{18}$
2. $\frac{a^2 - x^2}{x + 2} \cdot \frac{x^2 + 4x + 4}{x^2 + 2ax + a^2}$
3. $\frac{a + b}{(a - b)^2} \cdot \frac{a^2 - b^2}{b + 1} \cdot \frac{1 - b^2}{3a + 3b}$
4. $\frac{14x^2yz}{3ab^2} \div \frac{42yz^2}{a^2b^2}$
5. $\frac{x^4 - y^4}{x + y} \div \frac{3x^2 - 3y^2}{x^2 + 2xy + y^2}$
6. Divide $\frac{a^2 + 5a + 6}{a^2 - 6a + 9}$ by $\frac{a^2 + 4a + 4}{a^2 - 9}$.

5.2 LEAST COMMON MULTIPLES

7. Determine the LCM of the indicated integers.

   (a) 4, 8 (b) 24, 32 (c) 6, 10, 15

8. Determine an LCM of the indicated polynomials.

   (a) $x^2y,\ xy^2$ (b) $x^2 - a^2,\ (x - a)^2$

   (c) $x^2 + 4x + 4,\ x^2 - 4,\ x^2 - 4x + 4$

9. Determine the LCD of the indicated fractions. Then determine equivalent fractions with this LCD as the denominator.

(a) $\frac{3}{4}, \frac{-5}{8}$ (b) $\frac{1}{48}, \frac{-1}{36}$ (c) $\frac{1}{20}, \frac{1}{30}, \frac{3}{40}$

10. Determine the LCD of the indicated rational expressions. Then determine equivalent rational expressions with this LCD as the denominator.

(a) $\frac{1}{x^2}, \frac{-1}{x^3}$ (b) $\frac{x}{x^2 - 4}, \frac{1}{x^2 + 2x}$

(c) $\frac{1}{x^2 + 3x + 2}, \frac{2}{x^2 + 2x + 1}, \frac{-1}{x + 2}$

5.3 ADDITION AND SUBTRACTION

In Exercises 11–18 add or subtract as indicated. Simplify the result. Rational expressions, other than fractions, should be left in factored form.

11. $\frac{4}{9} + \frac{2}{9}$ 12. $\frac{5}{8} - \frac{3}{8}$ 13. $\frac{2}{25} - \frac{1}{50} + \frac{1}{75}$

14. $\frac{1}{x} + \frac{2}{x}$ 15. $\frac{1}{x^2} + \frac{2}{x}$ 16. $\frac{1}{x^2yz^2} + \frac{y}{x^2z^2} - \frac{1}{x^2yz}$

17. $\frac{a}{x^2 - a^2} + \frac{1}{x - a}$ 18. $\frac{1}{x^2 - 9} - \left(\frac{1}{x + 3} - \frac{1}{3 - x}\right)$

19. Combine, as indicated: $\frac{x}{x + 1}\left(\frac{1}{x^2 - 1} - \frac{1}{x^2 + x}\right)$

5.4 COMPLEX EXPRESSIONS

In Exercises 20–24 simplify each expression:

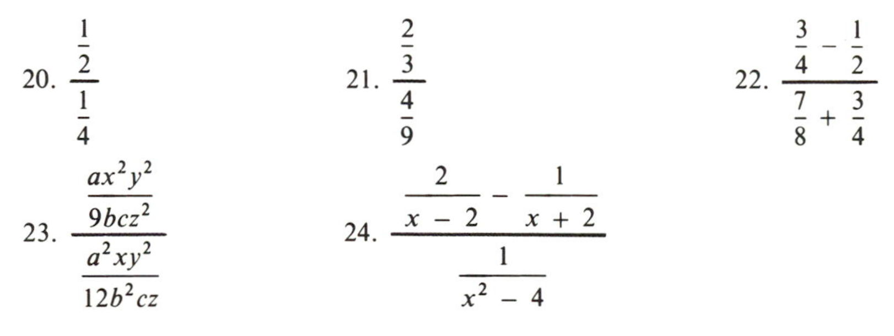

20. $\dfrac{\frac{1}{2}}{\frac{1}{4}}$ 21. $\dfrac{\frac{2}{3}}{\frac{4}{9}}$ 22. $\dfrac{\frac{3}{4} - \frac{1}{2}}{\frac{7}{8} + \frac{3}{4}}$

23. $\dfrac{\frac{ax^2y^2}{9bcz^2}}{\frac{a^2xy^2}{12b^2cz}}$ 24. $\dfrac{\frac{2}{x - 2} - \frac{1}{x + 2}}{\frac{1}{x^2 - 4}}$

5.5 CALCULATING WITH DECIMALS

25. Determine the sum:

$$\begin{array}{r} .439 \\ .83 \\ .7 \\ \underline{.52} \end{array}$$

26. Determine the product:

$.25 \times .006$

27. Find the value of $2x^3 + 3x^2 + 5x$ when $x = .2$.

TEST YOURSELF ON CHAPTER 5.

1. Multiply: $\frac{3}{8} \cdot \frac{4}{9}$

2. Multiply: $\frac{a^2(x + a)}{x^2 - 1} \cdot \frac{1 - x}{ax + a^2}$

3. Divide: $\frac{5a^2b^2}{12a^2 - 48} \div \frac{25ab^3}{18 - 9a}$

4. Determine the LCM of the indicated integers.

   (a) 8, 12 (b) 9, 15, 30

5. Determine the LCD of the indicated rational expressions. Then determine rational expressions with this LCD as the denominator.

   (a) $\frac{a}{y^2}, \frac{-1}{y^3}$ (b) $\frac{1}{ax - 5a}, \frac{2}{a^2}, \frac{-1}{x^2 - 25}$

6. Determine: $\frac{3}{100} - \frac{1}{20} + \frac{7}{50}$

7. Determine: $\frac{1}{x^2 - 1} - \left(\frac{1}{x + 1} + \frac{2}{1 - x}\right)$

8. Determine: $\frac{x^2y}{3}\left(\frac{3x}{4y} - \frac{y}{x^2}\right)$

9. Simplify: $\dfrac{\dfrac{20xy^2}{21abc}}{\dfrac{36x^3y}{35a^2b}}$

10. Simplify: $\dfrac{1 - \dfrac{1}{a}}{a - \dfrac{1}{a}}$

6

first-degree equations

6.1 ROOTS OF EQUATIONS

DEFINITION. An **equation** is a statement of equality (or equivalence). An equation has the form:

$$\square = \square$$

Here $\square$ is known as the **left side of the equation;** $\square$ is the **right side.**

You will study equations involving constants and variables.

EXAMPLE 1.

(a) $$\frac{2}{6} = \frac{1}{3}$$

is an equation expressing the equivalence of fractions. $\frac{2}{6}$ is the left side and $\frac{1}{3}$ the right side. There are no variables present in this equation.

The following equations involve variables.

(b) $$x + 1 = 6$$

This is an example of a **first-degree equation.** The highest-degree term present is a first-degree term. The left side of the equation is the first-degree polynomial $x + 1$; the right side is the constant 6.

(c) $$3y - 7 = 8$$

is also a first-degree equation.

(d) $$x^2 + 4x + 1 = 0$$

involves a second-degree term, and is known as a **second-degree equation** or a **quadratic equation.** Such equations will be considered in Chapter 13. ■

In the first three sections of this chapter, you will be primarily concerned with first-degree equations in a single variable. First-degree equations (in the variable x) can be written in the form

$$Ax + B = 0$$

where A and B are constants and $A \neq 0$. For example, in Section 6.2 you will see that

$$x + 1 = 6$$

can be rewritten as

$$x - 5 = 0.$$

Thus $A = 1$ and $B = -5$

Equations involving only constants are either *true* or *false* statements.

EXAMPLE 2.

(a) $$2 + 2 = 4 \quad \text{and} \quad \frac{3}{9} = \frac{2}{6}$$

are each true.

(b) $$2 \cdot 3 = 5 \quad \text{and} \quad \frac{1}{2} = \frac{3}{4}$$

are each false. ■

In Section 2.2 you substituted numbers for the variable(s) of a polynomial. You can also substitute a number for the variable of an equation. If the variable occurs more than once in an equation, substitute the same number for *each occurrence.*

DEFINITION. A **root** or **solution** of an equation (in a single variable) is a number that when substituted in the equation yields a true statement. A root of an equation is said to **satisfy the equation.**

EXAMPLE 3.

(a) 5 is a root of the equation

$$2x - 3 = 7.$$

For, substitute 5 for x. Then

$$2 \cdot 5 - 3 = 7$$

is a true statement. Thus 5 satisfies the given equation.

(b) -4 is a root of the equation

$$5x + 2 = -22 - x.$$

Here the variable x occurs twice. Substitute -4 for each occurrence of x, and obtain

$$5(-4) + 2 = -22 - (-4)$$

or

$$-18 = -18.$$ ■

This is true.

EXAMPLE 4. -3 is not a root of the equation

$$7 - 3x = 9 + x.$$

For,

$$7 - 3(-3) \neq 9 + (-3)$$

$\neq$ means "is not equal to".

or

$$16 \neq 6$$ ■

It is one thing to **solve an equation,** that is, to determine its roots. It is often a far simpler matter to **check** whether a given number is, indeed, a root of an equation. In Example 3, you *checked* that 5 is a root of the equation $2x - 3 = 7$ and that -4 is a root of the equation $5x + 2 = -22 - x$. In Example 4, the *check* indicated that -3 is not a root of the equation $7 - 3x = 9 + x$. Hereafter, when checking a root, place a question mark over the equality sign,

$$\overset{?}{=},$$

until you determine the truth (or falsehood) of the statement in question. Then write

$$\overset{\checkmark}{=}$$

if the statement (of equality) is true,

$$\overset{\text{x}}{=},$$

if the statement is false.

EXAMPLE 5. Check that $\frac{1}{2}$ is a root of the equation

$$\frac{5}{y} = 9 + 2y.$$

SOLUTION. Substitute $\frac{1}{2}$ for each occurrence of y.

$$\frac{5}{\frac{1}{2}} \stackrel{?}{=} 9 + 2\left(\frac{1}{2}\right)$$

$$5 \cdot 2 \stackrel{?}{=} 9 + 1$$

$$10 \stackrel{\checkmark}{=} 10$$

■

In the next section you will learn how to *solve* some equations.

EXERCISES

In Exercises 1–12: does the statement involve only constants?

i. If so, is it true or is it false?

ii. If not, is it a first-degree equation in a single variable?

1. $3x = 9$
2. $2y + 4 = 3y - 8$
3. $9 + 8 = 20 - 3$
4. $9 - 3 = 5 + 2$
5. $14 = 14$
6. $x = x$
7. $x = 2x$
8. $x = y$
9. $x^2 + 1 = 3x$
10. $x + y = 2$
11. $xy = 2$
12. $\frac{x}{2} = \frac{2x}{4}$

In Exercises 13–34 check whether the number in color is a root of the given equation.

13. $x + 5 = 10,\ 5$
14. $x - 4 = 9,\ 13$
15. $3y = 18,\ 6$
16. $4x = -16,\ 4$
17. $2x + 5 = x - 9,\ 3$
18. $u - 7 = 6u + 2,\ -1$
19. $10x + 1 = 0,\ \frac{1}{10}$
20. $17x = 34,\ 2$
21. $\frac{1}{x} = 6,\ \frac{1}{6}$
22. $\frac{1}{x} = \frac{1}{6},\ \frac{1}{6}$
23. $\frac{1}{-x} = \frac{1}{6},\ -6$
24. $\frac{1}{x} + \frac{1}{2} = 10,\ \frac{2}{19}$
25. $\frac{1}{x} - \frac{1}{3} = 3x,\ 6$
26. $x - 3 = \frac{x + 8}{2},\ 11$

27. $\frac{1}{5}x = x + 5, \quad 15$

28. $5(x - 1) = 2, \quad \frac{2}{5}$

29. $\frac{x+7}{3} + \frac{x-1}{2} = 0, \quad \frac{-11}{5}$

30. $\frac{x+1}{2} + \frac{x}{4} = 1, \quad 2$

31. $x = x, \quad -1$

32. $y = 2y, \quad 0$

33. $z = 2(z + 1), \quad 0$

34. $\frac{u}{2} = u + 2, \quad -2$

6.2 SOLVING EQUATIONS

You know how to *check* whether a given number is a root of an equation. It remains to determine how to find that root, that is, how to solve the equation in the first place!

DEFINITION. **Equivalent equations** are equations with exactly the same roots.

EXAMPLE 1.

$$5x - 52 = 2x + 8 \qquad \text{and} \qquad 3x = 60$$

are equivalent equations. Each has the root 20. For, in the first equation,

$$5 \cdot 20 - 52 = 2 \cdot 20 + 8$$

$$48 = 48$$

Also, in the second equation,

$$3 \cdot 20 = 60$$

Thus 20 is a root of each equation.

Later you will see that 20 is the *only* root of each equation. Clearly the second equation

$$3x = 60$$

is the *simpler* one. ■

An equation can be thought of as a balanced scale. *In order to preserve the balance, what is done to one side of the equation must be done to the other side.* You solve an equation by successively transforming it into simpler equivalent equations, until finally, you obtain the equation

$$x = c$$

with (the constant) root c. (Note that c is the *only root* of this last equation.)

The **Addition Principle** *enables you to add the same quantity to both sides of an equation.*

ADDITION PRINCIPLE

Let *R* be any rational expression (possibly a polynomial or even a constant). Then

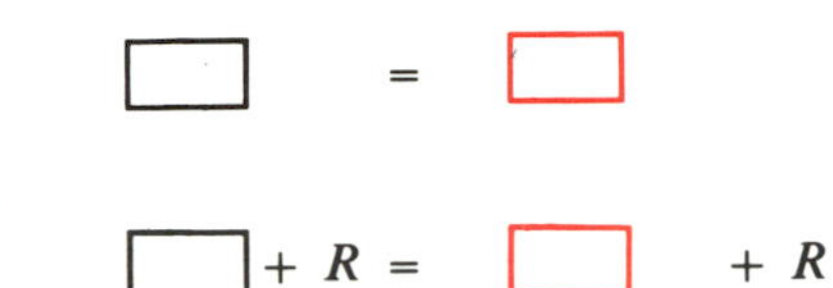

and

are equivalent equations.

Most frequently, the expression R to be added will be relatively simple. (See Exercise 43 for a restriction on the Addition Principle.) The symbol

R △—△ R

will indicate that the expression R is to be added to both sides of an equation.

EXAMPLE 2. Solve: $y - 5 = 0$

SOLUTION. In order to solve this equation, use the Addition Principle. Add 5 to both sides of the equation.

$$\begin{array}{rcr} y - 5 & = & 0 \\ \underline{+\ 5} & △—△ & \underline{+5} \\ y & = & 5 \end{array}$$

■

Recall that $a - b = a + (-b)$; thus *you can also subtract by means of the Addition Principle.*

EXAMPLE 3. Solve: $x + 7 = 19$

SOLUTION. Subtract 7 from both sides of the given equation.

$$\begin{array}{rcr} x + 7 & = & 19 \\ \underline{-\ 7} & △—△ & \underline{-7} \\ x & = & 12 \end{array}$$

You can check that 12 is, indeed, a root of the equation.

CHECK.

$$12 + 7 \stackrel{?}{=} 19$$

$$19 \stackrel{\checkmark}{=} 19$$

■

(In most of these examples the check will *not* be given.)

In order to simplify an equation, *bring expressions involving variables to one side; constants to the other side.* It does not matter which side of the simplified equation contains the variables. In Examples 2 and 3 the variables were brought to the left side and the constants to the right side. In Example 4, the reverse will be true.

EXAMPLE 4. Solve: $20 - x = 13$

SOLUTION. In order to bring constants to one side, variables to the other, add $-13 + x$ to both sides.

$$\begin{array}{rcl} 20 - x & = & 13 \\ \underline{-13 + x} & \triangle\!\!-\!\!\triangle & \underline{-13 + x} \\ 7 & = & x \end{array}$$

■

The second major tool in solving equations is the **Multiplication Principle.**

MULTIPLICATION PRINCIPLE

Let *c* be a *nonzero* constant. Then

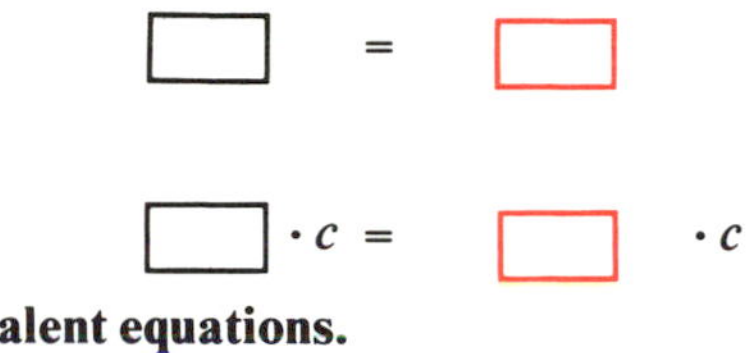

are equivalent equations.

EXAMPLE 5. Solve: $\frac{x}{2} = 3$

SOLUTION. Multiply both sides by 2.

$$\frac{x}{2} \cdot 2 = 3 \cdot 2$$

$$x = 6$$

■

Recall that $\frac{a}{b} = a \cdot \frac{1}{b}$. Thus *the Multiplication Principle enables you to multiply or divide both sides of an equation by the same nonzero constant.*

EXAMPLE 6. Solve: $5x = 28$

SOLUTION. Divide both sides by 5.

$$\frac{5x}{5} = \frac{28}{5}$$

$$x = \frac{28}{5}$$

■

The remaining examples use both the Addition and Multiplication Principles. Which principle do you apply first when both are used in solving an equation? Consider the equation

$$3x + 5 = 15,$$

as opposed to the equation

$$3(x + 5) = 15.$$

To solve

$$3x + 5 = 15,$$

first add -5 to both sides:

$$\begin{aligned} 3x + 5 &= 15 \\ -5 \quad &\quad -5 \\ \hline 3x &= 10 \\ \frac{3x}{3} &= \frac{10}{3} \\ x &= \frac{10}{3} \end{aligned}$$

But if you are given

$$3(x + 5) = 15,$$

first divide both sides by 3:

$$\begin{aligned} \frac{3(x + 5)}{3} &= \frac{15}{3} \\ x + 5 &= 5 \\ -5 \quad &\quad -5 \\ \hline x &= 0 \end{aligned}$$

EXAMPLE 7. Show that 20 is the one and only root of each of the equations of Example 1:

$$5x - 52 = 2x + 8 \qquad \text{and} \qquad 3x = 60$$

SOLUTION. According to the Addition and Multiplication Principles, the following are equivalent equations:

$$\begin{aligned} 5x - 52 &= 2x + 8 \\ -2x + 52 \quad &\quad -2x + 52 \\ \hline 3x &= 60 \\ \frac{3x}{3} &= \frac{60}{3} \\ x &= 20 \end{aligned}$$

Clearly, the last equation has the single root 20. Because all of these equations are equivalent, each equation has 20 as its one and only root.

EXAMPLE 8. Solve the equation:

$$\frac{1}{2}[y - (3y - 4)] = 1$$

SOLUTION.

$$\frac{1}{2}[y - (3y - 4)] = 1$$ First, multiply both sides by 2.

$$y - (3y - 4) = 2$$ Next, remove parentheses.

$$\underbrace{y - 3y} + 4 = 2$$

$$-4 \qquad -4$$

$$-2y = -2$$

$$\frac{-2y}{-2} = \frac{-2}{-2}$$

$$y = 1$$

CHECK.

$$\frac{1}{2}[1 - (3 \cdot 1 - 4)] \stackrel{?}{=} 1$$

$$\frac{1}{2}[1 - (-1)] \stackrel{?}{=} 1$$

$$\frac{1}{2} \cdot 2 \stackrel{?}{=} 1$$

$$1 \stackrel{\checkmark}{=} 1$$

■

In the final example, an equation containing second-degree terms reduces to a first-degree equation.

EXAMPLE 9. Solve: $x^2 + 7 = (x - 1)^2$

SOLUTION.

$$x^2 + 7 = (x - 1)^2$$ On the right side square $x - 1$.

$$x^2 + 7 = x^2 - 2x + 1$$

$$-x^2 - 7 + 2x \qquad -x^2 + 2x - 7$$

$$2x = -6$$

$$x = -3$$

CHECK.

$$(-3)^2 + 7 \stackrel{?}{=} (-3 - 1)^2$$

$$9 + 7 \stackrel{?}{=} (-4)^2$$

$$16 \stackrel{\checkmark}{=} 16$$

■

EXERCISES

In Exercises 1–42 solve each equation. Check the ones so indicated.

1. $x + 1 = 11$ (Check.)
2. $x - 2 = 3$
3. $y + 4 = -1$
4. $x + 16 = 14$
5. $z + 5 = -5$ (Check.)
6. $u + 13 = 13$
7. $4x = 8$
8. $-3x = 9$ (Check.)
9. $7x = -56$
10. $\frac{x}{2} = 5$
11. $\frac{1}{3}x = -4$ (Check.)
12. $3x = \frac{1}{3}$
13. $\frac{x}{2} = \frac{1}{5}$
14. $.7x = .28$ (Check.)
15. $2x + 1 = 7$
16. $3u - 5 = -2$
17. $8 - 5x = -7$
18. $9 + \frac{1}{4}x = 10$ (Check.)
19. $2x + 5 = 3x$
20. $5x - 2 = 7x + 2$
21. $-3x + 7 = x + 3$
22. $5x - 9 = 3x + 9$ (Check.)
23. $10x + 62 = x - 1$
24. $13x + 1 = -10x - 45$
25. $y - (2 - 3y) = 18$ (Check.)
26. $3(z - 7) = -9$
27. $10\left(z - \frac{1}{4}\right) = \frac{3}{4}$ (Check.)
28. $4(x - 7) = -5x - 1$
29. $-7(x + 5) = x - 3$
30. $3x - 1 = 5 - (x - 12)$
31. $x - [2 - (x - 3)] = 7 - x$ (Check.)
32. $(3x - 2) - (6 + 5x) = 1$
33. $5x - 6 - (4x + 16) = 0$
34. $1 - (1 - [x - (1 - x)]) = -1$
35. $x^2 + 5x = x^2 - 10$ (Check.)
36. $-5x + 2x^2 = 9x^2 + 35 - 7x^2$
37. $x^2 + 25 = (x - 5)^2$
38. $(x - 2)^2 + 3 = (x + 1)^2$ (Check.)
39. $(x + 5)(x - 3) = (x + 5)^2$
40. $(x + 1)(x - 2) = (x + 3)(x + 2)$

41. $2x - x^2 = 4x - (x + 2)^2$

42. $(x - 8)^2 + 6 = (x - 7)^2 + 1$ (Check.)

43. Are the equations

$$x + 1 = 1$$

and

$$x + 1 + \frac{1}{x} = 1 + \frac{1}{x}$$

equivalent?

44. Are the equations

$$x + 1 = x + 2$$

and

$$(x + 1)0 = (x + 2)0$$

equivalent?

6.3 EQUATIONS WITH RATIONAL EXPRESSIONS

When an equation involves rational expressions (or fractions), multiply both sides of the equation by the LCD of these expressions. Then solve the resulting equations as before. In the first two examples the denominators are constants.

EXAMPLE 1. Solve: $\frac{x - 2}{3} = \frac{2x}{5}$

SOLUTION.

$$\text{LCD}\left(\frac{x - 2}{3}, \frac{2x}{5}\right) = 15$$

$$\begin{aligned} \frac{x - 2}{3} &= \frac{2x}{5} \\ 5(x - 2) &= 3 \cdot 2x \\ 5x - 10 &= 6x \\ -5x \quad & \quad -5x \\ \hline -10 &= x \end{aligned}$$

First multiply both sides by 15 $\left(\text{or cross-multiply: } \frac{x - 2}{3} \times \frac{2x}{5}\right)$.

CHECK.

$$\frac{-10 - 2}{3} \stackrel{?}{=} \frac{2(-10)}{5}$$

$$\frac{-12}{3} \stackrel{?}{=} \frac{-20}{5}$$

$$-4 \stackrel{\checkmark}{=} -4$$

■

EXAMPLE 2. Solve: $\dfrac{3y - 1}{4} - \dfrac{y - 3}{6} = 1$

SOLUTION.

$$\text{LCD}\left(\frac{3y - 1}{4}, \frac{y - 3}{6}, 1\right) = 12$$

$$\frac{3y - 1}{4} - \frac{y - 3}{6} = 1$$ First multiply both sides by 12.

$$3(3y - 1) - 2(y - 3) = 12$$ Now simplify the left side.

$$9y - 3 - 2y + 6 = 12$$

$$7y + 3 = 12$$

$$7y = 9$$ Finally, divide both sides by 7.

$$y = \frac{9}{7}$$

■

EXAMPLE 3. Solve: $\dfrac{x}{x + 1} = 2$

SOLUTION. Multiply both sides by $x + 1$ $\left(\text{or cross-multiply: } \dfrac{x}{x + 1} \times \dfrac{2}{1}\right)$.

$$x = 2(x + 1)$$

Now simplify:

$$\begin{array}{rcl} x & = & 2x + 2 \\ -x - 2 & & -x - 2 \\ \hline -2 & = & x \end{array}$$

■

EXAMPLE 4. Solve: $\dfrac{1}{x + 3} = \dfrac{2}{x}$

SOLUTION. Multiply both sides of the equation by $x(x + 3)$ (or cross-multiply).

$$x = 2(x + 3)$$

Now simplify:

$$\begin{array}{rcl} x & = & 2x + 6 \\ -x - 6 & & -x - 6 \\ \hline -6 & = & x \end{array}$$

CHECK.

$$\frac{1}{-6 + 3} \stackrel{?}{=} \frac{2}{-6}$$

$$\frac{1}{-3} \stackrel{\checkmark}{=} \frac{-1}{3}$$

■

Note that in Example 3, you multiplied both sides of the given equation by $x + 1$, which is $\text{LCD}\left(\frac{x}{x + 1}, 2\right)$. In Example 4, you multiplied both sides by $x(x + 3)$, which is $\text{LCD}\left(\frac{1}{x + 3}, \frac{2}{x}\right)$.

EXAMPLE 5. Solve: $\frac{5}{x} + \frac{4}{x - 2} = \frac{10}{x}$

SOLUTION.

$$\text{LCD}\left(\frac{5}{x}, \frac{4}{x - 2}, \frac{10}{x}\right) = x(x - 2)$$

Multiply both sides by $x(x - 2)$.

$$\left(\frac{5}{x} + \frac{4}{x - 2}\right) \cdot x(x - 2) = \frac{10}{x} \cdot x(x - 2)$$

Simplify:

$$\begin{array}{rcl} 5(x - 2) + 4x & = & 10(x - 2) \\ 5x - 10 + 4x & = & 10x - 20 \\ -9x + 20 & & -9x + 20 \\ \hline 10 & = & x \end{array}$$

■

EXAMPLE 6. Solve: $\frac{1}{x^2 - 1} + \frac{1}{x - 1} = \frac{4}{x + 1}$

SOLUTION.

$$x^2 - 1 = (x + 1)(x - 1)$$

$$\text{LCD}\left(\frac{1}{x^2 - 1}, \frac{1}{x - 1}, \frac{4}{x + 1}\right) = (x + 1)(x - 1)$$

$$\left[\frac{1}{x^2 - 1} + \frac{1}{x - 1}\right] \cdot (x + 1)(x - 1) = \frac{4}{x + 1} \cdot (x + 1)(x - 1)$$

$$1 + (x + 1) = 4(x - 1)$$

$$2 + x = 4x - 4$$

$$\underline{+4 - x} \quad \triangle \!\!-\!\!-\!\! \triangle \quad \underline{-x + 4}$$

$$6 = 3x$$

$$2 = x$$

■

In several of these examples you multiplied both sides of an equation by an expression containing a *variable*. See Section 13.6 to find out what problems this sometimes entails.

EXERCISES

Solve each equation. Check the ones so indicated.

1. $\frac{y}{9} = -2$ (Check.)
2. $\frac{3x}{4} = 18$
3. $\frac{u}{10} = \frac{2}{5}$
4. $\frac{-2x}{27} = \frac{2}{3}$
5. $\frac{x}{2} = \frac{-2}{3}$
6. $\frac{2y}{5} = \frac{1}{12}$ (Check.)
7. $\frac{x}{5} = \frac{x + 4}{6}$
8. $\frac{3x + 1}{2} = \frac{7x + 1}{3}$
9. $\frac{z - 1}{7} = \frac{2 - z}{5}$
10. $\frac{3x - 7}{14} = -x$
11. $\frac{x}{8} + \frac{x}{4} = 3$
12. $\frac{x}{3} + \frac{2x}{9} = 5$
13. $\frac{t}{4} - \frac{3t}{5} = -7$
14. $\frac{2u}{3} - \frac{5}{6} = \frac{3}{4} - \frac{u}{8}$
15. $\frac{5}{6}x + \frac{3x}{4} = \frac{5x}{3} - 1$ (Check.)
16. $\frac{5t}{3} - \frac{3t}{4} = \frac{2t}{3} + \frac{1}{2}$
17. $\frac{7z}{5} - \frac{3z}{2} = \frac{1}{10} - \frac{1}{4}$
18. $\frac{2x}{9} - \frac{x}{12} = \frac{5x}{6} + \frac{x}{8}$ (Check.)

19. $\frac{35}{x} = 5$ (Check.)

20. $\frac{-4}{x} = 1$

21. $\frac{7}{1 - x} = -1$

22. $\frac{9}{x + 6} = 3$ (Check.)

23. $\frac{-2}{x + 8} = -1$

24. $\frac{x}{x + 4} = \frac{1}{2}$

25. $\frac{x}{x + 1} = -1$ (Check.)

26. $\frac{5x}{x + 3} = 2$ (Check.)

27. $\frac{15}{y} = \frac{20}{y + 1}$

28. $\frac{18}{x + 5} = \frac{8}{x}$

29. $\frac{27}{x - 2} = \frac{33}{x}$

30. $\frac{-12}{u - 3} = \frac{3}{u - 8}$ (Check.)

31. $\frac{x}{x + 1} = \frac{5}{6}$

32. $\frac{z + 5}{z} = -4$

33. $\frac{z + 7}{z + 2} = -4$ (Check.)

34. $\frac{y - 10}{3y} = 0$

35. $\frac{t - 2}{t + 5} = \frac{2}{3}$

36. $\frac{1 + t}{-2t} = -1$

37. $\frac{2}{x} + \frac{6}{x} = 2$ (Check.)

38. $\frac{5}{x} - \frac{2}{x} = 1$

39. $3x - \frac{1}{2} = \frac{1}{4}$

40. $\frac{1}{x} + \frac{1}{3} = \frac{1}{2}$

41. $\frac{3}{x} - \frac{1}{3} = \frac{2}{5}$

42. $\frac{5}{2x} + \frac{1}{2} = \frac{3}{2}$

43. $\frac{7}{x} - \frac{5}{2x} = \frac{3}{2}$

44. $\frac{1}{u} + \frac{5}{u - 2} = \frac{4}{u - 2}$

45. $\frac{5}{y + 1} - \frac{18}{y + 2} = \frac{-10}{y + 1}$

46. $\frac{25}{t + 1} - \frac{24}{t + 4} = \frac{10}{t + 1}$ (Check.)

47. $\frac{1}{u - 2} + \frac{5}{u + 4} = \frac{30}{(u - 2)(u + 4)}$

48. $\frac{18}{(z + 2)(z - 1)} + \frac{3}{z + 2} = \frac{4}{z - 1}$

49. $\frac{4}{x - 1} + \frac{3}{x + 1} = \frac{9}{x^2 - 1}$

50. $\frac{12}{y} - \frac{8}{y + 2} = \frac{48}{y^2 + 2y}$ (Check.)

51. $\frac{4}{z + 1} + \frac{3}{z + 2} - \frac{21}{z^2 + 3z + 2} = 0$

52. $\frac{8}{x + 5} + \frac{4}{x^2 + 7x + 10} = \frac{3}{x + 2}$

6.4 LITERAL EQUATIONS

DEFINITION. A **literal equation** is one involving at least one letter other than the variable for which you are solving.

Sometimes these letters all represent constants. The constant may be a complicated number, such as -96.004. It is easier to designate it by a *literal* constant, such as c. At other times several variables may be present. Thus in the equation

$$x + y = 20$$

there are two variables. You may sometimes want to solve this for x, and at other times for y. The methods already developed for solving equations apply here as well. Bring terms containing *the variable for which you are solving* to one side. Bring all other terms to the other side.

EXAMPLE 1. Let a be a constant. Solve

$$2x - 1 = a$$

for x.

SOLUTION. The term containing x remains on the left side: the other terms go to the right side. Thus

$$2x - 1 = a$$

$$2x = a + 1$$

$$x = \frac{a + 1}{2}$$

CHECK. Substitute $\frac{a+1}{2}$ for x in the original equation.

$$2\left(\frac{a + 1}{2}\right) - 1 \stackrel{?}{=} a$$

$$(a + 1) - 1 \stackrel{?}{=} a$$

$$a \stackrel{\checkmark}{=} a$$

■

EXAMPLE 2. Let a, b, c be constants. Solve

$$3x + a = b - cx$$

for x.

SOLUTION. Bring the terms containing x to the left side, the other terms to the right side.

$$
\begin{aligned}
3x + a &= b - cx \\
\underline{+cx - a} &\quad \underline{+ cx - a} \\
3x + cx &= b - a \\
(3 + c)x &= b - a \\
x &= \frac{b - a}{3 + c}
\end{aligned}
$$

Finally, divide both sides by $3 + c$.

CHECK. Substitute $\dfrac{b-a}{3+c}$ for x in the original equation.

$$3\left(\frac{b-a}{3+c}\right) + a \stackrel{?}{=} b - c\left(\frac{b-a}{3+c}\right)$$

$$\frac{3(b-a)}{3+c} + \frac{a(3+c)}{3+c} \stackrel{?}{=} \frac{b(3+c)}{3+c} - \frac{c(b-a)}{3+c}$$

Now multiply both sides by $3 + c$, and simplify.

$$3b - \cancel{3a} + \cancel{3a} + ac \stackrel{?}{=} 3b + \cancel{bc} - \cancel{bc} + ac$$

$$3b + ac \stackrel{\checkmark}{=} 3b + ac$$

■

In Example 2, when you divide both sides of the equation by $3 + c$, you are assuming $c \neq -3$. Otherwise, $3 + c = 3 + -3 = 0$. (Recall that division by 0 is undefined.) *In this section assume the constants and variables are such that division is defined.*

EXAMPLE 3. Suppose x, y, and z are variables. Solve

$$x = 2z - y$$

for (a) y, (b) z.

SOLUTION.

(a) Add $y - x$ to both sides, and obtain

$$y = 2z - x.$$

(b) Consider the given equation:

$$x = 2z - y$$

First add y to both sides. The only term in which z occurs remains on the right.

$$x + y = 2z$$

$$\frac{x + y}{2} = z$$

■

EXAMPLE 4. Suppose x, y, and z are variables. Solve

$$x = \frac{2y - 3z}{2 + z}$$

for (a) y, (b) z.

SOLUTION.

(a) Leave the term in which y occurs on the right side; bring all other terms to the left side.

$$x = \frac{2y - 3z}{2 + z}$$

First multiply both sides by $2 + z$.

$$x(2 + z) = 2y - 3z$$

$$x(2 + z) + 3z = 2y$$

$$\frac{x(2 + z) + 3z}{2} = y$$

(b) Bring terms in which z occurs to the left side; bring all other terms to the right side.

$$x = \frac{2y - 3z}{2 + z}$$

$$(2 + z)x = 2y - 3z$$

$$2x + zx = 2y - 3z$$

$$zx + 3z = 2y - 2x$$

Now isolate the common factor z on the left side.

$$z(x + 3) = 2y - 2x$$

$$z = \frac{2y - 2x}{x + 3}$$

■

EXAMPLE 5. Solve $\frac{1}{x} + \frac{1}{y} = \frac{1}{z}$ for y.

SOLUTION.

$$\text{LCD}\left(\frac{1}{x}, \frac{1}{y}, \frac{1}{z}\right) = xyz$$

$$\frac{1}{x} + \frac{1}{y} = \frac{1}{z}$$

First multiply both sides by xyz.

$$yz + xz = xy$$

$$yz - xy = -xz$$

Now isolate the common factor y on the left side.

$$y(z - x) = -xz$$

$$y = \frac{-xz}{z - x}$$

or

$$y = \frac{xz}{x - z}$$

■

EXAMPLE 6. The relationship between energy E and mass M is given by

$$E = Mc^2$$

where the constant c is the speed of light. Solve for M.

SOLUTION.

$$E = Mc^2$$

$$\frac{E}{c^2} = M$$

■

EXAMPLE 7.

$$F = \frac{9}{5}C + 32$$

expresses degrees Fahrenheit F in terms of degrees Centigrade (or Celsius) C.
(a) Solve for C. (b) Find the value of C when F = 98.6 .

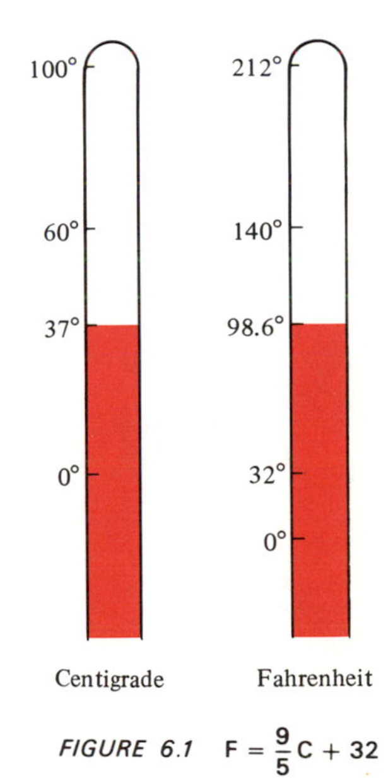

FIGURE 6.1 $F = \frac{9}{5}C + 32$

SOLUTION.

(a)

$$F = \frac{9}{5}C + 32$$

$$F - 32 = \frac{9}{5}C$$

Now multiply both sides by $\frac{5}{9}$.

$$\frac{5}{9}(F - 32) = C$$

(b) Let F = 98.6.

$$C = \frac{5}{9}(98.6 - 32)$$

$$= \frac{5}{9}(66.6)$$

$$= 37$$

■

EXERCISES

Assume a, b, c, K are constants. In Exercises 1–20 solve for x. Check the ones so indicated.

1. $5x = c$ (Check.)
2. $x - 7 = a$
3. $x + 13 = b - c$
4. $6x + c = b - x$
5. $ax + 7 = 3x - 2$ (Check.)
6. $2(x + a) - 1 = 5x + b$
7. $ax + 3 = (1 - a)x - b$
8. $ax - b(c + x) = 7x - a$
9. $\frac{3x - a}{b} = 9$
10. $\frac{a - 7bx}{2} = 1$
11. $\frac{4b - 3x}{6} = 5x - c$
12. $\frac{2 - 3b}{5x} = a - bc$
13. $\frac{4}{x} = a + b$
14. $\frac{1}{3 - x} = a - 2c$
15. $\frac{x}{x + a} = bc + 1$ (Check.)
16. $a^2x + a^2 = 1 - 2bc$
17. $a^2bx - 4 = 5 - 3a^2bx$
18. $\frac{c}{1 - x} = \frac{3}{4}$
19. $\frac{a}{4b - bx} = 7 - a$
20. $\frac{x}{2a - 3x} = -4$
21. Solve $a - 2t = 4 + b$ for t.
22. Solve $5t + 7 = abt - 1$ for t.
23. Solve $\frac{3y + 1}{a + b} = \frac{y}{3}$ for y.
24. Solve $\frac{1}{K} = \frac{2}{z} - \frac{1}{5}$ for z.

In Exercises 25–36 assume x, y, z, t are the variables. Solve for the indicated variables.

25. $2x + y = 1 + t$
   (a) for y, (b) for x
26. $3x - 2y = z + 1$
   (a) for x, (b) for y
27. $ax + by = cz$
   (a) for x, (b) for y
28. $3at - x = 1 + cz$
   (a) for t, (b) for z
29. $5ay - 3xz = 1$
   (a) for y, (b) for z
30. $\frac{1}{x} - \frac{1}{y} = 3z$
   (a) for x, (b) for y
31. $(x + y)(y + z) = 9$
   (a) for x, (b) for z
32. $xt - 3y + 3xy = 2t$
   (a) for t, (b) for y
33. $a(a - x) + 1 = b(t - b)$
   (a) for x, (b) for t
34. $\frac{4}{x} - \frac{1}{3t} = y$
   (a) for x, (b) for t

35. $\dfrac{K - 5t}{b} = \dfrac{3(x - 1)}{z}$
(a) for t, (b) for z

36. $\dfrac{K}{t} + \dfrac{3}{xt} = -1$
(a) for x, (b) for t

37. $C = 2\pi r$ gives the circumference C of a circle in terms of the radius length r. Solve for r. [See Figure 6.2 .]

38. $A = \dfrac{bh}{2}$ gives the area of a triangle in terms of the base length b and the altitude length h.
(a) Solve for b. (b) Solve for h. [See Figure 6.3 .]

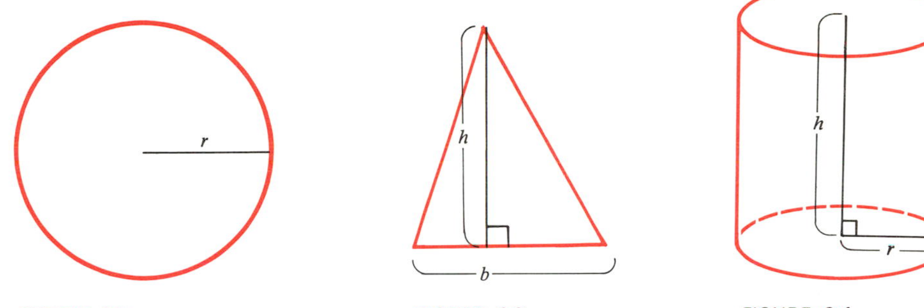

FIGURE 6.2

FIGURE 6.3

FIGURE 6.4

39. $V = \pi r^2 h$ gives the volume of a right circular cylinder in terms of r, the length of the radius of the base and the altitude length h. Solve for h. [See Figure 6.4 .]

40. (a) Solve $S = 2\pi r(r + h)$ for h.
(b) Determine h when $r = 5$ and $S = 100\pi$.

41. (a) Solve $h = \dfrac{v^2}{2g} + \dfrac{p}{c}$ for p.
(b) Determine p when $h = 2, g = 32, v = 10, c = 5$.

42. (a) Solve $S = \dfrac{a - rl}{1 - r}$ for r.
(b) Determine r when $S = 40, a = 10, l = 2$.

*6.5 OTHER TYPES OF EQUATIONS

You have seen equations that have a single root. In fact, every first-degree equation

$$Ax + B = 0, \qquad A \neq 0,$$

*Optional topic.

has a single root given by

$$x = \frac{-B}{A}.$$

For example,

$$3x + 2 = 0$$

has the single root $\frac{-2}{3}$.

There are also equations that have several roots.

EXAMPLE 1. The equation

$$|x| = 1$$

has both 1 and -1 as its roots. For, $|1| = 1$ and $|-1| = 1$. This type of equation will be discussed in Chapter 14. ■

EXAMPLE 2. The equation

$$x^2 = 4$$

has both 2 and -2 as its roots. For, $2^2 = 4$ and $(-2)^2 = 4$. This type of equation will be discussed in Chapter 13. ■

There are equations with any number of roots. To show this, recall that

if a product is zero, then one of its factors must be zero,

and

if a factor is zero, the product is zero.

Thus

$$ab = 0$$

means

$$a = 0 \quad \text{or} \quad b = 0$$

and

$$abcde = 0$$

means

$$a = 0 \quad \text{or} \quad b = 0 \quad \text{or} \quad c = 0 \quad \text{or}$$
$$d = 0 \quad \text{or} \quad e = 0.$$

The equation

$$(x - 1)(x - 2)(x - 3) = 0$$

has exactly 3 roots. For surely, 1, 2, and 3 are roots. For example, 2 is a root because

$$(2 - 1)(2 - 2)(2 - 3) = 1 \cdot 0 \cdot (-1) = 0$$

Also, if c were a root, then

$$(c - 1)(c - 2)(c - 3) = 0$$

By the above,

$$c - 1 = 0 \qquad \text{or} \qquad c - 2 = 0 \qquad \text{or} \qquad c - 3 = 0$$

Thus

$$c = 1 \qquad \text{or} \qquad c = 2 \qquad \text{or} \qquad c = 3$$

EXAMPLE 3. The equation

$$(x - 1)(x - 2)(x - 3)(x - 4)(x - 5) = 0$$

has exactly 5 roots: 1, 2, 3, 4, and 5; the equation

$$(x - 1)(x - 2)(x - 3)\ldots(x - 10) = 0$$

has exactly 10 roots; the equation

$$(x - 1)(x - 2)(x - 3)\cdots(x - 127) = 0$$

has exactly 127 roots. ■

There even are equations that have no roots.

EXAMPLE 4. The equation

$$x = x + 1$$

has no roots. For you can subtract x from both sides and obtain the equivalent *false* statement

$$0 = 1.$$

For any number you substitute for x in the given equation, you obtain a false statement. ■

When you simplify an equation and obtain a false statement, such as

$$0 = 1,$$

the original equation has no root.

EXAMPLE 5. Show that

$$2(x + 2) = 2(x + 4) - 1$$

has no root.

SOLUTION.

$$2(x+2) \stackrel{?}{=} 2(x+4) - 1$$

$$2x + 4 \stackrel{?}{=} 2x + 8 - 1$$

$$4 \stackrel{\text{x}}{=} 7$$

$\stackrel{\text{x}}{=}$ indicates that the statement

$4 = 7$

is false.

The given equation has no root. ■

DEFINITION. An **identity** is an equation for which *every* real number is a root.

EXAMPLE 6. The equation

$$x = 5x + 3x - 7x$$

is an identity. For by combining terms on the right side, you obtain the equivalent equation:

$$x = x.$$

Clearly *every* real number is a root of this equation. ■

To verify that an equation is an identity, simplify it and obtain an equivalent equation, such as

$$x = x,$$

that is true no matter what number is substituted for the variable. You may, in fact, obtain a true statement (without variables), such as

$$0 = 0.$$

(See Example 7.)

EXAMPLE 7. Show that

$$x^2 + 19 = (x+3)(x+4) + 7(1-x)$$

is an identity.

SOLUTION.

$$x^2 + 19 \stackrel{?}{=} (x+3)(x+4) + 7(1-x)$$ First simplify the right side.

$$x^2 + 19 \stackrel{?}{=} x^2 + \cancel{7x} + 12 + 7 - \cancel{7x}$$

$$\underline{-x^2 - 19} \quad \triangle \quad \underline{-x^2 \qquad - 19}$$

$$0 \stackrel{\checkmark}{=} 0$$

Here, the equation simplifies to the true statement, $0 = 0$. ■

EXERCISES

In Exercises 1–10 determine the number of (real) roots of the given equation.

1. $2x + 7 = x - 7$
2. $|x| = 9$
3. $x^2 = 25$
4. $x = x - 5$
5. $x^2 = 0$
6. $|x| = 0$
7. $x^2 = -1$
8. $|x| = -1$
9. $(x + 4)(x - 5) = 0$
10. $(x + 2)(x - 7)(x + 13)(x - 9)(x + 10) = 0$
11. Determine an equation with exactly 8 roots.
12. Determine an equation whose roots are 2, 5, and 7.

In Exercises 13–20 show that each equation has no root.

13. $x + 10 = x - 3$
14. $2y - (12 + y) = y + 3$
15. $3(u - 3) + 1 - u = 2(u + 5)$
16. $(y - 1)^2 = y(y - 2)$
17. $(x + 1)(x - 1) = (x + 1)^2 - 2x$
18. $(z + 3)(z + 4) + 3z = (z + 5)^2$
19. $x(x - 5) = (x - 2)^2 + 3 - x$
20. $t^2(t + 1) = t^3 + (t + 1)(t - 1)$

In Exercises 21–28 show that each equation is an identity.

21. $5x - 6 = 3(x - 2) + 2x$
22. $7u - (5 - 2u) = 9u - 5$
23. $y - (2 - 3y) = 4(y - 1) + 2$
24. $2t - [1 - (3 - t)] = 2 + t$
25. $(t - 4)^2 + 16t = (t + 4)^2$
26. $(z - 2)(z + 3) + 15 = (z - 3)^2 + 7z$
27. $\dfrac{y + 4}{2} + 1 - y = \dfrac{6 - y}{2}$
28. $(x + 2)^2 - (x + 1)^2 = 2x + 3$

In Exercises 29–40 determine (a) which equations have a single root, (b) which have no roots, (c) which are identities.

29. $5x + 2 = 3x - 4$
30. $7 - 3x = 4(1 - x) + x$
31. $6(2 + x) = 3(2x + 4)$
32. $(u + 5)^2 - (u - 5)^2 = 20u$
33. $(x + 3)^2 + (x + 4)^2 = 2(x + 5)^2$
34. $\dfrac{7 - 3y}{4} + \dfrac{2 - y}{2} = \dfrac{5 - y}{4}$
35. $3 - [8 - (2 - z)] = 4 - [2z - (3z - 1)]$
36. $5y - [3 + (2 - y)] = 6 - [3 - (2 - y)]$
37. $\dfrac{2x - 3}{2} + 2 = x - \dfrac{1}{4}$
38. $u^2 - (u + 4)^2 = -8(u + 2)$
39. $(y + 5)^2 - (y + 4)^2 = 2(y + 1)$
40. $\dfrac{6u}{5} - \dfrac{2u}{2} = \dfrac{u - 3}{3}$

REVIEW EXERCISES FOR CHAPTER 6

6.1 ROOTS OF EQUATIONS

In Exercises 1–4 does the statement involve only constants?

i. If so, is it true or is it false?

ii. If not, is it a first-degree equation in a single variable?

1. $4 + 2 = 6$
2. $5x = 20$
3. $x + y = 10$
4. $9 - 5 = 2 + 3$

In Exercises 5–8 check whether the number in color is a root of the given equation.

5. $x - 7 = 3,\quad 10$
6. $5x = -20,\quad 4$
7. $2x - 1 = 5 - 2x,\quad \frac{3}{2}$
8. $\frac{3}{x} = \frac{1}{5},\quad 15$

In Exercises 9–20 solve each equation. Check the ones so indicated.

6.2 SOLVING EQUATIONS

9. $x - 10 = 10$ (Check.)
10. $7u = 35$
11. $\frac{x}{3} = -5$
12. $2u + 1 = 5 - 2u$ (Check.)
13. $\frac{x}{4} = \frac{3}{5}$
14. $3(y - 2) = 1 - (4 - 2y)$

6.3 EQUATIONS WITH RATIONAL EXPRESSIONS

15. $\frac{x}{6} = \frac{2}{3}$
16. $\frac{x}{10} = \frac{x + 2}{15}$
17. $\frac{u + 3}{8} = \frac{u - 2}{3}$ (Check.)
18. $\frac{3}{y + 2} = \frac{6}{y + 7}$
19. $\frac{3}{x} + \frac{1}{3} = \frac{5}{6}$
20. $\frac{2}{x + 2} + \frac{1}{x - 2} = \frac{10}{x^2 - 4}$

6.4 LITERAL EQUATIONS

In Exercises 21–24 solve for the indicated variable.

21. $3x - c = 1 + c$ (for x)
22. $\frac{5u + a}{b} = 3$ (for u)
23. $y = \frac{3x + 5}{10}$ (for x)
24. $\frac{4}{y} + \frac{1}{2x} = z$ (for y)
25. Solve $5x + 3(y - z) = 2t + 3$
    (a) for x, (b) for y.
26. Solve $\frac{A}{t} + \frac{1}{st} = -2$
    (a) for s, (b) for t.

6.5 OTHER TYPES OF EQUATIONS

In Exercises 27–29 determine the number of roots of the given equation.

27. $x^2 = 16$ 28. $|x| = 10$ 29. $(x + 4)(x - 1)(x + 2) = 0$

In Exercises 30–32, determine

(a) which equations have a single root,

(b) which have no root,

(c) which are identities.

30. $|x| + 1 = 0$ 31. $5x - 4 = 4 - 5x$ 32. $5x - 4 = 4 + 5x$

TEST YOURSELF ON CHAPTER 6.

1. Check whether the number in color is a root of the given equation.

   (a) $\frac{x}{5} = 5,\quad 1$ (b) $3x + 2 = 17,\quad 5$

In Problems 2–7 solve each equation. Check the ones so indicated.

2. $4x + 2 = 1 - 3x$ (Check.)
3. $\frac{x}{10} = \frac{-6}{5}$
4. $9 - 2(y + 1) = -3(1 - 2y)$
5. $\frac{4}{x} = \frac{3}{x - 4}$ (Check.)
6. $\frac{z - 2}{4} + \frac{1}{2} = 1 + \frac{z - 2}{6}$
7. $\frac{2}{x + 1} + \frac{1}{x - 1} = \frac{4}{x^2 - 1}$
8. Solve for x:

$$5x + a - b = c + 4 - x$$

9. Solve

$$\frac{1}{y} - 3x = 4 + 2x$$

   (a) for x, (b) for y.

10. Which of the following equations have a single root?

    (a) $(x - 4)(x + 5) = 0$ (b) $|x| = 0$

    (c) $2x^2 = 2$ (d) $(x - 1)^2 + 2x = 1$

7
word problems

7.1 INTEGER AND AGE PROBLEMS

Algebraic methods enable you to solve problems that arise in various situations. Often a problem is stated in words. Your first task is to translate the problem into mathematical symbols. Each problem to be considered can be formulated in terms of an equation. You solve the equation in order to solve the original problem.

The most commonly used numbers are the integers:

$$\ldots -3, -2, -1, 0, 1, 2, 3, \ldots$$

You may not be directly interested in integers. But you use integers to describe money, distance, time, weight, and so on. Also, integer problems help develop your understanding of how to apply algebraic techniques to other types of problems.

EXAMPLE 1. Let x represent an integer. Express each of the following in algebraic symbols:

(a) two more than the integer
(b) five less than the integer
(c) half the integer
(d) six less than twice the integer
(e) half of six less than the integer

SOLUTION.

(a) $\underbrace{\text{two more than}}_{2\,+}\ \underbrace{\text{the integer}}_{x}$

or, more accurately,

$$x + 2$$

(b) $\underbrace{\text{five less than}}_{-5\,+}\ \underbrace{\text{the integer}}_{x}$

or

$$x - 5$$

(c) $\underbrace{\text{half the integer}}_{\frac{x}{2}}$

(d) $\underbrace{\text{six less than}}_{-6\,+}\ \underbrace{\text{twice the integer}}_{2x}$

or

$$2x - 6$$

Here first consider twice x; then subtract 6.

(e) $\underbrace{\text{half of}}_{\frac{1}{2}}\ \underbrace{\text{six less than the integer}}_{(-6\,+\,x)}$

or

$$\frac{x - 6}{2}$$

Here first subtract 6 from x; then take half of this. ■

Consecutive integers are 1 unit apart. Thus 5, 6, and 7 are three consecutive integers.

EXAMPLE 2. Determine two consecutive integers whose sum is 21.

SOLUTION. You are concerned with two consecutive integers. Let x be the smaller integer. The next consecutive integer is then $x + 1$.

Translate the problem:

The sum of two consecutive integers is 21.

$$x + (x + 1) = 21$$

Solve the equation:

$$2x + 1 = 21$$
$$2x = 20$$
$$x = 10$$
$$x + 1 = 11$$

Thus the two integers are 10 and 11. ■

EXAMPLE 3. The sum of three consecutive *odd* integers is 45. Determine these integers.

SOLUTION. The **odd integers** are

1, 3, 5, 7, and so on,

-1, -3, -5, -7, and so on.

[See Figure 7.1.] *Consecutive odd integers* are *two* units apart.

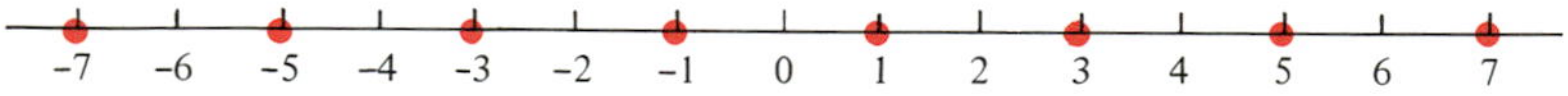

−7 −6 −5 −4 −3 −2 −1 0 1 2 3 4 5 6 7

FIGURE 7.1 The odd integers are 1, 3, 5, 7, and so on, −1, −3, −5, −7, and so on. The even integers are 0, 2, 4, 6, and so on, −2, −4, −6, and so on.

Let x be the smallest of these odd integers. Then $x + 2$ is the next odd integer, and $x + 4$ is the largest of these.

Translate the problem:

The sum of three consecutive odd integers is 45.

$$x + (x + 2) + (x + 4) = 45$$

Solve the equation.

$$3x + 6 = 45$$
$$3x = 39$$
$$x = 13$$
$$x + 2 = 15$$
$$x + 4 = 17$$

The three integers are 13, 15, 17.

CHECK.

$$13 + 15 + 17 \stackrel{?}{=} 45$$

$$45 \stackrel{\checkmark}{=} 45$$ ■

EXAMPLE 4. Eight more than an integer is three times this integer. Determine the integer.

SOLUTION. Let x be this integer.
Translate the problem:

$$\underbrace{\text{Eight more than an integer}}_{x + 8} \ \underset{=}{\text{is}} \ \underbrace{\text{three times the integer.}}_{3x}$$

Solve the above equation:

$$8 = 2x$$

$$4 = x$$

The integer in question is 4.

CHECK.

$$4 + 8 \stackrel{?}{=} 3 \cdot 4$$

$$12 \stackrel{\checkmark}{=} 12$$ ■

A related type of problem concerns determining *age, which is assumed to be an integer*. In such problems, a person's age may be given at various times—for example, now, three years ago, five years from now.

EXAMPLE 5. Let x represent a man's age *now*.

(a) *Three years ago* his age was

$$\underbrace{\text{three less than}}_{-3 +} \ \underbrace{\text{his present age}}_{x}$$

or

$$x - 3.$$

(b) *In seven years* his age will be

$$\underbrace{\text{seven more than}}_{7 +} \ \underbrace{\text{his present age.}}_{x}$$

or

$$x + 7.$$

(c) Suppose his wife is

$$\underbrace{\text{four years younger than}}_{-4\ +}\ \underbrace{\text{he is.}}_{x}$$

His wife's age (*now*) is

$$x - 4.$$

(d) *In two years* his son will be half as old as the father.

In two years the father's age will be $x + 2$. The son's age will be $\frac{x+2}{2}$ (*in two years*). The son's *present age* is $\frac{x+2}{2} - 2$, which simplifies as follows:

$$\frac{x+2}{2} - 2 = \frac{x + 2 - 4}{2}$$

$$= \frac{x-2}{2}$$

■

EXAMPLE 6. A man is now four times as old as his son. Six years ago he was ten times as old as his son. Determine the son's age now.

SOLUTION. Let x be the son's age *now*. Then $4x$ is the man's age *now*.

Six years ago, the son's age was $x - 6$. *Six years ago*, the man's age was $4x - 6$. Translate the problem:

(The equation is given in the second sentence of the problem, and can be reworded as follows:)

$$\underbrace{\text{The man's age six years ago}}_{4x\ -\ 6}\ \underset{=}{\text{was}}\ \underbrace{\text{ten times the son's age six years ago.}}_{10(x\ -\ 6)}$$

Solve the equation:

$$4x - 6 = 10x - 60$$

$$54 = 6x$$

$$9 = x$$

The son is now 9 years old and the father is 36 years old (4 times as old). Six years ago, the son was 3 and the man was 30 (10 times as old). ■

EXAMPLE 7. Two brothers are ten years apart in age. In two years the older brother will be twice as old as the younger. How old is each now?

SOLUTION. Let x be the younger brother's age *now*. The older brother's age is $x + 10$. *In two years* the younger brother's age will be $x + 2$ and the older brother's age will be $(x + 10) + 2$, or

$$x + 12.$$

Translate the problem:

The older brother's age in two years	will be	twice the younger brother's age (in two years)
$x + 12$	$=$	$2(x + 2)$

Solve the equation.

$$x + 12 = 2x + 4$$

$$8 = x \qquad \text{(the younger brother's age } now\text{)}$$

$$18 = x + 10 \qquad \text{(the older brother's age } now\text{)}$$

EXERCISES

1. Let x represent an integer. Express each of the following in terms of x.
   (a) five more than the integer
   (b) eight less than the integer
   (c) three times the integer
   (d) one-fourth of the integer
   (e) The integer is increased by three.
   (f) The integer is decreased by ten.
2. Let x represent an integer. Express each of the following in terms of x.
   (a) one more than half of the integer
   (b) half of one more than the integer
   (c) two less than half of the integer
   (d) half of two less than the integer
3. Let x represent an integer.
   (a) Express the next consecutive integer.
   (b) Express the sum of x and the next two consecutive integers.
   (c) Suppose x is odd. What are the next two consecutive odd integers?
4. Suppose the sum of two integers is 100. If x represents one of these integers, express the other in terms of x.
5. Determine two consecutive integers whose sum is 37.
6. Determine three consecutive integers whose sum is 18.

7. Determine two consecutive odd integers whose sum is 52.
8. Determine three consecutive even integers whose sum is 66.
9. Nine more than twice an integer is 43. Determine the integer.
10. Three less than five times an integer is 17. Determine the integer.
11. Half of an integer plus one-third of the next consecutive integer equals 7 . Determine the integers.
12. The sum of two integers is 60. The larger is six more than the smaller. Determine the integers.
13. Suppose x represents a woman's age now. Express her age in five years.
14. Jim is five years older than Jerry. If x represents Jerry's age, express Jim's age.
15. Rosita is three years older than Carmen. If x represents Rosita's age, express Carmen's age.
16. Five years ago a boy was twice as old as his sister. If x is the boy's present age, express his sister's present age.
17. A father is five times as old as his daughter. In five years he will be three times as old as the daughter. How old is each now?
18. Bill's grandfather is six times as old as Bill. In six years the grandfather will be four times as old as Bill. How old is Bill?
19. A man is five years older than his wife. Thirteen years ago he was twice her age. How old is the man?
20. A 35-year-old man has a 9-year-old daughter. In how many years will his age be double hers?
21. Harry is five years older than Don. Three years ago Don was two-thirds as old as Harry. How old is Harry?
22. A woman has a son and daughter. The woman is six times as old as her son, who is two years older than his sister. In six years the woman will be four times as old as her daughter. How old is the woman?

7.2 COIN AND MERCHANDISING PROBLEMS

In coin problems you evaluate the total value of various coins. For example, a nickel is worth 5¢; thus 6 nickels are worth 30¢. A quarter is worth 25¢; 7 quarters are worth $1.75 .

EXAMPLE 1. Ed finds three quarters, two dimes, six nickels, and two pennies in his pocket. How much money does he have?

SOLUTION. Table 7.1 is useful.

	Cents per Coin $\cdot$	Number of Coins $=$	Total Amount in Cents
quarters	25	3	75
dimes	10	2	20
nickels	5	6	30
pennies	1	2	2
			127

TABLE 7.1

According to the last column of the table, Ed has 127 cents (or $1.27). ■

EXAMPLE 2. Anita has $1.70 in nickels and dimes. If she has 22 coins in all, how many nickels has she?

SOLUTION. Let x be the number of nickels. Because there are 22 coins altogether, the number of dimes she has is $22 - x$. [See Table 7.2 .]

	Cents per Coin $\cdot$	Number of Coins $=$	Total Amount in Cents
nickels	5	x	$5x$
dimes	10	$22 - x$	$10(22 - x)$

TABLE 7.2

$$\underbrace{5x + 10(22 - x)}_{\text{The total value of nickels and dimes (in cents)}} \overset{\text{equals}}{=} \underbrace{170}_{\text{170 cents.}}$$

$$5x + 220 - 10x = 170$$

$$50 = 5x$$

$$10 = x$$

Anita has 10 nickels (and 12 dimes). ■

EXAMPLE 3. A man has two more dimes than quarters and three times as many nickels as dimes. Altogether he has $2.50 . How many of each coin does he have?

SOLUTION. Let x be the number of quarters. Then $x + 2$ is the number of dimes and $3(x + 2)$ is the number of nickels. [See Table 7.3 .]

	Cents per Coin	Number of Coins	= Total Amount in Cents
quarters	25	x	$25x$
dimes	10	$x + 2$	$10(x + 2)$
nickels	5	$3(x + 2)$	$15(x + 2)$

TABLE 7.3

The total value of quarters, of dimes, and of nickels — equals — 250 (cents).

$$\begin{aligned}
25x + 10(x + 2) + 15(x + 2) &= 250 \\
50x + 50 &= 250 \\
50x &= 200 \\
x &= 4 \\
x + 2 &= 6 \\
3(x + 2) &= 18
\end{aligned}$$

He has 4 quarters, 6 dimes, and 18 nickels. ■

There are merchandising problems closely related to the above coin problems.

EXAMPLE 4. A pushcart owner buys grapefruit for 10¢ apiece and sells them for 25¢ apiece. After selling all of them, he makes a profit of \$6.60 on this item. How many grapefruit does he buy?

SOLUTION. Let x be the number of grapefruit *bought*. Then x is also the number of grapefruit *sold*. Now use Table 7.4 .

	Cents per Grapefruit	Number of Grapefruit	= Total Amount in Cents
bought	10	x	$10x$
sold	25	x	$25x$

TABLE 7.4

The amount of money received — minus — the amount of money paid — equals — his profit.

$$\begin{aligned}
25x - 10x &= 660 \\
15x &= 660 \\
x &= 44
\end{aligned}$$

He buys 44 grapefruit. ■

EXAMPLE 5. Chocolates at $1.50 per pound are mixed with chocolates at $1.00 per pound. How much of each must be used in order to have a 40-pound mixture that sells at $1.20 per pound?

SOLUTION. Let x be the number of pounds of the $1.00 per pound chocolates. Then $(40 - x)$ pounds of the $1.50 per pound chocolates are used. [Use Table 7.5 .]

	Cents per Pound	· Number of Pounds	= Total Amount in Cents
$1.00 per pound chocolates	100	x	$100x$
$1.50 per pound chocolates	150	$40 - x$	$150(40 - x)$
Mixture ($1.20 per pound)	120	40	4800

TABLE 7.5.

The value of the $1.00 per pound chocolates + the value of the $1.50 per pound chocolates = the value of the mixture.

$$100x + 150(40 - x) = 4800$$

$$100x + 6000 - 150x = 4800$$

$$1200 = 50x$$

$$24 = x$$

$$16 = 40 - x$$

24 pounds of $1.00 chocolates must be mixed with 16 pounds of $1.50 chocolates.

■

EXERCISES

1. Determine the value (in dollars and cents) of eight quarters, three nickels, and twelve pennies.
2. Determine the value of eight half-dollars, six quarters, three dimes, and fourteen nickels.
3. Frank has five more quarters than nickels. Altogether he has $4.25. How many of each coin does he have?
4. Linda has $5.55 in nickels, dimes, and quarters, She has ten more dimes than nickels and nine more nickels than quarters. How many of each coin does she have?
5. The 34 coins Phil has are worth $6.10 . If they are just quarters and dimes, how many of each does he have?
6. Helen has five more pennies than nickels and three more nickels than dimes. Altogether her coins are worth $2.95. How many of each coin does she have?

7. I have the same number of nickels and quarters in my pocket. If I had two more nickels and three fewer quarters, I would have \$4.75. How many quarters have I?
8. Pat has a total of 21 coins in her purse in quarters, dimes, and nickels. If she had two more quarters, one more dime, and five more nickels, she would have \$3.55. She has the same number of nickels as quarters. How many of each coin does she have?
9. A man has five more pennies than nickels and one more nickel than dimes. Altogether he has \$1.39. How many pennies does he have?
10. A store buys candy bars for 5¢ apiece and sells them for 10¢ apiece. A profit of \$7.20 is made on this item after all have been sold. How many candy bars were there?
11. Jerry bought ice cream bars at 14¢ apiece. He ate two and sold the rest at 25¢ apiece. He made a profit of \$6.32 for his efforts. How many ice cream bars did he purchase?
12. A candy store owner buys tootsie rolls at three for a dime and sells them at four for a quarter. After selling all of them, his profit is \$4.20. How many tootsie rolls were there?
13. Coffee at 75 cents per pound is blended with coffee at \$1.25 per pound. How much of each must be used to make 50 pounds of a mixture at 95 cents per pound?
14. Ten pounds of hazel nuts at 75 cents per pound and twelve pounds of pecans at 80 cents per pound are to be combined with walnuts at 90 cents per pound to obtain a mixture that will sell at 85 cents per pound. How many pounds of walnuts must be used?
15. Nine pounds of chocolate chip cookies at \$2.10 per pound are mixed with six pounds of nut cookies at \$1.80 per pound and with ten pounds of ginger snaps at \$1.53 per pound. How much should the assorted cookies cost per pound?

7.3 DISTANCE PROBLEMS

Problems concerning motion at a *constant rate* can be handled with these algebraic methods. *The distance that an object travels at a constant rate of speed equals its rate multiplied by the time in transit.* Let

d = distance,

r = rate,

t = time.

Then

$$d = r \cdot t$$

Thus if an automobile travels at the constant rate of 60 miles per hour, in 5 hours it goes 300 miles.

You can use other forms of the distance formula. For example, to determine time, use

$$t = \frac{d}{r};$$

to determine the rate, use

$$r = \frac{d}{t}.$$

Throughout this section all rates are assumed to be constant.

EXAMPLE 1. An airplane traveling at the rate of 400 miles per hour flies 1800 miles. How long does this take?

SOLUTION. Let t be the number of hours.

r	$t = \frac{d}{r}$	d
400	t	1800

TABLE 7.6

$$t = \frac{d}{r}$$

$$t = \frac{1800}{400}$$

$$t = 4\frac{1}{2}$$

The trip takes $4\frac{1}{2}$ hours. ■

EXAMPLE 2. Two cars leave a gas station at the same time, one traveling eastward, the other westward. The eastward-bound car travels at 50 miles per hour; the westward-bound car speeds at 65 miles per hour. How far apart are they after 3 hours?

SOLUTION.

	r ·	t =	d
eastward	50	3	150
westward	65	3	195

TABLE 7.7

The cars are traveling in *opposite* directions. Their distance apart d is the *sum* of the distances each has traveled in 3 hours:

$$d = 150 + 195$$

$$d = 345$$

The distance apart is 345 miles. Note that each hour the cars move 65 + 50, or 115, miles apart, and $345 = 3 \cdot 115$. [See Figure 7.2(a).] ■

In Example 2 the cars were traveling in *opposite* directions, and their distance apart was the *sum* of the distances each had traveled. Had they left the gas station at the same time but traveled in the *same* direction, their distance apart would have been the distance the faster car traveled *minus* the distance the slower car traveled. [See Figure 7.2(b).]

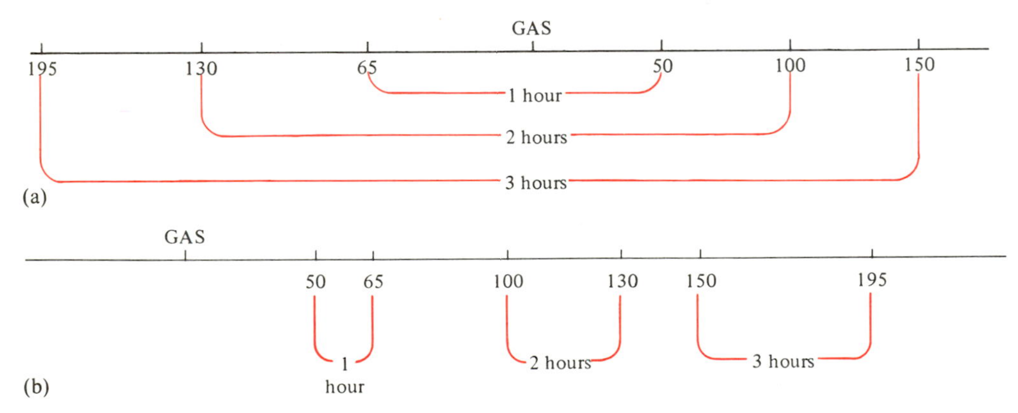

FIGURE 7.2 (a) The cars separate at the rate of 65 + 50 (or 115) miles per hour. After 3 hours they are $3 \cdot 115$ (or 345) miles apart.
(b) The cars separate at the rate of 65–50 (or 15) miles per hour. After 3 hours they are $3 \cdot 15$ (or 45) miles apart.

EXAMPLE 3. A bandit leaves Cheyenne, Wyoming in a stolen car at 4 a.m., traveling eastward at the constant rate of 60 miles per hour. At 6 a.m., a sheriff leaves Cheyenne and pursues him at the constant rate of 75 miles per hour.

(a) At what time does the sheriff overtake the bandit?
(b) How far from Cheyenne is the scene of capture?

SOLUTION. (a) Let t be the time (in hours) the bandit travels before being captured. The sheriff, who leaves Cheyenne 2 hours later, travels for $(t - 2)$ hours.

	r ·	t =	d
bandit	60	t	$60t$
sheriff	75	$t - 2$	$75(t - 2)$

TABLE *7.8*

Here, the bandit and the sheriff, although traveling over *different periods of time, travel the same distance.* Thus

$$\underbrace{60t}_{\text{the distance the bandit travels}} \overset{\text{equals}}{=} \underbrace{75(t-2)}_{\text{the distance the sheriff travels.}}$$

$$60t = 75t - 150$$

$$150 = 15t$$

$$10 = t$$

The bandit, who starts out at 4 a.m., travels for 10 hours. The capture occurs at 2 p.m.

(b) $60t = 60 \cdot 10 = 600$

Thus the capture takes place 600 miles east of Cheyenne. ■

EXAMPLE 4. Natty Bumppo swims straight across a lake at the rate of 25 yards per minute to the spot where he has left his canoe. He then paddles back to his starting point along a straight course at the rate of 75 yards per minute. If the round trip takes an hour and 20 minutes, how far is it from shore to shore?

SOLUTION.

	r	$\cdot$ t	$=$ d
swimming	25	t	$25t$
canoeing	75	$80 - t$	$75(80 - t)$

TABLE 7.9

Here, t represents the number of minutes spent swimming. Altogether the round trip takes 80 minutes; thus he spends $(80 - t)$ minutes canoeing. *The distance is the same both ways.* To determine the distance d, first determine t. Then $d = 25t$.

$$\underbrace{25t}_{\text{The distance (across the lake) he swims}} \overset{\text{equals}}{=} \underbrace{75(80-t)}_{\text{the distance he canoes.}}$$

$$t = 3(80 - t)$$

$$t = 240 - 3t$$

$$4t = 240$$

$$t = 60$$

$$25t = 1500$$

The distance across the lake is 1500 yards. ■

EXERCISES

(*All rates are assumed to be constant.*)

1. How long will it take to walk to class if you walk 3 miles per hour along a straight path, and if the classroom is a mile and a half from where you are?
2. An automobile headed westward travels for 2 hours at 70 miles per hour. It then slows down and continues westward at 60 miles per hour for the next hour and a quarter. How far does it travel?
3. Pensive Pete travels 60 miles per hour for 3 hours before realizing that he left his airplane tickets at his motel room. He returns at a rate of 80 miles per hour. How long does the return trip take?
4. Two cars leave the center of town at the same time traveling in opposite directions. One car is going 48 miles per hour; the other is going 62 miles per hour. How far apart are they after 6 hours?
5. Two trains approach one another along (straight) parallel tracks. They start out at the same time from stations 90 miles apart. One train travels at 100 miles per hour, the second train at 80 miles per hour. How far has the faster train gone when they pass each other?
6. Two bicyclists leave from the same place and at the same time traveling in the same direction along a straight road. One goes at 18 miles per hour, the other at 14 miles per hour. How far apart are they after an hour and a quarter?
7. A Mercedes and a Volkswagen leave a toll area at the same time traveling in the same direction along a straight road. The Volkswagen travels at two-thirds the rate of the Mercedes. At the end of 4 hours they are 100 miles apart. How fast is the Mercedes traveling?
8. A rowboat goes upstream at the rate of 4 miles per hour. It returns downstream at the rate of 8 miles per hour. If the round trip takes 3 hours, how far upstream did it go?
9. A hitchhiker begins to walk along a straight road at the rate of 3 miles per hour to the next town, which is 18 miles away. After 10 minutes a car picks him up, and in 15 more minutes he is in town. How fast was the car traveling?
10. Two cars leave the Hollywood Bowl at the same time and head for San Francisco. The first car takes 7 hours along the shorter route, which measures 420 miles. The other car, which travels 4 miles per hour faster, arrives a half hour later. How long is the second route?
11. A skier walks 3000 feet up a mountain road, rests for 10 minutes, and then skis down the road. He returns a half hour after he started. If he skis four times as fast as he walks, at what rate does he ski?
12. A cross-country runner can average 10 miles per hour over level ground and 6 miles per hour over hilly ground. Altogether it takes him an hour and forty minutes to cover 12 miles. How many of these miles are hilly?

7.4 INTEREST PROBLEMS

The **interest earned** *for one year* equals the yearly **interest rate** times the amount of money (or **principal**) invested. Let

I be the interest earned,

R be the interest rate,

P be the principal.

Then

$$I = R \cdot P$$

Percent means hundredths. For example, 6% means the same as .06 . Thus $100 invested for one year at 6% earns

.06(100) dollars

or $6 in one year.

EXAMPLE 1. A man deposits $800 in a bank whose annual (once a year) interest rate is 5%. How much interest does he earn if he leaves all the money, including the interest, in for 2 years?

SOLUTION. Let P_1 be his original principal (of $800). The interest rate is 5%, or .05.

R ·	P_1 =	I
.05	800	40

TABLE 7.10

After one year he earns $40 interest. He leaves this in the bank and therefore has $840 after one year. This is his principal, P_2, for the second year. The interest rate is the same.

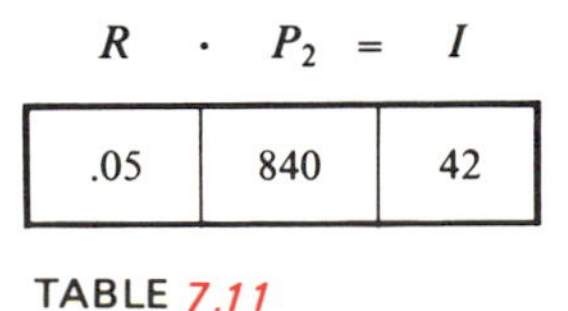

R ·	P_2 =	I
.05	840	42

TABLE 7.11

He earns $42 interest the second year. Thus in two years he earns $82 interest. ■

EXAMPLE 2. Part of "Dollar" Bill's \$4000 is invested in a bank at 6% and the other part is in 8%-interest-yielding bonds. His annual interest income from the two sources is \$260. How much money has Bill in the bank?

SOLUTION. Let x be the amount of dollars in the bank. Then $4000 - x$ is the amount in bonds.

	R	$\cdot$ P	$= I$
in bank	.06	x	$.06x$
in bonds	.08	$4000 - x$	$.08(4000 - x)$

TABLE 7.12

His annual interest income from the two sources is \$260. Thus

the bank interest	plus	the bond interest	equals	\$260.
$.06x$	$+$	$.08(4000 - x)$	$=$	260

To eliminate decimals, multiply both sides by 100.

$$6x + 8(4000 - x) = 26\,000$$

$$6x + 32\,000 - 8x = 26\,000$$

$$6000 = 2x$$

$$3000 = x$$

Bill has \$3000 in the bank. ■

EXAMPLE 3. Smart Sam invests \$1000 more at 9% than in a safer investment at 6%. He earns the same as if the entire amount were invested at $8\frac{1}{2}$%. How much money has Sam invested at 6%?

SOLUTION. Let x be the amount of dollars invested at 6%. Then $x + 1000$ is the amount invested at 9%. The entire amount, which he could invest at $8\frac{1}{2}$%, is $x + (x + 1000)$, or $2x + 1000$.

	R	$\cdot$ P	$= I$
at 6%	.06	x	$.06x$
at 9%	.09	$x + 1000$	$.09(x + 1000)$
at $8\frac{1}{2}$%	.085	$2x + 1000$	$.085(2x + 1000)$

TABLE 7.13

Note that $\frac{1}{2}\% = \frac{.010}{2} = .005$; hence

$$8\frac{1}{2}\% = .08 + .005 = .085.$$

$$\underbrace{.06x + .09(x + 1000)}_{\text{He earns}} \underbrace{=}_{\text{the same as if}} \underbrace{(2x + 1000)}_{\text{the entire amount}} \underbrace{(.085)}_{\text{were at } 8\frac{1}{2}\%}$$

Multiply both sides by 1000 to clear decimals.

$$60x + 90(x + 1000) = (2x + 1000)85$$

$$60x + 90x + 90\,000 = 170x + 85\,000$$

$$5000 = 20x$$

$$250 = x$$

Sam has \$250 invested at 6%. ■

EXERCISES

1. How much interest is earned in one year on a principal of \$560 if the annual interest rate is 6%?
2. What is the annual interest rate if \$88 interest is paid in a year on a principal of \$2200?
3. How much money must be invested for a year at 7% in order to earn \$249.90 interest?
4. A sum of \$2500 is left for two years in a bank that pays an annual interest rate of 5%. How much interest is earned?
5. A sum of \$3600 is left for three years in a bank that pays an annual interest rate of 6%. How much interest is earned (to the nearest cent)?
6. Part of a sum of \$1600 is invested at 5% and the remainder at 6%. The combined annual interest from this money is \$92. How much money is invested at 5%?
7. Part of a sum of \$5400 is invested in 8% bonds and the remainder in $6\frac{1}{2}\%$ tax-free bonds. The combined annual interest from these bonds is \$375. How much money is invested in 8% bonds?
8. If you have twice as much invested at 7% as at 5% and if your annual interest income from these two investments is \$950, how much have you invested at each rate?
9. A man invests \$2500 more at 6% than at 7%, and receives an annual interest income from these two investments of \$1060. How much is invested at each rate?
10. A man invests \$500 more at 8% than at 5.5%. He earns the same as if the entire sum were invested at 7%. How much money is invested at 8%?
11. A company invests \$12 000 at 5% and \$18 000 at 6%. At what rate should it invest its remaining \$5000 in order to receive a combined interest of \$2030?

7.5 MIXTURE PROBLEMS

Several ingredients are combined to form a mixture. Your task is to determine the amount of a specific ingredient in the mixture.

For example, if equal parts of alcohol and water are mixed to form 30 gallons of a solution, the amount of alcohol in the solution is $\frac{1}{2}$ (or 50%) of the solution, that is, 15 gallons.

In general, *the fraction of the substance times the amount of the mixture equals the amount of the substance in the mixture.*

Fraction of the Substance × Amount of Mixture

= Amount of the Substance

In the preceding paragraph, alcohol is the substance. Thus

$$\frac{1}{2}\text{ (fraction of the substance)} \times \begin{array}{c}\text{30 gallons of solution}\\ \text{(the mixture)}\end{array} = \begin{array}{c}\text{15 gallons of}\\ \text{the substance}\end{array}$$

EXAMPLE 1. How many gallons of a 12% salt solution should be combined with 10 gallons of an 18% salt solution to obtain a 16% solution?

SOLUTION. Salt is the substance. Let x be the number of gallons of 12% solution. Then there are $(x + 10)$ gallons in the final mixture (the 16% solution).

	Fraction of Salt ·	Amount of Mixture	= Amount of Salt
12% solution	.12	x	$.12x$
18% solution	.18	10	1.8
16% solution	.16	$x + 10$	$.16(x + 10)$

TABLE 7.14

$$\underbrace{\begin{array}{c}\text{Amount of Salt in}\\ \text{12\% Solution}\end{array}}_{} + \underbrace{\begin{array}{c}\text{Amount of Salt in}\\ \text{18\% Solution}\end{array}}_{} = \underbrace{\begin{array}{c}\text{Amount of Salt in}\\ \text{16\% Solution}\end{array}}_{}$$

$$\begin{aligned} .12x + 1.8 &= .16(x + 10) \\ 12x + 180 &= 16(x + 10) \\ 12x + 180 &= 16x + 160 \\ 20 &= 4x \\ 5 &= x \end{aligned}$$

Five gallons of 12% solution must be used. ■

EXAMPLE 2. How many ounces of an alloy containing 30% gold must be melted with an alloy containing 40% gold to obtain 100 ounces of an alloy containing 36% gold?

SOLUTION. Gold is the substance. Let x be the number of ounces of the 30% alloy. Then $(100 - x)$ ounces of the 40% alloy are used.

	Fraction of Gold ·	Ounces of Alloy =	Ounces of Gold
30% alloy	.3	x	$.3x$
40% alloy	.4	$100 - x$	$.4(100 - x)$
36% alloy	.36	100	36

TABLE 7.15

$$\underbrace{.3x}_{\text{Ounces of Gold in 30\% Alloy}} + \underbrace{.4(100 - x)}_{\text{Ounces of Gold in 40\% Alloy}} = \underbrace{36}_{\text{Ounces of Gold in 36\% Alloy}}$$

$$3x + 4(100 - x) = 360$$

$$3x + 400 - 4x = 360$$

$$40 = x$$

There must be 40 ounces of the 30% alloy. ■

EXAMPLE 3. Alloy A is 1 part copper to 3 parts tin. Alloy B is 1 part copper to 4 parts tin. How much of alloy B should be added to 24 pounds of alloy A to obtain an alloy that is 2 parts copper to 7 parts tin?

SOLUTION. Let copper be the substance. Let x be the number of pounds of alloy B.

	Fraction of Copper ·	Pounds of Alloy =	Pounds of Copper
A: 1 part copper, 3 parts tin	$\frac{1}{4}$	24	6
B: 1 part copper, 4 parts tin	$\frac{1}{5}$	x	$\frac{x}{5}$
C: 2 parts copper, 7 parts tin	$\frac{2}{9}$	$24 + x$	$\frac{2}{9}(24 + x)$

TABLE 7.16

Pounds of Copper in alloy B	+	Pounds of Copper in Alloy A	=	Pounds of Copper in Alloy C	
$\frac{x}{5}$	+	6	=	$\frac{2}{9}(24 + x)$	Multiply both sides by 45.
$9x$	+	270	=	$10(24 + x)$	
$9x$	+	270	=	$240 + 10x$	
30			=	x	

Thirty pounds of alloy B must be used. ■

EXERCISES

1. How many gallons of a 20% salt solution should be combined with 18 gallons of a 28% salt solution to obtain a 22% solution?
2. How many gallons of a 15% salt solution should be combined with a 25% salt solution to obtain 40 gallons of a 21% solution?
3. How much pure acid should be added to 5 gallons of 40% acid solution to obtain a 50% acid solution?
4. How many liters of a chemical that is 44% sulphuric acid must be combined with 12 liters of a chemical that is 56% sulphuric acid to obtain a 47% sulphuric acid mixture?
5. A martini mix contains 4 parts gin to 1 part vermouth. A drier mix contains 6 parts gin to 1 part vermouth. How much of the drier mix must be used for 15 ounces of a mixture that is 5 parts gin to 1 part vermouth?
6. How many ounces of an alloy containing 60% silver must be added to 50 pounds of an alloy containing 45% silver to obtain an alloy containing 51% silver?
7. How much cream that contains 33% butterfat must be blended with milk that contains 3% butterfat to obtain 10 gallons of half-and-half that contains 8% butterfat?
8. Ten gallons of chocolate ice cream that contains 18% butterfat is mixed with 8 gallons of vanilla ice cream that contains 12% butterfat to make chocolate ripple ice cream. What percent butterfat does the chocolate ripple contain?
9. An alloy is 2 parts zinc to 1 part tin. A second alloy is 3 parts zinc to 1 part tin. How much of each alloy should be used to make 200 tons of an alloy that is 7 parts zinc to 3 parts tin?
10. A grocer mixes 6 pounds of coffee that is 60% Colombian with one that is 80% Colombian. How much of the second type should be used to obtain a mixture that is 78% Colombian?

7.6 WORK PROBLEMS

Suppose a job is done by several people, each working at a constant rate. You sometimes have to consider the fraction of work done by each person *working alone*. For example, if a man can paint a room in 4 hours, he paints $\frac{1}{4}$ of the room per hour. If he stops after 3 hours, he has done $\frac{3}{4}$ of the paint job. There may be a second worker involved. This will affect the time to complete the painting.

In work problems,

the fraction of work done in 1 time unit · time = the fraction of work done

EXAMPLE 1. A father can paint a room in 4 hours and his son can paint it in 6 hours. How long does it take them to paint the room together?

SOLUTION. Let x be the number of hours it takes them to paint the room together. In 1 hour the father does $\frac{1}{4}$ of the work and the son does $\frac{1}{6}$ of the work.

	Fraction of Work Done in 1 Hour	· Hours	= Fraction of Work Done
Father	$\frac{1}{4}$	x	$\frac{x}{4}$
Son	$\frac{1}{6}$	x	$\frac{x}{6}$

TABLE 7.17

There is 1 job to be done—namely, to paint the room. Each does a part. Thus

$$\underbrace{\text{the fraction done by the father}} + \underbrace{\text{the fraction done by the son}} = \underbrace{1 \text{ (total job)}}$$

$$\frac{x}{4} + \frac{x}{6} = 1 \qquad \text{Multiply both sides by 12.}$$

$$3x + 2x = 12$$

$$5x = 12$$

$$x = \frac{12}{5}$$

It takes $\frac{12}{5}$ hours (2 hours and 24 minutes) to complete the paint job together. ■

EXAMPLE 2. A gardener can mow a lawn in 9 hours. When his assistant aids him, it takes the two together only 6 hours. How long does it take the assistant alone to mow the lawn?

SOLUTION. The gardener alone does $\frac{1}{9}$ of the work in 1 hour. Let x be the number of hours it takes the assistant alone to mow the lawn. The assistant does $\frac{1}{x}$ of the work in 1 hour.

	Fraction of Work Done in one Hour	· Hours	= Fractions of Work Done
Gardener	$\frac{1}{9}$	6	$\frac{2}{3}$
Assistant	$\frac{1}{x}$	6	$\frac{6}{x}$

TABLE *7.18*

$$\underbrace{\text{Fraction of work done by gardener}} + \underbrace{\text{Fraction of work done by assistant}} = \underbrace{1 \text{ (total job)}}$$

$$\frac{2}{3} + \frac{6}{x} = 1 \qquad \text{Multiply both sides by } 3x.$$

$$2x + 18 = 3x$$

$$18 = x$$

The assistant alone mows the lawn in 18 hours. ■

EXAMPLE 3. One pipe can *fill* a tank in 15 minutes and another pipe can *empty* the tank in 45 minutes. If the tank is half-full when both pipes are opened, how long will it take to fill the tank? [See Figure 7.3.]

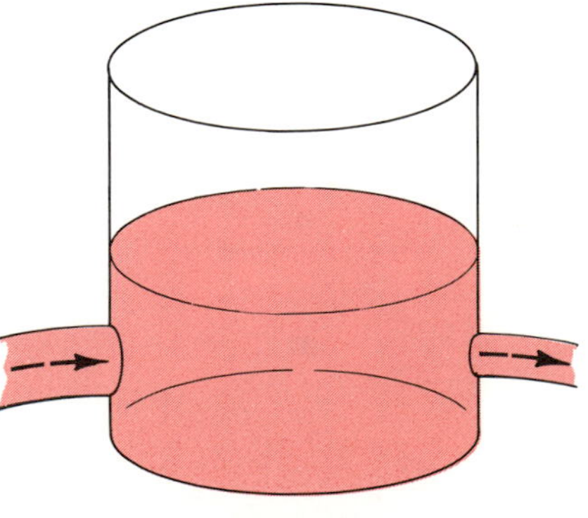

FIGURE 7.3

SOLUTION. Let x be the number of minutes necessary to fill the *empty* tank when both pipes are open. In 1 minute, the first pipe *fills* $\frac{1}{15}$ of the tank. The second pipe *empties* $\frac{1}{45}$ of the tank; its contribution in that minute is $\frac{-1}{45}$ (of the tank).

	Fraction of Work Done in 1 Minute	· Minutes	= Fraction of Work Done
First Pipe	$\frac{1}{15}$	x	$\frac{x}{15}$
Second Pipe	$\frac{-1}{45}$	x	$\frac{-x}{45}$

TABLE 7.19

The tank is half-filled to begin with. Only half the job of filling it remains. Thus

$$\underbrace{\text{Fraction of work done by first pipe}} + \underbrace{\text{Fraction of work done by second pipe}} = \underbrace{\frac{1}{2} \text{ (of total job)}}$$

$$\frac{x}{15} + \frac{-x}{45} = \frac{1}{2} \qquad \text{Multiply both sides by 90.}$$

$$6x - 2x = 45$$

$$4x = 45$$

$$x = \frac{45}{4}$$

It takes $\frac{45}{4}$ $\left(\text{or } 11\frac{1}{4}\right)$ minutes to fill the second half of the tank. ■

EXERCISES

1. Bob can paint a barn in 10 hours. Joe can paint it in 12 hours. How long does it take them together to paint the barn?
2. Two typists address a batch of envelopes. Tillie can complete the job by herself in 9 hours. Terri can do the job by herself in 8 hours. How long does it take the two together?
3. Judd and Judy can clean their apartment together in 3 hours. They each work at the same speed. How long would it take Judy to clean it by herself?
4. Two privates sweep the mess hall together in 4 hours. One works twice as fast as the other. How long does it take the faster private to do the job alone?
5. A carpenter must saw several beams of wood. It takes him 2 days if he works with his assistant and 3 days if he works alone. How long does it take the assistant alone to saw these beams?
6. Three workers set out to plaster a ceiling. One can do the job alone in 2 hours, one in 3 hours, and the last one in 4 hours. How long does it take them together to plaster the ceiling?
7. A father and his two sons gather the autumn leaves. The father works twice as fast as one son and three times as fast as the other. If it takes the father 6 hours to gather the leaves by himself, how long does it take all three of them to do the job?

8. In Exercise 7, how long would it take the two boys to gather the leaves without aid from their father?

9. One pipe fills a tank in 50 minutes, another pipe in 75 minutes. How long does it take for both pipes together to fill the tank?

10. Two equal-sized pipes fill a tank in an hour. How long does it take for one of these pipes to fill the tank?

11. One pipe can fill a tank in 15 minutes. A second pipe can empty a full tank in 18 minutes. If both pipes are open, how long does it take to fill the tank, when empty?

12. The hot water faucet fills a tub in 12 minutes, the cold water faucet in 8 minutes. When the stopper is lifted, a filled tub empties in 10 minutes. Both faucets are open and the stopper is lifted. How long does it take to fill two-thirds of the tub?

REVIEW EXERCISES FOR CHAPTER 7

7.1 INTEGER AND AGE PROBLEMS

1. The sum of three consecutive odd integers is 81. Determine these integers.

2. A woman is six years younger than her husband. Eighteen years ago she was two-thirds his age. How old is she now?

7.2 COIN AND MERCHANDIZING PROBLEMS

3. Dottie has \$2.40 in nickels, dimes, and quarters. She has twice as many nickels as quarters, and twice as many quarters as dimes. How many of each does she have?

4. Almonds at 80 cents per pound are mixed with Brazil nuts at \$1.20 per pound. How many pounds of each must be used to make 80 pounds of a mixture at 90 cents per pound?

7.3 DISTANCE PROBLEMS

5. Suppose you drive for 3 and a half hours at 60 miles per hour to visit your Aunt Tillie. If the return trip takes 3 hours, at what rate do you return?

6. Two trains approach each other along (straight) parallel tracks. One train travels at 100 miles per hour and goes twice as fast as the other. If their distance apart is 225 miles, in how many hours will they pass each other?

7.4 INTEREST PROBLEMS

7. How much money must be invested for a year at 6% in order to earn \$216 interest?

8. If you have twice as much invested at 8% as at 5% and if your annual interest income from these two investments is \$315, how much is invested at each rate?

7.5 MIXTURE PROBLEMS

9. How many gallons of a 32% salt solution must be combined with 12 gallons of a 20% salt solution to obtain a 29% solution?
10. Alloy A is 1 part copper to 4 parts tin. Alloy B is 1 part copper to 7 parts tin. How much of alloy B should be added to 30 pounds of alloy A to obtain an alloy that is 1 part copper to 5 parts tin?

7.6 WORK PROBLEMS

11. Bob can paint a room in 6 hours and Ed can paint it in 8 hours. How long does it take them to paint the room together?
12. One pipe fills a tank in 30 minutes, another pipe in 40 minutes. How long does it take for both pipes together to fill the tank?

TEST YOURSELF ON CHAPTER 7.

1. The sum of two integers is 90. The larger is eight more than the smaller. Determine the integers.
2. Fred has three times as many quarters as nickels. Altogether he has $8.00. How many quarters does he have?
3. A Thunderbird and an MG leave a gas station at the same time traveling in opposite directions. The Thunderbird travels at 55 miles per hour. After 3 hours the cars are 375 miles apart. How fast is the MG traveling?
4. How many gallons of a 12% salt solution must be combined with a 42% salt solution to obtain 30 gallons of an 18% solution?

8
functions and graphs

8.1 SETS AND FUNCTIONS

The notion of a "function" is crucial in all scientific studies. A function indicates how one variable corresponds to another. In order to understand what a function is, you must first know what a set is.

DEFINITION. Any collection of distinct objects is known as a **set.**

There are sets of people, sets of cities, sets of books.

EXAMPLE 1. The set of states in the United States consists of 50 objects—the 50 states. Among these are California, Idaho, and Virginia. ■

Often the objects of a set are indicated by means of **braces:**

$$\{ \quad \}$$

The objects are listed between the braces. *The order in which the objects are listed does not matter.*

EXAMPLE 2. The set consisting of Anne, Bill, Jerry, and Peggy is denoted by

{Anne, Bill, Jerry, Peggy}

or by

{Bill, Peggy, Jerry, Anne}. ■

In mathematics sets of numbers are important.

EXAMPLE 3. The set consisting of the first eight positive integers is

$$\{1, 2, 3, 4, 5, 6, 7, 8\}.$$ ■

The sets thus far considered have all been **finite,** that is, there is a specific number of objects—4 or 8 or 50 or 0. In contrast, here are some **infinite** sets with which you are familiar.

EXAMPLE 4. Each of the following sets is infinite:

(a) the set of all real numbers
(b) the set of all rational numbers
(c) the set of all integers
(d) the set of all positive integers ■

You are now going to consider *correspondences between two sets.* In other words, you will match the objects of one set with those of another set. At first, you will only consider match-ups between finite sets.

EXAMPLE 5. Suppose that Anne, Bill, Jerry, and Peggy all took a math quiz. Their grades were as follows:

Anne: 76, Bill: 92, Jerry: 79, Peggy: 97

You can consider the set of students,

{Anne, Bill, Jerry, Peggy},

and the set of their exam scores,

$$\{76, 92, 79, 97\}.$$

The correspondence identifying each student with his or her exam score can be given by:

Anne → 76

Bill → 92

Jerry → 79

Peggy → 97

The key point here is that *to each student there corresponds exactly one exam score.* ■

DEFINITION. A **function** is a correspondence between two sets, called the **domain** and **range,** such that to each object in the domain there corresponds *exactly one* object in the range. And every object in the range corresponds to *at least one* object in the domain.

In Example 5, the *domain* was {Anne, Bill, Jerry, Peggy} and the *range* was {76, 79, 92, 97}.

EXAMPLE 6. Consider the correspondence given by:

$$1 \to 2$$
$$2 \to 4$$
$$3 \to 6$$
$$4 \to 8$$
$$5 \to 10$$

To every object of the first set, $\{1, 2, 3, 4, 5\}$, there corresponds exactly one object of the second set, $\{2, 4, 6, 8, 10\}$. Thus this correspondence represents a function with *domain* $\{1, 2, 3, 4, 5\}$ and *range* $\{2, 4, 6, 8, 10\}$. ■

EXAMPLE 7. The correspondence indicated by

$$1 \to 1$$
$$2 \to \frac{1}{2}$$
$$3 \to \frac{1}{3}$$
$$4 \to \frac{1}{4}$$

represents a function with *domain* $\{1, 2, 3, 4\}$ and range $\left\{1, \frac{1}{2}, \frac{1}{3}, \frac{1}{4}\right\}$. Each number in the domain is identified with its reciprocal in the range. ■

EXAMPLE 8. The correspondence indicated by

$$5 \to 2$$
$$10 \to 3$$
$$15 \to 2$$

represents a function with domain $\{5, 10, 15\}$ and range $\{2, 3\}$. Observe that *two different objects of the domain,* 5 and 15, *correspond to the same object, 2, of the range.* This is allowed in the definition of a function. For all that is required is that

> *to each object of the domain there corresponds exactly one object of the range.* ■

EXAMPLE 9. The correspondence

$$2 \to 5$$
$$2 \searrow 15$$
$$3 \to 10$$

does *not* represent a function with domain {2, 3} and range {5, 10, 15}. For *there correspond two distinct objects,* 5 and 15, *of the "range" to the object* 2 *of the "domain".* ■

EXERCISES

In Exercises 1–10 use braces to indicate the following sets:

1. the set consisting of Bob and Carol and Ted and Alice
2. the set of colors of the rainbow
3. the set consisting of the New England States
4. the set consisting of the last three American Presidents
5. the set of outfield positions on a baseball team
6. the set consisting of the Five Books of Moses
7. the set of the first ten positive integers
8. the set of even positive integers < 10
9. the set of odd negative integers > -10
10. the set of real numbers that are neither positive nor negative

In Exercises 11–20 which of the following correspondences represent functions?

11. John → Jackie
Bob → Ethyl
Ted → Joan

12. John → Jackie
Bob → Jackie
Ted → Jackie

13. John → Jackie

14. John → Jackie
John → Ethyl
Bob → Joan

15. Jackie → John
Ethyl → Bob
Joan → Ted

16. John → Jackie
Bob → Ethyl
Ted → Joan
John → Joan

17. 1 → 3
2 → 6
3 → 9
4 → 12

18. 1 → 3
1 → 6
1 → 9
1 → 12

19. 1 → 3
2 → 3
3 → 3
4 → 3

20. 1 → 3
2 → 6
3 → 3
4 → 6

In Exercises 21–26 determine (a) the domain and (b) the range of the indicated function.

21. 1 → 10
2 → 9
3 → 8
4 → 7

22. 5 → 5
6 → 6
7 → 7

23. 100 → 1
200 → 2
300 → 3
400 → 4

24. $1 \to 1$
$2 \to 1$
$3 \to 1$
$4 \to 1$
$5 \to 1$

25. $-1 \to 0$
$-2 \to 0$
$-3 \to 0$
$-4 \to 1$
$-5 \to 1$
$-6 \to 1$

26. $1 \to 0$
$2 \to 4$
$3 \to 0$
$4 \to 1$
$5 \to 4$
$6 \to 0$
$7 \to 1$
$8 \to 1$
$9 \to 0$
$10 \to 1$

27. Determine two different functions with domain $\{1, 2\}$ and range $\{1, 2\}$.

28. Determine six different functions with domain $\{1, 2, 3\}$ and range $\{1, 2\}$.

29. How many functions are there with domain $\{1, 2, 3\}$ and range $\{1, 2, 3\}$?

8.2 EVALUATING FUNCTIONS

The functions considered up to now have involved relatively few match-ups. When there are a large number of or infinitely many match-ups, there is a need for better notation.

The letters f, g, h, F, G, H will stand for functions. Recall that a function is a correspondence between two sets—the domain and the range—such that to every object in the domain there corresponds exactly one object in the range.

DEFINITION. Let f be a function with domain A and range B. The objects in A are called the **arguments of** f. The object b in B that corresponds to a particular argument a is called the **value of** f **at** a.

Write

$$f(a) = b$$

and read this as

f of a equals b.

EXAMPLE 1. Suppose f is the function defined by the correspondence:

$$1 \to 10$$

$$2 \to 20$$

$$3 \to 30$$

$$4 \to 40$$

The domain is $\{1, 2, 3, 4\}$; the arguments of f are 1, 2, 3, and 4. The range is $\{10, 20, 30, 40\}$.

The value of f at 1 is 10; $f(1) = 10$.

The value of f at 2 is 20; $f(2) = 20$.

The value of f at 3 is 30; $f(3) = 30$.

The value of f at 4 is 40; $f(4) = 40$. ■

A function f is often defined by means of an equation. The equation specifies the value of f at an argument x. The left side of the equation is $f(x)$; the right side is an algebraic expression, such as a polynomial or a rational expression. A function defined by means of a polynomial is called a **polynomial function,** and a function defined by means of a rational expression is called a **rational function.**

EXAMPLE 2. Let A be the set of integers. Let f be the polynomial function defined by

$$f(x) = 2x + 5$$

for every x in A. Determine the following values of f:

(a) $f(0)$, (b) $f(1)$, (c) $f(2)$, (d) $f(-1)$, (e) $f(-2)$

SOLUTION. To determine each value, substitute the specified argument for x in the polynomial $2x + 5$.

(a) $$f(0) = 2 \cdot 0 + 5 = 5$$

(b) $$f(1) = 2 \cdot 1 + 5 = 7$$

(c) $$f(2) = 2 \cdot 2 + 5 = 9$$

(d) $$f(-1) = 2(-1) + 5 = 3$$

(e) $$f(-2) = 2(-2) + 5 = 1$$ ■

EXAMPLE 3. Let $\mathcal{R}$ be the set of real numbers. Let g be the rational function defined by

$$g(x) = \frac{x^3 - 1}{x^2 + 1}$$

for every x in $\mathcal{R}$. Determine the following values of g:

(a) $g(1)$, (b) $g(3)$, (c) $g(10)$, (d) $g(.1)$, (e) $g(-1)$

SOLUTION.

(a) $$g(1) = \frac{1^3 - 1}{1^2 + 1} = 0$$

(b) $$g(3) = \frac{3^3 - 1}{3^2 + 1} = \frac{27 - 1}{9 + 1} = \frac{26}{10} = \frac{13}{5}$$

(c) $$g(10) = \frac{10^3 - 1}{10^2 + 1} = \frac{1000 - 1}{100 + 1} = \frac{999}{101}$$

(d) $$g(.1) = \frac{(.1)^3 - 1}{(.1)^2 + 1} = \frac{.001 - 1}{.01 + 1} = \frac{-.999}{1.01} = \frac{-999}{1010}$$

(e) $$g(-1) = \frac{(-1)^3 - 1}{(-1)^2 + 1} = \frac{-1 - 1}{1 + 1} = \frac{-2}{2} = -1$$ ■

A rational function is not defined when the denominator of the defining rational expression is zero. In Example 3, the denominator, $x^2 + 1$, is positive for all x. The problem of a zero denominator never arose. In Example 4, which follows, the denominator is 0 when $x = 0$. Thus 0 is not in the domain of the function. (Recall that division by 0 is not defined.)

EXAMPLE 4. Let A be the set of nonzero real numbers. Let F be the rational function defined by

$$F(x) = \frac{x^2 - 3x + 5}{x}$$

for every x in A. Determine the following values of F:

(a) $F(2)$, (b) $F(4)$, (c) $F(5)$, (d) $F\left(\frac{1}{2}\right)$, (e) $F(-3)$

SOLUTION.

(a) $$F(2) = \frac{2^2 - 3(2) + 5}{2} = \frac{3}{2}$$

(b) $$F(4) = \frac{4^2 - 3(4) + 5}{4} = \frac{9}{4}$$

(c) $$F(5) = \frac{5^2 - 3(5) + 5}{5} = \frac{15}{5} = 3$$

(d) $$F\left(\frac{1}{2}\right) = \frac{\left(\frac{1}{2}\right)^2 - 3\left(\frac{1}{2}\right) + 5}{\frac{1}{2}} = \frac{\frac{1}{4} - \frac{3}{2} + 5}{\frac{1}{2}}$$

$$= \frac{1 - 3 \cdot 2 + 5 \cdot 4}{4} \div \frac{1}{2}$$

$$= \frac{15}{4} \cdot \frac{2}{1}$$

$$= \frac{15}{2}$$

(e) $$F(-3) = \frac{(-3)^2 - 3(-3) + 5}{-3} = \frac{9 + 9 + 5}{-3} = \frac{-23}{3}$$ ■

Functions are often indicated by means of verbal descriptions. Such functions arise in geometry and in applications to such fields as engineering, biology, and economics. You will frequently want to convert the verbal description into an algebraic description.

EXAMPLE 5. Part of a prairie is to be enclosed by a fence so as to form a rectangular region. The length of this rectangle must be three times the width. Determine the area of the enclosed region as a function of the width. What will be the area if the width of the fence is (a) 50 feet, (b) 200 feet? [See Figure 8.1.]

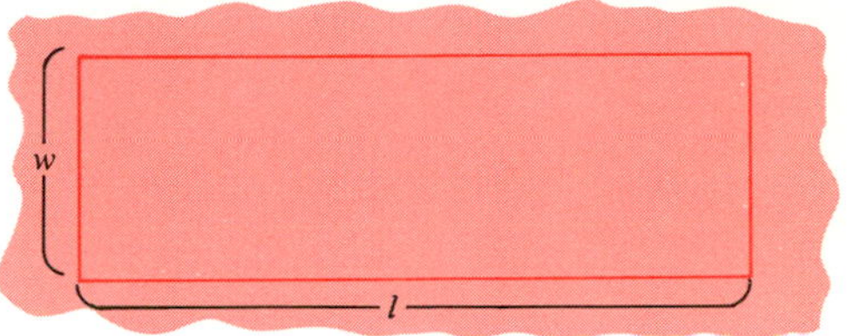

FIGURE 8.1 The length, l, of the fence is 3 times the width, w.

SOLUTION. The area A of a rectangle is given by

$$A = l \cdot w,$$

where l is the length and w the width. The length, l, of the rectangular fence is three times the width, w. Thus

$$l = 3w,$$

and

$$A = 3w \cdot w$$

Hence

$$A = 3w^2$$

(a) Let $w = 50$.

$$A = 3 \cdot 50^2 = 7500$$

The enclosed area is 7500 square feet.

(b) Let $w = 200$.

$$A = 3 \cdot 200^2 = 120\,000$$

The enclosed area is 120 000 square feet. ■

EXAMPLE 6. The profit derived from a novelty item is one-third of the cost, minus an initial expense of one hundred dollars. Determine the profit if the cost is (a) \$300, (b) \$3000, (c) \$21 000.

SOLUTION. Let c be the cost (in dollars). The profit, $f(c)$, is given by

$$f(c) = \frac{c}{3} - 100.$$

(a) $$f(300) = \frac{300}{3} - 100 = 0$$

(b) $$f(3000) = \frac{3000}{3} - 100 = 900$$

(c) $$f(21\,000) = \frac{21\,000}{3} - 100 = 6900$$

Thus a businessman would break even on a \$300 investment, and would derive profits of \$900 and \$6900 on investments of \$3000 and \$21 000, respectively. ■

EXERCISES

1. Suppose f is the function defined by the correspondence:

$$1 \to 0$$
$$2 \to 1$$
$$3 \to 0$$
$$4 \to 1$$

(a) What are the arguments of f? (b) Determine the value of f at 1. (c) Determine the value of f at 3.

2. Suppose g is the function defined by the correspondence:

$$\frac{1}{2} \to 1$$
$$\frac{2}{3} \to 2$$
$$\frac{3}{4} \to 3$$
$$\frac{4}{5} \to 4$$

(a) What are the arguments of g? (b) Determine the value of g at $\frac{2}{3}$. (c) Determine the value of g at $\frac{3}{4}$.

In Exercises 3–22 for each function determine the indicated function values.

3. $f(x) = 2x$

(a) $f(1)$ (b) $f(2)$ (c) $f(3)$ (d) $f(4)$

4. $g(x) = x + 2$
   (a) $g(2)$ (b) $g(5)$ (c) $g(10)$ (d) $g(18)$
5. $h(x) = x - 10$
   (a) $h(0)$ (b) $h(5)$ (c) $h(10)$ (d) $h(-10)$
6. $F(x) = 2x + 3$
   (a) $F(1)$ (b) $F(-1)$ (c) $F(3)$ (d) $F(-3)$
7. $G(x) = 5x - 1$
   (a) $G(0)$ (b) $G(5)$ (c) $G\left(\frac{1}{5}\right)$ (d) $G\left(\frac{2}{5}\right)$
8. $H(x) = 1 - 3x$
   (a) $H(1)$ (b) $H(0)$ (c) $H(-1)$ (d) $H\left(\frac{1}{9}\right)$
9. $f(x) = x^2 + 1$
   (a) $f(0)$ (b) $f(1)$ (c) $f(2)$ (d) $f(3)$
10. $g(x) = x^3 - 2$
   (a) $g(1)$ (b) $g(3)$ (c) $g(-1)$ (d) $g(-3)$
11. $h(t) = 10t + 4$
   (a) $h(5)$ (b) $h(10)$ (c) $h(100)$ (d) $h(-2)$
12. $F(u) = u^4 - u^2 + 1$
   (a) $F(0)$ (b) $F(1)$ (c) $F(-1)$ (d) $F(-2)$
13. $G(y) = y^3 - 2y^2 + 2y$
   (a) $G(0)$ (b) $G(-1)$ (c) $G\left(\frac{1}{2}\right)$ (d) $G\left(\frac{1}{4}\right)$
14. $H(z) = z^3 + 4z^2 - 3z + 5$
   (a) $H(1)$ (b) $H(.1)$ (c) $H(.2)$ (d) $H(10)$
15. $f(t) = \frac{1}{t}$
   (a) $f(1)$ (b) $f(4)$ (c) $f\left(\frac{1}{4}\right)$ (d) $f(.1)$
16. $f(x) = x^2 - \frac{1}{x^2}$
   (a) $f(1)$ (b) $f(2)$ (c) $f(-2)$ (d) $f\left(\frac{1}{2}\right)$
17. $g(u) = \frac{1}{u^2 - 2u - 5}$
   (a) $g(0)$ (b) $g(1)$ (c) $g(-1)$ (d) $g(5)$

18. $H(x) = \dfrac{x+1}{x-1}$

(a) $H(2)$ (b) $H(10)$ (c) $H(99)$ (d) $H(-1)$

19. $f(x) = \dfrac{x^4 - 3x^3}{x^2 + 5}$

(a) $f(0)$ (b) $f(1)$ (c) $f(-1)$ (d) $f(-2)$

20. $H(x) = 4$

(a) $H(1)$ (b) $H(2)$ (c) $H(4)$ (d) $H(-4)$

21. $f(z) = \dfrac{-1}{2}$

(a) $f\left(\dfrac{1}{2}\right)$ (b) $f(0)$ (c) $f\left(\dfrac{-1}{2}\right)$ (d) $f(100)$

22. $g(t) = (t-1)(t-4)$

(a) $g(0)$ (b) $g(1)$ (c) $g(-1)$ (d) $g(4)$

23. (a) Determine the area of a square as a function of s, the length of a side.

(b) Determine the area if $s = 9$.

(c) Determine the area if $s = \dfrac{1}{9}$.

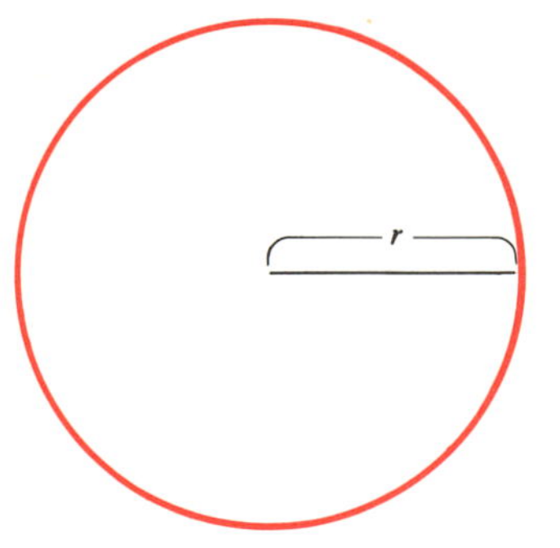

FIGURE 8.2

24. The area of a circle is π times the square of r, the length of the radius. [See Figure 8.2.]

(a) Determine the area of a circle as a function of r.

(b) Determine the area if $r = 1$.

(c) Determine the area if $r = 7$.

25. The area of a triangle is one-half b, the length of the base, times h, the length of the corresponding altitude. Suppose the base length is two-thirds that of the altitude. Determine the area of the triangle as a function of

(a) h, (b) b.

(c) Determine the area when $h = 18$.

(d) Determine the area when $b = 18$.

26. Water is pouring into a reservoir at the rate of 100 cubic feet per minute.

(a) Determine the volume of the water as a function of t, the number of minutes.

(b) What is the volume of the water after 10 minutes?

(c) What is the volume of the water after an hour?

27. Harry is five years older than his sister Harriet.

    (a) Determine Harry's age as a function of Harriet's age, h.

    (b) Determine Harriet's age as a function of Harry's age, H.

28. An airplane rises at the rate of $\frac{t^2}{5}$ feet per second.

    (a) How high is the plane after 5 seconds?

    (b) How high is the plane after half a minute?

29. An automobile salesman sells 6 cars his first week on the job. Each week thereafter, he sells 4 cars. Let $t = 1, 2, \ldots$ be the number of weeks he has been working.

    (a) Determine the number of cars he has sold as a function of t.

    (b) How many cars has he sold after 52 weeks?

30. A ball is thrown directly upward. Its height, in feet, after t seconds is given by $128t - 16t^2$. Determine the height after

    (a) 2 seconds, (b) 4 seconds, (c) 6 seconds,

    (d) 8 seconds. (e) What should be the domain of this function?

In Exercises 31–34 determine the largest possible domain of the indicated functions. [*Hint:* First find the values for which the denominator is 0.]

31. $f(x) = \dfrac{x + 3}{x - 1}$ 32. $g(x) = \dfrac{x + 2}{x + 1}$

33. $h(x) = \dfrac{1}{x^2}$ 34. $F(t) = \dfrac{1}{t^2 + 4}$

8.3 CARTESIAN COORDINATES

In order to picture functions geometrically, it is necessary to introduce a **Cartesian** (or **rectangular**) **coordinate system** in the plane.

In Chapter 1 you saw how real numbers correspond to points on a horizontal line L. Through the origin on L draw another line perpendicular to L. This second line is thus vertical. [See Figure 8.3.] From now on, call the horizontal line the **x-axis** and the vertical line the **y-axis.** Together the two lines are called the **coordinate axes.** The **origin,** which is the intersection of the coordinate axes, will also represent 0 on the y-axis.

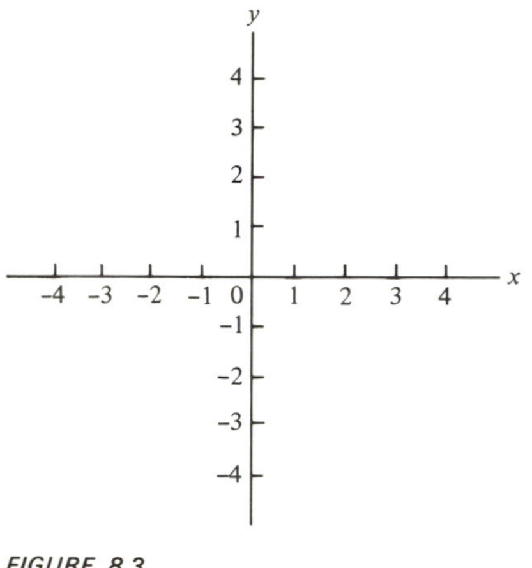

FIGURE 8.3

There is already a distance unit on the x-axis in terms of which positive numbers are marked off to the right of 0 and negative numbers are marked off to the left of 0. For convenience, choose the same distance unit on the y-axis. Mark off positive numbers *upward from* 0 and negative numbers *downward,* as in Figure 8.3.

Every point on the x-axis represents a real number. So too, every point on the y-axis represents a real number. Moreover, every real number corresponds to a point on each of these coordinate axes. Just as you identified points on the x-axis with the numbers they represent, so too, will you identify points on the y-axis with the numbers they represent.

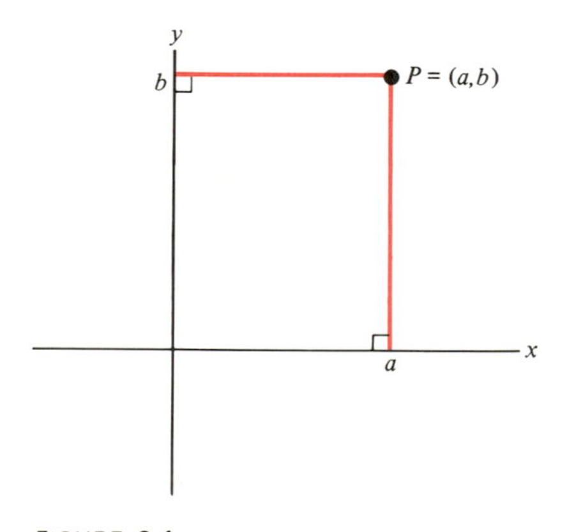

FIGURE 8.4

Points in the plane can be associated with ***pairs*** of real numbers, ***in a definite order***. Let P be a point in the plane. Draw a perpendicular from P to each of the coordinate axes (as in Figure 8.4). Let the vertical line from P intersect the x-axis where $x = a$, and let the horizontal line from P intersect the y-axis where $y = b$. The point P is thereby associated with the numbers a and b, *in this order*. To indicate the ordering of a and b, write

$$(a, b).$$

Call (a, b) an **ordered pair.** Thus an ordered pair (a, b) indicates two numbers a and b, *in the order written.* The number a is called the **first coordinate** or **x-coordinate** of P; b is called the **second coordinate** or **y-coordinate** of P. You say that the coordinates of P are (a, b) and write

$$P = (a, b).$$

[See Figure 8.4.]

Starting with an ordered pair (c, d) of numbers, you can also associate a unique point Q on the plane. Locate c on the x-axis; locate d on the y-axis. Draw perpendiculars through each of these points. The intersection of these perpendiculars is the point Q associated with the ordered pair (c, d).

EXAMPLE 1. Several points of the plane are represented in Figure 8.5. Their coordinates are indicated below Figure 8.5. ■

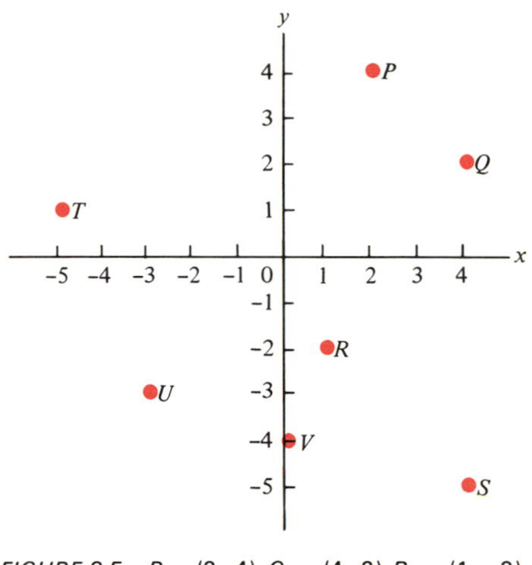

FIGURE 8.5 $P = (2, 4)$, $Q = (4, 2)$ $R = (1, -2)$, $S = (4, -5)$, $T = (-5, 1)$, $U = (-3, -3)$, $V = (0, -4)$

EXAMPLE 2. Plot the following points on a Cartesian coordinate system in the plane:

(a) $P = (4, 6)$ (b) $Q = \left(\frac{1}{2}, -3\right)$, (c) $R = \left(0, \frac{-1}{2}\right)$,

(d) $S = (-6.5, -1.5)$, (e) $T = (-5, 0)$

SOLUTION. See Figure 8.6. ■

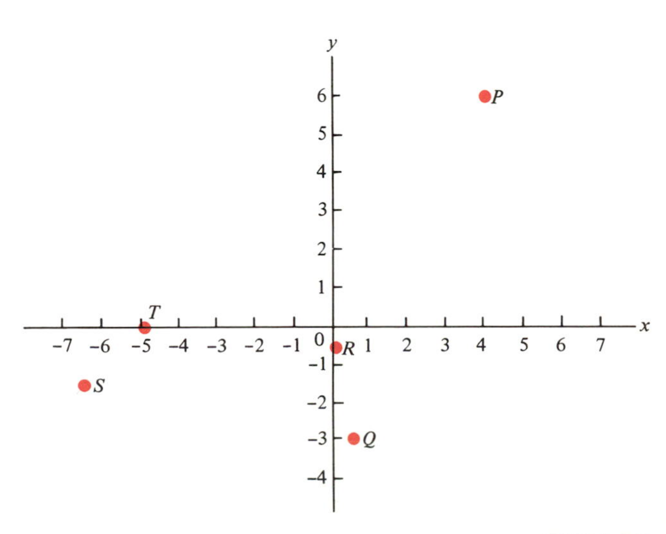

FIGURE 8.6

It is important to observe that *two ordered pairs are equal only when they agree in both coordinates.* Agreement at a single coordinate is not enough. For example,

$$(4, 5) \neq (4, 6)$$

These ordered pairs represent different points of the plane. [See Figure 8.7 (a).] Also

$$(3, 1) \neq (4, 1);$$

these ordered pairs represent different points. [See Figure 8.7 (b).] Finally,

$$(2, 3) \neq (3, 2);$$

the coordinates are reversed. The first coordinates do not agree ($2 \neq 3$); nor do the second coordinates agree. [See Figure 8.7 (c).]

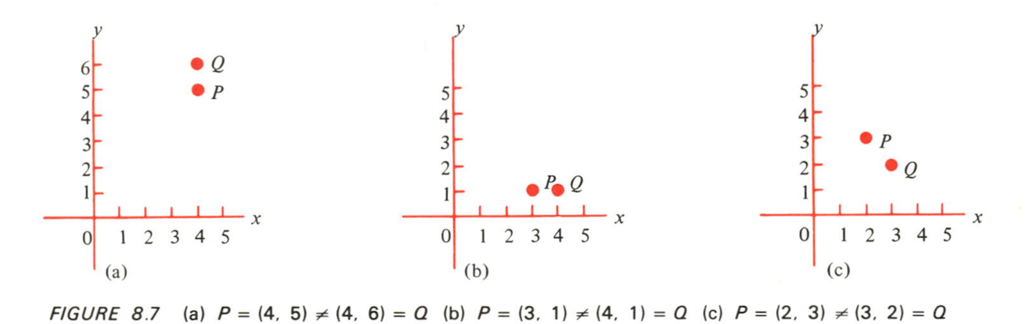

FIGURE 8.7 (a) $P = (4, 5) \neq (4, 6) = Q$ (b) $P = (3, 1) \neq (4, 1) = Q$ (c) $P = (2, 3) \neq (3, 2) = Q$

The coordinate axes divide the plane into 4 different regions, called **quadrants.** These quadrants are numbered in a counterclockwise direction, as indicated in Figure 8.8. *Every point that is not on a coordinate axis* (that is, for which *neither* coordinate is zero) *lies in exactly one quadrant.* The signs of the coordinates determine the quadrants, as indicated in Table 8.1.

Quadrant	x	y
I	+	+
II	−	+
III	−	−
IV	+	−

TABLE *8.1*

You will see references to "the first quadrant" (meaning quadrant I), "the second quadrant" (quadrant II), and so on.

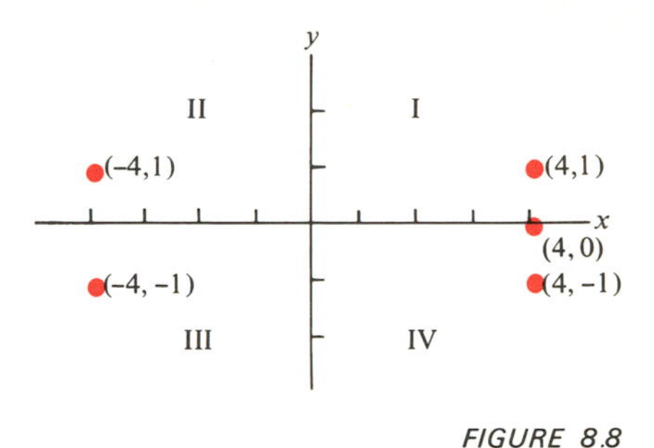

FIGURE 8.8

EXAMPLE 3. [Refer to Figure 8.8.]

(a) (4, 1) lies in the first quadrant. Both coordinates are positive.
(b) (−4, 1) lies in the second quadrant. Here the x-coordinate is negative and the y-coordinate is positive.
(c) (−4, −1) lies in the third quadrant. Both coordinates are negative.
(d) (4, −1) lies in the fourth quadrant. The x-coordinate is positive and the y-coordinate is negative.
(e) (4, 0) does not lie in any quadrant, for one of its coordinates is zero. This point is on the x-axis. ■

EXERCISES

1. In Figure 8.9, determine the coordinates of each point.

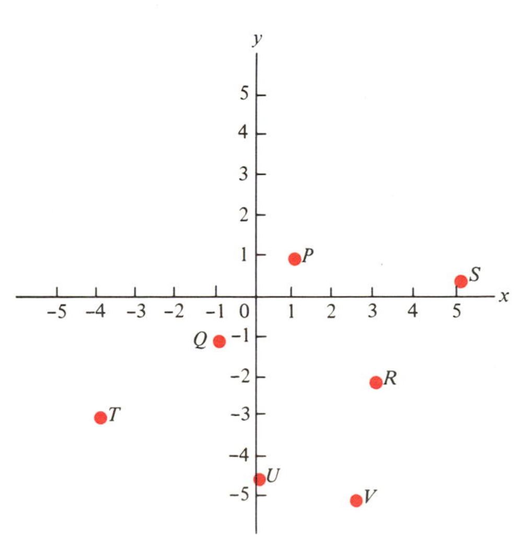

FIGURE 8.9

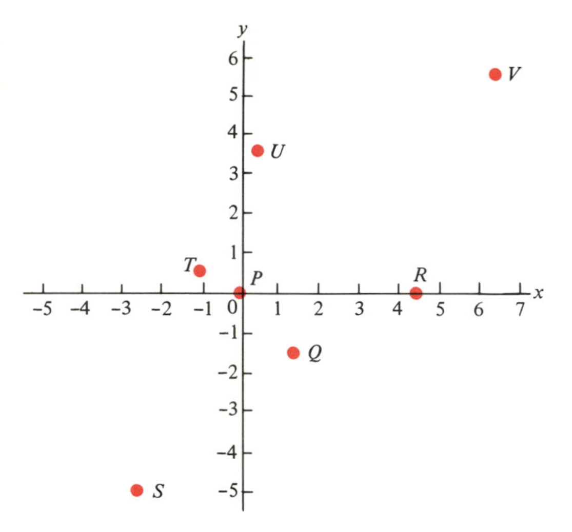

FIGURE 8.10

2. In Figure 8.10, determine the coordinates of each point.

On a coordinate system plot the following points:

3. $(3, 1)$
4. $(5, 4)$
5. $(-4, 2)$
6. $(0, 4)$
7. $(1, -3)$
8. $(-2, 1)$
9. $(-4, -3)$
10. $(6, -1)$
11. $(-3.5, 0)$
12. $\left(\frac{1}{2}, \frac{-1}{2}\right)$
13. $(-1, 0)$
14. $(0, 0)$
15. $(-4.5, 4.5)$
16. $(-3, -1.5)$
17. $(1.5, -2.5)$
18. $\left(\frac{-3}{2}, \frac{-1}{2}\right)$
19. $(-6, -3.5)$
20. $(4.5, -.5)$

In Exercises 21–34 determine the quadrant of each point, or indicate that the point lies on a coordinate axis.

21. $(1, -1)$
22. $(-2, 3)$
23. $(0, 4)$
24. $(10, -1)$
25. $(-3, -7)$
26. $(-1.3, 1.5)$
27. $(6, 0)$
28. $\left(0, \frac{1}{2}\right)$
29. $\left(\frac{1}{4}, \frac{-1}{3}\right)$
30. $(0, 0)$
31. $\left(\frac{1}{2}, \frac{2}{5}\right)$
32. $(-7, -4)$
33. $(4, 7)$
34. $(-3, .8)$

8.4 GRAPHS

Recall that a function is a correspondence between two sets, called the domain and range, such that to each object in the domain there corresponds exactly one object in the range. For example,

$$1 \rightarrow 4$$

$$2 \rightarrow 6$$

$$3 \rightarrow 10$$

describes a function whose domain is $\{1, 2, 3\}$ and whose range is $\{4, 6, 10\}$. You can regard this correspondence as yielding the ordered pairs (1, 4), (2, 6), and (3, 10). Thus this particular function may be described by the *set* of ordered pairs

$$\{(1, 4), (2, 6), (3, 10)\}.$$

Ordered pairs of numbers correspond to points of the plane. You can represent a function geometrically by locating the points corresponding to the ordered pairs of the function.

DEFINITION. The **graph of a function** f is the pictorial representation of the function on the plane. The graph consists of all points (x, y) such that $f(x) = y$.

If a function consists of finitely many ordered pairs, to graph the function, simply plot the corresponding points of the plane.

EXAMPLE 1. Graph the function given by:

$$\{(1, 2), (2, 4), (3, 1), (4, 5)\}$$

SOLUTION. See Figure 8.11. ■

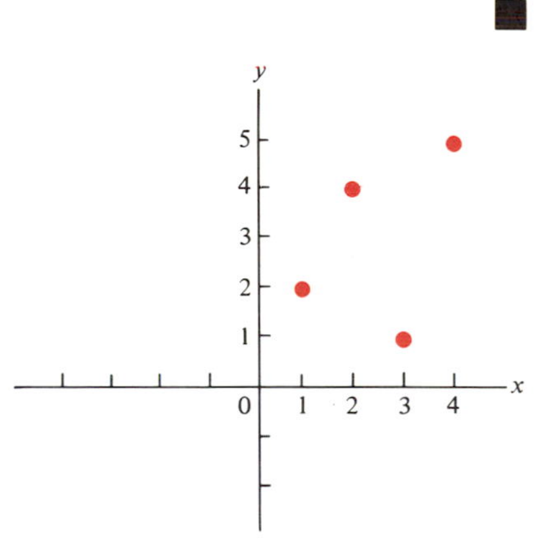

FIGURE 8.11

A function can also be given by an equation in two variables,

$$y = f(x).$$

When this is the case:

1. Assign several numerical values to x.
2. Compute the corresponding values of y.
3. Plot the points (x, y).
4. Join these points by a smooth curve. (This curve may, in fact, be a line, as in Examples 2 and 3.)

This procedure will yield the graph of the function provided the function is "relatively simple" and the numerical values in step 1 are "well-chosen". The next three examples illustrate the procedure.

In later mathematics courses, particularly in calculus, you will study more advanced methods of graphing functions.

When nothing further is stated, assume the domain is the set of all real numbers.

EXAMPLE 2. Graph the function given by:

$$f(x) = 3x + 2$$

SOLUTION. Let

$$y = f(x) = 3x + 2.$$

The numerical values chosen for x and the corresponding values of y are given in Table 8.2.

x	-2	-1	0	1	2
$y = 3x + 2$	-4	-1	2	5	8

TABLE *8.2*

These points (x, y) are plotted in Figure 8.12 (a); the "smooth curve" that joins them, in Figure 8.12 (b), is a (straight) line. ■

The graph of a function defined by an equation

$$\text{(A)} \qquad y = mx + b,$$

where m and b are any real numbers, is always a line. In Example 2,

$$m = 3, \qquad b = 2$$

(The significance of m and b will be explained in the next chapter, where you will study lines.) *A line is determined by two points.* Thus when a function of the form (A) is given, you will recognize that its graph is a line. You need only plot two points and draw the line that passes through them. (You may also wish to plot a third point as a check. If your plotting is accurate, a straight line passes through the three points.)

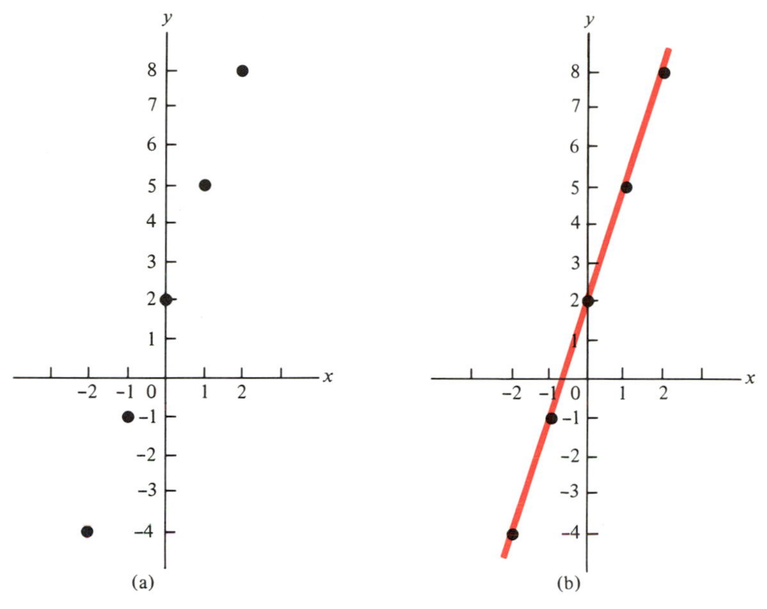

FIGURE 8.12 (b) The graph of $y = 3x + 2$

EXAMPLE 3. Graph the function g given by:

$$g(x) = 3 - x$$

SOLUTION. Let $y = g(x)$. The defining equation for g can be rewritten as

$$y = -x + 3.$$

This is the equation of a line; here

$$m = -1, \qquad b = 3.$$

You need only plot two points; however, a third point is given as a check. [See Table 8.3.]

x	0	1	2
$y = 3 - x$	3	2	1

TABLE *8.3*

These points are plotted in Figure 8.13 (a) on page 242; the line connecting them is shown in Figure 8.13 (b). ■

EXAMPLE 4. Graph the function F given by

$$F(x) = x^2.$$

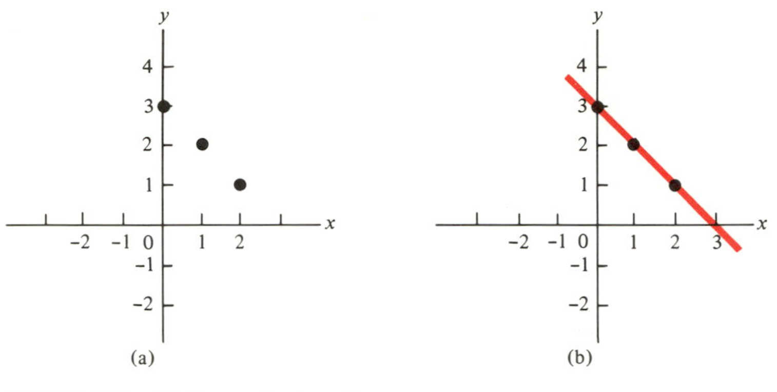

FIGURE 8.13 (b) The graph of $y = 3 - x$

SOLUTION. Let $y = F(x)$. The defining equation is *not* of the form $y = mx + b$. You must plot several points to determine the graph. It will be important to consider fractional values of x as well as integral values. (See Table 8.4.). Note that

$$\left(\frac{1}{4}\right)^2 = \frac{1}{4} \cdot \frac{1}{4} = \frac{1}{16}, \qquad \left(\frac{-1}{4}\right)^2 = \frac{(-1)}{4} \cdot \frac{(-1)}{4} = \frac{1}{16},$$

$$\left(\frac{1}{2}\right)^2 = \frac{1}{2} \cdot \frac{1}{2} = \frac{1}{4}, \qquad \left(\frac{3}{4}\right)^2 = \frac{3}{4} \cdot \frac{3}{4} = \frac{9}{16}.$$

x	0	$\frac{1}{4}$	$\frac{-1}{4}$	$\frac{1}{2}$	$\frac{-1}{2}$	$\frac{3}{4}$	$\frac{-3}{4}$	1	-1	2	-2
$y = x^2$	0	$\frac{1}{16}$	$\frac{1}{16}$	$\frac{1}{4}$	$\frac{1}{4}$	$\frac{9}{16}$	$\frac{9}{16}$	1	1	4	4

TABLE *8.4*

These points are plotted in Figure 8.14 (a); a smooth curve is drawn connecting them in Figure 8.14 (b).

As you see, $F(1) = F(-1)$ and $F(2) = F(-2)$. In general,

$$\begin{aligned} F(-x) &= (-x)^2 \\ &= (-x)(-x) \\ &= x^2 \\ &= F(x) \end{aligned}$$

Thus whenever (x, y) is on the graph, so is $(-x, y)$. The graph is symmetric with respect to the y-axis. This curve is known as a **parabola.** The lowest point on this parabola, (0, 0), is called the **vertex** of the parabola. You will learn more about parabolas in Chapter 15.

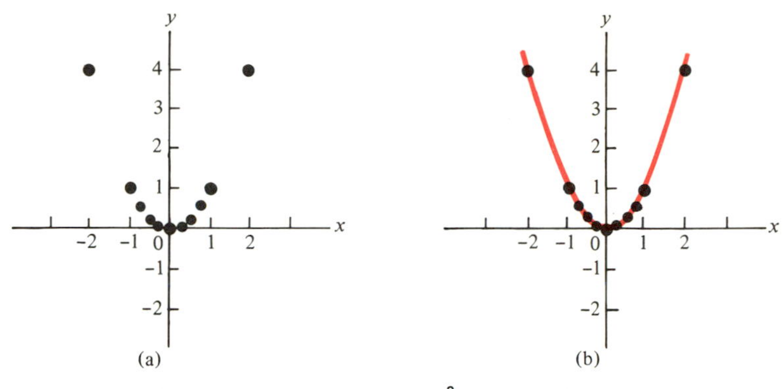

FIGURE 8.14 (b) The parabola given by $y = x^2$ with vertex (0, 0)

The definition of a function requires that

to each argument there corresponds exactly one value.

A single argument a cannot have two different values corresponding to it; you cannot have both

$$a \rightarrow b$$

$$a \rightarrow c,$$

that is, you cannot have both

$$(a, b)$$

and

$$(a, c)$$

on the graph.

Thus, two points with the same first coordinate cannot lie on the graph of a function. Because two points with the same first coordinate determine a vertical line, it follows that *a vertical line cannot intersect the graph of a function more than once.*

EXAMPLE 5.

$$\{(1, 4), (2, 2), (1, 5)\}$$

does not represent a function because two distinct ordered pairs have the same first coordinate. In Figure 8.15, the vertical line V intersects the graph twice—at P and at R. ■

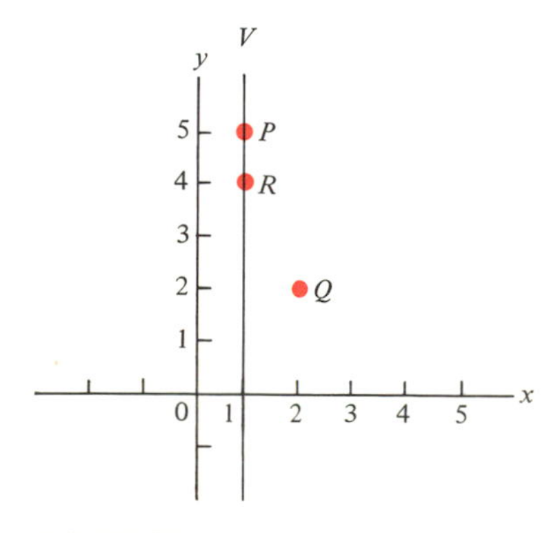

FIGURE 8.15

You can immediately tell whether a curve on the plane represents the graph of *some* function.

1. If *every* vertical line intersects the curve *at most once*, the curve is the graph of a function.

2. If there is *at least one* vertical line that intersects the curve *more than once*, the curve is *not* the graph of a function.

EXAMPLE 6. In Figure 8.16, curves (a), (b), and (c) are graphs of functions; curves (d), (e), and (f) are not. In each of the graphs (a), (b), and (c), every vertical line intersects the curve at most once. In each of the graphs (d), (e), and (f), there is at least one vertical line that intersects the curve more than once. Notice that in (e) the curve is itself a vertical line, V. Thus a *vertical line is not the graph of a function.* However a *horizontal line* [Figure 8.16(b)] is *the graph of a function.*

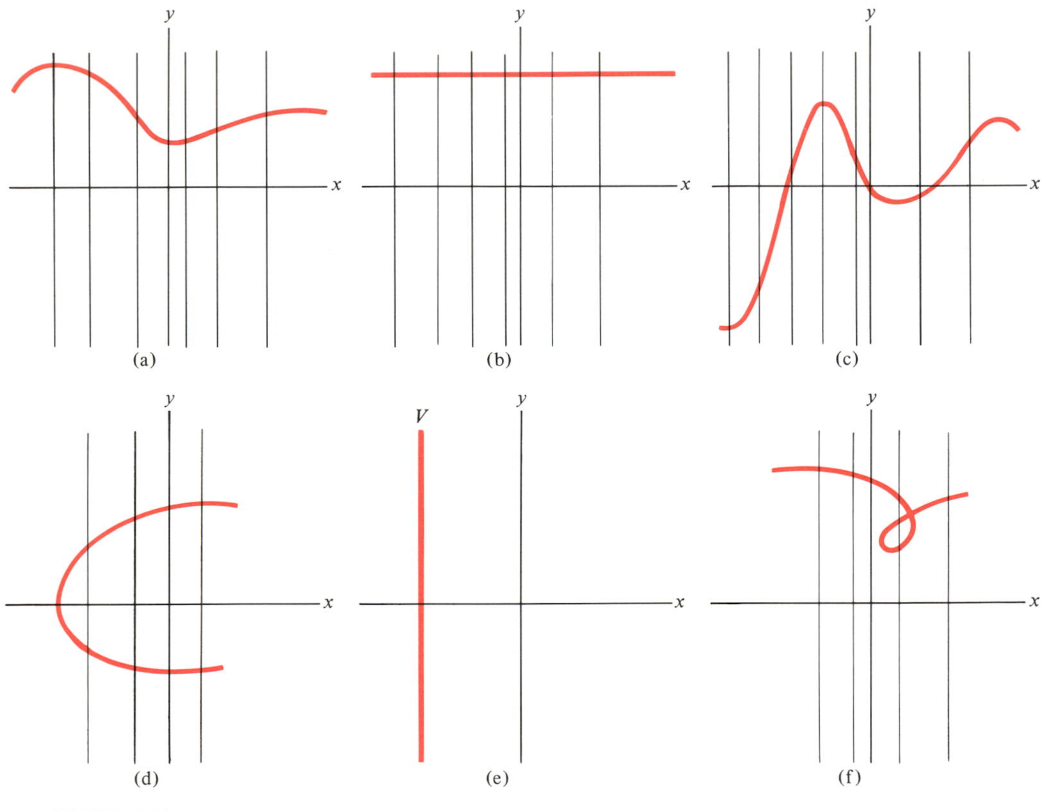

FIGURE 8.16

EXERCISES

Graph each of the functions indicated in Exercises 1–10.

1. $\{(1, 3), (2, 4), (3, 5), (4, 6)\}$
2. $\{(1, 1), (2, 2), (3, 3), (4, 4), (5, 5)\}$

3. $\{(1, -1), (2, -2), (3, -3), (4, -4), (5, -5)\}$
4. $\{(-1, 1), (-2, 2), (-3, 3), (-4, 4), (-5, 5)\}$
5. $\{(-1, -1), (-2, -2), (-3, -3), (-4, -4), (-5, -5)\}$
6. $\{(1, 6), (2, 5), (3, 4), (4, 3), (5, 2), (6, 1)\}$
7. $\{(1, 5), (2, 4), (3, 3), (4, 2), (5, 1), (6, 0), (7, -1), (8, -2), (9, -3), (10, -4)\}$
8. $\{(-3, 2), (0, 1), (1, 4), (4, 0)\}$
9. $\{(2, 6), (4, 2), (6, -1)\}$
10. $\{(3, -2)\}$

In Exercises 11–17 (Figure. 8.17) determine which diagrams represent graphs of functions.

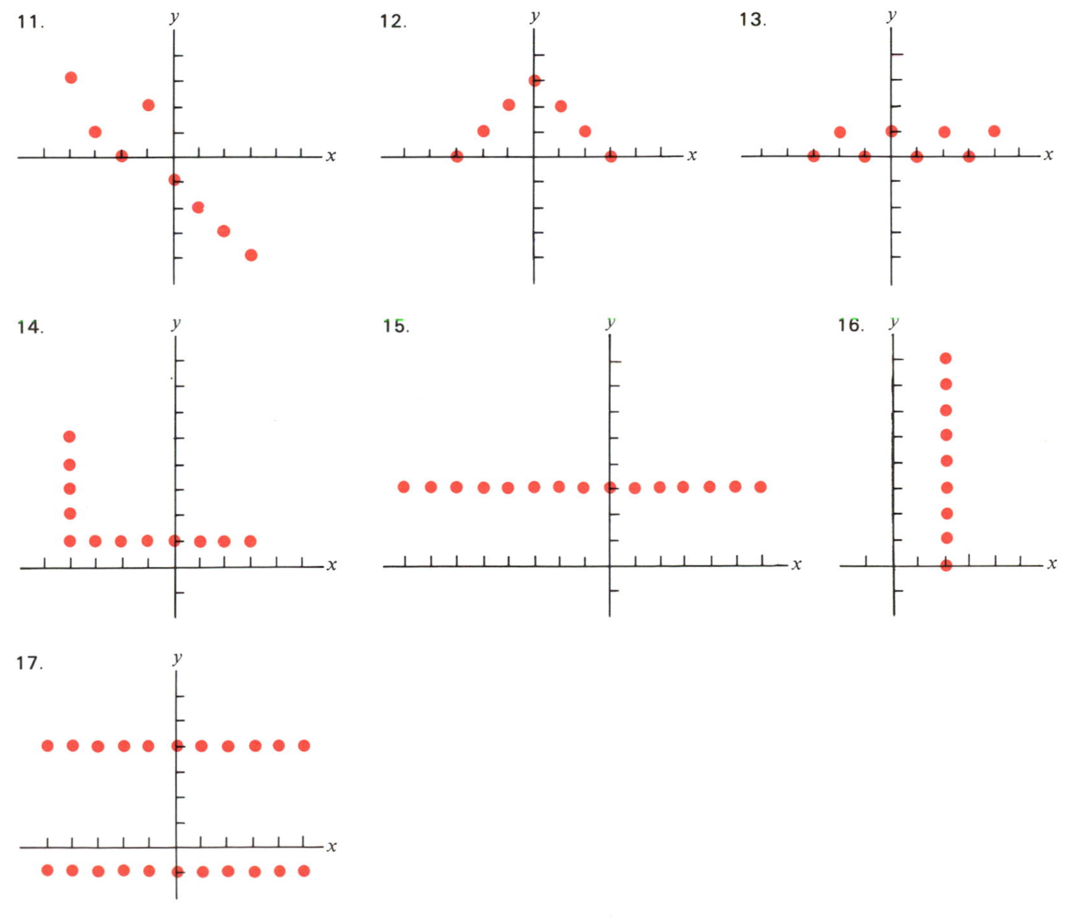

FIGURE 8.17

In Exercises 18–36 graph the function defined by the given equation.

18. $y = 5x$
19. $y = -x$
20. $y = -3x$
21. $y = x + 1$
22. $y = x + 4$
23. $y = x - 2$
24. $y = 2x - 1$
25. $y = 3x + 4$
26. $y = 5 - 2x$
27. $y = 1 - 4x$
28. $y = 6$
29. $y = 0$
30. $y = -2$
31. $y = 2x^2$
32. $y = x^2 + 5$
33. $y = 2 - x^2$
34. $y = 3x^2 - 5$
35. $y = (x - 1)^2$
36. $y = (x + 2)^2 + 1$

In Exercises 37–48 (Figure 8.18) determine which diagrams represent graphs of functions.

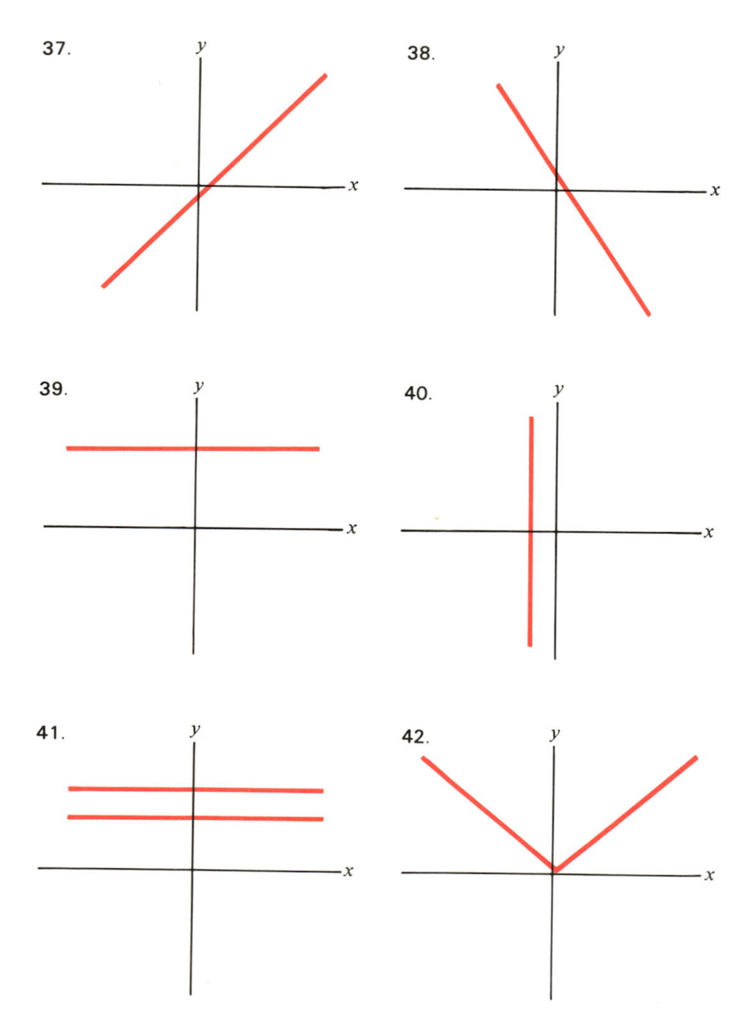

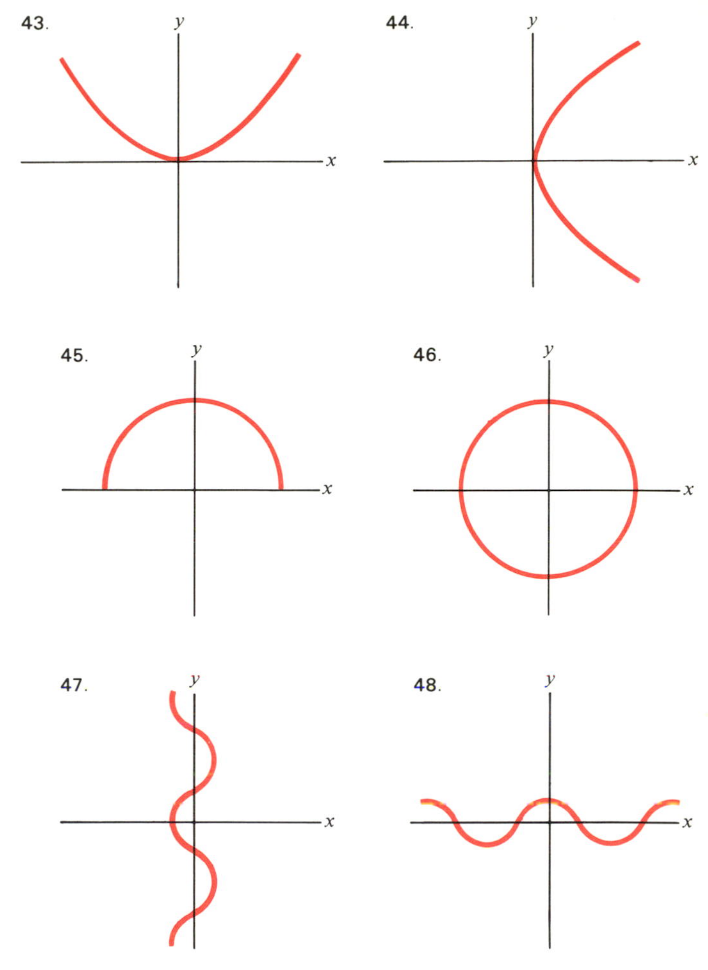

FIGURE 8.18

*8.5 INVERSE OF A FUNCTION

When a function is represented by a set of ordered pairs, distinct ordered pairs have distinct *first* coordinates. Every *vertical* line in the plane intersects the graph of a function at most once.

DEFINITION. A function is said to be **one–one** if its distinct ordered pairs have distinct *second* coordinates (as well as distinct first coordinates).

Thus a function is one–one if distinct arguments (first coordinates) have distinct function values (second coordinates).

*Optional topic.

Geometrically, every *horizontal* line in the plane (as well as every vertical line) intersects the graph of a one–one function at most once. [See Figure 8.19.]

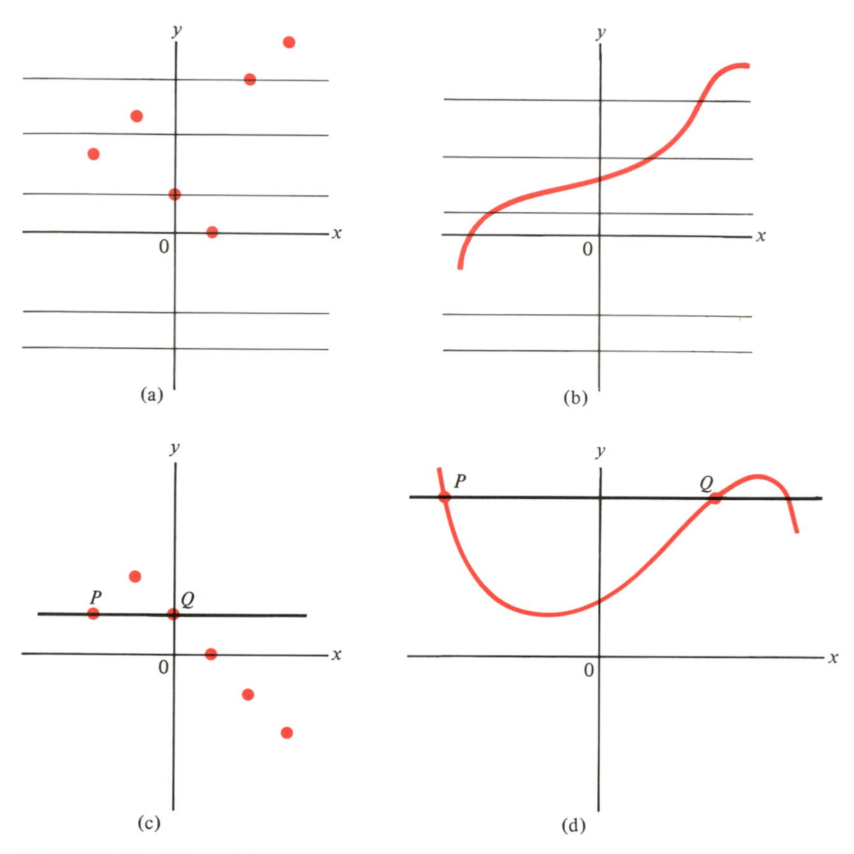

FIGURE 8.19 (a) and (b) each depict *one-one functions*. In each case distinct points have distinct second coordinates. Every horizontal line cuts the graph at most once. (c) and (d) each represent functions that are *not* one-one. In each graph there are 2 points P and Q with the same second coordinate. There is at least one horizontal line that cuts the graph more than once.

EXAMPLE 1. Let F be given by:

$$\{(1, 4), (2, 3), (3, 2), (4, 1)\}$$

(See Figure 8.20.) Distinct ordered pairs have distinct *first* coordinates. Thus *F is a function*. Furthermore, distinct ordered pairs have distinct *second* coordinates. Thus *the function F is one–one*. ■

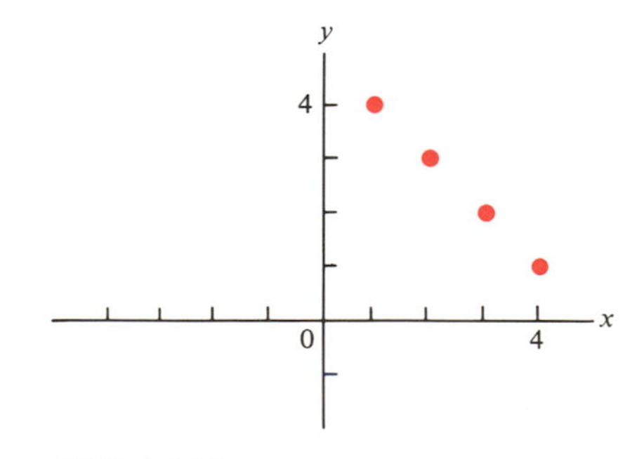

FIGURE 8.20

EXAMPLE 2. Let G be given by:

$$\{(1, 5), (2, 0), (3, 5)\}$$

(See Figure 8.21.) Distinct ordered pairs have distinct *first* coordinates. Thus *G is a function.* However, (1, 5) and (3, 5) have the same *second* coordinate, 5. Therefore, *the function G is not one–one.* ■

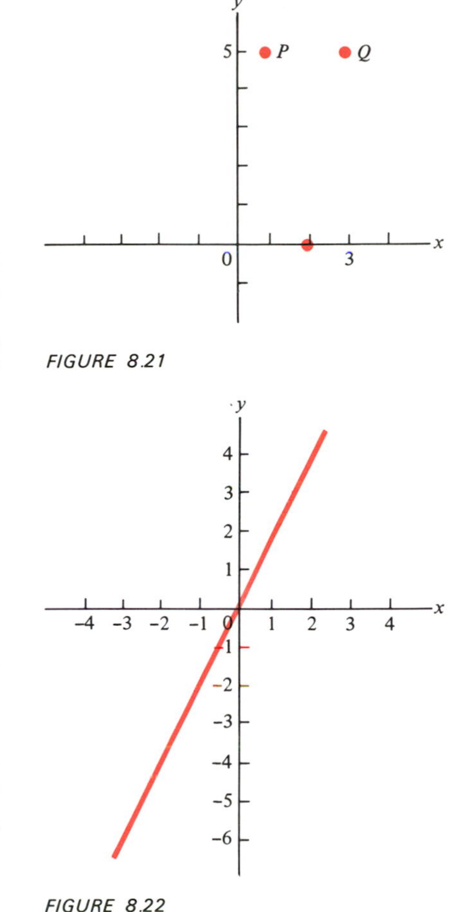

FIGURE 8.21

FIGURE 8.22

EXAMPLE 3. Let f be given by:

$$f(x) = 2x$$

(See Figure 8.22.) Show that the function f is one–one.

SOLUTION. Let x_1 and x_2 be *distinct* values of x. Can $(x_1, f(x_1))$ and $(x_2, f(x_2))$ have the same second coordinates, that is, can

$$f(x_1) = f(x_2)?$$

Here

$$f(x) = 2x$$

In this case, if

$$f(x_1) = f(x_2),$$

then

$$2x_1 = 2x_2$$

Divide both sides by 2.

and

$$x_1 = x_2$$

But x_1 and x_2 are *distinct*. $[x_1 \neq x_2]$ This is a contradiction! Therefore

$$f(x_1) \neq f(x_2),$$

and distinct ordered pairs,

$$(x_1, f(x_1)) \quad \text{and} \quad (x_2, f(x_2)),$$

have distinct second coordinates.

The notion of a one–one function is closely related to the concept of the "inverse of a function".

EXAMPLE 4. Let g be given by:

$$\{(1, 4),\ (3, 2),\ (5, 10)\}$$

Interchange the coordinates of each ordered pair of g.

Instead of:	write:
(1, 4)	(4, 1)
(3, 2)	(2, 3)
(5, 10)	(10, 5)

Note that

$$\{(4, 1),\ (2, 3),\ (10, 5)\}$$

inverts the original correspondence:

from $1 \rightarrow 4$, $3 \rightarrow 2$, $5 \rightarrow 10$ to $1 \leftarrow 4$, $3 \leftarrow 2$, $5 \leftarrow 10$ ■

DEFINITION. Let f be a one–one function. Interchange the coordinates of each ordered pair of f. The resulting correspondence is called the **inverse of** f.

Write

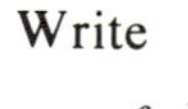

$$f^{-1}$$

for "the inverse of f".

If

$$(a, b)$$

is an ordered pair of f, then

$$(b, a)$$

is an ordered pair of f^{-1}.

Geometrically, *the point* (b, a) *is the reflection of the point* (a, b) *in the graph of the line* $y = x$. [See Figure 8.23.]

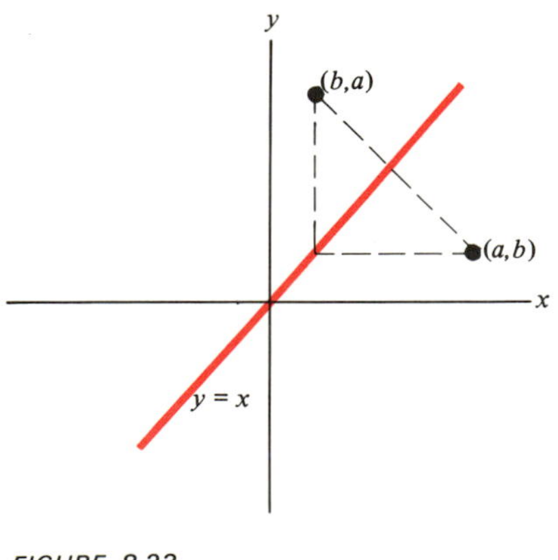

FIGURE 8.23

If a one–one function is given by an equation in x and y, to find its inverse:

1. Interchange the roles of x and y.
2. Solve for y in terms of x.

(In the remaining examples the verifications that the functions are one–one are omitted.)

EXAMPLE 5. Let f be given by:

$$f(x) = 3x + 2$$

Determine f^{-1}.

SOLUTION. Let $y = f(x)$. Then

$$y = 3x + 2$$

First interchange x and y in the defining equation:

$$x = 3y + 2$$

Now solve for y in terms of x:

$$x - 2 = 3y$$

$$\frac{x - 2}{3} = y$$

Thus f^{-1} is given by

$$f^{-1}(x) = \frac{x - 2}{3}.$$

■

It can be shown that *the inverse of a one–one function is itself a one–one function*. Moreover,

the inverse of the inverse of a function is the function itself:

$$(f^{-1})^{-1} = f$$

Observe that if you interchange the ordered pairs of the function twice, you get back the original ordered pairs:

$$(a, b) \rightarrow (b, a) \rightarrow (a, b)$$

EXAMPLE 6. Let f be given by:

$$\{(1, 6), (3, 9), (5, 0), (7, 7)\}$$

Then f^{-1} is given by:

$$\{(6, 1), (9, 3), (0, 5), (7, 7)\}$$

and $(f^{-1})^{-1}$ is given by:

$$\{(1, 6), (3, 9), (5, 0), (7, 7)\}$$

Thus

$$(f^{-1})^{-1} = f$$

■

EXERCISES

In Exercises 1–8 (Figure 8.24) which diagrams represent one–one functions?

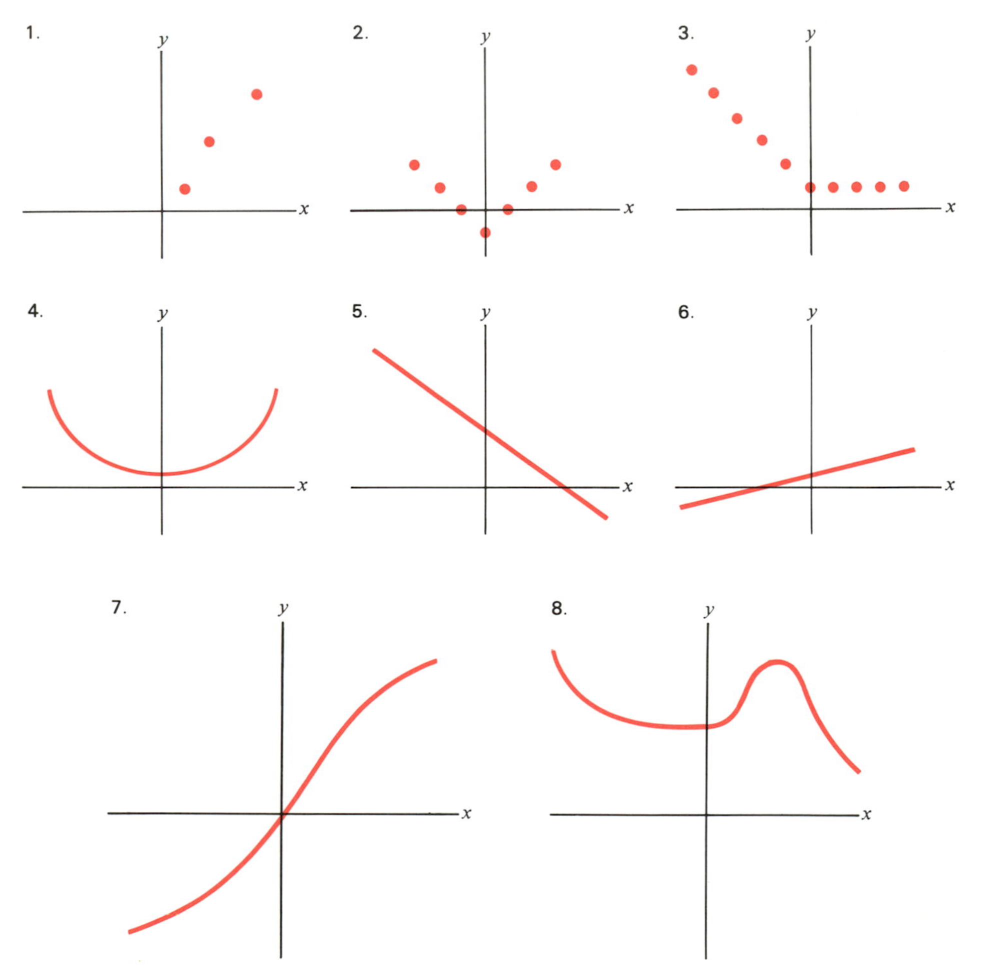

FIGURE 8.24

In Exercises 9–14 which of the indicated functions are one–one?

9. $\{(1, 4), (2, 5), (3, 6), (4, 7), (5, 8)\}$
10. $\{(1, 10), (2, 8), (3, 6), (4, 2), (5, 0)\}$
11. $\{(1, 1), (2, 2), (3, 3), (4, 4), (5, 5)\}$
12. $\{(1, 0), (2, 1), (3, 1), (4, 2)\}$

13. $\{(0, 1), (10, 2), (20, 1), (30, 2)\}$

14. $\{(-6, 2), (-5, 2), (-4, 3), (-3, 12), (-2, 0), (-1, 17), (0, 10), (1, 30)\}$

In Exercises 15–18 show that the function defined by each equation is *not* one–one. The domain of each function is the set of all real numbers. [*Hint:* For each function, find two arguments that have the same function values.]

15. $f(x) = 4$ 16. $f(x) = x^2$ 17. $f(x) = -2x^2$ 18. $f(x) = 3x^2 + 1$

19. In Figure 8.25, match the graphs of functions that are inverses of one another.

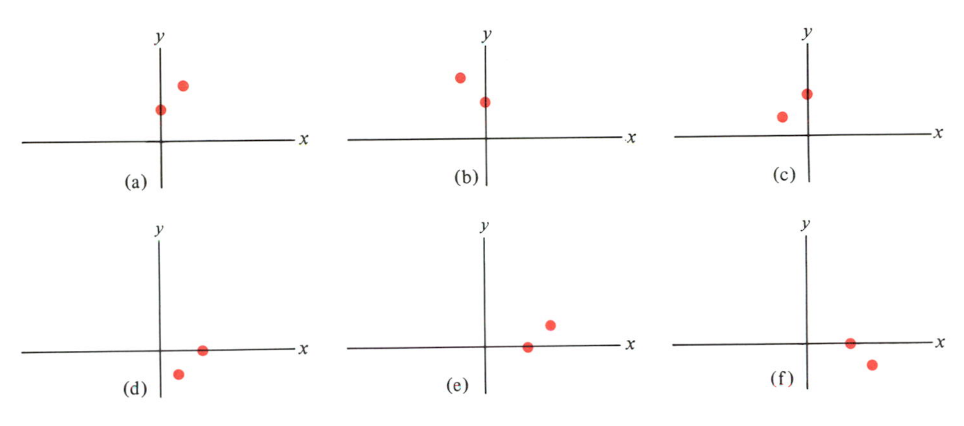

FIGURE 8.25

20. In Figure 8.26, match the graphs of functions that are inverses of one another.

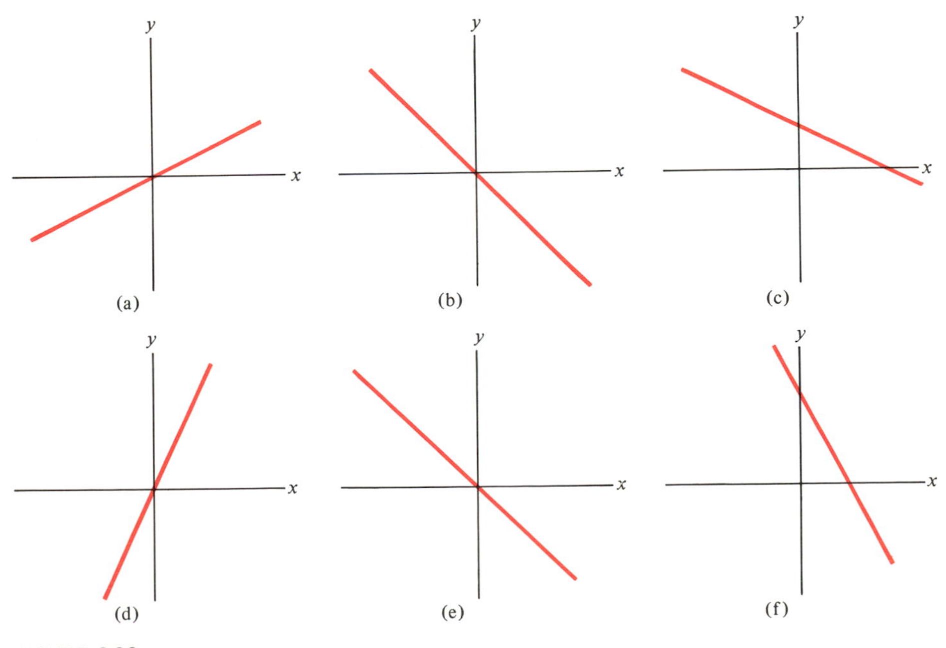

FIGURE 8.26

In Exercises 21–38 determine the inverse of the indicated one–one function.

21. $\{(1, 2), (2, 4), (3, 8)\}$

22. $\{(1, 10), (2, 9), (3, 8), (4, 7), (5, 6)\}$

23. $\{(1, 1), (2, 2), (3, 3), (4, 4), (5, 5)\}$

24. $\{(1, -1), (2, -2), (3, -3), (4, -4), (5, -5)\}$

25. $\{(2, 4)\}$

26. $\{(3, 3)\}$

27. $y = 2x$ 28. $y = -3x$ 29. $y = x + 8$ 30. $y = x - 1$

31. $y = 9 - x$ 32. $y = \frac{1}{2}x + \frac{1}{4}$ 33. $y = 3x + 7$ 34. $y = -2x + 5$

35. $y = 7 - 2x$ 36. $y = \frac{1}{3}(x - 3)$ 37. $y = \frac{1}{4}x + \frac{1}{2}$ 38. $y = \frac{2x - 3}{4}$

REVIEW EXERCISES FOR CHAPTER 8

8.1 SETS AND FUNCTIONS

1. Use braces to indicate the following sets:
   (a) the set of colors of the U. S. flag
   (b) the set consisting of the first six positive integers
   (c) the set consisting of the first seven even positive integers
2. Which of the following correspondences represent functions?

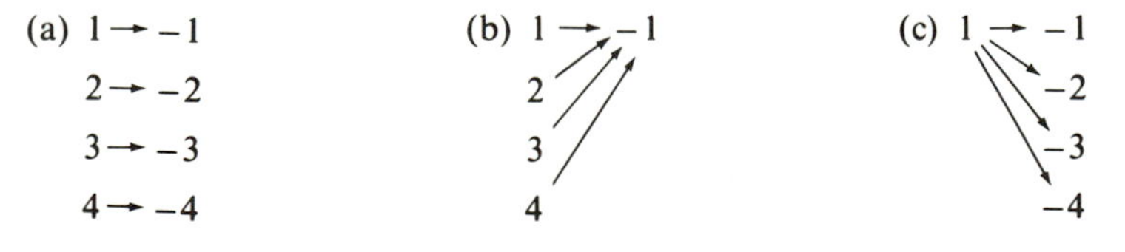

3. Determine:
   (a) the domain, (b) the range
   of the function defined by the correspondence:

$5 \to 1$

$10 \to 2$

$15 \to 1$

$20 \to 2$

$25 \to 3$

4. Determine two different functions with domain $\{1, 2\}$ and range $\{3, 4\}$.

8.2 EVALUATING FUNCTIONS

5. Suppose f is the function defined by the correspondence:

$$2 \to 1$$
$$4 \to 2$$
$$6 \to 3$$
$$8 \to 1$$

   (a) What are the arguments of f?
   (b) Determine the value of f at 4.
   (c) Determine the value of f at 8.

6. Suppose $f(x) = 3x$. Determine the following function values.
   (a) $f(1)$ (b) $f(4)$ (c) $f(0)$ (d) $f(-2)$

7. Suppose $g(x) = 2x + 5$. Determine the following function values.
   (a) $g(2)$ (b) $g(10)$ (c) $g\left(\frac{1}{2}\right)$ (d) $g\left(\frac{-1}{2}\right)$

8. Suppose $h(x) = \frac{x - 1}{x + 1}$. Determine the following function values.
   (a) $h(1)$ (b) $h(-2)$ (c) $h\left(\frac{1}{4}\right)$ (d) $h(.1)$

9. (a) Determine the volume of a cube as a function of the length of a side, s.
   (b) Determine the volume if the side is of length 5 inches.
   (c) Determine the volume if the side is of length .2 inch.

10. Determine the largest possible domain of the function $f(x) = \frac{x + 5}{x - 4}$.

8.3 CARTESIAN COORDINATES

11. In Figure 8.27 determine the coordinates of each point.

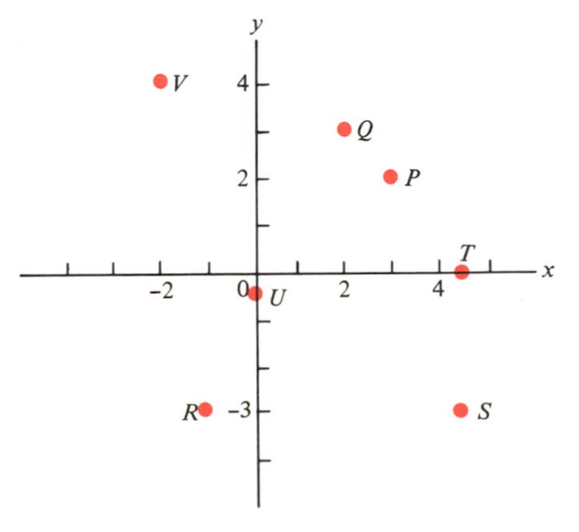

FIGURE 8.27

12. On a coordinate system plot the following ordered pairs:
    (a) (1, 5) (b) (3, −1) (c) (0, 4)
    (d) $\left(\frac{1}{2}, \frac{-1}{2}\right)$ (e) (−3, −4) (f) (−2, −2.5)

13. Determine the quadrant of each of the following points, or indicate that the point lies on a coordinate axis.
    (a) (4, 3) (b) (−4, 3) (c) (−2, −5)
    (d) (0, 6) (e) (1, −4) (f) $\left(\frac{1}{2}, \frac{3}{2}\right)$

8.4 GRAPHS

14. Graph the indicated function:

$$\{(1, 2), (2, 1), (3, 2), (4, 1), (5, 2), (6, 1)\}$$

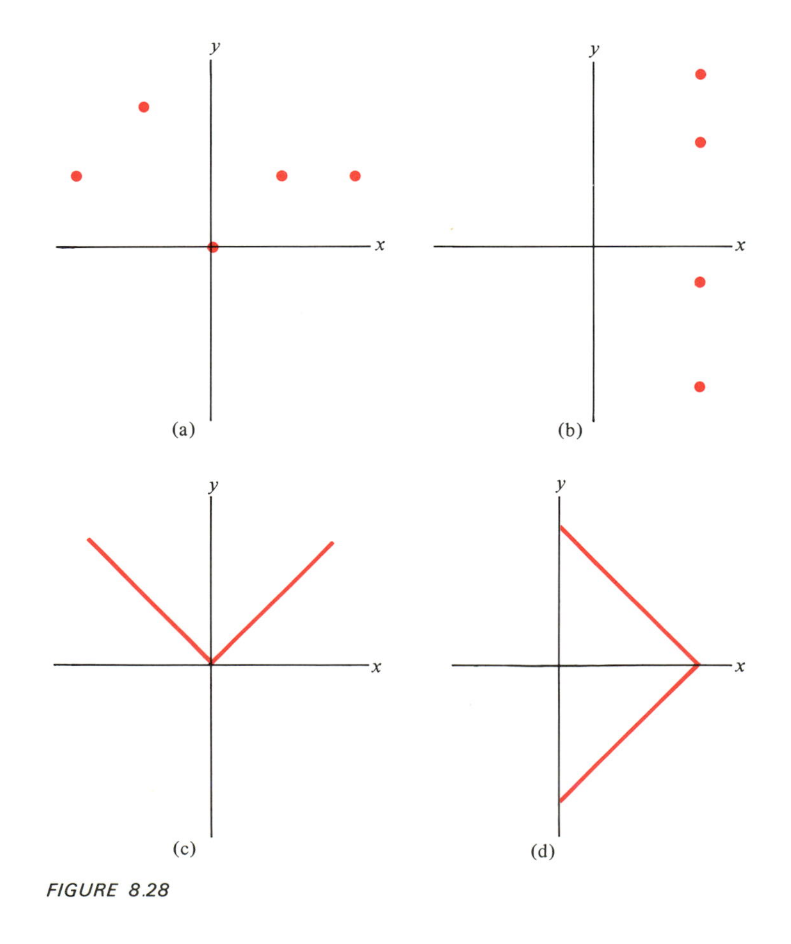

FIGURE 8.28

15. Graph the function defined by the equation:

$$y = -2x$$

16. Graph the function defined by the equation:

$$y = 3 - 2x$$

17. Which of the graphs in Figure 8.28 (on page 256) represent functions?

8.5 INVERSE OF A FUNCTION

18. Which of the graphs in Figure 8.29 represent one–one functions?
19. Which of the indicated functions are one–one?
    (a) $\{(1, 4), (2, 3), (3, 4), (4, 2)\}$
    (b) $\{(1, 1), (2, 3), (3, 5), (4, 7), (5, 9)\}$
    (c) $\{(-3, 3), (-2, 2), (-1, 1), (0, 0), (1, -1), (2, -2), (3, -3)\}$

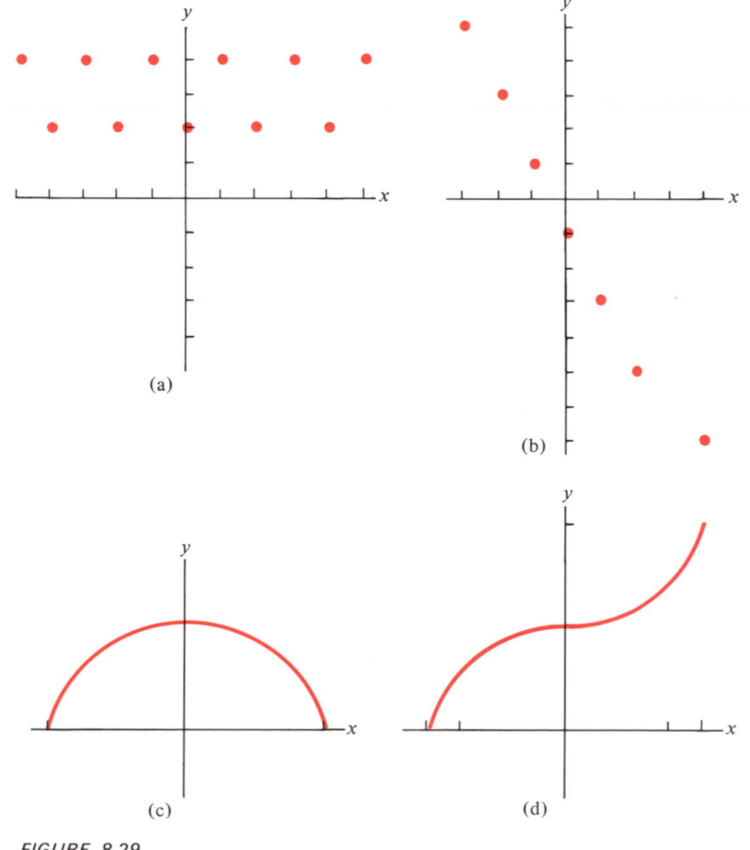

FIGURE 8.29

20. Determine the inverse of the function indicated by

$$\{(1, 7), (2, 4), (3, 2), (4, 0), (5, 6)\}.$$

21. Determine the inverse of the function defined by the equation

$$y = 3x - 4.$$

22. Determine the inverse of the function defined by the equation

$$y = 2(x + 5).$$

TEST YOURSELF ON CHAPTER 8.

1. Which of the following correspondences represent functions?

(a) $2 \to 1$
$3 \to 2$
$4 \to 1$
$5 \to 2$

(b) $1 \to 2$
$2 \to 3$
$1 \to 4$
$2 \to 5$

2. Determine the domain and range of the function defined by the correspondence:

$2 \to 3$
$4 \to 5$
$6 \to 7$
$8 \to 9$
$10 \to 9$

3. Suppose $f(x) = 1 - 4x$. Determine the following function values:

(a) $f(1)$ (b) $f(-1)$ (c) $f\left(\frac{1}{4}\right)$ (d) $f(-10)$

4. Suppose $g(x) = x^3 + 2x - 4$. Determine the following function values:

(a) $g(0)$ (b) $g(1)$ (c) $g(-1)$ (d) $g(2)$

5. (a) Describe the circumference of a circle as a function of the length of the radius r.

   (b) Determine the circumference if the radius is of length 5.

6. On a coordinate system, represent the following ordered pairs:

(a) $(4, 3)$ (b) $(-2, 3)$ (c) $(3, -2)$ (d) $(-1, -6)$

7. Graph the function defined by the equation

$$y = 2x + 1.$$

8. Which of the graphs in Figure 8.30 represent functions?

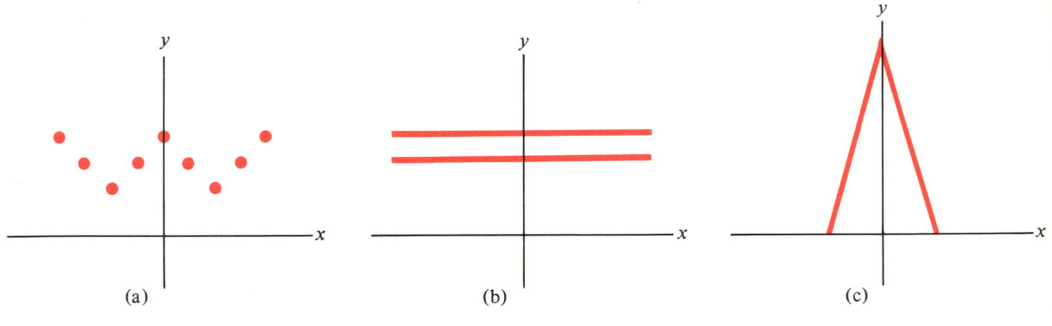

FIGURE 8.30

9

lines

9.1 PROPORTION

DEFINITION. Let a and b ($\neq 0$) be numbers. The **ratio of a to b** is the quotient

$$\frac{a}{b}.$$

EXAMPLE 1. Determine the ratio of y to x if

$$2y = 3x, \qquad x \neq 0.$$

SOLUTION. The ratio of y to x is simply the quotient

$$\frac{y}{x}, \qquad x \neq 0.$$

You are given the equation

$$2y = 3x.$$

Thus when $x \neq 0$, divide both sides of the equation by $2x$, and obtain

$$\frac{y}{x} = \frac{3}{2}.$$

■

DEFINITION. A **proportion** is a statement that two ratios are equal.

The proportion

$$\frac{a}{b} = \frac{c}{d}$$

is often read

a is to b and c is to d.

The equivalence of fractions is a proportion. For example,

$$\frac{3}{5} = \frac{6}{10}, \qquad \frac{1}{2} = \frac{4}{8}, \qquad \text{and} \qquad \frac{-5}{2} = \frac{10}{-4}$$

are each proportions.

EXAMPLE 2. Find the value of a in the proportion

$$\frac{a}{9} = \frac{20}{36}.$$

SOLUTION. Multiply both sides of the equation by 9.

$$a = \frac{\overset{5}{\cancel{20}} \cdot \cancel{9}}{\underset{1}{\cancel{\underset{4}{\cancel{36}}}}}$$

$$a = 5$$

■

EXAMPLE 3. Find the value of d in the proportion

$$\frac{-3}{7} = \frac{12}{d}.$$

SOLUTION. Multiply both sides of the equation by $7d$ (or cross-multiply):

$$-3d = 12 \cdot 7$$

Thus

$$d = -4 \cdot 7$$

$$d = -28$$

■

EXAMPLE 4. Find all possible values of b in the proportion

$$\frac{1}{b} = \frac{b}{4}.$$

SOLUTION. Cross-multiply:

$$b^2 = 4$$

Thus

$$b = 2 \qquad \text{or} \qquad b = -2$$

■

Many problems are solved by setting up proportions.

EXAMPLE 5. A psychologist needs 9 pounds of cheese to feed 100 rats. After some time the rat population increases to 150. How much cheese is needed to feed them?

SOLUTION. Let x be the number of pounds of cheese needed to feed 150 rats. Set up the proportion:

$$\frac{9}{100} = \frac{x}{150}$$

Now multiply both sides by 300, the LCD.

$$\frac{9}{\underset{1}{\cancel{100}}} \cdot \overset{3}{\cancel{300}} = \frac{x}{\underset{1}{\cancel{150}}} \cdot \overset{2}{\cancel{300}}$$

$$27 = 2x$$

$$\frac{27}{2} = x$$

Thus $13\frac{1}{2}$ pounds are needed. ■

DEFINITION. **Similar triangles** are triangles in which *the lengths of corresponding sides* are *proportional.*

Alternatively,

Similar triangles are those in which *corresponding angles* are *equal.*

(If two triangles satisfy either condition, they also satisfy the other.)

In Figure 9.1 (a), the triangles are similar. Corresponding sides are marked with the same number of bars. The sides of lengths a, b, c of one triangle correspond, respectively, to the sides of lengths a', b', c' of the other. Corresponding angles are equal.

$$\frac{a}{a'} = \frac{b}{b'} = \frac{c}{c'}$$

The triangles of Figure 9.1 (b) are also similar.

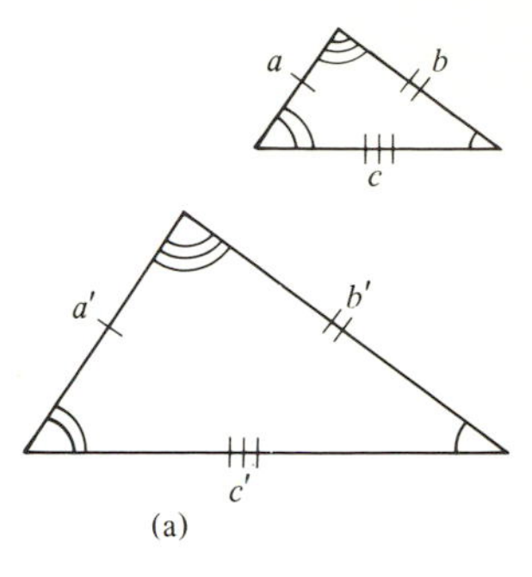

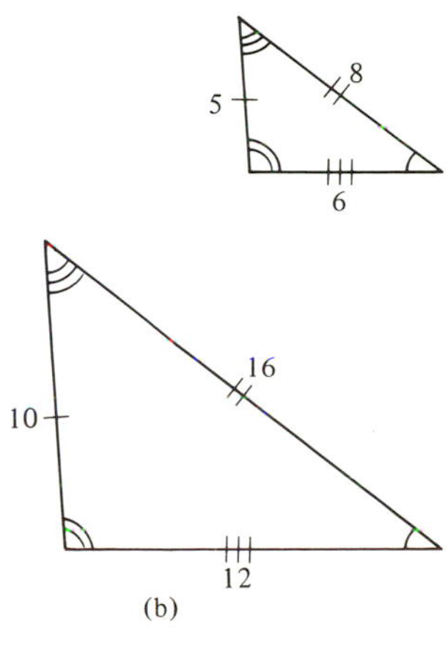

FIGURE 9.1 (a) Similar triangles: $\frac{a}{a'} = \frac{b}{b'} = \frac{c}{c'}$ (b) Similar triangles: $\frac{5}{10} = \frac{8}{16} = \frac{6}{12}$

Note that the equation

$$\frac{a}{a'} = \frac{b}{b'}$$

is equivalent to the equation

$$\frac{a}{b} = \frac{a'}{b'}.$$

For, multiply both sides of the first equation by $\frac{a'}{b}$ to obtain the second equation. Also, if you multiply both sides of the second equation by $\frac{b}{a'}$, you obtain the first. When the lengths of corresponding sides of a triangle are proportional, both equations hold.

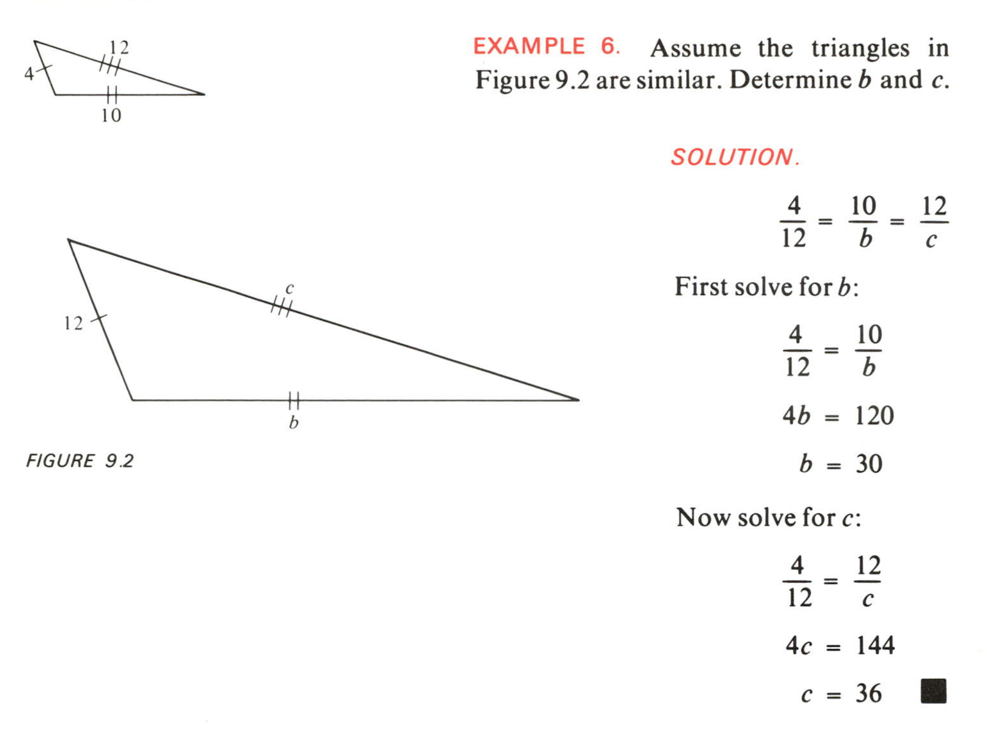

FIGURE 9.2

EXAMPLE 6. Assume the triangles in Figure 9.2 are similar. Determine b and c.

SOLUTION.

$$\frac{4}{12} = \frac{10}{b} = \frac{12}{c}$$

First solve for b:

$$\frac{4}{12} = \frac{10}{b}$$

$$4b = 120$$

$$b = 30$$

Now solve for c:

$$\frac{4}{12} = \frac{12}{c}$$

$$4c = 144$$

$$c = 36 \quad \blacksquare$$

In the next section, you will use the notion of similar triangles to obtain an important property of lines.

EXERCISES

In Exercises 1–4 determine the ratio of y to x, $x \neq 0$, in the given equation.

1. $y = 6x$
2. $x = 6y$
3. $5y = 7x$
4. $3y = -8x$

In Exercises 5–24 find all possible values of a, b, c, or d in each proportion.

5. $\frac{a}{5} = \frac{4}{2}$
6. $\frac{a}{7} = \frac{3}{1}$
7. $\frac{2}{b} = \frac{5}{15}$
8. $\frac{-1}{-2} = \frac{c}{12}$
9. $\frac{9}{6} = \frac{3}{d}$
10. $\frac{-7}{2} = \frac{c}{-4}$
11. $\frac{a}{3} = \frac{40}{15}$
12. $\frac{5}{9} = \frac{45}{d}$
13. $\frac{2}{b} = \frac{-2}{7}$
14. $\frac{-3}{5} = \frac{6}{d}$
15. $\frac{\frac{1}{2}}{2} = \frac{c}{4}$
16. $\frac{\frac{1}{3}}{\frac{2}{3}} = \frac{\frac{3}{7}}{d}$
17. $\frac{5}{6} = \frac{7}{d}$
18. $\frac{3}{5} = \frac{c}{4}$
19. $\frac{-7}{3} = \frac{1}{d}$
20. $\frac{3}{b} = \frac{b}{3}$
21. $\frac{a}{8} = \frac{2}{a}$
22. $\frac{.4}{.6} = \frac{12}{d}$
23. $\frac{.01}{.1} = \frac{c}{2}$
24. $\frac{.63}{.21} = \frac{.9}{d}$

25. A dozen eggs cost 96¢. How much do 3 eggs cost?
26. Bel Paese sells for \$2.40 per pound. How much does a 6-ounce slice cost?
27. If 20 pounds of fertilizer costs \$4, how much do 90 pounds cost?
28. If 2 pounds of potatoes cost 25 cents, how much do 6 pounds cost?
29. Oranges sell for a dollar a dozen. How much do 15 oranges cost?
30. If 24 hungry freshmen can eat 96 hamburgers, how many hamburgers should be prepared for 43 hungry freshmen?
31. Tom earns \$12.56 for 4 hours of work. At this rate how much will he earn for 10 hours of work?
32. A pint of cider costs 45¢. At this rate how much would you pay for 2 quarts of cider?
33. On a highway map, $\frac{1}{3}$ of an inch represents 10 miles. How many miles are represented by an inch and a half?
34. Bob owned a $\frac{3}{4}$ interest in some property. When it was sold, Bob received \$27 000 for his share. For how much was the property sold?
35. A 27-foot rope is cut into 2 pieces that are in the ratio of $\frac{7}{2}$. How long is each piece?
36. A batter has 2 hits for every 7 times at bat. If he keeps up this rate, how many hits will he have after 350 at-bats?
37. Suppose Vida Blue wins 3 games for every game he loses. How many games does he win in a season in which he loses 8 games?

In each of Exercises 38–42 the triangles are similar. Corresponding sides are marked with the same number of bars. Determine the lengths of the indicated sides.

38.

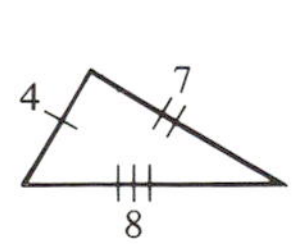

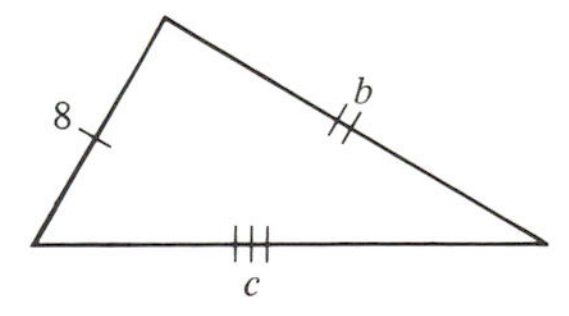

FIGURE 9.3

39.

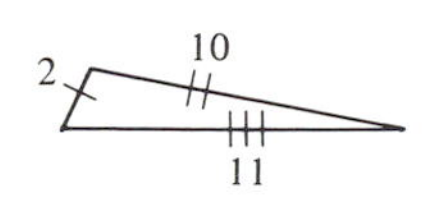

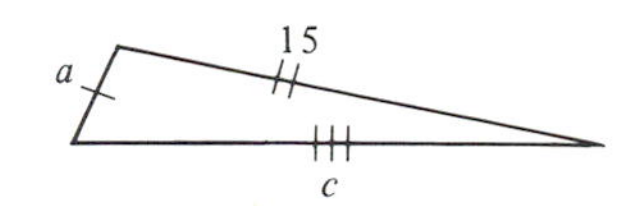

FIGURE 9.4

40.

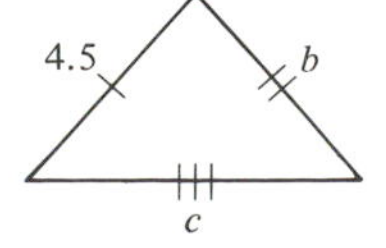

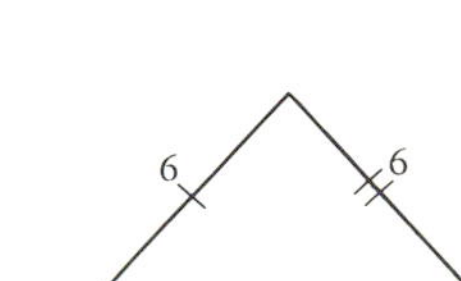
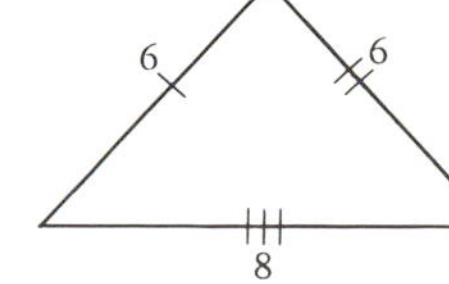

FIGURE 9.5

41.

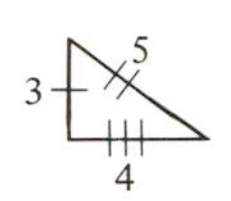

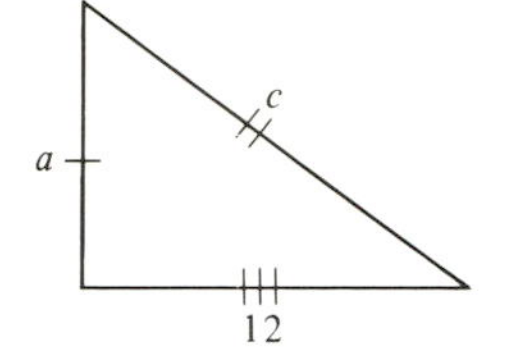

FIGURE 9.6

42.

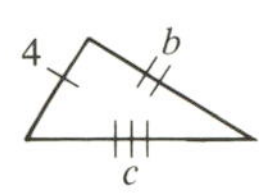

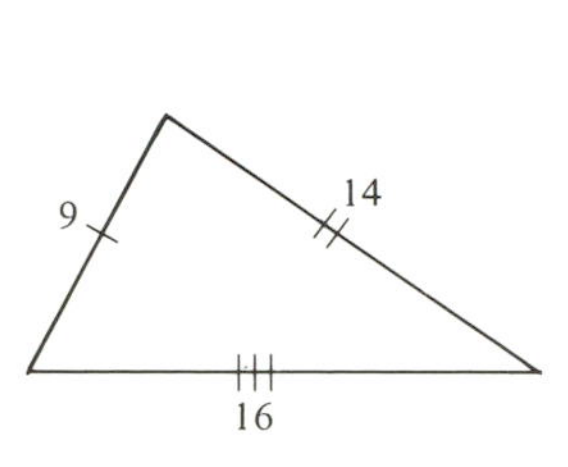

FIGURE 9.7

9.2 POINT-SLOPE EQUATION OF A LINE

Consider an equation in two variables, such as

$$5x - y = 9 - x.$$

An ordered pair (x_1, y_1) **satisfies** such an equation, or is a **solution** of the equation, if a ***true*** statement results when x_1 replaces x and $\mathbf{y_1}$ replaces $\mathbf{y}$ in the equation. Thus (2, 3) satisfies the above equation because

$$5 \cdot 2 - \mathbf{3} = 9 - 2$$

or

$$7 = 7.$$

As you have seen in Chapter 8, a line can be given by means of an equation in the two variables x and y. The ordered pairs (x_1, y_1) corresponding to points on the line (and only these ordered pairs) satisfy the equation of the line.

When you are given such an equation, you can graph the line by plotting two points and drawing the line through them.

It is one thing to draw a line; it is another matter to determine its equation. If you have sufficient information about a line, you can determine its equation.

The "slope" of a line expresses its slant or inclination. How fast do the y-values change along the line as the x-values change? Do the y-values increase or decrease as the x-values increase?

Let P_1 and P_2 be two different points on a line L. Suppose $P_1 = (x_1, y_1)$ and $P_2 = (x_2, y_2)$.

The **rise** or the **change in** y (along L from P_1 to P_2) is defined to be

$$y_2 - y_1.$$

$$\text{rise} \quad = \quad \text{change in } y \quad = \quad y_2 - y_1$$

The **run** or the **change in** x (along L from P_1 to P_2) is defined to be

$$x_2 - x_1.$$

$$\text{run} = \text{change in } x = x_2 - x_1$$

In Figure 9.8(a), the line slants *upward to the right*. The rise, $y_2 - y_1$, and the run, $x_2 - x_1$, are *both positive*. In Figure 9.8(b), the line slants *upward to the left*. The rise, $y_2 - y_1$, is *negative* although the run, $x_2 - x_1$, is still *positive*.

In the following, assume that the lines L discussed are not vertical. Then distinct points on each line L will have distinct x-values. Thus, if (x_1, y_1) and (x_2, y_2) are on L, the run, $x_2 - x_1$, is nonzero. Division by $x_2 - x_1$ is therefore defined.

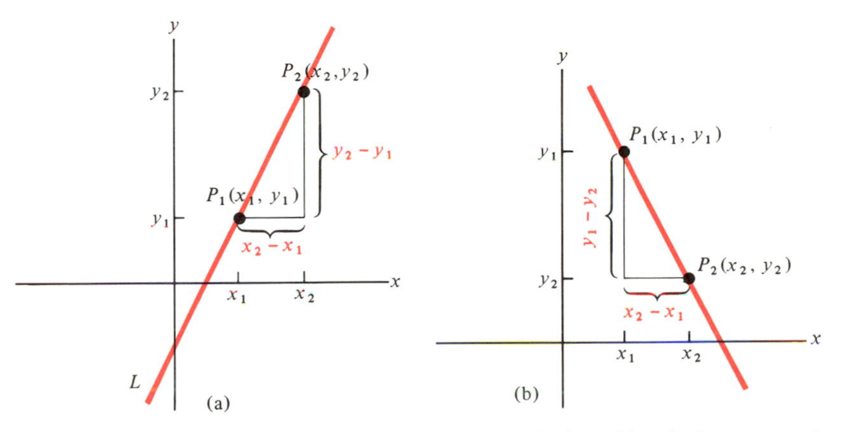

FIGURE 9.8 (a) The rise, $y_2 - y_1$, and run, $x_2 - x_1$, are both positive. L slopes upward to the right. (b) The rise, $y_2 - y_1$ or $-(y_1 - y_2)$, is negative. The run, $x_2 - x_1$, is positive. The line slopes upward to the left.

DEFINITION. Let L be a nonvertical line and let P_1 and P_2 be *any* two distinct points on L. Suppose $P_1 = (x_1, y_1)$ and $P_2 = (x_2, y_2)$. The slope of L is defined to be

$$\frac{y_2 - y_1}{x_2 - x_1}.$$

It is important to recognize that *no matter which two distinct points are chosen on L, this ratio is always the same.* In Figure 9.9 on page 268, the slope of the line is positive. Consider the rise and run from P_1 to P_2, as well as the rise and run from P_3 to P_4. Note that the right triangles P_1P_2Q and P_3P_4R are similar because corresponding angles are equal. The lengths of corresponding sides of similar triangles are proportional. Thus

$$\frac{y_2 - y_1}{x_2 - x_1} = \frac{y_4 - y_3}{x_4 - x_3}$$

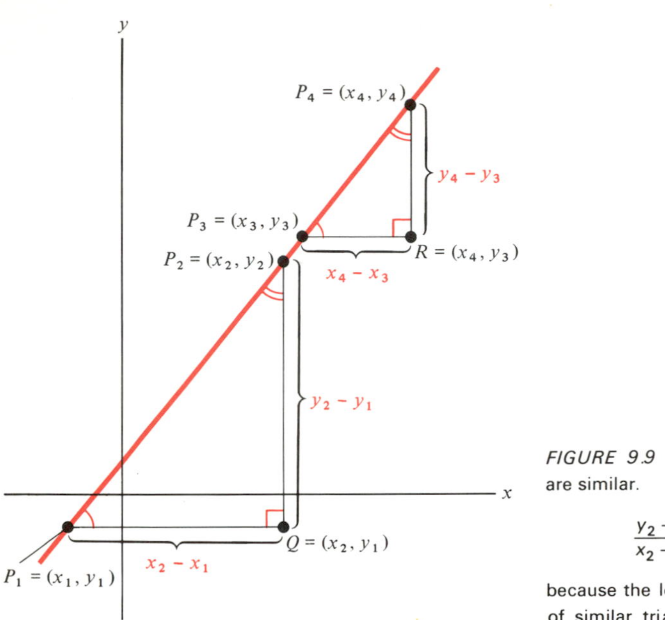

FIGURE 9.9 Triangles P_1P_2Q and P_3P_4R are similar.

$$\frac{y_2 - y_1}{x_2 - x_1} = \frac{y_4 - y_3}{x_4 - x_3}$$

because the lengths of corresponding sides of similar triangles are proportional.

Either of these ratios defines the slope of the line L.

Now you know that the ratio of rise to run does not depend on the points chosen on L.

In determining the slope of L, it does not matter whether you go from P_1 to P_2 or from P_2 to P_1 because

$$\frac{y_2 - y_1}{x_2 - x_1} = \frac{(-1)(y_1 - y_2)}{(-1)(x_1 - x_2)} = \frac{y_1 - y_2}{x_1 - x_2}.$$

EXAMPLE 1. Determine the slope of the line through the points (1, 2) and (2, 5).

SOLUTION. Let $(x_1, y_1) = (1, 2)$ and $(x_2, y_2) = (2, 5)$.

$$\text{Slope} = \frac{\text{rise}}{\text{run}} = \frac{y_2 - y_1}{x_2 - x_1} = \frac{5 - 2}{2 - 1} = 3$$

The slope is positive; the line slopes upward to the right. [See Figure 9.10(a) on page 269.] Note that

$$\frac{y_1 - y_2}{x_1 - x_2} = \frac{2 - 5}{1 - 2} = \frac{-3}{-1} = 3.$$

■

EXAMPLE 2. Determine the slope of the line through the points (1, 4) and (4, 1).

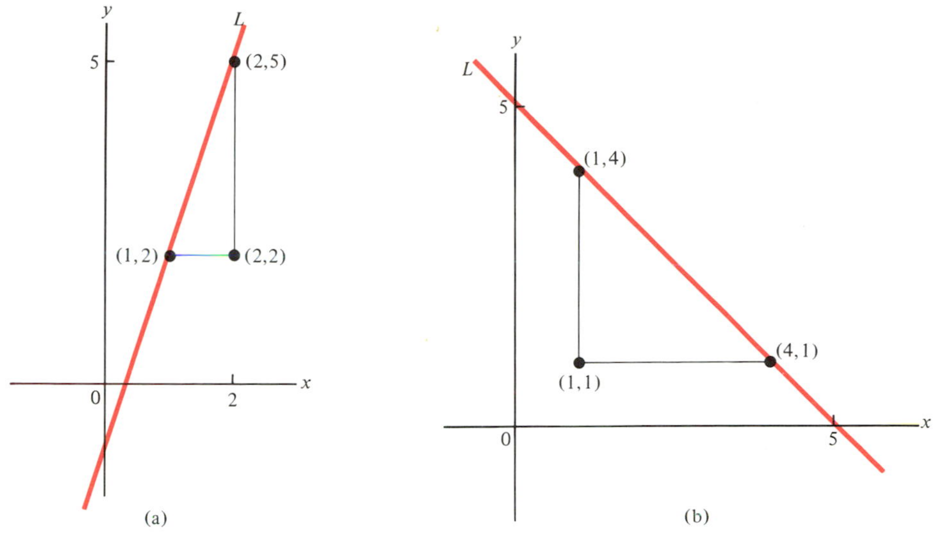

FIGURE 9.10 (a) slope $L = \frac{\text{rise}}{\text{run}} = \frac{5-2}{2-1} = 3$ (b) slope $L = \frac{\text{rise}}{\text{run}} = \frac{1-4}{4-1} = -1$

SOLUTION. Let $(x_1, y_1) = (1, 4)$ and $(x_2, y_2) = (4, 1)$.

$$\text{Slope} = \frac{\text{rise}}{\text{run}} = \frac{y_2 - y_1}{x_2 - x_1} = \frac{1-4}{4-1} = \frac{-3}{3} = -1$$

The slope is negative; the line slopes upward to the left. [See Figure 9.10(b).] ■

EXAMPLE 3. Consider the line L through $(-1, 2)$ and $(1, 4)$. Determine the y-coordinate of the point on L with x-coordinate 2.

SOLUTION. Let $(x_1, y_1) = (-1, 2)$ and $(x_2, y_2) = (1, 4)$. [See Figure 9.11 on page 270.]

$$\text{Slope } L = \frac{\text{rise}}{\text{run}} = \frac{y_2 - y_1}{x_2 - x_1} = \frac{4-2}{1-(-1)} = \frac{2}{2} = 1$$

Let (x_3, y_3) be the point on L with x-coordinate 2. Thus $x_3 = 2$, and you must determine y_3. The slope of L can be expressed in terms of the point (x_3, y_3) and one other point—say (x_2, y_2). You already know that the slope of L is 1. Thus

$$\frac{y_3 - y_2}{x_3 - x_2} = \frac{y_3 - 4}{2-1} = \text{slope } L = 1$$

$$y_3 - 4 = 1$$

$$y_3 = 5$$

■

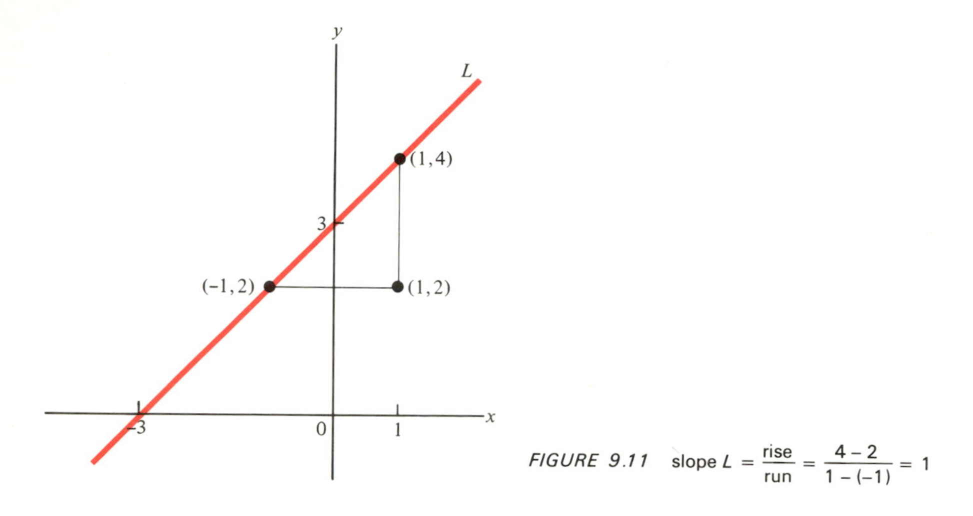

FIGURE 9.11 slope $L = \frac{\text{rise}}{\text{run}} = \frac{4-2}{1-(-1)} = 1$

The fact that the slope of a line L can be obtained from *any* two distinct points on L suggests the equation of L. Let (x_1, y_1) be a fixed point on L and let (x, y) be an arbitrary point on L other than (x_1, y_1). Let m be the slope of L. Then

$$\frac{y - y_1}{x - x_1} = m$$

or

$$y - y_1 = m(x - x_1)$$

The second equation is known as the **point-slope form of the equation of a line.** If you know a point (x_1, y_1) on L as well as the slope m of L, you can determine the equation of L.

As you will see, there are various forms in which the equation of a line can be written. Any two equations of a line L are equivalent because ordered pairs corresponding to the points of L (and only these ordered pairs) satisfy each of these equations. Thus you speak of *the* equation of a line.

EXAMPLE 4. Determine the equation of the line L through $(1, 3)$ with slope $\frac{1}{2}$.

SOLUTION. Here $(x_1, y_1) = (1, 3)$; $m = \frac{1}{2}$. Thus, substituting the indicated numbers in the equation

$$y - y_1 = m(x - x_1),$$

you obtain

$$y - 3 = \frac{1}{2}(x - 1).$$

This is the equation of L. Alternatively, multiply both sides by 2 to eliminate fractions, and obtain

$$2(y - 3) = x - 1$$

or

$$2y = x + 5.$$

[See Figure 9.12.] ■

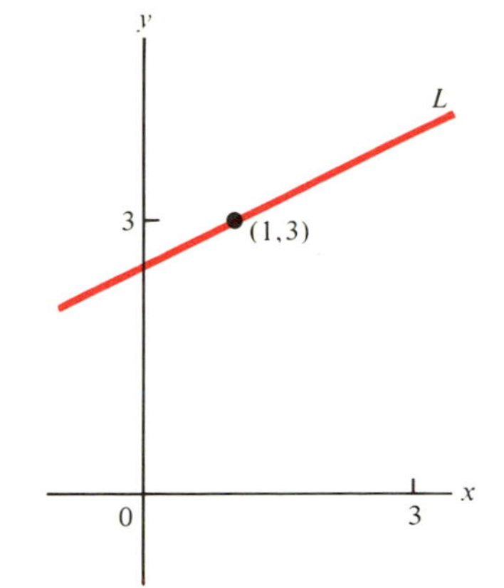

FIGURE 9.12 The line L given by $2(y - 3) = x - 1$

EXAMPLE 5. (a) Determine the equation of the line L through (0, 4) with slope -2.
(b) Determine the x-coordinate of the point on L with y-coordinate -2.
(c) Graph L.

SOLUTION. (a) Here $(x_1, y_1) = (0, 4)$, $m = -2$. Thus

$$y - y_1 = m(x - x_1)$$

$$y - 4 = -2(x - 0)$$

$$y - 4 = -2x$$

(b) Replace y by -2 in the preceding equation.

$$-2 - 4 = -2x$$

$$-6 = -2x$$

$$3 = x$$

(c) Two distinct points determine the graph of L. Use the points

$$P_1 = (0, 4) \text{ and } P_2 = (3, -2).$$

[See Figure 9.13.] ■

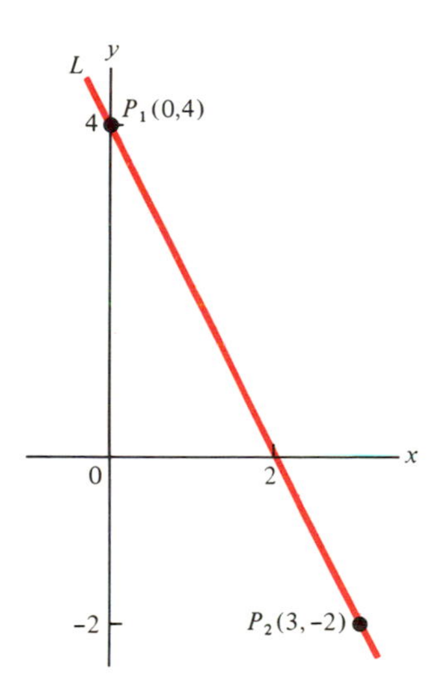

FIGURE 9.13 The line L given by $y - 4 = -2x$

On a *horizontal* line L, all of the y-values are the same. For any two distinct points (x_1, y_1) and (x_2, y_2) on L,

$$y_1 = y_2$$

Thus

$$\text{rise } L = y_2 - y_1 = 0$$

Therefore

$$m = \text{slope } L = \frac{\text{rise}}{\text{run}} = \frac{y_2 - y_1}{x_2 - x_1} = \frac{0}{x_2 - x_1} = 0$$

In other words, ***a horizontal line has slope*** **0**. The equation of a horizontal line is

$$y - y_1 = m(x_2 - x_1),$$

that is,

$$y - y_1 = 0,$$

or

$$y = y_1.$$

Thus the equation of a *horizontal* line is determined by a *single y-value*, y_1. [See Figure 9.14.]

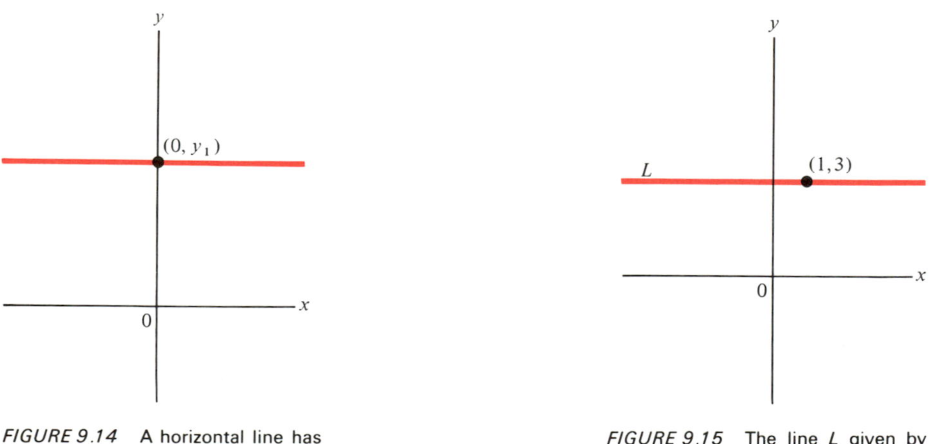

FIGURE 9.14 A horizontal line has the equation $y = y_1$.

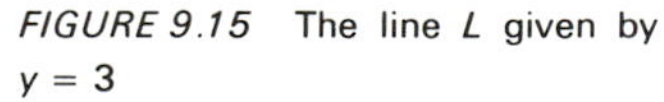

FIGURE 9.15 The line L given by $y = 3$

EXAMPLE 6. Determine the equation of the horizontal line L through (1, 3).

SOLUTION. Here $(x_1, y_1) = (1, 3)$. The equation of L is

$$y = 3.$$

[See Figure 9.15.] ■

What about a *vertical* line? So far, vertical lines have been excluded from the discussion because *all x-values are the same*. Thus a vertical line does not represent a *function*. The slope of a vertical line is undefined because the run, $x_2 - x_1$, is 0. The ratio of rise to run does not exist because division by 0 is undefined. However, the fact that all x-values are the same immediately yields the *equation* of a vertical line. Determine one x-value, x_1, and you know all

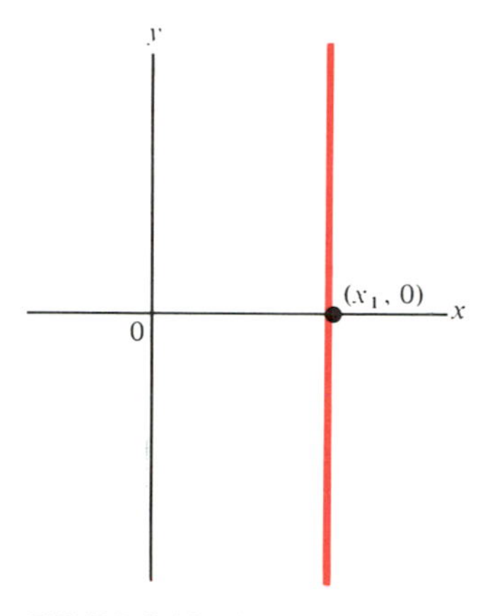

FIGURE 9.16 A vertical line has the equation $x = x_1$.

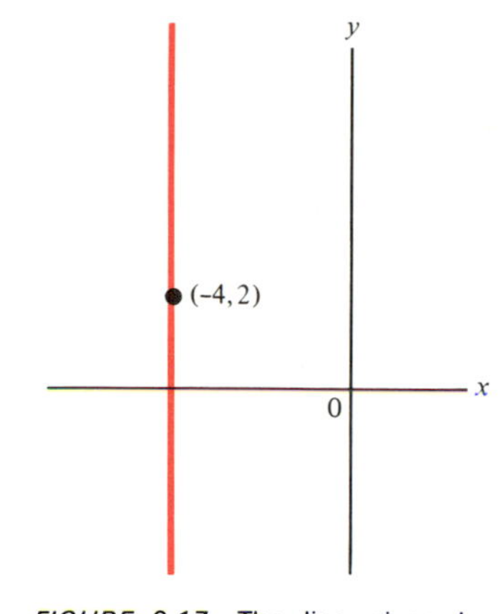

FIGURE 9.17 The line given by $x = -4$

x-values. The equation of a vertical line is then

$$x = x_1.$$

[See Figure 9.16.]

EXAMPLE 7. Determine the equation of the vertical line through $(-4, 2)$.

SOLUTION. Here $(x_1, y_1) = (-4, 2)$. The equation of L is

$$x = -4.$$

[See Figure 9.17.] ■

EXERCISES

In Exercises 1–14 determine the slope of the line through the given points.

1. $(1, 1)$ and $(3, 3)$
2. $(2, 1)$ and $(4, 2)$
3. $(8, 3)$ and $(2, 2)$
4. $(1, 6)$ and $(-1, -6)$
5. $(6, 3)$ and $(0, 4)$
6. $(2, 8)$ and $(10, 2)$
7. $(0, 8)$ and $(1, 1)$
8. $(1, 3)$ and $(5, -3)$
9. $(1, 6)$ and $(9, 10)$
10. $(8, 5)$ and $(2, 5)$
11. $(3, 2)$ and $(-7, -4)$
12. $(3, -2)$ and $(0, -1)$
13. $\left(-3, \frac{1}{2}\right)$ and $\left(0, \frac{1}{4}\right)$
14. $(.6, 1)$ and $(.2, -.2)$
15. A line L has slope 2 and passes through $(2, 4)$. Determine the x-coordinate of the point on L with y-coordinate 1.
16. A line L has slope -1 and passes through $(0, 4)$. Determine the x-coordinate of the point on L with y-coordinate -2.

17. A line L has slope 5 and passes through $(-1, 3)$. Determine the y-coordinate of the point on L with x-coordinate 1.

18. A line L has slope $\frac{1}{3}$ and passes through the origin. Determine the y-coordinate of the point on L with x-coordinate 9.

19. A line L passes through $(1, 3)$ and $(2, 5)$. Determine the x-coordinate of the point on L with y-coordinate -1.

20. A line L passes through $(-1, -2)$ and $(0, 6)$. Determine the x-coordinate of the point on L with y-coordinate 2.

21. A line L passes through $(1, -3)$ and the origin. Determine the y-coordinate of the point on L with x-coordinate 5.

22. A line L passes through $(-2, -4)$ and $(-1, -1)$. Determine the y-coordinate of the point on L with x-coordinate 1.

In Exercises 23–36 determine the equation of the line that has slope m and that passes through the point P.

23. $m = 3, P = (0, 4)$
24. $m = -1, P = (2, 2)$
25. $m = 2, P = (-1, 3)$
26. $m = 4, P = (0, 0)$
27. $m = \frac{1}{2}, P = (1, 2)$
28. $m = \frac{-1}{2}, P = (4, 0)$
29. $m = 10, P = (1, 1)$
30. $m = 5, P = (2, 4)$
31. $m = -3, P = (-1, -1)$
32. $m = \frac{-1}{4}, P = (1, 4)$
33. $m = 0, P = (6, 3)$
34. $m = 20, P = (0, 0)$
35. $m = \frac{2}{3}, P = (1, -1)$
36. $m = \frac{-3}{4}, P = (-1, 5)$

In Exercises 37–42: (a) Determine the equation of the line that has slope m and that passes through the point P. (b) Graph this line.

37. $m = 2, P = (3, 3)$
38. $m = -4, P = (1, 0)$
39. $m = 4, P = (1, 5)$
40. $m = 0, P = (1, 4)$
41. $m = \frac{-1}{2}, P = (4, 2)$
42. $m = \frac{1}{3}, P = (3, 6)$

43. Which of the following lines are horizontal? Which are vertical? Which are neither horizontal nor vertical?

(a) $y = x - 4$ (b) $y = 2x + 1$ (c) $y = 1$

(d) $x = 3$ (e) $y = 0$ (f) $x = -2$

44. Determine the equation of the vertical line through $(5, 5)$.

45. Determine the equation of the horizontal line through $(5, 5)$.

46. A line passes through $(2, 8)$ and $(-2, 8)$. Determine its equation.

47. A line passes through (1, −4) and (1, 0). Determine its equation.
48. Determine the equation of the vertical line through (0, −2).

9.3 ALTERNATE FORMS OF LINEAR EQUATIONS

The x-axis has the equation

$$y = 0;$$

the y-axis has the equation

$$x = 0.$$

[See Figure 9.18.]

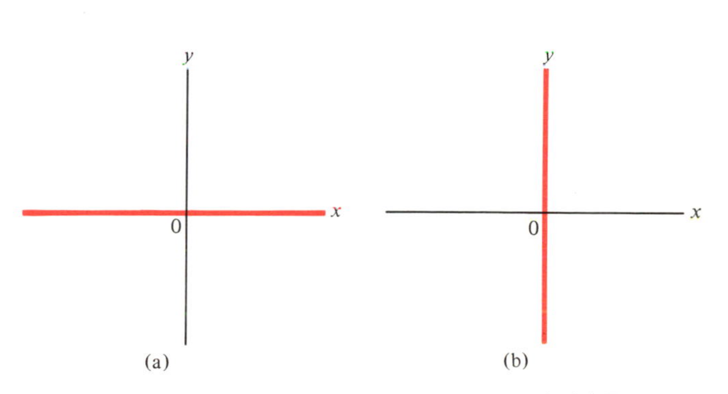

FIGURE 9.18 (a) The x-axis has the equation $y = 0$. (b) The y-axis has the equation $x = 0$.

A horizontal line intersects the y-axis once; a horizontal line other than the x-axis does not intersect the x-axis. A vertical line intersects the x-axis once; a vertical line other than the y-axis does not intersect the y-axis. Every other line in the plane intersects both coordinate axes exactly once. If $(a, 0)$ is the only point of intersection of a line L with the x-axis, then a is called the **x-intercept** of L. Similarly, if $(0, b)$ is the only point of intersection of L with the y-axis, then b is called the **y-intercept** of L. [See Figure 9.19.]

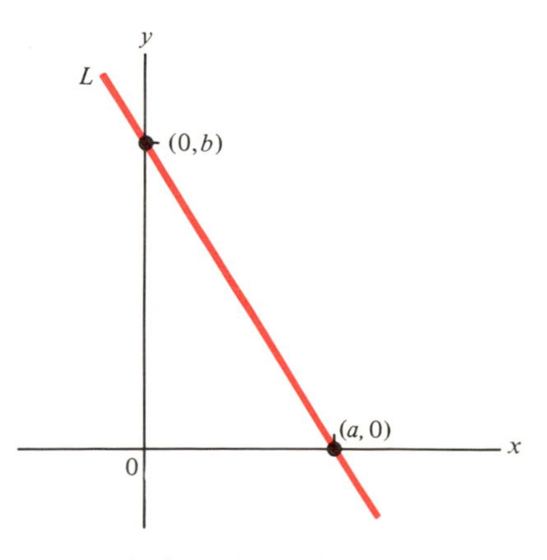

FIGURE 9.19 L has x-intercept a and y-intercept b.

Note that these intercepts are *numbers* rather than points. The x-intercept is the *x-coordinate* of the intersection of L with the x-axis. The y-intercept is the *y-coordinate* of the intersection of L with the y-axis.

To determine the x-intercept of a line L, set $y = 0$ in any equation of L; to determine the y-intercept, set $x = 0$.

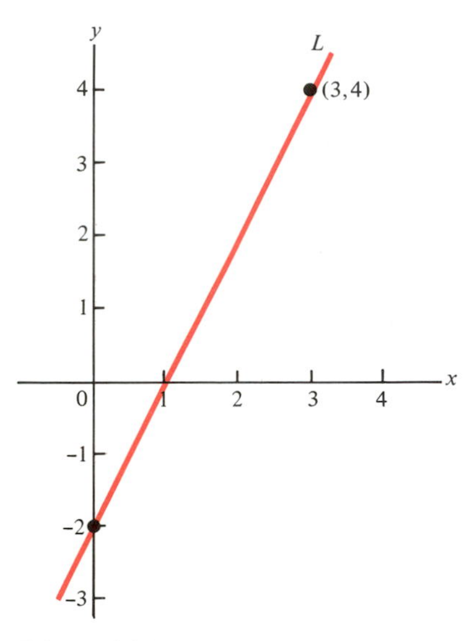

FIGURE 9.20

EXAMPLE 1. Let L be the line given by the equation:

$$y - 4 = 2(x - 3)$$

(a) Determine the x-intercept of L.
(b) Determine the y-intercept of L.

SOLUTION. (a) Set $y = 0$ in the given equation of L, and solve for x.

$$-4 = 2(x - 3)$$
$$-2 = x - 3$$
$$1 = x$$

The x-intercept is 1.

(b) Set $x = 0$ in the given equation of L, and solve for y.

$$y - 4 = 2(-3)$$
$$y - 4 = -6$$
$$y = -2$$

The y-intercept is -2. ■

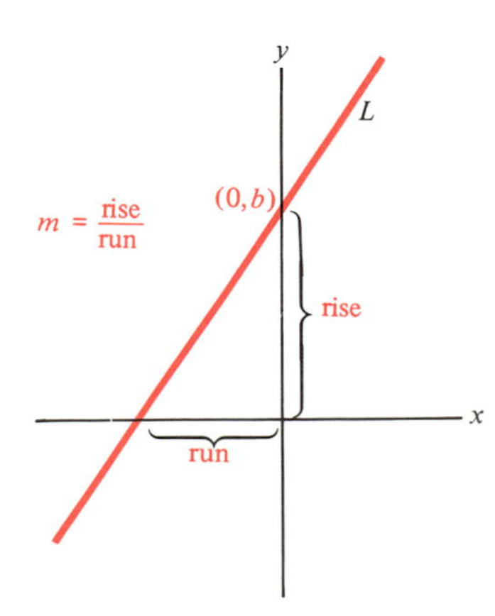

FIGURE 9.21 The slope-intercept form of the equation of L: $y = mx + b$

The point-slope form of the equation of L requires knowing one point on L in addition to the slope, m. Let this point be $(0, b)$, where b is the y-intercept. Then the equation of L is

$$y - b = m(x - 0)$$

or

$$\mathbf{y = mx + b.}$$

This second equation is in **slope-intercept form.** Thus the slope and y-intercept of a (nonvertical) line L determine the slope-intercept form of the equation of L. [See Figure 9.21.]

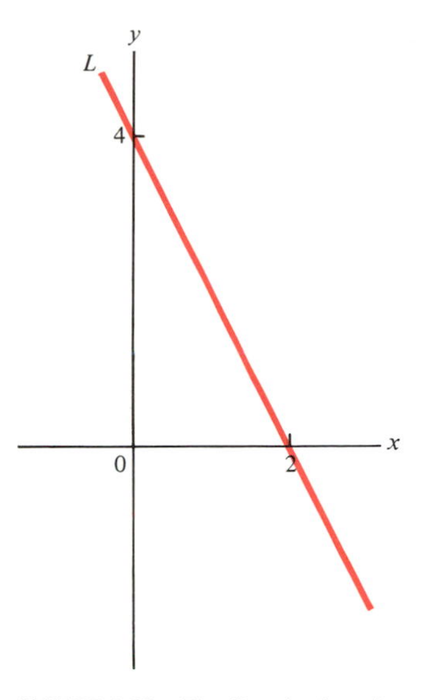

FIGURE 9.22 The line L given by $y = -2x + 4$

EXAMPLE 2. Let L have slope -2 and y-intercept 4. Determine the slope-intercept form of the equation of L.

SOLUTION. Here $m = -2$ and $b = 4$. The slope-intercept form of the equation of L is

$$y = -2x + 4. \qquad \blacksquare$$

Two distinct points on a line determine that line. To begin with, two distinct points (x_1, y_1) and (x_2, y_2) determine the slope m of a nonvertical line L:

$$m = \frac{y_2 - y_1}{x_2 - x_1}$$

Insert this in the point-slope form of the equation of L,

$$y - y_1 = m(x - x_1),$$

and obtain

$$\mathbf{y - y_1 = \frac{y_2 - y_1}{x_2 - x_1}(x - x_1),}$$

the 2-point form. [See Figure 9.23.]

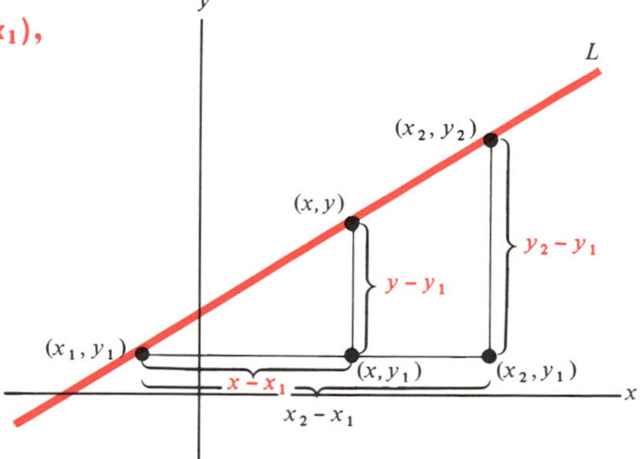

FIGURE 9.23 The 2-point form of the equation of L:

$$y - y_1 = \frac{y_2 - y_1}{x_2 - x_1}(x - x_1)$$

Note the similar triangles.

EXAMPLE 3. Determine the 2-point form of the equation of the line through (2, 5) and (1, 9).

SOLUTION. Either of these points can be (x_1, y_1). Arbitrarily, let $(x_1, y_1) = (2, 5)$ and $(x_2, y_2) = (1, 9)$.

$$y - y_1 = \frac{y_2 - y_1}{x_2 - x_1}(x - x_1)$$

$$y - 5 = \frac{9 - 5}{1 - 2}(x - 2)$$

$$y - 5 = -4(x - 2)$$

Had you let $(x_1, y_1) = (1, 9)$, the resulting equation would have been different in form, although equivalent. (See Exercise 47.) ■

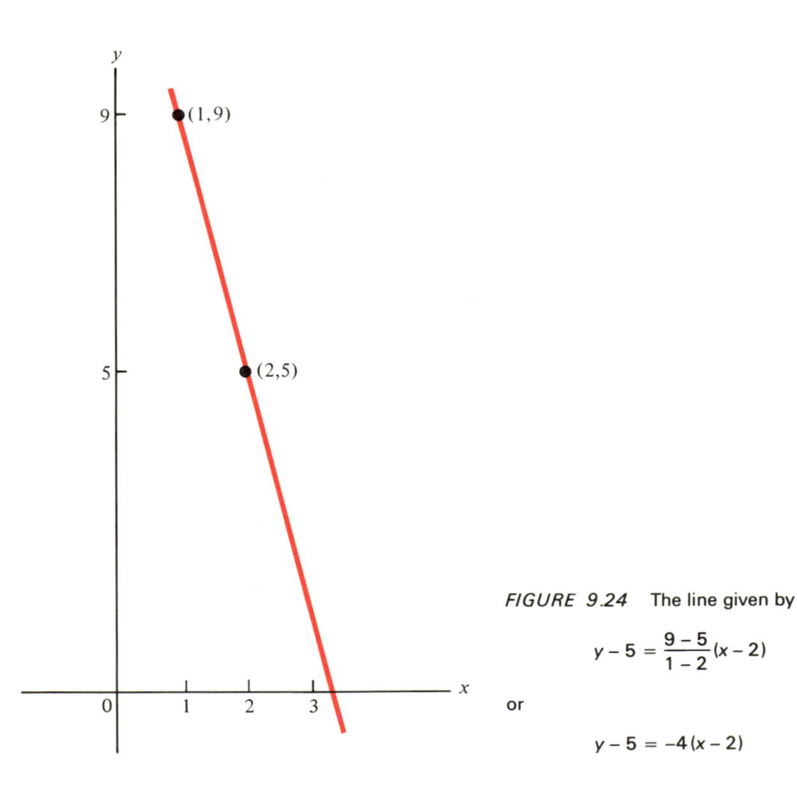

FIGURE 9.24 The line given by

$$y - 5 = \frac{9 - 5}{1 - 2}(x - 2)$$

or

$$y - 5 = -4(x - 2)$$

There is one more important form of the equation of a line, known as the **general form:** The equation of a line can be written as

(A) $$\boldsymbol{Ax + By + C = 0,}$$

where A and B are not *both* 0.

EXAMPLE 4. Determine the general form of the line given by

$$y - 7 = -3(x + 5).$$

SOLUTION. Add $3(x + 5)$ to both sides, simplify, and arrange the terms, as in Equation (A).

$$y - 7 = -3(x + 5)$$

$$y - 7 + 3(x + 5) = 0$$

$$y - 7 + 3x + 15 = 0$$

$$3x + y + 8 = 0$$

This is the general form of the equation. Here $A = 3, B = 1, C = 8$. ■

EXAMPLE 5.

(a) Determine the slope-intercept form of the equation

$$2x - 3y + 1 = 0$$

of a line L.

(b) Determine the x-coordinate of the point on L with y-coordinate -6.

(c) Graph L.

SOLUTION.

(a) $$2x + 1 = 3y$$

Change sides and divide by 3:

$$y = \frac{1}{3}(2x + 1)$$

$$y = \frac{2}{3}x + \frac{1}{3}$$

This is the slope-intercept form. The slope is $\frac{2}{3}$ and the y-intercept is $\frac{1}{3}$.

(b) Let $y = -6$ in the original equation:

$$2x - 3(-6) + 1 = 0$$
$$2x + 18 + 1 = 0$$
$$2x = -19$$
$$x = \frac{-19}{2}$$

(c) You now know two points on the line:

$$\left(0, \frac{1}{3}\right) \quad \text{and} \quad \left(\frac{-19}{2}, -6\right).$$

Draw the line through these two points, as in Figure 9.25. ■

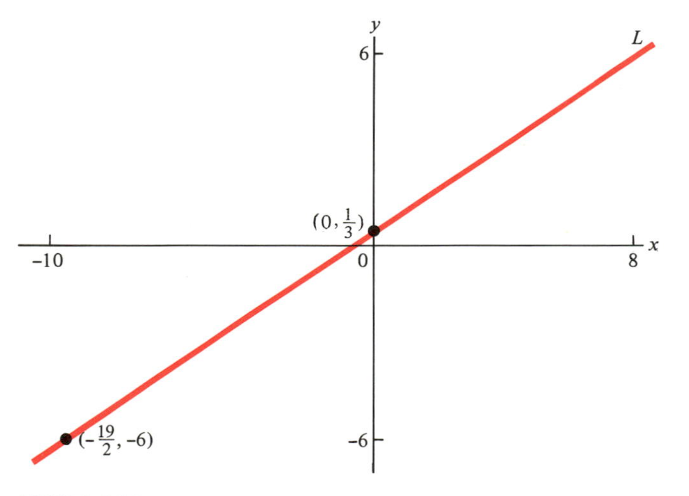

FIGURE 9.25

Note that the slope m of a line can be obtained from the equation

$$Ax + By + C = 0$$

as follows. If $B = 0$, the equation is

$$Ax + C = 0,$$

from which you obtain

$$x = \frac{-C}{A}.$$

In this case the line is vertical and its slope is not defined. [See Figure 9.26 (a).]

On the other hand, if $B \neq 0$, you can solve

$$Ax + By + C = 0$$

for y in terms of x:

$$By = -Ax - C$$

$$y = \frac{-A}{B}x - \frac{C}{B}$$

In this case the slope of the line is $\frac{-A}{B}$. [See Figure 9.26 (b).]

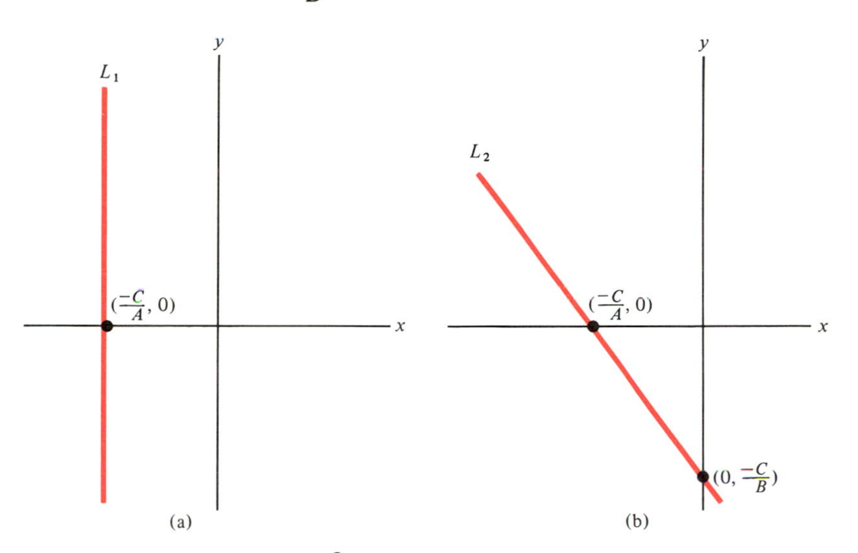

FIGURE 9.26 (a) The line L_1 is given by $x = \frac{-C}{A}$. (b) The line L_2 is given by $Ax + By + C = 0$.

(Assume $A > 0, B > 0, C > 0$.) $m = \dfrac{\frac{-C}{B}}{\frac{C}{A}} = \frac{-A}{B}$

EXERCISES

In Exercises 1–8 determine (a) the x-intercept and (b) the y-intercept of each of the indicated lines.

1. $y - 4 = x - 2$
2. $y - 1 = 5(x + 3)$
3. $y = 2(x + 7)$
4. $y + 1 = -4(x - 3)$
5. $y = 2x - 5$
6. $y = -3x + 1$
7. $2x + 3y + 5 = 0$
8. $5x - y = 0$

In Exercises 9–16 determine the slope-intercept form of the equation of the indicated line L.

9. L: $y - 5 = 2(x - 1)$
10. L: $y + 1 = 3(x + 4)$
11. L: $y = 2(x + 9)$
12. L: $y + \frac{1}{2} = x - \frac{3}{4}$

13. L has slope 3 and y-intercept 1.
14. L has slope 2 and x-intercept 2.
15. L passes through $(8, 1)$ and $(3, 4)$.
16. L has x-intercept 4 and y-intercept -3.

In Exercises 17–22 determine the 2-point form of the equation of the indicated line L.

17. L passes through $(1, 4)$ and $(3, 8)$.
18. L passes through $(4, 2)$ and $(1, 6)$.
19. L has x-intercept 1 and y-intercept 5.
20. L has x-intercept -2 and y-intercept -3.
21. L passes through $(-4, -2)$ and has x-intercept -3.
22. L passes through the origin and $(6, -5)$.

In Exercises 23–30 determine the general form of the equation of the indicated line L.

23. $L\colon y - 2 = x + 8$
24. $L\colon y - 7 = 2(x - 5)$
25. $L\colon y = -3x + 8$
26. L passes through $(1, 7)$ and $(-2, 4)$.
27. L passes through $(1, 8)$ and the origin.
28. L is the vertical line through $(2, 5)$.
29. L is the horizontal line through $(2, 5)$.
30. L has slope 4 and y-intercept -5.

In Exercises 31–38, (a) determine the slope of the indicated line L; (b) graph L.

31. $L\colon y + 1 = 5(x - 4)$
32. $L\colon 3x + 4y - 7 = 0$
33. $L\colon 2x - y - 2 = 0$
34. $L\colon 3y + 5 = 0$
35. L passes through the origin and $(3, 8)$.
36. L passes through $(-1, 5)$ and $(-2, -1)$.
37. L has x-intercept 5 and y-intercept 1.
38. L has x-intercept -2 and y-intercept -4.

In Exercises 39–46 determine the equation of the line L in any form.

39. L passes through $(-1, -1)$ and $(4, 2)$.
40. L passes through $(2, 4)$ and has x-intercept 5.
41. L passes through $(2, -3)$ and has y-intercept 5.
42. L passes through $(3, -1)$ and has x-intercept 3.
43. L passes through $(7, -2)$ and has y-intercept -2.
44. L passes through the origin and has slope -3.
45. L passes through $(1, 5)$ and has no x-intercept.

46. L passes through $(-2, 5)$ and has no y-intercept.

47. (a) In Example 3, p. 278, determine the 2-point form of the equation of the line by letting $(x_1, y_1) = (1, 9)$.

(b) Show that this equation is equivalent to the one obtained on p. 278:

$$y - 5 = -4(x - 2)$$

9.4 PARALLEL AND PERPENDICULAR LINES

In this section and in Chapter 10, you will be concerned with two or three lines at a time. Lines that are "parallel" or "perpendicular" are of particular interest.

Recall that *lines* (as opposed to line segments) extend indefinitely. [See Figure 9.27.]

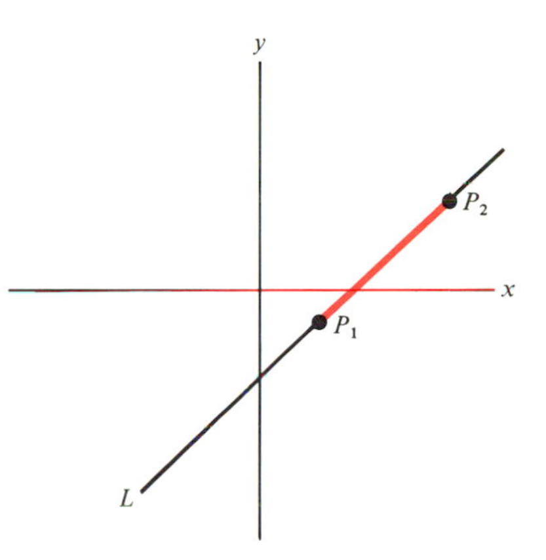

FIGURE 9.27 The line L extends indefinitely. $\overline{P_1 P_2}$ is a line segment on L.

DEFINITION. **Parallel lines** on the plane are lines that do not intersect.

If L_1 and L_2 are parallel lines, write

$$L_1 \parallel L_2.$$

In this case L_1 is also said to be **parallel to** L_2 (and L_2 **parallel to** L_1).

Two vertical lines

$$x = x_1 \quad \text{and} \quad x = x_2,$$

$x_1 \neq x_2$, are parallel.

A vertical line intersects a nonvertical line; hence these lines are not parallel. [See Figure 9.28.]

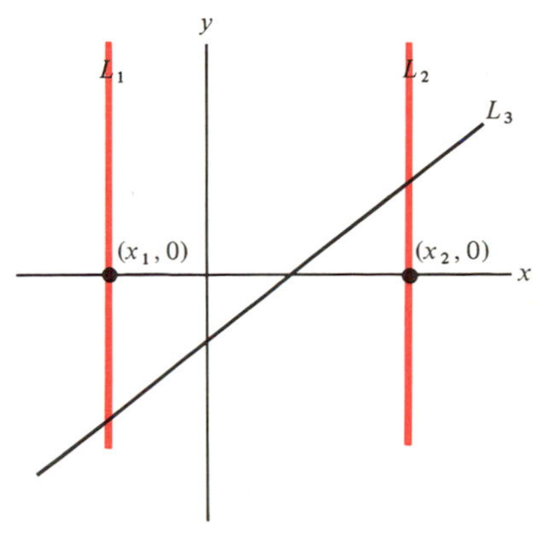

FIGURE 9.28 $L_1 \parallel L_2$
Each of the vertical lines L_1 and L_2 intersects the nonvertical line L_3. Thus neither is parallel to L_3.

Let L_1 and L_2 be distinct nonvertical lines. Then it can be shown that *if they are parallel, their slopes are equal, and if their slopes are equal, the lines are parallel.* Thus let

$$m_1 = \text{slope } L_1 \qquad \text{and}$$
$$m_2 = \text{slope } L_2.$$

If $L_1 \parallel L_2$, then $m_1 = m_2$. And if $m_1 = m_2$, then $L_1 \parallel L_2$. [See Figure 9.29.]

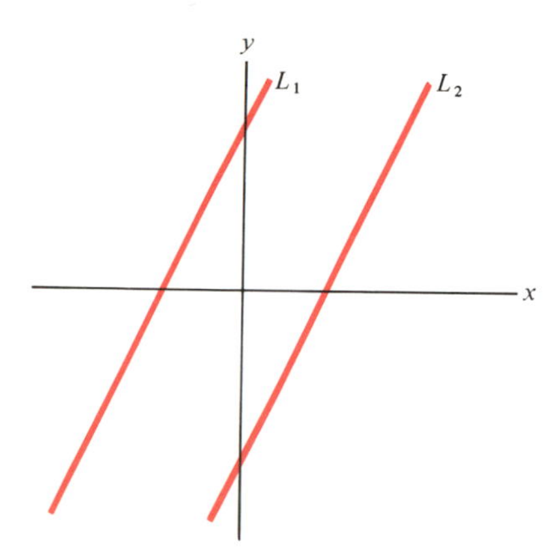

FIGURE 9.29 $L_1 \parallel L_2$

$m_1 = \text{slope } L_1 = \text{slope } L_2 = m_2$

EXAMPLE 1.

(a) The lines given by

$$y = 4x + 3 \qquad \text{and} \qquad y = 4x - 5$$

are different because the first has y-intercept 3 and the second has y-intercept -5. These lines are parallel because each has slope 4. [See Figure 9.30 (a).]

(b) The lines given by

$$L_1\colon y = 2x + 3 \qquad \text{and} \qquad L_2\colon y = -2x + 3$$

are not parallel because L_1 has slope 2, whereas L_2 has slope -2. [See Figure 9.30 (b).] ■

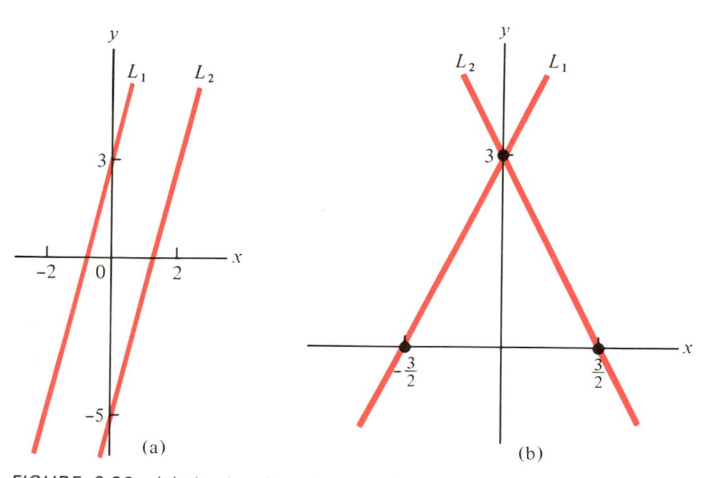

FIGURE 9.30 (a) L_1 is given by $y = 4x + 3$ and L_2 by $y = 4x - 5$. slope $L_1 = 4 =$ slope L_2. Thus L_1 and L_2 are parallel. (b) L_1 is given by $y = 2x + 3$ and L_2 by $y = -2x + 3$. slope $L_1 = 2$, slope $L_2 = -2$. Thus L_1 and L_2 are not parallel.

EXAMPLE 2. The lines given by

$$L_1: 2x + 6y + 1 = 0 \quad \text{and}$$

$$L_2: 3(y - 4) = 5 - x$$

are parallel. Here

$$m_1 = \text{slope } L_1 = \frac{-2}{6} = \frac{-1}{3},$$

$$m_2 = \text{slope } L_2 = \frac{-1}{3}.$$

(Check these computations! Also check that L_1 and L_2 are different lines.) ■

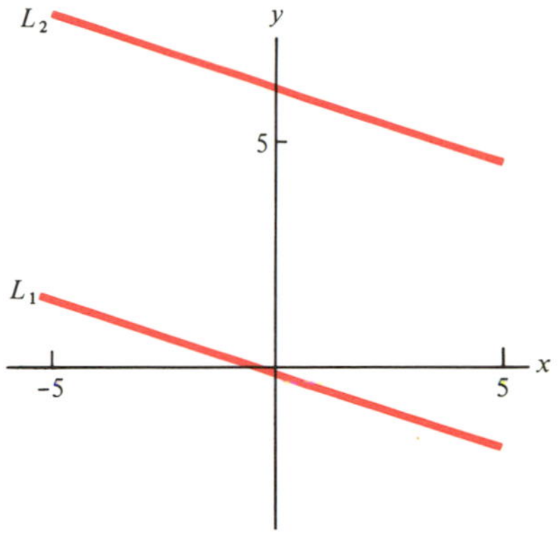

FIGURE 9.31

Let L be a given line and let P be a point that is *not* on L. *There is exactly one line that passes through P and is parallel to L.* [See Figure 9.32.]

EXAMPLE 3. Determine the equation of the line through (1, 4) that is parallel to the line given by

$$y = 2x - 5.$$

SOLUTION. The line given by

$$y = 2x - 5$$

has slope 2. Therefore, write the equation of the line that has slope 2 and that passes through (1, 4). The point-slope form of the equation of this line is

$$y - 4 = 2(x - 1).$$

[See Figure 9.33.] ■

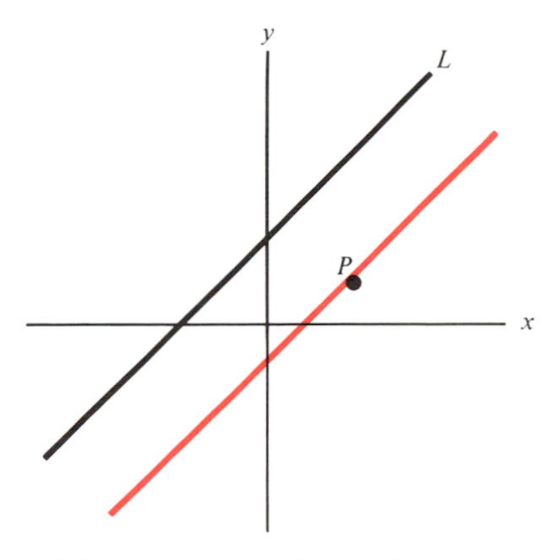

FIGURE 9.32 There is exactly one line that passes through P and is parallel to L.

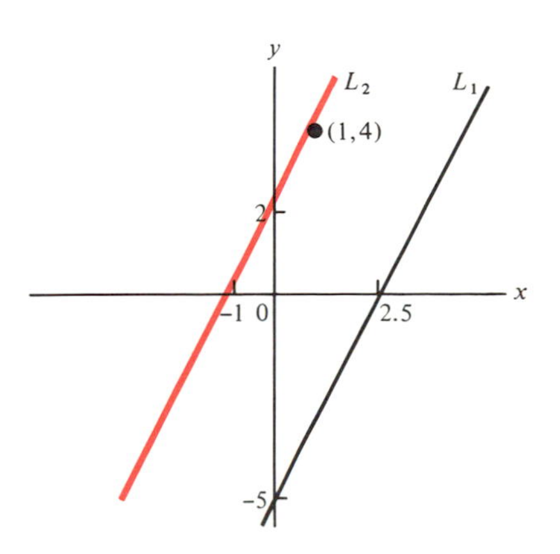

FIGURE 9.33 L_1 is given by $y = 2x - 5$. $L_2 \| L_1$ and L_2 passes through (1, 4). Thus L_2 is given by $y - 4 = 2(x - 1)$.

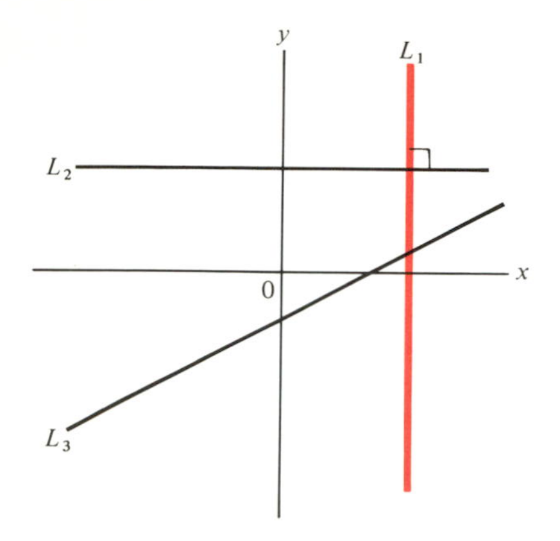

FIGURE 9.34 L_1 is vertical; L_2 is horizontal; $L_1 \perp L_2$. L_3 is not horizontal. L_1 is not perpendicular to L_3.

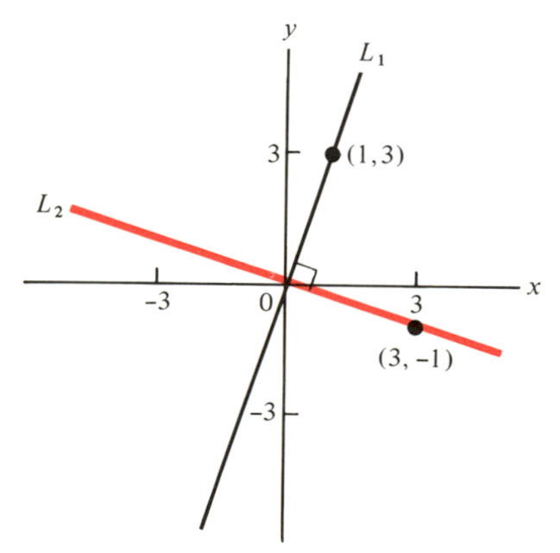

FIGURE 9.35 Slope $L_1 = \frac{3}{1} = 3$, slope $L_2 = \frac{-1}{3}$

DEFINITION. Two lines are **perpendicular** if they intersect at right angles.

If lines L_1 and L_2 are perpendicular, write

$$L_1 \perp L_2.$$

A vertical line and a horizontal line are perpendicular. A vertical line is not perpendicular to any nonhorizontal line. [See Figure 9.34.]

Draw a line through (0, 0) and (1, 3). Now draw another line through (0, 0), but perpendicular to the given line. Observe that the second line appears to pass through (3, −1). [See Figure 9.35.] Compute the slopes of both lines, and observe that these slopes, 3 and $\frac{-1}{3}$, are "negative reciprocals".

Let L_1 and L_2 be distinct nonvertical lines. Then it can be shown that *if L_1 and L_2 are perpendicular, their slopes are negative reciprocals, and if their slopes are negative reciprocals, the lines are perpendicular.* Thus let

$$m_1 = \text{slope } L_1$$

and

$$m_2 = \text{slope } L_2.$$

If $L_1 \perp L_2$, then $m_1 = \frac{-1}{m_2}$. Also, if $m_1 = \frac{-1}{m_2}$, then $L_1 \perp L_2$. Note that

$$\text{if} \quad m_1 = \frac{-1}{m_2},$$

$$\text{then} \quad m_1 \neq 0, \qquad m_2 = \frac{-1}{m_1}, \qquad \text{and} \qquad m_1 m_2 = -1.$$

EXAMPLE 4.

(a) The lines given by

$$L_1\colon y = 2x + 3 \qquad \text{and} \qquad L_2\colon y = \frac{-1}{2}x$$

are perpendicular. Here

$$m_1 = \text{slope } L_1 = 2 \qquad \text{and} \qquad m_2 = \text{slope } L_2 = \frac{-1}{2}$$

[See Figure 9.36 (a).]

(b) The lines given by

$$L_3\colon y = 4x + 1 \qquad \text{and} \qquad L_4\colon y = \frac{x}{4}$$

are *not* perpendicular. Here

$$m_3 = \text{slope } L_3 = 4 \qquad \text{and} \qquad m_4 = \text{slope } L_4 = \frac{1}{4}$$

But the *negative reciprocal* of 4 is $\frac{-1}{4}$.

[See Figure 9.36(b).]

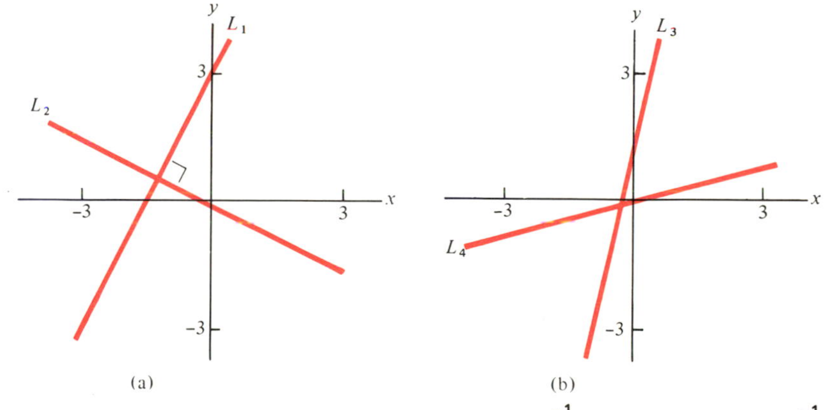

FIGURE 9.36 (a) L_1 is given by $y = 2x + 3$ and L_2 by $y = \frac{-1}{2}x$. Here $m_1 = 2$, $m_2 = \frac{-1}{2}$.

$L_1 \perp L_2$

(b) L_3 is given by $y = 4x + 1$ and L_4 is given by $y = \frac{x}{4}$. Here $m_3 = 4$, $m_4 = \frac{1}{4}$.

Thus L_3 and L_4 are not perpendicular.

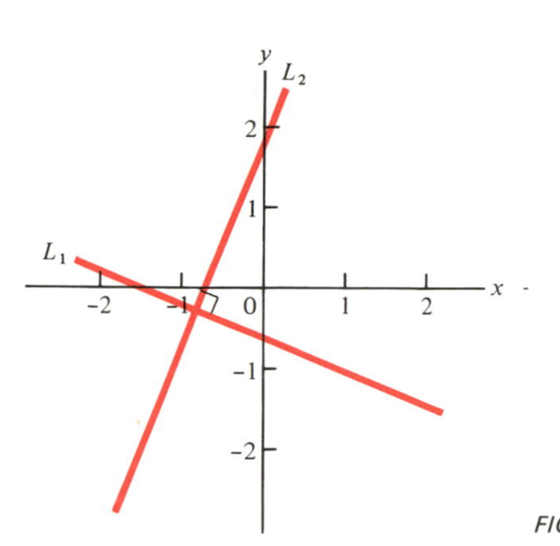

FIGURE 9.37

EXAMPLE 5. The lines

$$L_1\colon 2x + 5y + 3 = 0 \qquad \text{and}$$

$$L_2\colon 2(y - 1) = 5x + 2$$

are perpendicular. Here

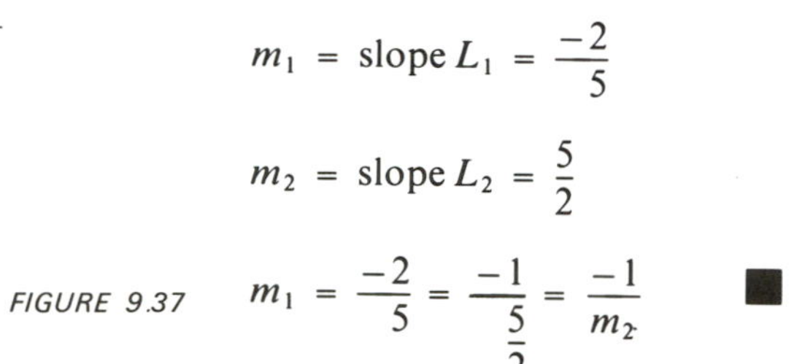

$$m_1 = \text{slope } L_1 = \frac{-2}{5}$$

$$m_2 = \text{slope } L_2 = \frac{5}{2}$$

$$m_1 = \frac{-2}{5} = \frac{-1}{\frac{5}{2}} = \frac{-1}{m_2}$$ ■

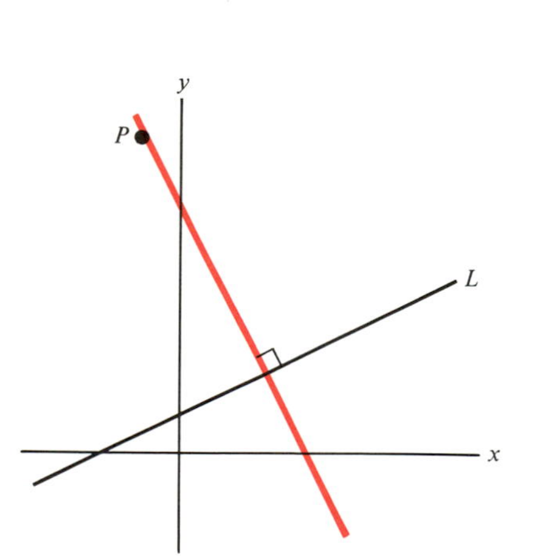

FIGURE 9.38

Let L be a given line and let P be a point that is *not* on L. *There is exactly one line that passes through P and is perpendicular to L.* [See Figure 9.38.]

EXAMPLE 6. Determine the equation of the line through $(-2, -2)$ that is perpendicular to the line

$$2y = x + 1.$$

SOLUTION. The equation of the given line can be written in the form

$$y = \frac{x}{2} + \frac{1}{2}.$$

This line has slope $\frac{1}{2}$. Now find the line that has slope -2 $\left(\text{that is, } \frac{-1}{\frac{1}{2}}\right)$ and that passes through $(-2, -2)$. The point-slope form of the equation of this line is then

$$y - (-2) = -2(x - (-2)),$$

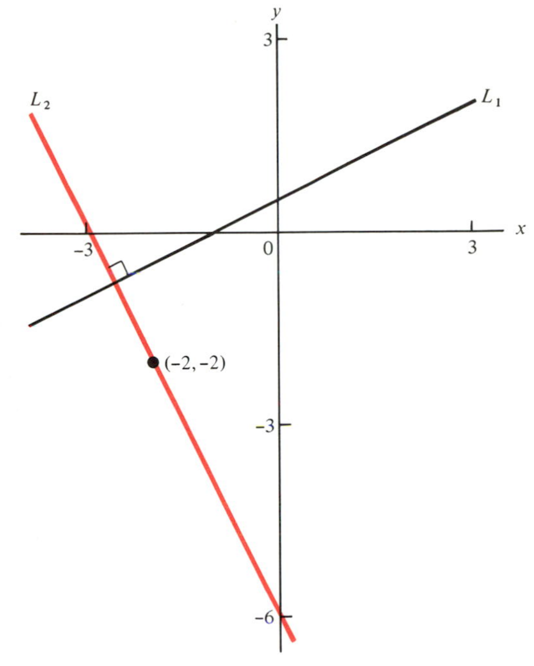

FIGURE 9.39 L_1 is given by $2y = x + 1$ and L_2 by $y + 2 = -2(x + 2)$. Here $m_1 = \frac{1}{2}$, $m_2 = -2$. Thus $L_1 \perp L_2$

or

$$y + 2 = -2(x + 2).$$

[See Figure 9.39.] ■

EXAMPLE 7. Consider the lines:

$$L_1: y = 4$$

$$L_2: x = 4$$

$$L_3: y = -4$$

$$L_4: x = -2$$

$$L_5: y = 4x - 2$$

Which pairs of lines are (a) parallel, (b) perpendicular?

SOLUTION. L_1 and L_3 are horizontal; L_2 and L_4 are vertical. L_5 is neither horizontal nor vertical.

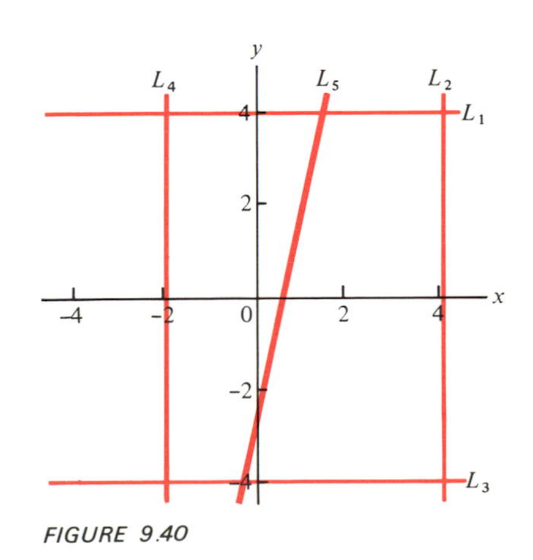

FIGURE 9.40

(a) Two horizontal lines are parallel. (Each has slope 0.) Two vertical lines are parallel. A $\begin{Bmatrix}\text{horizontal}\\ \text{vertical}\end{Bmatrix}$ line is not parallel to a $\begin{Bmatrix}\text{nonhorizontal}\\ \text{nonvertical}\end{Bmatrix}$ line. Thus,

$$L_1 \parallel L_3; \qquad L_2 \parallel L_4$$

(b) A $\begin{Bmatrix}\text{vertical}\\ \text{horizontal}\end{Bmatrix}$ line is perpendicular to a $\begin{Bmatrix}\text{horizontal}\\ \text{vertical}\end{Bmatrix}$ line, and to no other type of line. Thus,

$$L_1 \perp L_2, \qquad L_1 \perp L_4,$$

$$L_3 \perp L_2, \qquad L_3 \perp L_4$$

■

EXAMPLE 8. Show that the lines L_1 through (6, 2) and (2, 6) and L_2 through the origin and (5, 5) are perpendicular.

SOLUTION. Obtain the 2-point forms of the equations of these lines:

$$L_1: y - 2 = \frac{6 - 2}{2 - 6}(x - 6)$$

or

$$y - 2 = -(x - 6)$$

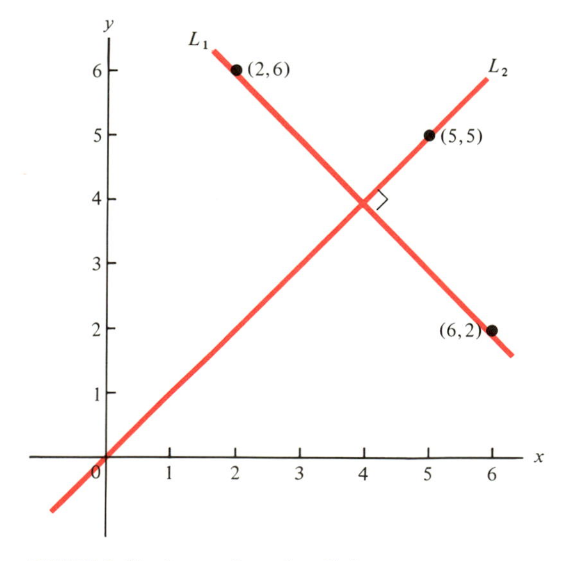

FIGURE 9.41 $L_1: y - 2 = -(x - 6); L_2: y = x$

Here $m_1 = -1, m_2 = 1, m_1 = \dfrac{-1}{m_2}$. Thus $L_1 \perp L_2$

$$m_1 = \text{slope } L_1 = -1$$

$$L_2\colon y - 0 = \frac{5-0}{5-0}(x - 0)$$

or

$$y = x$$

$$m_2 = \text{slope } L_2 = 1$$

$$m_1 = \frac{-1}{m_2}$$

$$L_1 \perp L_2$$

[See Figure 9.41.] ■

EXERCISES

In Exercises 1–20 determine whether the given lines are (a) parallel, (b) perpendicular, or (c) neither parallel nor perpendicular.

1. $y = 2x + 1$ and $y = 2x + 5$
2. $y = -x + 5$ and $y = -x - 3$
3. $y = 3x + 1$ and $y = \frac{-1}{3}x + 4$
4. $y = 2x + 1$ and $y = 3x - 1$
5. $y - 2 = 3(x + 1)$ and $y + 5 = 3(x - 1)$
6. $y + 1 = x - 1$ and $y + 4 = 1 - x$
7. $4x + 7y + 3 = 0$ and $7x + 4y + 3 = 0$
8. $x + y + 1 = 0$ and $5x - 5y + 2 = 0$
9. $y - 4 = \frac{1}{2}(x - 2)$ and $3x - 6y = 4$
10. $y = 2$ and $x = 2$
11. $y - 3 = 0$ and $y = 4$
12. $x + 7 = 0$ and $x - 7 = 0$
13. $2x + 8y = 3$ and $y = 4(3 - x)$
14. $y = 6x$ and $y = \frac{x}{6}$
15. $y = 3x$ and $x = \frac{-1}{3}y$
16. $y = 2x$ and $x = -2y$
17. $y = 3$ and $y = 3x$
18. $y = 0$ and $y = x$
19. $x + y + 1 = 0$ and $y = x$
20. $3x - 3y = 9$ and $y = x - 3$

In Exercises 21–28 determine whether the lines through P and Q and through R and S are (a) parallel, (b) perpendicular, or (c) neither parallel nor perpendicular.

21. $P = (3, 3), Q = (-4, -4); R = (2, -2), S = (-1, 1)$
22. $P = (3, 2), Q = (5, 3); R = (1, 4), S = (5, 6)$
23. $P = (3, 9), Q = (2, 8); R = (5, 3), S = (-2, 1)$
24. $P = (-1, -2), Q = (3, 0); R = (5, 6), S = (1, 4)$
25. $P = (3, 7), Q = (1, 4); R = (6, 0), S = (1, 5)$
26. $P = (3, 0), Q = (-1, -1); R = (1, 2), S = (2, -2)$
27. $P = (3, 5), Q = (3, -2); R = (4, 3), S = (4, 5)$
28. $P = (1, 7), Q = (-2, 7); R = (5, 9), S = (5, 0)$

In Exercises 29–38 determine the equations of the lines that pass through P and are (a) parallel to, (b) perpendicular to, the given line L.

29. $P = (1, 1), L: y = 2x + 8$
30. $P = (4, 0), L: y = 5 - x$
31. $P = (0, -3), L: 2x + y + 1 = 0$
32. $P = (0, 0), L: 4(y - 3) = 3(x + 2)$
33. $P = (2, 5), L: x + y = 2$
34. $P = (-3, 1), L: x - 2y = 4$
35. $P = (2, 8), L: y = -2$
36. $P = (3, 4), L: x + 3 = 0$
37. $P = (0, 0), L: x = 9$
38. $P = (2, 0), L: y = 2$

In Exercises 39–42 determine all pairs of (a) parallel lines, (b) perpendicular lines:

39. $L_1: y = 3x$
    $L_2: y = -3x$
    $L_3: y = \dfrac{x}{3}$
    $L_4: y = 3(x - 2)$
    $L_5: 3y = 2 - x$

40. $L_1: 2x + 3y = 5$
    $L_2: 9y = 5 - 6x$
    $L_3: 3x - 2y = 7$
    $L_4: 3x + 2y = 1$
    $L_5: y = \dfrac{3x}{2}$

41. $L_1: x + y = 4$
    $L_2: x - y = 7$
    $L_3: y = 3 - x$
    $L_4: y = x$
    $L_5: y = -x$

42. $L_1: x = 5$
    $L_2: y = -5$
    $L_3: x = \dfrac{-1}{5}$
    L_4: the x-axis
    L_5: the y-axis

In Exercises 43–45 let L_1, L_2, and L_3 be lines in the plane.

43. Suppose $L_1 \parallel L_2$ and $L_2 \parallel L_3$. Is $L_1 \parallel L_3$?
44. (a) Let L_1, L_2, and L_3 be *distinct* lines. Suppose $L_1 \perp L_2$ and $L_2 \perp L_3$. What can be said about L_1 and L_3?

    (b) In (a) if the word "distinct" is omitted, what can be said about L_1 and L_3?
45. Suppose $L_1 \perp L_2$ and $L_2 \parallel L_3$. What can be said about L_1 and L_3?

REVIEW EXERCISES FOR CHAPTER 9.

9.1 PROPORTION

1. Determine the ratio of y to x, $x \neq 0$, if $5y = -9x$.
2. Find the values of a, b, and c in the following proportions:

   (a) $\frac{a}{3} = \frac{3}{9}$ (b) $\frac{5}{b} = \frac{-10}{14}$ (c) $\frac{\frac{1}{3}}{\frac{2}{9}} = \frac{c}{4}$

3. Lemons sell for 75 cents a dozen. How much do 4 lemons cost?
4. The triangles in Figure 9.42 are similar. Corresponding sides are marked with the same number of bars. Determine the lengths of sides a and b.

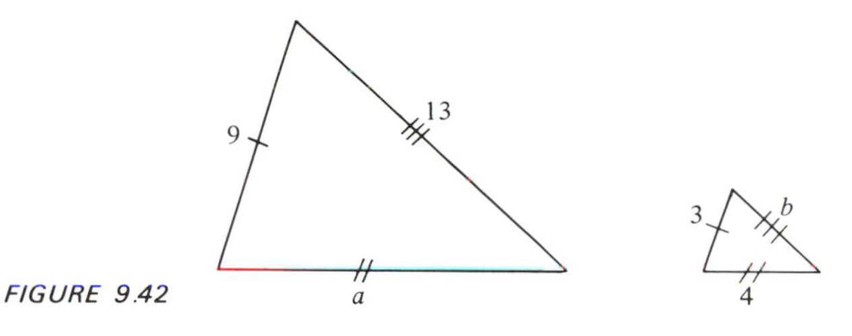

FIGURE 9.42

9.2 POINT-SLOPE EQUATION OF A LINE

5. Determine the slope of the line through (4, 2) and (6, 3).
6. Determine the slope of the line through (−2, 1) and (2, −2).
7. A line L has slope 3 and passes through (2, 5). Determine the x-coordinate of the point on L with y-coordinate −4.
8. (a) Determine the equation of the line that has slope 2 and that passes through the point (3, −1).

   (b) Graph this line.
9. (a) Determine the equation of the line that has slope $\frac{-1}{2}$ and that passes through the point (1, 2).

   (b) Graph this line.
10. Which of the following lines are horizontal? Which are vertical?

    (a) $y = -1$ (b) $x = 2$ (c) $y = x$
11. Determine the equation of the vertical line through $\left(1, \frac{-1}{2}\right)$.

9.3 ALTERNATE FORMS OF LINEAR EQUATIONS

12. Determine

    (a) the x-intercept and (b) the y-intercept of the line

    $$y + 2 = 2(x + 2).$$

13. Determine the slope-intercept form of the equation of the line that passes through (2, 5) and (5, 2).

14. Determine the 2-point form of the equation of the line that has x-intercept 2 and y-intercept -4.

15. Determine the general form of the equation of the line that passes through (2, 6) and $(-1, 1)$.

16. Determine (any form of) the equation of the line that passes through $(-1, -4)$ and has x-intercept -2.

9.4 PARALLEL AND PERPENDICULAR LINES

In Exercises 17–19 determine whether the given lines are (a) parallel, (b) perpendicular, or (c) neither parallel nor perpendicular.

17. $y = 4x - 1$ and $y = 1 + 4x$

18. $y = \frac{x}{2}$ and $y = 1 - 2x$

19. $y = 1 - x$ and $y = 1 + x$

20. Determine whether the line through (3, 2) and (2, 3) is

    (a) parallel to,

    (b) perpendicular to, or

    (c) neither parallel nor perpendicular to the line through (1, 4), and $(-1, -4)$.

21. Determine the equations of the lines that pass through (2, 5) and are (a) parallel to, (b) perpendicular to the line given by

$$y = 2x - 1.$$

22. Determine all pairs of (a) parallel lines, (b) perpendicular lines:

$$L_1: y = 2x \qquad L_2: y = \frac{-x}{2} \qquad L_3: y = 4 - 2x$$

$$L_4: y = 1 - \frac{x}{2} \qquad L_5: y = 1 + \frac{x}{2}$$

TEST YOURSELF ON CHAPTER 9.

1. Find the value of a and b in the following proportions:

(a) $\frac{15}{a} = \frac{25}{10}$ (b) $\frac{\frac{1}{2}}{\frac{3}{4}} = \frac{b}{9}$

2. Determine the slope of the line that passes through (2, 4) and $(-1, 0)$.

3. Determine the equation of the line that has slope 5 and that passes through the point $(2, -2)$.

4. Determine the equation of the horizontal line that passes through the point $(-1, -2)$.
5. Determine (a) the x-intercept and (b) the y-intercept of the line given by

$$y - 1 = -(x + 4).$$

6. Determine the general form of the equation of the line that passes through $(2, -1)$ and $(-1, 4)$.
7. Determine any form of the equation of the line that passes through the points $(2, 6)$ and $(1, 1)$.
8. Are the lines given by

$$y = 1 + 3x \qquad \text{and} \qquad y = \frac{4 - x}{3}$$

   (a) parallel, (b) perpendicular, or (c) neither parallel nor perpendicular?
9. Determine the equation of the line that passes through $(2, -3)$ and that is parallel to the line given by

$$y - 2 = 4x + 2.$$

10. Determine all pairs of (a) parallel lines, (b) perpendicular lines:

$$L_1: y = 4 - x \qquad L_2: y + 1 = x \qquad L_3: y = x - 4$$

$$L_4: y = 2x - 4 \qquad L_5: 2y = 4 - x$$

10

systems of linear equations

10.1 LINEAR SYSTEMS IN TWO VARIABLES

There are three possibilities for the intersection of two lines in the plane. Let L_1 and L_2 be these lines.

CASE 1. L_1 and L_2 do not intersect. In this case $L_1 \parallel L_2$ [See Figure 10.1(a).]

CASE 2. L_1 and L_2 intersect at a single point. [See Figure 10.1(b).]

CASE 3. L_1 and L_2 are identical. [See Figure 10.1(c).]

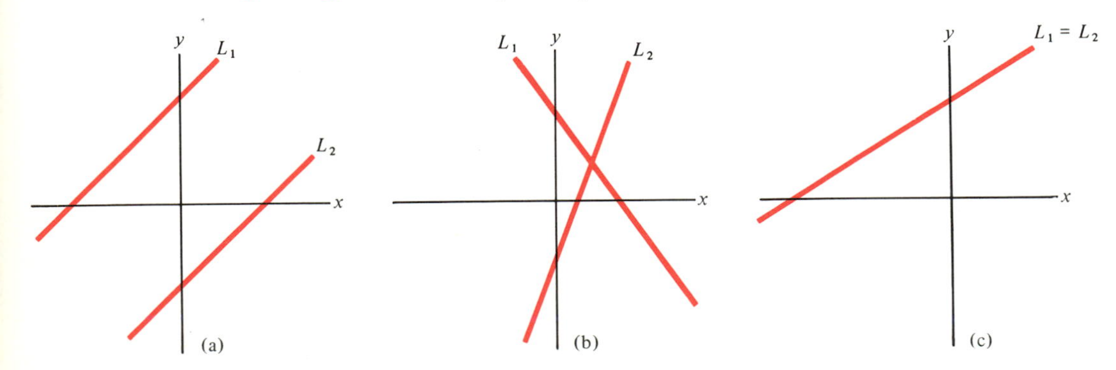

FIGURE 10.1 (a) L_1 and L_2 do not intersect. $L_1 \parallel L_2$ (b) L_1 and L_2 intersect at a single point. (c) L_1 and L_2 are identical.

Thus, *distinct* lines (Cases 1 and 2) intersect at a single point or not at all. In this section, you will consider Case 2, in which there is a single point of intersection. Case 1, in which there is no intersection (parallel lines) was discussed in Section 9.4.

Suppose neither L_1 nor L_2 is vertical. Let

$$m_1 = \text{slope } L_1; \qquad m_2 = \text{slope } L_2.$$

In Cases 1 and 3,

$$m_1 = m_2$$

Case 2 is characterized by

$$m_1 \neq m_2.$$

One way of determining the *approximate* intersection of two lines is by graphing each of the lines on the same coordinate system to see where they intersect.

EXAMPLE 1. Determine the intersection of

$$L_1\text{: } 2x + 3y = 8$$

and

$$L_2\text{: } 3x - y = 1$$

graphically.

SOLUTION.

$$m_1 = \text{slope } L_1 = \frac{-2}{3}$$

$$m_2 = \text{slope } L_2 = 3$$

There is a single point of intersection because $m_1 \neq m_2$.

Determine two points (for convenience, the x- and y- intercepts) on each line.

L_1:	x	0	4
	y	$\frac{8}{3}$	0

L_2:	x	0	$\frac{1}{3}$
	y	-1	0

TABLE 10.1

Draw L_1 through $\left(0, \frac{8}{3}\right)$ and $(4, 0)$ and L_2 through $(0, -1)$ and $\left(\frac{1}{3}, 0\right)$. The point of intersection can be read from the graph as $(1, 2)$. [See Figure 10.2, page 298.]

■

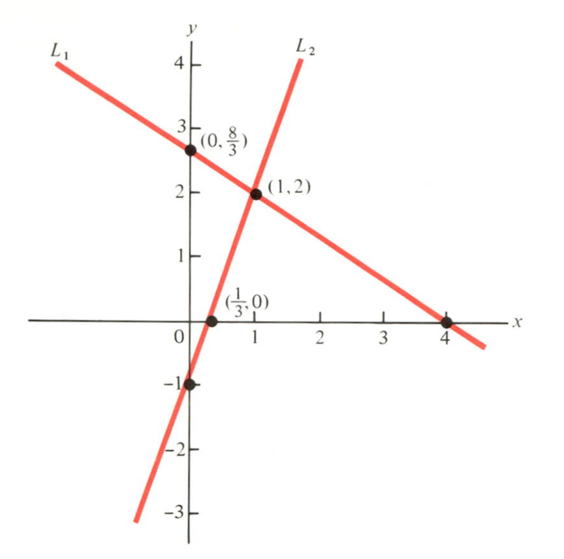

FIGURE 10.2 The single point of intersection is (1, 2).

The above graphical method tends to be inaccurate, particularly when the x- or y-coordinate of the intersection point is not an integer. There are also algebraic techniques for determining the intersection point of two given lines L_1 and L_2. The algebraic methods, which give the *exact* intersection, are preferable.

Recall that the general form of the equation of a line is

$$Ax + By + C = 0,$$

where A and B are not *both* 0. You can also write

$$Ax + By = -C.$$

Therefore the equations of lines L_1 and L_2 can be written as

$$a_1x + b_1y = k_1$$

$$a_2x + b_2y = k_2$$

(where a_1 and b_1 are not both 0, and a_2 and b_2 are not both 0).

Together, the equations of L_1 and L_2 form a **system of two linear equations in two variables.** (The word "linear" indicates that each equation represents a line.)

DEFINITION. The ordered pair (x_0, y_0) is called a **solution of the system** if (x_0, y_0) is a solution of (or satisfies) *both* equations of the system.

Solving the system is the algebraic equivalent of graphing these lines to find their point of intersection. In other words, *a solution of the system corresponds to a point of intersection of the lines.*

Thus in Example 1, you considered the system:

$$2x + 3y = 8$$
$$3x - y = 1$$

The solution of this system represents the point of intersection, (1, 2), of the corresponding lines.

In the **substitution method** for solving two equations:

1. Solve one equation for one variable—say y—in terms of the other, x (as you did when you solved literal equations in Section 6.4).
2. Then substitute this expression for y in the other equation and solve for x.
3. Replace x by this value in the expression for y, and solve for y.

EXAMPLE 2. Solve the system:

$$2x + y = 12$$
$$3x - 2y = 11$$

SOLUTION. From the first equation,

$$y = 12 - 2x$$

Substitute $12 - 2x$ for y in the second equation:

$$3x - 2(12 - 2x) = 11$$
$$3x - 24 + 4x = 11$$
$$7x = 35$$
$$x = 5$$

Replace x by 5 in the expression for y:

$$y = 12 - 2x$$

Thus

$$y = 12 - 2(5)$$
$$y = 2$$

The solution of the system of equations is (5, 2).

CHECK. Replace x by 5 and y by $\mathbf{2}$ in each of the original equations:

$2(5) + \mathbf{2} \stackrel{?}{=} 12$	$3(5) - 2(\mathbf{2}) \stackrel{?}{=} 11$
$12 \stackrel{\checkmark}{=} 12$	$11 \stackrel{\checkmark}{=} 11$

■

EXAMPLE 3. Solve the system:

$$7x - 3y = 7$$
$$x + 8y = 1$$

SOLUTION. From the second equation,

$$x = 1 - 8y$$

Substitute $1 - 8y$ for x in the first equation.

$$7(1 - 8y) - 3y = 7$$
$$7 - 56y - 3y = 7$$
$$-59y = 0$$
$$y = 0$$

Replace y by 0 in the expression for x:

$$x = 1 - 8y$$

Thus

$$x = 1 - 8(0)$$
$$x = 1$$

The solution of the system is $(1, 0)$. ■

You can also eliminate a variable by **"adding equations"** (or really, adding the corresponding *sides* of the equations). For if

$$\square = \square$$

and

$$\square = \square,$$

then

$$\square + \square = \square + \square$$

EXAMPLE 4. Solve the system:

$$4x + 3y = -1$$
$$2x - 3y = 13$$

SOLUTION. Note that $+3y$ occurs in the first equation and $-3y$ in the second. Adding, you obtain:

$$\begin{array}{l} 4x + 3y = -1 \\ \underline{2x - 3y = \ \ 13} \\ 6x \qquad\ \ = \ \ 12 \end{array}$$

Therefore

$$x = 2$$

Replace x by 2 in, say, the first equation:

$$\begin{aligned} 4(2) + 3y &= -1 \\ 8 + 3y &= -1 \\ 3y &= -9 \\ y &= -3 \end{aligned}$$

(Had you replaced x by 2 in the second equation, you would have also obtained $y = -3$.) The solution is $(2, -3)$. ■

In order to use this method effectively, you first may have to transform one or both of the equations into an equivalent equation.

EXAMPLE 5. Solve the system:

$$\begin{aligned} 5x + 4y &= 8 \\ 3x + 2y &= 6 \end{aligned}$$

SOLUTION. Note that $4y$ occurs in the first equation and $2y$ in the second. Multiply both sides of the second equation by -2:

$$-6x - 4y = -12$$

Now perform the addition to obtain:

$$\begin{array}{rcr} 5x + 4y &=& 8 \\ -6x - 4y &=& -12 \\ \hline -\ x \qquad &=& -\ 4 \end{array}$$

or

$$x \qquad = \quad 4$$

Replace x by 4 in the first equation:

$$\begin{aligned} 5(4) + 4y &= 8 \\ 20 + 4y &= 8 \\ 4y &= -12 \\ y &= -3 \end{aligned}$$

The solution is $(4, -3)$. ■

Both equations may have to be transformed so that upon addition one of the variables is eliminated.

EXAMPLE 6. Solve the system:

$$2x - 3y = 3$$
$$3x - 2y = 7$$

SOLUTION. Multiply, as indicated, in order to eliminate y.

Multiply both sides by 2. → $2x - 3y = 3$

$3x - 2y = 7$ ← Multiply both sides by -3.

$$\begin{aligned} 4x - 6y &= 6 \\ -9x + 6y &= -21 \\ \hline -5x \qquad &= -15 \end{aligned}$$

Add.

Thus

$$x = 3$$

Replace x by 3 in the first equation:

$$\begin{aligned} 2(3) - 3y &= 3 \\ 6 - 3y &= 3 \\ -3y &= -3 \\ y &= 1 \end{aligned}$$

The solution is (3, 1). ■

EXAMPLE 7. Three times a number is two less than another number. Seven times the first number is one more than twice the second number. Find these two numbers.

SOLUTION. Let x represent the first number and y the second. Set up a system of linear equations and solve.

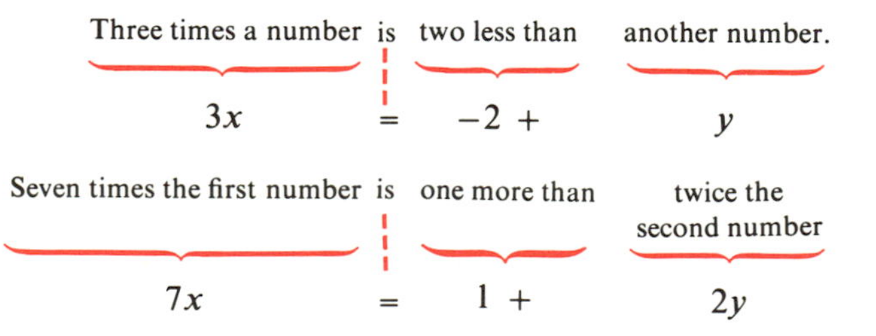

From the first equation,

$$3x + 2 = y$$

Replace y by $3x + 2$ in the second equation.

$$7x = 1 + 2(3x + 2)$$

$$7x = 1 + 6x + 4$$

$$x = 5$$

$$y = 3x + 2$$

$$y = 17$$

The numbers are 5 and 17. ■

EXAMPLE 8. Determine a and b so that the line given by

$$ax + by + 1 = 0$$

goes through the points $(2, -5)$ and $(-1, 1)$.

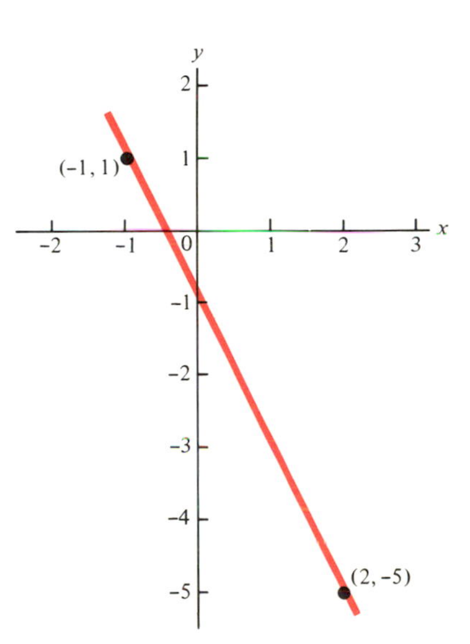

FIGURE 10.3 The line given by $2x + y + 1 = 0$

SOLUTION. Because the *point* $(2, -5)$ is on the line, the *ordered pair* $(2, -5)$ satisfies the equation of the line. Thus replace x by 2 and y by -5.
Therefore one equation *in the variables a and b* is

$$a(2) + b(-5) + 1 = 0$$

or

$$2a - 5b = -1.$$

Similarly, replace x by -1 and y by 1 in the equation of the line, and obtain a second equation in a and b:

$$-a + b + 1 = 0$$

or

$$b = a - 1$$

You now have two equations in a and b:

$$2a - 5b = -1$$

$$b = a - 1$$

Replace b by $a - 1$ in the first equation,

and obtain:

$$2a - 5(a - 1) = -1$$

Thus

$$2a - 5a + 5 = -1$$
$$-3a = -6$$
$$a = 2$$

Also,

$$b = a - 1$$
$$b = 2 - 1$$
$$b = 1$$

Thus, $a = 2$ and $b = 1$; the equation of the line is then

$$2x + y + 1 = 0.$$

■

EXERCISES

In Exercises 1–10 determine the intersection of L_1 and L_2 graphically.

1. L_1: $4x - 3y = 1$
   L_2: $x + y = 2$
2. L_1: $y = 3x$
   L_2: $y + 1 = 0$
3. L_1: $3x + y = 1$
   L_2: $x + y = 1$
4. L_1: $2x - y = 3$
   L_2: $x + 2y = -1$
5. L_1: $4x + 2y = 5$
   L_2: $x - 2y = 0$
6. L_1: $x - 2y = 1$
   L_2: $-x + 3y = -3$
7. L_1: $x - y = 5$
   L_2: $3x - 5y = 5$
8. L_1: $x + 2y = 4$
   L_2: $2x + 5y = 10$
9. L_1: $5x - 2y = 1$
   L_2: $x - 3y = 8$
10. L_1: $x + y = 2$
    L_2: $2x - y = 10$

In Exercises 11–18 solve each system by the substitution method.

11. $2x + y = 4$
    $3x - y = 1$
12. $3x - 4y = 0$
    $x + y = 7$
13. $x + y = 7$
    $3x - 5y = 5$
14. $x - 3y = 8$
    $3x - 2y = 10$
15. $4u + v = 24$
    $u - 3v = 6$
16. $5s + t = 3$
    $s - 2t = 5$
17. $4x - 8y = 0$
    $x + 2y = 1$
18. $7x - 2y = 1$
    $2x + 4y = 14$

In Exercises 19–26 solve each system by "adding equations".

19. $2x + y = 9$
    $3x - y = 1$
20. $5x - 2y = 3$
    $3x + 2y = 21$
21. $7x - 3y = 4$
    $-7x + 7y = 0$

22. $2x + 5y = 22$
    $4x - 5y = 14$

23. $u + 2v = 3$
    $u - 2v = 15$

24. $5x + 4y = -3$
    $6x - 2y = 10$

25. $9x - 7y = 15$
    $3x + 2y = 18$

26. $3x + 2y = 10$
    $2x + 3y = 10$

In Exercises 27–38 solve by either substituting or adding.

27. $2x - 7y = 0$
    $x + 4y = 0$

28. $x + 3y = 1$
    $5x + 14y = 8$

29. $5x + 3y = 24$
    $3x + 2y = 14$

30. $10x + 100y = 2$
    $100x - 200y = 8$

31. $2s + 5t = -3$
    $5s + 7t = 9$

32. $x + y + 2 = 0$
    $3x - 2y + 1 = 0$

33. $2x + 3y = -6$
    $7x + 4y = 5$

34. $9x - 8y = 26$
    $4x - 3y = 16$

35. $3x + 2y = 5$
    $12x + 7y = 25$

36. $-5x + 4y = 1$
    $3x - 2y = 0$

37. $3x - 2y = 9$
    $5x - 4y = 7$

38. $6x - 3y = 6$
    $7x - 4y = -2$

In Exercises 39–49 set up a system of linear equations and solve. (Note that Exercises 39–41 can easily be solved by the methods of Chapter 7.)

39. The sum of two numbers is 30. The larger number is 4 more than the smaller. Find these numbers.

40. Alice has \$3.40 in dimes and quarters. If the number of dimes and quarters were reversed, she would have \$4.30. How many dimes does she have?

41. Ted is twice as old as his brother. Four years ago he was four times as old as his brother. How old is each boy?

42. Twice a number is one less than three times another number. Five times the first number is one more than seven times the second. Find these two numbers.

43. A bookstore shelves books vertically. A 2-foot shelf holds 12 copies of a dictionary and 3 copies of an encyclopedia. A 3-foot shelf holds 4 copies of the dictionary and 15 copies of the encyclopedia. How thick is each book?

44. The perimeter of an isosceles triangle is 20 inches. If the length of each of the equal sides were increased by 3 inches and the base length were doubled, the perimeter would be 34 inches. Find the length of each side.

45. Determine a and b so that the line given by $ax + by - 5 = 0$ passes through the points $(2, 1)$ and $(5, 5)$.

46. Determine a and b so that the line given by $ax + by - 2 = 0$ passes through the points $(1, 1)$ and $(4, 6)$.

47. Determine a and b so that the line given by $ax + by - 2 = 0$ passes through the points $(1, 1)$ and $(4, -2)$.

48. Determine a and b so that the line given by $ax + by - 10 = 0$ has x-intercept 2 and y-intercept 5.

49. Determine a and b so that the line given by $ax + by + 2 = 0$ passes through the point $(-1, -1)$ and has slope -1.

10.2 LINEAR SYSTEMS IN THREE VARIABLES

The equation

$$2x + y - 3z = 10$$

is in *three* variables. Set $x = 4, y = 5, z = 1$, and obtain:

$$2 \cdot 4 + 5 - 3 \cdot 1 = 10$$

or

$$10 = 10$$

Thus the numbers

$$4, 5, 1 \qquad \text{(in this order)}$$

serve as a "solution" of the given equation.

DEFINITION. Let x_0, y_0, and z_0 be three real numbers. The symbol

$$(x_0, y_0, z_0)$$

is called an **ordered triple,** and represents the given numbers in the order written. The ordered triple (x_0, y_0, z_0) is called a **solution** of an equation

$$Ax + By + Cz = D, \qquad A, B, C \text{ not all } 0,$$

if

$$Ax_0 + By_0 + Cz_0 = D.$$

The equations

$$a_1x + b_1y + c_1z = d_1$$

$$a_2x + b_2y + c_2z = d_2$$

$$a_3x + b_3y + c_3z = d_3$$

constitute a **system of three linear equations in three variables** (provided a_1, b_1, and c_1 are not *all* 0; a_2, b_2, and c_2 are not *all* 0; and a_3, b_3, and c_3 are not *all* 0).

The ordered triple (x_0, y_0, z_0) is called a **solution of the** above **system** if (x_0, y_0, z_0) is a solution of *all three* equations of the system.

To solve a system of three linear equations in three variables, either substitute or "add equations" to obtain a system of two linear equations in two variables. Again use one of these methods to solve the resulting system.

EXAMPLE 1. Solve the system:

$$x - y - z = 0$$

$$x - y + z = 4$$

$$x + y + z = 10$$

SOLUTION. From the first equation,

(A) $\quad x = y + z$

Replace x by $y + z$ in the other equations, and obtain the system of two equations in two variables:

$$(y + z) - y + z = 4$$

(B) $\quad (y + z) + y + z = 10$

The first of these becomes

$$2z = 4$$

or

$$z = 2.$$

Replace z by 2 in Equation (B), and obtain:

$$2y + 4 = 10$$

$$2y = 6$$

$$y = 3$$

Replace y by 3 and **z** by **2** in Equation (A), and obtain:

$$x = 3 + \mathbf{2}$$

$$x = 5$$

The solution is (5, 3, 2).

CHECK. Replace x by 5, **y** by **3**, and **z** by **2** in each of the original equations.

$x - y - z \stackrel{?}{=} 0$	$x - y + z \stackrel{?}{=} 4$	$x + y + z \stackrel{?}{=} 10$
$5 - \mathbf{3} - \mathbf{2} \stackrel{?}{=} 0$	$5 - \mathbf{3} + \mathbf{2} \stackrel{?}{=} 4$	$5 + \mathbf{3} + \mathbf{2} \stackrel{?}{=} 10$
$0 \stackrel{\checkmark}{=} 0$	$4 \stackrel{\checkmark}{=} 4$	$10 \stackrel{\checkmark}{=} 10$

■

EXAMPLE 2. Solve the system:

(A) $\quad x + y \quad = 3$

(B) $\quad y + z = 5$

(C) $\quad x \quad + z = 4$

SOLUTION. From the first equation you obtain:

(A′) $\quad y = 3 - x$

Replace y by $3 - x$ in Equation (B), and consider this result together with Equation (C):

(B′) $\quad (3 - x) + z = 5$

or

(B″) $\quad -x + z = 2$
(C) $\quad x + z = 4$

"Add" these two equations in the variables x and z.

$$2z = 6$$
$$z = 3$$

Replace z by 3 in Equation (C):

$$x + 3 = 4$$
$$x = 1$$

Replace x by 1 in Equation (A′):

$$y = 3 - 1$$
$$y = 2$$

The solution is (1, 2, 3).

EXAMPLE 3. Solve the system:

(A) $\quad 4x + 3y - 3z = 2$

(B) $\quad 5x - 3y + 2z = 10$

(C) $\quad 2x - 2y + 3z = 14$

SOLUTION. Eliminate y so as to obtain two equations in x and z. First use Equations (A) and (B)

$$4x + 3y - 3z = 2$$
$$5x - 3y + 2z = 10$$

to obtain:

(D) $\quad 9x - z = 12$

or

(D′) $\quad 9x - 12 = z$

Next, multiply both sides of (B) by 2 and both sides of (C) by -3. Then add:

(B′) $\quad 10x - 6y + 4z = 20$

(C′) $\quad -6x + 6y - 9z = -42$
(E) $\quad 4x - 5z = -22$

You now have two equations in x and z:

(D′) $9x - 12 = z$

(E) $4x - 5z = -22$

From (D′), replace z by $9x - 12$ in (E):

$$4x - 5(9x - 12) = -22$$
$$4x - 45x + 60 = -22$$
$$-41x = -82 \qquad \text{Divide by } -41.$$
$$x = 2$$

Replace x by 2 in (D′):

$$z = 9(2) - 12$$
$$z = 6$$

Replace x by 2 and z by 6 in (A);

$$4(2) + 3y - 3(6) = 2$$
$$8 + 3y - 18 = 2$$
$$3y = 12$$
$$y = 4$$

The solution is (2, 4, 6). ■

EXAMPLE 4. The sum of three numbers is 21. Six times the smallest number equals the sum of the others. The largest number is one less than the sum of the others. Find these numbers.

SOLUTION. Let x be the smallest of these numbers, let y be the next smallest, and let z be the largest.

The sum of the three numbers is 21.

(A) $x + y + z = 21$

Six times the smallest number equals the sum of the others.

(B) $6x = y + z$

The largest number is one less than the sum of the others.

(C) $z = -1 + (x + y)$

Equations (A), (B), and (C) are three equations in the three variables x, y, and z. From (C), replace z by $x + y - 1$ in Equations (A) and (B):

$$(A') \qquad x + y + (x + y - 1) = 21$$

$$(B') \qquad 6x = y + (x + y - 1)$$

You now have two equations in the two variables x and y. Simplify (A′):

$$(A'') \qquad 2x + 2y = 22$$

$$x + y = 11$$

$$(A''') \qquad x = 11 - y$$

Simplify (B′):

$$(B'') \qquad 5x = 2y - 1$$

From (A‴), replace x by $11 - y$ in (B″).

$$5(11 - y) = 2y - 1$$

$$55 - 5y = 2y - 1$$

$$56 = 7y$$

$$8 = y$$

Equation (A‴) becomes

$$x = 11 - 8$$

or

$$x = 3.$$

Replace x by 3 and y by **8** in (C):

$$z = -1 + (3 + \mathbf{8})$$

$$z = 10$$

The three numbers are 3, 8, and 10. ■

In later mathematics courses you will learn that, geometrically, a linear equation in three variables

$$ax + by + cz = k$$

(a, b, and c not *all* 0) represents a plane (in space). [See Figure 10.4.]

A solution of a system of three linear equations in three variables represents a point of intersection of three planes. (These observations will not be used here.)

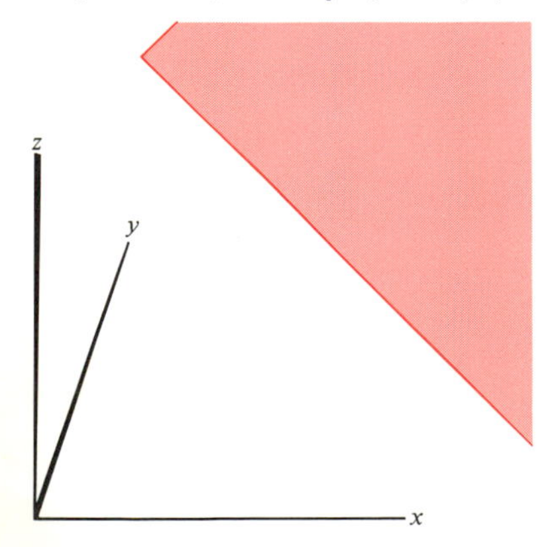

FIGURE 10.4 The equation $ax + by + cz = k$ (a, b, c not all 0) represents a plane in 3-dimensional space. Note that for 3-dimensional considerations a third coordinate axis (the z-axis) is added.

EXERCISES

In Exercises 1–20 solve each system. You may use any method.

1. $\begin{aligned} x + y \quad\quad &= 2 \\ x \quad + z &= 5 \\ y + z &= 5 \end{aligned}$

2. $\begin{aligned} x + y \quad\quad &= 6 \\ y + z &= 5 \\ x + \quad z &= 7 \end{aligned}$

3. $\begin{aligned} x + y \quad\quad &= 0 \\ y + 2z &= 5 \\ x \quad + z &= 4 \end{aligned}$

4. $\begin{aligned} x + 3y \quad\quad &= -2 \\ 2x \quad + z &= 0 \\ x + y + 2z &= 6 \end{aligned}$

5. $\begin{aligned} x + y \quad\quad &= 5 \\ x \quad + 2z &= 2 \\ x + y + z &= 3 \end{aligned}$

6. $\begin{aligned} x + y + z &= 2 \\ 5x \quad - 6z &= 9 \\ 4y + 10z &= 2 \end{aligned}$

7. $\begin{aligned} x - y + z &= 9 \\ 3x - 5y - 4z &= 2 \\ y + z &= 5 \end{aligned}$

8. $\begin{aligned} x + y + 3z &= 10 \\ 2x \quad + 5z &= 3 \\ -y - 3z &= -1 \end{aligned}$

9. $\begin{aligned} 2x + 3y - z &= 6 \\ x + 2y - z &= 1 \\ 4x + y - z &= 0 \end{aligned}$

10. $\begin{aligned} x + 5y - z &= 2 \\ 3x + 3y - 2z &= 4 \\ x - 7y + 5z &= -10 \end{aligned}$

11. $\begin{aligned} 2x - y + z &= 3 \\ 3x + y - 6z &= 7 \\ 5x - 2y - z &= 4 \end{aligned}$

12. $\begin{aligned} \frac{x}{2} + \frac{y}{4} + \frac{z}{4} &= 5 \\ \frac{x}{4} + \frac{y}{2} + \frac{z}{2} &= 4 \\ \frac{x}{4} + \frac{y}{4} + \frac{z}{2} &= 3 \end{aligned}$

13. $\begin{aligned} x + 5y - 10z &= 5 \\ x - 3y - 4z &= 9 \\ x - 7y + z &= 15 \end{aligned}$

14. $\begin{aligned} x + 3y + z &= 0 \\ 5x + y + 3z &= 0 \\ 2x + y + z &= 2 \end{aligned}$

15. $\begin{aligned} 5x - y + z &= -6 \\ 2x \quad + z &= 0 \\ 3x + y + 6z &= 0 \end{aligned}$

16. $\begin{aligned} 4x - 5y - z &= 5 \\ 2x + y - z &= 3 \\ 6x - 9y - 2z &= 13 \end{aligned}$

17. $\begin{aligned} 3x - 10y + 2z &= -4 \\ 2x + 3y - 5z &= 13 \\ 5x - 2y - 2z &= 0 \end{aligned}$

18. $\begin{aligned} 7x - 2y + 2z &= 19 \\ 3x + y - 4z &= -30 \\ 2x + 3y - 3z &= -16 \end{aligned}$

19. $\begin{aligned} x - 2y + z &= 0 \\ 3x + 4y + 3z &= 5 \\ 2x - 2y - z &= 4 \end{aligned}$

20. $\begin{aligned} 2x + 2y + z &= 1 \\ x + 8y + z &= 2 \\ 3x + 2y - 2z &= 3 \end{aligned}$

In Exercises 21–24 set up a system of three linear equations in three variables, and solve.

21. The sum of three numbers is 12. Five times the smallest number equals the sum of the other two. The largest number equals the sum of the other two. Find these numbers.

22. The sum of three numbers is 20. Four times the smallest number is the sum of the other two. Three times the largest number is five more than twice the sum of the other two. Find these numbers.

23. Helen has $2.60 in nickels, dimes, and quarters. The number of nickels and dimes together is one more than the number of quarters. The value of the nickels is that of the dimes. How many of each coin does she have?

24. With your two dollars you can buy 3 hot dogs, an order of french fries, and two sodas. You can save a dime by having 2 of each item. Or you can save a nickel by ordering 4 hot dogs and 1 soda. How much does a hot dog cost?

*10.3 2 x 2 DETERMINANTS AND CRAMER'S RULE

"Determinants" enable you to solve systems of linear equations mechanically.

DEFINITION. Let a, b, c, and d be any four numbers. The symbol

$$\begin{vmatrix} a & c \\ b & d \end{vmatrix}$$

is called a 2-**by**-2 (2×2) **determinant** with **entries** a, b, c, d; its **value** is

$$ad - bc.$$

EXAMPLE 1. Evaluate each 2×2 determinant:

(a) $\begin{vmatrix} 4 & 1 \\ 1 & 2 \end{vmatrix}$ (b) $\begin{vmatrix} 5 & 0 \\ 2 & -1 \end{vmatrix}$ (c) $\begin{vmatrix} 3 & 6 \\ 4 & 8 \end{vmatrix}$

SOLUTION.

(a) Here $a = 4, b = c = 1, d = 2$. Thus

$$\begin{vmatrix} 4 & 1 \\ 1 & 2 \end{vmatrix} = 4(2) - 1(1) = 7$$

(b) $$\begin{vmatrix} 5 & 0 \\ 2 & -1 \end{vmatrix} = 5(-1) - 2(0) = -5$$

(c) $$\begin{vmatrix} 3 & 6 \\ 4 & 8 \end{vmatrix} = 3(8) - 4(6) = 0$$

■

EXAMPLE 2. Determine b if $\begin{vmatrix} 5 & b \\ 3 & 4 \end{vmatrix} = 23.$

SOLUTION.

$$\begin{vmatrix} 5 & b \\ 3 & 4 \end{vmatrix} = 5(4) - 3b = 20 - 3b$$

*Optional topic.

Thus solve

$$20 - 3b = 23,$$

and obtain

$$-3b = 3,$$
$$b = -1.$$ ■

The following example will be used to determine a mechanical procedure for solving a system of two linear equations in two variables.

EXAMPLE 3.

(a) $$\begin{vmatrix} a_1 & b_1 \\ a_2 & b_2 \end{vmatrix} = a_1b_2 - a_2b_1$$

(b) $$\begin{vmatrix} k_1 & b_1 \\ k_2 & b_2 \end{vmatrix} = k_1b_2 - k_2b_1$$

(c) $$\begin{vmatrix} a_1 & k_1 \\ a_2 & k_2 \end{vmatrix} = a_1k_2 - a_2k_1$$ ■

Now consider two linear equations in two variables:

(A) $a_1x + b_1y = k_1$

(B) $a_2x + b_2y = k_2$

To eliminate y, multiply both sides of (A) by b_2, and both sides of (B) by $-b_1$.

$$a_1b_2x + b_1b_2y = k_1b_2$$
$$-a_2b_1x - b_1b_2y = -k_2b_1$$

Add:

$$(a_1b_2 - a_2b_1)x = k_1b_2 - k_2b_1$$

According to Example 3, parts (a) and (b), this equation can be expressed in terms of determinants:

$$\begin{vmatrix} a_1 & b_1 \\ a_2 & b_2 \end{vmatrix} x = \begin{vmatrix} k_1 & b_1 \\ k_2 & b_2 \end{vmatrix}$$

If the left-hand determinant is not zero,

$$x = \frac{\begin{vmatrix} k_1 & b_1 \\ k_2 & b_2 \end{vmatrix}}{\begin{vmatrix} a_1 & b_1 \\ a_2 & b_2 \end{vmatrix}}$$

Observe that the determinant in the *denominator* has as its entries the coefficients of x and y as they appear on the left side of the original equations. Call this determinant D:

$$\begin{matrix} a_1 x + b_1 y \\ a_2 x + b_2 y \end{matrix} \rightarrow \begin{vmatrix} a_1 & b_1 \\ a_2 & b_2 \end{vmatrix} = D$$

The determinant of the *numerator,* called D_x, is obtained from D by replacing a's (the coefficients of x) by k's:

$$D_x = \begin{vmatrix} k_1 & b_1 \\ k_2 & b_2 \end{vmatrix}$$

$$\begin{matrix} + b_1 y = k_1 \\ + b_2 y = k_2 \end{matrix} \rightarrow \begin{vmatrix} k_1 & b_1 \\ k_2 & b_2 \end{vmatrix} = D_x$$

Thus

$$x = \frac{D_x}{D}, \qquad \text{if } D \neq 0$$

Let D_y be the 2×2 determinant obtained from D by replacing b's (the coefficients of y) by k's. Thus

$$D_y = \begin{vmatrix} a_1 & k_1 \\ a_2 & k_2 \end{vmatrix}$$

$$\begin{matrix} a_1 x + \quad = k_1 \\ a_2 x + \quad = k_2 \end{matrix} \rightarrow \begin{vmatrix} a_1 & k_1 \\ a_2 & k_2 \end{vmatrix} = D_y$$

Similarly, it can be shown that

$$y = \frac{\begin{vmatrix} a_1 & k_1 \\ a_2 & k_2 \end{vmatrix}}{\begin{vmatrix} a_1 & b_1 \\ a_2 & b_2 \end{vmatrix}},$$

provided the determinant in the denominator, D, is not zero. The determinant in the numerator is D_y. Thus, when $D \neq 0$,

$$y = \frac{D_y}{D}$$

In this case *the system has the unique solution* (x, y), namely,

$$x = \frac{D_x}{D}, \qquad y = \frac{D_y}{D}$$

This solution by determinants is known as **Cramer's Rule.**

If $D = 0$, then the system has either no solution (geometrically, parallel lines) or infinitely many solutions (both equations represent the same line). Cramer's Rule, which involves division by D, does not apply.

EXAMPLE 4. Solve by Cramer's Rule:

$$2x + y = 3$$
$$3x + 2y = 5$$

SOLUTION.

$$D = \begin{vmatrix} 2 & 1 \\ 3 & 2 \end{vmatrix} = 2(2) - 3(1) = 1$$

$$D_x = \begin{vmatrix} 3 & 1 \\ 5 & 2 \end{vmatrix} = 3(2) - 5(1) = 1$$

$$D_y = \begin{vmatrix} 2 & 3 \\ 3 & 5 \end{vmatrix} = 2(5) - 3(3) = 1$$

Observe that $D \neq 0$; hence Cramer's Rule applies:

$$x = \frac{D_x}{D} = \frac{1}{1} = 1$$

$$y = \frac{D_y}{D} = \frac{1}{1} = 1$$

The unique solution is $(1, 1)$. ■

EXAMPLE 5. Solve by Cramer's Rule:

$$10x - 7y = 12$$
$$3x - 2y = 5$$

SOLUTION.

$$D = \begin{vmatrix} 10 & -7 \\ 3 & -2 \end{vmatrix} = 10(-2) - 3(-7) = 1$$

$$D_x = \begin{vmatrix} 12 & -7 \\ 5 & -2 \end{vmatrix} = 12(-2) - 5(-7) = 11$$

$$D_y = \begin{vmatrix} 10 & 12 \\ 3 & 5 \end{vmatrix} = 10(5) - 3(12) = 14$$

$D \neq 0$; hence, Cramer's Rule applies:

$$x = \frac{D_x}{D} = \frac{11}{1} = 11$$

$$y = \frac{D_y}{D} = \frac{14}{1} = 14$$

The unique solution is (11, 14). ■

EXAMPLE 6. Show that Cramer's Rule does not apply to the system:

$$3x + 5y = 1$$
$$6x + 10y = 4$$

SOLUTION.

$$D = \begin{vmatrix} 3 & 5 \\ 6 & 10 \end{vmatrix} = 3(10) - 6(5) = 0$$

Cramer's Rule only applies when $D \neq 0$. ■

In fact, the two lines represented by equations

(A) $3x + 5y = 1$

and

(B) $6x + 10y = 4$

have the same slope, $\frac{-3}{5}$, but are not identical. Note that the point $(2, -1)$ is on the line given by (A) because

$$3 \cdot 2 + 5(-1) = 1;$$

however, $(2, -1)$ is not on the line given by (B) because

$$6 \cdot 2 + 10(-1) \neq 4.$$

Thus the lines given by Equations (A) and (B) are parallel. [See Figure 10.5.]

Note that the left side of Equation (B) "is a multiple of" the left side of Equation (A). In general, the lines given by the equations

$$ax + by = c$$

and

$$kax + kby = kc$$

(a and b not both 0, and $k \neq 0$) are identical. For example,

$$3x + 5y = 1$$

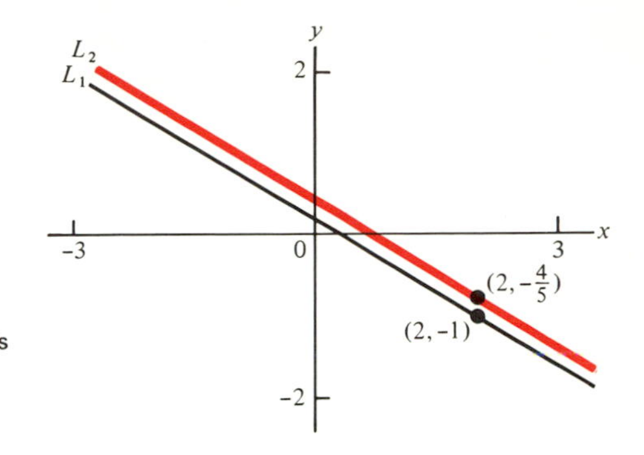

FIGURE 10.5 L_1 is given by $3x + 5y = 1$. L_2 is given by $6x + 10y = 4$.

$L_1 \parallel L_2$

and

$$6x + 10y = 2$$

represent the same line. But the lines represented by the equations

$$ax + by = c$$

and

$$kax + kby = lc, \qquad k \neq l,$$

are parallel. Thus

$$3x + 5y = 1$$

and

$$6x + 10y = 4$$

represent parallel lines.

EXERCISES

In Exercises 1–14 evaluate each determinant.

1. $\begin{vmatrix} 1 & 2 \\ 1 & 1 \end{vmatrix}$

2. $\begin{vmatrix} 3 & 1 \\ 4 & 1 \end{vmatrix}$

3. $\begin{vmatrix} 2 & 0 \\ 1 & 3 \end{vmatrix}$

4. $\begin{vmatrix} 5 & 2 \\ 1 & 0 \end{vmatrix}$

5. $\begin{vmatrix} 0 & 0 \\ 0 & 1 \end{vmatrix}$

6. $\begin{vmatrix} 8 & 3 \\ 2 & 1 \end{vmatrix}$

7. $\begin{vmatrix} 1 & -2 \\ 1 & 3 \end{vmatrix}$

8. $\begin{vmatrix} 2 & 4 \\ 4 & -1 \end{vmatrix}$

9. $\begin{vmatrix} 1 & -3 \\ -2 & 4 \end{vmatrix}$

10. $\begin{vmatrix} 5 & -9 \\ -7 & -4 \end{vmatrix}$

11. $\begin{vmatrix} -2 & -9 \\ 9 & \frac{1}{2} \end{vmatrix}$

12. $\begin{vmatrix} 6 & \frac{1}{2} \\ 8 & \frac{1}{3} \end{vmatrix}$

13. $\begin{vmatrix} \frac{1}{2} & \frac{2}{5} \\ \frac{1}{4} & \frac{-3}{5} \end{vmatrix}$

14. $\begin{vmatrix} .1 & .02 \\ -.4 & .01 \end{vmatrix}$

In Exercises 15–22 determine x.

15. $\begin{vmatrix} x & 5 \\ 2 & 1 \end{vmatrix} = 0$

16. $\begin{vmatrix} 3 & x \\ 8 & 1 \end{vmatrix} = -5$

17. $\begin{vmatrix} 2 & x \\ 9 & 1 \end{vmatrix} = 11$

18. $\begin{vmatrix} 4 & 3 \\ x & 2 \end{vmatrix} = -13$

19. $\begin{vmatrix} -1 & 2 \\ 4 & x \end{vmatrix} = -12$

20. $\begin{vmatrix} 4 & 9 \\ 8 & x \end{vmatrix} = 0$

21. $\begin{vmatrix} 5 & 2 \\ x & x \end{vmatrix} = 27$

22. $\begin{vmatrix} x & -x \\ -4 & 1 \end{vmatrix} = 5$

23. Evaluate: $\begin{vmatrix} a & a \\ b & b \end{vmatrix}$

24. Evaluate: $\begin{vmatrix} a & c \\ ka & kc \end{vmatrix}$

25. Compare $\begin{vmatrix} a & c \\ b & d \end{vmatrix}$ with $\begin{vmatrix} c & a \\ d & b \end{vmatrix}$.

26. Compare $\begin{vmatrix} a & c \\ b & d \end{vmatrix}$ with $\begin{vmatrix} b & d \\ a & c \end{vmatrix}$.

In Exercises 27–40: (a) Determine whether Cramer's Rule applies. (b) If it applies, use Cramer's Rule to find the solution.

27. $\begin{aligned} x + y &= 5 \\ x - y &= 1 \end{aligned}$

28. $\begin{aligned} 2x + y &= 1 \\ 3x + y &= 2 \end{aligned}$

29. $\begin{aligned} x - y &= 0 \\ x + 2y &= 1 \end{aligned}$

30. $\begin{aligned} 6x - 3y &= 0 \\ x + 2y &= 5 \end{aligned}$

31. $\begin{aligned} x + 3y &= 1 \\ 3x + 9y &= 2 \end{aligned}$

32. $\begin{aligned} 5x - 3y &= 10 \\ x - y &= -4 \end{aligned}$

33. $\begin{aligned} 2x + 3y &= -10 \\ x + 6y &= 1 \end{aligned}$

34. $\begin{aligned} 4x - 6y &= 8 \\ -2x + 3y &= -4 \end{aligned}$

35. $\begin{aligned} 5x + 3y &= 3 \\ 3x + 5y &= 5 \end{aligned}$

36. $\begin{aligned} 2x - 7y &= 3 \\ 9x + 2y &= 47 \end{aligned}$

37. $\begin{aligned} 15x + 10y &= 25 \\ 3x \qquad &= 10 \end{aligned}$

38. $\begin{aligned} 9x - 2y &= 3 \\ -4y &= 1 \end{aligned}$

39. $\begin{aligned} \frac{x}{2} + \frac{y}{3} &= \frac{1}{6} \\ 2x + 3y &= 4 \end{aligned}$

40. $\begin{aligned} \frac{x}{4} + \frac{y}{2} &= \frac{1}{2} \\ \frac{x}{2} - \frac{y}{4} &= \frac{1}{2} \end{aligned}$

*10.4 3 x 3 DETERMINANTS AND CRAMER'S RULE

3-by-3 (3×3) determinants are defined in terms of 2×2 determinants.

DEFINITION. Let $a_1, a_2, a_3, b_1, b_2, b_3, c_1, c_2$, and c_3 be any nine numbers. The symbol

$$\begin{vmatrix} a_1 & b_1 & c_1 \\ a_2 & b_2 & c_2 \\ a_3 & b_3 & c_3 \end{vmatrix}$$

is called a **3-by-3** (3×3) **determinant;** its **entries** are the above nine numbers; its **value** is

$$a_1\begin{vmatrix} b_2 & c_2 \\ b_3 & c_3 \end{vmatrix} - a_2\begin{vmatrix} b_1 & c_1 \\ b_3 & c_3 \end{vmatrix} + a_3\begin{vmatrix} b_1 & c_1 \\ b_2 & c_2 \end{vmatrix}.$$

The **rows** and **columns** of a 3×3 determinant are as indicated:

first row → $a_1 \quad b_1 \quad c_1$
second row → $a_2 \quad b_2 \quad c_2$
third row → $a_3 \quad b_3 \quad c_3$

first column ↓ second column ↓ third column ↓

$$\begin{vmatrix} a_1 & b_1 & c_1 \\ a_2 & b_2 & c_2 \\ a_3 & b_3 & c_3 \end{vmatrix}$$

By definition, a 3×3 determinant can be evaluated in terms of 2×2 determinants, as follows:

$$\begin{vmatrix} a_1 & b_1 & c_1 \\ a_2 & b_2 & c_2 \\ a_3 & b_3 & c_3 \end{vmatrix} = a_1\underbrace{\begin{vmatrix} b_2 & c_2 \\ b_3 & c_3 \end{vmatrix}}_{\text{Drop first row, first column.}} - a_2\underbrace{\begin{vmatrix} b_1 & c_1 \\ b_3 & c_3 \end{vmatrix}}_{\text{Drop second row, first column.}} + a_3\underbrace{\begin{vmatrix} b_1 & c_1 \\ b_2 & c_2 \end{vmatrix}}_{\text{Drop third row, first column.}}$$

$$= a_1(b_2c_3 - b_3c_2) - a_2(b_1c_3 - b_3c_1) + a_3(b_1c_2 - b_2c_1)$$

*Optional topic.

Because the value of each 2 × 2 determinant is a number, the resulting expression is a number.

Note that the signs preceding the 2 × 2 determinants alternate:

$$+ \quad - \quad +$$

Also, each 2 × 2 determinant is obtained from the 3 × 3 determinant by dropping the appropriate row along with the first column. Thus, to obtain the *second* 2 × 2 determinant

$$\begin{vmatrix} b_1 & c_1 \\ b_3 & c_3 \end{vmatrix},$$

drop the *second* row and first column:

$$\begin{vmatrix} a_1 & b_1 & c_1 \\ a_2 & b_2 & c_2 \\ a_3 & b_3 & c_3 \end{vmatrix} \rightarrow \begin{vmatrix} b_1 & c_1 \\ b_3 & c_3 \end{vmatrix}$$

EXAMPLE 1. Evaluate the 3 × 3 determinant:

$$\begin{vmatrix} 2 & 1 & 4 \\ 6 & 1 & 2 \\ 1 & 2 & 1 \end{vmatrix}$$

SOLUTION.

$$\begin{aligned}
\begin{vmatrix} 2 & 1 & 4 \\ 6 & 1 & 2 \\ 1 & 2 & 1 \end{vmatrix} &= 2\begin{vmatrix} 2 & 1 & 4 \\ 6 & 1 & 2 \\ 1 & 2 & 1 \end{vmatrix} - 6\begin{vmatrix} 2 & 1 & 4 \\ 6 & 1 & 2 \\ 1 & 2 & 1 \end{vmatrix} + 1\begin{vmatrix} 2 & 1 & 4 \\ 6 & 1 & 2 \\ 1 & 2 & 1 \end{vmatrix} \\
&= 2\begin{vmatrix} 1 & 2 \\ 2 & 1 \end{vmatrix} - 6\begin{vmatrix} 1 & 4 \\ 2 & 1 \end{vmatrix} + 1\begin{vmatrix} 1 & 4 \\ 1 & 2 \end{vmatrix} \\
&= 2[1(1) - 2(2)] - 6[1(1) - 2(4)] + 1[1(2) - 1(4)] \\
&= 2(-3) - 6(-7) + (-2) \\
&= 34
\end{aligned}$$

■

EXAMPLE 2. Evaluate the 3 × 3 determinant:

$$\begin{vmatrix} 4 & 1 & 0 \\ 0 & 2 & -8 \\ -2 & -1 & -1 \end{vmatrix}$$

SOLUTION.

$$\begin{vmatrix} 4 & 1 & 0 \\ 0 & 2 & -8 \\ -2 & -1 & -1 \end{vmatrix} = 4\begin{vmatrix} 2 & -8 \\ -1 & -1 \end{vmatrix} - \underbrace{0\begin{vmatrix} 1 & 0 \\ -1 & -1 \end{vmatrix}}_{=0} + (-2)\begin{vmatrix} 1 & 0 \\ 2 & -8 \end{vmatrix}$$

$$= 4[2(-1) - (-1)(-8)] - 2[1(-8) - 2(0)]$$

$$= 4(-10) - 2(-8)$$

$$= -24$$

■

EXAMPLE 3. Evaluate the 3 × 3 determinant:

$$\begin{vmatrix} 3 & 4 & 5 \\ 2 & 0 & 2 \\ 4 & -1 & 0 \end{vmatrix}$$

SOLUTION.

$$\begin{vmatrix} 3 & 4 & 5 \\ 2 & 0 & 2 \\ 4 & -1 & 0 \end{vmatrix} = 3\begin{vmatrix} 0 & 2 \\ -1 & 0 \end{vmatrix} - 2\begin{vmatrix} 4 & 5 \\ -1 & 0 \end{vmatrix} + 4\begin{vmatrix} 4 & 5 \\ 0 & 2 \end{vmatrix}$$

$$= 3(0 + 2) - 2(0 + 5) + 4(8 - 0)$$

$$= 28$$

■

EXAMPLE 4. Determine x:

$$\begin{vmatrix} 2 & 5 & 0 \\ x & 1 & 4 \\ 3 & -1 & 1 \end{vmatrix} = 100$$

SOLUTION.

$$\begin{vmatrix} 2 & 5 & 0 \\ x & 1 & 4 \\ 3 & -1 & 1 \end{vmatrix} = 2\begin{vmatrix} 1 & 4 \\ -1 & 1 \end{vmatrix} - x\begin{vmatrix} 5 & 0 \\ -1 & 1 \end{vmatrix} + 3\begin{vmatrix} 5 & 0 \\ 1 & 4 \end{vmatrix}$$

$$= 2(1 + 4) - x(5 + 0) + 3(20 - 0)$$

$$= 10 - 5x + 60$$

$$= 70 - 5x$$

Thus solve the equation

$$70 - 5x = 100$$

and obtain

$$-5x = 30,$$
$$x = -6. \qquad \blacksquare$$

Just as 2×2 determinants were used to solve systems of two equations, 3×3 determinants are used for systems of three linear equations in three variables.

Consider the system:

$$\begin{aligned} a_1x + b_1y + c_1z &= k_1 \\ a_2x + b_2y + c_2z &= k_2 \\ a_3x + b_3y + c_3z &= k_3 \end{aligned}$$

Let D be the 3×3 determinant whose entries are the coefficients of x, y, and z as they appear on the left sides of these equations:

$$D = \begin{vmatrix} a_1 & b_1 & c_1 \\ a_2 & b_2 & c_2 \\ a_3 & b_3 & c_3 \end{vmatrix}$$

Let D_x be the determinant obtained from D by replacing a's (the coefficients of x) by k's:

$$D_x = \begin{vmatrix} k_1 & b_1 & c_1 \\ k_2 & b_2 & c_2 \\ k_3 & b_3 & c_3 \end{vmatrix}$$

Similarly, replace b's by k's in D to obtain D_y, and replace c's by k's in D to obtain D_z:

$$D_y = \begin{vmatrix} a_1 & k_1 & c_1 \\ a_2 & k_2 & c_2 \\ a_3 & k_3 & c_3 \end{vmatrix}, \qquad D_z = \begin{vmatrix} a_1 & b_1 & k_1 \\ a_2 & b_2 & k_2 \\ a_3 & b_3 & k_3 \end{vmatrix}$$

Note that here D, D_x, D_y, and D_z are all 3×3 determinants.

As with systems of two linear equations, formulas for x, y, and z can be obtained in terms of determinants. **Cramer's Rule** *for systems of three linear equations* states that *if* $D \neq 0$, *then the system has a unique solution* (x, y, z), in which

$$x = \frac{D_x}{D}, \qquad y = \frac{D_y}{D}, \qquad z = \frac{D_z}{D}.$$

If $D = 0$, the system does not have a *unique* solution. In fact, in this case either there are no solutions or there are infinitely many.

EXAMPLE 5. Solve by Cramer's Rule:

$$\begin{aligned} 2x + 4y - 3z &= 1 \\ 5x - y + z &= 6 \\ y + z &= 5 \end{aligned}$$

SOLUTION.

$$D = \begin{vmatrix} 2 & 4 & -3 \\ 5 & -1 & 1 \\ 0 & 1 & 1 \end{vmatrix}$$

$$= 2\begin{vmatrix} -1 & 1 \\ 1 & 1 \end{vmatrix} - 5\begin{vmatrix} 4 & -3 \\ 1 & 1 \end{vmatrix} + 0\begin{vmatrix} 4 & -3 \\ -1 & 1 \end{vmatrix}$$

$$= 2[-1(1) - 1(1)] - 5[4(1) - 1(-3)] + 0$$

$$= 2(-2) - 5(7)$$

$$= -39$$

Because $D \neq 0$, Cramer's Rule applies.

$$D_x = \begin{vmatrix} 1 & 4 & -3 \\ 6 & -1 & 1 \\ 5 & 1 & 1 \end{vmatrix}$$

$$= 1\begin{vmatrix} -1 & 1 \\ 1 & 1 \end{vmatrix} - 6\begin{vmatrix} 4 & -3 \\ 1 & 1 \end{vmatrix} + 5\begin{vmatrix} 4 & -3 \\ -1 & 1 \end{vmatrix}$$

$$= 1[-1(1) - 1(1)] - 6[4(1) - 1(-3)] + 5[4(1) - (-1)(-3)]$$

$$= 1(-2) - 6(7) + 5(1)$$

$$= -39$$

$$D_y = \begin{vmatrix} 2 & 1 & -3 \\ 5 & 6 & 1 \\ 0 & 5 & 1 \end{vmatrix}$$

$$= 2\begin{vmatrix} 6 & 1 \\ 5 & 1 \end{vmatrix} - 5\begin{vmatrix} 1 & -3 \\ 5 & 1 \end{vmatrix} + 0\begin{vmatrix} 1 & -3 \\ 6 & 1 \end{vmatrix}$$

$$= 2[6(1) - 5(1)] - 5[1(1) - 5(-3)] + 0$$

$$= 2(1) - 5(16)$$

$$= -78$$

$$D_z = \begin{vmatrix} 2 & 4 & 1 \\ 5 & -1 & 6 \\ 0 & 1 & 5 \end{vmatrix}$$

$$= 2\begin{vmatrix} -1 & 6 \\ 1 & 5 \end{vmatrix} - 5\begin{vmatrix} 4 & 1 \\ 1 & 5 \end{vmatrix} + 0\begin{vmatrix} 4 & 1 \\ -1 & 6 \end{vmatrix}$$

$$= 2[-1(5) - 1(6)] - 5[4(5) - 1(1)] + 0$$

$$= 2(-11) - 5(19)$$

$$= -117$$

Thus by Cramer's Rule,

$$x = \frac{D_x}{D} = \frac{-39}{-39} = 1$$

$$y = \frac{D_y}{D} = \frac{-78}{-39} = 2$$

$$z = \frac{D_z}{D} = \frac{-117}{-39} = 3$$

The solution is (1, 2, 3).

CHECK. Replace x by 1, y by 2, and z by 3 in the original equations:

$2(1) + 4(2) - 3(3) \stackrel{?}{=} 1$	$5(1) - 2 + 3 \stackrel{?}{=} 6$	$2 + 3 \stackrel{?}{=} 5$
$1 \stackrel{\checkmark}{=} 1$	$6 \stackrel{\checkmark}{=} 6$	$5 \stackrel{\checkmark}{=} 5$

All three statements are true; the solution checks. ■

EXAMPLE 6. Show that the system

$$\begin{aligned} 2x - y + 3z &= 1 \\ 5x + 2y - z &= 0 \\ 7x + y + 2z &= 4 \end{aligned}$$

does *not* have a unique solution.

SOLUTION.

$$D = \begin{vmatrix} 2 & -1 & 3 \\ 5 & 2 & -1 \\ 7 & 1 & 2 \end{vmatrix}$$

$$= 2\begin{vmatrix} 2 & -1 \\ 1 & 2 \end{vmatrix} - 5\begin{vmatrix} -1 & 3 \\ 1 & 2 \end{vmatrix} + 7\begin{vmatrix} -1 & 3 \\ 2 & -1 \end{vmatrix}$$

$$= 2[2(2) - 1(-1)] - 5[-1(2) - 1(3)] + 7[-1(-1) - 2(3)]$$

$$= 2(5) - 5(-5) + 7(-5)$$

$$= 0$$

When $D = 0$, either no solution exists or the solution is not unique.

As you might expect, 4×4 determinants can be defined in terms of 3×3 determinants, just as 3×3 determinants have been defined in terms of 2×2's. Then 5×5 determinants can be defined in terms of 4×4's, 6×6's in terms of 5×5's, and so on. And Cramer's Rule can be extended to systems of four (or more) linear equations in the same number of variables.

EXERCISES

In Exercises 1–12 evaluate each determinant.

1. $\begin{vmatrix} 5 & 1 & 1 \\ 2 & 0 & 1 \\ 1 & 0 & 1 \end{vmatrix}$

2. $\begin{vmatrix} 3 & 8 & 1 \\ 0 & 1 & 0 \\ 1 & 0 & 1 \end{vmatrix}$

3. $\begin{vmatrix} 1 & -2 & 1 \\ 2 & 1 & 1 \\ 1 & 0 & 1 \end{vmatrix}$

4. $\begin{vmatrix} 1 & 3 & 1 \\ 2 & 1 & 2 \\ -1 & 1 & 0 \end{vmatrix}$

5. $\begin{vmatrix} 6 & 1 & 0 \\ 0 & 1 & 0 \\ 1 & 2 & 0 \end{vmatrix}$

6. $\begin{vmatrix} 0 & 0 & 9 \\ 1 & 0 & 2 \\ 1 & 1 & 3 \end{vmatrix}$

7. $\begin{vmatrix} 2 & 1 & 4 \\ 1 & 3 & 1 \\ 1 & -2 & 3 \end{vmatrix}$

8. $\begin{vmatrix} 1 & -1 & 1 \\ -1 & 1 & -1 \\ 1 & -1 & 1 \end{vmatrix}$

9. $\begin{vmatrix} 0 & 2 & 4 \\ 0 & 6 & 2 \\ 3 & 1 & 0 \end{vmatrix}$

10. $\begin{vmatrix} 2 & 5 & 3 \\ 1 & 8 & -4 \\ -3 & 2 & 3 \end{vmatrix}$

11. $\begin{vmatrix} 1 & .6 & .2 \\ 0 & .9 & .1 \\ 1 & .5 & .1 \end{vmatrix}$

12. $\begin{vmatrix} \frac{1}{2} & \frac{1}{2} & 0 \\ 1 & -1 & \frac{1}{4} \\ 0 & 1 & \frac{1}{2} \end{vmatrix}$

In Exercises 13–16 determine x.

13. $\begin{vmatrix} x & 0 & 0 \\ 0 & 2 & 0 \\ 0 & 0 & 3 \end{vmatrix} = 12$

14. $\begin{vmatrix} x & 7 & -1 \\ 0 & 5 & 1 \\ 0 & 0 & 2 \end{vmatrix} = 40$

15. $\begin{vmatrix} x & 0 & 0 \\ 4 & 2 & 1 \\ 5 & 2 & 2 \end{vmatrix} = 10$

16. $\begin{vmatrix} 0 & 7 & 2 \\ x & 3 & 6 \\ 0 & 4 & 1 \end{vmatrix} = 12$

17. Evaluate: $\begin{vmatrix} a_1 & b_1 & b_1 \\ a_2 & b_2 & b_2 \\ a_3 & b_3 & b_3 \end{vmatrix}$

18. Evaluate: $\begin{vmatrix} a_1 & b_1 & c_1 \\ 0 & 0 & 0 \\ a_3 & b_3 & c_3 \end{vmatrix}$

19. Compare $\begin{vmatrix} ka_1 & b_1 & c_1 \\ ka_2 & b_2 & c_2 \\ ka_3 & b_3 & c_3 \end{vmatrix}$ with $\begin{vmatrix} a_1 & b_1 & c_1 \\ a_2 & b_2 & c_2 \\ a_3 & b_3 & c_3 \end{vmatrix}$.

20. Compare $\begin{vmatrix} a_1 & c_1 & b_1 \\ a_2 & c_2 & b_2 \\ a_3 & c_3 & b_3 \end{vmatrix}$ with $\begin{vmatrix} a_1 & b_1 & c_1 \\ a_2 & b_2 & c_2 \\ a_3 & b_3 & c_3 \end{vmatrix}$.

In Exercises 21–32: (a) Determine whether Cramer's Rule applies. (b) If it applies, use Cramer's Rule to find the solution.

21. $\begin{aligned} x + y \qquad &= 2 \\ x \qquad - z &= 0 \\ 2x \qquad + z &= 3 \end{aligned}$

22. $\begin{aligned} 5x \qquad + z &= 6 \\ y + z &= 1 \\ x - 3y + z &= 2 \end{aligned}$

23. $\begin{aligned} x + 2y \qquad &= 3 \\ y - z &= 2 \\ 5x + 3y \qquad &= 1 \end{aligned}$

24. $\begin{aligned} 3x - y \qquad &= 10 \\ 2y + z &= 9 \\ x + y - z &= 1 \end{aligned}$

25. $\begin{aligned} x \qquad + z &= 0 \\ 2x - 3y + z &= 6 \\ x - y - z &= 12 \end{aligned}$

26. $\begin{aligned} 3x + 2y - 3z &= 7 \\ 2x - 5y - 5z &= 5 \\ y + z &= 1 \end{aligned}$

27. $\begin{aligned} 2x + 7y + z &= 0 \\ 5x + 6y \qquad &= 1 \\ 3x - y - z &= 0 \end{aligned}$

28. $\begin{aligned} 5x - y + 2z &= 5 \\ 3y - 4z &= 0 \\ 2x \qquad + 4z &= 2 \end{aligned}$

29. $\begin{aligned} x + 2y + z &= 1 \\ 5x - y \qquad &= 2 \\ 4x - 3y - z &= 2 \end{aligned}$

30. $\begin{aligned} x + y + 2z &= 0 \\ 3y - z &= 0 \\ 2x - y + 5z &= 0 \end{aligned}$

31. $\begin{aligned} x - y \qquad &= 0 \\ 2x - y - z &= 0 \\ x - 3y - 2z &= 4 \end{aligned}$

32. $\begin{aligned} 5x + y - 2z &= 4 \\ -x \qquad + z &= -1 \\ -10x - 2y + 4z &= 0 \end{aligned}$

REVIEW EXERCISES FOR CHAPTER 10

10.1 LINEAR SYSTEMS IN TWO VARIABLES

1. Determine the intersection of L_1 and L_2 graphically:

$$L_1: 2x - 3y = -4$$

$$L_2: x + y = 3$$

2. Solve by substituting:

$$5x - y = 1$$

$$2x + 3y = 14$$

3. Solve by adding:

$$5s - t = 2$$

$$2s + t = 5$$

In Exercises 4–6 solve by either substituting or adding:

4. $\begin{aligned} 4x + 3y &= 1 \\ x + 2y &= -1 \end{aligned}$

5. $\begin{aligned} 3x + 5y &= 4 \\ 2x + 3y &= 2 \end{aligned}$

6. $\begin{aligned} 3x + y &= 4 \\ 3x - 2y &= -8 \end{aligned}$

7. Determine a and b so that the line given by $ax + by + 1 = 0$ passes through the point $(1, 1)$ and has slope $\frac{2}{3}$.

10.2 LINEAR SYSTEMS IN THREE VARIABLES

In Exercises 8–10 solve the given system. You may use any methods.

8. $\begin{aligned} x \quad\quad + z &= 0 \\ 2x - y \quad\quad &= 0 \\ x + y + z &= 2 \end{aligned}$

9. $\begin{aligned} 2x - y + 4z &= 5 \\ x + y - z &= 4 \\ 3x \quad\quad - z &= 1 \end{aligned}$

10. $\begin{aligned} 7x + y + z &= 2 \\ 3x - 4y + 2z &= 3 \\ 6x + 3y + z &= 7 \end{aligned}$

11. The sum of three numbers is 31. The largest of these numbers is 6 more than the smallest. The sum of the two smaller numbers is 5 more than the largest number. Find these numbers.

10.3 2 x 2 DETERMINANTS AND CRAMER'S RULE

In Exercises 12–14 evaluate each determinant.

12. $\begin{vmatrix} 2 & 0 \\ 1 & 1 \end{vmatrix}$

13. $\begin{vmatrix} 4 & -1 \\ 2 & 3 \end{vmatrix}$

14. $\begin{vmatrix} 1 & -5 \\ -2 & -1 \end{vmatrix}$

15. Determine x: $\begin{vmatrix} 2 & 1 \\ 1 & x \end{vmatrix} = 3$

In Exercises 16–18: (a) Determine whether Cramer's Rule applies. (b) If it applies, use Cramer's Rule to find the solution.

16. $\begin{aligned} 3x + y &= 4 \\ x - 2y &= -8 \end{aligned}$

17. $\begin{aligned} 3x + 2y &= 5 \\ 6x + 4y &= 9 \end{aligned}$

18. $\begin{aligned} 2x - y &= 1 \\ x + y &= 2 \end{aligned}$

10.4 3 x 3 DETERMINANTS AND CRAMER'S RULE

In Exercises 19–20 evaluate each determinant.

19. $\begin{vmatrix} 2 & 0 & 1 \\ 0 & 1 & 1 \\ 0 & -1 & 0 \end{vmatrix}$

20. $\begin{vmatrix} 1 & 3 & 0 \\ 1 & 2 & -1 \\ -1 & 1 & 2 \end{vmatrix}$

21. Determine x: $\begin{vmatrix} 0 & 0 & x \\ 7 & 1 & 2 \\ 1 & 1 & -1 \end{vmatrix} = 30$

In Exercises 22–23: (a) Determine whether Cramer's Rule applies. (b) If it applies, use Cramer's Rule to find the solution.

22. $$\begin{aligned} x + y + z &= 1 \\ 2x \quad\quad + z &= 3 \\ y - z &= 1 \end{aligned}$$

23. $$\begin{aligned} x + y \quad\quad &= 5 \\ y - z &= 0 \\ 2x + y + z &= 10 \end{aligned}$$

TEST YOURSELF ON CHAPTER 10.

Choose any 7 of these. Problems 3 and 6 are based on Section 10.3.

1. Solve by substituting:

$$\begin{aligned} 2x + y &= 7 \\ 3x - 2y &= 0 \end{aligned}$$

2. Solve by adding:

$$\begin{aligned} 3x - 2y &= 7 \\ 9x - 10y &= 5 \end{aligned}$$

3. Use Cramer's Rule to determine the solution:

$$\begin{aligned} x + 3y &= 3 \\ x + 5y &= 1 \end{aligned}$$

In Exercises 4 and 5 solve by any method:

4. $$\begin{aligned} 3x + 4y &= 4 \\ x - y &= 6 \end{aligned}$$

5. $$\begin{aligned} 10x - 3y &= 4 \\ 3x - 2y &= -1 \end{aligned}$$

6. (a) Evaluate: $\begin{vmatrix} 2 & 4 \\ 3 & 2 \end{vmatrix}$

(b) Determine x so that $\begin{vmatrix} 2 & x \\ 1 & 5 \end{vmatrix} = 0.$

In Exercises 7–9 solve by any method:

7. $$\begin{aligned} 3x \quad\quad + z &= 8 \\ x + y \quad\quad &= 1 \\ x + y + z &= 6 \end{aligned}$$

8. $$\begin{aligned} 2x + y + 3z &= 9 \\ x - y + z &= 0 \\ 5x + 3y + 2z &= 7 \end{aligned}$$

9. $$\begin{aligned} x + y + z &= 8 \\ x - 3y + z &= 0 \\ -x - y + 2z &= 7 \end{aligned}$$

11

exponents, roots, and radicals

11.1 ZERO AND NEGATIVE EXPONENTS

You are already familiar with *positive integral exponents.* Thus in the expression

$$a^7 \qquad \text{(the 7th } power \text{ of } a\text{)}$$

a is the *base* and 7 the *exponent*. It will be useful to define *zero and negative integral exponents,* as well.

Recall that

$$2^4 = 2 \times 2 \times 2 \times 2 = 16,$$

$$2^5 = 2 \times 2 \times 2 \times 2 \times 2 = 32.$$

In order to see how zero and negative exponents should be defined, fill in the blanks in the right-hand column of Tables 11.1 and 11.2 by observing the patterns:

$2^5 =$	32
$2^4 =$	16
$2^3 =$	8
$2^2 =$	4
$2^1 =$	2
$2^0 =$	
$2^{-1} =$	
$2^{-2} =$	
$2^{-3} =$	
$2^{-4} =$	
$2^{-5} =$	

TABLE *11.1*

$10^5 =$	100 000
$10^4 =$	10 000
$10^3 =$	1000
$10^2 =$	100
$10^1 =$	10
$10^0 =$	
$10^{-1} =$	
$10^{-2} =$	
$10^{-3} =$	
$10^{-4} =$	
$10^{-5} =$	

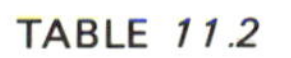

TABLE *11.2*

In Table 11.1 observe that every time the exponent is decreased by 1, the number in the right-hand column is *divided by* 2. And in Table 11.2 every time the exponent is decreased by 1, the number in the right-hand column is *divided by* 10. It seems natural to complete these tables as follows:

$$2^1 = 2 \qquad 10^1 = 10$$

$$2^0 = 1 \qquad 10^0 = 1$$

$$2^{-1} = \frac{1}{2} = \frac{1}{2^1} \qquad 10^{-1} = \frac{1}{10} = \frac{1}{10^1}$$

$$2^{-2} = \frac{1}{4} = \frac{1}{2^2} \qquad 10^{-2} = \frac{1}{100} = \frac{1}{10^2}$$

$$2^{-3} = \frac{1}{8} = \frac{1}{2^3} \qquad 10^{-3} = \frac{1}{1000} = \frac{1}{10^3}$$

$$2^{-4} = \frac{1}{16} = \frac{1}{2^4} \qquad 10^{-4} = \frac{1}{10\,000} = \frac{1}{10^4}$$

$$2^{-5} = \frac{1}{32} = \frac{1}{2^5} \qquad 10^{-5} = \frac{1}{100\,000} = \frac{1}{10^5}$$

These patterns suggest the following definition.

DEFINITION. Let $a \neq 0$ and let m be a positive integer. (Then $-m$ is a negative integer.) Define

$$a^{-m} = \frac{1}{a^m}$$

and

$$a^0 = 1.$$

As you learned in Section 2.5,

$$\frac{a^5}{a^2} = \frac{\not{a} \cdot \not{a} \cdot a \cdot a \cdot a}{\not{a} \cdot \not{a}} = a^3,$$

$$\frac{a^2}{a^2} = \frac{\not{a} \cdot \not{a}}{\not{a} \cdot \not{a}} = 1,$$

and

$$\frac{a^2}{a^5} = \frac{\not{a} \cdot \not{a}}{\not{a} \cdot \not{a} \cdot a \cdot a \cdot a} = \frac{1}{a^3}.$$

You also learned that *when m and n are positive integers and m > n, then*

$$\frac{a^m}{a^n} = a^{m-n}.$$

Thus

$$\frac{a^5}{a^2} = a^3 = a^{5-2}$$

Now observe that with this new definition,

$$\frac{a^m}{a^n} = a^{m-n} \qquad \textit{for all integers } m \textit{ and } n$$

(*positive, negative, or zero*). For example,

$$\frac{a^2}{a^2} = 1 = a^0 = a^{2-2}$$

and

$$\frac{a^2}{a^5} = \frac{1}{a^3} = a^{-3} = a^{2-5}$$

Also,

$$\begin{aligned}\frac{a^{-3}}{a^2} &= \frac{\frac{1}{a^3}}{a^2} \\ &= \frac{1}{a^3} \div a^2 \\ &= \frac{1}{a^3} \cdot \frac{1}{a^2} \\ &= \frac{1}{a^5} \\ &= a^{-5}\end{aligned}$$

Thus

$$\frac{a^{-3}}{a^2} = a^{-3-2}$$

Similarly,

$$\begin{aligned}\frac{a^{-3}}{a^{-2}} &= \frac{\frac{1}{a^3}}{\frac{1}{a^2}} \\ &= \frac{1}{a^3} \cdot \frac{a^2}{1} \\ &= \frac{1}{a} \\ &= a^{-1}\end{aligned}$$

Thus

$$\frac{a^{-3}}{a^{-2}} = a^{-3-(-2)}$$

Observe that 0^m is defined only for positive m. For if you allowed such expressions as 0^0 or 0^{-2}, the above rule would yield

$$0^0 = \frac{0^2}{0^2} \qquad \text{and} \qquad 0^{-2} = \frac{0^1}{0^3}.$$

In each case, the denominator on the right side of the equation is zero, and thus division is undefined.

Here are some examples illustrating these definitions. First, *any nonzero number to the* 0th *power is* 1.

EXAMPLE 1. Let $a \neq 0, b \neq 0, c \neq 0$.

(a) $5^0 = 1$ (b) $a^0 = 1$ (c) $(5bc)^0 = 1$ ■

Next,

$$a^{-m} = \frac{1}{a^m}, \qquad a \neq 0, \qquad \text{for any integer } m$$

EXAMPLE 2.

(a) $3^{-1} = \frac{1}{3}$ (b) $3^{-2} = \frac{1}{3^2} = \frac{1}{9}$

(c) $3^{-3} = \frac{1}{3^3} = \frac{1}{27}$ (d) $3^{-4} = \frac{1}{3^4} = \frac{1}{81}$ ■

The base a can be a fraction.

EXAMPLE 3.

(a) $\left(\frac{2}{7}\right)^{-1} = \frac{1}{\frac{2}{7}} = \frac{7}{2}$ (b) $\left(\frac{2}{7}\right)^{-2} = \frac{1}{\left(\frac{2}{7}\right)^2} = \frac{1}{\frac{4}{49}} = \frac{49}{4}$ ■

Now, consider the law:

$$\frac{a^m}{a^n} = a^{m-n}, \qquad \text{for } a \neq 0, \qquad \text{for all integers } m \text{ and } n$$

EXAMPLE 4.

(a) $\frac{5^2}{5^2} = 5^0$

$\left(\text{Note that } \frac{5^2}{5^2} = \frac{25}{25} = 1 = 5^0.\right)$

(b) $\frac{5^1}{5^3} = 5^{1-3} = 5^{-2}$

$\left(\text{Note that } \frac{5^1}{5^3} = \frac{1}{5^2} = 5^{-2}.\right)$

(c) $\dfrac{5^7}{5^{13}} = 5^{7-13} = 5^{-6}$ ■

EXAMPLE 5. Let $a \neq 0, b \neq 0, c \neq 0$.

(a) $\dfrac{a^4}{a^7} = a^{4-7} = a^{-3}$ (b) $\dfrac{a^0}{a^9} = a^{0-9} = a^{-9}$

(c) $\dfrac{a^3 \cdot b^5}{a^6 \cdot b^6} = \dfrac{a^3}{a^6} \cdot \dfrac{b^5}{b^6} = a^{3-6}b^{5-6} = a^{-3}b^{-1}$

(d) $\dfrac{a^7 \cdot b^4 \cdot c^6}{a^{10} \cdot b^4 \cdot c^4} = a^{7-10}b^{4-4}c^{6-4} = a^{-3}b^0c^2 = a^{-3}c^2$ ■

Finally, note that

$$a^{-n} = \frac{1}{a^n}$$

for every integer n—positive, negative, or zero. For example, if $n = -2$,

$$a^{-(-2)} = a^2$$

and

$$\frac{1}{a^{-2}} = \frac{1}{\frac{1}{a^2}}$$

$$= 1 \div \frac{1}{a^2}$$

$$= 1 \cdot \frac{a^2}{1}$$

$$= a^2$$

Therefore

$$a^{-(-2)} = \frac{1}{a^{-2}}$$

Also,

$$\frac{1}{a^{-n}} = \frac{1}{\frac{1}{a^n}} = a^n$$

and

$$\frac{a^{-m}}{a^{-n}} = \frac{\frac{1}{a^m}}{\frac{1}{a^n}} = \frac{1}{a^m} \div \frac{1}{a^n} = \frac{1}{a^m} \cdot \frac{a^n}{1} = \frac{a^n}{a^m}$$

Thus

$$\frac{a^{-m}}{a^{-n}} = \frac{a^n}{a^m}$$

EXERCISES

Assume $a \neq 0, b \neq 0, x \neq 0, y \neq 0$.
In Exercises 1–26 evaluate each expression.

1. 10^{-1}
2. 3^{-2}
3. 6^0
4. 2^{-3}
5. 2^{-5}
6. 7^{-2}
7. 9^0
8. 8^{-2}
9. 12^{-2}
10. 3^{-3}
11. 10^{-4}
12. 10^{-6}
13. $(-5)^{-1}$
14. $(-9)^0$
15. $(-2)^{-3}$
16. $(-5)^{-2}$
17. $\left(\frac{1}{2}\right)^{-1}$
18. $\left(\frac{2}{3}\right)^{-2}$
19. $\left(\frac{2}{5}\right)^{-1}$
20. $\left(\frac{5}{2}\right)^{-2}$
21. $\left(\frac{4}{7}\right)^{0}$
22. $\frac{2^{-2}}{3^{-3}}$
23. $\frac{2^{-2}}{3}$
24. $\frac{(2 \cdot 5)^{-1}}{3^{-2}}$
25. $(2^{-2})^{-3}$
26. $\frac{(2^4)^{-1}}{2^{-2}}$

In Exercises 27–38 determine the exponent m.

27. $\frac{1}{b^1} = b^m$
28. $\frac{1}{a^2} = a^m$
29. $\frac{1}{2} = 2^m$
30. $\frac{1}{8} = 2^m$
31. $1 = 2^m$
32. $2 = \left(\frac{1}{2}\right)^m$
33. $\frac{1}{25} = 5^m$
34. $\frac{1}{27} = 3^m$
35. $\frac{3}{2} = \left(\frac{2}{3}\right)^m$
36. $\frac{4}{9} = \left(\frac{3}{2}\right)^m$
37. $\frac{1}{32x^5} = (2x)^m$
38. $\frac{2}{x^5} = 2x^m$

In Exercises 39–48 simplify and express your answer without writing a fraction. Use negative exponents, if necessary.

39. $\frac{a^4b^2}{a^6b}$ *Answer:* $a^{-2}b$
40. $\frac{x^5}{x^8}$
41. $\frac{y^4}{y^{10}}$
42. $\frac{a^7}{a^7}$
43. $\frac{x^5y^8}{x^3y^{10}}$
44. $\frac{a^3b^3}{a^6b^0}$
45. $\frac{a^5b^8}{a^7b^9}$
46. $\frac{a^2x^3y^4}{a^4x^3y^2}$
47. $\frac{(x+y)^4}{(x+y)^5}$
48. $\frac{a^3(x-y)^2}{a^6(x-y)^6}$

11.2 THE RULES FOR EXPONENTS

Clearly,

$$
\begin{aligned}
4^2 \cdot 4^3 &= (4 \cdot 4)(4 \cdot 4 \cdot 4) \\
&= 4 \cdot 4 \cdot 4 \cdot 4 \cdot 4 \\
&= 4^5
\end{aligned}
$$

Thus

$$4^2 \cdot 4^3 = 4^{2+3}$$

Now let a be any number, and let m and n be *positive* integers. Then recall that

$$
\begin{aligned}
a^m \cdot a^n &= \underbrace{(a \cdot a \cdot a \ldots a)}_{m \text{ times}}\underbrace{(a \cdot a \cdot a \ldots a)}_{n \text{ times}} \\
&= \underbrace{a \cdot a \cdot a \ldots a}_{m+n \text{ times}} \\
&= a^{m+n}.
\end{aligned}
$$

Therefore

(A) $\quad a^m \cdot a^n = a^{m+n}$

Thus, *to multiply powers of the same base, add the exponents.*

Observe that (for $a \neq 0$)

$$
\begin{aligned}
a^m \cdot a^0 &= a^m \cdot 1 \\
&= a^m \\
&= a^{m+0}.
\end{aligned}
$$

Thus

$$a^m \cdot a^0 = a^{m+0}$$

(Here m can be any integer—positive, negative, or zero.)

In Section 11.1, you learned that

(B) $\quad \dfrac{a^m}{a^n} = a^{m-n}$

for $a \neq 0$ and for *all* integers m and n.

Let $a \neq 0$. Rule (B) can be used to show that Rule (A),

$$a^m \cdot a^n = a^{m+n},$$

holds for *all* integers m and n. For example, let $m = 6$ and $n = -9$. Then

$$\begin{aligned} a^6 \cdot a^{-9} &= a^6 \cdot \frac{1}{a^9} \\ &= \frac{a^6}{a^9} \\ &= a^{6-9} \\ &= a^{6+(-9)} \end{aligned}$$

For the remainder of Section 11.2, *assume* $a \neq 0$ *and* $b \neq 0$.

EXAMPLE 1.

(a) $a^8 \cdot a^5 = a^{8+5} = a^{13}$
(b) $7^4 \cdot 7^2 = 7^{4+2} = 7^6$
(c) $b^6 \cdot b^{-10} = b^{6-10} = b^{-4} = \dfrac{1}{b^4}$
(d) $8^{-3} \cdot 8^{-2} = 8^{-3+(-2)} = 8^{-5} = \dfrac{1}{8^5}$ ■

Similarly, for all integers m, n, and r,

$$a^m \cdot a^n \cdot a^r = a^{m+n+r}$$

Thus

$$a^5 \cdot a^3 \cdot a^2 = a^{5+3+2} = a^{10}$$

and

$$b^{-7} \cdot b^2 \cdot b^{-3} = b^{-7+2+(-3)} = b^{-8} = \frac{1}{b^8}$$

Next, what is the cube of the square of a?

$$(a^2)^3 = a^2 \cdot a^2 \cdot a^2 = a^{2+2+2} = a^6 = a^{2 \cdot 3}$$

Thus

$$(a^2)^3 = a^{2 \cdot 3}$$

Now, what is the nth power of the mth power of a?
Suppose m and n are positive integers. Then

$$\begin{aligned} (a^m)^n &= \underbrace{a^m \cdot a^m \cdot a^m \cdots a^m}_{n \text{ times}} \\ &= \underbrace{\underbrace{a \cdot a \cdot a \cdots a}_{m \text{ times}} \cdot \underbrace{a \cdot a \cdot a \cdots a}_{m \text{ times}} \cdots \underbrace{a \cdot a \cdot a \cdots a}_{m \text{ times}}}_{n \text{ of these}} \\ &= \underbrace{a \cdot a \cdot a \cdots a}_{mn \text{ times}} \\ &= a^{mn} \end{aligned}$$

Does this result hold when n is a negative integer? Consider:

$$(a^2)^{-3} = \frac{1}{(a^2)^3} = \frac{1}{a^{2 \cdot 3}} = \frac{1}{a^6} = a^{-6} = a^{2(-3)}$$

Thus

$$(a^2)^{-3} = a^{2(-3)}$$

In fact, for all integers m and n,

(C) $$(a^m)^n = a^{mn}$$

(*To find a power of a power, multiply the corresponding exponents.*)

EXAMPLE 2.

(a) $(5^3)^4 = 5^{3 \cdot 4} = 5^{12}$

(b) $(9^2)^{-5} = 9^{2(-5)} = 9^{-10} = \dfrac{1}{9^{10}}$

(c) $(7^{-2})^{-4} = 7^{(-2)(-4)} = 7^8$ ■

EXAMPLE 3.

(a) $(a^4)^2 = a^{4 \cdot 2} = a^8$ (b) $a^{(4^2)} = a^{16}$

Thus

$$(a^4)^2 \neq a^{(4^2)}$$

Observe that

$$(a^4)^2 = a^4 \cdot a^4 = a^{4+4},$$

whereas

$$a^{(4^2)} = a^{4 \cdot 4}.$$ ■

How do you find the cube of a product? By the Commutative and Associative Laws,

$$\begin{aligned}(ab)^3 &= ab \cdot ab \cdot ab \\ &= (aaa)(bbb) \\ &= a^3b^3\end{aligned}$$

In general, for any integer m,

(D) $$(ab)^m = a^m b^m$$

(*The* mth *power of a product is the product of the* mth *powers.*)

EXAMPLE 4.

(a) $(xy)^7 = x^7y^7$

(b) $(2a)^5 = 2^5a^5 = 32a^5$

(c) $(2 \cdot 5)^2 = 2^2 \cdot 5^2 = 4 \cdot 25 = 100$
Note further that

$$(2 \cdot 5)^2 = 10^2 = 100.$$

(d) $(3xyz)^3 = 3^3x^3y^3z^3 = 27x^3y^3z^3$ ■

How do you find the cube of a quotient? Assume $b \neq 0$.

$$\left(\frac{a}{b}\right)^3 = \frac{a}{b} \cdot \frac{a}{b} \cdot \frac{a}{b}$$

$$= \frac{a \cdot a \cdot a}{b \cdot b \cdot b}$$

$$= \frac{a^3}{b^3}$$

In general, for any integer m,

(E) $$\left(\frac{a}{b}\right)^m = \frac{a^m}{b^m}, \quad b \neq 0$$

(*The* m*th power of a quotient is the quotient of the* m*th powers.*)

EXAMPLE 5. Assume $a \neq 0, x \neq 0, y \neq 0$.

(a) $\left(\frac{x}{y}\right)^4 = \frac{x^4}{y^4}$

(b) $\left(\frac{3}{4}\right)^2 = \frac{3^2}{4^2} = \frac{9}{16}$

(c) $\left(\frac{10}{3}\right)^3 = \frac{10^3}{3^3} = \frac{1000}{27}$

(d) $\left(\frac{3a^{-2}b^2}{5a}\right)^3 = \left(\frac{3b^2}{5a^3}\right)^3 = \frac{27b^6}{125a^9}$

(e) $\left(\frac{x}{y}\right)^{-2} = \frac{x^{-2}}{y^{-2}} = \frac{\frac{1}{x^2}}{\frac{1}{y^2}} = \frac{1}{x^2} \cdot \frac{y^2}{1} = \frac{y^2}{x^2}$

EXERCISES

Assume $a \neq 0, b \neq 0, x \neq 0, y \neq 0$.
In Exercises 1–34 simplify and write your answer using only positive exponents.

1. $a^2 \cdot a$
2. $a^6 \cdot a^2$
3. $a^4 \cdot a^5$
4. $a^4 \cdot a^{-3}$
5. $a^4 \cdot a^{-5}$
6. $\frac{a^8}{a^4}$
7. $\frac{a^{-3}}{a^7}$
8. $\frac{a^{-3}}{a^{-7}}$

9. $(a^2)^5$
10. $(a^{-3})^2$
11. $(a^2x)^2$
12. $(a^3xy)^{-1}$
13. $\dfrac{10^{-2}}{a^{-3}}$
14. $\dfrac{a^0}{a^{-7}b}$
15. $\dfrac{(a^2)^3(b^3)^2}{a^6b^4}$
16. $\dfrac{3^2(xy^2)^3}{(9x)^2}$
17. $\left(\dfrac{2x^2y^3}{xy^2}\right)^2$
18. $\dfrac{(3x^4y)^{-2}}{(9xy^2)^{-1}}$
19. $\dfrac{(2a^2)^3b}{4a^5b^{-1}}$
20. $\left(\dfrac{10^2xy}{10xy^0}\right)^{-3}$
21. $\dfrac{(2a^{-3}b)^{-2}}{(8ab^{-4})^{-1}}$
22. $\left(\dfrac{25ax^2y^3}{5a^4x^{-2}y^0}\right)^{-2}$
23. $\dfrac{a^4 \cdot a^3 \cdot a^9}{a^{-2}}$
24. $a^2 \cdot (a^5)^2 \cdot a^{-3}$
25. $(x + y)^{-1}$
26. $\dfrac{(a + b)^{-2}}{(a + b)^{-4}}$
27. $3^{-1} + 3$
28. $2^{-1} + 2^{-2}$
29. $\dfrac{a^{-1}}{b^{-1}} - \dfrac{b}{a}$
30. $\dfrac{a^{-1}}{b^{-1}} + \dfrac{b^{-1}}{a^{-1}}$
31. $\dfrac{(x^2)^{-1}}{y^2} + \dfrac{2}{3x^2y^2}$
32. $(a^{-1} + b^{-1})^{-1}$
33. $(a^{-1} - b^{-1})ab$
34. $\dfrac{x^{-1} - y^{-1}}{x^{-2} - y^{-2}}$
35. (a) Determine the square of the cube of 2.
    (b) Determine the cube of the square of 2.
36. What power of 2 is $\dfrac{1}{64}$?

11.3 SCIENTIFIC NOTATION

Scientific notation is a convenient way of expressing large numbers, such as 676 000 000, and small positive numbers, such as .000 000 004 . It is commonly used in fields such as chemistry, biology, engineering, physics, and astronomy, where such numbers frequently occur. You will use scientific notation in your study of logarithms (Chapter 12).

The powers of 10 play a key role in scientific notation. Observe the pattern:

$$10^1 = 10 \qquad 10^0 = 1 \qquad 10^{-1} = \frac{1}{10} = .1$$

$$10^2 = 100 \qquad\qquad 10^{-2} = \frac{1}{100} = .01$$

$$10^3 = 1000 \qquad\qquad 10^{-3} = \frac{1}{1000} = .001$$

$$10^4 = 10\,000 \qquad\qquad 10^{-4} = \frac{1}{10\,000} = .0001$$

$$10^5 = 100\,000 \qquad 10^{-5} = \frac{1}{100\,000} = .000\,01$$

$$10^6 = 1\,000\,000 \qquad 10^{-6} = \frac{1}{1\,000\,000} = .000\,001$$

Every *positive* number N can be expressed as the product of

(a) a number between 1 and 10

and

(b) a power of 10.

More precisely,

$$N = M \cdot 10^m,$$

where $1 \le M < 10$ and m is an integer. This form of writing a number is called **scientific notation.**

EXAMPLE 1. To express 4730 in scientific notation, note that

$$4730 = \frac{4730}{1000} \times 1000 = 4.730 \times 10^3.$$

Observe that *division by* 1000 (or 10^3) is accomplished by *moving the decimal point 3 places to the left* (to obtain 4.730). Balance this by multiplying by 10^3. ■

EXAMPLE 2. To express .000 56 in scientific notation, note that

$$.000\,56 = (.000\,56 \times 10\,000) \cdot \frac{1}{10\,000} = 5.6 \times 10^{-4}.$$

Here, *multiplication by* 10 000 (or 10^4) is accomplished by *moving the decimal point 4 places to the right* (to obtain 5.6). Balance this by multiplying by 10^{-4}. ■

Let N be a positive number, expressed as a decimal. To obtain the scientific notation

$$M \cdot 10^m$$

for N:

1. Place the decimal point after the first nonzero digit to obtain M.

2a. If you moved the decimal point k places to the *left*, you *divided by* 10^k. To balance this, multiply by 10^k. Thus let $m = k$.

2b. If you moved the decimal point k places to the *right*, you *multiplied by* 10^k. To balance this, multiply by 10^{-k}. Thus let $m = -k$.

2c. If $1 \le N < 10$, then $m = 0$

EXAMPLE 3. Express in scientific notation:

(a) 763 (b) .008 406 (c) 9.058 (d) 34.81

SOLUTION.

(a) $N = 763$
Place the decimal point after the 7. Thus

$$M = 7.63$$

You moved the decimal point 2 places to the *left:*

763. to 7.63

Therefore $m = 2$ and

$$763 = 7.63 \times 10^2$$

(b) $N = .008\,406$
The first *nonzero* digit is 8. Thus

$$M = 8.406$$

(The zeros to the left of 8 can be omitted.) You moved the decimal point 3 places to the *right:*

.008 406 to ~~00~~8.406

Therefore $m = -3$ and

$$.008\,406 = 8.406 \times 10^{-3}$$

(c) $1 \leq 9.058 < 10.$
Thus $m = 0$ and

$$9.058 = 9.058 \times 10^0$$

If N lies between 1 and 10, as is the case here, the factor $10^0(= 1)$ is usually omitted. Thus the scientific notation of 9.058 is

$$9.058$$

itself.

(d) $34.81 = 3.481 \times 10^1 = 3.481 \times 10$ Write "10^1" as "10". ■

EXAMPLE 4. Express in scientific notation:

(a) 6 380 000 (b) .000 000 53 (c) .4 (d) $\frac{1}{4}$ (e) 7

SOLUTION.

(a) $6\,380\,000 = 6.38 \times 10^6$

(b) $.000\,000\,53 = 5.3 \times 10^{-7}$

(c) $.4 = 4 \times 10^{-1}$

(d) $\frac{1}{4} = .25 = 2.5 \times 10^{-1}$

(e) $1 \leq 7 < 10$. Thus the scientific notation is 7 itself. ■

To convert back from scientific notation

$$M \cdot 10^m, \qquad \text{where } m \neq 0,$$

to the usual decimal notation:

1. Move the decimal point m places to the *right* if $m > 0$. (Here you multiply by a *positive* power of 10.)
2. Move the decimal point m places to the *left* if $m < 0$. (Here you multiply by a *negative* power of 10, or *divide* by a *positive* power of 10.)

Note that this *reverses* the former process.

EXAMPLE 5.

(a) $1.48 \times 10^4 = 14\,800$

Here $M = 1.48$ and $m = 4 > 0$. The decimal point is moved to the *right* 4 places.

(b) $6.04 \times 10^{-3} = .006\,04$

Here $M = 6.04$ and $m = -3 < 0$. The decimal point is moved to the *left* 3 places. ■

Computations are often simplified by expressing numbers in terms of powers of 10. Instead of using scientific notation in the computation, it is usually best to express a number as the product of an *integer* and a power of 10. The final result is then written in scientific notation.

EXAMPLE 6. Find the value of:

$$\frac{25\,000 \times 160\,000}{800\,000 \times .0125}$$

Express your answer in scientific notation.

SOLUTION.

$$\frac{25\,000 \times 160\,000}{800\,000 \times .0125} = \frac{\overset{}{\cancel{25}} \times 10^3 \times \overset{2}{\cancel{16}} \times 10^4}{\cancel{8} \times 10^5 \times \underset{5}{\cancel{125}} \times 10^{-4}}$$

$$= \frac{2}{5} \times 10^{3+4-(5-4)}$$

$$= .4 \times 10^6$$

$$= 4 \times 10^5$$

EXERCISES

In Exercises 1–22 express each number in scientific notation.

1. 14.9
2. 362
3. 5084
4. 7.82
5. .695
6. .008 02
7. .9
8. 9.99
9. 10
10. 8036.2
11. 248 000
12. 3 062 000 000
13. .000 820 001
14. .000 000 009
15. 8
16. 843×10^4
17. 27.9×10^{-2}
18. 9.36×10^{-5}
19. $\dfrac{3 \times 4 \times 5}{100}$
20. .01
21. $5 \times 80 \times 10^6$
22. .000 03

In Exercises 23–40 convert scientific notation back to the usual decimal notation.

23. 8.34×10^1
24. 6.06×10^4
25. 4.44×10^8
26. 3.09×10^{-2}
27. 5.15×10^0
28. 6.092×10^{-4}
29. 8.181×10^{-1}
30. 3.59×10^{-6}
31. 4.74×10^5
32. 6×10^7
33. 6.2×10^{-2}
34. 3.87×10^{-5}
35. 1.11×10^1
36. 3.93×10^{-7}
37. 5.87×10^{-3}
38. 2.024×10^1
39. $3.896\,24 \times 10^2$
40. 3.796×10^{-8}

In Exercises 41–48 find the value of each number. Express your answer in scientific notation.

41. $10^8 \times 10^4 \times 10^{-2}$
42. $\dfrac{10^7 \times 10^9}{10^3 \times 10^8}$
43. $60\,000 \times 5000 \times 20\,000$
44. $9000 \times .0004$
45. $25\,000 \times .002 \times 500 \times .000\,04$
46. $\dfrac{270 \times 8000}{.002 \times 300\,000}$
47. $\dfrac{200 \times .0004 \times 64\,000}{4000 \times .032 \times .08}$
48. $\dfrac{144\,000 \times .000\,36}{.0072 \times 960\,000}$
49. The human eye is normally capable of responding to light waves whose lengths are in the range of 3.8×10^{-5} to 7.6×10^{-5} centimeters. Which of the following centimeter readings lie within this range?

    (a) .000 45 (b) .000 045 (c) 60×10^{-6} (d) $.039 \times 10^{-3}$
50. The coefficient of linear expansion indicates the fractional change in the length of a rod per degree change of temperature. For tungsten the coefficient is 4.4×10^{-6} per degree Centigrade. What is the fractional change in the length of a tungsten rod when the temperature rises from 10° to 30° Centigrade?
51. A light year is the distance over which light can travel in a year's time. One light year is approximately 6×10^{12} miles. If a star is 250 light years away from earth, how far away is it in miles?

11.4 RATIONAL EXPONENTS

Up to now, you have considered only *integers as exponents*. How should you define an expression such as

$$9^{\frac{1}{2}},$$

with the *rational exponent* $\frac{1}{2}$?

Recall that for $a \neq 0$,

$$(a^m)^n = a^{mn}$$

for all *integers* m and n. If this rule is to apply to rational exponents, then

$$(9^{\frac{1}{2}})^2 = 9^{\frac{1}{2}\cdot 2} = 9^1 = 9.$$

Thus

$$9^{\frac{1}{2}} \cdot 9^{\frac{1}{2}} = 9,$$

and

$9^{\frac{1}{2}}$ *multiplied by itself is* 9.

Note that

$$3 \cdot 3 = 9 \qquad \text{and} \qquad (-3)(-3) = 9.$$

It is customary to let

$$9^{\frac{1}{2}} = 3.$$

Also, because it is difficult to read in print, from now on

$9^{\frac{1}{2}}$ will be written as $9^{1/2}$.

Thus

$$9^{1/2} = 3$$

✱ *DEFINITION.* Let $a \geq 0$. Then b is **the square root of** a if $b^2 = a$ and $b \geq 0$.

Write

$$a^{1/2} = b.$$

(The notation $\sqrt{a}$ is also used for the square root of a, and will be discussed later.)

EXAMPLE 1.

(a) $4^{1/2} = 2$ because $2^2 = 4$ and $2 \geq 0$.
(b) $25^{1/2} = 5$ because $5^2 = 25$ (and $5 \geq 0$)
(c) $100^{1/2} = 10$
(d) $0^{1/2} = 0$ ■

EXAMPLE 2. The **Pythagorean Theorem** states that the square of the length of the hypotenuse of a right triangle is given by $a^2 + b^2$, where a and b are the lengths of the other two sides. In Figure 11.1, c is the length of the hypotenuse. Thus

$$c^2 = a^2 + b^2$$

Find c if $a = 3$ and $b = 4$.

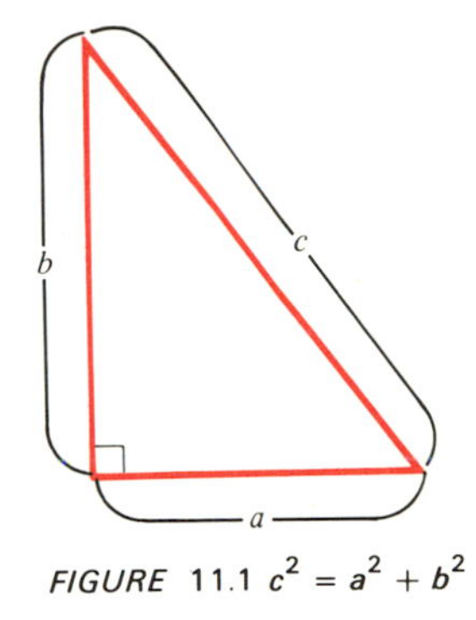

FIGURE 11.1 $c^2 = a^2 + b^2$

SOLUTION.

$$\begin{aligned} c^2 &= a^2 + b^2 \\ &= 3^2 + 4^2 \\ &= 9 + 16 \\ &= 25 \end{aligned}$$

Thus

$$c = 25^{1/2} = 5$$ ■

Every nonnegative number has a square root. But these square roots may not be rational. For example, $2^{1/2}$ and $3^{1/2}$ are irrational numbers. You can *approximate* $2^{1/2}$ to 3 places by 1.414 and $3^{1/2}$ by 1.732. However, *these are not the exact values.* Except in applications, it is best to leave these square roots in the form

$$2^{1/2}, \qquad 3^{1/2}, \qquad \text{and so on.}$$

The square of a real number b is at least 0.

$$b^2 \geq 0$$

Thus a negative number, such as -4, cannot be the square of a real number. Therefore $(-4)^{1/2}$ is not defined *in the real number system.* However, because

$$4^{1/2} = 2,$$

you can write

$$-(4^{1/2}) = -2.$$

When you write

$$-4^{1/2}, \qquad \text{you mean } -(4^{1/2}).$$

Similarly,

$$-25^{1/2} = -(25^{1/2}) = -5$$

DEFINITION. The real number b is the **cube root of** a if $b^3 = a$.

Write

$$b = a^{1/3}.$$

EXAMPLE 3.

(a) $8^{1/3} = 2$ because $2^3 = 8$
(b) $(-8)^{1/3} = -2$ because $(-2)^3 = -8$
(c) $27^{1/3} = 3$
(d) $0^{1/3} = 0$ ■

Every real number—positive, negative, or zero—has a cube root.

In general, you can consider the nth roots of numbers, for $n = 1, 2, 3, 4, 5, \ldots$.

DEFINITION. Let n be an *even* positive integer: $2, 4, 6, \ldots$, and let $a \geq 0$. Then b is the **nth root of** a if $b^n = a$, and $b \geq 0$.

Let n be an *odd* positive integer: $1, 3, 5, \ldots$, and let a be any real number. Then b is the **nth root of** a if $b^n = a$.

Write

$$b = a^{1/n}.$$

Only nonnegative numbers have (real) nth roots for n $= 2, 4, 6, \ldots$. *But every real number—positive, negative, or zero—has an nth root for n* $= 1, 3, 5, 7, \ldots$.

EXAMPLE 4.

(a) $81^{1/4} = 3$ because $3^4 = 81$ and $3 \geq 0$
(b) $32^{1/5} = 2$ because $2^5 = 32$
(c) $(-32)^{1/5} = -2$ because $(-2)^5 = -32$
(d) -5 has no 6th root because negative numbers have nth roots only for odd n.
(e) $1^{1/n} = 1$ for every positive integer n because $1^n = 1$
(f) $(-1)^{1/n} = -1$ for every *odd* positive integer n because *odd* powers of -1 equal -1. ■

For *rational numbers* $\frac{m}{n}$, rational powers $a^{m/n}$ can now be defined so as to extend the law

$$(a^m)^n = a^{mn}.$$

DEFINITION. Let m be a nonzero integer, let n be a positive integer, and let a be a real number (with $a \geq 0$ if n is even, and $a \neq 0$ if $m < 0$). Define

$$a^{m/n} = (a^{1/n})^m.$$

In other words, *the* $\left(\frac{m}{n}\right)$th *power of a is the* mth *power of the* nth *root of a.*

Thus

$$8^{2/3} = (8^{1/3})^2 = 2^2 = 4$$

It can be shown that

$$(a^{1/n})^m = (a^m)^{1/n}.$$

In fact,

$$8^{2/3} = (8^2)^{1/3} = 64^{1/3} = 4$$

Thus $a^{m/n}$ *is also the* nth *root of the* mth *power of* a. However, in practice, it is often easier to first take roots, then powers, for you will usually be working with smaller and more easily recognized numbers. For example, consider $25^{3/2}$. First,

$$25^{3/2} = (25^{1/2})^3 = 5^3 = 125$$

Next,

$$25^{3/2} = (25^3)^{1/2} = 15\,625^{1/2} = 125$$

Note the difficulty in the second method.

EXAMPLE 5. Determine:

(a) $16^{3/4}$ (b) $4^{3/2}$ (c) $(-27)^{2/3}$ (d) $25^{-1/2}$

SOLUTION.

(a) $16^{3/4} = (16^{1/4})^3 = 2^3 = 8$
(b) $4^{3/2} = (4^{1/2})^3 = 2^3 = 8$
(c) $(-27)^{2/3} = [(-27)^{1/3}]^2 = (-3)^2 = 9$
(d) Here the exponent is $\frac{-1}{2}$.

$$25^{-1/2} = (25^{1/2})^{-1} = 5^{-1} = \frac{1}{5}$$

■

If $a < 0$ and m and n are both even, then $(a^m)^{1/n}$ is defined, but $a^{1/n}$ and, hence, $(a^{1/n})^m$ are not. For example, suppose $a = -2, m = 2$, and $n = 2$. Then

$$(a^m)^{1/n} = [(-2)^2]^{1/2} = 4^{1/2} = 2,$$

but

$$(a^{1/n})^m \qquad \text{or} \qquad [(-2)^{1/2}]^2$$

would involve "the square root of a negative number".

Recall that

$$\frac{-1}{2} = -\frac{1}{2} = \frac{1}{-2}.$$

So far, only $\frac{-1}{2}$ has been defined as an exponent. Thus

$$25^{-1/2} = \frac{1}{5}$$

You will write

$$25^{-(1/2)}$$

when the exponent is

$$-\frac{1}{2},$$

according to the following definition.

DEFINITION. Let $a \neq 0$ and let m and n be positive integers (with a positive if n is even). Define

$$a^{-(m/n)} \text{ to be } a^{-m/n} \text{ and } a^{m/-n} \text{ to be } a^{-m/n}.$$

Thus, by definition,

$$a^{-(m/n)} = a^{m/-n} = a^{-m/n}$$

EXAMPLE 6. Determine:

(a) $8^{-(1/3)}$ (b) $8^{2/-3}$ (c) $(-32)^{-3/5}$

SOLUTION.

(a) $8^{-(1/3)} = 8^{-1/3} = (8^{1/3})^{-1} = 2^{-1} = \frac{1}{2}$

(b) $8^{2/-3} = 8^{-2/3} = (8^{1/3})^{-2} = 2^{-2} = \frac{1}{2^2} = \frac{1}{4}$

(c) $(-32)^{-3/5} = [(-32)^{1/5}]^{-3} = (-2)^{-3} = \frac{1}{(-2)^3} = \frac{1}{-8} = \frac{-1}{8}$ ■

EXERCISES

In Exercises 1–36 simplify.

1. $16^{1/2}$
2. $36^{1/2}$
3. $1^{1/2}$
4. $49^{1/2}$
5. $81^{1/2}$
6. $64^{1/2}$
7. $121^{1/2}$
8. $144^{1/2}$
9. $8^{1/3}$
10. $(-8)^{1/3}$
11. $125^{1/3}$
12. $-125^{1/3}$
13. $16^{1/4}$
14. $81^{1/4}$
15. $64^{2/3}$
16. $(-8)^{2/3}$
17. $(-1)^{4/3}$
18. $16^{5/4}$
19. $32^{2/5}$
20. $9^{-1/2}$
21. $16^{-1/2}$
22. $81^{-1/2}$
23. $100^{-1/2}$
24. $1000^{-1/3}$
25. $8^{-1/3}$
26. $8^{-2/3}$
27. $16^{-1/4}$
28. $16^{-3/4}$

29. $81^{-1/4}$
30. $81^{-3/4}$
31. $9^{-3/2}$
32. $125^{-2/3}$
33. $10\,000^{-3/4}$
34. $100^{-5/2}$
35. $36^{-(1/2)}$
36. $49^{1/-2}$

In Exercises 37–39 find the length of the hypotenuse of a right triangle if the lengths of the other sides are as indicated.

37. 12 and 16
38. 5 and 12
39. 7 and 24
40. One side of a right triangle is of length 8 and the hypotenuse is of length 10. Find the length of the remaining side.
41. The area of a square tablecloth is 3600 square inches. Find the length of a side.
42. The volume of a cubic box is 64 cubic feet. Find the length of a side.
43. The area of a circle is 144π. Find the length of the radius.
44. The area of a circle is $10\,000\pi$. Find the length of the radius.
45. Determine the square root of the square of -1.
46. Determine the cube root of the cube of -1.
47. Determine the square of the cube root of -1.
48. Determine the cube root of the square of -1.

11.5 THE RULES FOR RATIONAL EXPONENTS

The rules for integral exponents apply also to *rational* (or *fractional*) exponents. Thus, let r and s be *rational numbers* and suppose that a^r and a^s are both defined.

(A) $$a^r \cdot a^s = a^{r+s}$$

(B) $$\frac{a^r}{a^s} = a^{r-s}, \qquad a \neq 0$$

(C) $$(a^r)^s = a^{rs}$$

Also, suppose b^r is defined.

(D) $$(ab)^r = a^r b^r$$

(E) $$\left(\frac{a}{b}\right)^r = \frac{a^r}{b^r}, \qquad b \neq 0$$

The requirement that a^r and a^s are both defined is crucial. For example, in (C), suppose that $a = -2, r = 2, s = \frac{1}{2}$. Then

$$(a^r)^s = [(-2)^2]^{1/2} = 4^{1/2} = 2$$

But

$$a^{rs} = (-2)^{2 \cdot 1/2} = (-2)^1 = -2$$

Thus here,

$$(a^r)^s \neq a^{rs}$$

because a^s is not defined.

To apply Rules (A), (B), and (C), combine the fractions r and s.

EXAMPLE 1. Let $x > 0$.

(a) $x^{1/4} \cdot x^{3/4} = x^{1/4+3/4} = x^1 = x$
(b) $x^{1/2} \cdot x^{1/4} = x^{1/2+1/4} = x^{(2+1)/4} = x^{3/4}$
(c) $x^{1/2} \cdot x^{1/3} = x^{1/2+1/3} = x^{(3+2)/6} = x^{5/6}$
(d) $x^{5/6} \cdot x^{-1/4} = x^{(10-3)/12} = x^{7/12}$ ■

EXAMPLE 2. Let $a > 0$.

(a) $\dfrac{a^{2/5}}{a^{1/5}} = a^{2/5-1/5} = a^{1/5}$

(b) $\dfrac{a^{1/2}}{a^{1/4}} = a^{1/2-1/4} = a^{(2-1)/4} = a^{1/4}$

(c) $\dfrac{a^{2/3}}{a^{1/5}} = a^{2/3-1/5} = a^{(10-3)/15} = a^{7/15}$ ■

EXAMPLE 3. Let $b > 0$.

(a) $(b^{1/2})^{1/3} = b^{(1/2)(1/3)} = b^{1/6}$
(b) $(b^{2/5})^{3/2} = b^{(2/5)(3/2)} = b^{3/5}$
(c) $(b^{-3/4})^{-2/9} = b^{(-3/4)(-2/9)} = b^{1/6}$ ■

Now consider Rules (D) and (E).

EXAMPLE 4. Let $x > 0, y > 0$.

(a) $(xy)^{3/4} = x^{3/4}y^{3/4}$

(b) $(32x)^{-2/5} = 32^{-2/5}x^{-2/5} = \dfrac{1}{4x^{2/5}}$

(c) $\left(\dfrac{8}{27}\right)^{2/3} = \dfrac{8^{2/3}}{27^{2/3}} = \dfrac{4}{9}$ ■

EXAMPLE 5. Let $a > 0, b \neq 0, c > 0$.

(a) $(a^{1/2}b^{2/3})^{12} = a^{(1/2)(12)}b^{(2/3)(12)} = a^6b^8$

(b) $\dfrac{(a^2bc^{1/4})^4}{(a^{1/2}b^3c^2)^2} = \dfrac{a^8b^4c}{ab^6c^4} = \dfrac{a^7}{b^2c^3}$ ■

Note that in part (b) each variable appears only once and with only *positive* exponents in the *simplified* form

$$\frac{a^7}{b^2c^3}.$$

To simplify a numerical expression such as $8^{1/2}$, separate all square factors of 8, and use Rule (D). The simplified form will involve the rational power of a smaller integer.

EXAMPLE 6. Simplify:

(a) $8^{1/2}$

(b) $\left(\frac{1}{32}\right)^{1/4}$

SOLUTION.

(a)
$$\begin{aligned} 8^{1/2} &= (4 \cdot 2)^{1/2} \\ &= 4^{1/2} \cdot 2^{1/2} && \text{by Rule (D)} \\ &= 2 \cdot 2^{1/2} \end{aligned}$$

(b)
$$\begin{aligned} \left(\frac{1}{32}\right)^{1/4} &= \frac{1^{1/4}}{32^{1/4}} && \text{by Rule (E)} \\ &= \frac{1}{(16 \cdot 2)^{1/4}} \\ &= \frac{1}{16^{1/4} \cdot 2^{1/4}} && \text{by Rule (D)} \\ &= \frac{1}{2} \cdot \frac{1}{2^{1/4}} \end{aligned}$$

EXERCISES

In Exercises 1–42 let $a > 0$, $b > 0$, $c > 0$, $x > 0$, $y > 0$, $z > 0$. Simplify and express your answer using only positive exponents.

1. $x^{1/2} \cdot x^{1/2}$
2. $x \cdot x^{1/2}$
3. $x^2 \cdot x^{1/4}$
4. $a^{3/4} \cdot a^{5/4}$
5. $\frac{y^{3/4}}{y^{1/4}}$
6. $\frac{b^2}{b^{1/2}}$
7. $(a^{2/3})^6$
8. $(x^{1/2})^{1/4}$
9. $(a^{2/3})^{3/2}$
10. $\frac{(b^{5/6})^{3/2}}{b^{1/4}}$
11. $x^{1/4} \cdot x^{1/2}$
12. $a^{2/3} \cdot a^{1/6}$
13. $\frac{y^{3/4}}{y^{1/2}}$
14. $\frac{z^{3/5}}{z^{1/10}}$
15. $x^{1/2} \cdot x^{1/3}$
16. $a^{2/5} \cdot a^{1/2}$
17. $z^{3/4} \cdot z^{1/3}$
18. $b^{3/8} \cdot b^{1/6}$
19. $x^{7/10} \cdot x^{-1/5}$
20. $y^{2/3} \cdot y^{-1/9}$
21. $\frac{x^{3/4}}{x^{1/3}}$
22. $\frac{a^{3/5}}{a^{1/2}}$
23. $\frac{b^{1/6}}{b^{-1/3}}$
24. $\frac{y^{4/5}}{y^{-3/10}}$
25. $a^{3/4} \cdot a^{1/4} \cdot a^{1/2}$
26. $x^{1/2} \cdot x^{5/8} \cdot x^{-1/4}$
27. $\frac{y^{2/3}y^{1/4}}{y^{1/2}}$

28. $\frac{z \cdot z^{3/4}}{z^{1/2} \cdot z^0}$

29. $\frac{(x^{1/4}y^{2/3})^2}{x^{1/2}y^{1/3}}$

30. $\frac{(a^4b^6)^{-1/2}}{a^{1/2}b^{-1}}$

31. $\frac{(ab^2c^3)^{1/6}}{(a^{-1}b^{-2})^{-1/2}}$

32. $\frac{(27xyz)^{1/3}}{(25x^2y^{1/2})^{1/2}}$

33. $(x^{1/2}y^{1/3})^6 \cdot \left(\frac{x}{y^2}\right)^{1/3}$

34. $\left[\left(\frac{xy^{1/2}}{z}\right)^{3/2}\right]^4$

35. $\left(\frac{x^2y^{1/3}}{z^{1/9}}\right)^3 \left(\frac{x^{-1/2}z^{1/4}}{y^{1/2}}\right)^2$

36. $\frac{(9a^4b^{4/3}c^{2/3})^{-1/2}}{(125a^{3/5}b^9)^{-1/3}}$

37. $27^{1/2}$

38. $\left(\frac{1}{8}\right)^{1/2}$

39. $(32a^4)^{1/2}$

40. $(-8a^6b^9)^{2/3}$

41. $(75a^2b^5)^{-1/2}$

42. $\frac{(63a^4)^{3/2}}{(28b^6)^{1/2}}$

In Exercises 43–52, simplify.

43. $(16x)^{1/2}, \quad x > 0$

44. $\left(\frac{27}{8}\right)^{1/3}$

45. $\left(\frac{4}{9}\right)^{3/2}$

46. $\left(\frac{100}{49}\right)^{-1/2}$

47. $\left(\frac{-1}{125}\right)^{-2/3}$

48. $\left(\frac{16}{81}\right)^{-3/4}$

49. $3^{1/3} \cdot 3^{2/3}$

50. $9^{1/2} \cdot 16^{3/4}$

51. $\frac{25^{3/2}}{5}$

52. $27^{-1/3} \cdot 9^{3/2}$

11.6 RADICAL NOTATION

DEFINITION. Let n be a positive integer: 1, 2, 3, . . . , and let a be a real number (with $a \geq 0$ if n is even). Define

$$\sqrt[n]{a} = a^{1/n}.$$

For the case of $n = 2$, write

$$\sqrt{a} \quad \text{in place of} \quad \sqrt[2]{a}.$$

Thus $\sqrt{a}$ indicates the square root of a; $\sqrt[n]{a}$ indicates the nth root of a.

EXAMPLE 1.

(a) $\sqrt{9} = 9^{1/2} = 3$

(b) $\sqrt[3]{8} = 8^{1/3} = 2$

(c) $\sqrt[3]{-8} = (-8)^{1/3} = -2$

(d) $\sqrt[4]{16} = 16^{1/4} = 2$

(e) $\sqrt[4]{-16}$ is not defined because $-16 < 0$ and 4 is *even*. ■

$$\sqrt[n]{a}$$

is called **radical notation;** the symbol

$$\sqrt{}$$

is a **radical.**

Radical notation is widely used. The purpose of this section is simply to reconsider the material on rational exponents in terms of radicals.

Recall that for positive integers m and n and for any real number a,

$$a^{m/n} \qquad \text{was defined as} \qquad (a^{1/n})^m,$$

provided that $a \geq 0$ if n is even. You also learned that

$$(a^{1/n})^m = (a^m)^{1/n}.$$

In radical notation,

$$(a^{1/n})^m = (\sqrt[n]{a})^m$$

and

$$(a^m)^{1/n} = \sqrt[n]{a^m}$$

Thus

$$a^{m/n} = (\sqrt[n]{a})^m = \sqrt[n]{a^m}$$

and

$$8^{2/3} = (\sqrt[3]{8})^2 = 2^2 = 4$$

EXAMPLE 2. Let $a \geq 0, x \neq 0$. Express in radical notation:

(a) $a^{3/4}$ (b) $2^{2/3}$ (c) $\dfrac{1}{x^{3/5}}$

SOLUTION. (There are several possible forms for the final answer.)

(a) $a^{3/4} = (\sqrt[4]{a})^3 = \sqrt[4]{a^3}$

(b) $2^{2/3} = (\sqrt[3]{2})^2 = \sqrt[3]{2^2} = \sqrt[3]{4}$

(c) $\dfrac{1}{x^{3/5}} = \dfrac{1}{(\sqrt[5]{x})^3} = \dfrac{1}{\sqrt[5]{x^3}}$ ■

EXAMPLE 3. Let $a \geq 0$. Express in terms of rational exponents:

(a) $\sqrt[5]{17}$, (b) $\sqrt[3]{a^2}$, (c) $(\sqrt[7]{a})^4$

SOLUTION.

(a) $\sqrt[5]{17} = 17^{1/5}$ (b) $\sqrt[3]{a^2} = a^{2/3}$ (c) $\sqrt[7]{a^4} = a^{4/7}$ ■

Rules (D) and (E) for rational exponents $\frac{1}{n}$ can now be restated in radical notation. Let m and n be rational numbers and suppose that $\sqrt[n]{a}$ and $\sqrt[n]{b}$ are defined.

(D) $\quad \sqrt[n]{ab} = \sqrt[n]{a} \cdot \sqrt[n]{b}$

(E) $\quad \sqrt[n]{\frac{a}{b}} = \frac{\sqrt[n]{a}}{\sqrt[n]{b}}, \quad b \neq 0$

EXAMPLE 4. Let $a \geq 0$. Simplify the following:

(a) $\sqrt{9\,000\,000}$ (b) $\sqrt[3]{56}$ (c) $\sqrt[4]{\frac{16}{81}}$

(d) $\sqrt{9a^2b^4}$ (e) $\sqrt[3]{\frac{a^6b^9c}{27}}$ (f) $\sqrt[3]{.008}$

SOLUTION.

(a) $\sqrt{9\,000\,000} = \sqrt{9 \times 10^6} = \sqrt{9}\sqrt{10^6} = 3(10^3) = 3000$

(b) $\sqrt[3]{56} = \sqrt[3]{8 \cdot 7} = \sqrt[3]{8} \cdot \sqrt[3]{7} = 2\sqrt[3]{7}$

(c) $\sqrt[4]{\frac{16}{81}} = \frac{\sqrt[4]{16}}{\sqrt[4]{81}} = \frac{2}{3}$

(d) $\sqrt{9a^2b^4} = \sqrt{9}\sqrt{a^2}\sqrt{b^4} = 3ab^{4/2} = 3ab^2$

(e) $\sqrt[3]{\frac{a^6b^9c}{27}} = \frac{\sqrt[3]{a^6}\sqrt[3]{b^9}\sqrt[3]{c}}{\sqrt[3]{27}} = \frac{a^2b^3\sqrt[3]{c}}{3}$

(f) $\sqrt[3]{.008} = \sqrt[3]{8 \times 10^{-3}} = \sqrt[3]{8}\sqrt[3]{10^{-3}} = 2 \times 10^{-1} = 2(.1) = .2$ ■

When applying Rules (A), (B), and (C), first convert to rational exponents.

EXAMPLE 5. Let $x > 0$. Express in terms of rational exponents:

(a) $\sqrt{x} \cdot \sqrt[4]{x}$ (b) $\frac{x}{\sqrt{x}}$ (c) $\sqrt[3]{x^9}$

(d) $\sqrt{x + 5}$ (e) $\sqrt{x} + 5$

SOLUTION.

(a) $\sqrt{x} \cdot \sqrt[4]{x} = x^{1/2} \cdot x^{1/4} = x^{3/4}$ (b) $\frac{x}{\sqrt{x}} = \frac{x}{x^{1/2}} = x^{1-1/2} = x^{1/2}$

(c) $\sqrt[3]{x^9} = (x^9)^{1/3} = x^3$ (d) $\sqrt{x + 5} = (x + 5)^{1/2}$

(e) $\sqrt{x} + 5 = x^{1/2} + 5$ ■

EXERCISES

In Exercises 1–12 simplify.

1. $\sqrt{25}$
2. $\sqrt{100}$
3. $\sqrt[3]{27}$
4. $\sqrt[3]{-27}$
5. $\sqrt[4]{16}$
6. $\sqrt[5]{-1}$
7. $\sqrt[5]{32}$
8. $\sqrt[4]{10\,000}$
9. $(\sqrt[3]{8})^2$
10. $(\sqrt[3]{-8})^2$
11. $\sqrt{(-1)^2}$
12. $(\sqrt[4]{16})^3$

In Exercises 13–26 express in radical notation. Assume $a > 0, b > 0, c > 0$.

13. $a^{1/5}$
14. $b^{2/3}$
15. $c^{4/3}$
16. $c^{-1/2}$
17. $c^{-3/4}$
18. $a^{1/2}b^{1/3}$
19. $3(ab)^{2/5}$
20. $3ab^{2/5}$
21. $(a + b)^{1/2}$
22. $a + b^{1/2}$
23. $(a + 5b)^{-1/4}$
24. $(a + b + c)^{-2/7}$
25. $a^{2/3}b^{3/4}$
26. $5a^{3/2}b^{1/2}$

In Exercises 27–40 express in terms of rational exponents. Assume $x > 0$, $y > 0$, $z > 0$.

27. $\sqrt{x}$
28. $\sqrt[4]{y}$
29. $\sqrt[3]{xy}$
30. $\sqrt[3]{x}y$
31. $\sqrt[3]{x}\sqrt[4]{z}$
32. $\sqrt[5]{3xy}$
33. $(\sqrt[3]{5x^2y})^2$
34. $\sqrt[3]{(xyz)^2}$
35. $\sqrt{2 + x}$
36. $2 + \sqrt{x}$
37. $\sqrt{x} + \sqrt{y}$
38. $\sqrt{x + y}$
39. $\dfrac{3}{\sqrt{x}}$
40. $\dfrac{\sqrt{x}}{\sqrt[3]{yz}}$

In Exercises 41–68 simplify, if possible. Assume $a > 0, b > 0, c > 0$.

41. $\sqrt{25a^2}$
42. $\sqrt[3]{125b^3}$
43. $\sqrt[4]{a^4b^4c^8}$
44. $\sqrt{\dfrac{a^6b^2}{c^2}}$
45. $\sqrt[3]{\dfrac{8a^9b^3}{c^{12}}}$
46. $\sqrt[3]{8}\sqrt{a^2b^4}$
47. $\dfrac{\sqrt[5]{a^{10}b^{15}}}{\sqrt[4]{16c^4b^8}}$
48. $\sqrt{a^2 + b^2}$
49. $\sqrt{a^2} + \sqrt{b^2}$
50. $\sqrt{a^2} - \sqrt{b^2} + \sqrt{a^2 - b^2}$, $a \geq b$
51. $\sqrt{.01}$
52. $\sqrt{.0004}$
53. $\sqrt[4]{160\,000a^{24}}$
54. $\sqrt[4]{a^5b^8c^9}$
55. $\sqrt{12a^4c^7}$
56. $-\sqrt{\dfrac{8}{a^4b^6}}$
57. $\dfrac{-a^3b}{\sqrt{ab}}$
58. $\dfrac{\sqrt{4a^2}}{\sqrt{a}}$
59. $\left(\sqrt{\dfrac{1}{25a^4}}\right)^3$
60. $\sqrt{ab^2}\sqrt[3]{a^9b^6}$
61. $\sqrt[4]{32a^7b^8}$
62. $\sqrt[5]{-32a^6bc^5}$
63. $\sqrt[3]{\dfrac{81a^4b^9c^{12}}{24a}}$
64. $\sqrt[4]{80a^4b^6c^8}$
65. $\sqrt[4]{81a^8}$
66. $\sqrt[3]{(ab)^6c^9}$
67. $\sqrt[3]{a^3 + b^9}$
68. $\sqrt[3]{a^3 + a^6b^3}$

11.7 COMBINING RADICALS

Expressions with radicals (or rational exponents) can be added, subtracted, multiplied, factored, or divided. The *Distributive Laws,*

$$a(b + c) = ab + ac,$$

$$(b + c)a = ba + ca,$$

play an important role in these combinations.

EXAMPLE 1. Assume $x > 0$. Combine, as indicated:

(a) $\sqrt[4]{5} + 3\sqrt[4]{5}$

(b) $7(10^{1/3}) - 4(10^{1/3}) + 10^{1/3}$

(c) $\sqrt{25x^3} + \sqrt{64x^3} - \sqrt{x^3}$

(d) $\dfrac{\sqrt{4x^3y^2}}{3} + \sqrt{\dfrac{x^3y^2}{4}}$

SOLUTION.

(a) $\sqrt[4]{5} + 3\sqrt[4]{5} = 1 \cdot \sqrt[4]{5} + 3\sqrt[4]{5} = (1 + 3)\sqrt[4]{5} = 4\sqrt[4]{5}$

(b) $7(10^{1/3}) - 4(10^{1/3}) + 10^{1/3} = (7 - 4 + 1)(10^{1/3}) = 4(10^{1/3})$

(c) First note that

$$\sqrt{25x^3} = \sqrt{25}\sqrt{x^2}\sqrt{x} = 5x\sqrt{x},$$

$$\sqrt{64x^3} = \sqrt{64}\sqrt{x^2}\sqrt{x} = 8x\sqrt{x},$$

and

$$\sqrt{x^3} = \sqrt{x^2}\sqrt{x} = x\sqrt{x}.$$

Thus

$$\sqrt{25x^3} + \sqrt{64x^3} - \sqrt{x^3} = 5x\sqrt{x} + 8x\sqrt{x} - x\sqrt{x} = 12x\sqrt{x}$$

(d)
$$\frac{\sqrt{4x^3y^2}}{3} = \frac{\sqrt{4}\sqrt{x^2}\sqrt{x}\sqrt{y^2}}{3} = \frac{2xy\sqrt{x}}{3}$$

Here 3 is not under the radical sign. However, in the following, 4 is under the radical sign:

$$\sqrt{\frac{x^3y^2}{4}} = \frac{\sqrt{x^2}\sqrt{x}\sqrt{y^2}}{\sqrt{4}} = \frac{xy\sqrt{x}}{2}$$

Thus

$$\frac{\sqrt{4x^3y^2}}{3} + \sqrt{\frac{x^3y^2}{4}} = \frac{2xy\sqrt{x}}{3} + \frac{xy\sqrt{x}}{2}$$

$$= \frac{4xy\sqrt{x} + 3xy\sqrt{x}}{6}$$

$$= \frac{7xy\sqrt{x}}{6}$$

■

In general,

$$\sqrt{a} + \sqrt{b} \neq \sqrt{a + b}$$

For example,

$$\sqrt{4} + \sqrt{9} = 2 + 3 = 5$$

But

$$\sqrt{4 + 9} = \sqrt{13}$$

Thus

$$\sqrt{4} + \sqrt{9} \neq \sqrt{4 + 9}$$

EXAMPLE 2. Multiply:

(a) $\sqrt{2}(3 - \sqrt{3})$ (b) $\sqrt{2}(\sqrt{2} + 1)$ (c) $(\sqrt{x} + \sqrt{y})(\sqrt{x} - \sqrt{y})$

SOLUTION.

(a) $\sqrt{2}(3 - \sqrt{3}) = \sqrt{2} \cdot 3 - \sqrt{2}\sqrt{3} = 3\sqrt{2} - \sqrt{6}$
(b) $\sqrt{2}(\sqrt{2} + 1) = \sqrt{2}\sqrt{2} + \sqrt{2} \cdot 1 = 2 + \sqrt{2}$
(c) $(\sqrt{x} + \sqrt{y})(\sqrt{x} - \sqrt{y}) = \sqrt{x}\sqrt{x} + \cancel{\sqrt{y}\sqrt{x}} - \cancel{\sqrt{x}\sqrt{y}} - \sqrt{y}\sqrt{y}$
$= x - y$ ■

EXAMPLE 3. Multiply:

(a) $x^{1/2}(x^{1/3} + x^{1/2})$ (b) $(ab^{1/2} + 1)(a^{1/2}b^{1/2} - 2)$

SOLUTION.

(a) $x^{1/2}(x^{1/3} + x^{1/2}) = x^{1/2}x^{1/3} + x^{1/2}x^{1/2}$

$= x^{1/2+1/3} + x^{1/2+1/2}$

$= x^{5/6} + x$

(b) $(ab^{1/2} + 1)(a^{1/2}b^{1/2} - 2) = ab^{1/2}a^{1/2}b^{1/2} + 1 \cdot a^{1/2}b^{1/2} - 2ab^{1/2} - 2$

$= a^{3/2}b + a^{1/2}b^{1/2} - 2ab^{1/2} - 2$ ■

You can factor irrational numbers and algebraic expressions containing radicals. For example,

$$\sqrt{20} = \sqrt{4 \cdot 5} = \sqrt{4} \cdot \sqrt{5} = 2\sqrt{5}$$

$$x^{3/2} + x^{1/2}y = x^{1/2}x + x^{1/2}y = x^{1/2}(x + y)$$

Factoring expressions with radicals may lead to further simplification.

EXAMPLE 4.

(a) Factor: $2 + \sqrt{12}$

(b) Simplify: $\dfrac{2 + \sqrt{12}}{4}$

SOLUTION.

(a) $$\begin{aligned} 2 + \sqrt{12} &= 2 + \sqrt{4 \cdot 3} \\ &= 2 + \sqrt{4}\sqrt{3} \\ &= 2 + 2\sqrt{3} \\ &= 2(1 + \sqrt{3}) \end{aligned}$$

(b) $$\begin{aligned} \frac{2 + \sqrt{12}}{4} &= \frac{2(1 + \sqrt{3})}{4} \quad \text{by part (a)} \\ &= \frac{1 + \sqrt{3}}{2} \end{aligned}$$ ■

EXAMPLE 5.

(a) Factor: $\sqrt[3]{x^3y^4} + \sqrt{x^6y^8}$

(b) Simplify: $\dfrac{\sqrt[3]{x^3y^4} + \sqrt{x^6y^8}}{x^2y}$

SOLUTION.

(a) $$\begin{aligned} \sqrt[3]{x^3y^4} + \sqrt{x^6y^8} &= \sqrt[3]{x^3}\sqrt[3]{y^3}\sqrt[3]{y} + \sqrt{x^6}\sqrt{y^8} \\ &= xy\sqrt[3]{y} + x^3y^4 \\ &= xy(\sqrt[3]{y} + x^2y^3) \end{aligned}$$

(b) Use part (a) to factor the numerator.

$$\begin{aligned} \frac{\sqrt[3]{x^3y^4} + \sqrt{x^6y^8}}{x^2y} &= \frac{xy(\sqrt[3]{y} + x^2y^3)}{x^2y} \\ &= \frac{\sqrt[3]{y} + x^2y^3}{x} \end{aligned}$$ ■

EXAMPLE 6. Simplify:

(a) $5\sqrt{18} - 2\sqrt{8} + \sqrt{50}$ (b) $\sqrt{x^5y^4z} - 2\sqrt{xy^2z^3}$ (c) $81^{1/3} - 24^{1/3}$

SOLUTION.

(a) Although the radicals are all different, they have a common factor, $\sqrt{2}$. Thus

$$\begin{aligned} \sqrt{18} &= \sqrt{9}\sqrt{2} = 3\sqrt{2} \\ \sqrt{8} &= \sqrt{4}\sqrt{2} = 2\sqrt{2} \\ \sqrt{50} &= \sqrt{25}\sqrt{2} = 5\sqrt{2} \end{aligned}$$

$$5\sqrt{18} - 2\sqrt{8} + \sqrt{50} = 5\cdot 3\sqrt{2} - 2\cdot 2\sqrt{2} + 5\sqrt{2}$$
$$= 15\sqrt{2} - 4\sqrt{2} + 5\sqrt{2}$$
$$= 16\sqrt{2}$$

(b) $$\sqrt{x^5y^4z} - 2\sqrt{xy^2z^3} = x^2y^2\sqrt{xz} - 2yz\sqrt{xz}$$
$$= (x^2y - 2z)y\sqrt{xz}$$

(c) $$81^{1/3} - 24^{1/3} = (27\cdot 3)^{1/3} - (8\cdot 3)^{1/3}$$
$$= 27^{1/3}\cdot 3^{1/3} - 8^{1/3}\cdot 3^{1/3}$$
$$= 3(3^{1/3}) - 2(3^{1/3})$$
$$= 3^{1/3}$$

■

EXERCISES

Assume all letters represent positive numbers.
In Exercises 1–20 combine, as indicated.

1. $4\sqrt{3} + \sqrt{3}$
2. $\sqrt[3]{5} + 6\sqrt[3]{5}$
3. $2\sqrt{7} - \sqrt{7}$
4. $3\sqrt[4]{5} + 2\sqrt[4]{5} - \sqrt[4]{5}$
5. $\sqrt{x} + 4\sqrt{x}$
6. $\sqrt{xy} + 2\sqrt{xy} - 3\sqrt{xy}$
7. $\sqrt{9a} + \sqrt{4a}$
8. $\sqrt{16x} + \sqrt{25x} - \sqrt{x}$
9. $\sqrt{a^2b^3} - 4\sqrt{a^2b^3}$
10. $5\sqrt{x^3y^3} - \sqrt{4x^3y^3} + 4\sqrt{x^3y^3}$
11. $\dfrac{\sqrt{7}}{4} + \dfrac{\sqrt{7}}{4}$
12. $\sqrt{3} - \dfrac{\sqrt{3}}{2}$
13. $\dfrac{\sqrt[4]{5}}{2} + 2\sqrt[4]{5}$
14. $\sqrt{\dfrac{13}{4}} + \dfrac{\sqrt{13}}{2}$
15. $\dfrac{\sqrt{11}}{3} + \dfrac{\sqrt{11}}{2}$
16. $\dfrac{2\sqrt{7}}{5} + \dfrac{\sqrt{7}}{2}$
17. $\dfrac{\sqrt{3}}{4} - \left(\dfrac{\sqrt{3}}{2} + \dfrac{\sqrt{3}}{5}\right)$
18. $\dfrac{\sqrt{12}}{5} - \dfrac{3\sqrt{12}}{4} + \dfrac{\sqrt{12}}{2}$
19. $5^{1/3} + 2(5^{1/3})$
20. $\dfrac{x^{1/2}y^{1/3}}{5} - \dfrac{2x^{1/2}y^{1/3}}{3}$

In Exercises 21–36 multiply as indicated.

21. $3(\sqrt{2} + \sqrt{7})$
22. $\sqrt{2}(\sqrt{3} + \sqrt{5})$
23. $\sqrt{3}(3 + \sqrt{3} - \sqrt{2})$
24. $(\sqrt{a} + \sqrt{b})(\sqrt{a} - \sqrt{b})$
25. $(\sqrt{x} + \sqrt{a})(\sqrt{2x} + \sqrt{a})$
26. $(x + 3\sqrt{y})(\sqrt{x} + \sqrt{y})$
27. $(\sqrt{x} + 2\sqrt{y})^2$
28. $(\sqrt{x+y} - \sqrt{y})(\sqrt{x+y} + \sqrt{y})$
29. $x^{1/2}(x^{1/2} + 1)$
30. $x^{1/4}(x^{1/2} + x^{5/2})$
31. $a^{3/4}(a^{1/4} - a^{1/2})$
32. $y^{1/6}(y - y^{1/2} + y^{1/3})$

33. $(x^{1/2} - 1)(x^{1/2} + 1)$

34. $(a^{1/2} - b^{1/2})(a^{1/2} + b^{1/2})$

35. $(x^{1/4} - y^{1/4})^2$

36. $(x^{2/3} + x^{1/3})(x^{1/2} - x^{1/3})$

In Exercises 37–46, factor.

37. $3\sqrt{18}$

38. $5 - \sqrt{50}$

39. $\sqrt{6} + \sqrt{12} - \sqrt{60}$

40. $24 - 2\sqrt{8} + \sqrt{16}$

41. $\sqrt{a^2b} + \sqrt{9b}$

42. $\sqrt{a^2b^4c} - \sqrt{a^4b^2c^3}$

43. $\sqrt{9x^2y^4z} + \sqrt{x^4y^4z}$

44. $\sqrt{36x^3} - \sqrt{16x^5}$

45. $\sqrt{24x^3y^3} - \sqrt{54xy} + \sqrt{150x^3y}$

46. $a^{1/2}b^{3/2} - a^{3/2}b^{1/2}$

In Exercises 47–56, simplify.

47. $\dfrac{8 + 10\sqrt{3}}{2}$

48. $\dfrac{8 - \sqrt{12}}{6}$

49. $\dfrac{\sqrt{45} + \sqrt{10}}{\sqrt{5}}$

50. $\dfrac{3a\sqrt{x} + 6b\sqrt{x}}{3\sqrt{x}}$

51. $\dfrac{\sqrt{x} - \sqrt{xy}}{\sqrt{x}}$

52. $\dfrac{\sqrt[3]{27y^2} - 2\sqrt[3]{y^2}}{\sqrt[3]{y}}$

53. $\dfrac{\sqrt{18a^2b^3} - \sqrt{9a^4b^3}}{3a^4b}$

54. $\dfrac{\sqrt{125x^2y^4} - \sqrt{49x^3y^2}}{x^2y}$

55. $\dfrac{8a^{1/2}b^{3/2} - 2a^{3/2}b^{1/2}}{4a^{1/2}b^{1/2}}$

56. $\dfrac{(9ab^3)^{1/2} - (36ab^3)^{1/2}}{3a^{1/2}b^{3/2}}$

In Exercises 57–68 combine, as indicated.

57. $\sqrt{2} + \sqrt{8}$

58. $3\sqrt{5} - \sqrt{20}$

59. $5\sqrt{7} + \sqrt{63} - \sqrt{28}$

60. $7\sqrt{300} + 2\sqrt{12} - \sqrt{27}$

61. $5\sqrt{24} - (\sqrt{54} - 3\sqrt{150})$

62. $3\sqrt{98} - (\sqrt{32} + 4\sqrt{18})$

63. $\sqrt{a^5b^2} - a\sqrt{a^3b^2}$

64. $\sqrt[4]{16x^5} + \sqrt[4]{x}$

65. $\sqrt[3]{54x^3y^3} + \sqrt[3]{2x^6y^9}$

66. $\sqrt{100x^2y^4z^5} - \sqrt{25y^4z}$

67. $\dfrac{\sqrt{25a^2b^4c}}{2} - \dfrac{\sqrt{9a^2b^2c}}{3}$

68. $\dfrac{\sqrt{xyz^2}}{5} + \dfrac{\sqrt{4xyz^4}}{4}$

11.8 RATIONALIZING THE DENOMINATOR

Fractions are usually easiest to work with when their denominators are as simple as possible. For example, $\frac{4}{6}$ can be reduced to $\frac{2}{3}$. Similarly, it is better to write $\frac{4}{-3}$ in the standard form $\frac{-4}{3}$ (with *positive* denominator).

When the denominator of a fraction involves radicals (or fractional powers), convert to an equivalent fraction whose denominator is clear of radicals. The same remark applies to rational expressions. This process is called **rationalizing the denominator.**

Consider, for example, the fraction

$$\frac{1}{\sqrt{2}}$$

with a radical in the denominator. By the Fundamental Principle of Fractions,

$$\frac{1}{\sqrt{2}} = \frac{1 \cdot \sqrt{2}}{\sqrt{2} \cdot \sqrt{2}} = \frac{\sqrt{2}}{2}$$

Thus

$$\frac{1}{\sqrt{2}} = \frac{\sqrt{2}}{2}$$

The denominator of $\frac{\sqrt{2}}{2}$ is rational. To see that this is the simpler form, consider the *rational approximation* of $\sqrt{2}$:

$$\sqrt{2} \approx 1.414$$

Read: $\sqrt{2}$ is approximately equal to 1.414.

Thus

$$\frac{\sqrt{2}}{2} \approx \frac{1.414}{2} = .707,$$

whereas

$$\frac{1}{\sqrt{2}} \approx \frac{1}{1.414} = \frac{1000}{1414}$$

It is much harder to convert $\frac{1000}{1414}$ to a decimal than it was to convert $\frac{1.414}{2}$.

EXAMPLE 1. Let a be any number, let $x \neq 0$, and let $y > 0$. Rationalize the denominator of each:

(a) $\frac{3}{\sqrt{7}}$ (b) $\frac{a}{x^{1/3}}$ (c) $\frac{-1}{y^{5/4}}$

SOLUTION.

(a) Multiply the numerator and denominator by $\sqrt{7}$:

$$\frac{3}{\sqrt{7}} = \frac{3 \cdot \sqrt{7}}{\sqrt{7} \cdot \sqrt{7}} = \frac{3\sqrt{7}}{7}$$

(b) To clear the denominator of fractional powers, multiply the numerator and denominator by $x^{2/3}$. Thus

$$\frac{a}{x^{1/3}} = \frac{a \cdot x^{2/3}}{x^{1/3} \cdot x^{2/3}} = \frac{ax^{2/3}}{x}$$

(c) To obtain an integral power of y in the denominator, multiply the numerator and denominator by $y^{3/4}$:

$$\frac{-1}{y^{5/4}} = \frac{-1 \cdot y^{3/4}}{y^{5/4} \cdot y^{3/4}} = \frac{-y^{3/4}}{y^2}$$ ■

Your next goal is to rationalize the denominator of an expression such as

$$\frac{5}{1 - \sqrt{x}}.$$

Suppose you tried to multiply numerator and denominator by $\sqrt{x}$; then

$$\frac{5}{1 - \sqrt{x}} = \frac{5 \cdot \sqrt{x}}{(1 - \sqrt{x}) \cdot \sqrt{x}} = \frac{5\sqrt{x}}{\sqrt{x} - x}$$

The denominator is still not clear of radicals. Another method must be found. To this end, the expressions

$$A + \sqrt{B} \qquad \text{and} \qquad A - \sqrt{B}, \qquad B \geq 0,$$

are called **conjugates.** Each is **the conjugate of** the other. (A may or may not involve a radical.)

You will be concerned with conjugates of the forms:

$$c + \sqrt{b} \qquad \text{and} \qquad c - \sqrt{b}$$

$$\sqrt{a} + \sqrt{b} \qquad \text{and} \qquad \sqrt{a} - \sqrt{b}$$

$$c\sqrt{a} + d\sqrt{b} \qquad \text{and} \qquad c\sqrt{a} - d\sqrt{b}$$

Here c and d do not involve radicals.

EXAMPLE 2.

(a) $1 + \sqrt{3}$ and $1 - \sqrt{3}$ are conjugates.
(b) $1 - x^{1/2}$ and $1 + x^{1/2}$ are conjugates.
(c) $\sqrt{x} + \sqrt{y}$ and $\sqrt{x} - \sqrt{y}$ are conjugates.
(d) $x + 2\sqrt{a}$ and $x - 2\sqrt{a}$ are conjugates. ■

When you multiply conjugates of the above forms, radicals are cleared. The cross-terms are eliminated because of the difference of signs.

EXAMPLE 3.

(a) $(1 + \sqrt{3})(1 - \sqrt{3}) = 1 + \cancel{\sqrt{3}} - \cancel{\sqrt{3}} - \sqrt{3}\,\sqrt{3} = 1 - 3 = -2$
(b) $(1 - x^{1/2})(1 + x^{1/2}) = 1 - x^{1/2}x^{1/2} = 1 - x$
(c) $(\sqrt{x} + \sqrt{y})(\sqrt{x} - \sqrt{y}) = x - y$
(d) $(x + 2\sqrt{a})(x - 2\sqrt{a}) = x^2 - 4a$ ■

To rationalize a denominator of the form $c + \sqrt{b}$, $\sqrt{a} - \sqrt{b}$, and so on, *multiply the numerator and denominator by the conjugate of the denominator.*

EXAMPLE 4. Rationalize the denominator of each expression:

(a) $\dfrac{5}{1 - \sqrt{x}}$ (b) $\dfrac{a}{\sqrt{x} + \sqrt{y}}$ (c) $\dfrac{\sqrt{3} + \sqrt{2}}{\sqrt{3} - \sqrt{2}}$

SOLUTION.

(a) $$\frac{5}{1 - \sqrt{x}} = \frac{5\cdot(1 + \sqrt{x})}{(1 - \sqrt{x})\cdot(1 + \sqrt{x})} = \frac{5(1 + \sqrt{x})}{1 - x}$$

(b) $$\frac{a}{\sqrt{x} + \sqrt{y}} = \frac{a\cdot(\sqrt{x} - \sqrt{y})}{(\sqrt{x} + \sqrt{y})\cdot(\sqrt{x} - \sqrt{y})} = \frac{a(\sqrt{x} - \sqrt{y})}{x - y}$$

(c) $$\frac{\sqrt{3} + \sqrt{2}}{\sqrt{3} - \sqrt{2}} = \frac{(\sqrt{3} + \sqrt{2})\cdot(\sqrt{3} + \sqrt{2})}{(\sqrt{3} - \sqrt{2})\cdot(\sqrt{3} + \sqrt{2})} = \frac{3 + 2\sqrt{6} + 2}{3 - 2} = 5 + 2\sqrt{6}$$

Observe that in (a) and (b), the numerator is left in factored form. There was nothing to be gained by multiplying out. In part (c) the numerator is simpler after multiplying out. ■

EXERCISES

Rationalize the denominator. Assume that all letters represent positive numbers and that all denominators are nonzero.

1. $\dfrac{2}{\sqrt{3}}$
2. $\dfrac{5}{\sqrt{2}}$
3. $\dfrac{7}{5^{1/2}}$
4. $\dfrac{-1}{\sqrt{6}}$
5. $\dfrac{3}{2\sqrt{2}}$
6. $\dfrac{-2}{3\sqrt{5}}$
7. $\dfrac{\sqrt{5}}{3\sqrt{2}}$
8. $\dfrac{-3\sqrt{7}}{2\sqrt{5}}$
9. $\dfrac{1}{\sqrt{a}}$
10. $\dfrac{-1}{\sqrt{2x}}$
11. $\dfrac{x}{\sqrt{ab}}$
12. $\dfrac{3}{\sqrt{x-1}}$
13. $\dfrac{-5}{2\sqrt{x}}$
14. $\dfrac{-1}{\sqrt{x}\sqrt{y}}$
15. $\dfrac{1 + 2x}{x^{1/2}}$
16. $\dfrac{\sqrt{x} + \sqrt{y}}{\sqrt{x^3 y}}$
17. $\dfrac{1 - \sqrt{2y}}{\sqrt{4y}}$
18. $\dfrac{\sqrt{a} - \sqrt{b}}{\sqrt{8a^3b^3}}$
19. $\dfrac{1}{3^{1/3}}$
20. $\dfrac{-2}{5^{1/4}}$
21. $\dfrac{3}{\sqrt[3]{x^2}}$
22. $\dfrac{-7}{a^{4/5}}$
23. $\dfrac{x^{1/2}}{(x - a)^{1/3}}$
24. $\dfrac{2}{y^{7/5}}$
25. $\dfrac{1}{1 + \sqrt{2}}$
26. $\dfrac{-2}{2 + \sqrt{3}}$
27. $\dfrac{5}{3 - \sqrt{2}}$
28. $\dfrac{-3}{7 - \sqrt{3}}$
29. $\dfrac{4}{1 + \sqrt{a}}$
30. $\dfrac{-5}{2 - \sqrt{b}}$
31. $\dfrac{\sqrt{x}}{1 + \sqrt{x}}$
32. $\dfrac{-x\sqrt{x}}{2 - \sqrt{x}}$
33. $\dfrac{-3}{\sqrt{x} + 1}$
34. $\dfrac{-5}{\sqrt{x} - y}$
35. $\dfrac{-5}{\sqrt{x - y}}$
36. $\dfrac{-5}{\sqrt{x} - \sqrt{y}}$

37. $\dfrac{3}{a - \sqrt{b+1}}$
38. $\dfrac{-2ab}{\sqrt{a} - \sqrt{b}}$
39. $\dfrac{\sqrt{x}}{\sqrt{x} + \sqrt{y}}$
40. $\dfrac{1 + \sqrt{x}}{\sqrt{3} + \sqrt{x}}$
41. $\dfrac{6}{3 + 2\sqrt{x}}$
42. $\dfrac{\sqrt{3}}{5 - 7\sqrt{a}}$
43. $\dfrac{-2}{5 + 7\sqrt{2}}$
44. $\dfrac{9}{3 - \sqrt{3x}}$
45. $\dfrac{2}{7\sqrt{2} + 5\sqrt{3}}$
46. $\dfrac{-1}{a\sqrt{x} + b\sqrt{y}}$
47. $\dfrac{a + b}{a\sqrt{x} - b\sqrt{y}}$
48. $\dfrac{7 - 3x}{3\sqrt{x} - 2\sqrt{3}}$

11.9 COMPLEX NUMBERS

The square roots of some integers are integers. For example,

$$\sqrt{4} = 2, \qquad \sqrt{25} = 5, \qquad \sqrt{100} = 10$$

You cannot go very far in algebra without meeting square roots of other integers. You consider expressions such as

$$\sqrt{3}, \qquad \sqrt{5}, \qquad \sqrt{8} \text{ (or } 2\sqrt{2}\text{)}.$$

These square roots are not integers; in fact, they cannot be expressed as quotients of integers. Thus you consider *irrational numbers. Rational and irrational numbers together form the real numbers.* Up until now, real numbers were all you needed to know about.

In order to solve many problems in advanced mathematics, engineering, and science you have to consider *square roots of negative numbers.* As you know, negative numbers have no square roots *within the real number system* (because for every real number x, $x^2 \geq 0$). To get around this difficulty, it is necessary to consider a *new type of number.*

DEFINITION. Call i "the square root of -1".

Write

$$i = \sqrt{-1}.$$

Note that i is a *new number.*

Because $i = \sqrt{-1}$,

i^2 will have to be -1.

You will also have to consider such square roots as

$$\sqrt{-3} \qquad \text{and} \qquad \sqrt{-4}.$$

These can be defined in terms of i by extending the rule

$$\sqrt{ab} = \sqrt{a}\,\sqrt{b}.$$

Multiplication in the new number system will be defined so that

$$\sqrt{-3} = \sqrt{3(-1)} = \sqrt{3}\,\sqrt{-1} = \sqrt{3}i,$$
$$\sqrt{-4} = \sqrt{4(-1)} = \sqrt{4}\,\sqrt{-1} = 2i.$$

Now you will be able to find a root of a second-degree equation, such as

$$x^2 = -3 \qquad \text{and} \qquad x^2 = -4.$$

Extend the rule

$$(ab)^n = a^n b^n$$

to obtain

$$(\sqrt{3}i)^2 = (\sqrt{3})^2 i^2 = 3(-1) = -3.$$

Thus $\sqrt{3}i$ is a root of the equation

$$x^2 = -3.$$

Similarly,

$$\begin{aligned}(2i)^2 &= 2^2 i^2\\ &= 4(-1)\\ &= -4\end{aligned}$$

Therefore $2i$ is a root of the equation

$$x^2 = -4.$$

As you have seen, you must consider numbers of the form

$$bi, \qquad \text{where } b \text{ is real.}$$

It turns out that you must also work with numbers of the form

$$a + bi, \qquad \text{where } a \text{ and } b \text{ are both real.}$$

All "polynomial equations"

$$c_n x^n + \cdots + c_2 x^2 + c_1 x + c_0 = 0$$

can be solved using these numbers, regardless of the degree, n, of the polynomial.

DEFINITION. A **complex number** is an expression of the form

$$a + bi,$$

where a and b are real numbers.

EXAMPLE 1. Each of the following is a complex number:

(a) $2 + 3i$
(b) $3 - 2i$ (or $3 + (-2)i$)

(c) $7 + \frac{1}{4}i$

(d) $\frac{3 + i}{4}$ $\left(\text{or } \frac{3}{4} + \frac{1}{4}i\right)$

(e) $1 + \sqrt{2}i$ ■

DEFINITION. The complex number

$$a + 0i$$

(with $b = 0$) is considered to be the same as the real number a. Write

$$a \qquad \text{instead of} \qquad a + 0i.$$

Complex numbers of the form

$$0 + bi$$

(with $a = 0$) are called **imaginary.** Write

$$bi \qquad \text{instead of} \qquad 0 + bi.$$

EXAMPLE 2.

(a) Write

$$6 \qquad \text{for} \qquad 6 + 0i.$$

(b) Write

$$\sqrt{5} \qquad \text{for} \qquad \sqrt{5} + 0i.$$

(c) Write

$$7i \qquad \text{for} \qquad 0 + 7i.$$

$7i$ is an imaginary number.

(d) Write

$$-7i \qquad \text{for} \qquad 0 + (-7)i.$$

$-7i$ is an imaginary number. ■

DEFINITION. Let $a + bi$ be a complex number. Then a is called the **real part** and b the **imaginary part of** $a + bi$.

Note that *both the real and the imaginary parts of a complex number are real numbers.*

EXAMPLE 3.

(a) The real part of $3 + 5i$ is 3; the imaginary part is 5.

(b) The real part of $\frac{1}{2} - \sqrt{2}i$ is $\frac{1}{2}$; the imaginary part is $-\sqrt{2}$.

(c) The real part of $-3i$ is 0; the imaginary part is -3. ■

Addition, subtraction, multiplication, and division are defined for complex numbers. Simply treat complex numbers

$$a + bi$$

as you would treat binomials

$$a + bx,$$

and set

$$i^2 = -1.$$

To add or subtract complex numbers, add or subtract their real parts and their imaginary parts, separately.

EXAMPLE 4. Compute the following sum and difference:

(a) $(3 + 4i) + (7 - 2i)$ (b) $(3 + 4i) - (7 - 2i)$

SOLUTION.

(a) Method 1: $(3 + 4i) + (7 - 2i) = (3 + 7) + [4 + (-2)]i$
$= 10 + 2i$

Method 2: Add:

$$\begin{array}{r} 3 + 4i \\ \underline{7 - 2i} \\ 10 + 2i \end{array}$$

(b) Method 1: $(3 + 4i) - (7 - 2i) = (3 - 7) + [4 - (-2)]i$
$= -4 + 6i$

Method 2: *Subtract:*

$$\begin{array}{r} 3 + 4i \\ - \quad + \\ \underline{+ 7 - 2i} \\ -4 + 6i \end{array}$$

■

EXAMPLE 5. Compute the following sum and difference:

(a) $(8 - \sqrt{2}i) + (8 + \sqrt{2}i)$
(b) $(8 - \sqrt{2}i) - (8 + \sqrt{2}i)$

SOLUTION.

(a) Add:

$$\begin{array}{l} 8 - \sqrt{2}i \\ \underline{8 + \sqrt{2}i} \\ 16 \end{array}$$

(b) *Subtract:*

$$\begin{array}{r} 8 - \sqrt{2}i \\ - \quad - \\ \underline{+ 8 + \sqrt{2}i} \\ -2\sqrt{2}i \end{array}$$

■

To multiply complex numbers, extend the Associative, Commutative, and Distributive Laws from real numbers to complex.

EXAMPLE 6. Find the following products:

(a) $5(2 + 5i)$
(b) $2i(3 + 2i)$
(c) $(6 + 3i)(2 + i)$
(d) $(3 - 2i)(1 - i)$

SOLUTION.

(a) $5(2 + 5i) = 5 \cdot 2 + 5 \cdot 5i = 10 + 25i$
(b) $2i(3 + 2i) = (2i)3 + (2i)(2i)$
$= 6i + 4i^2$
$= 6i - 4$ $i^2 = -1$
$= -4 + 6i$

(c) *Multiply:*

$$\begin{array}{r} 6 + 3i \\ 2 + i \\ \hline 12 + 6i \\ 6i - 3 \\ \hline 9 + 12i \end{array}$$

$(3i)i = 3i^2 = -3$

Thus

$$(6 + 3i)(2 + i) = 9 + 12i$$

(d) *Multiply:*

$$\begin{array}{r} 3 - 2i \\ 1 - i \\ \hline 3 - 2i \\ - 3i - 2 \\ \hline 1 - 5i \end{array}$$

$(-2i)(-i) = 2i^2 = -2$

Thus

$$(3 - 2i)(1 - i) = 1 - 5i$$

■

DEFINITION. The **conjugate of a complex number** $a + bi$ is $a - bi$.

EXAMPLE 7.

(a) The conjugate of $3 + 2i$ is $3 - 2i$.
(b) The conjugate of $3 - 2i$ is $3 - (-2)i$, or $3 + 2i$.
(c) The conjugate of $\sqrt{2}i$ is $-\sqrt{2}i$ (because $\sqrt{2}i = 0 + \sqrt{2}i$).
(d) The conjugate of 7 is 7 (because $7 = 7 + 0i = 7 - 0i$).

The product of a complex number and its conjugate is a real number.

EXAMPLE 8. Find the product of $2 + 5i$ and its conjugate.

SOLUTION. The conjugate of $2 + 5i$ is $2 - 5i$. *Multiply:*

$$\begin{array}{l} 2 + 5i \\ \underline{2 - 5i} \\ 4 + 10i \\ \underline{ -10i + 25} \\ 29 \end{array}$$

$(5i)(-5i) = -25i^2 = 25$

Thus

$$(2 + 5i)(2 - 5i) = 29$$

Note that the product is a real number. ■

To divide two complex numbers, first express the quotient as a ratio or "fraction". Then multiply both numerator and denominator by the conjugate of the denominator. This will yield a fraction whose denominator is real. (Recall a similar use of "conjugates" in rationalizing a denominator.)

EXAMPLE 9.

$$\frac{6 + 5i}{3 + 2i} = \frac{(6 + 5i) \cdot (3 - 2i)}{(3 + 2i) \cdot (3 - 2i)}$$

$$= \frac{28 + 3i}{3^2 + 2^2}$$

$$= \frac{28}{13} + \frac{3}{13}i$$

$$\begin{array}{r} 6 + 5i \\ \underline{3 - 2i} \\ 18 + 15i \\ \underline{-12i + 10} \\ 28 + 3i \end{array}$$

■

Note that when the denominator of the fraction obtained is an integer, as in Example 9, the given complex number can easily be put in the form $a + bi$. In Example 9, $a = \frac{28}{13}$, $b = \frac{3}{13}$.

Powers of complex numbers are defined as are powers of real numbers. Thus

$$(2 + i)^2 = (2 + i)(2 + i) = 3 + 4i$$

$$(2 + i)^{10} = \underbrace{(2 + i)(2 + i) \cdots (2 + i)}_{\text{10 times}}$$

$$\begin{array}{r} 2 + i \\ \underline{2 + i} \\ 4 + 2i \\ \underline{2i - 1} \\ 3 + 4i \end{array}$$

$$(2 + i)^{-1} = \frac{1}{2 + i}$$

$$= \frac{1}{2 + i} \cdot \frac{2 - i}{2 - i}$$

$$= \frac{2 - i}{4 + 1}$$

$$= \frac{2}{5} - \frac{1}{5}i$$

The powers of i are particularly important. But after the 4th power, they repeat:

$$i^1 = i \qquad i^5 = i^4 \cdot i = 1 \cdot i = i$$

$$i^2 = -1 \qquad i^6 = i^4 \cdot i^2 = i^2 = -1$$

$$i^3 = i^2 \cdot i = -i \qquad i^7 = i^4 \cdot i^3 = i^3 = -i$$

$$i^4 = i^2 \cdot i^2 = (-1)(-1) = 1 \qquad i^8 = i^4 \cdot i^4 = 1$$

and so on. Thus

$$i^{10} = i^4 \cdot i^4 \cdot i^2 = -1$$

$$i^{16} = i^4 \cdot i^4 \cdot i^4 \cdot i^4 = 1$$

For *positive* real numbers a and b,

$$\sqrt{-a} = \sqrt{a}i \qquad \text{and} \qquad \sqrt{-ab} = \sqrt{a}\sqrt{b}i$$

Thus

$$\sqrt{-4} = \sqrt{4}i \quad (\text{or } 2i) \qquad \text{and}$$

$$\sqrt{-(4)(9)} = \sqrt{4}\sqrt{9}i (= 2 \cdot 3i = 6i)$$

For the most part, the rules that hold for real numbers are also true for complex numbers. However, there is one important exception. The rule

$$\sqrt{a}\sqrt{b} = \sqrt{ab}$$

is true when $a \geq 0$ and $b \geq 0$. However, what happens when both a and b are negative? Let $a = -4$ and $b = -4$. Both $\sqrt{a}$ and $\sqrt{b}$ are complex numbers. Then

$$\sqrt{a}\sqrt{b} = \sqrt{-4} \cdot \sqrt{-4} = 2i \cdot 2i = 4i^2 = -4$$

but

$$\sqrt{ab} = \sqrt{(-4)(-4)} = \sqrt{16} = 4$$

Thus

$$\sqrt{-4} \cdot \sqrt{-4} \neq \sqrt{(-4)(-4)}$$

The rule

$$\sqrt{a}\sqrt{b} = \sqrt{ab},$$

which is true for $a \geq 0$ and $b \geq 0$, is *false* when both a and b are negative.

Finally, complex numbers enable you to "factor" previously irreducible polynomials. As a simple example,

$$x^2 + 1 = (x + i)(x - i)$$

EXERCISES

In Exercises 1–8 write each expression in the form *bi*.

1. $\sqrt{-9}$
2. $\sqrt{-100}$
3. $\sqrt{-8}$
4. $\sqrt{-7}$
5. $\sqrt{-\frac{1}{4}}$
6. $\sqrt{-\frac{1}{16}}$
7. $\sqrt{\frac{-4}{9}}$
8. $\sqrt{\frac{-2}{5}}$

In Exercises 9–20 determine (a) the real part, (b) the imaginary part, and (c) the conjugate of the given complex number.

9. $1 + i$
10. $2 + 5i$
11. $3 + 3i$
12. $1 - i$
13. $-2 + 4i$
14. 2
15. $2i$
16. $-2i$
17. $\frac{1}{2} + \sqrt{2}i$
18. $\frac{3 - 5i}{2}$
19. $-\sqrt{2} + \sqrt{3}i$
20. 0

In Exercises 21–58 determine the indicated complex number in the form $a + bi$.

21. $(2 + 4i) + (1 + 2i)$
22. $(6 + 3i) + (8 - 2i)$
23. $(5 - 2i) + (-2 + 2i)$
24. $(2 - 3i) + (5 - 9i)$
25. $6 + (3 + 7i)$
26. $(4 + 2i) + 3i$
27. $\left(\frac{1}{2} + 2i\right) + \left(\frac{1}{4} - \frac{1}{2}i\right)$
28. $\left(\frac{1}{3} - \frac{1}{2}i\right) + \left(\frac{1}{6} - \frac{3}{4}i\right)$
29. $(4 + 7i) - (3 + 4i)$
30. $(8 + 2i) - (-6 + i)$
31. $(6 + 5i) - (7 - i)$
32. $i - (4 - 2i)$
33. $\left(\frac{1}{2} + \frac{1}{4}i\right) - \left(\frac{1}{4} + \frac{1}{2}i\right)$
34. $\left(\frac{2}{5} + \frac{1}{3}i\right) - \left(\frac{3}{5} - \frac{1}{6}i\right)$
35. $(2i)(3i)$
36. $(5i)(-4i)$
37. $-4(2 + 5i)$
38. $3i(7 - 2i)$
39. $(1 + i)(1 - i)$
40. $(2 + 3i)(1 - i)$
41. $(4 + 3i)(2 + i)$
42. $(1 + 2i)(3 - i)$
43. $(5 - 5i)(1 - 2i)$
44. $(10 - 3i)(6 + i)$
45. $(3i)^2$
46. $(-5i)^2$
47. $(1 + 2i)^2$
48. $(1 + 2i)^3$
49. $(3 - i)^2$
50. $(3 - i)^3$
51. $\frac{1 + i}{1 - i}$
52. $\frac{2 + i}{1 + i}$
53. $\frac{3 - 5i}{i}$
54. $\frac{4 - 3i}{2}$
55. $\frac{2 + 5i}{1 - 2i}$
56. $\frac{3 + i}{3 - i}$
57. $(5 + i)^{-1}$
58. $(5 + i)^{-2}$

In Exercises 59–67, simplify:

59. i^7
60. i^9
61. i^{12}
62. $i^{34} + i^{35}$
63. $i^{61} - i^{62}$
64. i^{-1}
65. i^{-2}
66. i^{-3}
67. i^{-4}
68. Is $\sqrt{(-1)(-1)} = \sqrt{-1}\,\sqrt{-1}$?

REVIEW EXERCISES FOR CHAPTER 11

In Exercises 1–36, assume $a > 0, b > 0, c > 0, x > 0, y > 0, z > 0$.

11.1 ZERO AND NEGATIVE EXPONENTS

In Exercises 1–3, evaluate:

1. 4^{-2}
2. $\left(\frac{3}{5}\right)^{-1}$
3. $\frac{2^{-4}}{7^{-1}}$

In Exercises 4–6 determine the exponent m.

4. $\frac{1}{a^3} = a^m$
5. $\frac{1}{36} = 6^m$
6. $\frac{1}{16x^4} = (2x)^m$

11.2 THE RULES FOR EXPONENTS

In Exercises 7–9 simplify and write your answers using only positive exponents.

7. $(a^{-2})^3$
8. $\frac{a^4a^5}{a^{-2}a^{10}}$
9. $\frac{(2a^{-1}b)^2}{(4ab^{-2})^{-2}}$

11.3 SCIENTIFIC NOTATION

In Exercises 10–12 express each in scientific notation.

10. 6049.4
11. .1018
12. .000 639

In Exercises 13–15 convert scientific notation back to the usual decimal notation.

13. 7.19×10^2
14. 6.38×10^{-1}
15. 1.21×10^{-4}

11.4 RATIONAL EXPONENTS

In Exercises 16–18 simplify:

16. $25^{1/2}$
17. $32^{3/5}$
18. $100^{-3/2}$

11.5 THE RULES FOR RATIONAL EXPONENTS

In Exercises 19–22 simplify and express your answer using only positive exponents.

19. $a^{3/4} \cdot a^{1/2}$
20. $\frac{a^{1/4}b^{1/2}}{a^{3/4}b^{3/4}}$
21. $(64a^{1/4}bc^2)^{1/2}$
22. $\frac{(2a^{1/3}b^{1/6}c)^6}{8a^2bc^3}$

11.6 RADICAL NOTATION

23. Simplify:
    (a) $\sqrt[3]{125}$
    (b) $(\sqrt[5]{-32})^2$

24. Express in radical notation:

    (a) $x^{1/2}y^{1/3}$ (b) $(x + y)^{3/4}$

25. Express in terms of rational exponents:

    (a) $\sqrt[5]{(xyz)^3}$ (b) $\dfrac{\sqrt[3]{x + y}}{\sqrt{x}}$

In Exercises 26–28 simplify, if possible:

26. $\sqrt{64a^2b^4}$ 27. $\sqrt[3]{27a^6}\sqrt{b^2c^8}$ 28. $\sqrt{3^2 + 4^2} - \sqrt{a^2} - \sqrt{4c^2}$

11.7 COMBINING RADICALS

29. Combine, as indicated:

    (a) $4\sqrt{5} + \sqrt{5} - 2\sqrt{5}$ (b) $\dfrac{\sqrt{12}}{2} + \dfrac{\sqrt{12}}{3}$

30. Multiply, as indicated:

    (a) $\sqrt{x + y}\sqrt{x - y}$ (b) $x^{1/4}(x + x^{1/2} - x^{3/4})$

31. Factor:

    (a) $4 - \sqrt{32}$ (b) $\sqrt{x^2y^3z^5} + \sqrt{9x^4yz}$

32. Simplify:

    (a) $\dfrac{\sqrt{8} + \sqrt{20}}{2}$ (b) $\dfrac{\sqrt{9xy} - \sqrt{25x^3y} + \sqrt{2x^3y^5}}{\sqrt{xy}}$

33. Combine, as indicated:

    (a) $\sqrt{50} + \sqrt{72} - \sqrt{18}$ (b) $\dfrac{\sqrt{100x^2y^2z^4} + x\sqrt{81y^3}}{xy}$

11.8 RATIONALIZING THE DENOMINATOR

In Exercises 34–36 rationalize the denominator:

34. $\dfrac{5}{\sqrt{3x}}$ 35. $\dfrac{-1}{1 + \sqrt{3}}$ 36. $\dfrac{x}{\sqrt{x} - 2\sqrt{y}}$

11.9 COMPLEX NUMBERS

37. Write each expression in the form bi:

    (a) $\sqrt{-81}$ (b) $\sqrt{\dfrac{-1}{9}}$

38. Determine (i) the real part, (ii) the imaginary part, and (iii) the conjugate of:

    (a) $2 - 3i$ (b) $\sqrt{2}$ (c) $7i$

In Exercises 39–41 determine the indicated complex number in the form $a + bi$.

39. $(7 + 2i) + (-3 + 5i)$

40. $(1 - 2i)(2 + 3i)$

41. $\dfrac{4 + i}{1 + i}$

42. Simplify:
    (a) i^{13}
    (b) $i^{16} - i^{-1}$

TEST YOURSELF ON CHAPTER 11.

In Exercises 1–8, assume $a > 0, b > 0$.

1. Simplify, and write your answers using positive exponents only:
   (a) $\dfrac{(a^{-2})^{-1}}{a^{-1}}$
   (b) $\dfrac{5^{-3}}{3^0 2^{-2}}$
2. (a) Express .949 in scientific notation.
   (b) Convert 3.14×10^4 back to the usual decimal notation.
3. Simplify, and express your answer using positive exponents only:
   $$\frac{a^{3/2} \cdot a^{1/2} b}{(a^{1/2} b)^2}$$
4. Simplify:
   (a) $(\sqrt[4]{16})^3$
   (b) $\sqrt{a^2 b^4 (a + b)^2}$
5. Combine:
   $$\frac{\sqrt{18}}{2} + \sqrt{\frac{18}{2}}$$
6. Multiply: $a^{1/3} b^{2/3} (3a^{1/3} b^{2/3} + a^{-1/3} b^{1/3})$
7. Simplify: $\dfrac{\sqrt{8a^3 b} - \sqrt{ab}}{2\sqrt{a}}$
8. Rationalize the denominator of $\dfrac{3}{1 - \sqrt{a}}$.
9. Determine (a) the real part, (b) the imaginary part, and (c) the conjugate of $\sqrt{5} - 2i$.
10. Determine $(3 - i)(2 + 4i)$ in the form $a + bi$.

12
logarithms

12.1 DEFINITION OF LOGARITHMS

Logarithms, which you will be learning about in this chapter, have important uses in such fields as physics, chemistry, and economics, as well as in advanced mathematics. Some of these applications will be illustrated at the end of this chapter. You will also see how a knowledge of logarithms simplifies difficult computations in arithmetic.

Logarithms are closely related to exponents. When you say

$$b^m = n,$$

b is called the **base,** m is the **exponent,** and n is the **mth power of b**. Often you are asked:

What power of b is n?

The emphasis here is on the *exponent* m, and you want to feature m (alone) on the right side of an equation. For this purpose, logarithmic notation is useful.

DEFINITION. Let b and n be positive numbers with $b \neq 1$. Then if

$$b^m = n,$$

say that m is **the logarithm of n to the base b**, and write

$$\log_b n = m.$$

Thus

$$\log_b n = m \qquad \text{means} \qquad b^m = n.$$

(You often speak of "the log of n" instead of "the logarithm of n".)

EXAMPLE 1.

(a) $\log_2 8 = 3$

base power exponent

(the log of 8 to the base 2 is 3) because

$2^3 = 8$

base exponent power

(b) $\log_{10} 10\,000 = 4$

base power exponent

(the log of 10 000 to the base 10 is 4) because

$10^4 = 10\,000$

base exponent power

(c) $\log_{10} .01 = -2$

base power exponent

(the log of .01 to the base 10 is -2) because

$$10^{-2} = \frac{1}{10^2} = \frac{1}{100} = .01$$

base exponent power

■

Observe that $\log_b n$ *is an exponent* because

$$\log_b n = m \qquad \text{means} \qquad b^m = n.$$

Also, if

$$\log_b n = m,$$

then

n is the $(\log_b n)$th power of b

because ‖

n is the mth power of b.

All powers n of a positive number b are positive. Thus $\log_b n$ *is defined only for positive n.* For example, suppose $b = 4$.

$$4^2 = 16 \qquad (\text{or } \log_4 16 = 2)$$

$$4^{1/2} = 2 \qquad \left(\text{or } \log_4 2 = \frac{1}{2}\right)$$

$$4^0 = 1 \qquad (\text{or } \log_4 1 = 0)$$

$$4^{-1} = \frac{1}{4} \qquad \left(\text{or } \log_4 \frac{1}{4} = -1\right)$$

$$4^{-3} = \frac{1}{64} \qquad \left(\text{or } \log_4 \frac{1}{64} = -3\right)$$

The various powers n of 4 are all positive. Thus $\log_4 n$ is defined only for positive n. On the other hand, logarithms, which are exponents, can be fractions or negative numbers.

EXAMPLE 2.

(a) $\log_4 16 = 2$ because $4^2 = 16$

(b) $\log_{16} 4 = \frac{1}{2}$ because $16^{1/2} = 4$

(c) $\log_{16} \frac{1}{4} = \frac{-1}{2}$ because $16^{-1/2} = \frac{1}{16^{1/2}} = \frac{1}{4}$ ■

EXAMPLE 3.

(a) $\log_4 8 = \frac{3}{2}$ because $4^{3/2} = (4^{1/2})^3 = 2^3 = 8$

(b) $\log_{16} 8 = \frac{3}{4}$ because $16^{3/4} = (16^{1/4})^3 = 2^3 = 8$

(c) $\log_{1/2} 8 = -3$ because $\left(\frac{1}{2}\right)^{-3} = \frac{1}{\left(\frac{1}{2}\right)^3} = \frac{1}{\frac{1}{8}} = 8$

Note that $\log_{1/2} 8$ stands for $\log_{\frac{1}{2}} 8$, which is difficult to read in print. ■

EXAMPLE 4. Find the value of:

(a) $\log_3 27$ (b) $\log_6 1$ (c) $\log_{1/8} \frac{1}{2}$

SOLUTION.

(a) What power of 3 is 27? ***Answer:*** The 3rd power

$$3^3 = 27$$

Therefore

$$\log_3 27 = 3$$

(b) What power of a number, such as 6, is 1? ***Answer:*** The 0th power

$$6^0 = 1$$

Therefore

$$\log_6 1 = 0$$

(c) $$\left(\frac{1}{8}\right)^{1/3} = \frac{1}{8^{1/3}} = \frac{1}{2}$$

Therefore

$$\log_{1/8} \frac{1}{2} = \frac{1}{3}$$ ■

EXAMPLE 5.

(a) Find the value of b:

$$\log_b 125 = 3$$

(b) Find the value of n:

$$\log_9 n = -2$$

SOLUTION.

(a) Solve the equation:

$$b^3 = 125$$

Answer: $b = 5$. For,

$$5^3 = 125$$

(b) Solve the equation:

$$9^{-2} = n$$

Answer: $n = 9^{-2} = \frac{1}{9^2} = \frac{1}{81}$ ■

In the definition of $\log_b n$, $b \neq 1$. This is because every power of 1 is 1. Thus the base 1 is uninteresting.

EXERCISES

In Exercises 1–20 express in logarithmic notation.

1. $2^4 = 16$
2. $5^3 = 125$
3. $10^6 = 1\,000\,000$
4. $7^3 = 243$
5. $10^{-1} = \frac{1}{10}$
6. $9^{-2} = \frac{1}{81}$
7. $4^{1/2} = 2$
8. $27^{1/3} = 3$
9. $16^{3/4} = 8$
10. $125^{2/3} = 25$
11. $9^{-1/2} = \frac{1}{3}$
12. $64^{-1/3} = \frac{1}{4}$
13. $\left(\frac{1}{3}\right)^{-2} = 9$
14. $\left(\frac{2}{5}\right)^3 = \frac{8}{125}$
15. $\left(\frac{4}{7}\right)^{-1} = \frac{7}{4}$
16. $\left(\frac{5}{3}\right)^{-2} = \frac{9}{25}$
17. $(.1)^4 = .0001$
18. $(.2)^3 = .008$
19. $.01^{1/2} = .1$
20. $.0144^{1/2} = .12$

In Exercises 21–32 express in exponential notation.

21. $\log_2 16 = 4$
22. $\log_3 27 = 3$
23. $\log_8 2 = \frac{1}{3}$
24. $\log_{100} 10 = \frac{1}{2}$
25. $\log_7 \frac{1}{7} = -1$
26. $\log_{16} 8 = \frac{3}{4}$
27. $\log_{49} 7 = \frac{1}{2}$
28. $\log_7 \frac{1}{49} = -2$
29. $\log_{10} .01 = -2$
30. $\log_{.2} .04 = 2$
31. $\log_{81} 27 = \frac{3}{4}$
32. $\log_{2/3} \frac{8}{27} = 3$

In Exercises 33–48 find the value of each logarithm.

33. $\log_5 25$
34. $\log_{11} 11$
35. $\log_{10} 10\,000$
36. $\log_2 64$
37. $\log_4 64$
38. $\log_8 64$
39. $\log_{16} 64$
40. $\log_{64} 8$
41. $\log_2 \frac{1}{64}$
42. $\log_{1/2} 64$
43. $\log_2 1$
44. $\log_{.3} .09$
45. $\log_{1/3} \frac{1}{27}$
46. $\log_{3/2} \frac{27}{8}$
47. $\log_{2/3} \frac{27}{8}$
48. $\log_{1/27} 81$

In Exercises 49–64 find the value of x.

49. $\log_5 x = 2$
50. $\log_x 8 = 3$
51. $\log_7 7 = x$
52. $\log_x 29 = 1$
53. $\log_x \frac{1}{3} = -2$
54. $\log_5 \sqrt{5} = x$

55. $\log_{10} 1 = x$ 56. $\log_x 121 = 2$ 57. $\log_x \frac{1}{4} = -2$

58. $\log_{2/5} \frac{4}{25} = x$ 59. $\log_x .64 = 2$ 60. $\log_x .64 = -2$

61. $\log_x .0016 = 4$ 62. $\log_{.09} x = 2$ 63. $\log_x 100\,000 = -5$

64. $\log_{32} 64 = x$

12.2 PROPERTIES OF LOGARITHMS

Logarithms have been defined in terms of exponents. Thus it is not surprising that the properties of logarithms are derived from properties of exponents.

Recall the laws of exponents:

(A) $b^{m_1+m_2} = b^{m_1}b^{m_2}$

(B) $b^{m_1-m_2} = \dfrac{b^{m_1}}{b^{m_2}}, \quad b \neq 0$

(C) $(b^m)^r = b^{mr}$

The corresponding properties of logarithms are as follows:

(A) $\log_b n_1 n_2 = \log_b n_1 + \log_b n_2$

(B) $\log_b \dfrac{n_1}{n_2} = \log_b n_1 - \log_b n_2$

(C) $\log_b n^r = r \log_b n$

Observe that *the base b is held constant in each equation.* Also, b ($\neq 1$), n, n_1, n_2 are positive real numbers, but r can be any rational number. Property **(A)** asserts:

The log of a product is the sum of the logs.

Property **(B)** asserts:

The log of a quotient is the difference of the logs.

Property **(C)** asserts:

The log of the rth *power of n is r times the log of n.*

Before proving these properties, consider some examples.

EXAMPLE 1. $\log_2 (8 \cdot 4) = \log_2 8 + \log_2 4$, according to **(A)**. Consider the left side of this equation:

$$\log_2 (8 \cdot 4) = \log_2 32 = 5$$

Now consider the right side and show this too equals 5:

$$\underbrace{\log_2 8}_{3} + \underbrace{\log_2 4}_{2} = 3 + 2 = 5$$

■

EXAMPLE 2. $\log_3 \frac{27}{9} = \log_3 27 - \log_3 9$, according to (**B**).

$$\log_3 \frac{27}{9} = \log_3 3 = 1$$

$$\underbrace{\log_3 27}_{3} - \underbrace{\log_3 9}_{2} = 3 - 2 = 1$$

■

EXAMPLE 3. $\log_{10} 100^3 = 3 \log_{10} 100$, according to (**C**).

$$\log_{10} 100^3 = \log_{10} 1\,000\,000 = 6$$

because $10^6 = 1\,000\,000$

$$3 \underbrace{\log_{10} 100}_{2} = 3 \cdot 2 = 6$$

■

These examples illustrate the laws, *but they do not prove them.* Perhaps, you might think these rules hold only for specially chosen numbers. You will now see that, in fact, these properties are true for all numbers (for which exponents or logarithms have been defined).

Here once again are the properties of exponents.

(A) $$b^{m_1+m_2} = b^{m_1}b^{m_2}$$

(B) $$b^{m_1-m_2} = \frac{b^{m_1}}{b^{m_2}}, \qquad b \neq 0$$

(C) $$(b^m)^r = b^{mr}$$

These properties will be used to prove the corresponding properties (**A**), (**B**), (**C**) of logarithms.

(**A**) $$\boxed{\log_b n_1 n_2 = \log_b n_1 + \log_b n_2}$$

Proof: Let

$$\log_b n_1 = m_1, \qquad \log_b n_2 = m_2.$$

By the definition of logarithms,

$$b^{m_1} = n_1, \qquad b^{m_2} = n_2$$

By property (A) of exponents,

$$b^{m_1+m_2} = b^{m_1} \cdot b^{m_2} = n_1 n_2$$

Again by the definition of logarithms,

$$\log_b n_1 n_2 = m_1 + m_2 = \log_b n_1 + \log_b n_2 \qquad \blacksquare$$

Property (**C**) will be verified next.

(**C**) $\boxed{\log_b n^r = r \log_b n}$

Proof: Let

$$\log_b n = m.$$

By the definition of logarithms,

$$b^m = n$$

Consider the rth power of each side.

$$(b^m)^r = n^r$$

Apply property (C) of exponents to the left side.

$$b^{mr} = (b^m)^r = n^r$$

Logarithmic notation yields

$$\log_b n^r = mr = rm = r \log_b n. \qquad \blacksquare$$

Properties (**A**) and (**C**) will be used to prove (**B**).

(**B**) $\boxed{\log_b \frac{n_1}{n_2} = \log_b n_1 - \log_b n_2}$

Proof:

$$\log_b \frac{n_1}{n_2} = \log_b \left(n_1 \cdot \frac{1}{n_2}\right)$$

$$= \log_b n_1 + \log_b \frac{1}{n_2} \qquad \text{by (A)}$$

$$= \log_b n_1 + \log_b n_2^{-1}$$

$$= \log_b n_1 + (-1) \log_b n_2 \qquad \text{by (C)}$$

$$= \log_b n_1 - \log_b n_2 \qquad \blacksquare$$

Here are some further examples illustrating the properties of logarithms. (Assume all letters represent positive numbers.)

EXAMPLE 4. Express $\log_b rst$ as a sum of logarithms.

SOLUTION.

$$\begin{aligned}\log_b rst &= \log_b (rs)t \\ &= \log_b rs + \log_b t \\ &= \log_b r + \log_b s + \log_b t\end{aligned}$$

■

EXAMPLE 5. Express $\log_2 \dfrac{x^3}{yz}$ in terms of simpler logarithms.

SOLUTION.

$$\begin{aligned}\log_2 \frac{x^3}{yz} &= \log_2 x^3 - \log_2 yz \\ &= 3\log_2 x - (\log_2 y + \log_2 z) \\ &= 3\log_2 x - \log_2 y - \log_2 z\end{aligned}$$

■

EXAMPLE 6. Express $\log_{10} \dfrac{\sqrt{ax^3}}{\sqrt[3]{5y^2}}$ in terms of simpler logarithms.

SOLUTION.

$$\begin{aligned}\log_{10} \frac{\sqrt{ax^3}}{\sqrt[3]{5y^2}} &= \log_{10} \sqrt{ax^3} - \log_{10} \sqrt[3]{5y^2} \\ &= \log_{10} (ax^3)^{1/2} - \log_{10} (5y^2)^{1/3} \\ &= \frac{1}{2}\log_{10}(ax^3) - \frac{1}{3}\log_{10}(5y^2) \\ &= \frac{1}{2}(\log_{10} a + \log_{10} x^3) - \frac{1}{3}(\log_{10} 5 + \log_{10} y^2) \\ &= \frac{1}{2}(\log_{10} a + 3\log_{10} x) - \frac{1}{3}(\log_{10} 5 + 2\log_{10} y)\end{aligned}$$

or

$$= \frac{1}{2}\log_{10} a + \frac{3}{2}\log_{10} x - \frac{1}{3}\log_{10} 5 - \frac{2}{3}\log_{10} y$$

■

EXAMPLE 7. Express $\log_b 48$ in terms of $\log_b 2$ and $\log_b 3$.

SOLUTION.

$$
\begin{aligned}
48 &= 2^4 \cdot 3 \\
\log_b 48 &= \log_b (2^4 \cdot 3) \\
&= \log_b 2^4 + \log_b 3 \\
&= 4 \log_b 2 + \log_b 3
\end{aligned}
$$

■

EXERCISES

Assume all variables represent positive numbers.

In Exercises 1–10: (a) Verify the following equations by evaluating both sides. (b) Indicate which properties of logarithms are illustrated.

1. $\log_5 125 = \log_5 25 + \log_5 5$ *Answer:* (a) $3 = 2 + 1$ (b) $\log_b n_1 n_2 = \log_b n_1 + \log_b n_2$ Here $b = 5, n_1 = 25, n_2 = 5$
2. $\log_3 81 = \log_3 3 + \log_3 27$
3. $\log_2 64 = \log_2 4 + \log_2 16$
4. $\log_3 9 = \log_3 27 - \log_3 3$
5. $\log_{10} 1000 = \log_{10} 100\,000 - \log_{10} 100$
6. $\log_4 1 = \log_4 64 - \log_4 64$
7. $\log_2 64 = 2 \log_2 8$
8. $\log_{10} 100\,000\,000 = 4 \log_{10} 100$
9. $\log_3 81 = 2 \log_3 9$
10. $\log_2 128 = \log_2 4 + 2 \log_2 8 - \log_2 2$

In Exercises 11–30 express each in terms of simpler logarithms.

11. $\log_b 10x$ *Answer:* $\log_b 10 + \log_b x$
12. $\log_b xyz$
13. $\log_b \dfrac{x}{y}$
14. $\log_b \dfrac{x}{3}$
15. $\log_{10} \dfrac{ab}{c}$
16. $\log_{10} \dfrac{a}{bc}$
17. $\log_{10} x^3$
18. $\log_2 y^{1/2}$
19. $\log_5 xy^4$
20. $\log_b (xy)^4$
21. $\log_3 \dfrac{x^2 y^{1/2}}{z}$
22. $\log_b \sqrt{xyz}$
23. $\log_b \sqrt{\dfrac{5x}{3}}$
24. $\log_b \dfrac{\sqrt{5x}}{3}$
25. $\log_{10} \dfrac{a^4 b^2}{c^3}$
26. $\log_{10} \dfrac{19a^7 b^2}{cd^4}$
27. $\log_b x^{1/2} y^{-1/3} z^{3/4}$
28. $\log_b \dfrac{1}{\sqrt{x^3 y^4}}$
29. $\log_b \left(\dfrac{x^{1/2} y^2}{z^3}\right)^{-5}$
30. $\log_b \dfrac{(x^3 y^{1/3})^7}{(wz^2)^3}$

In Exercises 31–42 express each in terms of $\log_b 2$, $\log_b 3$, and $\log_b 5$.

31. $\log_b 6$ *Answer:* $\log_b 2 + \log_b 3$

32. $\log_b 15$

33. $\log_b 25$

34. $\log_b 81$

35. $\log_b \frac{2}{3}$

36. $\log_b \frac{3}{4}$

37. $\log_b 12$

38. $\log_b 24$

39. $\log_b \frac{48}{25}$

40. $\log_b \frac{15}{64}$

41. $\log_b \frac{1}{3}$

42. $\log_b \frac{1}{20}$

In Exercises 43–56 express each as a single logarithm.

43. $\log_b s + \log_b t$

44. $\log_b x - \log_b y$

45. $\log_b r + \log_b s + \log_b t$

46. $\log_b 10 - \log_b a + \log_b x$

47. $\log_b 10 - (\log_b a + \log_b x)$

48. $\log_b 10 - (\log_b a - \log_b x)$

49. $2 \log_b x$

50. $3 \log_b m + \frac{1}{2} \log_b n$

51. $\frac{1}{4} \log_b x - \frac{1}{3} \log_b y^2 + \log_b z$

52. $\frac{1}{5}\left(\log_b r - 2 \log_b s + \frac{1}{3} \log_b t\right)$

53. $\dfrac{\log_b 10 - \log_b 7}{3}$

54. $\dfrac{\log_{10} a + 3 \log_{10} b}{5} + \dfrac{\log_{10} c - \log_{10} d}{3}$

55. $2\left(\log_3 7 + \frac{1}{3} \log_3 5\right) - \log_3 17$

56. $3\left[2 \log_3 9 + \frac{1}{2}(\log_3 5 - \log_3 10)\right]$

12.3 COMMON LOGARITHMS

Logarithms to the base 10 are known as **common logarithms.** As you will see in Section 12.6, common logarithms simplify arithmetic computations.

When you work with common logs you are considering powers of 10. Thus

$$\log_{10} 100 = 2$$

because

$$10^2 = 100$$

You can immediately find logs of powers of 10. In fact,

$$\log_{10} 10 = 1$$

because

$$10^1 = 10$$

Furthermore, by property (**C**) of logarithms,

$$\begin{aligned}\log_{10} 10^r &= r \log_{10} 10 \\ &= r \cdot 1 \\ &= r\end{aligned}$$

Thus

$$\begin{aligned}\log_{10} 100 &= \log_{10} 10^2 = 2 \\ \log_{10} 1000 &= \log_{10} 10^3 = 3 \\ \log_{10} 10\,000 &= \log_{10} 10^4 = 4 \\ \log_{10} 100\,000 &= \log_{10} 10^5 = 5\end{aligned}$$

and

$$\begin{aligned}\log_{10} 1 & & &= \log_{10} 10^0 = 0 \\ \log_{10} .1 &= \log_{10} \frac{1}{10} &&= \log_{10} 10^{-1} = -1 \\ \log_{10} .01 &= \log_{10} \frac{1}{100} &&= \log_{10} 10^{-2} = -2 \\ \log_{10} .001 &= \log_{10} \frac{1}{1000} &&= \log_{10} 10^{-3} = -3 \\ \log_{10} .0001 &= \log_{10} \frac{1}{10\,000} &&= \log_{10} 10^{-4} = -4 \\ \log_{10} .000\,01 &= \log_{10} \frac{1}{100\,000} &&= \log_{10} 10^{-5} = -5\end{aligned}$$

You could also determine the logs of *rational powers of* 10, as above. For example,

$$\begin{aligned}\log_{10} 10^{1/2} &= \frac{1}{2} \log_{10} 10 = \frac{1}{2} \\ \log_{10} 10^{2/3} &= \frac{2}{3} \log_{10} 10 = \frac{2}{3}\end{aligned}$$

The trouble is that numbers such as $10^{1/2}$ and $10^{2/3}$ are *irrational.* Numbers such as 5, $\frac{4}{3}$, 1.32 are not rational powers of 10. Recall that $\log_{10} n$ is defined for all positive n. *Except for special values of n* (such as 10, 100, $10^{1/2}$) $\log_{10} n$ *is irrational.*

It would be a hopeless task to determine, directly, the common log of most positive numbers (rational or irrational). For instance, to find $\log_{10} 1.32$, you would have to find the power of 10 that yields 1.32. Fortunately, tables have been compiled (by advanced methods) that indicate the *approximate* log of any number between 1 and 10. Such a table of common logs is found inside the back cover. As you will see, this table will enable you to determine the log of every positive number.

Let $1 \leq n < 10$. Furthermore, suppose n is a 3-digit number. For example, consider:

$$n = 1.32$$

The first 2 digits (together with the decimal point) are indicated in the first column in Table 12.1. The logs of all 3-digit numbers beginning with 1.3 are given in the (horizontal) *row* of 1.3. Thus $\log_{10} 1.32$ is in the (vertical) *column* headed by 2.

n	0	1	2	3	4	5	6	7	8	9
1.0	.0000	.0043	.0086	.0128	.0170	.0212	.0253	.0294	.0334	.0374
1.1	.0414	.0453	.0492	.0531	.0569	.0607	.0645	.0682	.0719	.0755
1.2	.0792	.0828	.0864	.0899	.0934	.0969	.1004	.1038	.1072	.1106
1.3	.1139	.1173	.1206	.1239	.1271	.1303	.1335	.1367	.1399	.1430
1.4	.1461	.1492	.1523	.1553	.1584	.1614	.1644	.1673	.1703	.1732
1.5	.1761	.1790	.1818	.1847	.1875	.1903	.1931	.1959	.1987	.2014
1.6	.2041	.2068	.2095	.2122	.2148	.2175	.2201	.2227	.2253	.2279
1.7	.2304	.2330	.2355	.2380	.2405	.2430	.2455	.2480	.2504	.2529
1.8	.2553	.2577	.2601	.2625	.2648	.2672	.2695	.2718	.2742	.2765
1.9	.2788	.2810	.2833	.2856	.2878	.2900	.2923	.2945	.2967	.2989
2.0	.3010	.3032	.3054	.3075	.3096	.3118	.3139	.3160	.3181	.3201
2.1	.3222	.3243	.3263	.3284	.3304	.3324	.3345	.3365	.3385	.3404
2.2	.3424	.3444	.3464	.3483	.3502	.3522	.3541	.3560	.3579	.3598
2.3	.3617	.3636	.3655	.3674	.3692	.3711	.3729	.3747	.3766	.3784
2.4	.3802	.3820	.3838	.3856	.3874	.3892	.3909	.3927	.3945	.3962

TABLE *12.1* $\log_{10} 1.32 = .1206$

(Strictly speaking, .1206 is only the *approximate* value of log 1.32.) From now on, the base 10 will usually be omitted in the notation. Thus write:

$$\log 1.32 = .1206$$

EXAMPLE 1. Find the value of: log 4.89

SOLUTION. Consider the row of 4.8 and the column of 9 in Table 12.2 on page 388.

n	0	1	2	3	4	5	6	7	8	9
4.0	.6021	.6031	.6042	.6053	.6064	.6075	.6085	.6096	.6107	.6117
4.1	.6128	.6138	.6149	.6160	.6170	.6180	.6191	.6201	.6212	.6222
4.2	.6232	.6243	.6253	.6263	.6274	.6284	.6294	.6304	.6314	.6325
4.3	.6335	.6345	.6355	.6365	.6375	.6385	.6395	.6405	.6415	.6425
4.4	.6435	.6444	.6454	.6464	.6474	.6484	.6493	.6503	.6513	.6522
4.5	.6532	.6542	.6551	.6561	.6571	.6580	.6590	.6599	.6609	.6618
4.6	.6628	.6637	.6646	.6656	.6665	.6675	.6684	.6693	.6702	.6712
4.7	.6721	.6730	.6739	.6749	.6758	.6767	.6776	.6785	.6794	.6803
4.8	.6812	.6821	.6830	.6839	.6848	.6857	.6866	.6875	.6884	.6893
4.9	.6902	.6911	.6920	.6928	.6937	.6946	.6955	.6964	.6972	.6981
5.0	.6990	.6998	.7007	.7016	.7024	.7033	.7042	.7050	.7059	.7067
5.1	.7076	.7084	.7093	.7101	.7110	.7118	.7126	.7135	.7143	.7152
5.2	.7160	.7168	.7177	.7185	.7193	.7202	.7210	.7218	.7226	.7235
5.3	.7243	.7251	.7259	.7267	.7275	.7284	.7292	.7300	.7308	.7316
5.4	.7324	.7332	.7340	.7348	.7356	.7364	.7372	.7380	.7388	.7396

TABLE *12.2* log 4.89 = .6893 ■

EXAMPLE 2. Find the value of: log 6.70

SOLUTION. This logarithm is found on the second page of the table of logs, inside the back cover. [Also, see Table 12.3 .] Consider the row of 6.7 and the column headed by 0.

n	0	1	2	3	4	5	6	7	8	9
5.5	.7404	.7412	.7419	.7427	.7435	.7443	.7451	.7459	.7466	.7474
5.6	.7482	.7490	.7497	.7505	.7513	.7520	.7528	.7536	.7543	.7551
5.7	.7559	.7566	.7574	.7582	.7589	.7597	.7604	.7612	.7619	.7627
5.8	.7634	.7642	.7649	.7657	.7664	.7672	.7679	.7686	.7694	.7701
5.9	.7709	.7716	.7723	.7731	.7738	.7745	.7752	.7760	.7767	.7774
6.0	.7782	.7789	.7796	.7803	.7810	.7818	.7825	.7832	.7839	.7846
6.1	.7853	.7860	.7868	.7875	.7882	.7889	.7896	.7903	.7910	.7917
6.2	.7924	.7931	.7938	.7945	.7952	.7959	.7966	.7973	.7980	.7987
6.3	.7993	.8000	.8007	.8014	.8021	.8028	.8035	.8041	.8048	.8055
6.4	.8062	.8069	.8075	.8082	.8089	.8096	.8102	.8109	.8116	.8122
6.5	.8129	.8136	.8142	.8149	.8156	.8162	.8169	.8176	.8182	.8189
6.6	.8195	.8202	.8209	.8215	.8222	.8228	.8235	.8241	.8248	.8254
6.7	.8261	.8267	.8274	.8280	.8287	.8293	.8299	.8306	.8312	.8319
6.8	.8325	.8331	.8338	.8344	.8351	.8357	.8363	.8370	.8376	.8382
6.9	.8388	.8395	.8401	.8407	.8414	.8420	.8426	.8432	.8439	.8445

TABLE *12.3* log 6.70 = .8261 ■

To find the log of a number n bigger than 10 or the log of a decimal, use **scientific notation** (Section 11.3). Thus write n as the product of a number between 1 and 10 and a power of 10. In Example 3, some numbers are converted to scientific notation. The logarithms of the first two of these numbers are determined in Examples 4 and 6.

EXAMPLE 3.

(a) $743 = 7.43 \times 10^2$ (b) $.008\,36 = 8.36 \times 10^{-3}$

(c) 6.95 is already between 1 and 10. Thus

$$6.95 = 6.95 \times 10^0$$

You can consider both 6.95 and 6.95×10^0 as being in scientific notation.

Note that in part (a) to obtain

743 from 7.43

move the decimal point 2 places to the right. Thus multiply by

$$10^2.$$

In part (b) to obtain

.008 36 from 8.36

move the decimal point 3 places to the left (and add zeros). Here multiply by a negative power of 10,

$$10^{-3}.$$ ■

The properties of logarithms enable you to determine the logarithm of any (3-digit) positive number n. Recall that

(**A**) $\log n_1 n_2 = \log n_1 + \log n_2$

and

(**C**) $\log_{10} 10^r = r$

EXAMPLE 4. Find the value of: log 743

SOLUTION.

$$\begin{aligned} 743 &= 7.43 \times 10^2 \\ \log 743 &= \log(7.43 \times 10^2) \\ &= \log 7.43 + \log 10^2 \\ &= \log 7.43 + 2 \end{aligned}$$

Determine log 7.43. [See Table 12.4.]

n	0	1	2	3	4	5	6	7	8	9
7.0	.8451	.8457	.8463	.8470	.8476	.8482	.8488	.8494	.8500	.8506
7.1	.8513	.8519	.8525	.8531	.8537	.8543	.8549	.8555	.8561	.8567
7.2	.8573	.8579	.8585	.8591	.8597	.8603	.8609	.8615	.8621	.8627
7.3	.8633	.8639	.8645	.8651	.8657	.8663	.8669	.8675	.8681	.8686
7.4	.8692	.8698	.8704	.8710	.8716	.8722	.8727	.8733	.8739	.8745

TABLE *12.4* log 7.43 = .8710

Thus

$$\log 743 = \underbrace{\log 7.43}_{.8710} + 2$$

$$= 2.8710$$

■

The log of a number consists of an integral part, called the **characteristic** and a decimal part, called the **mantissa.** In Example 4,

$$\log 743 = 2.8710$$

integral part or characteristic — decimal part or mantissa

Thus here

the characteristic is 2; the mantissa is .8710 .

Observe that *the characteristic equals the power of 10 when the original number is written in scientific notation.* For example,

$$743 = 7.43 \times 10^2$$

(characteristic)

and the characteristic of log 743 is 2.

The characteristic is always an integer—positive, negative, or zero. The mantissa is the part of the logarithm read from the table. It is important to note that *the mantissa is always written as a positive decimal—never as a negative decimal.*

EXAMPLE 5. Find the value of: log 936 000

SOLUTION. Write 936 000 = 9.36×10^5. Now log 9.36 can be read from the table. Thus

$$\log 936\,000 = \log(9.36 \times 10^5)$$
$$= \log 9.36 + \log 10^5$$
$$= \underbrace{\log 9.36}_{\text{mantissa}} + \underbrace{5}_{\text{characteristic}}$$

Use Table 12.5 to find log 9.36.

9.0	.9542	.9547	.9552	.9557	.9562	.9566	.9571	.9576	.9581	.9586
9.1	.9590	.9595	.9600	.9605	.9609	.9614	.9619	.9624	.9628	.9633
9.2	.9638	.9643	.9647	.9652	.9657	.9661	.9666	.9671	.9675	.9680
9.3	.9685	.9689	.9694	.9699	.9703	.9708	.9713	.9717	.9722	.9727
9.4	.9731	.9736	.9741	.9745	.9750	.9754	.9759	.9763	.9768	.9773
9.5	.9777	.9782	.9786	.9791	.9795	.9800	.9805	.9809	.9814	.9818
9.6	.9823	.9827	.9832	.9836	.9841	.9845	.9850	.9854	.9859	.9863
9.7	.9868	.9872	.9877	.9881	.9886	.9890	.9894	.9899	.9903	.9908
9.8	.9912	.9917	.9921	.9926	.9930	.9934	.9939	.9943	.9948	.9952
9.9	.9956	.9961	.9965	.9969	.9974	.9978	.9983	.9987	.9991	.9996
n	0	1	2	3	4	5	6	7	8	9

TABLE *12.5* $\log 9.36 = \underbrace{.9713}_{\text{mantissa}}$

Thus

$$\log 936\,000 = 5.9713$$

■

EXAMPLE 6. Find the value of: log .008 36

SOLUTION.

$$.008\,36 = 8.36 \times 10^{-3}$$
$$\log .008\,36 = \log(8.36 \times 10^{-3})$$
$$= \log 8.36 + \log 10^{-3}$$
$$= \underbrace{(\log 8.36)}_{\text{mantissa}} - \underbrace{3}_{\text{characteristic}}$$

Next, log 8.36 can be read from Table 12.6 on page 392.

n	0	1	2	3	4	5	6	7	8	9
8.0	.9031	.9036	.9042	.9047	.9053	.9058	.9063	.9069	.9074	.9079
8.1	.9085	.9090	.9096	.9101	.9106	.9112	.9117	.9122	.9128	.9133
8.2	.9138	.9143	.9149	.9154	.9159	.9165	.9170	.9175	.9180	.9186
8.3	.9191	.9196	.9201	.9206	.9212	.9217	.9222	.9227	.9232	.9238
8.4	.9243	.9248	.9253	.9258	.9263	.9269	.9274	.9279	.9284	.9289

TABLE *12.6* $\log 8.36 = \underbrace{.9222}_{\text{mantissa}}$

Therefore

$$\log .008\ 36 = \underbrace{\log 8.36}_{.9222} - 3$$

$$= \underbrace{.9222}_{\text{mantissa}} \underbrace{- 3}_{\text{characteristic}}$$

If you added the positive mantissa .9222 to the negative characteristic -3 you would obtain

$$\begin{array}{r} -3.0000 \\ +\ \ .9222 \\ \hline -2.0778 \end{array} = -2 - .0778$$

This would change the mantissa from .9222 to the *negative decimal* $-.0778$, and you would no longer be able to locate it in the table. ■

EXAMPLE 7. Find the value of: log .0233

SOLUTION.

$$.0233 = 2.33 \times 10^{-2}$$

$$\log .0233 = \log(2.33 \times 10^{-2})$$

$$= \log 2.33 - 2$$

Read log 2.33 from Table 12.7 on page 393.

n	0	1	2	3	4	5	6	7	8	9
2.0	.3010	.3032	.3054	.3075	.3096	.3118	.3139	.3160	.3181	.3201
2.1	.3222	.3243	.3263	.3284	.3304	.3324	.3345	.3365	.3385	.3404
2.2	.3424	.3444	.3464	.3483	.3502	.3522	.3541	.3560	.3579	.3598
2.3	.3617	.3636	.3655	.3674	.3692	.3711	.3729	.3747	.3766	.3784
2.4	.3802	.3820	.3838	.3856	.3874	.3892	.3909	.3927	.3945	.3962

TABLE *12.7* $\log 2.33 = \underbrace{.3674}_{\text{mantissa}}$

$$\log .0233 = \underbrace{.3674}_{\text{mantissa}} \underbrace{- \ 2}_{\text{characteristic}}$$

EXERCISES

In Exercises 1–14 find the value of the indicated common log.

1. log 1.03
2. log 1.92
3. log 4.31
4. log 9.77
5. log 3.00
6. log 8.07
7. log 9.40
8. log 9.99
9. log 6.03
10. log 5.11
11. log 7.28
12. log 6.46
13. log 1.09
14. log 9.85

In Exercises 15–30 determine the characteristic of the indicated common log.

15. log 27.4
16. log 382
17. log 68 700
18. log 9.14
19. log .0384
20. log .101
21. log .000 724
22. log .0101
23. log 8 760 000
24. log .000 087 6
25. $\log (4.82 \times 10^{8})$
26. $\log (2.84 \times 10^{-7})$
27. $\log (485 \times 10^{4})$
28. $\log (.283 \times 10^{-3})$
29. $\log (30 \times 200)$
30. $\log (30 \times 40)$

In Exercises 31–50 find the value of the indicated common log.

31. log 872
32. log 9430
33. log 16.9
34. log 148 000
35. log 256 000 000
36. log 70 100
37. log .935
38. log .004 32
39. log .0136
40. log .652
41. log 6.39
42. log 86.5
43. log .000 832
44. log .0736
45. log 62
46. log .38
47. log .004 920
48. log 830 000
49. $\log (836 \times 10^{-5})$
50. $\log (.413 \times 10^{-2})$

12.4 ANTILOGS

In addition to determining the common log of a number, you will also want to find the number with a given logarithm. For instance, you may be told that

$$\log n = 2.5717,$$

and asked to find this number n. This can also be expressed by:

$$n = \text{antilog}\, 2.5717$$

DEFINITION. Let n be any real number. Define

$$\textbf{antilog}\, m = n$$

if

$$\log n = m.$$

(Do you see that antilog m is simply 10^m?)
Thus

$$\text{antilog}\, 3 = 1000 \qquad \text{because} \qquad \log 1000 = 3$$

$$(\text{or because } 10^3 = 1000)$$

To determine an antilog, reverse the process of finding a log. Before finding an antilog, recall how the log is found.

EXAMPLE 1.

$$\log 484 = \log (4.84 \times 10^2)$$

$$= \underbrace{2}_{\text{characteristic}} + \underbrace{\log 4.84}_{\text{mantissa}}$$

$$= 2.\underbrace{6848}_{\text{mantissa}}$$

because $\log 4.84 = .6848$

[See Table 12.8 on page 395.]
Suppose you are asked to determine:

$$\text{antilog}\, 2.6848$$

The mantissa came from the log table. Thus you now skim through the table to locate this mantissa, .6848 . You note the row 4.8 and the column 4 in which the mantissa .6848 is located. Thus this mantissa corresponds to 4.84, and you have determined the *digits* of the antilog. Now *use the characteristic to place the*

n	0	1	2	3	4	5	6	7	8	9
4.5	.6532	.6542	.6551	.6561	.6571	.6580	.6590	.6599	.6609	.6618
4.6	.6628	.6637	.6646	.6656	.6665	.6675	.6684	.6693	.6702	.6712
4.7	.6721	.6730	.6739	.6749	.6758	.6767	.6776	.6785	.6794	.6803
4.8	.6812	.6821	.6830	.6839	.6848	.6857	.6866	.6875	.6884	.6893
4.9	.6902	.6911	.6920	.6928	.6937	.6946	.6955	.6964	.6972	.6981
5.0	.6990	.6998	.7007	.7016	.7024	.7033	.7042	.7050	.7059	.7067
5.1	.7076	.7084	.7093	.7101	.7110	.7118	.7126	.7135	.7143	.7152
5.2	.7160	.7168	.7177	.7185	.7193	.7202	.7210	.7218	.7226	.7235
5.3	.7243	.7251	.7259	.7267	.7275	.7284	.7292	.7300	.7308	.7316
5.4	.7324	.7332	.7340	.7348	.7356	.7364	.7372	.7380	.7388	.7396

TABLE *12.8* log 4.84 = .6848 and thus antilog .6848 = 4.84

decimal point. Because the characteristic is 2, multiply by 10^2. This means you must move the decimal point 2 places to the *right.*

$$\text{antilog}\ .6848 = 4.84$$

$$\text{antilog}\ 2.6848 = 484$$

■

EXAMPLE 2. Find the value of: antilog 3.6665

SOLUTION. You are given that

$$\log n = \underbrace{3}_{\text{characteristic}}.\underbrace{6665}_{\text{mantissa}}$$

and you must find n.

First locate the mantissa .6665 in Table 12.9.

n	0	1	2	3	4	5	6	7	8	9
4.5	.6532	.6542	.6551	.6561	.6571	.6580	.6590	.6599	.6609	.6618
4.6	.6628	.6637	.6646	.6656	.6665	.6675	.6684	.6693	.6702	.6712
4.7	.6721	.6730	.6739	.6749	.6758	.6767	.6776	.6785	.6794	.6803
4.8	.6812	.6821	.6830	.6839	.6848	.6857	.6866	.6875	.6884	.6893
4.9	.6902	.6911	.6920	.6928	.6937	.6946	.6955	.6964	.6972	.6981
5.0	.6990	.6998	.7007	.7016	.7024	.7033	.7042	.7050	.7059	.7067
5.1	.7076	.7084	.7093	.7101	.7110	.7118	.7126	.7135	.7143	.7152
5.2	.7160	.7168	.7177	.7185	.7193	.7202	.7210	.7218	.7226	.7235
5.3	.7243	.7251	.7259	.7267	.7275	.7284	.7292	.7300	.7308	.7316
5.4	.7324	.7332	.7340	.7348	.7356	.7364	.7372	.7380	.7388	.7396

TABLE *12.9* log 4.64 = .6665

Now move the decimal point 3 places to the right because the characteristic is 3. You must add a 0 after the second 4.

$$\log 4640 = 3.6665$$

$$4640 = \text{antilog } 3.6665$$

■

EXAMPLE 3. Find the value of: antilog .9917

SOLUTION. The characteristic is 0. According to Table 12.10,

$$\log 9.81 = .9917$$

9.5	.9777	.9782	.9786	.9791	.9795	.9800	.9805	.9809	.9814	.9818
9.6	.9823	.9827	.9832	.9836	.9841	.9845	.9850	.9854	.9859	.9863
9.7	.9868	.9872	.9877	.9881	.9886	.9890	.9894	.9899	.9903	.9908
9.8	.9912	.9917	.9921	.9926	.9930	.9934	.9939	.9943	.9948	.9952
9.9	.9956	.9961	.9965	.9969	.9974	.9978	.9983	.9987	.9991	.9996
n	0	1	2	3	4	5	6	7	8	9

TABLE *12.10*

Thus

$$9.81 = \text{antilog } .9917$$

Because the characteristic is 0, this is the solution. ■

EXAMPLE 4. Find the value of: antilog (.4829 − 1)

SOLUTION. Locate the mantissa, .4829, in Table 12.11.

n	0	1	2	3	4	5	6	7	8	9
3.0	.4771	.4786	.4800	.4814	.4829	.4843	.4857	.4871	.4886	.4900
3.1	.4914	.4928	.4942	.4955	.4969	.4983	.4997	.5011	.5024	.5038
3.2	.5051	.5065	.5079	.5092	.5105	.5119	.5132	.5145	.5159	.5172
3.3	.5185	.5198	.5211	.5224	.5237	.5250	.5263	.5276	.5289	.5302
3.4	.5315	.5328	.5340	.5353	.5366	.5378	.5391	.5403	.5416	.5428

TABLE *12.11*

$$\log 3.04 = .4829$$

Because the characteristic is -1, move the decimal point 1 place to the left.

$$\log .304 = .4829 - 1$$
$$.304 = \text{antilog}\,(.4829 - 1)$$

EXAMPLE 5. Find the value of: antilog (.8182 − 4)

SOLUTION. Locate the mantissa, .8182, in Table 12.12.

n	0	1	2	3	4	5	6	7	8	9
6.5	.8129	.8136	.8142	.8149	.8156	.8162	.8169	.8176	.8182	.8189
6.6	.8195	.8202	.8209	.8215	.8222	.8228	.8235	.8241	.8248	.8254
6.7	.8261	.8267	.8274	.8280	.8287	.8293	.8299	.8306	.8312	.8319
6.8	.8325	.8331	.8338	.8344	.8351	.8357	.8363	.8370	.8376	.8382
6.9	.8388	.8395	.8401	.8407	.8414	.8420	.8426	.8432	.8439	.8445

TABLE *12.12.* $\log 6.58 = .8182$

Because the characteristic is -4, move the decimal point 4 places to the left. Add enough 0's.

$$\log .000\,658 = .8182 - 4$$
$$.000\,658 = \text{antilog}\,(.8182 - 4)$$ ■

In the next section you will learn how to approximate the antilog when the given mantissa is *not* located in the table of logs.

EXERCISES

Find the value of each antilog.

1. antilog .2945
2. antilog .9355
3. antilog .8716
4. antilog .8254
5. antilog .4346
6. antilog .2504
7. antilog .5623
8. antilog 1.5623
9. antilog 4.5623
10. antilog (.5623 − 1)
11. antilog (.5623 − 3)
12. antilog (.5623 − 5)
13. antilog .7846
14. antilog 2.7846
15. antilog 6.7846
16. antilog (.7846 − 1)
17. antilog (.7846 − 2)
18. antilog (.7846 − 6)
19. antilog 1.0374
20. antilog 4.9809
21. antilog 2.9805
22. antilog 6.9708
23. antilog (.8488 − 3)
24. antilog (.9818 − 1)
25. antilog (.9805 − 2)
26. antilog (.2227 − 4)
27. antilog .0294

28. antilog 5.6010
29. antilog .5623
30. antilog 4.7067
31. antilog 5.6493
32. antilog (.6493 − 5)
33. antilog 8.8089
34. antilog (.8089 − 8)

12.5 INTERPOLATING

The log of a 4-digit number, such as 2.215, does not appear in the log table. To obtain log 2.215, you *approximate* the value by a method known as **linear interpolation.**

EXAMPLE 1. Find the value of: log 2.215

SOLUTION. First observe that

$$2.210 < 2.215 < 2.220.$$

The logs of the 3-digit numbers

$$2.21\,(=2.210) \qquad \text{and} \qquad 2.22\,(=2.220)$$

are given in Table 12.13.

n	0	1	2	3	4	5	6	7	8	9
2.0	.3010	.3032	.3054	.3075	.3096	.3118	.3139	.3160	.3181	.3201
2.1	.3222	.3243	.3263	.3284	.3304	.3324	.3345	.3365	.3385	.3404
2.2	.3424	.3444	.3464	.3483	.3502	.3522	.3541	.3560	.3579	.3598
2.3	.3617	.3636	.3655	.3674	.3692	.3711	.3729	.3747	.3766	.3784
2.4	.3802	.3820	.3838	.3856	.3874	.3892	.3909	.3927	.3945	.3962

TABLE *12.13.* log 2.21 = .3444, log 2.22 = .3464

Clearly, 2.215 lies exactly halfway between 2.210 and 2.220, as you can see in Table 12.14 .

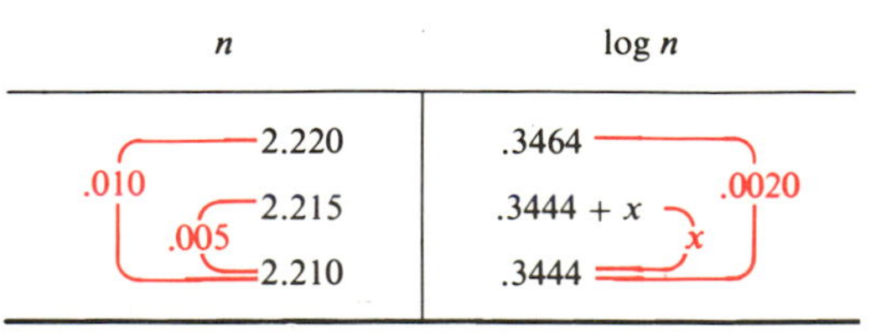

TABLE *12.14.*

In Table 12.14 the left-hand column indicates the values of n. The right-hand column indicates the values of log n. Observe the differences between the *largest* and *smallest* values in *each column*.

$$\begin{array}{r} 2.220 \\ -2.210 \\ \hline .010 \end{array} \qquad \begin{array}{r} .3464 \\ -.3444 \\ \hline .0020 \end{array}$$

You seek the log of the intermediate number, 2.215. Therefore, also note the differences:

$$\begin{array}{r} 2.215 \\ -2.210 \\ \hline .005 \end{array} \qquad \text{and} \qquad \begin{array}{r} .3444 + x \\ -.3444 \\ \hline x \end{array}$$

You need the *proportional increase*, x, on the right side. Thus set up the proportion:

$$\frac{.005}{.010} = \frac{x}{.0020}$$

Multiply the numerator and denominator on the left side by 1000, and reduce to lowest terms

$$\frac{5}{10} = \frac{x}{.0020}$$

$$\frac{1}{2} = \frac{x}{.0020}$$

$$\frac{.0020}{2} = x$$

$$.0010 = x$$

Thus add:

$$\begin{array}{r} .3444 \\ .0010 \\ \hline .3454 \end{array}$$

and obtain:

$$\log 2.215 = .3454$$

This is a reasonably good *approximation* to the actual value of log 2.215 .

In practice it is often unnecessary to be so careful about the decimal points in setting up the proportion. You could let d stand for the number of units (ten-thousandths) in the log column [See Table 12.15.]

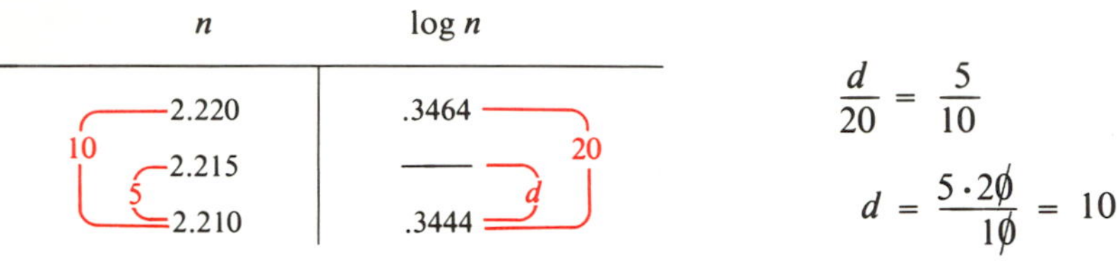

TABLE 12.15.

$$\frac{d}{20} = \frac{5}{10}$$

$$d = \frac{5 \cdot 2\not{0}}{1\not{0}} = 10$$

Thus add 10 "units" (.0010) to .3444 and obtain

$$\begin{array}{rl} & .3444 \\ & \underline{.0010} \\ \log 2.215 = & .3454, \end{array}$$

as above. ■

EXAMPLE 2. Find the value of: log 9564

SOLUTION.

$$\log 9564 = \log(9.564 \times 10^3)$$
$$= 3 + \log 9.564$$

Find log 9.56 and log 9.57 in the table inside the back cover. Then consider Table 12.16.

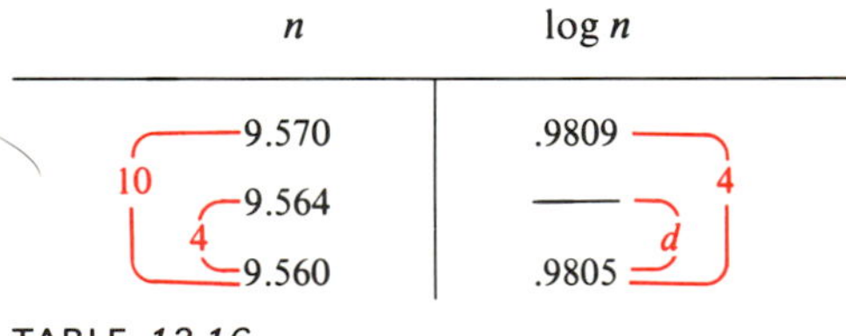

TABLE 12.16.

Because 9.564 is $\frac{4}{10}$ of the way from 9.560 to 9.570, log 9.564 should be about  $\frac{4}{10}$ of the way from .9805 to .9809.

Set up the proportion:

$$\frac{4}{10} = \frac{d}{4}$$

$$\frac{16}{10} = d$$

$$1.6 = d$$

Round off to the nearest integer. Because the digit to the right of decimal point is .5 or more, approximate d as 2. Therefore add:

$$\begin{array}{rr} & .9805 \\ & \underline{.0002} \\ \log 9.564 = & .9807 \end{array}$$

and

$$\log 9564 = 3.9807$$

■

EXAMPLE 3. Find the value of: log .089 11

SOLUTION.

$$\log .089\,11 = \log(8.911 \times 10^{-2})$$
$$= \log 8.911 - 2$$

[See Table 12.17.]

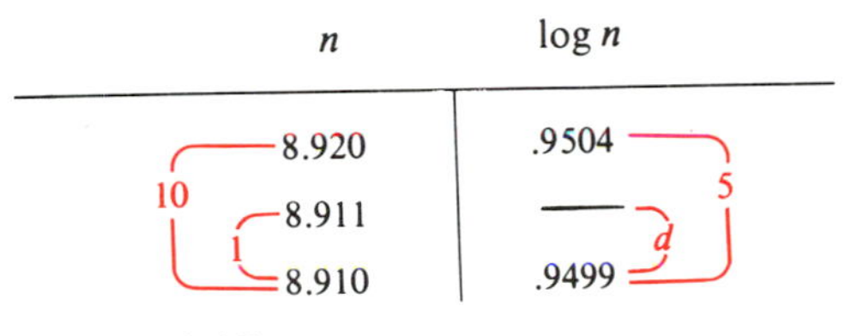

n	$\log n$
8.920	.9504
8.911	—
8.910	.9499

TABLE *12.17.*

Set up the proportion:

$$\frac{d}{5} = \frac{1}{10}$$

$$d = \frac{5}{10} = \frac{1}{2}$$

Because this is $\frac{1}{2}$ (or more), approximate d as 1. Add:

$$\begin{array}{rr} & .9499 \\ & \underline{.0001} \\ \log 8.911 = & .9500 \end{array}$$

Thus

$$\log .08911 = .9500 - 2$$

■

Interpolation can also be used to determine the antilog of a 4-placed decimal that is not in the log table.

EXAMPLE 4. Find the value of: antilog .9276

SOLUTION. .9276 is not found in the log table. Write down the closest numbers to .9276 that are in the table—one smaller, the other larger than .9276. These values are .9274 and .9279.

$$\log 8.46 = .9274 \qquad \text{and} \qquad \log 8.47 = .9279$$

Consider Table 12.18.

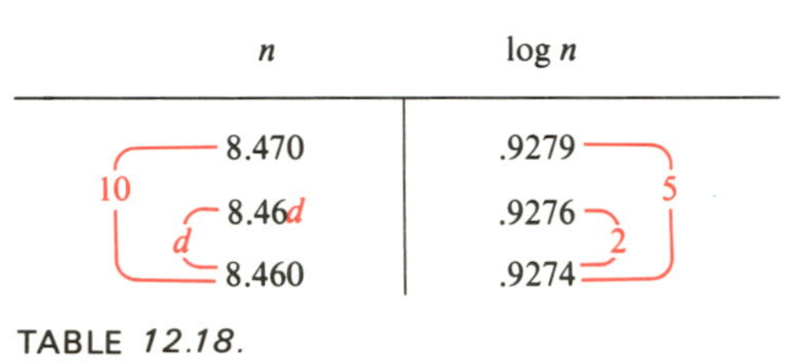

n	$\log n$
8.470	.9279
8.46d	.9276
8.460	.9274

TABLE *12.18.*

Because .9276 is $\frac{2}{5}$ of the way from .9274 to .9279, antilog .9276 should be about $\frac{2}{5}$ of the way from 8.460 to 8.470 .

Now d represents the increase in the *left-hand* column. Set up the proportion:

$$\frac{d}{10} = \frac{2}{5}$$

$$d = \frac{2 \cdot 10}{5}$$

$$d = 4$$

Add:

$$\begin{array}{r} 8.460 \\ \underline{.004} \\ \log\ 8.464 = .9276 \end{array}$$

Therefore

$$8.464 = \text{antilog } .9276$$ ■

EXAMPLE 5. Find the value of: antilog 3.5253

SOLUTION. As in the preceding section, first consider the mantissa, .5253 . Find n such that

$$\log n = .5253.$$

[See Table 12.19 on page 403.]

n	$\log n$
3.360	.5263
3.35d	.5253
3.350	.5250

(Brackets: 10 spans 3.360 to 3.350; d spans 3.35d to 3.350; 13 spans .5263 to .5250; 3 spans .5253 to .5250.)

TABLE *12.19.*

$$\frac{d}{10} = \frac{3}{13}$$

$$d = \frac{3 \cdot 10}{13} = \frac{30}{13} = 2\,\frac{4}{13}$$

Because

$$\frac{4}{13} < \frac{1}{2}$$

take

$$d = 2.$$

Therefore

$$3.352 = \text{antilog}\ .5253$$

To find antilog 3.5253 (with characteristic 3), move the decimal point 3 places to the right.

$$3\,352. = \text{antilog}\ 3.5253$$

■

EXERCISES

In Exercises 1–30 find the value of each common log by interpolating.

1. log 4.105
2. log 2.246
3. log 8.124
4. log 9.908
5. log 41.37
6. log 309.3
7. log .7035
8. log 99 480
9. log 103 600
10. log .085 52
11. log .007 699
12. log 9301
13. log 7.002
14. log 70 020
15. log .7002
16. log .070 02
17. log 3.589
18. log 35 890
19. log .003 589
20. log .000 035 89
21. log 8585
22. log 89.37
23. log .008 511
24. log .000 060 06
25. $\log (3.947 \times 10^{7})$
26. $\log (5.053 \times 10^{-4})$
27. $\log (8366 \times 10^{-8})$
28. $\log (.8888 \times 10^{6})$
29. $\log 98.01 \times 10^{8}$
30. $\log (.006\,329 \times 10^{-6})$

In Exercises 31–60 find the value of each antilog by interpolating.

31. antilog .6022
32. antilog .6587
33. antilog .9544
34. antilog .1538
35. antilog 1.9796
36. antilog 3.9215
37. antilog 4.9876
38. antilog (.9967 − 1)
39. antilog (.9984 − 2)
40. antilog 5.8390
41. antilog .8230
42. antilog 1.8230

43. antilog 4.8230
44. antilog (.8230 − 1)
45. antilog .8240
46. antilog 6.8240
47. antilog (.8240 − 2)
48. antilog (.8240 − 5)
49. antilog 6.0270
50. antilog (.3432 − 4)
51. antilog (.3931 − 1)
52. antilog (.3950 − 3)
53. antilog .1900
54. antilog .0005
55. antilog (.9999 − 1)
56. antilog 3.8247
57. antilog 6.6666
58. antilog 4.4343
59. antilog (.9224 − 2)
60. antilog (.9580 − 7)

12.6 COMPUTING WITH LOGARITHMS

Computations involving products, quotients, and powers (or roots) can be greatly simplified by the use of log tables. Of course, the results you obtain will usually be approximations.

In this era of the pocket calculator, you may wonder why it is necessary to worry about these computations. The answer is that computing with logarithms is probably the best way to come to grips with the properties of logarithms. And you will need to understand logarithms in your further studies of the natural sciences and mathematics.

First recall the basic properties of logarithms that will be used.

(A) $\log n_1 n_2 = \log n_1 + \log n_2$

(B) $\log \dfrac{n_1}{n_2} = \log n_1 - \log n_2$

(C) $\log n^r = r \log n$

Thus instead of multiplying or dividing, you now add or subtract. Instead of finding a power or root, you now multiply or divide. The operations you perform are therefore simpler.

The first example could easily be done directly. However, it illustrates the method.

EXAMPLE 1. Find the value of: (8.39) (.002 64)

SOLUTION. Let $n = (8.39)(.002\,64)$. Instead of finding n directly, consider $\log n$, and use Property (**A**).

$$\begin{aligned}\log n &= \log(8.39)(.002\,64)\\ &= \log(8.39) + \log(.002\,64)\\ &= \log 8.39 + \log(2.64 \times 10^{-3})\\ &= .9238 + .4216 - 3\end{aligned}$$

Thus add:

$$\begin{array}{r} .9238 \\ +\ \underline{.4216 - 3} \\ 1.3454 - 3 = .3454 - 2 \end{array}$$

$1 - 3 = -2$

Therefore

$$\log n = .3454 - 2$$

$$n = \text{antilog}\,(.3454 - 2)$$

In other words, to find the original product n, determine the indicated antilog. [See Table 12.20 .]

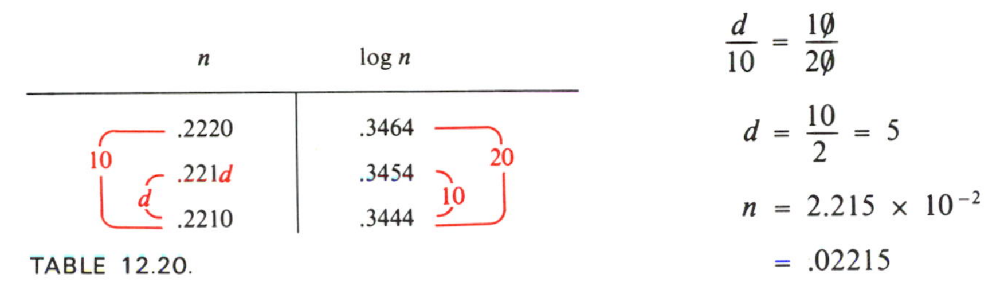

n	$\log n$
.2220	.3464
.221d	.3454
.2210	.3444

TABLE 12.20.

$$\frac{d}{10} = \frac{10}{20}$$

$$d = \frac{10}{2} = 5$$

$$n = 2.215 \times 10^{-2}$$

$$= .02215$$

Compare this with the *exact* answer you obtain by multiplying:

$$\begin{array}{r} 8.39 \\ \underline{.00264} \\ 3356 \\ 5034 \\ \underline{1678} \\ .0221496 \end{array}$$

If you round off the last 2 digits, you obtain .022 15, as in the logarithmic method. ■

EXAMPLE 2. Find the value of: $\dfrac{1.78}{9.35}$

SOLUTION. Let $n = \dfrac{1.78}{9.35}$. Then

$$\log n = \log\left(\frac{1.78}{9.35}\right)$$

$$= \log 1.78 - \log 9.35$$

$$= .2504 - .9708$$

If you subtracted directly, you would obtain the *negative mantissa*, $-.7204$, which you could not locate in your table of *positive matissas*. To avoid this, write

$$.2504 = 1.2504 - 1$$

Thus you obtain:

$$\begin{array}{r} 1.2504 - 1 \\ - \underline{.9708 \quad\;\;} \\ .2796 - 1 \end{array}$$

positive mantissa

$$\log n = .2796 - 1$$

$$n = \text{antilog}\,(.2796 - 1)$$

Interpolate as in Table 12.21.

n	$\log n$
1.910	.2810
1.90d	.2796
1.900	.2788

(Differences: 10 between 1.910 and 1.900, d between 1.90d and 1.900; 22 between .2810 and .2788, 8 between .2796 and .2788.)

TABLE *12.21.*

$$\frac{d}{10} = \frac{8}{22}$$

$$d = \frac{10 \cdot 8}{22} = \frac{80}{22} = 3\,\frac{14}{22} \approx 4$$

$$n = 1.904 \times 10^{-1}$$

$$= .1904$$

■

EXAMPLE 3. Find the value of: $\left(\dfrac{.009\ 35}{17.8}\right)^2$

SOLUTION. Let $n = \left(\dfrac{.009\ 35}{17.8}\right)^2$.

Consider $\log n$, and use properties (**B**) and (**C**):

$$\begin{aligned} \log n &= \log \left(\frac{.009\ 35}{17.8}\right)^2 \\ &= 2 \log \left(\frac{.009\ 35}{17.8}\right) \\ &= 2(\log .009\ 35 - \log 17.8) \\ &= 2[\log (9.35 \times 10^{-3}) - \log (1.78 \times 10^{1})] \\ &= 2[(.9708 - 3) - 1.2504] \end{aligned}$$

$$2[.9708 - 3 - .2504 - 1]$$

$$\begin{array}{r} .9708 - 3 \\ - \underline{.2504 - 1} \\ .7204 - 4 \\ \times \underline{\qquad\quad 2} \\ 1.4408 - 8 \end{array} = .4408 - 7$$

$1 - 8 = -7$

$$\log n = .4408 - 7$$
$$n = \text{antilog}\,(.4408 - 7)$$

Interpolate as in Table 12.22 .

n	$\log n$
2.760	.4409
2.75d	.4408
2.750	.4393

(Interpolation markings: 10 spans 2.750 to 2.760; d spans 2.750 to 2.75d; 16 spans .4393 to .4409; 15 spans .4393 to .4408.)

TABLE *12.22*

$$\frac{d}{10} = \frac{15}{16}$$
$$d = \frac{10 \cdot 15}{16} = \frac{150}{16} = 9\,\frac{6}{16} \approx 9.$$
$$n = 2.759 \times 10^{-7}$$
$$= .000\,000\,275\,9$$ ■

EXAMPLE 4. Find the value of: $\sqrt[5]{.308}$

SOLUTION. Let

$$n = \sqrt[5]{.308} = (.308)^{1/5}$$
$$\log n = \log\,(.308)^{1/5}$$
$$= \frac{1}{5}\log .308$$
$$= \frac{\log\,(3.08 \times 10^{-1})}{5}$$
$$= \frac{.4886 - 1}{5}$$

If you were to divide at this point, you would obtain

$$\frac{.4886 - 1}{5} = .9772 - \frac{1}{5}.$$

This is not in proper form because $-\frac{1}{5}$ *is not an integer*. Recall that *the characteristic must be an integer* in order to use the log table. When you divide by 5, you must obtain a negative integer as characteristic, in addition to a positive decimal mantissa. Thus write

$$.4886 - 1 = 4.4886 - 5,$$
$$4 - 5 = -1$$

and obtain:

$$\log n = \frac{4.4886 - 5}{5} = .8977 - 1$$

Now antilog $(.8977 - 1)$ can be determined from Table 12.23 on page 408.

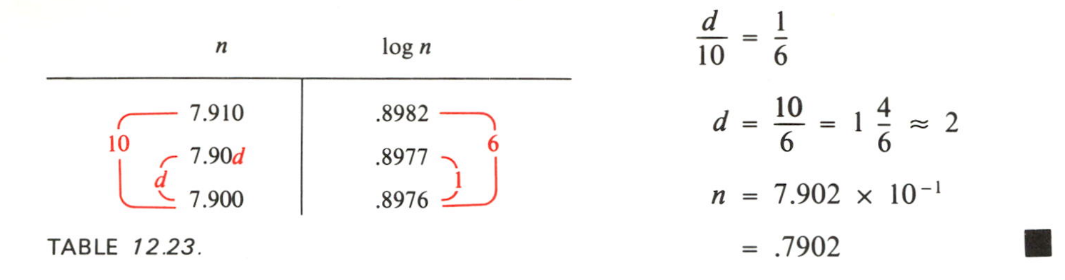

n	$\log n$
7.910	.8982
7.90*d*	.8977
7.900	.8976

TABLE *12.23.*

$$\frac{d}{10} = \frac{1}{6}$$

$$d = \frac{10}{6} = 1\frac{4}{6} \approx 2$$

$$n = 7.902 \times 10^{-1}$$

$$= .7902$$

■

EXAMPLE 5. Find the value of $\left(\dfrac{9.37^4 \times \sqrt{.008\,46}}{947}\right)^{3/4}$.

SOLUTION. Let $n = \left[\dfrac{(9.37)^4 \times (.008\,46)^{1/2}}{947}\right]^{3/4}$.

$$\log n = \log\left[\frac{(9.37)^4(.008\,46)^{1/2}}{947}\right]^{3/4}$$

$$= \frac{3}{4}\log\left[\frac{(9.37)^4(.008\,46)^{1/2}}{947}\right] \qquad \text{by property (C)}$$

$$= \frac{3}{4}[\log(9.37)^4 + \log(.008\,46)^{1/2} - \log 947] \qquad \text{by properties (A) and (B)}$$

$$= \frac{3}{4}\left[4\log 9.37 + \frac{1}{2}\log(.008\,46) - \log 947\right]$$

$$= 3\log 9.37 + \frac{3}{8}\log .008\,46 - \frac{3}{4}\log 947$$

$$= 3\log 9.37 + \frac{3}{8}\log(8.46 \times 10^{-3}) - \frac{3}{4}\log(9.47 \times 10^2)$$

$$= 3(.9717) + \frac{3}{8}(.9274 - 3) - \frac{3}{4}(2.9763)$$

$$= 2.9151 + \frac{3}{8}(.9274 - 3) - 2.2322$$

The middle expression requires further discussion.

$$\frac{3}{8}(.9274 - 3) = \frac{2.7822 - 9}{8}$$

$$= \frac{1.7822 - 8}{8}$$

Here $2 - 9 = 1 - 8 (= -7)$. But now the negative part, -8, *is divisible by* 8. Upon dividing you obtain

$$\underbrace{.2228}_{\text{mantissa}} \underbrace{- \ 1}_{\text{characteristic}}$$

(with an integer, -1, as the characteristic).

Thus

$$\begin{array}{rl} & 2.9151 \\ + & .2228 - 1 \\ \hline & 3.1379 - 1 \\ - & 2.2322 \\ \hline \end{array}$$

$$\log n = .9057 - 1$$

$$n = \text{antilog}\,(.9057 - 1)$$

Interpolate, as in Table 12.24 .

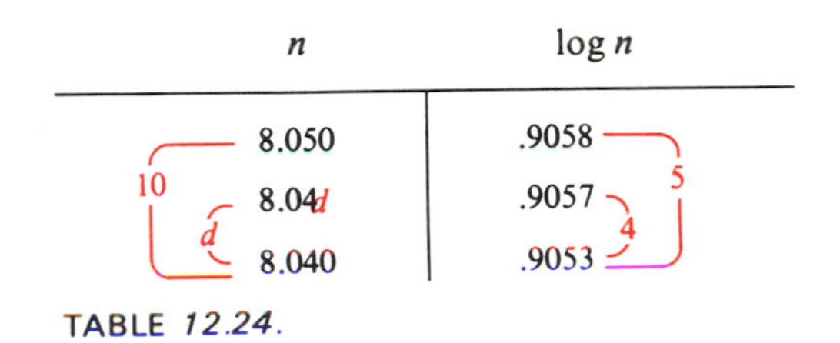

n	$\log n$
8.050	.9058
8.04d	.9057
8.040	.9053

TABLE 12.24.

$$\frac{d}{10} = \frac{4}{5}$$

$$d = \frac{10 \cdot 4}{5} = \frac{40}{5} = 8$$

$$n = 8.048 \times 10^{-1}$$

$$= .8048$$

EXAMPLE 6. Find the value of: $\dfrac{(2.805)^4(1.117)^{1/3}}{(8.083)^{1/2}}$

SOLUTION. Let $n = \dfrac{(2.805)^4(1.117)^{1/3}}{(8.083)^{1/2}}$.

$$\log n = \log \frac{(2.805)^4(1.117)^{1/3}}{(8.083)^{1/2}}$$

$$= 4 \log 2.805 + \frac{1}{3} \log 1.117 - \frac{1}{2} \log 8.083$$

Interpolate to find these logs. [See Tables 12.25, 12.26, and 12.27.]

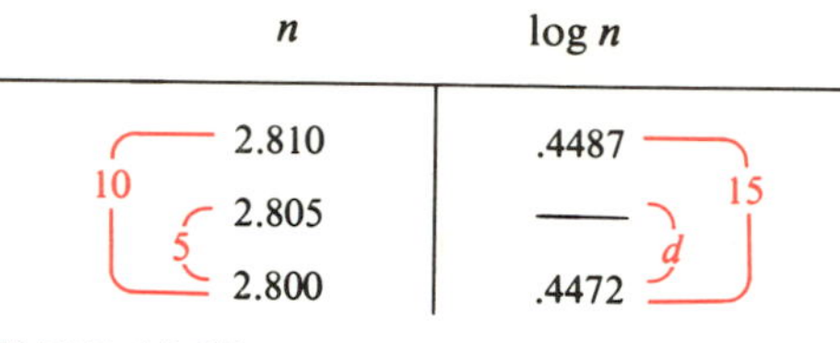

n	$\log n$
2.810	.4487
2.805	——
2.800	.4472

TABLE 12.25.

$$\frac{d}{15} = \frac{5}{10} = \frac{1}{2}$$

$$d = \frac{15}{2} = 7\frac{1}{2} \approx 8$$

$$\begin{aligned} \log 2.805 &= .4472 \\ &\underline{+.0008} \\ &.4480 \\ &\underline{\times \quad 4} \\ 4 \log 2.805 &= 1.7920 \end{aligned}$$

n	$\log n$
1.120	.0492
1.117	——
1.110	.0453

(10, 7, 39, d)

TABLE 12.26.

$$\frac{d}{39} = \frac{7}{10}$$

$$d = \frac{7 \cdot 39}{10} = \frac{273}{10} = 27.3 \approx 27$$

$$\begin{aligned} \log 1.117 &= .0453 \\ &\underline{+.0027} \\ &.0480 \end{aligned}$$

$$\frac{1}{3} \log 1.117 = \frac{.0480}{3} = .0160$$

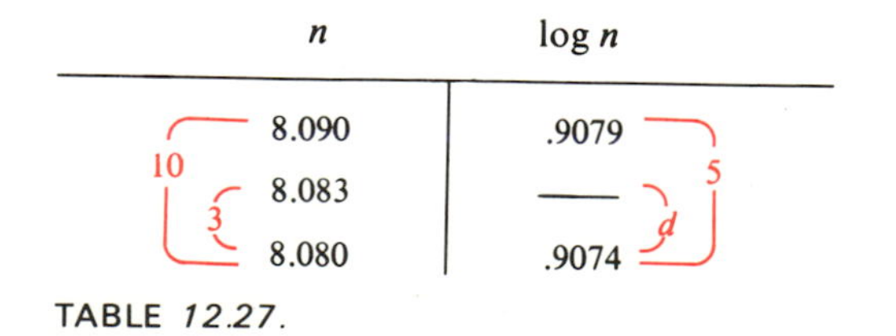

n	$\log n$
8.090	.9079
8.083	——
8.080	.9074

TABLE 12.27.

$$\frac{d}{5} = \frac{3}{10}$$

$$d = \frac{5 \cdot 3}{10} = \frac{15}{10} = 1.5 \approx 2$$

$$\begin{aligned} \log 8.083 &= .9074 \\ &\quad + .0002 \\ &\quad\ \ \overline{.9076} \end{aligned}$$

$$\frac{1}{2}\log 8.083 = \frac{.9076}{2} = .4538$$

$$\begin{aligned} 4\log 2.805 &= \quad 1.7920 \\ +\frac{1}{3}\log 1.117 &= +\ \ \underline{.0160} \\ &\quad\ \ 1.8080 \\ -\frac{1}{2}\log 8.083 &= -\ \ \underline{.4538} \\ \log n &= \quad 1.3542 \\ n &= \text{antilog } 1.3542 \end{aligned}$$

Interpolate as in Table 12.28.

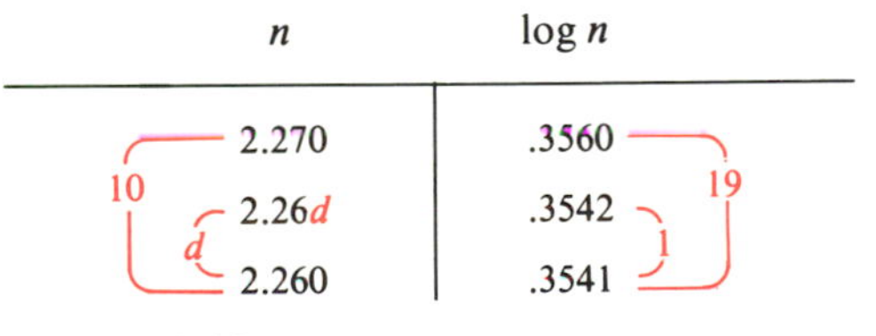

n	$\log n$
2.270	.3560
2.26d	.3542
2.260	.3541

TABLE 12.28

$$\frac{d}{10} = \frac{1}{19}$$

$$d = \frac{10}{19} \approx 1$$

$$\begin{aligned} n &= 2.261 \times 10^1 \\ &= 22.61 \end{aligned}$$

EXERCISES

In Exercises 1–32 compute using common logarithms.

1. (8.93) (6.16)
2. (107) (84.8)
3. (.0926) (9840)
4. (891) (6.32) (.007 41)
5. $\dfrac{842}{1.73}$
6. $\dfrac{1.95}{9.83}$

7. $\dfrac{80\,800}{1.78}$

8. $\dfrac{.009\,36}{.000\,74}$

9. $(34.3)^5$

10. $(.832)^{12}$

11. $107^{1/3}$

12. $\sqrt{99\,600}$

13. $\dfrac{(44.9)(8.85)}{73\,600}$

14. $\dfrac{(.003\,96)(.91)}{.004\,35}$

15. $\dfrac{.731}{(903)(.835)}$

16. $\dfrac{(.828)(6.14)}{(.003\,62)(.941)}$

17. $(9.84)^3(7.35)^5$

18. $\dfrac{(62.1)^7}{(.0386)^4}$

19. $\sqrt{184}\,\sqrt[3]{1840}$

20. $\sqrt{\dfrac{89\,200}{1.05}}$

21. $\dfrac{\sqrt{89.3}(996)}{804^5}$

22. $\dfrac{(692\,000^{1/10})(386\,000\,000^{1/5})}{(2.04)^{10}}$

23. $\sqrt{\dfrac{888(.333)}{77.7}}$

24. $\dfrac{\sqrt{888(.333)}}{77.7}$

25. $\dfrac{\sqrt{193}\ \sqrt[4]{8.83}}{(92.4)^5(.0083)^4}$

26. $\sqrt{\dfrac{(93.2)^5(.171)^{1/3}}{(8.09)^{3/4}}}$

27. $(9.935)(.083\,47)$

28. $\dfrac{83.82}{(.004\,962)(.6396)}$

29. $\sqrt{\dfrac{(9.936)(.8706)}{.024\,71}}$

30. $\dfrac{(1.444)^5(.083\,02)^3}{(.1628)^{3/4}}$

31. $\sqrt{\dfrac{(.9909)^5(.1131)^3}{7.707}}$

32. $\dfrac{(9413^4)(18\,820\,000^{1/2})}{(.007\,352)^9(.008\,214)^{1/3}}$

33. The amount A of money in a bank after n years is given by

$$A = P(1 + r)^n,$$

where P is the original principal and r is the annual interest rate. How much money do you have in the bank after 20 years if you deposit \$1000 at 5%?

34. In Exercise 33, how much money do you have if you deposit \$2500 at 6% for 12 years?

35. The kinetic energy K of a body is given by $K = \dfrac{1}{2}mv^2$, where m is the body's mass and v is its velocity. Determine the kinetic energy of a body when $m = 676\,000$ lb and $v = 1980$ ft/sec.

36. The mechanical efficiency e of a machine is defined by:

$$e = \frac{wh}{Fd}$$

Here a weight w is lifted to a height h by a force F that when applied to the machine, acts for a distance d. Determine e when:

$$w = 485 \text{ pounds} \qquad h = 52.8 \text{ feet}$$

$$F = 312 \text{ pounds} \qquad d = 101 \text{ feet}$$

*12.7 LOGARITHMIC AND EXPONENTIAL FUNCTIONS

When the base of a logarithmic equation is fixed, such as in

$$m = \log_{10} n,$$

or when base 10 is understood, as in

$$m = \log n,$$

a correspondence is set up between numbers n and m. *To each given* $n > 0$ *there corresponds exactly one value* m. For example,

$$\text{when } n = 100, \text{ then } m = \log 100 = 2$$

$$\text{when } n = 10\,000, \text{ then } m = \log 10\,000 = 4$$

and from the log table,

$$\text{when } n = 62.8, \text{ then } m = \log 62.8 = 1.7980$$

$$\text{when } n = .628, \text{ then } m = \log .628 = .7980 - 1$$

A function is thereby defined. You have been able to determine the log of a *positive* number with 3 or 4 digits by use of a table (and interpolation). Actually, the log of any *positive* number, irrational as well as rational, can be approximated by means of the techniques of calculus. Logs of "nearby" numbers are close to one another. For example, consider the irrational number $\sqrt{101}$. Because

$$\sqrt{101} \text{ is } \textit{approximately } \sqrt{100} \text{ (or 10)}$$

and because

$$\log 10 = 1,$$

$\log \sqrt{101}$ is close to (and slightly larger than) 1. Again, $\log x$ is defined for *all positive* numbers. A glance at the log table will convince you that

$\log x$ is an **increasing function.**

*Optional topic.

This means that if

$$0 < x_1 < x_2,$$

then

$$\log x_1 < \log x_2.$$

In words,

$$\log x \text{ gets bigger as } x \text{ gets bigger.}$$

Recall that a function is said to be *one–one* if distinct arguments have distinct function values. (See Section 8.5.) Consequently,

$$\log x \textit{ is one–one.}$$

For if $x_1 \neq x_2$, one of these numbers is smaller than the other. Suppose, for example, $x_1 < x_2$. Then $\log x_1 < \log x_2$. Therefore

$$\log x_1 \neq \log x_2$$

The graph of the common logarithmic function can now be drawn. Consider the sample values in Table 12.29.

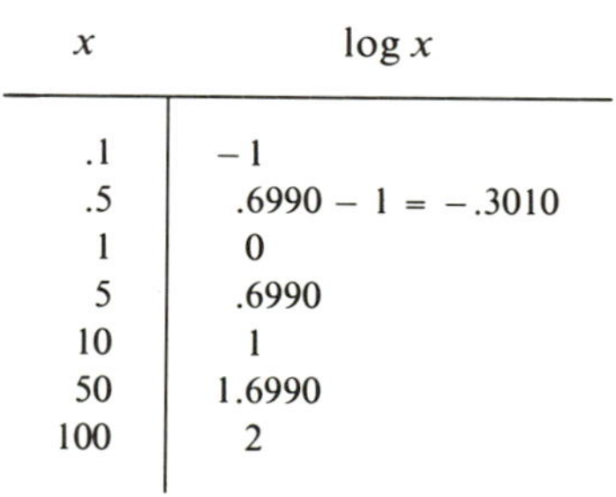

x	$\log x$
.1	-1
.5	$.6990 - 1 = -.3010$
1	0
5	.6990
10	1
50	1.6990
100	2

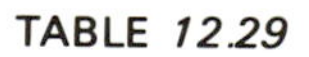

TABLE *12.29*

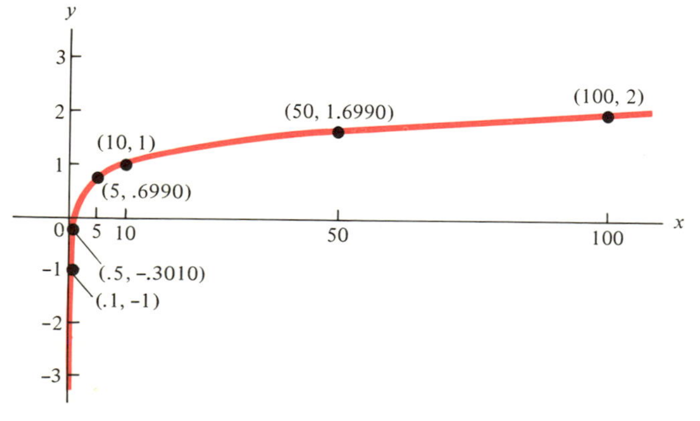

FIGURE 12.1 The graph of $y = \log x$

The graph of log x is shown in Figure 12.1. Note that different scales are used on the x- and y-axes. Log x is negative for $0 < x < 1$.

Recall that whenever a function is one–one, its inverse is defined and is also a one–one function. Use the symbol

$$\log^{-1}(x)$$

to denote the inverse function of log x. Geometrically, the graph of $\log^{-1}(x)$ is the reflection of log x in the line $y = x$ [Figure 12.2].

To obtain

$$\log^{-1}(x)$$

first interchange x and y in the defining equation,

$$y = \log_{10} x,$$

and obtain

$$x = \log_{10} y.$$

Next, solve for y in terms of x. Because of the definition of the logarithm to the base 10,

$$10^x = y$$

Therefore, $\log^{-1}(x)$ is given by

$$y = 10^x.$$

In other words, $\log^{-1}(x)$, which is an **exponential function,** expresses all powers

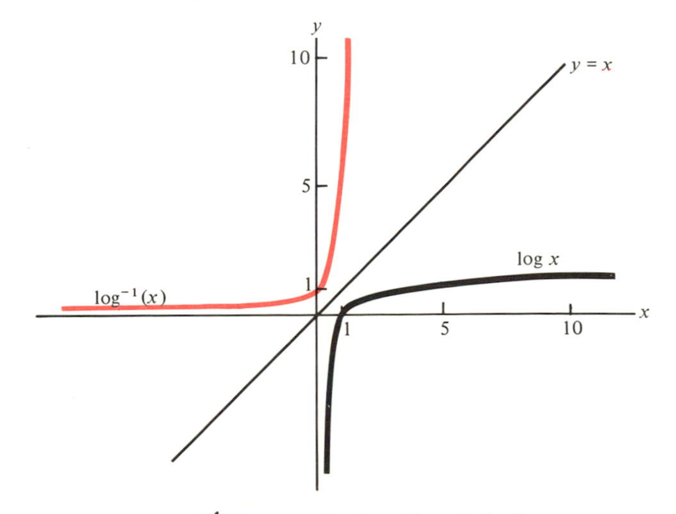

FIGURE 12.2 $\log^{-1}(x)$ is the reflection of log x in the line $y = x$. $\log^{-1}(x)$ is defined for all real numbers.

$$\log^{-1}(x) = 10^x$$

of 10. Note that the domain of $\log^{-1}(x)$ is the set of all real numbers. Also, $\log^{-1}(x)$ is an increasing function, just as $\log x$ is.

*12.8 EXPONENTIAL EQUATIONS; CHANGING THE BASE

In an **exponential equation** an exponent is a variable (or a nonconstant polynomial).

$$7^t = 15 \qquad \text{and} \qquad 3^{2x-1} = .145$$

are each exponential equations.

Property (**C**) of logarithms states:

$$\log n^r = r \log n$$

This property enables you to solve many exponential equations.

EXAMPLE 1. Solve for t:

$$7^t = 15$$

SOLUTION. Consider the log of each side of the given equation.

$$\log 7^t = \log 15$$

By property (**C**),

$$t \log 7 = \log 15$$

$$t = \frac{\log 15}{\log 7}$$

By use of the log table,

$$t = \frac{1.1761}{.8451} = 1.39 \text{ (to 2 decimal places)}$$

Write "$t \approx 1.39$"; this is to be read "t is approximately 1.39". ■

Make sure you understand that

$$\frac{\log 15}{\log 7} \neq \log 15 - \log 7.$$

$\left(\text{Rather, } \log \frac{15}{7} = \log 15 - \log 7. \text{ But this is } \textit{not} \text{ the case, above.}\right)$

*Optional topic.

EXAMPLE 2. Solve for x:

$$3^{2x-1} = .145$$

SOLUTION.

$$3^{2x-1} = .145$$

$$\log 3^{2x-1} = \log .145$$

$$(2x - 1)\log 3 = \log(1.45 \times 10^{-1})$$

$$2x - 1 = \frac{\log(1.45 \times 10^{-1})}{\log 3}$$

$$= \frac{.1614 - 1}{.4771}$$

$$= \frac{-.8386}{.4771}$$

Note that here a negative decimal is permissible. Only when the mantissa table is to be used, are negative decimals avoided.

$$2x - 1 \approx -1.76$$

$$2x \approx -.76$$

$$x \approx -.38$$

■

EXAMPLE 3. A chemical substance increases at a rate given by

$$y = 25 \cdot 10^t,$$

where y is the number of grams present after t hours. Determine how long it takes to obtain 1000 grams of this substance.

SOLUTION. Substitute 1000 for y in the given equation,

$$y = 25 \cdot 10^t.$$

$$1000 = 25 \cdot 10^t$$

$$\frac{1000}{25} = 10^t$$

$$40 = 10^t$$

$$\log 40 = \log 10^t$$

$$\log 40 = t \log 10$$

$$\frac{\log 40}{\log 10} = t$$

$$\frac{1.60}{1} \approx t$$

It takes about 1 hour and 36 minutes to obtain 1000 grams. ■

A **logarithmic equation** is an equation in which the logarithm of a variable (or of a nonconstant polynomial) appears.

EXAMPLE 4. Solve the following logarithmic equation for x:

$$\log 20x = 3$$

SOLUTION. The base 10 is assumed. Thus

$$\log_{10} 20x = 3$$

or

$$10^3 = 20x$$

$$1000 = 20x$$

$$50 = x \qquad \blacksquare$$

Properties of logarithms are often useful in solving logarithmic equations.

EXAMPLE 5. Solve for x:

$$\log 2 + \log x = 4$$

SOLUTION. The log of a product is the sum of the logs. Thus

$$\log_{10} 2 + \log_{10} x = \log_{10} 2x$$

and the given equation becomes

$$\log_{10} 2x = 4.$$

$$10^4 = 2x$$

$$10\,000 = 2x$$

$$5000 = x \qquad \blacksquare$$

A logarithmic equation enables you to change the base of a given logarithm.

CHANGE-OF-BASE FORMULA

Let $n > 0$, $a > 0$.

$$\log_a n = \frac{\log n}{\log a}$$

On the right side, the common log is assumed, although, as you will see, any (positive) base can be assumed. Before proving this formula, consider the following example.

EXAMPLE 6. Show that $\log_2 93 = \dfrac{\log 93}{\log 2}$.

SOLUTION. Let $y = \log_2 93$. Then

$$2^y = 93$$

$$\log 2^y = \log 93$$

$$y \log 2 = \log 93$$

$$y = \frac{\log 93}{\log 2}$$

The value of y can be obtained by dividing these common logs.

$$y = \frac{1.9685}{.3010} \approx 6.54$$

■

The proof of the change-of-base formula is similar to Example 6.

Proof: Let $n > 0$, $a > 0$, and let

$$y = \log_a n$$

Then

$$a^y = n$$

$$\log a^y = \log n$$

$$y \log a = \log n$$

$$y = \frac{\log n}{\log a}$$

Therefore because $y = \log_a n$,

$$\log_a n = \frac{\log n}{\log a}$$

■

Note that the base 10 was not essential in this proof. Thus the formula can be generalized.

GENERAL CHANGE-OF-BASE FORMULA

Let $n > 0$, $a > 0$, $b > 0$.

$$\log_a n = \frac{\log_b n}{\log_b a}$$

EXAMPLE 7. Use the General Change-of-Base Formula to express $\log_{17} 536$ in terms of base 3.

SOLUTION. In the General Change-of-Base Formula, let $n = 536$, $a = 17$, $b = 3$.

$$\log_{17} 536 = \frac{\log_3 536}{\log_3 17}$$ ■

EXERCISES

In Exercises 1–12 solve the indicated exponential equations to 2 places after the decimal point.

1. $5^x = 35$
2. $8^t = 100$
3. $7^{2y} = 93$
4. $3^{x+1} = 148$
5. $12^{5x} = 685$
6. $2^{-2t} = 100$
7. $14.8^{4t-1} = 17.3$
8. $9^{-x} = 73$
9. $6^{3-x} = 27$
10. $4^{1-2t} = 30$
11. $100^{3t+1} = .01$
12. $600^{1-3t} = 9000$

In Exercises 13–22 solve the indicated logarithmic equations. (Base 10 is understood.)

13. $\log x = 4$
14. $\log(5x) = 2$
15. $\log(-2x) = 1$
16. $\log(x + 1) = -3$
17. $\log(4x) = -1$
18. $\log(x + 1) = -3$
19. $\log\left(\dfrac{x}{2}\right) = 3$
20. $\log(4x + 1) = 0$
21. $\log 3 + \log x = 1$
22. $\log 2 + \log(x + 1) = 3$
23. A chemical substance increases at a rate given by

$$s = 10\,000 \cdot 5^t,$$

where s is the number of grams present after t hours. Determine how long it takes to obtain 5 000 000 grams of the substance.

24. A radioactive substance s decays at a rate given by

$$s = s_0(2.7)^{-t/4},$$

where s_0 is the amount originally present and t is the time in hours. After how many hours will half of the original substance be left?

25. The amount A of money in a bank after n years is given by

$$A = P(1 + r)^n,$$

where P is the original principal and r is the annual interest rate. If you invest \$2000 at 5%, how many years will it take for this to grow to over \$3000?

26. Use the formula of Exercise 25 to determine how long it will take to double your original principal

(a) at 4%, (b) at 8%.

In Exercises 27–34 find the value of the indicated logarithm to 2 decimal places.

27. $\log_2 3$
28. $\log_3 2$
29. $\log_5 19$
30. $\log_7 138$
31. $\log_6 84$
32. $\log_4 60\,000$
33. $\log_{100} 2800$
34. $\log_{.001} 934\,000$

In Exercises 35–42 use the General Change-of-Base Formula to express the given logarithm to the indicated base.

35. $\log_2 5$, base 3 Answer: $\dfrac{\log_3 5}{\log_3 2}$
36. $\log_3 10$, base 5
37. $\log_4 29$, base 2
38. $\log_7 364$, base 8
39. $\log_{10} 93$, base 2
40. $\log_{64} 843$, base 4
41. $\log_7 9.02$, base 12
42. $\log_5 .001$, base 3

In Exercises 43–44 assume $x > 0$. Simplify each expression.

43. $10^{\log_{10} x}$
44. $\log_{10} 10^x$

REVIEW EXERCISES FOR CHAPTER 12

12.1 DEFINITION OF LOGARITHMS

In Exercises 1–3 express in logarithmic notation.

1. $10^3 = 1000$
2. $\left(\dfrac{1}{2}\right)^2 = \dfrac{1}{4}$
3. $16^{1/2} = 4$

In Exercises 4–6 find the value of each indicated logarithm.

4. $\log_6 36$
5. $\log_3 81$
6. $\log_{.2} .04$

In Exercises 7–9 find the value of x.

7. $\log_3 9 = x$
8. $\log_x 25 = 2$
9. $\log_3 \sqrt{3} = x$

12.2 PROPERTIES OF LOGARITHMS

10. (a) Verify the following equation by evaluating both sides.
    (b) Indicate which property of logarithms is illustrated.

$$\log_2 64 = \log_2 16 + \log_2 4$$

In Exercises 11–14 assume $b > 0, x > 0, y > 0, z > 0$.
In Exercises 11–12 express each in terms of simpler logarithms.

11. $\log_b \dfrac{xy}{2}$
12. $\log_{10} \dfrac{(x^2y^3)^2}{z}$

13. Express $\log_b 60$ in terms of $\log_b 2$, $\log_b 3$, and $\log_b 5$.
14. Express $\log_b x + 3 \log_b y - \log_b z$ as a single logarithm.

In Exercises 15–31 common logs are understood.

12.3 COMMON LOGARITHMS

15. Find the value of: log 8.42
16. Determine the characteristic of the indicated logarithms:

(a) log 485 (b) log .926 (c) log .000 142

In Exercises 17–19 find the value of each.

17. log 614 18. log .808 19. log .0631

12.4 ANTILOGS

In Exercises 20–22 find the value of each indicated antilog.

20. antilog .9345 21. antilog 3.6618 22. antilog (.5955 − 1)

12.5 INTERPOLATING

In Exercises 23–24 find the value of each indicated log by interpolating.

23. log 8.206 24. log .073 14

In Exercises 25–26 find the value of each indicated antilog by interpolating.

25. antilog .6994 26. antilog 1.9797

12.6 COMPUTING WITH LOGARITHMS

In Exercises 27–29 compute using logarithms.

27. $(6.42)(9.09)$ 28. $\dfrac{396^{1/10}}{(1.71)^2}$ 29. $\dfrac{\sqrt{291(.444)}}{(28.4)^2}$

12.8 EXPONENTIAL EQUATIONS; CHANGING THE BASE

30. Solve to 2 places after the decimal point:

$$3^{x-1} = 259$$

31. Solve the logarithmic equation:

$$\log(5x) = 6$$

32. Find the value of $\log_3 5000$ to 2 decimal places.
33. Use the General Change-of-Base Formula to express $\log_2 7$ to the base 3.

TEST YOURSELF ON CHAPTER 12.

1. Find the value of $\log_3 \frac{1}{81}$.
2. Find the value of x: $\log_x 49 = 2$
3. Express $\log_3 40$ in terms of $\log_3 2$ and $\log_3 5$.

   In Exercises 4–7 common logs are understood.

4. Find the value of: $\log 848$
5. Find the value of: antilog 1.9552
6. Find the value of $\log 2.904$ by interpolating.
7. Compute $\frac{6.43^4}{\sqrt{1.01}}$ using logarithms.

13
quadratic equations

13.1 SOLVING BY FACTORING

A **quadratic equation** (in a single variable, x) is an equation that can be written in the form

$$Ax^2 + Bx + C = 0,$$

where $A \neq 0$. The quadratic equation

$$x^2 + 5x - 6 = 0$$

is already in this form; $A = 1, B = 5, C = -6$. The quadratic equation

$$7x^2 + 5x - 31 = 2x - x^2$$

can be transformed into

$$8x^2 + 3x - 31 = 0.$$

Here $A = 8, B = 3, C = -31$

In Chapter 6, you were able to solve an equation containing second-degree terms only if all such terms cancelled out. Such equations reduce to first-degree equations.

EXAMPLE 1. Solve:

$$3x^2 + x(x + 1) + 9 = 2x(5 + 2x)$$

SOLUTION.

$$3x^2 + x(x+1) + 9 = 2x(5+2x)$$

$$\underbrace{3x^2 + x^2}_{-4x^2} + x + 9 = 10x + 4x^2 \quad -4x^2$$

$$x + 9 = 10x$$

$$9 = 9x$$

$$x = 1$$

■

Some quadratic equations can be solved by means of a very simple principle.

If the product of two numbers is zero, then at least one of these numbers must be zero. And if at least one of the two factors a or b is zero, then the product ab is zero.

Thus you can solve an equation

$$ab = 0$$

by considering the two *simpler* equations

$$a = 0, \qquad b = 0$$

separately.

EXAMPLE 2. Solve the equation:

$$x(x-4) = 0$$

SOLUTION. This is of the form

$$ab = 0,$$

where $a = x$ and $b = x - 4$. You obtain two equations:

$x = 0$	$x - 4 = 0$
	$x = 4$

The roots of the given equation are 0 and 4. Note that the left side of this equation equals $x^2 - 4x$. Thus you have solved the quadratic equation

$$x^2 - 4x = 0$$

■

EXAMPLE 3. Solve the equation:

$$(t+7)(2t-3) = 0$$

SOLUTION. Obtain the two simpler equations:

$$t + 7 = 0, \qquad 2t - 3 = 0$$

Solve each of these equations separately.

$$t + 7 = 0 \qquad\qquad 2t - 3 = 0$$
$$t = -7 \qquad\qquad 2t = 3$$
$$t = \frac{3}{2}$$

The roots of the given equation are -7 and $\frac{3}{2}$. Note that

$$(t + 7)(2t - 3) = 2t^2 + 11t - 21.$$

Thus the roots of the quadratic equation

$$2t^2 + 11t - 21 = 0$$

are -7 and $\frac{3}{2}$. ■

Consider a quadratic equation

$$Ax^2 + Bx + C = 0, \qquad A \neq 0.$$

If you can factor the polynomial on the left side into two first-degree polynomials, apply the above method.

EXAMPLE 4. Solve:

$$x^2 - 6x = 0$$

SOLUTION.

$$x^2 - 6x = x(x - 6)$$

Thus

$$x(x - 6) = 0$$

$$x = 0 \qquad\qquad x - 6 = 0$$
$$x = 6$$

The roots are 0 and 6. ■

EXAMPLE 5. Solve:

$$x^2 - 7x + 10 = 0$$

SOLUTION.

$$x^2 - 7x + 10 = (x - 2)(x - 5)$$

$$(x - 2)(x - 5) = 0$$

$$x - 2 = 0 \quad \Big| \quad x - 5 = 0$$

$$x = 2 \quad \Big| \quad x = 5$$

The roots are 2 and 5. ■

EXAMPLE 6. Solve:

$$x^2 + 12x + 36 = 0$$

SOLUTION.

$$x^2 + 12x + 36 = (x + 6)^2$$

$$(x + 6)^2 = 0$$

If $a^2 = 0$, then $a = 0$, and if $a = 0$, then $a^2 = 0$. Thus

$$x + 6 = 0$$

$$x = -6$$

The only root is -6. ■

EXAMPLE 7. Solve:

$$3x^2 + x - 2 = 0$$

SOLUTION.

$$3x^2 + x - 2 = (3x - 2)(x + 1)$$

$$(3x - 2)(x + 1) = 0$$

$$3x - 2 = 0 \quad \Big| \quad x + 1 = 0$$

$$3x = 2 \quad \Big| \quad x = -1$$

$$x = \frac{2}{3} \quad \Big|$$

The roots are $\frac{2}{3}$ and -1. ■

In the next example, after you clear of fractions, a quadratic equation emerges. You will be able to solve this by factoring.

EXAMPLE 8. Solve:

$$\frac{4}{x+1} - \frac{1}{x} = 1$$

SOLUTION. Multiply both sides by the LCD, $x(x + 1)$.

$$4x - (x + 1) = x(x + 1)$$
$$3x - 1 = x^2 + x$$
$$x^2 - 2x + 1 = 0$$
$$(x - 1)^2 = 0$$
$$x = 1$$

■

The same methods apply when the product of more than two factors equals 0.

EXAMPLE 9. Solve:

$$x^3 - 4x^2 + 3x = 0$$

SOLUTION.

$$x^3 - 4x^2 + 3x = x(x^2 - 4x + 3)$$
$$= x(x - 1)(x - 3)$$

Thus

$$x(x - 1)(x - 3) = 0$$

You obtain three simpler equations:

$x = 0$	$x - 1 = 0$	$x - 3 = 0$
	$x = 1$	$x = 3$

The roots of the given equation are 0, 1, and 3. ■

EXAMPLE 10. Find a quadratic equation whose roots are 2 and 4.

SOLUTION. If 2 and 4 are roots of the quadratic equation $Ax^2 + Bx + C = 0$, then when you replace x by each of the numbers 2 and 4 in the quadratic equation, you obtain a true statement in each case. Thus both first-degree equations

$$x = 2 \quad \text{and} \quad x = 4$$

are true. Therefore

$$x - 2 = 0 \quad \text{and} \quad x - 4 = 0$$

Also,

$$(x - 2)(x - 4) = 0 \cdot 0 = 0$$

Consequently,

$$x^2 - 6x + 8 = 0$$

is a quadratic equation whose roots are 2 and 4. ■

EXERCISES

In Exercises 1–48 determine all roots of each equation. Assume $A \neq 0$.

1. $x(x - 2) = 0$
2. $3x(x + 5) = 0$
3. $\frac{x}{4}(x - 7) = 0$
4. $3x(2x + 1) = 0$
5. $(x - 1)(x - 2) = 0$
6. $(x + 5)(x - 3) = 0$
7. $\left(x + \frac{1}{2}\right)\left(x - \frac{1}{2}\right) = 0$
8. $\left(x + \frac{1}{4}\right)\left(x - \frac{1}{3}\right) = 0$
9. $(3x - 2)(2x - 4) = 0$
10. $(7x + 5)(2x - 3) = 0$
11. $(x - A)(x - B) = 0$
12. $Ax(x + B) = 0$
13. $(Ax - B)(Ax + B) = 0$
14. $(3Ax - C)\left(\frac{x}{A} + B\right) = 0$
15. $(x - 2)(x - 3)(x - 4) = 0$
16. $(x + 4)(x - 2)(2x + 1) = 0$
17. $x(x + 3)(2x + 1)(3x - 4) = 0$
18. $(x + 3)(x - 1)(2x + 1)(3x + 4)(2x - 5) = 0$
19. $x^2 - 3x = 0$
20. $x^2 + \frac{x}{2} = 0$
21. $4x^2 = 9x$
22. $Ax^2 = Bx$
23. $x^2 - 8x + 7 = 0$
24. $x^2 + 4x + 4 = 0$
25. $x^2 - 5x + 6 = 0$
26. $x^2 - 2x - 8 = 0$
27. $t^2 - 4t - 21 = 0$
28. $y^2 - 3y - 18 = 0$
29. $r^2 - 8r + 16 = 0$
30. $s^2 - s - 90 = 0$
31. $25x^2 = 36x$
32. $9x^2 = 4x$
33. $2x^2 + 5x - 3 = 0$
34. $4t^2 - 11t - 3 = 0$
35. $5u^2 - 4u - 1 = 0$
36. $6z^2 - 7z - 3 = 0$
37. $s^2 - \frac{s}{4} - \frac{1}{8} = 0$
38. $t^2 - \frac{2t}{3} + \frac{1}{9} = 0$
39. $1 + \frac{1}{u^2} = \frac{2}{u}$
40. $\frac{4}{x} + x = 5$

41. $\frac{2}{x^2} - \frac{3}{x} = -1$

42. $1 + \frac{6}{t^2} = \frac{5}{t}$

43. $\frac{2}{1 - u} + \frac{u}{2 + u} = 0$

44. $\frac{1}{y - 4} + \frac{1}{y + 2} = \frac{-1}{4}$

45. $x^3 + 6x^2 + 9x = 0$

46. $x^3 - 6x^2 + 5x = 0$

47. $x^3 + 2x^2 = 15x$

48. $x^3 + 20x = 9x^2$

In Exercises 49–56 determine a quadratic equation whose roots are as indicated.

49. 1 and 2

50. 1 and -2

51. 3 and -4

52. -3 and 4

53. $\frac{2}{3}$ and 1

54. 0 and $\frac{1}{2}$

55. only 3

56. only $\frac{-1}{4}$

13.2 EQUATIONS OF THE FORM $x^2 = a$

A quadratic equation of the form

$$x^2 = a$$

is easily handled.

EXAMPLE 1. Solve:

$$x^2 = 16$$

SOLUTION. Because $4^2 = 16$ and $(-4)^2 = 16$, the roots are 4 and -4. For brevity, write

± 4 for "plus and minus 4".

Thus the roots are given by $x = \pm 4$. ■

If a is positive, there are exactly two numbers whose square is a—namely,

$$\pm \sqrt{a}$$

You can also solve the equation

$$x^2 = a, \qquad a > 0$$

by factoring. Note that

$$x^2 - a = 0$$

and thus

$$x^2 - (\sqrt{a})^2 = 0.$$

$$(x - \sqrt{a})(x + \sqrt{a}) = 0$$

$(x - \sqrt{a}) = 0$	$(x + \sqrt{a}) = 0$
$x - \sqrt{a} = 0$	$x + \sqrt{a} = 0$
$x = \sqrt{a}$	$x = -\sqrt{a}$

$$x = \pm\sqrt{a}$$ ■

EXAMPLE 2. Solve:

$$x^2 = 17$$

SOLUTION. The roots are given by

$$x = \pm\sqrt{17}.$$

(The answer is generally left in this form, rather than approximated.) ■

EXAMPLE 3. Solve:

$$x^2 = 18$$

SOLUTION. The roots are given by

$$x = \pm\sqrt{18} = \pm\sqrt{9 \cdot 2} = \pm\sqrt{9} \cdot \sqrt{2} = \pm 3\sqrt{2}.$$ ■

EXAMPLE 4. Assume $a > 0$. Solve for x:

$$9x^2 - 25a^2b^4 = 0$$

SOLUTION.

$$9x^2 - 25a^2b^4 = 0$$

$$9x^2 = 25a^2b^4$$

$$x^2 = \frac{25a^2b^4}{9}$$

$$x^2 = \left(\frac{5ab^2}{3}\right)^2$$

$$x = \pm\frac{5ab^2}{3}$$ ■

EXAMPLE 5. Solve for y:

$$(2y + 1)^2 = 9$$

SOLUTION. This is of the form

$$x^2 = 9$$

with $x = 2y + 1$. Thus

$$2y + 1 = \pm 3$$

Two first-degree equations are obtained. Each of these is solved.

$2y + 1 = 3$	$2y + 1 = -3$
$2y = 2$	$2y = -4$
$y = 1$	$y = -2$

The roots are 1 and -2. ■

EXAMPLE 6. Let $A \neq 0$. Solve for x:

$$(Ax - B)^2 = 7$$

SOLUTION. $Ax - B = \pm\sqrt{7}$

$Ax - B = \sqrt{7}$	$Ax - B = -\sqrt{7}$
$Ax = B + \sqrt{7}$	$Ax = B - \sqrt{7}$
$x = \dfrac{B + \sqrt{7}}{A}$	$x = \dfrac{B - \sqrt{7}}{A}$

The 2 roots differ only in the sign of $\sqrt{7}$. Thus the roots can be written as

$$\frac{B \pm \sqrt{7}}{A}.$$ ■

The equation

$$x^2 = 0$$

has 0 *as its only root.*

EXAMPLE 7. Solve for y:

$$(3y - C)^2 = 0$$

SOLUTION. This is of the form

$$x^2 = 0$$

with $x = 3y - C$. The only root of $x^2 = 0$ is 0. Thus

$$3y - C = 0$$

$$3y = C$$

$$y = \frac{C}{3}$$

The given equation has only 1 root, $\frac{C}{3}$ ■

Equations of the form

$$x^2 = -a$$

where $a > 0$ *have complex roots. For*

$$x^2 = a(-1)$$
$$x = \pm\sqrt{a}\,\sqrt{-1}$$
$$= \pm\sqrt{a}\,i$$

EXAMPLE 8. Solve for x:

$$x^2 = -12$$

SOLUTION.

$$x = \pm\sqrt{12}\,i$$
$$x = \pm 2\sqrt{3}\,i$$

■

EXERCISES

Assume $A > 0$, $B \neq 0$, $C \neq 0$.
In Exercises 1–40 solve for x.

1. $x^2 = 1$
2. $x^2 = 64$
3. $x^2 = 11$
4. $x^2 = 19$
5. $x^2 = 8$
6. $x^2 = 24$
7. $5x^2 = 20$
8. $2x^2 = 72$
9. $2x^2 = 0$
10. $48 - 3x^2 = 0$
11. $9x^2 = 25$
12. $100x^2 = 1$
13. $\frac{x^2}{4} = 7$
14. $16x^2 = 21$
15. $8x^2 = 49$
16. $40x^2 = 3$
17. $x^2 = 16B^2$
18. $x^2 = 5A^4$
19. $x^2 = 81A^2B^4C^6$
20. $4x^2 = 9A^4B^4$
21. $(x + 3)^2 = 25$
22. $(3x)^2 = 100$
23. $(2x - 5)^2 = 1$
24. $\left(\frac{x}{3} + 1\right)^2 = 0$
25. $\left(\frac{x + 2}{2}\right)^2 = 4$
26. $\left(\frac{2x - 1}{3}\right)^2 = 3$
27. $x^2 = A^2$
28. $x^2 = A$
29. $A^2x^2 - B^2 = 0$
30. $A^2(x - B)^2 = C^2$
31. $4x^2 - 49A^2B^2 = 0$
32. $\left(\frac{Ax + B}{C}\right)^2 = 0$
33. $\frac{A^2B^2}{16}(x - C)^2 = 1$
34. $16(x - 1)^2 = 25$
35. $x^2 = -16$
36. $x^2 + 18 = 0$
37. $x^2 = -24$
38. $x^2 + A^2 = 0$
39. $(x - 2)^2 = -4$
40. $(x + 1)^2 = -3$

In Exercises 41–50 solve for the indicated variable.

41. $t^2 = 121$, for t

42. $9y^2 = 5$, for y

43. $(2z + 2)^2 = 4$, for z

44. $A^2u^2 = 5B^2$, for u

45. $(t - 1)^2 = 12$, for t

46. $3A^2y^2 = 27B^2$, for y

47. $A\left(\frac{y - B}{C}\right)^2 = A^3$, for y

48. $A^2\left(\frac{z + 5}{2}\right)^2 = 144$, for z

49. $u^2 + 64 = 0$, for u

50. $4(t - 3)^2 = -8$, for t

In Exercises 51–52 determine all possible values of x in each proportion.

51. $\frac{x}{2} = \frac{8}{x}$

52. $\frac{x}{5} = \frac{4}{x}$

13.3 COMPLETING THE SQUARE

Consider the equation:

$$x^2 + 4x - 9 = 0$$

You cannot factor the polynomial on the left side (at least not by the previous techniques).

Add 9 to both sides and obtain:

(A) $$x^2 + 4x = 9$$

The left side is not a square, as is. But perhaps by adding a constant to both sides, the equation

$$x^2 + 4x + \_\_\_\_\_ = 9 + \_\_\_\_\_$$

will be of the form

$$t^2 = a,$$

as in Section 13.2.

Consider $(x + 2)^2$:

$$\begin{array}{r} x + 2 \\ x + 2 \\ \hline x^2 + 2x \\ 2x + 4 \\ \hline x^2 + 4x + 4 \end{array}$$

Thus $(x + 2)^2$ is 4 more than $x^2 + 4x$.

Therefore add 4 to both sides of Equation (A), and obtain:

$$x^2 + 4x + 4 = 9 + 4$$
$$(x + 2)^2 = 13$$
$$x + 2 = \pm\sqrt{13}$$
$$x = -2 \pm \sqrt{13}$$

If you are given the equation

$$x^2 + 10x - 9 = 0,$$

in which the coefficient of x is 10 (instead of 4, as above) you would then consider:

$$(x + 5)^2 = x^2 + 10x + 25$$

Note that

$$5 = \frac{10}{2}$$

and

$$5^2 = \left(\frac{10}{2}\right)^2 = 25.$$

In general, for the equation

$$x^2 + Bx + C = 0$$

or

$$x^2 + Bx = -C,$$

add $\left(\frac{B}{2}\right)^2$ to both sides:

$$x^2 + Bx + \left(\frac{B}{2}\right)^2 = \underbrace{\left(\frac{B}{2}\right)^2 - C}_{D}.$$

$$\begin{array}{r} x + \frac{B}{2} \\ x + \frac{B}{2} \\ \hline x^2 + \frac{B}{2}x \\ \frac{B}{2}x + \left(\frac{B}{2}\right)^2 \\ \hline x^2 + Bx + \left(\frac{B}{2}\right)^2 \end{array}$$

$$\left(x + \frac{B}{2}\right)^2 = D$$

If $D \geq 0$, *then*

$$x + \frac{B}{2} = \pm\sqrt{D}$$

$$x = -\frac{B}{2} \pm \sqrt{D}$$

If $D < 0$, then $-D > 0$ and

$$x = -\frac{B}{2} \pm \sqrt{-D}\, i$$

EXAMPLE 1. Solve:

$$x^2 - 8x - 5 = 0$$

SOLUTION.

$$x^2 - 8x = 5$$

Here $B = -8, \frac{B}{2} = -4, \left(\frac{B}{2}\right)^2 = (-4)^2 = 16$

$$x^2 - 8x + 16 = 5 + 16$$

$$(x - 4)^2 = 21$$

$$x - 4 = \pm \sqrt{21}$$

$$x = 4 \pm \sqrt{21}$$ ■

EXAMPLE 2. Solve:

$$y^2 + 7y + 8 = 0$$

SOLUTION.

$$y^2 + 7y = -8$$

Here $B = 7, \frac{B}{2} = \frac{7}{2}, \left(\frac{B}{2}\right)^2 = \frac{49}{4}$

$$y^2 + 7y + \frac{49}{4} = -8 + \frac{49}{4}$$

$$= \frac{-32 + 49}{4}$$

$$= \frac{17}{4}$$

$$\left(y + \frac{7}{2}\right)^2 = \frac{17}{4}$$

$$y + \frac{7}{2} = \pm \frac{\sqrt{17}}{2}$$

$$y = \frac{-7}{2} \pm \frac{\sqrt{17}}{2}$$

$$y = \frac{-7 \pm \sqrt{17}}{2}$$ ■

EXAMPLE 3. Solve:

$$x^2 + 10x + 40 = 0$$

SOLUTION. $x^2 + 10x = -40$

$$B = 10, \qquad \left(\frac{B}{2}\right)^2 = 25$$

$$x^2 + 10x + 25 = 25 - 40$$

$$(x + 5)^2 = -15$$

$$x + 5 = \pm\sqrt{-15}$$

$$= \pm\sqrt{15}\,i$$

$$x = -5 \pm \sqrt{15}\,i.$$

■

Up to now, you have considered equations

$$x^2 + Bx + C = 0$$

with leading coefficient 1. For the more general equation

$$Ax^2 + Bx + C = 0,$$

where $A \neq 0$, divide both sides by A.

$$x^2 + \frac{B}{A}x + \frac{C}{A} = 0$$

This is now of the form

$$x^2 + bx + c = 0$$

$b = \dfrac{B}{A}, \ c = \dfrac{C}{A}$

that you have already considered.

EXAMPLE 4. Solve:

$$4x^2 + 3x - 1 = 0$$

SOLUTION. Divide both sides by the leading coefficient, 4.

$$x^2 + \frac{3}{4}x - \frac{1}{4} = 0$$

$$x^2 + \frac{3}{4}x = \frac{1}{4}$$

$$b = \frac{3}{4}, \qquad \frac{b}{2} = \frac{3}{8}, \qquad \left(\frac{b}{2}\right)^2 = \frac{9}{64}$$

$$x^2 + \frac{3}{4}x + \frac{9}{64} = \frac{1}{4} + \frac{9}{64}$$

$$= \frac{16 + 9}{64}$$

$$= \frac{25}{64}$$

$$\left(x + \frac{3}{8}\right)^2 = \left(\frac{5}{8}\right)^2$$

$$x + \frac{3}{8} = \pm\frac{5}{8}$$

$$x = \frac{-3}{8} + \frac{5}{8} = \frac{2}{8} = \frac{1}{4} \qquad \Bigg| \qquad x = \frac{-3}{8} - \frac{5}{8} = \frac{-8}{8} = -1$$

The roots are $\frac{1}{4}$ and -1. ■

EXERCISES

Solve each equation by completing the square.

1. $x^2 - 2x - 1 = 0$
2. $u^2 + 2u + 1 = 0$
3. $x^2 + 4x - 2 = 0$
4. $x^2 + 6x + 2 = 0$
5. $y^2 - 4y - 1 = 0$
6. $x^2 + 4x - 1 = 0$
7. $x^2 - 8x + 13 = 0$
8. $x^2 + 14x + 47 = 0$
9. $z^2 - 2z - 4 = 0$
10. $y^2 - y - 5 = 0$
11. $x^2 + x - 1 = 0$
12. $x^2 + 3x + 1 = 0$
13. $x^2 + 5x + 2 = 0$
14. $x^2 - 3x = 5$
15. $x^2 - 5x = -5$
16. $x^2 - 9x + 19 = 0$
17. $x^2 + \frac{1}{2}x - 1 = 0$
18. $x^2 + \frac{3}{2}x - \frac{1}{2} = 0$
19. $x^2 + \frac{2}{3}x - 2 = 0$
20. $x^2 + 4x + 10 = 0$
21. $t^2 + 4t + 2 = 0$
22. $y^2 + y + 1 = 0$
23. $x^2 + 2x + 2 = 0$
24. $x^2 + 4x + 1 = 0$
25. $x^2 - 3x + 3 = 0$
26. $x^2 + 3x + 3 = 0$
27. $2x^2 + 4x + 1 = 0$
28. $2x^2 + 4x - 1 = 0$
29. $2t^2 - 6t + 1 = 0$
30. $4x^2 + 8x + 1 = 0$

31. $4t^2 + 8t = 1$

32. $2x^2 + 5x - 1 = 0$

33. $4x^2 + 4x + 3 = 0$

34. $2x^2 + 3x + 1 = 0$

35. $\frac{x^2}{4} + \frac{x}{2} - \frac{1}{4} = 0$

36. $5x^2 + 2x + 5 = 0$

13.4 THE QUADRATIC FORMULA

The **quadratic formula** yields the roots of any quadratic equation. Write the equation in the form

$$Ax^2 + Bx + C = 0, \qquad A \neq 0.$$

Then the quadratic formula asserts that

$$\mathbf{x = \frac{-B \pm \sqrt{B^2 - 4AC}}{2A}}.$$

Before verifying the quadratic formula, first consider how it is used.

The quantity $B^2 - 4AC$ is known as the **discriminant** of the equation. The quadratic formula calls for the square root of the discriminant. Three cases occur, according to whether the discriminant is positive, zero, or negative.

CASE 1. $B^2 - 4AC > 0$. There are 2 distinct roots, and they are *real numbers.*

CASE 2. $B^2 - 4AC = 0$. There is exactly 1 root. It is the real number

$$\frac{-B}{2A}.$$

CASE 3. $B^2 - 4AC < 0$. There are 2 distinct roots. But they are *complex conjugates.* (Recall that complex conjugates are complex numbers of the form $a + bi$ and $a - bi$.)

EXAMPLE 1. Use the quadratic formula to solve:

$$2x^2 + 5x - 3 = 0$$

SOLUTION. $A = 2, B = 5, C = -3$

$$x = \frac{-B \pm \sqrt{B^2 - 4AC}}{2A}$$

$$= \frac{-5 \pm \sqrt{25 - 4(2)(-3)}}{2(2)}$$

$$= \frac{-5 \pm \sqrt{49}}{4}$$

$$= \frac{-5 \pm 7}{4}$$

$$x = \frac{-5 + 7}{4} \qquad\qquad x = \frac{-5 - 7}{4}$$

$$x = \frac{1}{2} \qquad\qquad x = -3$$

The roots are $\frac{1}{2}$ and -3. Note that the discriminant is positive:

$$B^2 - 4AC = 49 > 0.$$

There are 2 real roots. ■

EXAMPLE 2. Use the quadratic formula to solve:

$$4x^2 + 9 = 12x$$

SOLUTION. First write the equation in the form

$$Ax^2 + Bx + C = 0:$$

$$4x^2 - 12x + 9 = 0$$

Thus $A = 4, B = -12, C = 9$

$$x = \frac{12 \pm \sqrt{144 - 4(4)(9)}}{2(4)}$$

$$= \frac{12 \pm \sqrt{144 - 144}}{8}$$

$$= \frac{12}{8}$$

$$= \frac{3}{2}$$

Here the discriminant is zero: $B^2 - 4AC = 0$. There is exactly 1 root, and it is the real number $\frac{3}{2}$. ■

EXAMPLE 3. Use the quadratic equation to solve:

$$x^2 + x + 1 = 0$$

SOLUTION. $A = B = C = 1$

$$x = \frac{-1 \pm \sqrt{1 - 4(1)(1)}}{2(1)}$$

$$= \frac{-1 \pm \sqrt{-3}}{2}$$

The roots are $\frac{-1}{2} \pm \frac{\sqrt{3}}{2}i$. Note that they are complex conjugates. Here the discriminant is negative:

$$B^2 - 4AC = -3 < 0$$ ■

The quadratic formula is obtained by completing the square. Consider the quadratic equation

$$Ax^2 + Bx + C = 0.$$

Transform this as follows:

$$x^2 + \frac{B}{A}x + \frac{C}{A} = 0$$

$$x^2 + \frac{B}{A}x = -\frac{C}{A}$$

Complete the square. The coefficient of x is $\frac{B}{A}$.

$$x^2 + \frac{B}{A}x + \left(\frac{B}{2A}\right)^2 = \left(\frac{B}{2A}\right)^2 - \frac{C}{A}$$

$$\left(x + \frac{B}{2A}\right)^2 = \frac{B^2}{4A^2} - \frac{C}{A}$$

$$= \frac{B^2 - 4AC}{4A^2}$$

$$x + \frac{B}{2A} = \pm\sqrt{\frac{B^2 - 4AC}{(2A)^2}}$$

$$= \pm\frac{\sqrt{B^2 - 4AC}}{2A}$$

$$x = \frac{-B}{2A} \pm \frac{\sqrt{B^2 - 4AC}}{2A}$$

$$= \frac{-B \pm \sqrt{B^2 - 4AC}}{2A}$$ ■

EXERCISES

In Exercises 1–12: (a) Determine the discriminant. (b) Without solving, state whether the given equation has 2 distinct real roots, exactly 1 real root, or 2 complex conjugate roots.

1. $x^2 + 7x - 1 = 0$
2. $x^2 + 2x + 3 = 0$
3. $x^2 - x - 1 = 0$
4. $x^2 + x + 2 = 0$
5. $x^2 + 1 = 2x$
6. $x^2 + x + 5 = 0$
7. $2x^2 + 5x + 1 = 0$
8. $3x^2 + 1 = x$
9. $9x^2 + 6x + 1 = 0$
10. $2x^2 + 4x = 3$
11. $3x^2 + 10x - 1 = 0$
12. $4x^2 - 20x + 25 = 0$

In Exercises 13–34 solve by means of the quadratic formula.

13. $x^2 + x + 2 = 0$
14. $x^2 + 3x + 1 = 0$
15. $x^2 - 2x - 2 = 0$
16. $x^2 + 3x + 3 = 0$
17. $x^2 - 5x = 10$
18. $x^2 + 4 = 4x$
19. $2x^2 + 1 = 0$
20. $3x^2 + 4x = 0$
21. $2x^2 + 9x - 2 = 0$
22. $9x^2 - 24x + 16 = 0$
23. $5x^2 + 10x + 3 = 0$
24. $3x^2 - 7x + 3 = 0$
25. $\frac{x^2}{7} + 2x + 7 = 0$
26. $\frac{x^2}{2} - 3x = 5$
27. $x^2 + \frac{1}{4} = x$
28. $x^2 + .2x = .4$
29. $x^2 + 12x + 36 = 0$
30. $3x^2 + 10x - 10 = 0$
31. $\frac{x^2 + 3x}{2} + 1 = 0$
32. $\frac{x^2 - x}{4} + \frac{1}{2} = 1$
33. $2x^2 + Bx - B = 0$
34. $x^2 + Bx + B = 0$

In Exercises 35–39 determine B so that the given equation has just 1 root. [*Hint:* Consider the discriminant.]

35. $x^2 + Bx + 16 = 0$
36. $x^2 - Bx + 4 = 0$
37. $x^2 + Bx + 1 = 0$
38. $x^2 + Bx + 9 = 0$
39. $x^2 + Bx + 4 = 0$
40. Show that the equation

$$x^2 + Bx - 2 = 0$$

always has real roots (regardless of the coefficient B).

13.5 WORD PROBLEMS

Practical problems can often be reduced to quadratic equations. Whenever possible, solve the associated quadratic equation by factoring.

EXAMPLE 1. The sum of two numbers is 14 and the product is 48. Find these numbers.

SOLUTION. Let x be one of the numbers. Then $14 - x$ is the other because

$$x + (14 - x) = 14.$$

Also,

the product of these numbers is 48.

$$x(14 - x) = 48$$

$$14x - x^2 = 48$$

$$0 = x^2 - 14x + 48$$

$$0 = (x - 8)(x - 6)$$

$x - 8 = 0$	$x - 6 = 0$
$x = 8$	$x = 6$
$14 - x = 6$	$14 - x = 8$

In either case the two numbers are 6 and 8. ■

EXAMPLE 2. Find a *positive* number that is 2 more than its reciprocal.

SOLUTION. Let x be the number. Then $\frac{1}{x}$ is its reciprocal.

The number is 2 more than its reciprocal.

$$x = 2 + \frac{1}{x}$$

Multiply both sides by x. (See the beginning of Section 13.6 for what can happen when you multiply both sides of an equation by a variable.)

$$x^2 = 2x + 1$$

$$x^2 - 2x - 1 = 0$$

By the quadratic formula,

$$x = \frac{2 \pm \sqrt{(-2)^2 - 4(1)(-1)}}{2}$$

$$= \frac{2 \pm \sqrt{4 + 4}}{2}$$

$$= \frac{2 \pm 2\sqrt{2}}{2}$$

$$= 1 \pm \sqrt{2}$$

Because x must be positive and $\sqrt{2} \approx 1.4$, choose the root $1 + \sqrt{2}$. (The other root, $1 - \sqrt{2} \approx -.4$, is negative.)

CHECK. First note that

$$\frac{1}{1+\sqrt{2}} = \frac{1 \cdot (1-\sqrt{2})}{(1+\sqrt{2}) \cdot (1-\sqrt{2})} = \frac{1-\sqrt{2}}{1-2} = \sqrt{2} - 1.$$

Now check:

$$1 + \sqrt{2} \stackrel{?}{=} 2 + \frac{1}{1+\sqrt{2}}$$

$$1 + \sqrt{2} \stackrel{?}{=} 2 + (\sqrt{2} - 1)$$

$$1 + \sqrt{2} \stackrel{\checkmark}{=} 1 + \sqrt{2}$$

■

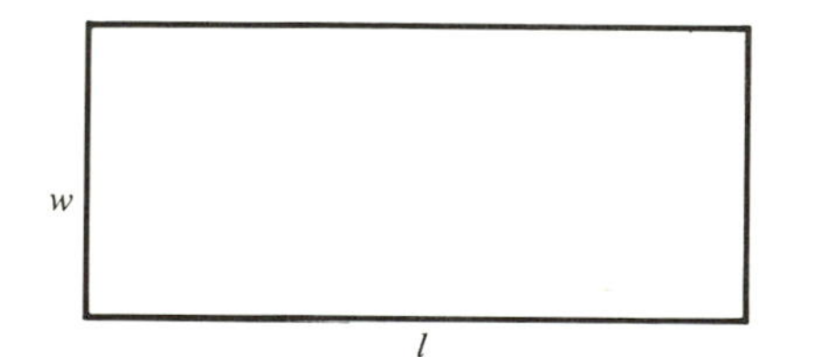

FIGURE 13.1

EXAMPLE 3. The perimeter (boundary) of a rectangle is 42 inches. The area of the rectangle is 108 square inches. Determine the dimensions of the rectangle.

SOLUTION.

Let l and w be the dimensions of the rectangle [See Figure 13.1.]

The perimeter is $2l + 2w$. Thus

$$2l + 2w = 42$$

$$l + w = 21$$

$$w = 21 - l$$

The area of the rectangle is $l \cdot w$ or $l(21 - l)$. Thus

$$l(21 - l) = 108$$

$$21l - l^2 = 108$$

$$0 = l^2 - 21l + 108$$

$$0 = (l - 9)(l - 12)$$

$l - 9 = 0$	$l - 12 = 0$
$l = 9$	$l = 12$
$w = 21 - 9 = 12$	$w = 21 - 12 = 9$

The dimensions of the rectangle are 12 by 9. ■

In distance problems (Section 7.3), *the distance that an object travels at a constant rate of speed equals its rate multiplied by the time in transit.* If

d = distance

r = rate

t = time,

then

$$d = r \cdot t$$

EXAMPLE 4. A canoeist paddles at a constant rate. He finds that it takes him 2 hours longer to make a 12-mile trip upstream than it does downstream. If the current is 3 miles per hour, how long would the trip take in still water?

SOLUTION. Let r be the rate (in hours) in still water. The rate is $r + 3$ downstream (with the current) and $r - 3$ upstream (against the current). The distance, 12 miles, is the same both ways. [See Table 13.1.] To find the time t, divide d by r.

	r	$t = \frac{d}{r}$	d
downstream	$r + 3$	$\frac{12}{r+3}$	12
upstream	$r - 3$	$\frac{12}{r-3}$	12

TABLE *13.1.*

The time upstream is 2 hours more than the time downstream.

$$\frac{12}{r-3} = 2 + \frac{12}{r+3}$$

Multiply by the LCD, $(r - 3)(r + 3)$:

$$12(r + 3) = 2(r - 3)(r + 3) + 12(r - 3)$$
$$6(r + 3) = (r - 3)(r + 3) + 6(r - 3)$$
$$\cancel{6r} + 18 = r^2 - 9 + \cancel{6r} - 18$$
$$0 = r^2 - 45$$
$$r^2 = 45$$
$$r = \pm\sqrt{45}$$
$$r = \pm 3\sqrt{5}$$

The rate is positive; thus $r = 3\sqrt{5}$ (approximately 6.7) miles per hour. ■

There are distance problems that involve an object traveling at a *variable rate* and that can be expressed in terms of quadratic equations. Because the rate is *not constant*, the formula $d = r \cdot t$ no longer applies.

EXAMPLE 5. A skier accelerates as he travels downhill. The distance s traveled in t seconds is given by

$$s = 10t^2 + 10t \text{(feet)}.$$

How long does it take him to go 560 feet downhill?

SOLUTION. Set $s = 560$.

$$560 = 10t^2 + 10t$$

$$56 = t^2 + t$$

$$t^2 + t - 56 = 0$$

$$(t + 8)(t - 7) = 0$$

$t + 8 = 0$	$t - 7 = 0$
$t = -8$	$t = 7$
(Reject, because t, the number of seconds, must be positive.)	

It takes the skier 7 seconds to travel 560 feet. ■

Work problems were discussed in Section 7.6. Recall that in work problems,

the fraction of work done in 1 time unit · time = the fraction of work done

EXAMPLE 6. A father and son working together can paint a house in 8 days. It takes the son, alone, 12 days longer than his father to paint the house. How long does it take the father, alone, to paint the house? Also, when working together, what fraction of the work does the father do?

SOLUTION. Let x be the number of days for the father to paint the house.

Then $x + 12$ is the number of days for the son to paint the house.

In 1 day the father paints $\frac{1}{x}$ of the house; the son paints $\frac{1}{x + 12}$ of the house. They work together 8 days. [See Table 13.2.]

	Fraction of Work Done in 1 Day	· Days	= Fraction of Work Done
Father	$\frac{1}{x}$	8	$\frac{8}{x}$
Son	$\frac{1}{x + 12}$	8	$\frac{8}{x + 12}$

TABLE *13.2.*

There is 1 job to be done—namely, to paint the house.

$$\underbrace{\text{The fraction of work done by the father}}_{\frac{8}{x}} + \underbrace{\text{the fraction of work done by the son}}_{\frac{8}{x+12}} = 1$$

Multiply both sides by the LCD, $x(x + 12)$.

$$8(x + 12) + 8x = x(x + 12)$$

$$8x + 96 + 8x = x^2 + 12x$$

$$x^2 - 4x - 96 = 0$$

$$(x - 12)(x + 8) = 0$$

$$x - 12 = 0 \qquad\qquad x + 8 = 0$$

$$x = 12 \qquad\qquad x = -8$$

(Reject.)

Clearly, x, the number of days, is positive. Thus $x = 12$ (and $x + 12 = 24$). The father paints $\frac{2}{3}$ of the house because the fraction of work done by him is $\frac{8}{12}$. ■

EXERCISES

1. The sum of two numbers is 12 and the product is 20. Find these numbers.
2. The sum of two numbers is 17 and the product is 72. Find these numbers.
3. The sum of two numbers is 0 and the product is -25. Find these numbers.
4. Twice a certain positive number is five less than its square. Find this number.
5. The square of a negative number is six more than five times the number. Find this number.
6. Show that no real number can equal two more than its square.
7. The sum of a number and its reciprocal is 4. Find all such numbers.
8. One more than the reciprocal of a negative number is twice that number. Find the number.
9. The product of two consecutive negative integers is 56. Determine these integers.
10. The sum of the squares of two consecutive positive odd integers is 74. Determine these integers.
11. The perimeter of a rectangle is 30 inches. The area of the rectangle is 50 square inches. Determine the dimensions of the rectangle.

12. The perimeter of a rectangle is 3 feet, 10 inches. The area of the rectangle is 130 square inches. Determine the dimensions of the rectangle.

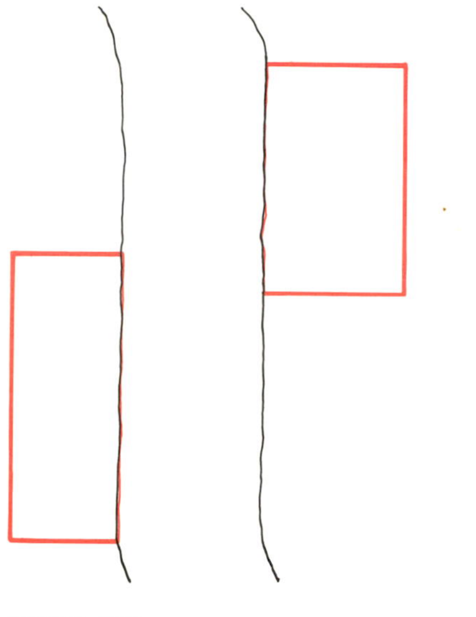

FIGURE 13.2

13. Two rectangular fields with different dimensions border on a river. [See Figure 13.2.] In each case the field is bounded along one of its longer sides by the river, whereas a 90-foot fence runs along the other three sides. The area of each field is 1000 square feet. Determine the dimensions of each field.

14. The area of a rectangular Persian rug is 100 square feet. If the length of the rug were decreased by 5 feet and the width increased by 5 feet, the area would be 150 square feet. Determine the dimensions of the rug.

15. The area of a rectangular wood-paneled floor is 160 square feet. If the length were increased by 5 feet and the width increased by 2 feet, the area would be 250 square feet. Determine the dimensions of the floor.

16. A canoe travels at a constant rate in still water. It takes 15 hours longer to make a 40-mile trip upstream than it does downstream. If the current is 3 miles per hour, how long does the trip take in still water?

17. Two cars, each traveling at a constant rate, leave Boston headed for Philadelphia, 300 miles away. One car travels 10 miles per hour faster and arrives 1 hour earlier than the other car. Determine the rate of the faster car.

18. To get to town Joe must jog 3 miles through the woods and then ride his scooter the remaining 5 miles. He rides 14 miles per hour faster than he jogs (at a constant rate). If it takes 45 minutes to make the entire trip, how fast does he jog?

19. An object falls from a window 400 feet high. Its distance above the ground, as it is falling, is given by

$$s = 400 - 16t^2 \text{(feet)},$$

where t is measured in seconds. How long does it take to hit the ground?

20. A ball is thrown straight up from ground level. Its height, h feet, after t seconds is given by

$$h = 128t - 16t^2.$$

(a) How long does it take to rise 112 feet?

(b) At what time t is it 112 feet above the ground on its way down?

21. If each works alone, a left-handed clerk takes 6 hours longer than a right-hander to stuff a batch of envelopes. Together they do the job in 4 hours. How long does it take the right-handed clerk, alone, to stuff the envelopes?

22. The hot-water faucet takes 30 minutes longer than the cold-water faucet to fill a tub. When both faucets are turned on, the tub is filled in 8 minutes. How long does it take the hot water faucet, alone, to fill the tub?
23. Sam and Suzy sip a soda together in 6 seconds. On separate sodas, Sam can outsip Suzy by 5 seconds. How long does it take Suzy to sip her soda?

13.6 EQUATIONS WITH SQUARE ROOTS

The Addition Principle (Section 6.2) asserts that you can add the same quantity to both sides of an equation. Similarly, the Multiplication Principle asserts that you can multiply both sides of an equation by the same *nonzero constant.* But if you multiply both sides of an equation by a *variable,* or if you *square* both sides of an equation, you may obtain an equation that has roots that are not solutions of the original equation. The following examples illustrate what can happen.

EXAMPLE 1. Consider the equation:

$$x = 4$$

Multiply both sides by the *variable* x:

$$x^2 = 4x$$

$$x^2 - 4x = 0$$

$$x(x - 4) = 0$$

$$x = 0 \quad \Big| \quad x - 4 = 0$$

$$x = 4$$

The new equation, $x^2 = 4x$, now has an additional root, 0, that is not a root of the original equation, $x = 4$. ■

EXAMPLE 2. Consider the equation:

$$x = -5$$

Square both sides and obtain:

$$x^2 = 25$$

$$x = \pm 5$$

The new equation has 2 roots, $+5$ and -5, whereas the original equation has the single root, -5. ■

Equations with square roots are often solved by squaring both sides. In so doing, you do not lose any roots. But Example 2 should convince you that you *may* gain extra roots. *You must check to see which roots are also roots of the original equation.*

EXAMPLE 3. Solve the equation:

$$\sqrt{3x - 14} = x - 6$$

SOLUTION. Square both sides.

$$3x - 14 = (x - 6)^2$$

$$3x - 14 = x^2 - 12x + 36$$

$$x^2 - 15x + 50 = 0$$

$$(x - 5)(x - 10) = 0$$

$$x = 5, 10$$

Now check each root to see if it satisfies the original equation

$$\sqrt{3x - 14} = x - 6.$$

For $x = 5$:

$$\sqrt{3(5) - 14} \stackrel{?}{=} 5 - 6$$

$$\sqrt{1} \stackrel{?}{=} -1$$

But $\sqrt{1}$ is *positive.*

$$1 \stackrel{\text{x}}{=} -1$$

This is *false,* so that 5 is *not* a root of the given equation.
For $x = 10$:

$$\sqrt{3(10) - 14} \stackrel{?}{=} 10 - 6$$

$$\sqrt{16} \stackrel{?}{=} 4$$

$$4 \stackrel{\checkmark}{=} 4$$

This is *true.* Therefore 10 is a root, and is the only root of

$$\sqrt{3x - 14} = x - 6.$$ ■

In part (a) of the next example, when you square, you get just the root of the original equation. In part (b) when you square, you get a root, although the original equation has no roots.

EXAMPLE 4. Solve:

(a) $\sqrt{x-3} = 5$ (b) $\sqrt{x-3} = -5$

SOLUTION.

(a)
$$\sqrt{x-3} = 5$$
$$x - 3 = 25$$
$$x = 28$$

CHECK.

$$\sqrt{28-3} \overset{?}{=} 5$$
$$\sqrt{25} \overset{?}{=} 5$$
$$5 \overset{\checkmark}{=} 5 \text{ (True)}$$

(b) $\sqrt{t} \geq 0$; thus $\sqrt{x-3} = -5$ has no root. Nevertheless, see what happens if you proceed as in (a).

$$\sqrt{x-3} = -5$$
$$x - 3 = 25$$
$$x = 28$$

CHECK.

$$\sqrt{28-3} \overset{?}{=} -5$$
$$\sqrt{25} \overset{?}{=} -5$$
$$5 \overset{\text{x}}{=} -5 \text{ (False)}$$

■

EXAMPLE 5. Solve:

$$x + \sqrt{x-2} = 8$$

SOLUTION. When only one radical occurs, isolate this on one side.

$$\sqrt{x-2} = 8 - x$$

Square both sides.

$$x - 2 = (8 - x)^2$$
$$x - 2 = 64 - 16x + x^2$$
$$x^2 - 17x + 66 = 0$$
$$(x-6)(x-11) = 0$$
$$x = 6, 11$$

Check to see if these are roots of the given equation.

$x = 6$:

$$6 + \sqrt{6-2} \overset{?}{=} 8$$
$$6 + \sqrt{4} \overset{?}{=} 8$$
$$6 + 2 \overset{\checkmark}{=} 8 \text{ (True)}$$

$x = 11$:

$$11 + \sqrt{11-2} \overset{?}{=} 8$$
$$11 + \sqrt{9} \overset{?}{=} 8$$
$$11 + 3 \overset{\text{x}}{=} 8 \text{ (False)}$$

Thus 6 is the only root of the given equation.

■

EXAMPLE 6. Solve:

$$\sqrt{1 + 4t} - 1 - \sqrt{2t} = 0$$

SOLUTION. Bring one of the radicals to the right side:

$$\sqrt{1 + 4t} - 1 = \sqrt{2t}$$

Square both sides.

$$1 + 4t - 2\sqrt{1 + 4t} + 1 = 2t$$

$$2t + 2 = 2\sqrt{1 + 4t}$$

$$t + 1 = \sqrt{1 + 4t}$$

This last equation contains one radical, as opposed to the original equation, which contained two. Again, square both sides to eliminate the radical.

$$(t + 1)^2 = 1 + 4t$$

$$t^2 + 2t + 1 = 1 + 4t$$

$$t^2 - 2t = 0$$

$$t(t - 2) = 0$$

$$t = 0, 2$$

CHECK. (the given equation).

$t = 0$:

$$\sqrt{1 + 4(0)} - 1 - \sqrt{2(0)} \stackrel{?}{=} 0$$

$$1 - 1 \stackrel{\checkmark}{=} 0 \text{ (True)}$$

$t = 2$:

$$\sqrt{1 + 4(2)} - 1 - \sqrt{2(2)} \stackrel{?}{=} 0$$

$$3 - 1 - 2 \stackrel{\checkmark}{=} 0 \text{ (True)}$$

Both 0 and 2 are roots of the given equation. ■

EXAMPLE 7. Solve:

$$\sqrt{3x}\sqrt{x + 1} = 6$$

SOLUTION. Square both sides and use $(ab)^2 = a^2b^2$

$$3x(x + 1) = 36$$

$$x(x + 1) = 12$$

$$x^2 + x - 12 = 0$$

$$(x + 4)(x - 3) = 0$$

$$x = -4, 3$$

CHECK. (the given equation)

$x = -4$:

$$\sqrt{3(-4)}\sqrt{-4+1} \stackrel{?}{=} 6$$

$$\sqrt{4(3)(-1)}\ \sqrt{3(-1)} \stackrel{?}{=} 6$$

$$2\sqrt{3}i\ \sqrt{3}i \stackrel{?}{=} 6$$

$$6i^2 \stackrel{?}{=} 6$$

$$-6 \stackrel{\text{x}}{=} 6 \text{ (False)}$$

$x = 3$:

$$\sqrt{3(3)}\sqrt{3+1} \stackrel{?}{=} 6$$

$$\sqrt{9}\ \sqrt{4} \stackrel{?}{=} 6$$

$$3(2) \stackrel{\checkmark}{=} 6 \text{ (True)}$$

Thus 3 is the only root of the given equation. ■

EXERCISES

Solve each equation and check the (possible) roots.

1. $\sqrt{x+2} = 2$
2. $\sqrt{x-1} = 3$
3. $\sqrt{2t+2} = 4$
4. $\sqrt{3t+1} = -4$
5. $\sqrt{x+8} - 5 = 0$
6. $\sqrt{2y-5} + 3 = 0$
7. $(13x - 3)^{1/2} - 6 = 0$
8. $\sqrt{x - \frac{3}{4}} = \frac{1}{2}$
9. $\sqrt{2x+2} = 2\sqrt{3}$
10. $\sqrt{5x+3} - 3\sqrt{2} = 0$
11. $x + \sqrt{x-1} = 7$
12. $3x + \sqrt{x-1} = 7$
13. $\sqrt{2u-1} - 8 = -u$
14. $\sqrt{6u+1} - 2u = 1 - u$
15. $\sqrt{1-2x} + x + 1 = 0$
16. $\sqrt{x} + \sqrt{x+5} = 5$
17. $\sqrt{2y} - \sqrt{y-1} = 1$
18. $\sqrt{8-y} + \sqrt{y+1} = 3$
19. $\sqrt{4y+1} + 3 = \sqrt{16+y}$
20. $\sqrt{t-3} - \sqrt{2t-5} + 1 = 0$
21. $\sqrt{1+3y} - 3 = \sqrt{y}$
22. $\sqrt{x+1} - \sqrt{x-2} = 1$
23. $\sqrt{x}\ \sqrt{x-3} = 2$
24. $\sqrt{x}\ \sqrt{x+7} = 12$
25. $\sqrt{z+3}\ \sqrt{z-9} = 8$
26. $\sqrt{3t+1}\ \sqrt{t-7} = 5$
27. $\sqrt{3u}\ \sqrt{9u+1} = 2$
28. $2\sqrt{4x+1} = 3\sqrt{2x}$
29. $5\sqrt{3x+6} = 3\sqrt{9x+10}$
30. $3\sqrt{2x} = 4\sqrt{x+1}$

REVIEW EXERCISES FOR CHAPTER 13

13.1 SOLVING BY FACTORING

In Exercises 1–4 determine all roots of each equation.

1. $(x-2)(x+3) = 0$
2. $x^2 - 9 = 0$

3. $x^2 + 5x + 4 = 0$

4. $1 - \frac{1}{t} = \frac{6}{t^2}$

5. Determine a quadratic equation whose roots are 4 and -5.

13.2 EQUATIONS OF THE FORM $x^2 = a$

In Exercises 6–9 solve for x.

6. $x^2 = 81$

7. $x^2 = 25A^2$

8. $\left(\frac{2x + 1}{3}\right)^2 = 4$

9. $x^2 + 48 = 0$

10. Suppose $A \neq 0$. Solve $4A^2u^2 = 64B^2$ for u.

13.3 COMPLETING THE SQUARE

In Exercises 11–14 solve each equation by completing the square.

11. $x^2 + 6x = 11$

12. $t^2 - 5t + 1 = 0$

13. $2y^2 + 8y + 3 = 0$

14. $x^2 + 3x + 3 = 0$

13.4 THE QUADRATIC FORMULA

In Exercises 15–17: (a) Determine the discriminant, (b) Without solving, state whether there are 2 distinct real roots, exactly 1 real root, or 2 complex conjugate roots.

15. $x^2 + 5x + 5 = 0$

16. $3y^2 - 7y + 5 = 0$

17. $9x^2 + 12x + 4 = 0$

In Exercises 18–21 solve by means of the quadratic formula.

18. $x^2 + 10x + 1 = 0$

19. $x^2 + 9 = 9x$

20. $y^2 - 5y + 5 = 0$

21. $2x^2 - 5x + 5 = 0$

13.5 WORD PROBLEMS

22. Determine all numbers that are such that the sum of the number and its reciprocal is 9.

23. The product of the two consecutive negative odd integers is 143. Determine these integers.

24. The perimeter of a rectangle is 32 inches. The area of the rectangle is 63 square inches. Determine the dimensions of the rectangle.

13.6 EQUATIONS WITH SQUARE ROOTS

In Exercises 25–28 solve each equation and check the (possible) roots.

25. $\sqrt{x + 3} - 3 = 0$

26. $\sqrt{x + 2} + 4 = 0$

27. $\sqrt{t + 1} - \sqrt{t} + 1 = 0$

28. $\sqrt{y}\,\sqrt{y + 3} = 2$

TEST YOURSELF ON CHAPTER 13.

In Problems 1–5 solve for x. You may use any method, except that in Problems 1 and 2, do not use the quadratic formula.

1. $x^2 + 3x - 18 = 0$
2. $x^2 - 45A^2 = 0$
3. $x^2 + 3x + 4 = 0$
4. $x^2 + 3x + 1 = 0$
5. $2x^2 + 6x + 1 = 0$
6. The sum of two numbers is 14 and the product is 45. Find these numbers.
7. Solve and check all (possible) roots:

$$\sqrt{y+3} - \sqrt{y} = 1$$

14

inequalities and absolute value

14.1 INTERVALS

Let a, b, c be real numbers. Recall that a is less than b,

$$a < b,$$

or equivalently, b is greater than a,

$$b > a,$$

if a lies to the *left* of b on the real line.

$$\text{If} \quad a < b \quad \text{and} \quad b < c, \quad \text{then} \quad a < c$$

For a lies to the left of b and b lies to the left of c. Therefore, a lies to the left of c [See Figure 14.1.]

FIGURE 14.1 $a < b$ and $b < c$. a lies to the left of c. Therefore $a < c$

For every two real numbers a and b, exactly one of the following holds:

$$a < b, \qquad a = b, \qquad a > b$$

You write

$$a \leq b$$

if either $a < b$ or $a = b$, and you say that a is less than or equal to b. Thus $3 \leq 4$ (because $3 < 4$) and $4 \leq 4$ (because $4 = 4$)

$$\text{If} \quad a \leq b \quad \text{and} \quad b \leq c, \quad \text{then} \quad a \leq c$$

On the other hand,

$$a \geq b$$

(a is greater than or equal to b) indicates that $a > b$ or $a = b$.

Instead of considering the *entire* real line, you may be interested only in the numbers "between a and b".

Let

$$a < x < b$$

stand for

$$\text{both} \quad a < x \quad \text{and} \quad x < b.$$

(In both cases, the symbol points to the *smaller* of the two numbers.) Similarly,

$$a \leq x \leq b \qquad \text{means} \qquad a \leq x \quad \text{and} \quad x \leq b.$$

DEFINITION. Let a and b be real numbers and let $a < b$. The **open interval**

$$(a, b)$$

consists of all real numbers x such that

$$a < x < b.$$

The **closed interval**

$$[a, b]$$

consists of all real numbers x such that

$$a \leq x \leq b.$$

a is called the **left end-point** and b the **right end-point** of each of these intervals.

Observe that *neither* end-point a nor b is in the *open* interval (a, b). *Both* end-points a and b are in the *closed* interval $[a, b]$.

Geometrically, the open interval (a, b) consists of all points that lie *both* to the right of a *and* to the left of b. [See Figure 14.2.]

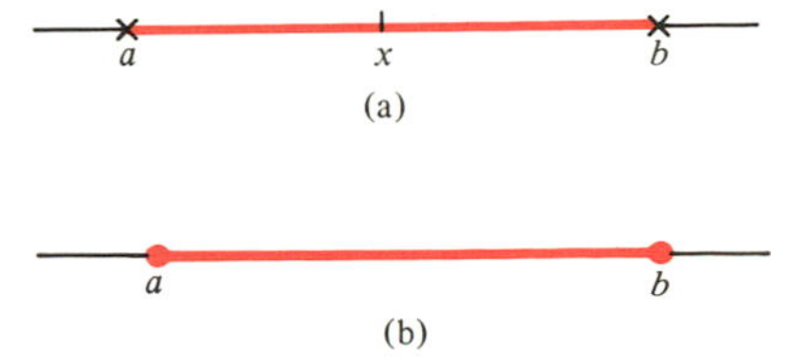

FIGURE 14.2 (a) The open interval (a, b) consists of all x that lie *both* to the right of a *and* to the left of b. The points a and b are excluded. (b) The closed interval $[a, b]$ consists of the numbers in (a, b) plus the end-points a and b.

EXAMPLE 1.

(a) The open interval $(3, 5)$ consists of all real numbers x such that

$$3 < x < 5.$$

Thus 3.1, 4, and $\frac{9}{2}$ are in this open interval. For example,

$$3 < 3.1 < 5$$

Neither 3 nor 5 is in this open interval. Also, such numbers as -4, 0, and 2 lie to the left of 3 and therefore are not in this interval. Similarly, 5.1, 6, and 10 lie to the right of 5, and are not in this interval.

(b) The closed interval [3, 5] consists of all the numbers of the open interval (3, 5) plus the two end-points 3 and 5. Thus 3, 3.7, 4.36, and 5 are in this closed interval; $\frac{1}{2}$, 2.9, and 5.01 are not. ■

EXAMPLE 2.

(a) Which of the numbers

$$-10, -7.5, -7, -6.6, -3, -2.1, -2, -1.9, 0, 4$$

are in $(-7, -2)$?

(b) Which of these numbers are in $[-7, -2]$?

SOLUTION.

(a) $(-7, -2)$ consists of all numbers x satisfying

$$-7 < x < -2.$$

Thus

$-6.6, -3$, and -2.1 are in this open interval;

$-10, -7.5, -7, -2, -1.9, 0$, and 4 are not.

(b) $[-7, -2]$ consists of all numbers x satisfying

$$-7 \leq x \leq -2.$$

Thus

$-7, -6.6, -3, -2.1$, and -2 are in this closed interval;

$-10, -7.5, -1.9, 0$, and 4 are not. ■

Frequently, the domain of a function is an interval. The defining equation applies *only to those x in this interval.*

EXAMPLE 3. Let $f(x)$ be defined by

$$f(x) = 2x \quad \text{and} \quad 0 \leq x \leq 5.$$

The defining equation

$$f(x) = 2x$$

applies only to numbers x that lie in the closed interval [0, 5]. Thus

$$f(0) = 2 \cdot 0 = 0$$

$$f(1) = 2 \cdot 1 = 2$$

$$f(2) = 2 \cdot 2 = 4$$

$$f\left(\frac{5}{2}\right) = 2 \cdot \frac{5}{2} = 5$$

$$f(5) = 2 \cdot 5 = 10$$

On the other hand,

$$f(-1), \qquad f(6), \qquad f(10)$$

are not defined because -1, 6, and 10 do not lie in the interval [0, 5]. [Thus you *cannot* say that $f(10) = 20$.] The graph of this function consists of the *line segment* depicted in Figure 14.3.

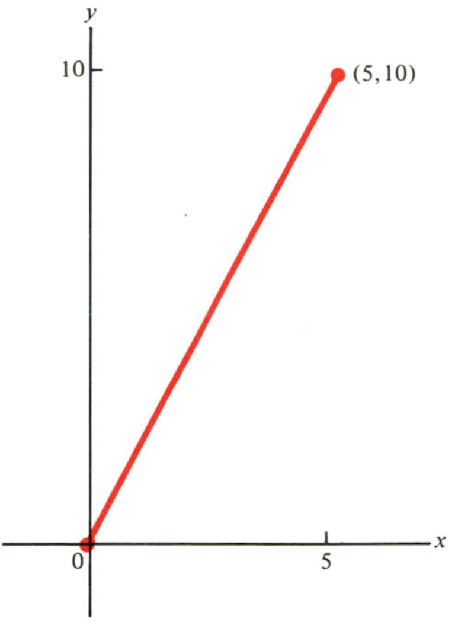

FIGURE 14.3

$f(x) = 2x, 0 \leq x \leq 5$

The graph of this function consists of the above line segment, including the points (0, 0) and (5, 10).

Sometimes you will want to speak about

$$\text{all } x > 2 \qquad \text{or} \qquad \text{all } x \leq 5.$$

DEFINITION. Let a and b be real numbers.

The **open right ray**

$$(a, \infty)$$

consists of all real numbers x such that

$$a < x.$$

The **closed right ray**

$$[a, \infty)$$

consists of all real numbers x such that

$$a \leq x.$$

The **open left ray**

$$(-\infty, b)$$

consists of all real numbers x such that

$$x < b.$$

The **closed left ray**

$$(-\infty, b]$$

consists of all real numbers x such that

$$x \leq b.$$

Geometrically, the open *right* ray

$$(a, \infty)$$

consists of all x that lie to the *right* of a; the closed *right* ray

$$[a, \infty)$$

includes the *left* end-point a. The open *left* ray

$$(-\infty, b)$$

consists of all x that lie to the *left* of b; the closed *left* ray

$$(-\infty, b]$$

includes the *right* end-point b. See Figure 14.4. [The symbol "∞" (infinity) is used to indicate that these lines extend indefinitely in one direction.]

FIGURE 14.4 (a) The open right ray (a, ∞). The point a is excluded. (b) The closed right ray $[a, \infty)$. The point a is included. (c) The open left ray $(-\infty, b)$. The point b is excluded. (d) The closed left ray $(-\infty, b]$. The point b is included.

EXAMPLE 4.

(a) The open right ray $(3, \infty)$ consists of all real numbers x that lie to the right of 3. Thus 4 is in this ray, but 3 is not.
(b) The closed right ray $[-4, \infty)$ consists of -4 together with all numbers that lie to the right of -4. Among these are -3, 0, and all positive numbers.
(c) The open left ray $(-\infty, 0)$ consists of all numbers x that lie to the left of 0. Thus $(-\infty, 0)$ is the set of negative numbers.
(d) The closed left ray $(-\infty, 2]$ consists of 2 together with all numbers less than 2. ■

EXERCISES

In Exercises 1–16 describe each set of numbers as either

(a) an open interval, (b) a closed interval,
(c) an open right ray, (d) a closed right ray,
(e) an open left ray, or (f) a closed left ray.

1. $(4, 7)$ 2. $[3, 6]$ 3. $(-2, -1)$ 4. $[4, \infty)$
5. $(-\infty, 6]$ 6. $(-3, 2)$ 7. $(-\infty, -1)$ 8. $[0, \infty)$
9. $[-4, -2]$ 10. $[-4, \infty)$ 11. $(-\infty, -4)$ 12. $[100, 200]$
13. $(100, \infty)$ 14. $(-\infty, 100]$ 15. $(-\infty, -100)$ 16. $[-100, \infty)$

In Exercises 17–30 answer "true" or "false".

17. 4 is in $[2, 7]$.
18. -3 is in $(-4, 4)$.
19. -2.1 is in $[-4, 1]$.
20. 0 is in $(0, 6)$.
21. $\frac{1}{2}$ is in $[-1, 1]$.
22. $\frac{-3}{2}$ is in $[-1, 0]$.
23. $\frac{7}{2}$ is in $[3, 4.5]$.
24. -2 is in $(-3, -2)$.
25. 4 is in $[4, \infty]$.
26. -2 is in $(-1, \infty)$.
27. 3 is in $(3, \infty)$.
28. 3 is in $(-\infty, -3)$.
29. $\frac{1}{2}$ is in $(-\infty, 0)$.
30. 0 is in $(-\infty, 0)$.

In Exercises 31–39 describe in one of the forms

$$[a, b], \quad (a, b), \quad (a, \infty), \quad [a, \infty), \quad (-\infty, b), \quad \text{or} \quad (-\infty, b]$$

the set of numbers satisfying each of the following:

31. $2 \leq x \leq 9$ *Answer:* $[2, 9]$
32. $0 < x < 1$
33. $-1 \leq x \leq 0$
34. $x > 6$
35. $x < -6$
36. $-100 < x < 100$
37. $x \geq 8$
38. $x \leq 8$
39. $-7 < x < -5$

In Exercises 40–46: (a) Graph the indicated function. (b) What is the domain of the function?

40. $y = x$ and $0 < x < 4$
41. $y = x + 1$ and $-2 \leq x \leq 2$
42. $y = x + 1$ and $-2 < x < 2$
43. $y = x + 2$ and $x > 2$
44. $y = x - 1$ and $x \geq -1$
45. $y = 2x + 1$ and $x < 1$
46. $y = x + 2$ and $x \leq -2$

14.2 ARITHMETIC OF INEQUALITIES

There are several important arithmetic properties of inequalities.

Let a, b, c, d be real numbers.

(A) **If**

$$a < b,$$

then

$$a + c < b + c$$

and

$$a - c < b - c$$

Property (A) asserts that *if you add or subtract the same number on both sides of an inequality the sense (or direction) of the inequality is preserved.*

To illustrate Property (A), let $a < b$. Then a lies to the left of b. Suppose, for example, $c > 0$. Then $a + c$ lies c units to the right of a, and $b + c$ lies *the same number of units* to the right of b. Thus $a + c$ lies to the left of $b + c$, and therefore

$$a + c < b + c$$

[See Figure 14.5.]

FIGURE 14.5 If $a < b$, then $a + c < b + c$. (The diagram is for the case of $c > 0$ and $a + c < b$.)

(B) If

$$a < b \qquad \text{and} \qquad c < d,$$

then

$$a + c < b + d$$

Thus inequalities may be added.

To verify Property (B), suppose $a < b$. Then $a + c < b + c$ (by Property A). If $c < d$, then $b + c < b + d$ (by Property A). Thus

$$a + c < b + c \qquad \text{and} \qquad b + c < b + d$$

Consequently,

$$a + c < b + d$$ ■

EXAMPLE 1.

$$2 < 4$$

By Property (A),

$$\underbrace{2 + 5}_{7} < \underbrace{4 + 5}_{9}$$

Also, by Property (A),

$$\underbrace{2 - 5}_{-3} < \underbrace{4 - 5}_{-1}$$

■

EXAMPLE 2.

(a) $3 < 6$ and $1 < 5$
By Property (B),

$$\underbrace{3 + 1}_{4} < \underbrace{6 + 5}_{11}$$

(b) $-7 < -3 \quad \text{and} \quad -4 < -1$

Again by Property (B),

$$\underbrace{(-7) + (-4)}_{-11} < \underbrace{(-3) + (-1)}_{-4}$$

■

Recall that c is *positive* if $c > 0$; c is *negative* if $c < 0$.

$(C)_1$ **If**

$$a < b \quad \text{and} \quad c > 0,$$

then

$$ac < bc$$

$(C)_2$ **If**

$$a < b \quad \text{and} \quad c < 0$$

then

$$ac > bc$$

Thus *multiplication by a positive number preserves the sense of an inequality. Multiplication by a* **negative** *number* **reverses** *the sense of an inequality.*

(The proofs are omitted.)

EXAMPLE 3.

$$2 < 3$$

(a) $$\underbrace{2(4)}_{8} < \underbrace{3(4)}_{12}$$

by Property $(C)_1$. [See Figure 14.6(a).]

(b) $\underbrace{2(-4)}_{-8} > \underbrace{3(-4)}_{-12}$

by Property $(C)_2$. [See Figure 14.6(b).] ■

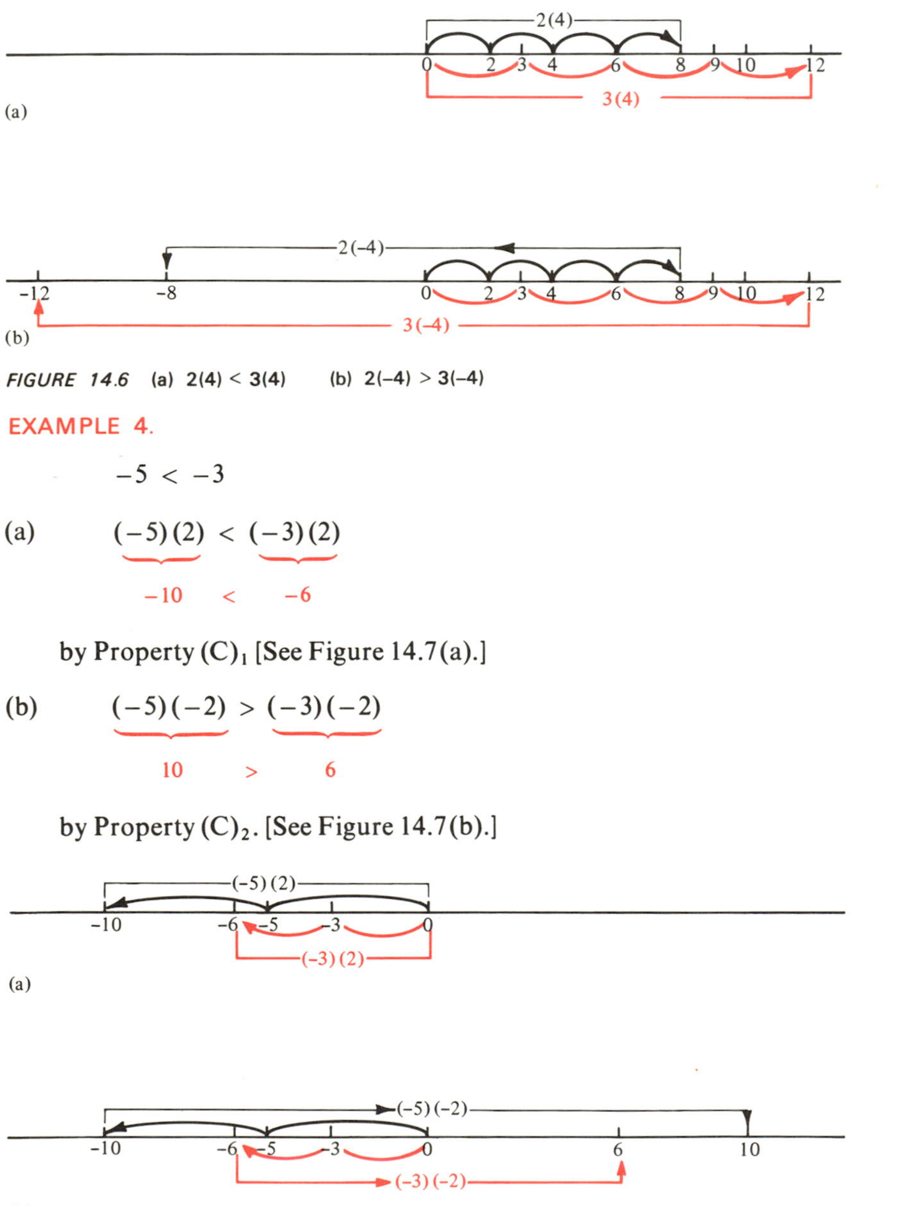

FIGURE 14.6 (a) $2(4) < 3(4)$ (b) $2(-4) > 3(-4)$

EXAMPLE 4.

$$-5 < -3$$

(a) $\underbrace{(-5)(2)}_{-10} < \underbrace{(-3)(2)}_{-6}$

by Property $(C)_1$ [See Figure 14.7(a).]

(b) $\underbrace{(-5)(-2)}_{10} > \underbrace{(-3)(-2)}_{6}$

by Property $(C)_2$. [See Figure 14.7(b).]

FIGURE 14.7 (a) $(-5)(2) < (-3)(2)$ (b) $(-5)(-2) > (-3)(-2)$

Division by c can be considered as multiplication by $\frac{1}{c}$. Because $c > 0$ is equivalent to $\frac{1}{c} > 0$, the following also hold.

(C)$_3$ If

$$a < b \quad \text{and} \quad c > 0,$$

then $\frac{a}{c} < \frac{b}{c}$

(C)$_4$ If

$$a < b \quad \text{and} \quad c < 0,$$

then

$$\frac{a}{c} > \frac{b}{c}$$

These properties can also be stated in terms of $\leq$. For example:

(A′) If

$$a \leq b,$$

then

$$a + c \leq b + c$$

and

$$a - c \leq b - c$$

Property (A) implies that

$$\text{if} \quad a < b, \qquad \text{then} \quad 0 < b - a$$

For, if

$$a < b,$$

then

$$a - a < b - a,$$

that is,

$$0 < b - a$$

Also, if

$$0 < b - a,$$

then by adding a to both sides,

$$a < b$$

■

Thus

$$a < b \quad \text{is equivalent to} \quad 0 < b - a.$$

EXERCISES

In exercises 1–22 fill in:

$$< \quad \text{or} \quad >$$

1. $5 + c \;\square\; 10 + c$
2. $-2 + c \;\square\; -1 + c$
3. $-4 + c \;\square\; 0 + c$
4. If $c > 0$, $-4c \;\square\; -7c$
5. $a + 3 \;\square\; a + 2$
6. $a - 6 \;\square\; a - 5$
7. $a - 0 \;\square\; a - 2$
8. $a - 4 \;\square\; 2 + a$
9. $-7 + -2 \;\square\; -4 + -1$
10. $-8 + -10 \;\square\; -9 + -7$
11. $5(8) \;\square\; 3(8)$
12. $-2(8) \;\square\; -1(8)$
13. $6(-3) \;\square\; 15(-3)$
14. $(-6)(-2) \;\square\; (-4)(-2)$
15. $(-2)(-7) \;\square\; (-2)(-5)$
16. $(-3)(4)(7) \;\square\; (-3)(3)(6)$
17. If $x < y$ and $z < 0$, then $zx \;\square\; zy$
18. If $x < y$, then $y - x \;\square\; 0$
19. $\frac{1}{6} \;\square\; \frac{1}{5}$
20. If $r < 0$, then $\frac{-4}{r} \;\square\; \frac{-3}{r}$
21. $\frac{1}{9} + \frac{1}{4} \;\square\; \frac{1}{3} + \frac{1}{4}$
22. $\frac{2}{3}\left(\frac{-2}{5}\right) \;\square\; \left(\frac{-3}{4}\right)\frac{2}{3}$
23. Show that if $x < 4$, then $x + 2 < 6$.
24. Show that if $y \leq 9$, then $y - 4 \leq 5$.
25. Show that if $a \leq 4$, then $2a \leq 8$.
26. Show that if $b < -2$, then $\frac{b}{2} < -1$.
27. Show that if $x \leq 5$, then $-2x \geq -10$.
28. Show that if $x < 3$, then $3x < 10$.
29. Show that if $2x < 3$, then $x < 2$.
30. Show that if $x < 10$, then $\frac{-x}{10} > -1$.

14.3 SOLVING INEQUALITIES

Recall that an equation is a statement of equality, and has the form:

$$\square = \square$$

DEFINITION. An **inequality** is a statement of one of the forms:

$$\square < \square$$
$$\square \leq \square$$
$$\square > \square$$
$$\square \geq \square$$

EXAMPLE 1. Each of the following is an inequality:

(a) $2 < 5$
(b) $x + 1 \leq 6$
(c) $2x - 3 \geq 4 - 3x$

A solution (or root) of an equation (in a single variable) is a number that when substituted in the equation yields a true statement. For most of the equations you have considered there were either one or two solutions.

DEFINITION. A **solution of an inequality** (in a single variable) is a number that when substituted in the inequality yields a true statement. The **solution set of an inequality** is the set of all solutions.

For many of the inequalities you will consider the solution set will be infinite. For example, the inequality

$$x < 0$$

has as its solution set, the set of negative numbers.

Equivalent equations are equations with exactly the same solutions (or roots).

DEFINITION. **Equivalent inequalities** are inequalities with exactly the same solution sets.

You solve an inequality by successively transforming it into simpler equivalent inequalities. The arithmetic properties you studied in the preceding section enable you to simplify inequalities. Thus adding or subtracting the same number, or multiplying or dividing by the same *positive* number on both sides of an inequality *preserves* the sense of the inequality. On the other hand, multiplying or dividing both sides of an inequality by a *negative* number *reverses* the sense of the inequality. Here is how you use these properties to simplify an inequality.

The **Addition Principle** asserts that you can add the same quantity to both sides of an inequality.

ADDITION PRINCIPLE (FOR INEQUALITIES).

Let *R* be any rational expression (possibly a polynomial or even a constant).

and

$$\square + R < \square + R$$

are equivalent inequalities.

Because

$$a - b = a + (-b),$$

you can also subtract by means of the Addition Principle.

EXAMPLE 2. Solve:

$$x + 5 < 10$$

SOLUTION. Use the Addition Principle as you would for the equation $x + 5 = 10$. Thus subtract 5 from (or add -5 to) both sides of the given inequality:

$$\begin{aligned} x + 5 &< 10 \\ -5 \quad & \quad -5 \\ x &< 5 \end{aligned}$$

The solution set is the open left ray

$$(-\infty, 5).$$

■

The **Multiplication Principle** asserts that *you can multiply both sides of an inequality by the same positive number.* But *if you multiply both sides by the same negative* number, the sense of inequality is reversed.

MULTIPLICATION PRINCIPLE (FOR INEQUALITIES)

(a) Let $p > 0$. Then

$$\square < \square$$

and

$$\square \cdot p < \square \cdot p$$

are equivalent inequalities.

(b) Let $n < 0$. Then

$$\square < \square$$

and

$$\square \cdot n > \square \cdot n$$

are equivalent inequalities.

Because

$$\frac{a}{b} = a \cdot \frac{1}{b}, \qquad b \neq 0,$$

the Multiplication Principle also enables you to divide both sides of an inequality by the same nonzero constant.

EXAMPLE 3. Solve:

$$4x > 24$$

SOLUTION. Divide both sides by the *positive* number 4 $\left(\text{or multiply by } \frac{1}{4}\right)$.

$$\frac{4x}{4} > \frac{24}{4}$$

$$x > 6$$

The solution set is the open right ray $(6, \infty)$. ■

EXAMPLE 4. Solve:

$$\frac{-x}{2} < 4$$

SOLUTION. Multiply both sides by the *negative* number -2. This *reverses* the sense of inequality.

$$\frac{-x}{2} < 4$$

$$x > -8$$

The solution set is the open right ray $(-8, \infty)$. ■

The Addition and Multiplication Principles also apply to inequalities involving $\leq$.

EXAMPLE 5. Solve:

$$3x + 6 \leq 9 - x$$

SOLUTION. Bring variables to one side; constants to the other.

$$\begin{array}{rcl} 3x + 6 & \leq & 9 - x \\ \underline{+x - 6} & \triangle\!\!-\!\!\triangle & \underline{-6 + x} \\ 4x & \leq & 3 \\ x & \leq & \frac{3}{4} \end{array}$$

The solution set is the closed left ray $\left(-\infty, \frac{3}{4}\right]$. ■

EXAMPLE 6. Solve:

$$\frac{x}{2} + \frac{1}{3} \geq \frac{3x + 1}{4}$$

SOLUTION. The LCD is 12. Thus multiply both sides of the inequality by 12.

$$\begin{aligned} 6x + 4 &\geq 3(3x + 1) \\ 6x + 4 &\geq 9x + 3 \\ \underline{-6x - 3} &\quad \underline{-6x - 3} \\ 1 &\geq 3x \\ \frac{1}{3} &\geq x \qquad \left(\text{or } x \leq \frac{1}{3}\right) \end{aligned}$$

The solution set is $\left(-\infty, \frac{1}{3}\right]$. ■

Sometimes, you will be given two inequalities to solve simultaneously.

EXAMPLE 7. Solve:

$$3 < x + 2 < 9$$

SOLUTION. You are given two inequalities that hold for x:

$$3 < x + 2 \qquad \text{and} \qquad x + 2 < 9$$

You can work with both at the same time by subtracting 2 from each expression.

$$\begin{aligned} 3 &< x + 2 < 9 \\ \underline{-2} &\quad \underline{-2} \quad \underline{-2} \\ 1 &< x < 7 \end{aligned}$$

The solution set is $(1, 7)$. ■

EXAMPLE 8. Solve:

$$-10 \leq 3x - 5 \leq 10$$

SOLUTION. Add 5 to each expression.

$$-5 \leq 3x \leq 15$$

Divide each expression by (the positive number) 3.

$$\frac{-5}{3} \leq x \leq 5$$

The solution set is $\left[\frac{-5}{3}, 5\right]$. ■

EXAMPLE 9. Solve:

$$4 < -x < 10$$

SOLUTION. Multiply each expression by -1. This *reverses* the senses of both inequalities.

$$-4 > x > -10$$

or

$$-10 < x < -4$$

The solution set is $(-10, -4)$. ■

EXERCISES

In Exercises 1–60 solve the indicated inequalities.

1. $x + 2 < 6$
2. $x + 3 < 0$
3. $x - 8 < 5$
4. $x + 4 < -2$
5. $t + 6 \leq 10$
6. $2 - y \leq 8$
7. $y + 2 > 3$
8. $x - 4 > -1$
9. $x + 10 \geq 10$
10. $t - \frac{1}{2} \geq 1$
11. $2x < 6$
12. $3x \leq -6$
13. $\frac{t}{2} > 4$
14. $3x \geq -9$
15. $5x < 0$
16. $\frac{x}{4} \geq 0$
17. $-x < 1$
18. $-3s < 5$
19. $-2y > 2$
20. $-4z \geq -2$
21. $-\frac{x}{3} \leq \frac{1}{6}$
22. $\frac{-4x}{3} > 2$
23. $x + 2 < 5x - 1$
24. $2x - 7 \geq x - 9$
25. $15x + 2 \leq 12x - 7$
26. $\frac{t}{2} + 1 \leq 3t + 2$
27. $2(x - 1) + 3(x + 2) \geq x - 5$
28. $2(3x - 2) \leq 4(1 - x)$
29. $7(2x + 1) + 3 < 5\left(2x - \frac{1}{5}\right)$
30. $4(x + 1) - 3 \leq 2(1 - 2x) - 1$
31. $\frac{x}{4} + \frac{1}{2} \leq \frac{3}{8}$
32. $\frac{x}{5} > \frac{1}{3}$
33. $\frac{3x}{7} \geq \frac{-2}{9}$
34. $\frac{-2t}{5} > \frac{3}{10}$
35. $\frac{1 + y}{2} < \frac{3}{4}$
36. $\frac{2x - 1}{3} > \frac{2}{5}$
37. $\frac{1 - 2x}{10} \leq \frac{x}{5}$
38. $\frac{3 - u}{3} < \frac{u}{4}$
39. $\frac{x}{3} - \frac{1}{6} \leq \frac{3x}{4}$

40. $\frac{5x - 1}{2} + \frac{1}{5} > \frac{x + 1}{10}$

41. $2 < x + 2 < 8$

42. $0 < x - 3 < 4$

43. $-1 \leq x + 8 \leq 5$

44. $\frac{1}{2} \leq x + \frac{1}{2} \leq 2$

45. $9 < 3x < 12$

46. $-10 \leq 2y \leq -6$

47. $-15 < 5z < 15$

48. $0 \leq \frac{x}{3} \leq 9$

49. $-1 < 2x + 1 < 8$

50. $-2 < 3x - 2 < 4$

51. $-9 < 5t - 1 < 9$

52. $-2 \leq \frac{2 + x}{3} \leq 2$

53. $-1 \leq -x \leq 2$

54. $-4 < -x < 4$

55. $0 < -2x < 8$

56. $-3 < 1 - 2x < 9$

57. $-2 < 3 - 5x < 4$

58. $-34 \leq 1 - 7x \leq 22$

59. $\frac{1}{2} < \frac{3 - 2x}{4} < 1$

60. $\frac{1}{2} \leq \frac{3}{4} - 2x \leq 1$

61. A highway has a minimum speed limit of 40 miles per hour and a maximum speed limit of 60 miles per hour. A car travels on this highway for three hours within the legal range. What interval describes the distance traveled?

62. On an April morning the temperature in Boston ranges between 50° and 59° Fahrenheit. Centigrade (C) can be determined from Fahrenheit (F) by the formula

$$C = \frac{5}{9}(F - 32).$$

On the Centigrade scale what is the range of temperature in Boston that morning?

63. In London one December morning the temperature ranges between 0° and 5° Centigrade. What is the range on the Fahrenheit scale? (See Exercise 62.)

64. During the preliminary phase of labor the expectant mother must take 6 to 9 deep breaths per minute. What is the range of breaths she should take during a 40-second period?

65. Suppose $y = 2x$, $1 \leq x \leq 4$. What is the range of y?

66. Suppose $y = 3x + 1$, $0 < x < 5$. What is the range of y?

67. Suppose $y = -x$, $0 < x$. What is the range of y?

14.4 GRAPHING INEQUALITIES

The graph of the *equation*

$$y = mx + b$$

is the line with slope m and y-intercept b. This line divides the plane into two regions. Points lying *below* this line satisfy the *inequality*

$$y < mx + b.$$

[See Figure 14.8(a).]

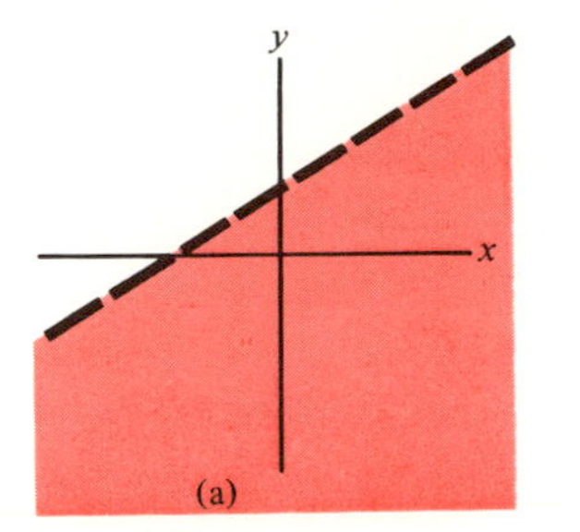

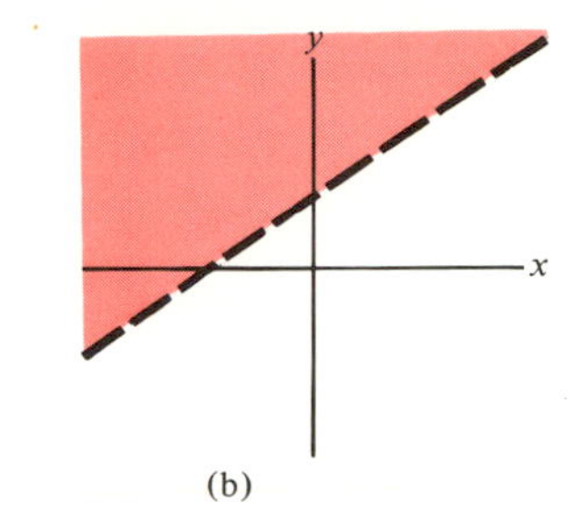

FIGURE 14.8 (a) The graph of the inequality $y < mx + b$. The line is excluded. (In the diagram $m > 0, b > 0$)
(b) The graph of the inequality $y > mx + b$. The line is excluded. (In the diagram $m > 0, b > 0$)

Points lying *above* the line satisfy the inequality

$$y > mx + b.$$

[See Figure 14.8(b).]

EXAMPLE 1. Consider the line given by the equation

$$y = x.$$

(a) The point (2, 1) lies *below* this line. Observe that the y-value is **1**, the x-value is 2, and

$$1 < 2.$$

Thus (2, 1) satisfies the inequality

$$y < x.$$

(b) The point (1, 2) lies above the line and

$$2 > 1.$$

Thus (1, 2) satisfies the inequality

$$y > x.$$

■

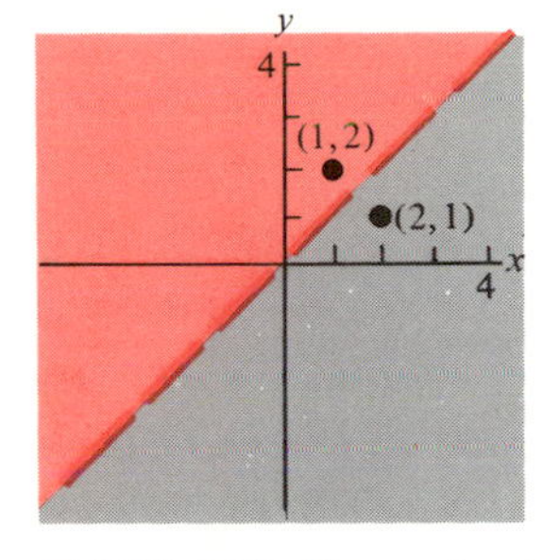

FIGURE 14.9 The graphs of $y < x$ (color) and $y > x$ (gray).

EXAMPLE 2. Graph the inequality:

$$y < 2 - x$$

SOLUTION. First draw the line given by

$$y = 2 - x.$$

The graph of the inequality

$$y < 2 - x$$

consists of all points lying *below* this line.

■

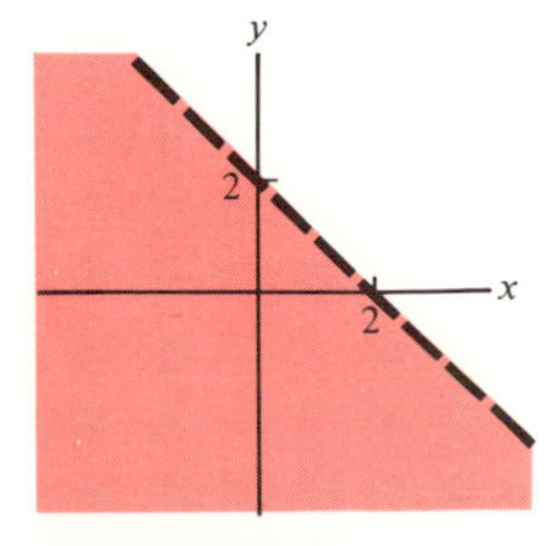

FIGURE 14.10

The graph of an inequality

$$y \leq mx + b$$

consists of all points lying *on or below* the line given by the equation

$$y = mx + b.$$

[See Figure 14.11(a).] The graph of an inequality

$$y \geq mx + b$$

consists of all points lying *on or above* the line given by

$$y = mx + b.$$

[See Figure 14.11(b).]

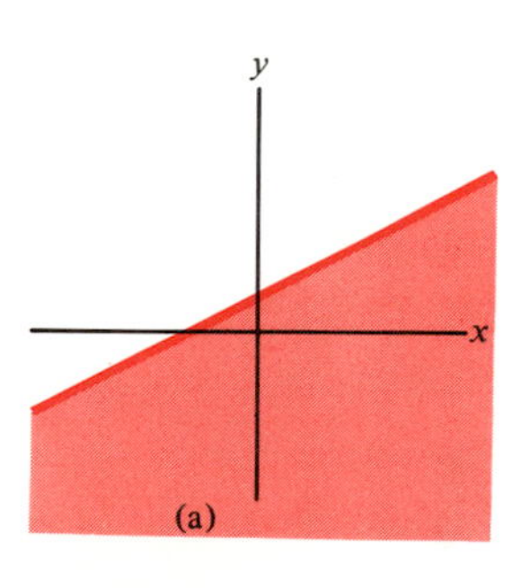

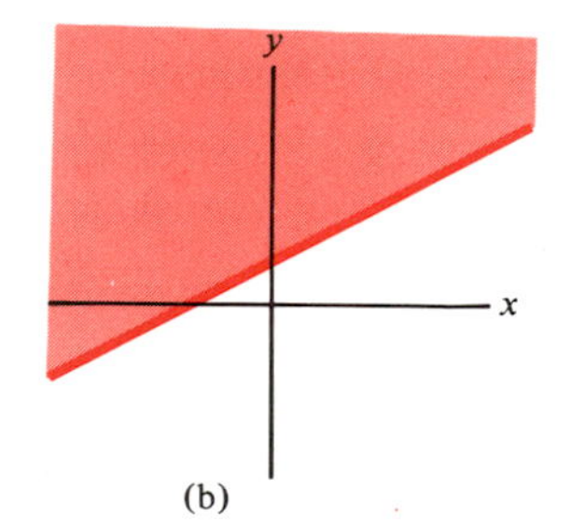

FIGURE 14.11 (a) The graph of the inequality $y \leq mx + b$. The line is included. (In the diagram $m > 0$, $b > 0$) (b) The graph of the inequality $y \geq mx + b$. The line is included. (In the diagram $m > 0, b > 0$)

EXAMPLE 3. Graph the inequality:

$$y \geq 3x - 1$$

SOLUTION. First draw the line given by

$$y = 3x - 1.$$

The graph of the inequality

$$y \geq 3x - 1$$

consists of all points lying *on or above* this line. [See Figure 14.12.] ■

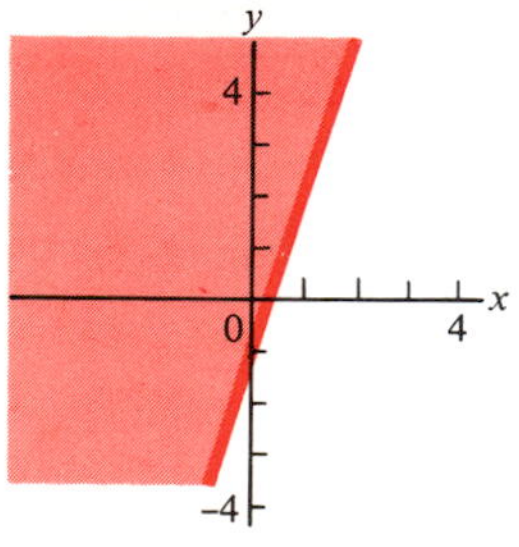

FIGURE 14.12

EXAMPLE 4. Graph the inequality:

$$2x - y \leq 4$$

SOLUTION. Rewrite the inequality in the form

$$2x - 4 \leq y$$

or

$$y \geq 2x - 4.$$

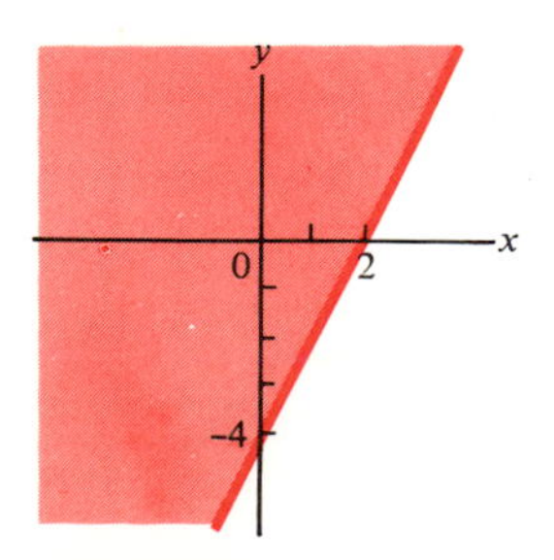

FIGURE 14.13

Now draw the line with equation

$$y = 2x - 4.$$

The graph of the given inequality consists of all points lying *on or above* this line. [See Figure 14.13.] ■

The graph of an inequality does *not* depict a function. For, the graph of a function is such that every vertical line intersects the graph at most once. In fact, every vertical line intersects the graph of each inequality described above infinitely many times. [See Figure 14.14.]

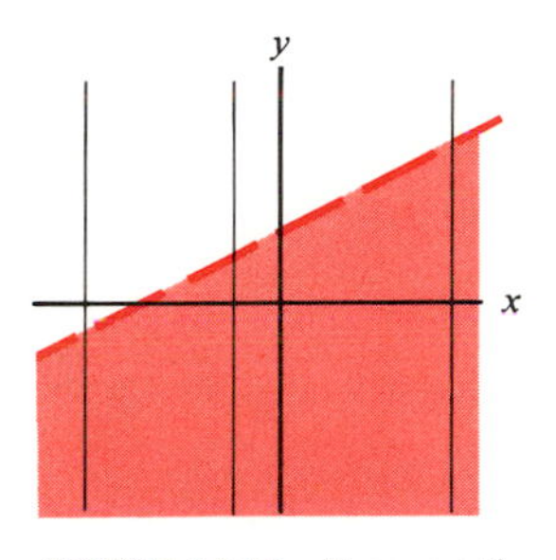

FIGURE 14.14 Every vertical line intersects the graph of this inequality infinitely many times.

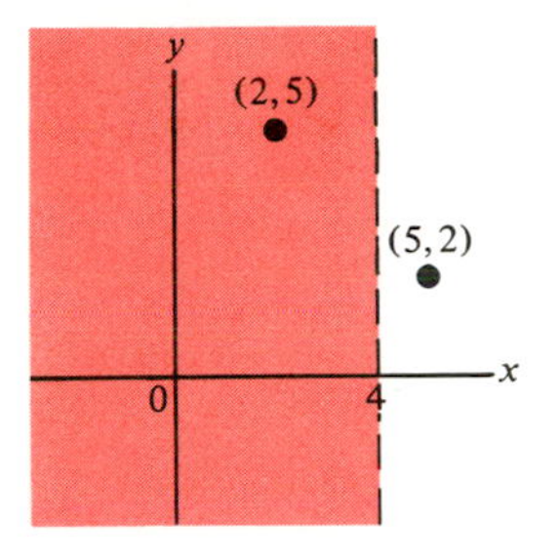

FIGURE 14.15 The graph of $x < 4$

A vertical line is given by an equation of the form

$$x = c,$$

where c is a real number. Thus the point (a, b) satisfies the inequality

$$x < c$$

if $a < c$. In this case the point lies to the *left of* the vertical line. For example, (2, 5) satisfies the inequality

$$x < 4,$$

but (5, 2) does not. (Again, a function is not defined.) [See Figure 14.15.]

EXERCISES

In Exercises 1–20 indicate which points satisfy the given inequality.

1. $y < 2x$

   (a) (1, 1) (b) (1, 2) (c) (1, 3) (d) (−1, −1)

2. $y < 5x$

   (a) (2, 0) (b) (2, 3) (c) (3, 12) (d) (0, 0)

3. $y < 1 - 3x$
   (a) $(0, 0)$ (b) $(1, 1)$ (c) $(1, -1)$ (d) $(-1, 3)$
4. $x < 2 + y$
   (a) $(0, 2)$ (b) $(-1, 0)$ (c) $(-2, -2)$ (d) $(2, 0)$
5. $y > x$
   (a) $(3, 5)$ (b) $(5, 3)$ (c) $(-3, -5)$ (d) $(-5, -3)$
6. $x > 1 - y$
   (a) $(2, 0)$ (b) $(0, 2)$ (c) $(-1, 2)$ (d) $(-1, 3)$
7. $y > 2x + 3$
   (a) $(0, 4)$ (b) $(1, 6)$ (c) $(-2, 0)$ (d) $(3, 9)$
8. $y > 5 - 3x$
   (a) $(1, 2)$ (b) $\left(\frac{1}{3}, 3\right)$ (c) $\left(\frac{1}{2}, 4\right)$ (d) $(2, 0)$
9. $y \leq 4x$
   (a) $(1, 4)$ (b) $(0, 0)$ (c) $(-1, -4)$ (d) $(1, -4)$
10. $y \leq 2x + 1$
    (a) $(1, 1)$ (b) $(2, 5)$ (c) $(-1, -1)$ (d) $(0, 1)$
11. $y \leq \frac{x + 1}{2}$
    (a) $\left(0, \frac{1}{2}\right)$ (b) $(1, 2)$ (c) $(1, 1)$ (d) $(2, 2)$
12. $y \leq x + \frac{1}{2}$
    (a) $(1, 1)$ (b) $(1, 2)$ (c) $\left(1, \frac{3}{2}\right)$ (d) $(-1, 0)$
13. $y \geq x + 1$
    (a) $(5, 6)$ (b) $\left(0, \frac{1}{2}\right)$ (c) $(3, 3.9)$ (d) $(-4, -3)$
14. $y \geq 2$
    (a) $(2, 0)$ (b) $(3, 1)$ (c) $(2, 2)$ (d) $(2, 3)$
15. $2y \geq x$
    (a) $(1, 1)$ (b) $(1, 2)$ (c) $(2, 1)$ (d) $(2, 2)$
16. $y + 1 \geq 2x$
    (a) $(1, 1)$ (b) $(1, 2)$ (c) $(2, 1)$ (d) $(2, 3)$
17. $x - 2y < 5$
    (a) $(0, 0)$ (b) $(2, 1)$ (c) $(3, -1)$ (d) $(5, 1)$
18. $2x + 3y \leq 1$
    (a) $(1, 1)$ (b) $\left(0, \frac{1}{3}\right)$ (c) $\left(\frac{1}{2}, 0\right)$ (d) $(-1, 1)$

19. $3x - 7y + 1 \leq 0$
    (a) (0, 0) (b) (3, 7) (c) (7, 3) (d) (2, 1)

20. $2x - 5 > 3y$
    (a) (1, 1) (b) (0, 0) (c) (1, 2) (d) (4, 1)

In Exercises 21–36 graph each inequality.

21. $y < 2x$
22. $y < x + 1$
23. $y < 2x - 1$
24. $y < 1 - 2x$
25. $y > x$
26. $y > x - 2$
27. $y > \frac{x}{2}$
28. $2x < 1$
29. $y \leq 3$
30. $y \leq 2x + 1$
31. $y \leq 4 - 2x$
32. $y \geq -3$
33. $y \geq x - 4$
34. $x \geq 10$
35. $2x + y < 1$
36. $x - 4y - 1 \geq 0$

14.5 QUADRATIC INEQUALITIES

In Chapter 13, you considered quadratic equations

$$Ax^2 + Bx + C = 0, \qquad A \neq 0.$$

A **quadratic inequality** is obtained if "=" is replaced by one of the symbols:

$$<, \quad \leq, \quad >, \quad \geq$$

Only quadratic inequalities in which the left side can easily be factored will be considered.

Recall that

$ab > 0$, if *both* factors are positive or *both* are negative;

$ab = 0$, if at least one factor is 0;

$ab < 0$, if one factor is positive and the other negative.

EXAMPLE 1. Solve:

$$x^2 + 3x > 0$$

SOLUTION. Factor the left side to obtain

$$x(x + 3) > 0.$$

Next, observe that

$$x(x + 3) = 0 \qquad \text{when } x = 0 \qquad \text{and also, when} \qquad x = -3.$$

The points 0 and -3 divide the x-axis into three regions. In each region $x(x + 3)$

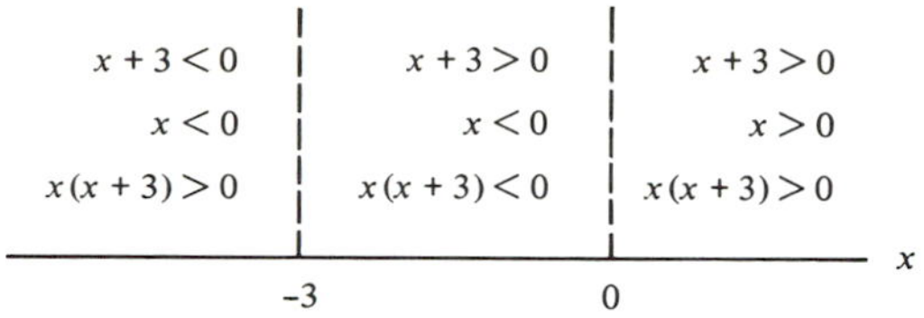

$x + 3 < 0$	$x + 3 > 0$	$x + 3 > 0$
$x < 0$	$x < 0$	$x > 0$
$x(x + 3) > 0$	$x(x + 3) < 0$	$x(x + 3) > 0$

$-3 \qquad 0 \qquad x$

FIGURE 14.16

is either positive throughout the region or else negative. Figure 14.16 indicates the signs of the factors x and $x + 3$ and, hence, of their product in each region. For instance, in the left-most region, that is, in $(-\infty, -3)$, x is less than -3, which, in turn, is less than 0:

$$x < -3 < 0$$

Thus

$$x < 0$$

Also,

$$x + 3 < -3 + 3 = 0$$

Therefore

$$x + 3 < 0$$

Each of the factors x and $x + 3$ is negative in $(-\infty, -3)$. Consequently,

$$x(x + 3) > 0$$

Instead of the above analysis, you may pick an arbitrary number in each region. This will indicate the sign of $x(x + 3)$ throughout the region. For example, in the middle region, that is, in $(-3, 0)$, replace x by -1 in the expression $x(x + 3)$:

$$(-1)\,(-1 + 3) = -2 < 0$$

Thus $x(x + 3) < 0$ in $(-3, 0)$

In the right-most region, $(0, \infty)$, both factors, x and $x + 3$, are positive, as, consequently, is the product, $x(x + 3)$.

The solution set of the given inequality consists of

all x in $\quad (-\infty, -3) \quad$ or in $\quad (0, \infty)$. ■

The notation

$A \cup B$ (read: A **union** B)

indicates the set of objects in *at least one* of the sets A or B. Thus the solution set in Example 1 is

$$(-\infty, -3) \cup (0, \infty).$$

[See Figure 14.17.]

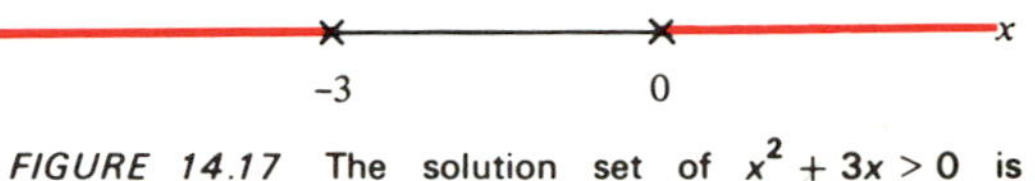

FIGURE 14.17 The solution set of $x^2 + 3x > 0$ is $(-\infty, -3) \cup (0, \infty)$.

EXAMPLE 2. Solve:

$$x^2 + x - 2 \le 0$$

SOLUTION. Factor the left side:

$$(x + 2)(x - 1) \le 0$$

Here the points -2 and 1 divide the x-axis into three regions. The possibilities for the factors are shown in Figure 14.18. Because $(x + 2)(x - 1) \le 0$

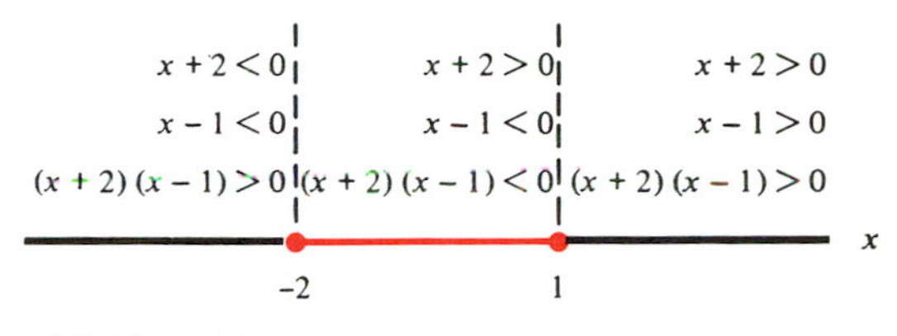

FIGURE 14.18

when $x = -2$ and when $x = 1$, the solution set for the inequality

$$(x + 2)(x - 1) \le 0,$$

and hence for the given inequality,

$$x^2 + x - 2 \le 0$$

is the closed interval $[-2, 1]$. ■

EXAMPLE 3. Solve:

$$x^2 + 25 > 10x$$

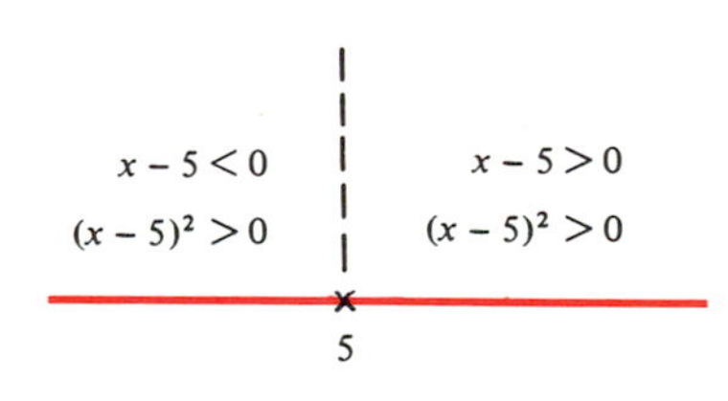

FIGURE 14.19

SOLUTION. Transform this as follows:

$$x^2 + 25 > 10x$$

$$x^2 - 10x + 25 > 0$$

$$(x - 5)^2 > 0$$

Next, $(x - 5)^2 = 0$ only when $x - 5 = 0$, that is, when $x = 5$. Thus there are only two regions in Figure 14.19. Observe that in both regions $(x - 5)^2 > 0$. The solution set of the given inequality consists of all real numbers other than 5.

■

You can solve certain inequalities by transforming them into quadratic inequalities.

EXAMPLE 4. Solve:

$$\frac{3}{x} < 5$$

SOLUTION. Note that $x \neq 0$ (because division by x is assumed). If you multiply both sides by x, you will have to consider two separate cases, depending on whether x is positive or negative. For, multiplication by a *negative* number *reverses* the sense of inequality. To avoid this difficulty, multiply both sides by x^2, which is *positive* (because $x \neq 0$).

$$\frac{3}{x} < 5$$

$$\frac{3}{x} \cdot x^2 < 5 \cdot x^2$$

$$3x < 5x^2$$

$$0 < 5x^2 - 3x$$

$$0 < x(5x - 3)$$

Note that $5x - 3 = 0$ when $x = \frac{3}{5}$. Figure 14.20 indicates that the solution set is $(-\infty, 0) \cup \left(\frac{3}{5}, \infty\right)$. ■

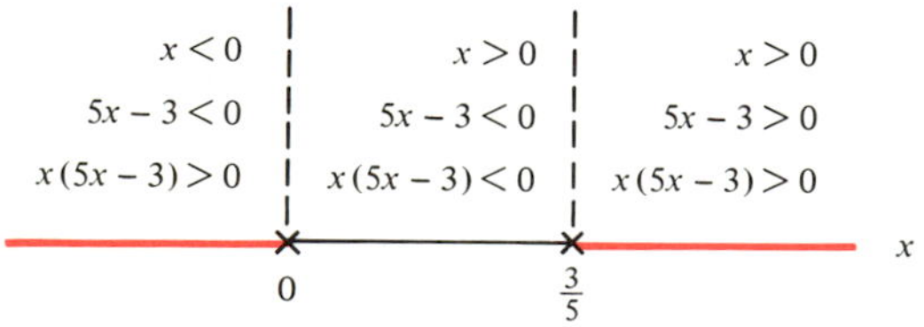

$x < 0$	$x > 0$	$x > 0$
$5x - 3 < 0$	$5x - 3 < 0$	$5x - 3 > 0$
$x(5x - 3) > 0$	$x(5x - 3) < 0$	$x(5x - 3) > 0$

$0 \qquad \frac{3}{5} \qquad x$

FIGURE 14.20

EXAMPLE 5. Solve:

$$\frac{x + 3}{x + 2} \geq 2$$

SOLUTION. Here $x \neq -2$ because division by $x + 2$ is assumed. If you multiply both sides by $x + 2$, you have to consider two cases, according to whether $x + 2 > 0$ or $x + 2 < 0$. Instead, multiply both sides by $(x + 2)^2$, which is positive.

$$\frac{x+3}{x+2}\cdot(x+2)^2 \geq 2\cdot(x+2)^2$$

$$(x+3)(x+2) \geq 2(x^2+4x+4)$$

$$x^2+5x+6 \geq 2x^2+8x+8$$

$$0 \geq x^2+3x+2$$

$$0 \geq (x+2)(x+1)$$

or

$$(x+2)(x+1) \leq 0$$

Note that $(x+2)(x+1) \leq 0$ when $x = -2$ and when $x = -1$. Figure 14.21 indicates that the solution set is the closed interval $[-2, -1]$. ■

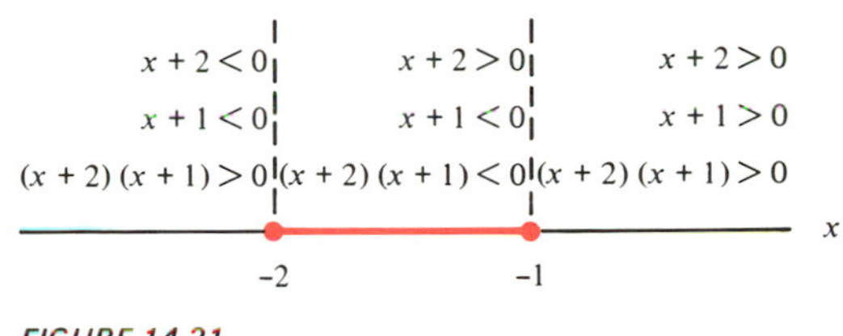

FIGURE 14.21

EXERCISES

In Exercises 1–40 solve each inequality.

1. $(x+7)(x-3) < 0$
2. $(x-5)(x-9) > 0$
3. $x^2 - 5x < 0$
4. $x^2 > 7x$
5. $t^2 - 4t \leq 0$
6. $y^2 + 10y \geq 0$
7. $x^2 + 2x + 1 < 0$
8. $x^2 - 2x > -1$
9. $x^2 + 4x + 4 \geq 0$
10. $x^2 + 9 \leq 6x$
11. $z^2 + 4z + 3 > 0$
12. $x^2 + 5x + 6 \leq 0$
13. $x^2 - 7x + 10 < 0$
14. $x^2 + 11x + 24 > 0$
15. $u^2 - 4u - 12 \leq 0$
16. $x^2 + 2x > 15$
17. $2x^2 + 3x + 1 < 0$
18. $3x^2 + 7x + 2 > 0$
19. $4t^2 - 8t + 3 > 0$
20. $25x^2 - 1 \leq 0$
21. $\frac{3}{x} < 1$
22. $\frac{-2}{x} < 3$
23. $\frac{4}{t} > -1$
24. $1 - \frac{5}{z} \geq \frac{1}{2}$
25. $\frac{1}{x+1} < 1$
26. $\frac{1}{x+5} > 1$
27. $\frac{1}{x+4} \leq 1$
28. $\frac{1}{x-2} < 1$
29. $\frac{x}{x+2} < 1$
30. $\frac{x}{x-3} \geq 1$
31. $\frac{x}{x+5} < 2$
32. $\frac{5x}{x+1} > 2$
33. $\frac{t+3}{t+2} < 1$
34. $\frac{u+2}{u+9} \geq 1$
35. $\frac{x+5}{x+2} < 2$
36. $\frac{x+1}{x-1} > 2$

37. $\dfrac{x+2}{x+4} < 3$ 38. $\dfrac{x+2}{x-1} < 2$ 39. $(x+2)^2 + 4 > 0$

40. $(x-3)^2 + 5 < 0$

41. Find the set of all real numbers whose square is less than twice the number.

42. Find the set of all real numbers whose square is at most three times the number.

14.6 ABSOLUTE VALUE IN EQUATIONS

Recall that the **absolute value of a number** n, written

$$|n|,$$

is its distance from the origin on the real line. Thus

$$|n| = \begin{cases} n, & \text{if } n \geq 0 \\ -n, & \text{if } n < 0 \end{cases}$$

Also,

$$|n| \geq 0 \text{ for every number } n$$

[See Figures 14.22 and 14.23.]

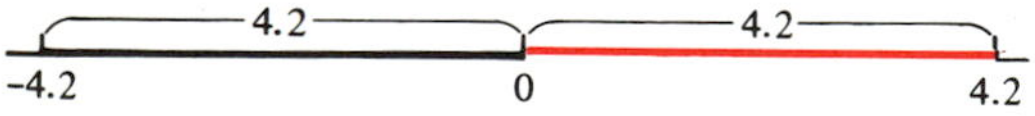

4.2 4.2

-4.2 0 4.2

FIGURE 14.22 4.2 is 4.2 units to the right of the origin. −4.2 is 4.2 units to the left of the origin. Therefore $|4.2| = |-4.2| = 4.2$

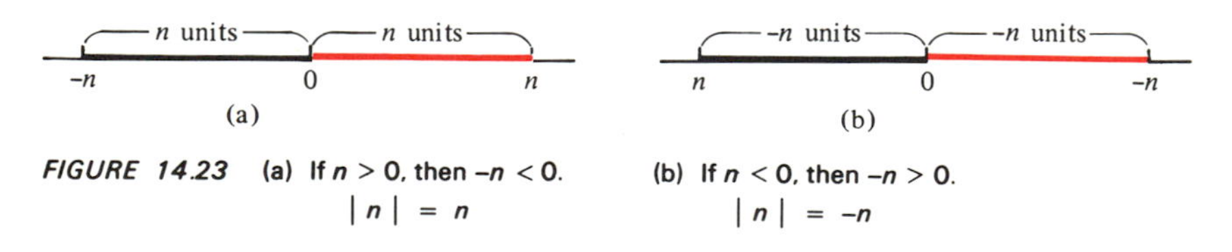

FIGURE 14.23 (a) If $n > 0$, then $-n < 0$. $|n| = n$ (b) If $n < 0$, then $-n > 0$. $|n| = -n$

Clearly,

$$|n - 0| = |n|$$

Thus

$$|n - 0| \text{ is the distance from } n \text{ to } 0 \text{ (the origin).}$$

[See Figure 14.23.]

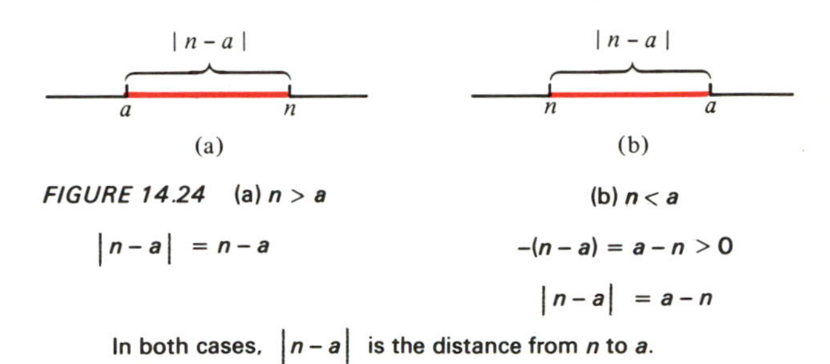

FIGURE 14.24 (a) $n > a$ (b) $n < a$

$|n - a| = n - a$ $\qquad$ $-(n-a) = a - n > 0$

$|n - a| = a - n$

In both cases, $|n - a|$ is the distance from n to a.

In general,

$|n - a|$ *is the distance from n to a.*

[See Figure 14.24.]

EXAMPLE 1.

(a) $|8 - 3|$ is the distance from 8 to 3.

$$|8 - 3| = 5$$

(b) $|-2 - 3|$ is the distance from -2 to 3.

$$|-2 - 3| = |-5| = 5$$

[See Figure 14.25.]

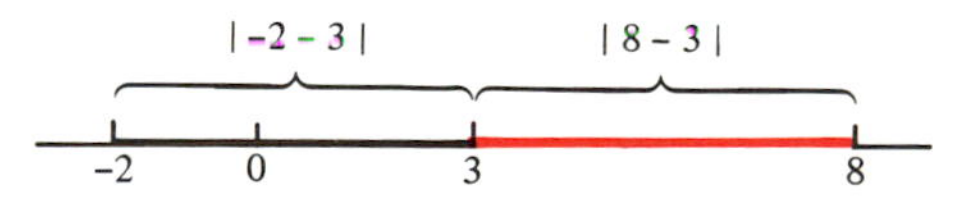

FIGURE 14.25 −2 and 8 are each 5 units from 3.

There is a systematic way of solving equations involving absolute value. From the formula

$$|x| = \begin{cases} x, & \text{if } x \geq 0, \\ -x, & \text{if } x < 0, \end{cases}$$

it follows that the equation

$$|x| = c, \qquad \text{where } c \geq 0,$$

has two solutions, given by

$$x = c, \qquad x = -c.$$

As usual, write these as

$$x = \pm c.$$

EXAMPLE 2. Solve:

$$|x| = 9$$

SOLUTION.

$$x = \pm 9$$ ■

Similarly, the equation

$$|x - a| = c, \qquad c \geq 0,$$

is transformed into

$$x - a = \pm c.$$

Thus

$$x = a \pm c$$ ■

EXAMPLE 3. Solve:

$$|x - 3| = 5$$

SOLUTION.

$$x - 3 = \pm 5$$

$$x = 3 \pm 5$$

$$x = 3 + 5 = 8 \qquad \bigg| \qquad x = 3 - 5 = -2$$

The roots are 8 and -2. ■

EXAMPLE 4. Solve:

$$|x + 4| = 4$$

SOLUTION.

$$|x - (-4)| = 4$$

$$x - (-4) = \pm 4$$

$$x = -4 \pm 4$$

$$x = -4 + 4 = 0 \qquad \bigg| \qquad x = -4 - 4 = -8$$

The roots are 0 and -8.

CHECK.

$\|0 + 4\| \stackrel{?}{=} \|4\|$	$\|-8 + 4\| \stackrel{?}{=} 4$
$4 \stackrel{\checkmark}{=} 4$	$\|-4\| \stackrel{?}{=} 4$
	$4 \stackrel{\checkmark}{=} 4$

■

EXAMPLE 5. Solve:

$$|2x + 1| = 6$$

SOLUTION.

$$|2x - (-1)| = 6$$
$$2x - (-1) = \pm 6$$
$$2x = -1 \pm 6$$
$$x = -\frac{1}{2} \pm 3$$

The roots are $\frac{5}{2}$ and $\frac{-7}{2}$. ■

EXAMPLE 6. Solve:

$$|x - 1| = -1$$

SOLUTION. $|n| \geq 0$. The left side of the given equation is *nonnegative*, whereas the right side is *negative*. There is *no* solution. ■

In an equation such as

$$|x - 4| = 2x,$$

in which a *variable* x appears without absolute value, *you must check the roots* you obtain by the above method. Because the left side is the absolute value of a number, the right side, too, must be at least 0.

EXAMPLE 7. Solve:

$$|x - 4| = 2x$$

SOLUTION. $x - 4 = \pm 2x$

$x - 4 = 2x$	$x - 4 = -2x$
$-4 = x$	$3x = 4$
(False, because $x \geq 0$)	$x = \frac{4}{3}$

CHECK.

$$\left|\frac{4}{3}-4\right| \stackrel{?}{=} 2\cdot\frac{4}{3}$$

$$\left|\frac{4}{3}-\frac{12}{3}\right| \stackrel{?}{=} \frac{8}{3}$$

$$\left|-\frac{8}{3}\right| \stackrel{?}{=} \frac{8}{3}$$

$$\frac{8}{3} \stackrel{\checkmark}{=} \frac{8}{3} \quad \text{(True)}$$

Thus $\frac{8}{3}$ is the only root. ■

EXAMPLE 8. Solve:

$$|x-3| = \frac{x}{2}$$

SOLUTION.

$$x-3 = \pm\frac{x}{2}$$

$x-3=\frac{x}{2}$	$x-3=-\frac{x}{2}$
$2x-6=x$	$2x-6=-x$
$x=6$	$3x=6$
	$x=2$
CHECK.	
$\lvert 6-3\rvert \stackrel{?}{=} \frac{6}{2}$	$\lvert 2-3\rvert \stackrel{?}{=} \frac{2}{2}$
$\lvert 3\rvert \stackrel{?}{=} 3$	$\lvert -1\rvert \stackrel{?}{=} 1$
$3=3$ (True)	$1=1$ (True)

Both 2 and 6 are roots of the given equation. ■

Finally, if

$$|a| = |b|,$$

then by considering all combinations of signs for a and b, you will see that

$$a = \pm b.$$

EXAMPLE 9. Solve:

$$|2x - 3| = |6x|$$

SOLUTION. $2x - 3 = \pm 6x$

$2x - 3 = 6x$	$2x - 3 = -6x$
$-3 = 4x$	$8x = 3$
$\frac{-3}{4} = x$	$x = \frac{3}{8}$

Thus $\frac{-3}{4}$ and $\frac{3}{8}$ are the roots. ■

EXERCISES

1. Which numbers are 2 units from 5?
2. Which numbers are 5 units from 2?
3. Which numbers are 3 units from 3?
4. Which numbers are 7 units from 1?
5. Which numbers are 4 units from -1?
6. Which numbers are 3 units from -5?
7. Which numbers are $\frac{1}{2}$ unit from 2?
8. Which numbers are $\frac{3}{2}$ units from -1?
9. Which numbers are $\frac{3}{4}$ unit from $\frac{1}{2}$?
10. Which numbers are 1.7 units from -1.2?

In Exercises 11–46 solve each equation. Check the ones so indicated.

11. $|x| = 7$
12. $|x| = \frac{1}{4}$
13. $|x| = 5.3$
14. $|x| = 0$
15. $|x| = -2$
16. $|3x| = 6$
17. $|-2x| = 4$
18. $|-4x| = -4$
19. $|x - 1| = 6$ (Check.)
20. $|x - 5| = 10$
21. $|x - 4| = 3$
22. $|x - 3| = 0$
23. $|x + 2| = 4$
24. $|x + 10| = 1$
25. $|2x - 1| = 9$ (Check.)

26. $|3x + 1| = 4$
27. $\left|\frac{x}{2} - 5\right| = 3$
28. $\left|\frac{x - 5}{2}\right| = 3$
29. $|5x + 1| = 0$
30. $|4x - 3| = -3$
31. $|2x + 1| = 4x$ (Check.)
32. $|3x - 2| = 2x - 3$ (Check.)
33. $|x + 1| = 5 - 2x$ (Check.)
34. $|9 - x| = 2x + 1$ (Check.)
35. $|x + 5| = 4x + 1$ (Check.)
36. $|6x - 1| = 9 - 3x$ (Check.)
37. $|x + 1| = x$ (Check.)
38. $|x + 1| = 2x$ (Check.)
39. $|x| = x$
40. $\left|\frac{1 - 6x}{2}\right| = x + 2$ (Check.)
41. $|2x| = |x + 1|$
42. $|3x - 1| = |1 + 5x|$ (Check.)
43. $|9 - x| = |6x|$
44. $|x + 4| = |8 - 2x|$
45. $|x + 3| = |x - 2|$
46. $|x + 4| = |1 - x|$

14.7 ABSOLUTE VALUE IN INEQUALITIES

As you know, $|x|$, or $|x - 0|$, is the distance from x to the origin. Thus $|x| < 4$ if its distance from the origin is less than 4. [See Figure 14.26.] The distance from 3.5 to 0 is 3.5, which is less than 4. Also, the distance from -3 to 0 is 3, which is less than 4.

$$|-3 - 0| = |-3| = 3$$

On the other hand, the distance from 5 to 0 is 5, which is greater than 4. And the distance from -6 to 0 is 6, which is greater than 4.

$$|-6 - 0| = |-6| = 6$$

Thus 3.5 and -3 are both solutions of $|x| < 4$. But 5 and -6 are not solutions of $|x| < 4$. In fact, the solutions of $|x| < 4$ are the numbers x with

$$-4 < x < 4.$$

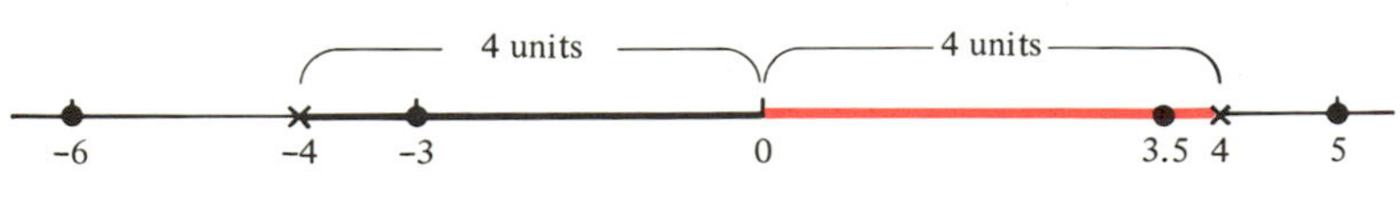

FIGURE 14.26 $|x| < 4$ means the same as $-4 < x < 4$.

Let $c > 0$. In general,

$$|x| < c \quad \text{means the same as} \quad -c < x < c.$$

[See Figure 14.27 on page 489.]

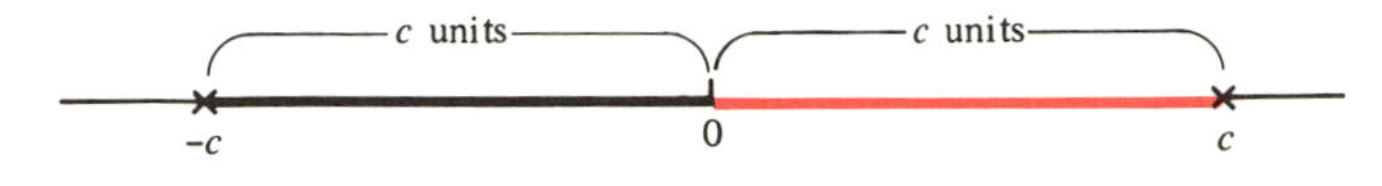

FIGURE 14.27 Let $c > 0$. Then $|x| < c$ means the same as $-c < x < c$.

EXAMPLE 1. Solve:

$$|x| < 3$$

SOLUTION.

$$|x| < 3 \qquad \text{means} \qquad -3 < x < 3.$$

The solution set is the *open* interval $(-3, 3)$. ■

Next, observe that $|x - 2|$ is the distance from x to 2 Thus

$$|x - 2| < 1$$

means that x lies within 1 unit of 2. [See Figure 14.28.] The distance from 2.8 to 2 is .8, which is less than 1. The distance from 1.5 to 2 is .5, which is less than 1.

$$|1.5 - 2| = |-.5| = .5$$

On the other hand, the distance between 3.1 to 2 is 1.1, which is greater than 1. And the distance from 0 to 2 is 2, which is greater than 1.

$$|0 - 2| = |-2| = 2$$

Thus 2.8 and 1.5 are solutions of $|x - 2| < 1$, but 3.1 and 0 are not. The solutions of $|x - 2| < 1$ are the numbers x with

$$2 - 1 < x < 2 + 1$$

or

$$1 < x < 3.$$

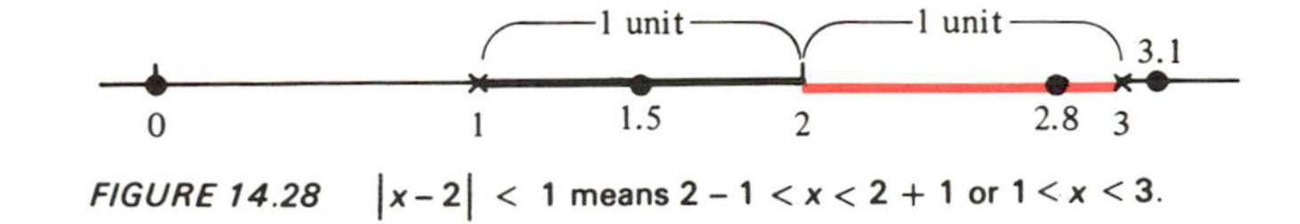

FIGURE 14.28 $|x - 2| < 1$ means $2 - 1 < x < 2 + 1$ or $1 < x < 3$.

For any number a, $|x - a|$ is the distance from x to a. Let $c > 0$. Then

$$|x - a| < c$$

means the same as

$$-c < x - a < c.$$

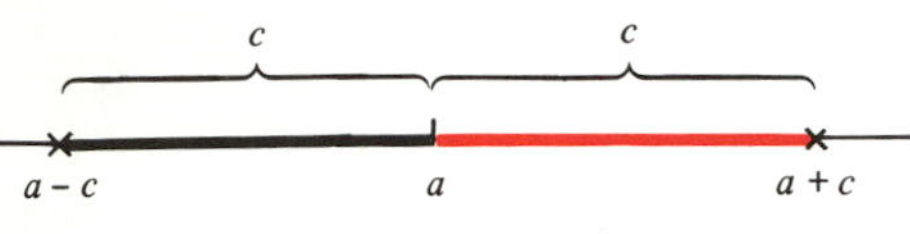

FIGURE 14.29 Let $c > 0$. Then $|x - a| < c$ means the same as $a - c < x < a + c$.

Add a to each expression and observe that this is equivalent to

$$a - c < x < a + c.$$

EXAMPLE 2. Solve:

$$|x - 6| < 4$$

SOLUTION.

$$-4 < x - 6 < 4$$

Add 6 to all three expressions.

$$2 < x < 10$$

The solution set is the *open* interval (2, 10). ■

Clearly

$$|x| \leq c \quad \text{means the same as} \quad -c \leq x \leq c,$$

and

$$|x - a| \leq c \quad \text{means the same as} \quad a - c \leq x \leq a + c.$$

EXAMPLE 3. Solve:

$$|5x - 2| \leq 18$$

SOLUTION.

$$-18 \leq 5x - 2 \leq 18$$

$$-16 \leq 5x \leq 20$$

$$\frac{-16}{5} \leq x \leq 4$$

The solution set is the *closed* interval $\left[\frac{-16}{5}, 4\right]$. ■

Multiplication by a negative number reverses the sense of an inequality. When two inequalities are involved, both senses are reversed.

EXAMPLE 4. Solve:

$$|3 - 2x| < 5$$

SOLUTION.

$$-5 < 3 - 2x < 5$$

$$\underline{-3} \quad \triangle\!\!-\!\!\triangle \quad \underline{-3 \qquad\quad} \quad \triangle\!\!-\!\!\triangle \quad \underline{-3}$$

$$-8 < -2x < 2$$

Multiply each expression by $\frac{-1}{2}$.

$$4 > x > -1$$

or

$$-1 < x < 4$$

The solution set is the *open* interval $(-1, 4)$. ■

Next observe that

$$|x| > 5$$

means that the distance from x to the origin is greater than 5. Also, if $c > 0$, then

$$|x| > c$$

means the distance from x to the origin is greater than c. As you see in Figure 14.30, if $|x| > c$, then

$$x < -c \quad \text{or} \quad x > c.$$

Thus x lies in $(-\infty, -c) \cup (c, \infty)$.

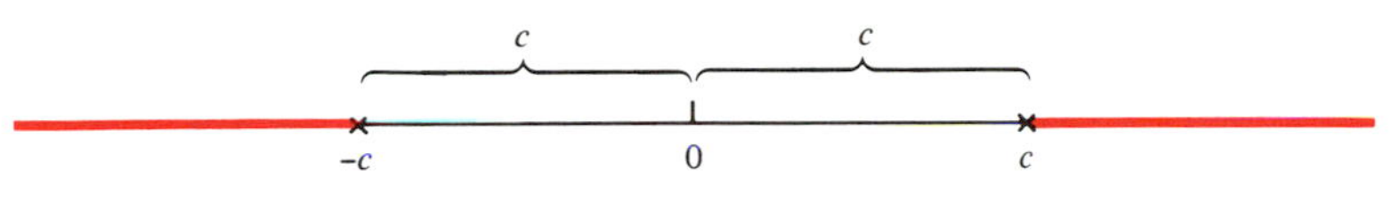

FIGURE 14.30 Let $c > 0$. Then $|x| > c$ means that x lies in $(-\infty, -c) \cup (c, \infty)$.

Similarly,

$$|x - a| > c$$

means that the distance from x to a is greater than c, and

$$\text{either} \quad x - a < -c \quad \text{or} \quad x - a > c$$

that is,

$$x < a - c \quad \text{or} \quad x > a + c$$

Thus if $|x - a| > c$, then x lies in $(-\infty, a - c) \cup (a + c, \infty)$. [See Figure 14.31.]

FIGURE 14.31 Let $c > 0$. Then $|x - a| > c$ means that x lies in $(-\infty, a - c) \cup (a + c, \infty)$.

EXAMPLE 5. Solve:

$$|x - 4| > 10$$

SOLUTION.

$$x - 4 < -10 \quad \text{or} \quad x - 4 > 10$$

$$x < -6 \quad \text{or} \quad x > 14$$

The solution set is $(-\infty, -6) \cup (14, \infty)$. ■

Clearly,

$$|x| \geq c \quad \text{means that } x \text{ lies in} \quad (-\infty, -c] \cup [c, \infty),$$

and

$$|x - a| \geq c \quad \text{means that } x \text{ lies in} \quad (-\infty, a - c] \cup [a + c, \infty).$$

EXAMPLE 6. Solve:

$$|2x + 5| \geq 9$$

SOLUTION.

$$\begin{array}{rcl} 2x + 5 \leq -9 & \text{or} & 2x + 5 \geq 9 \\ 2x \leq -14 & & 2x \geq 4 \\ x \leq -7 & & x \geq 2 \end{array}$$

The solution set is $(-\infty, -7] \cup [2, \infty)$. ■

Finally, observe that for $a > 0$, the inequality

$$x^2 < a^2 \quad \text{is equivalent to} \quad |x| < a, \quad \text{and hence to} \quad -a < x < a.$$

[See Figure 14.32.]

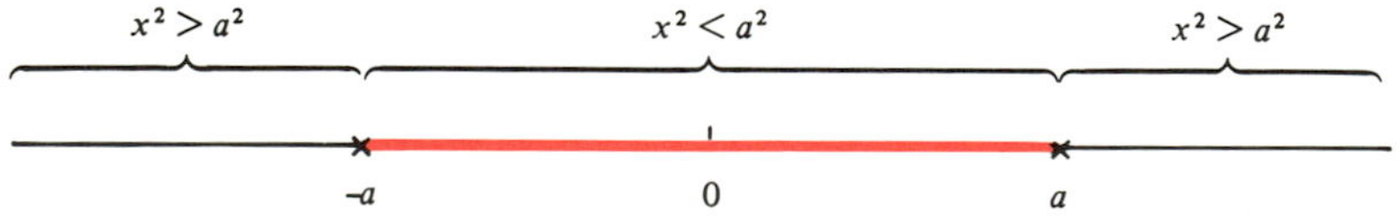

$x^2 > a^2$ $x^2 < a^2$ $x^2 > a^2$

$-a$ 0 a

FIGURE 14.32 Let $a > 0$. Then $x^2 < a^2$ is equivalent to $|x| < a$ and hence to $-a < x < a$.

For, if

$$|x| < a$$

then

$$x^2 = |x|^2 < a^2,$$

whereas if

$$|x| \geq a,$$

then

$$x^2 = |x|^2 \geq a^2$$ ■

Similarly,

$$x^2 \leq a^2, \quad |x| \leq a, \quad \text{and} \quad -a \leq x \leq a$$

are all equivalent.

EXAMPLE 7. If $x^2 < 9$, then

$$-3 < x < 3$$

■

EXERCISES

In Exercises 1–36 solve each inequality.

1. $|x| < 10$
2. $|x| < \frac{1}{2}$
3. $|x| \leq 4$
4. $|x| \leq \frac{5}{3}$
5. $|x - 1| < 7$
6. $|x - 2| < 4$
7. $|x + 3| < 8$
8. $\left|x + \frac{1}{2}\right| < \frac{1}{2}$
9. $\left|x - \frac{1}{2}\right| < 4$
10. $\left|x + \frac{3}{4}\right| < \frac{1}{2}$
11. $|2x - 1| < 5$
12. $|3x - 2| < 7$
13. $|5x + 1| < 6$
14. $\left|\frac{x + 3}{2}\right| < 4$
15. $|2x + 1| < 0$
16. $|3x + 1| \leq 0$
17. $\left|\frac{3x - 1}{4}\right| < 2$
18. $\left|\frac{x}{2} - \frac{1}{3}\right| < \frac{1}{6}$
19. $|x| > 5$
20. $|x| > \frac{1}{3}$
21. $|x| \geq 40$
22. $|y| \geq \frac{3}{4}$
23. $1 + |y| \geq 2$
24. $|y| \geq -2$
25. $|z - 1| > 6$
26. $|t + 2| \geq 2$
27. $|3x + 1| > 4$
28. $\left|\frac{x}{2} + 1\right| \geq 11$
29. $|4x - 3| > 15$
30. $|2x + 5| > 3$
31. $\left|\frac{x}{2} - \frac{4}{3}\right| > \frac{1}{3}$
32. $\left|\frac{5x}{8} - \frac{3}{4}\right| \geq \frac{1}{2}$
33. $|5x - 3| \geq 0$
34. $|5x - 3| > 0$
35. $|5x - 3| > -1$
36. $|5x - 3| > 1$

In Exercises 37–52 determine c.

37. $-5 < x < 5$ means the same as $|x| < c$.
38. $-2 \leq x \leq 2$ means the same as $|x| \leq c$.
39. x is in $(-\infty, -9) \cup (9, \infty)$ means the same as $|x| > c$.
40. Either $x \leq -1$ or $x \geq 1$ means the same as $|x| \geq c$.
41. $-4 < x < 6$ means the same as $|x - 1| < c$.
42. $-7 < x < 3$ means the same as $|x + 2| < c$.
43. $0 \leq x \leq 8$ means the same as $|x - c| \leq 4$.
44. $-9 < x < -5$ means the same as $|x + c| < 2$.
45. $-8 < x < -4$ means the same as $|x - c| < 2$.
46. $-2 < x < 3$ means that $|x - c| < \frac{5}{2}$.

47. $x^2 < 25$ means the same as $|x| < c$.
48. $x^2 \leq 100$ means the same as $|x| \leq c$.
49. $x^2 < \frac{1}{4}$ means the same as $-c < x < c$.
50. $x^2 \geq 1$ means the same as $|x| \geq c$.
51. $-6 < x < 6$ means the same as $x^2 < c$.
52. $-.1 \leq x \leq .1$ means the same as $x^2 \leq c$.

14.8 PROPERTIES OF ABSOLUTE VALUE

For every number n, $|n|$ has been defined as the distance from n to the origin. Moreover,

$$|n| = \begin{cases} n, & \text{if } n \geq 0 \\ -n, & \text{if } n < 0 \end{cases}$$

Clearly,

$$|n|^2 = |-n|^2 = n^2 = (-n)^2$$

Consequently,

$$|n| = |-n| = \sqrt{n^2} = \sqrt{(-n)^2}$$

EXAMPLE 1.

(a) Let $n = -5$. Then $|-5|^2 = 5^2 = (-5)^2 = 25$
(b) Let $n = 5$. Then $|5| = |-5| = \sqrt{5^2} = \sqrt{(-5)^2} = 5$ ■

Consider the **absolute value function,** defined by $f(x) = |x|$ for all real numbers x. Thus

$$f(x) = \begin{cases} x, & \text{if } x \geq 0 \\ -x, & \text{if } x < 0 \end{cases}$$

x	-4	-3	-2	-1	0	1	2	3	4
$f(x) = \|x\|$	4	3	2	1	0	1	2	3	4

TABLE *14.1.*

The graph consists of the line

$$y = x, \qquad \text{for } x \geq 0$$

and the line

$$y = -x, \qquad \text{for } x < 0.$$

Thus the graph consists of two straight lines meeting at the origin at a right angle. Observe that the absolute value function is not one–one. The same function value can correspond to different arguments. For example,

$$f(1) = f(-1) = 1$$

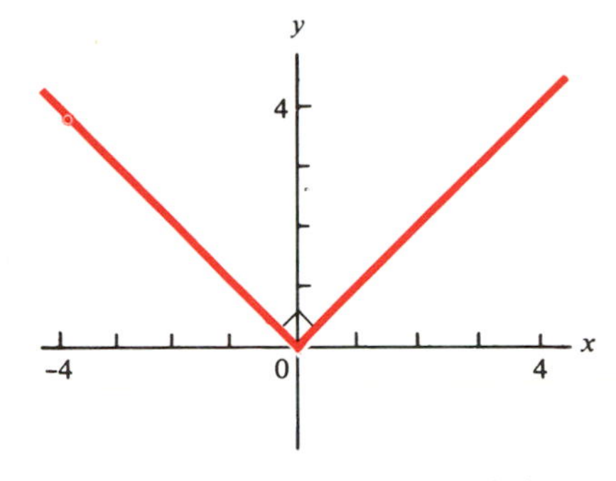

FIGURE 14.33 The graph of $f(x) = |x|$

EXAMPLE 2. Graph the function given by

$$f(x) = |x + 2|.$$

x	−5	−4	−3	−2	−1	0	1	2	3
$x+2$	−3	−2	−1	0	1	2	3	4	5
$f(x) = \|x+2\|$	3	2	1	0	1	2	3	4	5

TABLE *14.2.*

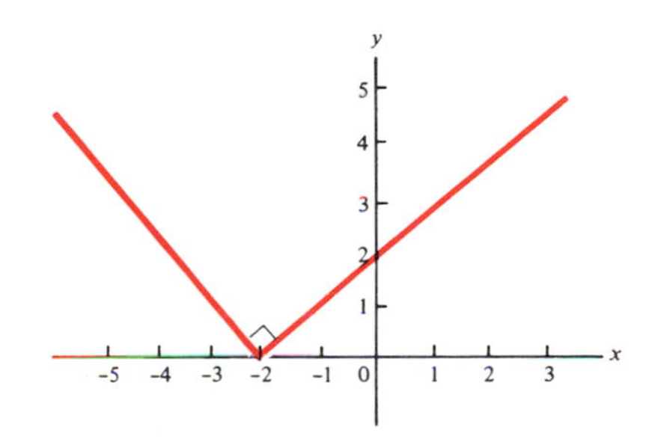

FIGURE 14.34 The graph of $f(x) = |x + 2|$

EXAMPLE 3. Graph the function given by

$$f(x) = |x| + 2.$$

SOLUTION.

x	−4	−3	−2	−1	0	1	2	3	4
$\|x\|$	4	3	2	1	0	1	2	3	4
$f(x) = \|x\| + 2$	6	5	4	3	2	3	4	5	6

TABLE *14.3.*

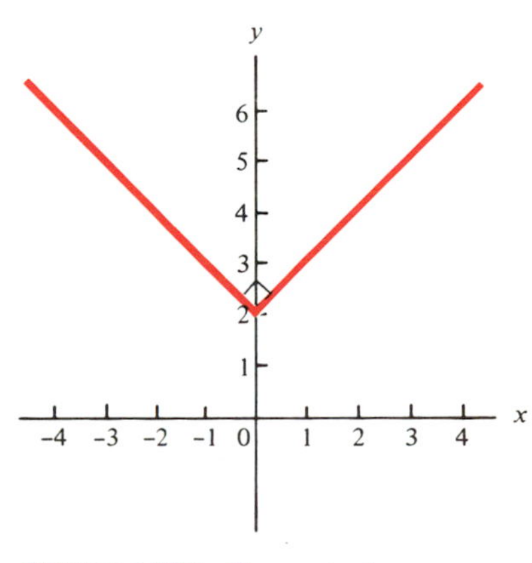

FIGURE 14.35 The graph of $f(x) = |x| + 2$

Note that

$|x + 2|$ means first add 2 to x; then take the absolute value. (See Example 2.)
$|x| + 2$ means first take the absolute value of x; then add 2 to this. (See Example 3.)

The absolute value of a product (or *quotient*) *equals the product* (or *quotient*) *of the absolute values.*

Thus let a and b be arbitrary real numbers.

$$|ab| = |a|\ |b|$$

and

$$\left|\frac{a}{b}\right| = \frac{|a|}{|b|}, \qquad \text{if } b \neq 0$$

EXAMPLE 4.

(a) Let $a = 5, b = -3$.

$$|ab| = |5(-3)| = |-15| = 15$$
$$|a|\ |b| = |5|\ |-3| = 5 \cdot 3 = 15$$

(b) Let $a = -9, b = -2$.

$$|ab| = |(-9)(-2)| = |18| = 18$$
$$|a|\ |b| = |-9|\ |-2| = 9 \cdot 2 = 18$$

(c) Let $a = -3, b = 4$.

$$\left|\frac{a}{b}\right| = \left|\frac{-3}{4}\right| = \frac{3}{4}, \qquad \frac{|a|}{|b|} = \frac{|-3|}{|4|} = \frac{3}{4}$$

■

EXAMPLE 5. Solve:

$$\left|\frac{3x}{x + 1}\right| = 2$$

SOLUTION. Use $\left|\frac{a}{b}\right| = \frac{|a|}{|b|}$.

$$\frac{|3x|}{|x + 1|} = 2$$

Multiply both sides by $|x + 1|$, which is positive.

$$|3x| = 2|x + 1|$$
$$3x = \pm 2(x + 1)$$

$3x = 2(x + 1)$	$3x = -2(x + 1)$
$3x = 2x + 2$	$3x = -2x - 2$
$x = 2$	$5x = -2$
	$x = \frac{-2}{5}$

■

The absolute value of a sum or difference is at most the sum of the absolute values:

$$|a + b| \leq |a| + |b|$$

$$|a - b| \leq |a| + |b|$$

These inequalities can be combined:

$$|a \pm b| \leq |a| + |b|$$

The signs of the numbers a and b determine which of the symbols

$<$ or $=$

holds in each case.

EXAMPLE 6.

(a) Let $a = 7, b = 5$.

$$|a + b| = \underbrace{|7 + 5|}_{12} = \underbrace{|7| + |5|}_{12} = |a| + |b|$$

Thus here, $|a + b| = |a| + |b|$

(b) Let $a = -1, b = 8$.

$$|a + b| = \underbrace{|-1 + 8|}_{7} < \underbrace{|-1| + |8|}_{9} = |a| + |b|$$

Thus here, $|a + b| < |a| + |b|$

(c) Let $a = 6, b = 3$.

$$|a - b| = \underbrace{|6 - 3|}_{3} < \underbrace{|6| + |3|}_{9} = |a| + |b|$$

Thus here, $|a - b| < |a| + |b|$

(d) Let $a = 6, b = -3$.

$$|a - b| = \underbrace{|6 - (-3)|}_{9} = \underbrace{|6| + |-3|}_{9} = |a| + |b|$$

Thus here, $|a - b| = |a| - |b|$ ■

EXERCISES

In Exercises 1–4 find the value of each sum or difference.

1. $|9| + |-9|$
2. $|-\pi| - |\pi|$
3. $\sqrt{(6.31)^2} - |-6.31|$
4. $|-4|^2 - \sqrt{(-4)^2}$

In Exercises 5–10 find the indicated function values.

5. Let $f(x) = |x + 5|$.
   (a) $f(0)$ (b) $f(5)$ (c) $f(-2)$ (d) $f(-5)$ (e) $f(-6)$
6. Let $f(x) = |x - 1|$.
   (a) $f(0)$ (b) $f(1)$ (c) $f(10)$ (d) $f(-1)$ (e) $f(-2)$
7. Let $g(x) = |2x|$.
   (a) $g(0)$ (b) $g(2)$ (c) $g(5)$ (d) $g(-2)$ (e) $g(-5)$
8. Let $f(t) = |2t + 1|$.
   (a) $f(0)$ (b) $f(1)$ (c) $f(3)$ (d) $f(-1)$ (e) $f\left(\frac{-1}{2}\right)$
9. Let $g(x) = \frac{|x|}{3}$.
   (a) $g(3)$ (b) $g(12)$ (c) $g(2)$ (d) $g(-3)$ (e) $g(-5)$
10. Let $f(x) = |x| + 1$.
   (a) $f(0)$ (b) $f(1)$ (c) $f(4)$ (d) $f(-1)$ (e) $f(-4)$

In Exercises 11–22 graph each function. Unless otherwise indicated, assume the domain is the set of all real numbers.

11. $f(x) = |x|$
12. $f(x) = |x|, \quad -2 < x < 3$
13. $f(x) = |x + 1|$
14. $f(x) = |x - 2|$
15. $f(x) = |x| + \frac{1}{2}$
16. $f(x) = |x| + 3$
17. $f(x) = |x| - 1$
18. $f(x) = |x| + \frac{1}{2}, \; -\frac{1}{2} \leq x \leq \frac{3}{2}$
19. $f(x) = 2|x|$
20. $f(x) = -2|x|$
21. $f(x) = \left|\frac{x}{2}\right|$
22. $f(x) = |x| - x$

In Exercises 23–32 answer "true" or "false".

23. For all a, $|3a| = 3|a|$
24. For all a, $\left|\frac{a}{10}\right| = \frac{|a|}{10}$
25. For all a, $|a + 6| = |a| + 6$
26. For all a, $|a - 4| = |a| - 4$
27. For all a and b, $|-ab| = |-a|\,|b|$
28. For all a and b, $b \neq 0$, $\left|\frac{-a}{-b}\right| = \frac{|-a|}{|-b|}$
29. For $a > 0$ and $b > 0$, $|a + b| = |a| + |b|$
30. For $a > 0$ and $b > 0$, $|a - b| = |a| + |b|$

31. For $a > 0$ and $b < 0$, $|a - b| = |a| + |b|$

32. For $a < 0$ and $b < 0$, $|a + b| = |a| + |b|$

In Exercises 33–44 determine all roots of each equation.

33. $\left|\dfrac{2x}{x + 2}\right| = 1$ 34. $\left|\dfrac{2x}{x - 1}\right| = 1$ 35. $\left|\dfrac{x}{3x + 5}\right| = 2$

36. $\left|\dfrac{4x + 1}{x - 2}\right| = 3$ 37. $\left|\dfrac{2x + 1}{x - 2}\right| = 4$ 38. $\left|\dfrac{2x + 5}{3x - 1}\right| = 2$

39. $\left|\dfrac{1 - x}{5x - 2}\right| = 2$ 40. $\left|\dfrac{2x + 1}{5 - 2x}\right| = \dfrac{1}{2}$ 41. $\left|\dfrac{3x - 7}{x + 2}\right| = \dfrac{3}{4}$

42. $\left|\dfrac{x + 2}{x - 1}\right| = 1$ 43. $\left|\dfrac{4x - 3}{x + 5}\right| = 0$ 44. $\left|\dfrac{2x - 1}{x + 3}\right| = -1$

REVIEW EXERCISES FOR CHAPTER 14

14.1 INTERVALS

In Exercises 1–4 describe each set of numbers as either

(a) an open interval, (b) a closed interval,
(c) an open right ray, (d) a closed right ray,
(e) an open left ray, or (f) a closed left ray.

1. $[2, 3]$ 2. $(0, \infty)$
3. $(-1, 0)$ 4. $(-\infty, -6]$

5. Answer "true" or "false".

(a) 2 is in $(2, 7)$. (b) -3 is in $[-5, -2]$.

(c) 0 is in $(-\infty, 0)$. (d) $\dfrac{1}{2}$ is in $[-1, 1]$.

6. Describe in one of the forms

$$[a, b],\ (a, b),\ (a, \infty),\ [a, \infty),\ (-\infty, b),\ \text{or}\ (-\infty, b]$$

the set of numbers satisfying each of the following:

(a) $5 \le x \le 7$ (b) $-2 < x < 3$
(c) $x \ge -1$ (d) $x < 3$

14.2 ARITHMETIC OF INEQUALITIES

In Exercises 7–9 fill in "<" or ">".

7. $x + 3 \;\square\; x - 4$

8. If $a < 0$, $-2a \;\square\; -4a$

9. If $a > 0$, $\frac{a}{3} \; \square \; \frac{a}{4}$

10. Show that if $x \leq 5$, then $x - 3 \leq 2$.

14.3 SOLVING INEQUALITIES

In Exercises 11–15 solve each inequality.

11. $5x \leq 15$
12. $x - 2 > 4x + 13$
13. $2(x + 1) + 3(x - 2) \geq 1$
14. $\frac{1 + t}{3} < \frac{t}{2}$
15. $-1 \leq 2t + 1 \leq 7$
16. Suppose $y = 5x - 2, 0 < x < 2$. What is the range of y?

14.4 GRAPHING INEQUALITIES

In Exercises 17–18 indicate which points satisfy the indicated inequality.

17. $y < 1 - x$

    (a) $(0, 1)$ (b) $(1, 0)$ (c) $(-2, 2)$ (d) $(2, -1)$

18. $2x - y \geq 1$

    (a) $(1, 1)$ (b) $(2, 2)$ (c) $(0, 0)$ (d) $(4, 6)$

In Exercises 19–20 graph each inequality.

19. $y > x - 3$
20. $x - y \leq 1$

14.5 QUADRATIC INEQUALITIES

In Exercises 21–24 solve each inequality.

21. $x^2 + 5x + 6 > 0$
22. $16t^2 \leq 1$
23. $\frac{1}{2 - x} < 1$
24. $\frac{x + 4}{x + 1} < 2$

14.6 ABSOLUTE VALUE IN EQUATIONS

25. Which numbers are 6 units from 4?

In Exercises 26–29 solve each equation. Check the ones so indicated.

26. $|4x| = 12$
27. $|x - 7| = 7$
28. $|2 - x| = 3x$ (Check.)
29. $|x + 1| = |x - 3|$ (Check.)

14.7 ABSOLUTE VALUE IN INEQUALITIES

In Exercises 30–32 solve each inequality.

30. $|x - 1| < 4$

31. $|y + 3| \geq 3$

32. $\left|\frac{x}{4} + \frac{1}{2}\right| \leq 0$

In Exercises 33–34 find the value of c.

33. $-3 \leq x \leq 3$ means the same as $|x| \leq c$.

34. $-1 < x < 7$ means the same as $|x - c| < 4$.

14.8 PROPERTIES OF ABSOLUTE VALUE

35. Let $f(x) = |x + 2|$. Determine:
    (a) $f(-2)$ (b) $f(-1)$
    (c) $f(1)$ (d) $f(2)$

36. Let $f(x) = |x| + 2$. Determine:
    (a) $f(-2)$ (b) $f(-1)$
    (c) $f(1)$ (d) $f(2)$

37. Graph the indicated function:

$$f(x) = |x| - 2, \qquad -2 \leq x \leq 2$$

38. Answer "true" or "false".
    (a) For all a and b, $|a + b| \geq |a| + |b|$
    (b) For $a < 0$ and $b > 0$, $|a + b| = |a| + |b|$

39. Determine all roots of the equation

$$\left|\frac{x}{x + 1}\right| = 1.$$

TEST YOURSELF ON CHAPTER 14.

1. Answer "true" or "false".
    (a) 0 is in $(-1, 1)$.
    (b) 2 is in $(-2, 2)$.
    (c) -1 is in $(-\infty, -1]$.

2. Fill in "<" or ">":
    (a) $x - 2 \; \square \; x - 1$
    (b) If $x < 0$, $\frac{x}{2} \; \square \; \frac{x}{3}$

3. Solve the inequality:

$$x + 2 \leq 3x - 6$$

4. Suppose $y = 3x + 1, 1 \leq x \leq 3$. What is the range of y?

5. Solve the inequality:

$$\frac{x - 1}{x + 2} < 1$$

6. Determine all roots of the equation:

$$|2x - 4| = 4$$

7. Solve the inequality:

$$|x + 1| > 5$$

8. Determine c:

$$-2 < x < 4 \qquad \text{means the same as} \qquad |x - c| < 3.$$

9. Let $f(x) = |x + 3|$. Determine:

   (a) $f(-3)$ (b) $f(3)$

10. Determine all roots of:

$$\left|\frac{x}{3}\right| = \frac{1}{2}$$

15

conic sections

15.1 DISTANCE

DEFINITION. The **distance between two points** P_1 and P_2 on the plane is defined to be the length of the line segment $\overline{P_1 P_2}$.

Write

$$\text{dist}(P_1, P_2)$$

for the distance between P_1 and P_2.

Because the length of the segment $\overline{P_1 P_2}$ is the same as that of $\overline{P_2 P_1}$,

$$\text{dist}(P_1, P_2) = \text{dist}(P_2, P_1).$$

First consider distance along horizontal and vertical lines. If the line segment $\overline{P_1 P_2}$ is *horizontal*, P_1 and P_2 have the same y-coordinate. Let $P_1 = (x_1, y_1)$ and $P_2 = (x_2, y_1)$. Consequently,

$$\text{dist}(P_1, P_2) = \textit{length } \overline{P_1 P_2} = |x_2 - x_1|$$

[If P_2 lies to the right of P_1, as in Figure 15.1(a), this length is $x_2 - x_1$. If P_2 lies to the left of P_1, as in Figure 15.1(b), this length is $x_1 - x_2$, which equals $-(x_2 - x_1)$. Note that $|x_2 - x_1|$ covers both cases.]

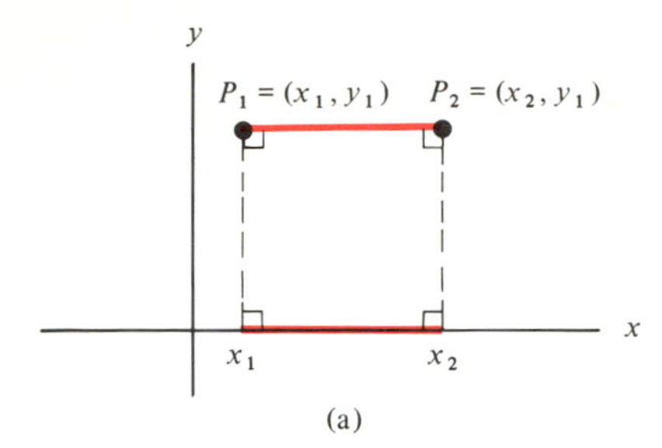

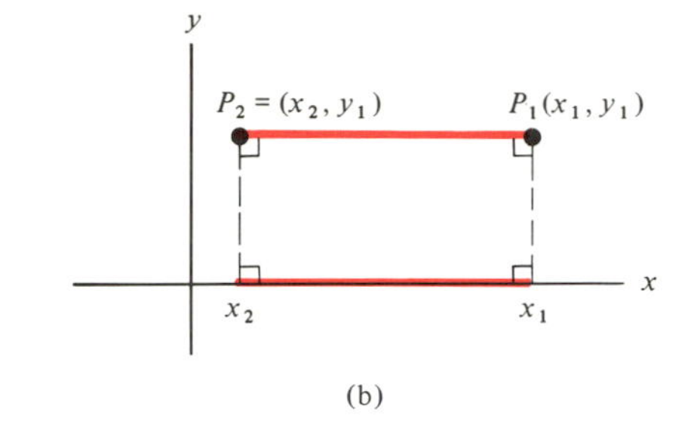

FIGURE 15.1 (a) $\text{dist}(P_1, P_2) = x_2 - x_1 = |x_2 - x_1|$ (b) $\text{dist}(P_1, P_2) = x_1 - x_2 = |x_2 - x_1|$

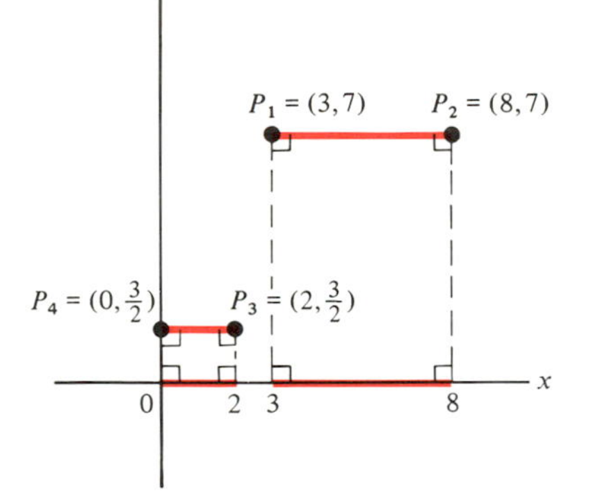

FIGURE 15.2

EXAMPLE 1.

(a) Let $P_1 = (3, 7)$, $P_2 = (8, 7)$. $\overline{P_1P_2}$ is horizontal. Both points have the same *y*-coordinate.

$$\text{dist}(P_1, P_2) = |8 - 3| = 5$$

(b) Let $P_3 = \left(2, \frac{3}{2}\right)$, $P_4 = \left(0, \frac{3}{2}\right)$. Then $\overline{P_3P_4}$ is horizontal and

$$\text{dist}(P_3, P_4) = |0 - 2| = 2.$$

■

If the line segment $\overline{P_1P_2}$ is *vertical*, P_1 and P_2 have the same *x*-coordinate. Let $P_1 = (x_1, y_1)$ and $P_2 = (x_1, y_2)$. Then

$$\text{dist}(P_1, P_2) = \text{length } \overline{P_1P_2} = |y_2 - y_1|$$

[See Figure 15.3.]

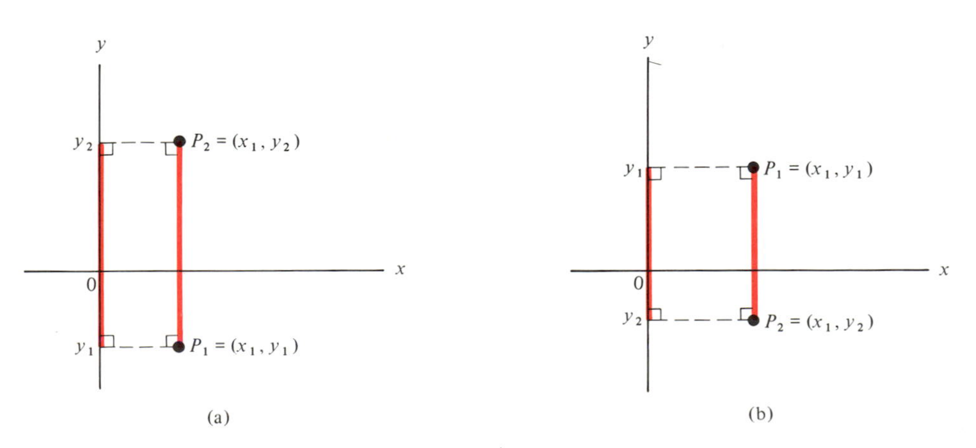

FIGURE 15.3 (a) $\text{dist}(P_1, P_2) = y_2 - y_1 = |y_2 - y_1|$ (b) $\text{dist}(P_1, P_2) = y_1 - y_2 = |y_2 - y_1|$

EXAMPLE 2.

(a) Let $P_1 = (-2, -4)$, $P_2 = (-2, 5)$. Then $\overline{P_1P_2}$ is vertical. Both points have the same x-coordinate.

$$\text{dist}(P_1, P_2) = |5 - (-4)| = 9$$

(b) Let $P_3 = (6, 9)$, $P_4 = (6, 6)$. Then $\overline{P_3P_4}$ is vertical and

$$\text{dist}(P_3, P_4) = |6 - 9| = 3.$$

■

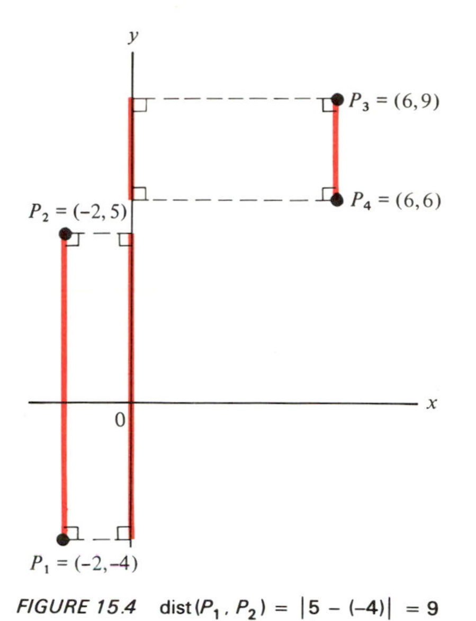

FIGURE 15.4 $\text{dist}(P_1, P_2) = |5 - (-4)| = 9$
$\text{dist}(P_3, P_4) = |6 - 9| = 3$

DEFINITION. The **midpoint** of a line segment $\overline{P_1P_2}$ is the point M on this segment such that

$$\text{dist}(P_1, M) = \text{dist}(P_2, M).$$

Thus, the midpoint divides the line segment into two segments of equal length.

For a *horizontal* line segment $\overline{P_1P_2}$, let $P_1 = (x_1, y_1)$ and $P_2 = (x_2, y_1)$. Then *the midpoint of* $\overline{P_1P_2}$ *is given by*

$$M = \left(\frac{x_1 + x_2}{2}, y_1\right).$$

For,

$$\begin{aligned} \text{dist}(P_1, M) &= \left|\frac{x_1 + x_2}{2} - x_1\right| \\ &= \left|\frac{x_1 + x_2 - 2x_1}{2}\right| \\ &= \frac{|x_2 - x_1|}{2} \end{aligned}$$

Also,

$$\begin{aligned} \text{dist}(P_2, M) &= \left|\frac{x_1 + x_2}{2} - x_2\right| \\ &= \frac{|x_1 + x_2 - 2x_2|}{2} \\ &= \frac{|x_1 - x_2|}{2} \\ &= \frac{|x_2 - x_1|}{2} \end{aligned}$$

Thus

$$\text{dist}(P_1, M) = \text{dist}(P_2, M) \qquad \blacksquare$$

[See Figure 15.5(a).]

Similarly, for a *vertical* line $\overline{P_1P_2}$, let $P_1 = (x_1, y_1)$ and $P_2 = (x_1, y_2)$. Then *the midpoint of* $\overline{P_1P_2}$ *is given by*

$$M = \left(x_1, \frac{y_1 + y_2}{2}\right).$$

[See Figure 15.5(b).]

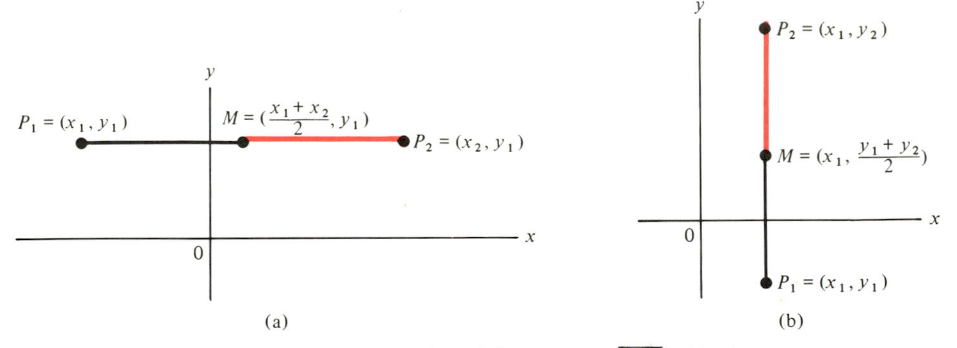

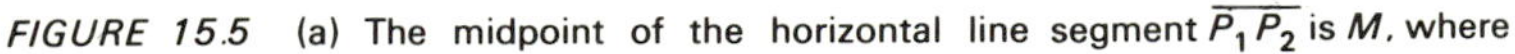

FIGURE 15.5 (a) The midpoint of the horizontal line segment $\overline{P_1P_2}$ is M, where

$$M = \left(\frac{x_1 + x_2}{2}, y_1\right)$$

(b) The midpoint of the vertical line segment $\overline{P_1P_2}$ is M, where

$$M = \left(x_1, \frac{y_1 + y_2}{2}\right)$$

EXAMPLE 3.

(a) Let $P_1 = (2, 4)$ and $P_2 = (8, 4)$. Then $\dfrac{2 + 8}{2} = 5$. Thus the midpoint of $\overline{P_1P_2}$ is $(5, 4)$. [See Figure 15.6(a) on page 507.]

(b) Let $P_3 = (1, -3)$ and $P_4 = (1, 1)$. Then $\dfrac{-3 + 1}{2} = -1$. Thus the midpoint of $\overline{P_3P_4}$ is $(1, -1)$. [See Figure 15.6(b).] $\blacksquare$

Sometimes, you will want to know the distance between a point and a horizontal or vertical line. (Of course, the point, here, is not on the line.) What is meant by distance in this case is the *distance along a perpendicular from the point P to the line L*. Denote this by

$$\text{dist}(P, L).$$

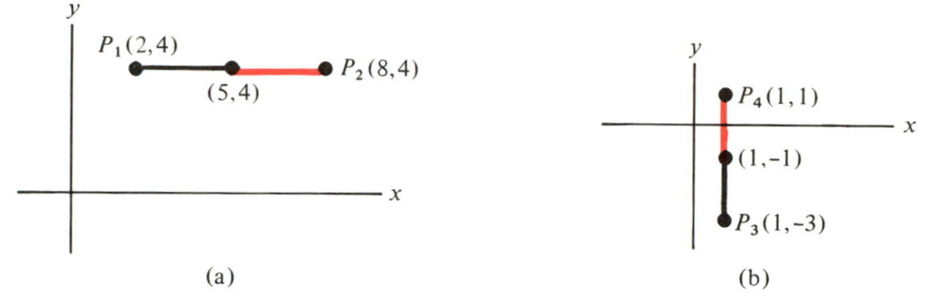

FIGURE 15.6 (a) The midpoint of $\overline{P_1P_2}$ is (5, 4). (b) The midpoint of $\overline{P_3P_4}$ is (1, −1).

Assume this perpendicular meets L at a point Q. Then the perpendicular line segment is given by $\overline{PQ}$.

If L is vertical, then $\overline{PQ}$ is horizontal. Let $P = (x_1, y_1)$ and let the equation of L be

$$x = x_2.$$

Then

$$Q = (x_2, y_1) \qquad \text{[See Figure 15.7(a).]}$$

And

$$\text{dist}(P, L) = \text{dist}(P, Q) = |x_2 - x_1|$$

Similarly, *if L is horizontal, then $\overline{PQ}$ is vertical.* Let $P = (x_1, y_1)$ and let the equation of L be

$$y = y_2.$$

Then

$$Q = (x_1, y_2) \qquad \text{[See Figure 15.7(b).]}$$

And

$$\text{dist}(P, L) = \text{dist}(P, Q) = |y_2 - y_1|$$

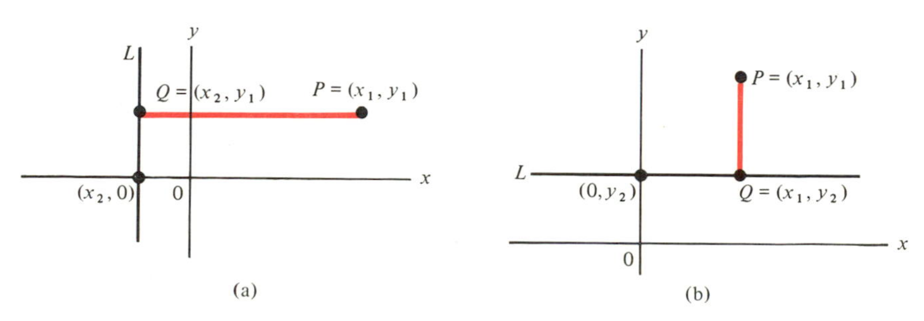

FIGURE 15.7 (a) If L is vertical, $\text{dist}(P, L) = \text{dist}(P, Q) = |x_2 - x_1|$ (b) If L is horizontal, $\text{dist}(P, L) = \text{dist}(P, Q) = |y_2 - y_1|$

EXAMPLE 4.

(a) Let $P_1 = (3, 4)$ and let L_1 be given by $x = -2$. Then L_1 is vertical and

$$\text{dist}(P_1, L_1) = |-2 - 3| = 5. \qquad \text{[See Figure 15.8(a).]}$$

(b) Let $P_2 = (2, -3)$ and let L_2 be given by $y = 5$. Then L_2 is horizontal and

$$\text{dist}(P_2, L_2) = |5 - (-3)| = 8. \qquad \text{[See Figure 15.8(b).]}$$ ■

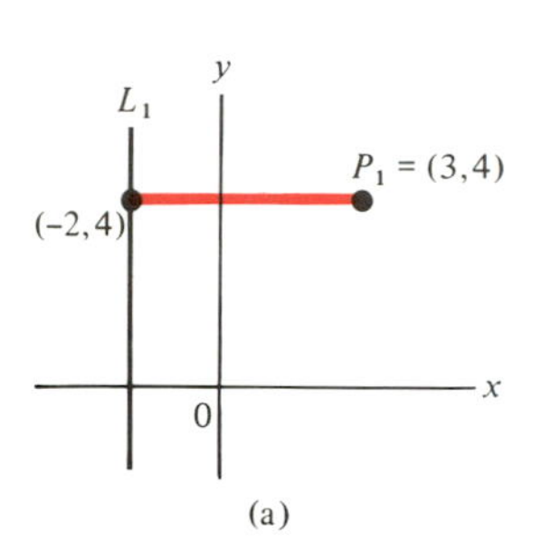

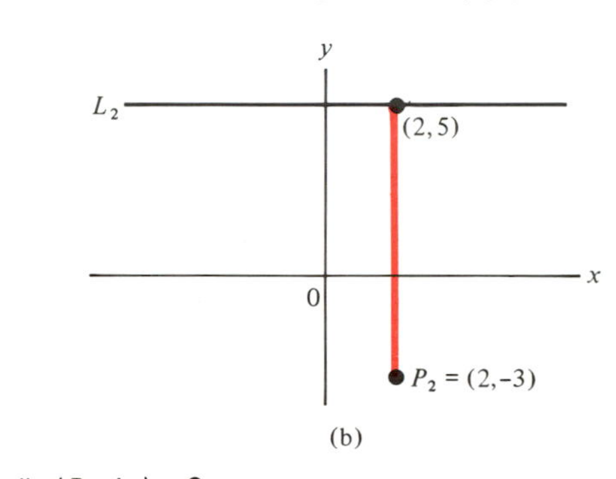

FIGURE 15.8 (a) $\text{dist}(P_1, L_1) = 5$ 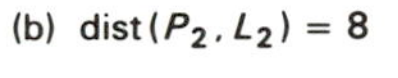 (b) $\text{dist}(P_2, L_2) = 8$

How do you express $\text{dist}(P_1, P_2)$ when the line segment $\overline{P_1P_2}$ is neither horizontal nor vertical? Let $P_1 = (x_1, y_1)$ and $P_2 = (x_2, y_2)$. Draw the perpendicular to the y-axis through P_1, and the perpendicular to the x-axis through P_2—as in Figure 15.9. Let Q be the intersection of these perpendiculars. Observe that

$$Q = (x_2, y_1).$$

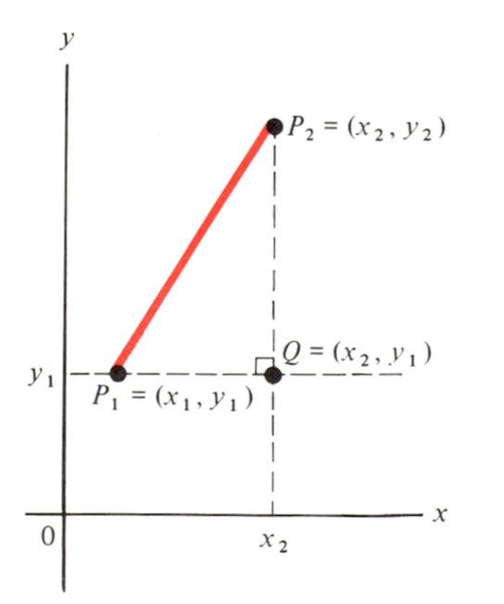

FIGURE 15.9 $\overline{P_1P_2}$ is neither horizontal nor vertical.
$\text{dist}(P_1, P_2) = \sqrt{(x_2 - x_1)^2 + (y_2 - y_1)^2}$

Also, $\overline{P_1P_2}$ is the *hypotenuse of the right triangle* P_1QP_2. According to the *Pythagorean Theorem,*

$$(\text{length } \overline{P_1P_2})^2 = (\text{length } \overline{P_1Q})^2 + (\text{length } \overline{P_2Q})^2$$

or

$$[\text{dist}(P_1, P_2)]^2 = [\text{dist}(P_1, Q)]^2 + [\text{dist}(P_2, Q)]^2$$

But $\overline{P_1Q}$ is horizontal and $\overline{P_2Q}$ is vertical. Therefore

$$[\text{dist}(P_1, P_2)]^2 = (x_2 - x_1)^2 + (y_2 - y_1)^2$$

and

$$\text{dist}(P_1, P_2) = \sqrt{(x_2 - x_1)^2 + (y_2 - y_1)^2}$$

Note that if $\overline{P_1P_2}$ is either horizontal or vertical, this formula still holds. For example, in the horizontal case,

$$y_2 = y_1$$

Hence,

$$y_2 - y_1 = 0, \qquad (y_2 - y_1)^2 = 0$$

and, as you saw in Section 14.8,

$$\begin{aligned}\text{dist}(P_1, P_2) &= |x_2 - x_1| = \sqrt{(x_2 - x_1)^2} = \sqrt{(x_2 - x_1)^2 + 0}\\ &= \sqrt{(x_2 - x_1)^2 + (y_2 - y_1)^2}\end{aligned}$$

EXAMPLE 5.

(a) Let $P_1 = (4, 5)$, $P_2 = (1, 9)$.

$$\begin{aligned}&\text{dist}(P_1, P_2)\\ &\quad= \sqrt{(1-4)^2 + (9-5)^2}\\ &\quad= \sqrt{9 + 16}\\ &\quad= \sqrt{25}\\ &\quad= 5\end{aligned}$$

(b) Let $P_3 = (-2, 1)$, $P_4 = (2, -1)$.

$$\begin{aligned}&\text{dist}(P_3, P_4)\\ &\quad= \sqrt{(2-(-2))^2 + (-1-1)^2}\\ &\quad= \sqrt{16 + 4}\\ &\quad= \sqrt{20}\\ &\quad= \sqrt{4}\sqrt{5}\\ &\quad= 2\sqrt{5}\end{aligned}$$

■

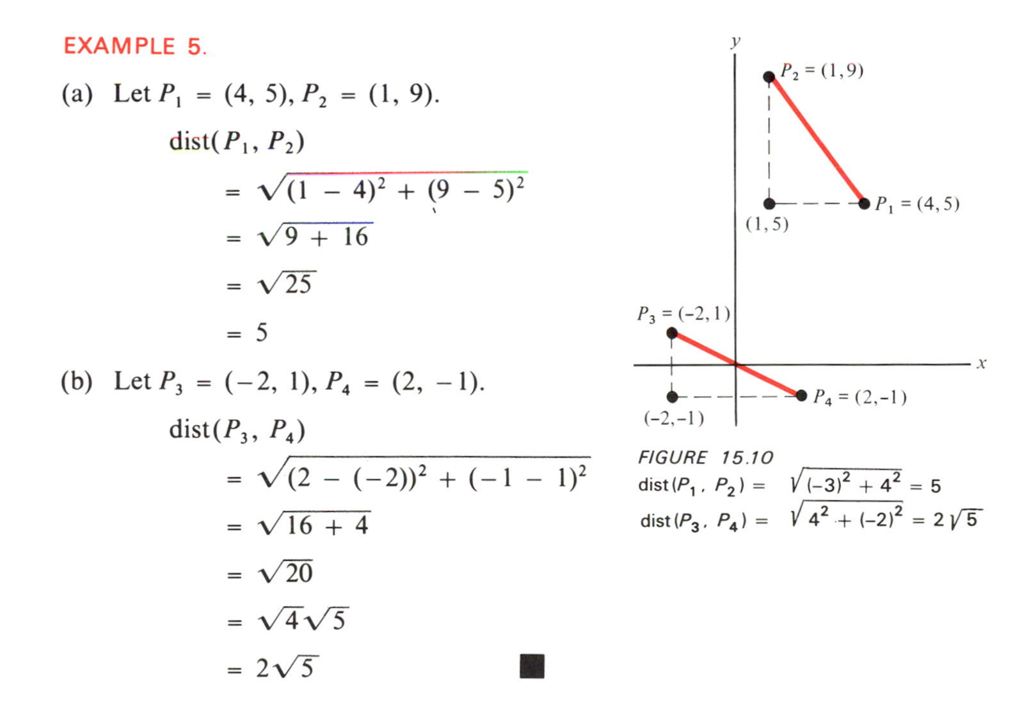

FIGURE 15.10
$\text{dist}(P_1, P_2) = \sqrt{(-3)^2 + 4^2} = 5$
$\text{dist}(P_3, P_4) = \sqrt{4^2 + (-2)^2} = 2\sqrt{5}$

EXERCISES

In Exercises 1–24 determine $\text{dist}(P_1, P_2)$.

1. $P_1 = (2, 5)$, $P_2 = (4, 5)$
2. $P_1 = (1, 2)$, $P_2 = (1, 6)$
3. $P_1 = (0, -3)$, $P_2 = (5, -3)$
4. $P_1 = (-8, -2)$, $P_2 = (-2, -2)$

5. $P_1 = (7, -1)$, $P_2 = (7, -4)$

6. $P_1 = (-3, -3)$, $P_2 = \left(\frac{-1}{2}, -3\right)$

7. $P_1 = (4, 0)$, $P_2 = (-8, 0)$

8. $P_1 = (-3, -5)$, $P_2 = (-4.4, -5)$

9. $P_1 = \left(\frac{1}{2}, \frac{1}{2}\right)$, $P_2 = \left(\frac{1}{4}, \frac{1}{2}\right)$

10. $P_1 = \left(\frac{1}{3}, \frac{2}{3}\right)$, $P_2 = \left(\frac{1}{3}, \frac{-1}{3}\right)$

11. $P_1 = (1, 1)$, $P_2 = (4, 5)$

12. $P_1 = (2, 2)$, $P_2 = (4, 4)$

13. $P_1 = (6, 3)$, $P_2 = (3, 6)$

14. $P_1 = (1, 4)$, $P_2 = (8, 5)$

15. $P_1 = (-1, 1)$, $P_2 = (1, -1)$

16. $P_1 = (4, 3)$, $P_2 = (-5, 1)$

17. $P_1 = (8, 2)$, $P_2 = (6, -2)$

18. $P_1 = (-4, -3)$, $P_2 = (-9, 2)$

19. $P_1 = (-10, -8)$, $P_2 = (-7, -12)$

20. $P_1 = \left(\frac{1}{2}, 1\right)$, $P_2 = \left(0, \frac{-1}{2}\right)$

21. $P_1 = (8, -3)$, $P_2 = (2, 1)$

22. $P_1 = (9, -5)$, $P_2 = (10, 0)$

23. $P_1 = (1, 1)$, $P_2 = (1.2, .8)$

24. $P_1 = \left(\frac{1}{4}, \frac{1}{2}\right)$, $P_2 = \left(\frac{3}{4}, 0\right)$

In Exercises 25–36 determine dist(P, L), where the point P and line L are as indicated.

25. $P = (1, 2)$ $L: x = 4$

26. $P = (3, 4)$ $L: y = 6$

27. $P = (4, 7)$ $L: x = -4$

28. $P = (3, -3)$ $L: y = -5$

29. $P = (7, 2)$ $L: y = 0$

30. $P = (-7, -3)$ L: the x-axis

31. $P = (-2, -5)$ $L: y = -2$

32. $P = (4, 1)$ $L: y = 100$

33. $P = (-7, 9)$ $L: x = -9$

34. $P = \left(-2, \frac{1}{2}\right)$ $L: x = \frac{-1}{4}$

35. $P = (3, 2)$ $L: y = \frac{1}{2}$

36. $P = (1.4, 2.7)$ $L: x = .7$

37. Describe the points of the plane that are the same distance from (4, 3) as from (−4, 3).

38. Find an equation for the points of the plane that are the same distance from (5, 5) as from (5, 1).

15.2 CIRCLES

Circles, parabolas, ellipses, and hyperbolas can be obtained as the intersections of planes and cones, as indicated in Figure 15.11. For this reason, these figures are known as **conic sections.**[1] As you will see, these figures can also be obtained in terms of distance.

[1]For further discussion, see the Instructor's Manual, pp. 11–15.

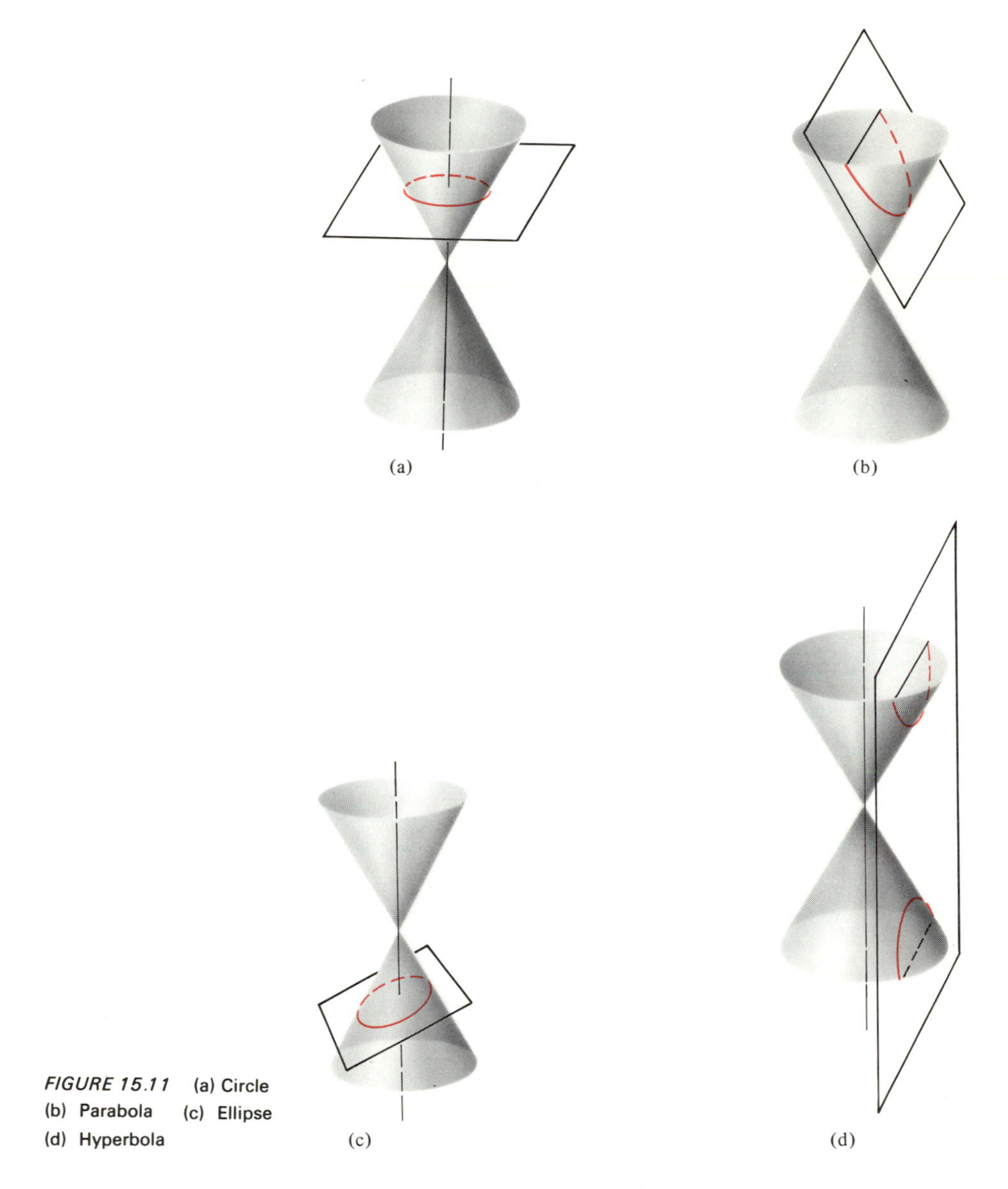

FIGURE 15.11 (a) Circle (b) Parabola (c) Ellipse (d) Hyperbola

DEFINITION. A **circle** consists of all points of the plane at a fixed distance from a given point. The given point is called the **center** of the circle. Each line segment from the center to a point on the circle is called a **radius** of the circle.

The length of each radius is the same.

EXAMPLE 1. Consider the circle centered at the origin and of radius length 5. This circle consists of all points 5 units away from the origin, P_0. Let $P = (x, y)$

and suppose P is on the circle. Then

$$\text{dist}(P, P_0) = 5$$

Therefore, because $P_0 = (0, 0)$,

$$\sqrt{(x - 0)^2 + (y - 0)^2} = 5$$

Square both sides. Thus

$$x^2 + y^2 = 25$$

This equation defines the circle. The set of (real) solutions of the equation is the set of points on the circle.

Some of the points on the circle are given in Table 15.1.

x	0	± 3	± 4	± 5
y	± 5	± 4	± 3	0

TABLE *15.1*

Thus, if $x = 3$, y can be 4 or -4. In fact,

$$3^2 + 4^2 = 9 + 16 = 25$$

and

$$3^2 + (-4)^2 = 9 + 16 = 25$$

Similarly, if $x = -3$, y can be 4 or -4. The circle is pictured in Figure 15.12. ■

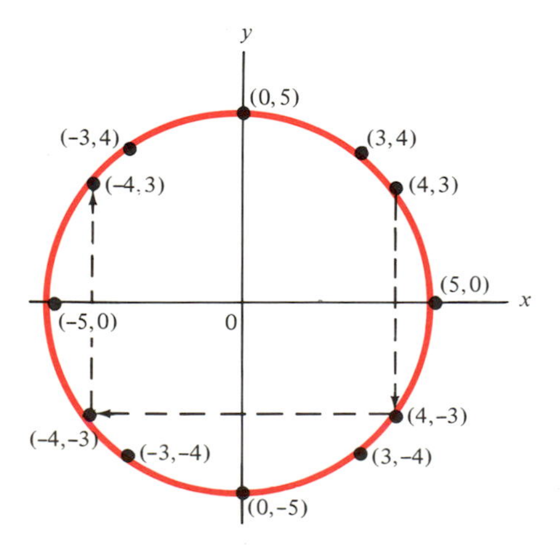

FIGURE 15.12 The graph of the circle with the equation $x^2 + y^2 = 25$

Both x^2 and y^2 appear in the equation $x^2 + y^2 = 25$, and there are no other occurrences of the variables x and y. Thus whenever (x, y) is on the graph,

$$(x, -y), \qquad (-x, y), \qquad \text{and} \qquad (-x, -y)$$

must also be on the graph. *The graph of a circle centered at the origin is symmetric with respect to both coordinate axes.*

For example, if (x, y) is on the graph, then

$$x^2 + y^2 = 25.$$

Therefore

$$x^2 + (-y)^2 = x^2 + y^2 = 25,$$

and $(x, -y)$ is also on the graph.

This shows that the graph is symmetric with respect to the x-axis. Similarly, it can be shown that the graph is symmetric with respect to the y-axis.

In Figure 15.12 observe that the point $(4, 3)$ is on the graph. By x-axis symmetry, the point $(4, -3)$ is also on the graph, and by y-axis symmetry, the point $(-4, -3)$ is on the graph. Finally, because $(-4, -3)$ is on the graph, by x-axis symmetry $(4, -3)$ is also on the graph.

If you "folded the plane" along the x-axis, the upper semicircle would lie on top of the lower semicircle. And if you folded the plane along the y-axis, the right semicircle would lie on top of the left semicircle.

Observe that for the circle centered at the origin and of radius length 5, the values of x and y are each in the closed interval $[-5, 5]$. Thus

$$-5 \le x \le 5 \qquad \text{and} \qquad -5 \le y \le 5$$

The circle of Example 1 was centered at the origin and had radius length 5. Its defining equation was

$$x^2 + y^2 = 5^2.$$

Now let $P_0 = (x_0, y_0)$ and let $r > 0$. Consider the circle centered at P_0 and of radius length r. A point P is on this circle if

$$\text{dist}(P, P_0) = r.$$

Thus, if $P = (x, y)$, then $\text{dist}(P, P_0) = \sqrt{(x - x_0)^2 + (y - y_0)^2}$, and

$$\sqrt{(x - x_0)^2 + (y - y_0)^2} = r$$

Square both sides to obtain the defining equation:

$$(x - x_0)^2 + (y - y_0)^2 = r^2, \qquad r > 0$$

This is the equation of a circle centered at (x_0, y_0) and of radius length r. See Figure 15.13 on page 514. [Observe that if $r = 0$, the only solution of the equation is (x_0, y_0).]

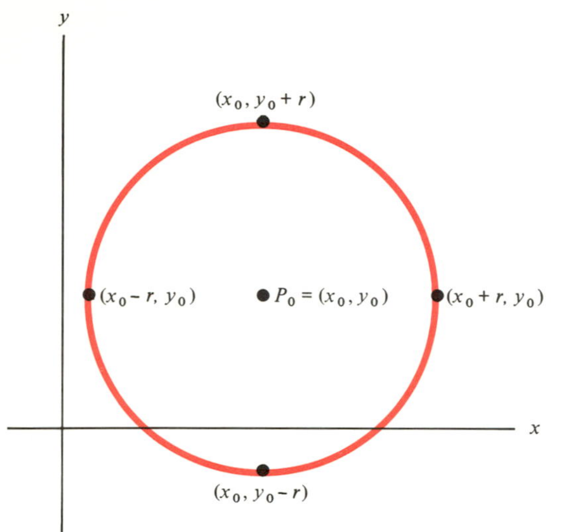

FIGURE 15.13 The circle centered at P_0 and of radius length r. Its equation is

$$(x - x_0)^2 + (y - y_0)^2 = r^2.$$

EXAMPLE 2. Determine the equation of the circle consisting of all points that are 3 units from (1, 2).

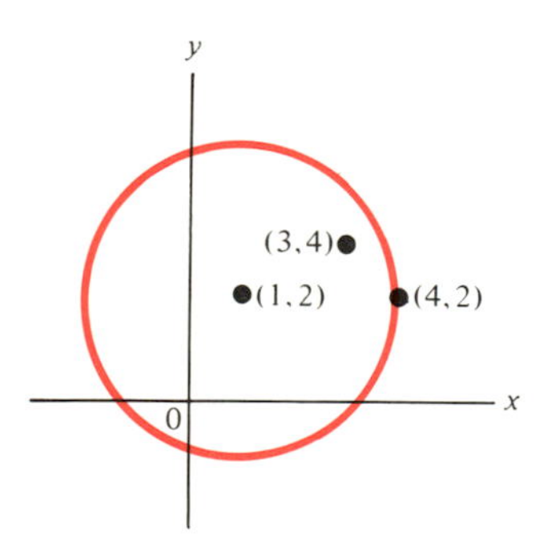

FIGURE 15.14

SOLUTION. Let $(x_0, y_0) = (1, 2)$, $r = 3$. [The center of the circle is (1, 2); the radius length is 3.] Therefore, the equation of the circle is

$$(x - 1)^2 + (y - 2)^2 = 3^2$$

or

$$(x - 1)^2 + (y - 2)^2 = 9.$$

The point (4, 2) is on this circle because

$$(4 - 1)^2 + (2 - 2)^2 =$$

$$3^2 + 0^2 = 9.$$

The point (3, 4) is *not* on the circle because

$$(3 - 1)^2 + (4 - 2)^2 =$$

$$2^2 + 2^2 = 8 \neq 9.$$ ■

EXAMPLE 3. Determine the center and radius length of the circle given by

$$(x + 2)^2 + (y - 9)^2 = 12.$$

SOLUTION. This equation is of the form

$$(x - x_0)^2 + (y - y_0)^2 = r^2.$$

Here $x_0 = -2$, $y_0 = 9$, $r^2 = 12$. Thus r is the square root of 12.

$$r = \sqrt{12} = \sqrt{4}\sqrt{3} = 2\sqrt{3}$$

The center is $(-2, 9)$; the radius length is $2\sqrt{3}$. ■

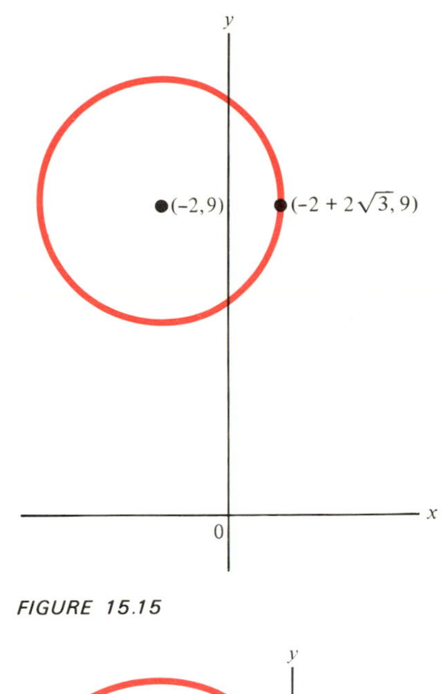

FIGURE 15.15

EXAMPLE 4.

(a) Show that the equation

$$x^2 + y^2 + 8x - 2y = 0$$

represents a circle.

(b) Determine the center and radius of this circle.

SOLUTION.

(a) Regroup terms to obtain

$$(x^2 + 8x) + (y^2 - 2y) = 0.$$

Next, complete the square within each group. (See Section 13.3.) To $x^2 + 8x$, add $\left(\frac{8}{2}\right)^2$ (or 16); to $y^2 - 2y$, add $\left(\frac{-2}{2}\right)^2$ (or 1). Thus

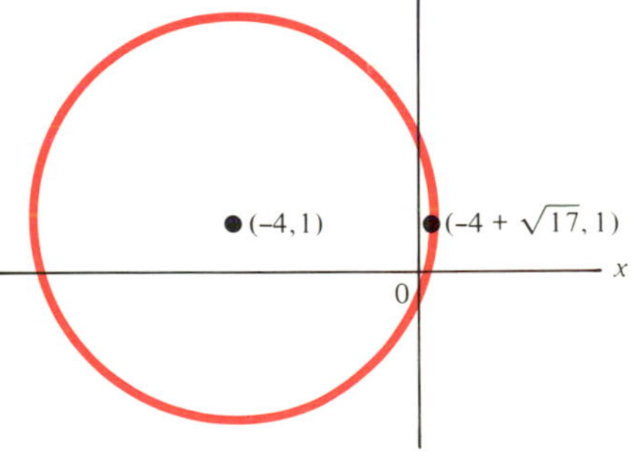

FIGURE 15.16

$$(x^2 + 8x + 16) + (y^2 - 2y + 1) = 16 + 1$$

and

$$(x + 4)^2 + (y - 1)^2 = 17$$

This is of the form

$$(x - x_0)^2 + (y - y_0)^2 = r^2,$$

where

$$x_0 = -4, \qquad y_0 = 1, \qquad r = \sqrt{17} \approx 4.12.$$

(b) The center is at $(-4, 1)$. The radius length is $\sqrt{17}$. ■

EXERCISES

In Exercises 1–16 determine (a) the center and (b) the radius length of the given circle.

1. $x^2 + y^2 = 9$
2. $x^2 + y^2 = 100$
3. $x^2 + y^2 = 8$
4. $x^2 + y^2 = 7$
5. $(x - 1)^2 + y^2 = 16$
6. $x^2 + (y - 3)^2 = 1$
7. $(x - 2)^2 + (y - 4)^2 = 49$
8. $(x + 2)^2 + (y + 4)^2 = 64$
9. $(x + 8)^2 + (y - 7)^2 = 144$
10. $\left(x - \frac{1}{2}\right)^2 + (y + 1)^2 = 20$
11. $\left(x + \frac{5}{3}\right)^2 + \left(y - \frac{1}{3}\right)^2 = 2$
12. $(x - 1.1)^2 + (y + .2)^2 = 25$
13. $\left(x - \frac{1}{4}\right)^2 + \left(y - \frac{1}{2}\right)^2 = \frac{1}{4}$
14. $\left(x + \frac{2}{5}\right)^2 + \left(y - \frac{4}{5}\right)^2 = \frac{9}{25}$
15. $(x + 2.7)^2 + (y - 1.7)^2 = .01$
16. $(x + \pi)^2 + (y - \pi)^2 = .09\pi^2$
17. Determine the equation of the circle consisting of all points that are 4 units from the origin.
18. Determine the equation of the circle consisting of all points that are 2 units from (2, 0).
19. Determine the equation of the circle consisting of all points that are 5 units from (2, 3).
20. Determine the equation of the circle consisting of all points that are 3 units from $(-3, -3)$.
21. Determine the equation of the circle consisting of all points that are $\frac{1}{2}$ unit from $\left(\frac{1}{4}, \frac{1}{2}\right)$.
22. Determine the equation of the circle consisting of all points that are .4 unit from $(1.2, -.8)$.

In Exercises 23–25 with the aid of a compass, graph the circle with the given center and radius length.

23. Center is the origin, radius length 6.
24. Center is (0, 4), radius length 4.
25. Center is $(-3, 0)$, radius length 3.

In Exercises 26–28 with the aid of a compass, graph the circle with the given equation.

26. $x^2 + y^2 = 16$
27. $(x - 3)^2 + y^2 = 9$
28. $(x - 3)^2 + (y - 7)^2 = 4$

In Exercises 29–40 complete the square to determine the center and radius length of the corresponding circle.

29. $x^2 + y^2 + 10x = 0$

30. $3x^2 + 3y^2 - 12y = 0$

31. $x^2 + y^2 + 2x + 6y = 0$

32. $x^2 + y^2 + 4x - 4y = 0$

33. $x^2 + y^2 + 6x + 18y = 0$

34. $x^2 + y^2 + 12x - 10y = 0$

35. $x^2 + y^2 + 6x + 5y = 0$

36. $x^2 + y^2 + 3x - 9y = 0$

37. $x^2 + y^2 + 4x - 3y = 1$

38. $x^2 + y^2 + 10x = y + 6$

39. $2x^2 + 2y^2 + 8x + 40y = 0$

40. $4x^2 + 4y^2 + 20x + 10y = 1$

15.3 PARABOLAS

EXAMPLE 1. Consider the equation

$$x^2 = 4y.$$

The second power of only one of the variables appears. (In the equation of the circle, the second powers of *both* variables appear.) Some corresponding values are given in Table 15.2. On the basis of these values a graph is drawn in Figure 15.17.

x	0	± 1	± 2	± 3	± 4	± 5	± 6
x^2	0	1	4	9	16	25	36
y	0	$\frac{1}{4}$	1	$2\frac{1}{4}$	4	$6\frac{1}{4}$	9

TABLE 15.2 ■

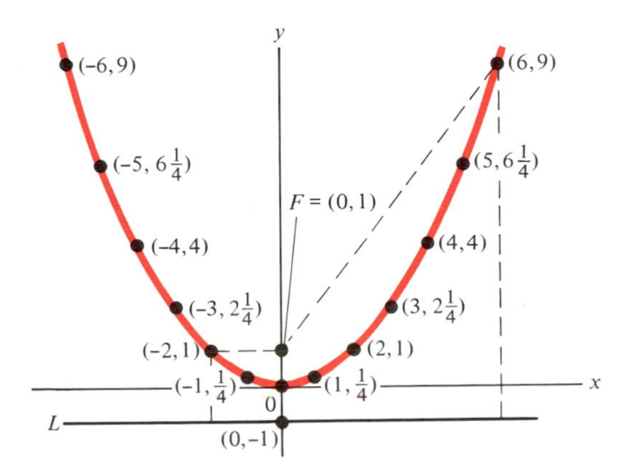

FIGURE 15.17 The parabola is given by $x^2 = 4y$. The point $F = (0, 1)$. The line L is given by $y = -1$.
Every point on the parabola is the same distance from F as from L.

This curve is known as a *parabola*. The "turning point" of the parabola—here, the origin—is called the *vertex*. Precise definitions will be given later.

In Figure 15.17 the point $F[=(0, 1)]$ and the horizontal line L, given by $y = -1$, are also indicated. It may surprise you to learn that *every point on the parabola is the same distance from F as from L*. Let us check this for several sample points on the parabola.

Let $P[=(x, y)]$ be on the parabola. Recall that

$$\text{dist}(P, F) = \sqrt{(x-0)^2 + (y-1)^2}$$

and that, because L is horizontal,

$$\text{dist}(P, L) = y - (-1) = y + 1.$$

[See Table 15.3.]

P	$\text{dist}(P, F) = \sqrt{(x-0)^2 + (y-1)^2}$	$\text{dist}(P, L) = y + 1$
(0, 0)	$\|0 - 1\| = 1$	$0 + 1 = 1$
(2, 1)	$\|2 - 0\| = 2$	$1 + 1 = 2$
(4, 4)	$\sqrt{(4-0)^2 + (4-1)^2} = \sqrt{25} = 5$	$4 + 1 = 5$
(−2, 1)	$\|-2 - 0\| = 2$	$1 + 1 = 2$
(−6, 9)	$\sqrt{(-6-0)^2 + (9-1)^2} = \sqrt{100} = 10$	$9 + 1 = 10$

TABLE *15.3*

Recall that the circle was defined in terms of distance. A circle consists of all points of the plane at a fixed distance from a given point.

DEFINITION. A **parabola** consists of all points of the plane that are the same distance from a given point F and a given line L. The point F is called the **focus** and the line L is called the **directrix.**

Thus, if P is on the parabola,

$$\text{dist}(P, F) = \text{dist}(P, L).$$

Throughout this section *p will represent a positive number.*

First consider the case where the focus F is $(0, p)$ and where the directrix L is the line with equation $y = -p$. Thus the focus is on the *positive y-axis*, and the directrix is a *horizontal line* lying *below* the focus. [See Figure 15.18.]

Let $P[=(x, y)]$ be on the parabola. Then

$$\text{dist}(P, F) = \text{dist}(P, L);$$

hence

$$\sqrt{(x - 0)^2 + (y - p)^2} = y + p$$

Square both sides.

$$x^2 + (y - p)^2 = (y + p)^2$$

$$x^2 + \cancel{y^2} - 2py + \cancel{p^2} = \cancel{y^2} + 2py + \cancel{p^2}$$

$$x^2 = 4py$$

This is the equation of a parabola with focus at $(0, p)$ and directrix given by $y = -p$.

Draw the line perpendicular to the directrix and through the focus. This line is called the **axis (of symmetry)** of the parabola. The intersection point of the axis with the parabola is called the **vertex.**

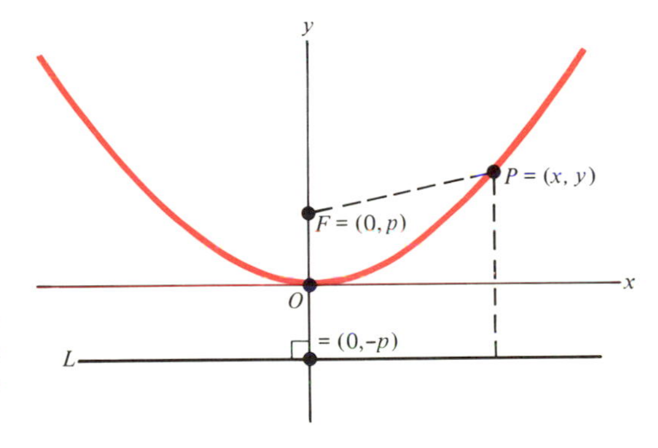

FIGURE 15.18. The parabola is given by $x^2 = 4py$. The focus F is $(0, p)$. The directrix L is given by $y = -p$. The y-axis is the axis of symmetry. The origin O is the vertex.

For a parabola given by

$$x^2 = 4py,$$

the axis of symmetry is the y-axis and the vertex is the origin. Whenever (x, y) is on the parabola, so is $(-x, y)$. Observe that x takes on all real values, whereas y takes on only nonnegative values. In symbols,

$$-\infty < x < \infty, \qquad y \geq 0$$

The parabola

$$x^2 = 4y$$

that you first considered is of this form with $p = 1$. Its focus is $(0, 1)$, its vertex is $(0, 0)$. Its directrix is the line with the equation $y = -1$, and its axis of symmetry is the y-axis.

The parabola with the equation

$$x^2 = 4py, \qquad p > 0$$

opens upward. [See Figure 15.19(a) on page 521.]

There are other possible positions for a parabola with vertex at the origin (and vertical or horizontal axis).

The parabola with the equation

$$x^2 = -4py, \qquad p > 0,$$

opens downward. Its focus is now $(0, -p)$ and its directrix is now the line given by $y = p$. Again the y-axis is the axis of symmetry. [See Figure 15.19 (b).]

The parabola with the equation

$$y^2 = 4px, \qquad p > 0,$$

opens to the right. $\left(\text{Observe that } x = \frac{y^2}{4p} \geq 0.\right)$ The focus is $(p, 0)$ and the directrix is the line given by $x = -p$. Now the x-axis is the axis of symmetry. [See Figure 15.19 (c).]

The parabola with the equation

$$y^2 = -4px, \qquad p > 0,$$

opens to the left. The focus is $(-p, 0)$ and the directrix is the line given by $x = p$. The x-axis is the axis of symmetry. [See Figure 15.19 (d).]

Table 15.4 summarizes information about the various parabolas.

equation	opens	focus	directrix	axis	values of variables
$x^2 = 4py$	upward	$(0, p)$	$y = -p$	y-axis	$-\infty < x < \infty$, $y \geq 0$
$x^2 = -4py$	downward	$(0, -p)$	$y = p$	y-axis	$-\infty < x < \infty$, $y \leq 0$
$y^2 = 4px$	to right	$(p, 0)$	$x = -p$	x-axis	$-\infty < y < \infty$, $x \geq 0$
$y^2 = -4px$	to left	$(-p, 0)$	$x = p$	x-axis	$-\infty < y < \infty$, $x \leq 0$

TABLE *15.4 POSSIBILITIES FOR A PARABOLA WITH VERTEX AT THE ORIGIN (AND VERTICAL OR HORIZONTAL AXIS)*

Note that if the *first-degree term* involves y, then the axis of the parabola is the y-axis. If the *first-degree term* involves x, then the axis of the parabola is the x-axis.

EXAMPLE 2.

(a) Determine the focus, directrix, and axis of the parabola with the equation:

$$x^2 = -8y$$

(b) Graph the parabola.

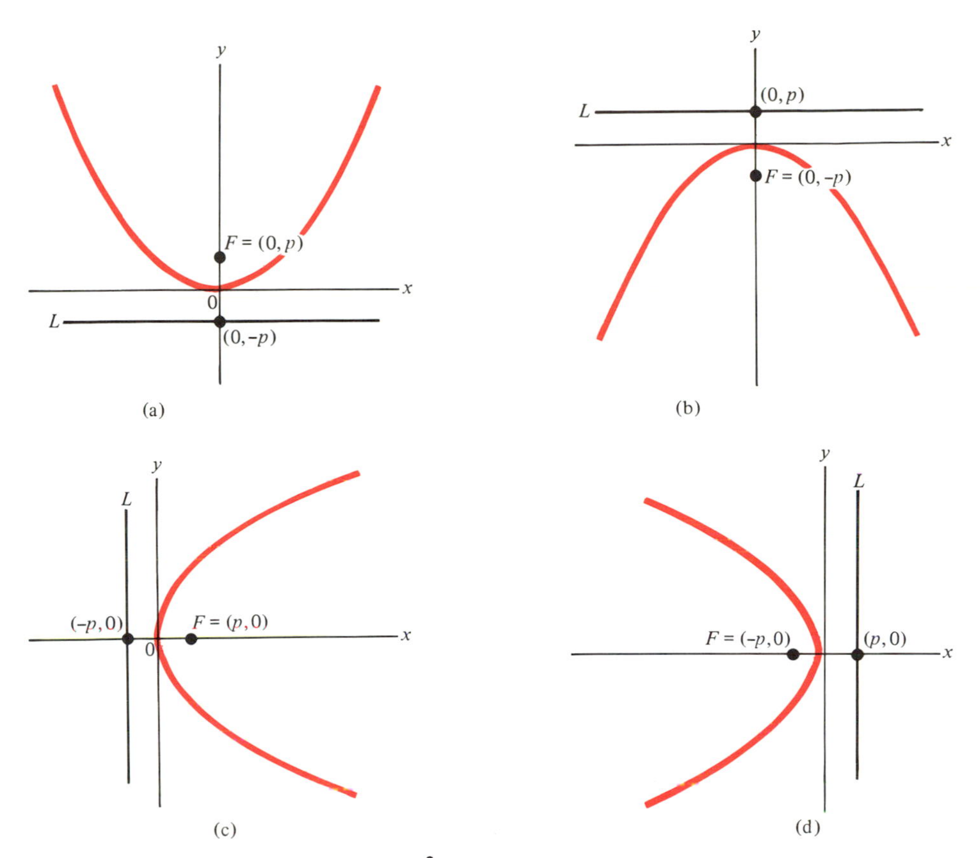

FIGURE 15.19 (a) The parabola given by $x^2 = 4py$ opens upward. The focus F is $(0, p)$. The directrix L is given by $y = -p$. (b) The parabola given by $x^2 = -4py$ opens downward. The focus F is $(0, -p)$. The directrix L is given by $y = p$. (c) The parabola given by $y^2 = 4px$ opens to the right. The focus F is $(p, 0)$. The directrix L is given by $x = -p$. (d) The parabola given by $y^2 = -4px$ opens to the left. The focus F is $(-p, 0)$. The directrix L is given by $x = p$.

SOLUTION.

(a) The equation of this parabola is of the form

$$x^2 = -4py,$$

with $p = 2$. Therefore the focus is $(0, -2)$, the directrix is the line given by $y = 2$, and the axis is the y-axis.

(b)

x	0	± 2	± 4	± 8
x^2	0	4	16	64
$y = \frac{-x^2}{8}$	0	$\frac{-1}{2}$	-2	-8

TABLE *15.5*

[See Figure 15.20 on page 522.] ■

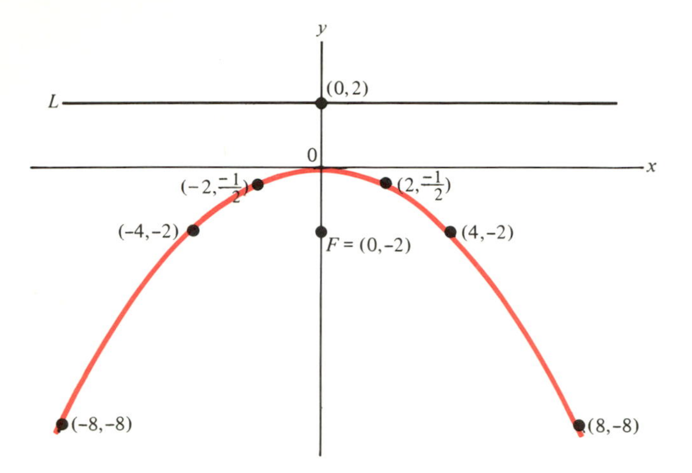

FIGURE 15.20 The parabola is given by $x^2 = -8y$. The focus F is $(0, -2)$. The directrix L is given by $y = 2$.

EXAMPLE 3.

(a) Determine the focus, directrix, and axis of the parabola with the equation:

$$y^2 = 2x$$

(b) Graph the parabola.

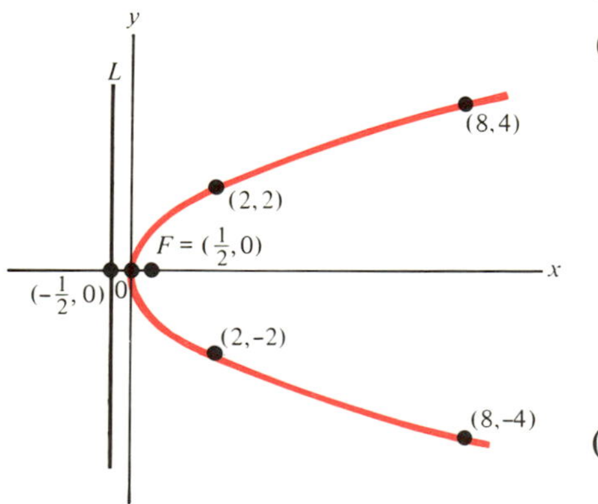

FIGURE 15.21 The parabola is given by $y^2 = 2x$. The focus F is $\left(\frac{1}{2}, 0\right)$ The directrix L is given by $x = \frac{-1}{2}$.

SOLUTION.

(a) The equation of the parabola is of the form

$$y^2 = 4px,$$

with $p = \frac{1}{2}$. Thus the focus is $\left(\frac{1}{2}, 0\right)$, the directrix is the line given by $x = \frac{-1}{2}$, and the axis is the x-axis.

(b) Note that y can be positive, negative, or 0; however, $x = \frac{y^2}{2} \geq 0$ because $y^2 \geq 0$. Also,

$$2x = y^2$$

$$\pm\sqrt{2x} = y$$

x	0	2	8
$2x$	0	4	16
$y = \pm\sqrt{2x}$	0	± 2	± 4

TABLE 15.6

EXAMPLE 4. Determine the equation of the parabola with focus $(-3, 0)$ and directrix given by $x = 3$.

SOLUTION. The axis is horizontal because the directrix (given by $x = 3$) is vertical. In fact, the axis is the x-axis because the focus, $(-3, 0)$, is on the x-axis. The vertex is at the origin, midway between $(-3, 0)$ and $(3, 0)$. Thus, $p = 3$. The focus is to the *left* of the directrix. The equation of the parabola is of the form

$$y^2 = -4px.$$

Because $p = 3$, the equation is

$$y^2 = -12x.$$

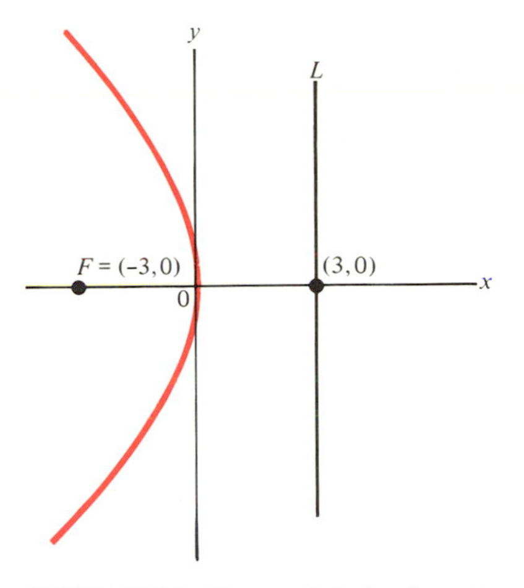

FIGURE 15.22 The parabola is given by $y^2 = -12x$. The focus F is $(-3, 0)$. The directrix L is given by $x = 3$.

The vertex of a parabola need not be the origin. Moreover, the axis of a parabola need be neither vertical nor horizontal. In these cases the equation of the parabola becomes somewhat more complicated.

There are many applications of parabolas in science. There are comets with parabolic paths. The cable of a suspension bridge with evenly distributed weight is parabolic. [See Figure 15.23.] Architects sometimes construct arches in the shape of parabolas. [See Figure 15.24 on page 524.]

Parabolic reflectors are often used. [See Figure 15.25.] Rays of light approaching the reflector along lines parallel to the axis of the parabolic reflector are reflected so that they pass through the focus. In an automobile headlight this reflection principle works in reverse. A cross section of the headlight is a

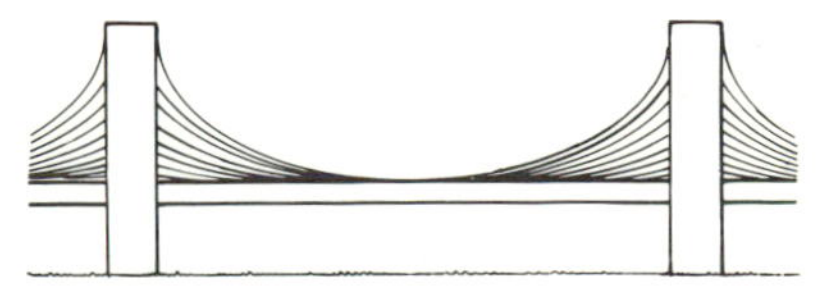

FIGURE 15.23 Parabolic suspension bridge

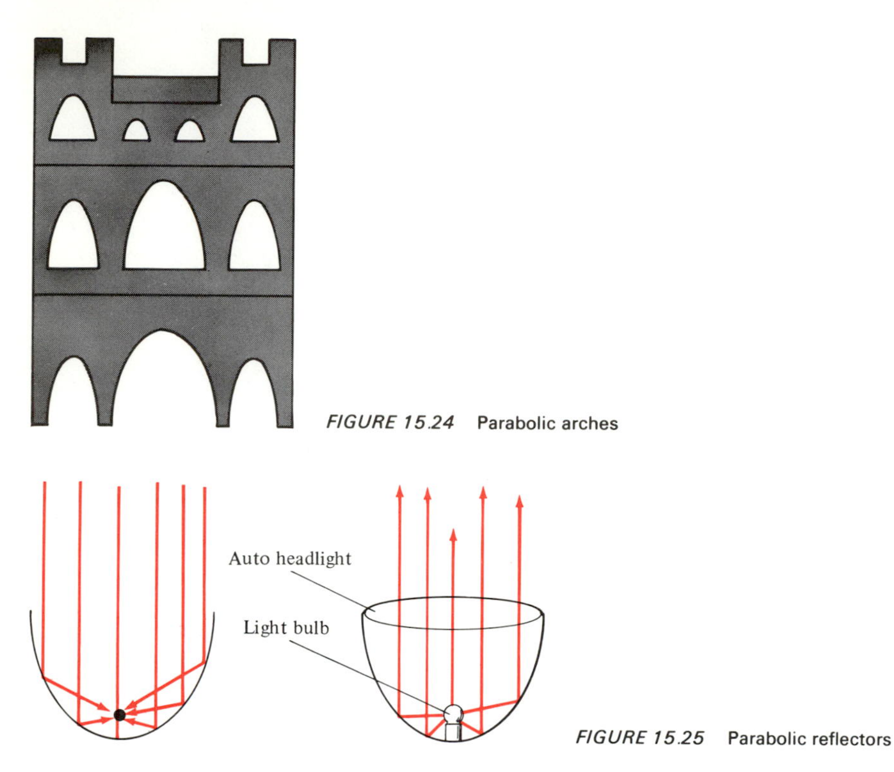

FIGURE 15.24 Parabolic arches

FIGURE 15.25 Parabolic reflectors

parabola with the bulb at the focus. The light is then reflected off the parabola along parallel rays.

EXERCISES

In Exercises 1–12 determine the vertex, axis, focus, and directrix of the parabola with the given equation. Graph the ones so indicated.

1. $x^2 = 8y$ (Graph.)
2. $x^2 = -4y$ (Graph.)
3. $x^2 = 12y$
4. $x^2 = -y$ (Graph.)
5. $y^2 = 4x$ (Graph.)
6. $y^2 = -16x$ (Graph.)
7. $y^2 = x$
8. $y^2 = -2x$ (Graph.)
9. $x^2 = 3y$
10. $x^2 + 9y = 0$
11. $3y^2 - 6x = 0$
12. $15x - 6y^2 = 0$

In Exercises 13–20 determine the equation of the parabola with focus and directrix as indicated.

13. focus $(3, 0)$, directrix: $x = -3$
14. focus $(-3, 0)$, directrix: $x = 3$
15. focus $(0, 3)$, directrix: $y = -3$
16. focus $(0, -3)$, directrix: $y = 3$

17. focus (5, 0), directrix: $x = -5$
18. focus $(-2, 0)$, directrix: $x = 2$
19. focus (0, 10), directrix: $y = -10$
20. focus $(0, -6)$, directrix: $y = 6$

In Exercises 21–28 determine the equation of the parabola with vertex at the origin and

21. directrix: $x = -8$
22. directrix: $x = -4$
23. focus $(-4, 0)$
24. focus (2, 0)
25. directrix: $y = -5$
26. directrix: $y = 1$
27. focus (0, 1)
28. focus $(0, -9)$
29. Determine the equation of the parabola that consists of all points that are the same distance from (6, 0) as from the line $x = -6$.
30. Determine the equation of the parabola that consists of all points that are the same distance from $(0, -4)$ as from the line $y = 4$.

*15.4 ELLIPSES AND HYPERBOLAS

EXAMPLE 1. Consider the equation

$$\frac{x^2}{25} + \frac{y^2}{9} = 1 .$$

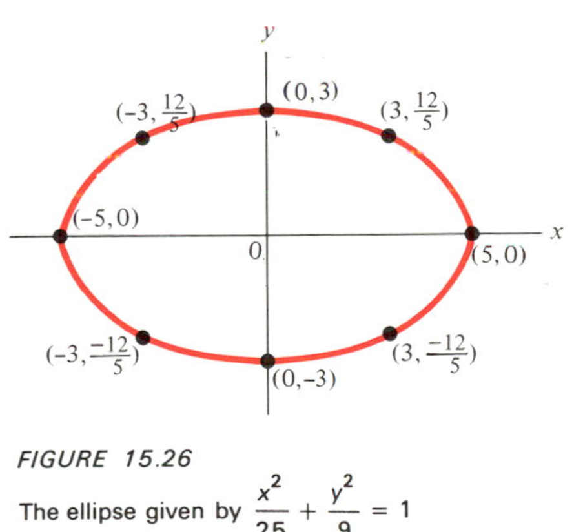

FIGURE 15.26

The ellipse given by $\frac{x^2}{25} + \frac{y^2}{9} = 1$

As in the equation of a circle, the second powers of both variables appear. But here, the (fractional) coefficients of x^2 and y^2 *differ*. Thus when you multiply by $225[= \text{LCD}(25, 9)]$ to clear of fractions, the coefficients of x^2 and y^2 in

(A) $$9x^2 + 25y^2 = 225$$

differ. Note finally that the x^2 and y^2 terms are separated by +.

To graph this curve, plot several key points. First, set $x = 0$ in Equation (A):

$$25y^2 = 225$$
$$y^2 = 9$$
$$y = \pm 3$$

Thus (0, 3) and $(0, -3)$ lie on the curve.

*Optional topic.

Substitute 0 for y in Equation (A):

$$9x^2 = 225$$
$$x^2 = 25$$
$$x = \pm 5$$

Thus (5, 0) and (−5, 0) lie on the curve.

Next, when $x = 3$ and when $x = -3$, $x^2 = 9$. Replace x^2 by 9 in Equation (A) and obtain:

$$9(9) + 25y^2 = 225$$
$$25y^2 = 144$$
$$y^2 = \frac{144}{25}$$
$$y = \frac{\pm 12}{5}$$

Thus $\left(3, \frac{12}{5}\right)$, $\left(3, \frac{-12}{5}\right)$, $\left(-3, \frac{12}{5}\right)$, and $\left(-3, \frac{-12}{5}\right)$ lie on the curve.

You now know that the points indicated in Table 15.7 are on the curve.

x	0	3	−3	5	−5
y	3 −3	$\frac{12}{5}$ $\frac{-12}{5}$	$\frac{12}{5}$ $\frac{-12}{5}$	0	0

TABLE *15.7* ■

This curve, which resembles a "flattened circle", is known as an *ellipse.*

Circles and parabolas were defined in terms of distance. The same sort of definition applies to ellipses.

In Example 1, let $F_1 = (-4, 0)$ and let $F_2 = (4, 0)$. Then *the sum*

$$\text{dist}(P, F_1) + \text{dist}(P, F_2)$$

is the same for every point P on the ellipse. Let us check this for several sample points. [See Table 15.8 on page 527.]

In fact, for every point P on this ellipse,

$$\text{dist}(P, F_1) + \text{dist}(P, F_2) = 10$$

DEFINITION. Let F_1 and F_2 be two given points, and let a be a number such that

$$\text{dist}(F_1, F_2) < 2a.$$

P	$\text{dist}(P, F_1) = \sqrt{(x+4)^2 + (y-0)^2}$	$\text{dist}(P, F_2) = \sqrt{(x-4)^2 + (y-0)^2}$	$\text{dist}(P, F_1) + \text{dist}(p, F_2)$
$(0, 3)$	$\sqrt{(0+4)^2+(3-0)^2}$	$\sqrt{(0-4)^2+(3-0)^2}$	
	$= \sqrt{25} = 5$	$= \sqrt{25} = 5$	10
$(0, -3)$	$\sqrt{(0+4)^2+(-3-0)^2}$	$\sqrt{(0-4)^2+(-3-0)^2}$	
	$= \sqrt{25} = 5$	$= \sqrt{25} = 5$	10
$(5, 0)$	$5 + 4 = 9$	$5 - 4 = 1$	10
$\left(3, \frac{12}{5}\right)$	$\sqrt{(3+4)^2+\left(\frac{12}{5}-0\right)^2}$	$\sqrt{(3-4)^2+\left(\frac{12}{5}-0\right)^2}$	
	$= \sqrt{49 + \frac{144}{25}}$	$= \sqrt{1 + \frac{144}{25}}$	
	$= \sqrt{\frac{1369}{25}}$	$= \sqrt{\frac{169}{25}}$	
	$= \frac{37}{5}$	$= \frac{13}{5}$	$\frac{50}{5} = 10$

TABLE 15.8

An **ellipse** consists of all points P of the plane such that

$$\text{dist}(P, F_1) + \text{dist}(P, F_2) = 2a.$$

The points F_1 and F_2 are called the **foci** of the ellipse. (Each is a **focus**.) The midpoint of the line segment F_1F_2 is called the **center** of the ellipse. [See Figure 15.27.]

A fairly accurate sketch of an ellipse can be obtained as follows. Stick tacks into a cardboard (representing the plane) at F_1 and F_2. Attach a string of length $2a$ to the tacks, as in Figure 15.28 . With a pencil pressed tightly against the string, trace out the ellipse.

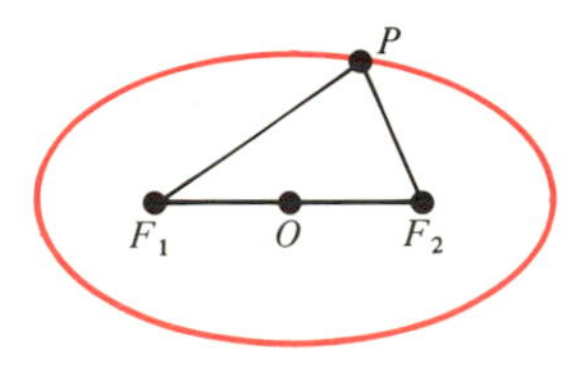

FIGURE 15.27 An ellipse with foci F_1 and F_2 and center O.

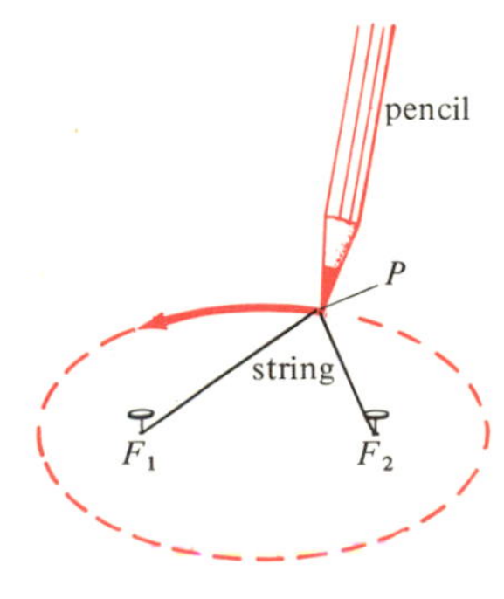

FIGURE 15.28

The equation of an ellipse centered at the origin and with foci on a coordinate axis can be written in the form

$$\frac{x^2}{a^2} + \frac{y^2}{b^2} = 1, \qquad \text{where } a > 0 \text{ and } b > 0.$$

Both x^2 and y^2 appear in this equation, and there are no other occurrences of the variables x and y. Thus whenever (x, y) is on the graph,

$$(x, -y), \qquad (-x, y), \qquad \text{and} \qquad (-x, -y)$$

must also be on the graph. *The graph of an ellipse is symmetric with respect to both coordinate axes.*

Observe that

$$\frac{x^2}{a^2} \leq \frac{x^2}{a^2} + \frac{y^2}{b^2} = 1 .$$

Thus

$$x^2 \leq a^2$$

and, as you saw in Section 14.7, it follows that

$$-a \leq x \leq a.$$

Similarly,

$$-b \leq y \leq b$$

Thus the values of x are in the closed interval $[-a, a]$ and the values of y are in the closed interval $[-b, b]$.

As you did in Example 1, first set $y = 0$ and then set $x = 0$ in the equation of an ellipse. You will see that the points

$$(a, 0), \qquad (-a, 0), \qquad (0, b), \qquad (0, -b)$$

are on the ellipse. They are called the **vertices** of the ellipse. Each **vertex** is on a coordinate axis. To graph an ellipse, first locate these vertices. Then draw a smooth curve through these points.

A ray of light that originates at one focus will bound off an elliptic reflector and pass through the other focus [See Figure 15.29.] The orbit of every planet is an ellipse with the sun at one focus [See Figure 15.30.] Periodic comets, which return to the sun at fixed intervals, have elliptical orbits.

Figure 15.31 shows a graph known as a *hyperbola.* Notice that it consists of two curves and that it crosses only one of the coordinate axes. Also, as a point moving along the hyperbola gets further away from the origin, the point gets closer and closer to one of two straight lines through the origin.

DEFINITION. Let F_1 and F_2 be two given points, and let a be a positive number

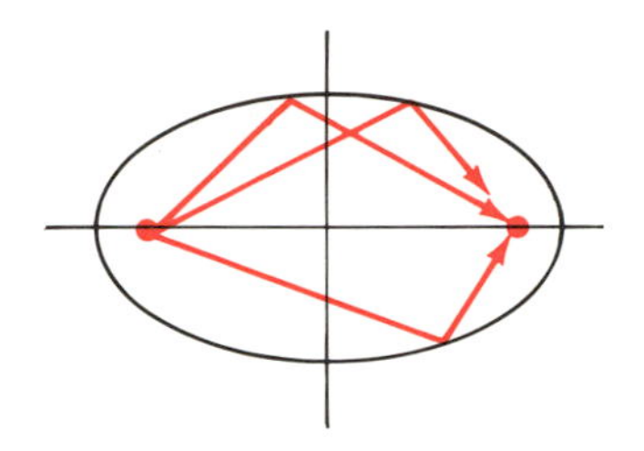

FIGURE 15.29 An elliptic reflector

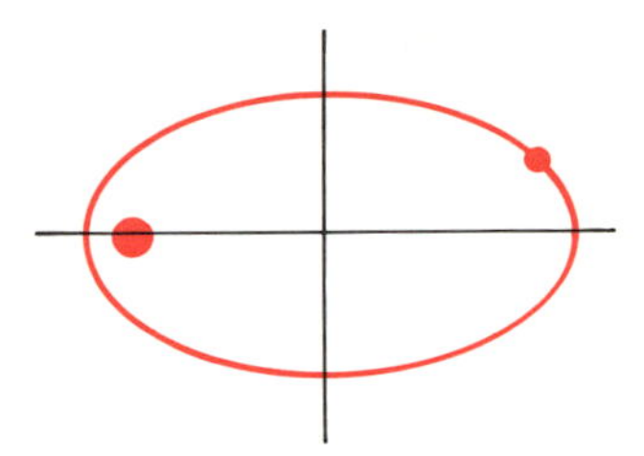

FIGURE 15.30 The orbit of every planet is an ellipse with the sun at one focus.

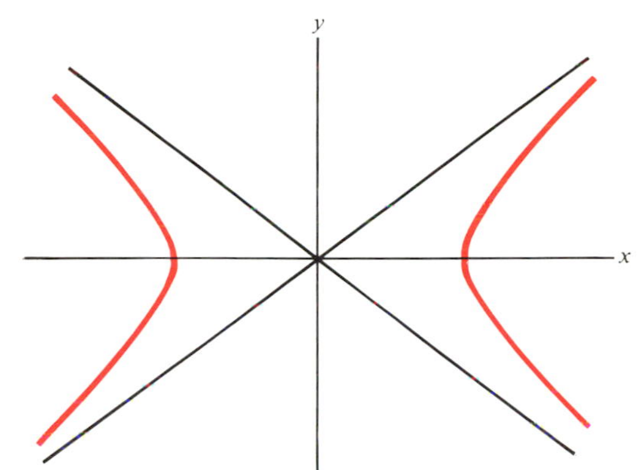

FIGURE 15.31 A hyperbola (in color)

such that

$$\text{dist}(F_1, F_2) > 2a.$$

A hyperbola consists of all points P of the plane such that

$$|\,\text{dist}(P, F_1) - \text{dist}(P, F_2)\,| = 2a.$$

The points F_1 and F_2 are called the **foci** of the hyperbola. (Each is a **focus.**) The midpoint of the line segment F_1F_2 is called the **center** of the hyperbola. [See Figure 15.32.]

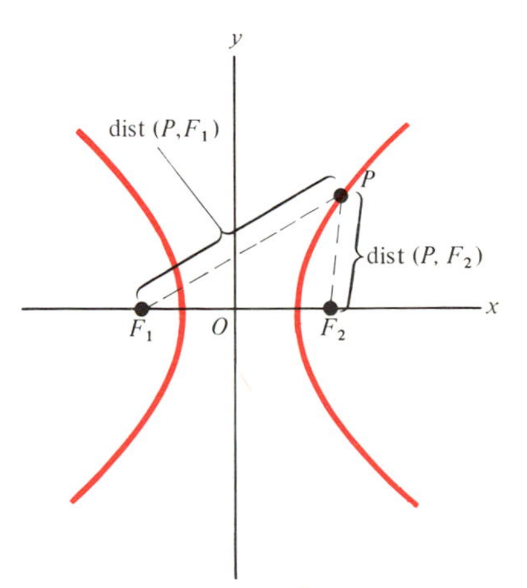

FIGURE 15.32 A hyperbola with foci F_1 and F_2 and center O. For points P on the hyperbola, $|\text{dist}(P, F_1) - \text{dist}(P, F_2)|$ is constant.

The distance condition involves absolute value. It can be rewritten in terms of two equations:

$$\text{dist}(P, F_1) - \text{dist}(P, F_2) = 2a$$

or

$$\text{dist}(P, F_2) - \text{dist}(P, F_1) = 2a$$

The equation of a hyperbola whose center is at the origin and whose foci are on a coordinate axis can be written in the form

$$\frac{x^2}{a^2} - \frac{y^2}{b^2} = 1$$

or

$$\frac{y^2}{a^2} - \frac{x^2}{b^2} = 1 .$$

Here

$$a > 0 \qquad \text{and} \qquad b > 0$$

Notice that both x^2 and y^2 appear in the equations of the hyperbola, and there are no other occurrences of the variables x and y. Thus whenever (x, y) is on the graph,

$$(x, -y), \qquad (-x, y), \qquad \text{and} \qquad (-x, -y)$$

must also be on the graph. *The graph is symmetric with respect to both coordinate axes.*

First, consider the equation

$$\frac{x^2}{a^2} - \frac{y^2}{b^2} = 1 .$$

Set $y = 0$ and obtain:

$$\frac{x^2}{a^2} = 1$$

$$x^2 = a^2$$

$$x = \pm a$$

Thus the points

$$(a, 0) \qquad \text{and} \qquad (-a, 0)$$

are on the hyperbola. They are called the **vertices** of the hyperbola. Each **vertex** is on the x-axis. However, when $x = 0$ you do *not* obtain any point on the hyperbola. For consider the equation

$$\frac{-y^2}{b^2} = 1$$

or

$$\left(\frac{y}{b}\right)^2 = -1 .$$

This has no (real) solution because a square is *nonnegative*. Therefore the y-axis ($x = 0$) lies between the two parts, or **branches**, of the hyperbola.

Observe that

$$\frac{x^2}{a^2} \geq \frac{x^2}{a^2} - \frac{y^2}{b^2} = 1 .$$

Thus

$$x^2 \geq a^2$$

Therefore

$$x \leq -a \qquad \text{or} \qquad x \geq a$$

However, y can take on all real values.

The lines through the origin

$$y = \frac{b}{a}x$$

and

$$y = \frac{-b}{a}x,$$

although not part of the hyperbola, nevertheless play a key role in graphing the hyperbola. These lines are called **asymptotes.** [See Figure 15.33.] A point moving along the hyperbola gets closer and closer to an asymptote as the point gets further and further from the origin.

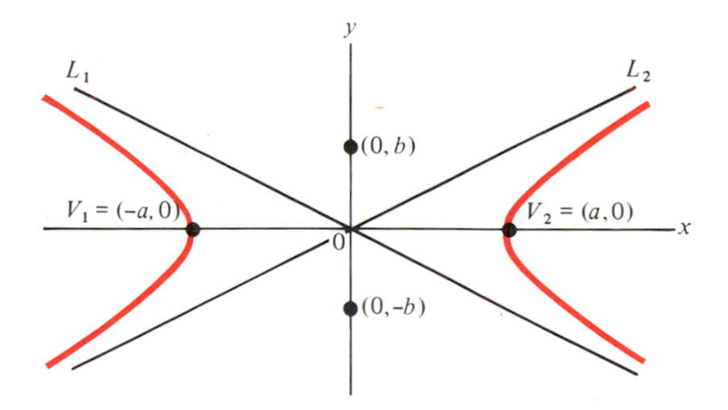

FIGURE 15.33 A hyperbola whose equation is of the form

$$\frac{x^2}{a^2} - \frac{y^2}{b^2} = 1.$$

Each curve (in color) is a branch of the hyperbola. The vertices are V_1 and V_2; the asymptotes are given by L_1: $y = \frac{-b}{a}x$ and L_2: $y = \frac{b}{a}x$.

To graph a hyperbola:

1. Plot the vertices.
2. Draw the asymptotes.
3. Sketch the hyperbola so that as $|x|$ gets larger, each curve approaches the asymptotes. *The hyperbola never crosses its asymptotes.*

EXAMPLE 2. Consider the hyperbola given by

$$\frac{x^2}{16} - \frac{y^2}{9} = 1 .$$

(a) Locate the vertices.
(b) What are the asymptotes?
(c) Sketch this hyperbola.

SOLUTION.

(a) The equation is of the form

$$\frac{x^2}{a^2} - \frac{y^2}{b^2} = 1,$$

where $a = 4$, $b = 3$. The vertices are $(\pm a, 0)$. Thus the vertices are $(4, 0)$ and $(-4, 0)$.

(b) The equations of the asymptotes are

$$y = \frac{3}{4}x \qquad \left(\text{because } \frac{b}{a} = \frac{3}{4}\right) \qquad \text{and} \qquad y = \frac{-3}{4}x.$$

(c) Plot the vertices. Then draw the asymptotes. Finally, sketch the hyperbola through the vertices so that each curve approaches the asymptotes. [See Figure 15.34 on page 533.] ■

For a hyperbola centered at the origin but with vertices on the y-axis, the equation is of the form

$$\frac{y^2}{a^2} - \frac{x^2}{b^2} = 1 .$$

Now the vertices are $(0, a)$ and $(0, -a)$. The asymptotes are given by

$$y = \frac{a}{b}x \qquad \text{and} \qquad y = \frac{-a}{b}x.$$

Also, $y \leq -a$ or $y \geq a$, whereas x takes on all real values.

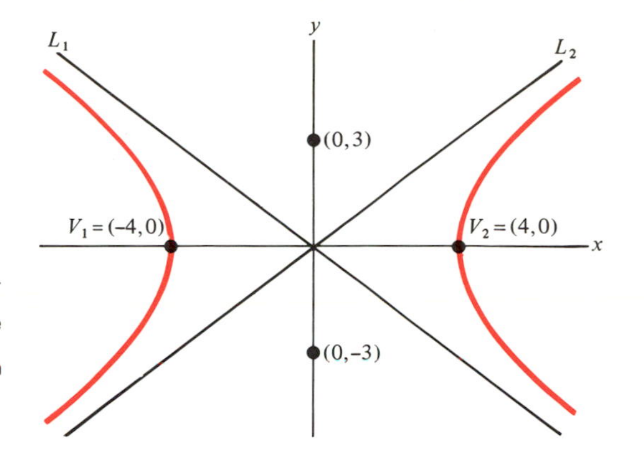

FIGURE 15.34 The hyperbola whose equation is $\frac{x^2}{16} - \frac{y^2}{9} = 1$. The vertices are (−4, 0) and (4, 0). The asymptotes are given by $L_1 : y = \frac{-3}{4}x$ and $L_2 : y = \frac{3}{4}x$.

EXAMPLE 3.

(a) Show that the equation

$$16y^2 - x^2 = 16$$

represents a hyperbola.

(b) Locate the vertices.

(c) What are the asymptotes?

(d) Sketch the graph.

SOLUTION.

(a) Divide both sides of the given equation by 16.

$$y^2 - \frac{x^2}{16} = 1$$

This is the equation of a hyperbola centered at the origin and with vertices on the y-axis.

(b) Set $x = 0$. The vertices are $(0, 1)$ and $(0, -1)$.

(c) $a = 1$ and $b = 4$. The asymptotes are now given by $y = \frac{a}{b}x$ and $y = \frac{-a}{b}x$.

Thus they are the lines whose equations are

$$y = \frac{x}{4} \qquad \text{and} \qquad y = \frac{-x}{4}.$$

(d) See Figure 15.35 on page 534. ■

There are comets whose paths are hyperbolic. Inverse variation, which you will study in the next chapter, is described by an equation

$$xy = k,$$

whose graph is a hyperbola.

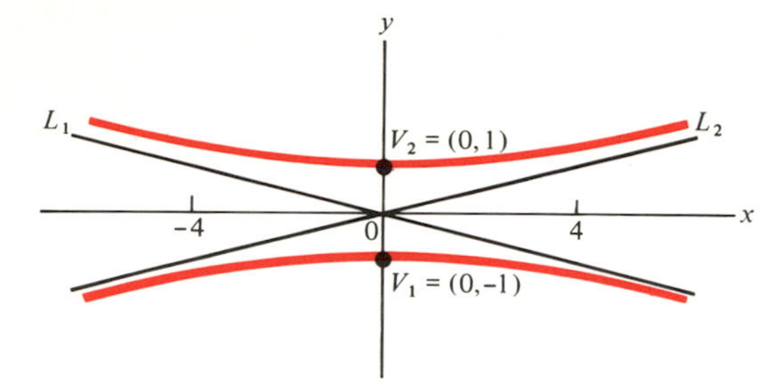

FIGURE 15.35 The hyperbola whose equation is $16y^2 - x^2 = 16$ or $y^2 - \frac{x^2}{16} = 1$. The vertices are V_1 and V_2. The asymptotes are given by $L_1 : y = \frac{-x}{4}$ and $L_2 : y = \frac{x}{4}$.

EXERCISES

In Exercises 1–16 an equation of an ellipse is given. Locate the vertices. Graph the ellipses so indicated.

1. $\frac{x^2}{25} + \frac{y^2}{16} = 1$ (Graph.)
2. $\frac{x^2}{16} + \frac{y^2}{25} = 1$ (Graph.)
3. $\frac{x^2}{9} + \frac{y^2}{4} = 1$
4. $\frac{x^2}{9} + \frac{y^2}{25} = 1$
5. $\frac{x^2}{169} + \frac{y^2}{144} = 1$ (Graph.)
6. $\frac{x^2}{169} + \frac{y^2}{25} = 1$ (Graph.)
7. $\frac{x^2}{100} + \frac{y^2}{36} = 1$ (Graph.)
8. $\frac{x^2}{25} + \frac{y^2}{169} = 1$ (Graph.)
9. $x^2 + \frac{y^2}{4} = 1$
10. $\frac{x^2}{100} + \frac{y^2}{81} = 1$
11. $\frac{x^2}{13} + \frac{y^2}{4} = 1$
12. $\frac{x^2}{64} + \frac{y^2}{15} = 1$
13. $9x^2 + 4y^2 = 36$
14. $25x^2 + y^2 = 25$
15. $9x^2 + 36y^2 = 36$
16. $16x^2 + 100y^2 = 400$

In Exercises 17–32 an equation of a hyperbola is given.

(a) Locate the vertices.

(b) What are the asymptotes?

(c) Graph the indicated hyperbolas.

17. $\frac{x^2}{9} - \frac{y^2}{16} = 1$ (Graph.)
18. $\frac{x^2}{64} - \frac{y^2}{36} = 1$ (Graph.)
19. $x^2 - \frac{y^2}{4} = 1$
20. $\frac{x^2}{9} - y^2 = 1$
21. $\frac{y^2}{16} - \frac{x^2}{9} = 1$ (Graph.)
22. $\frac{y^2}{25} - \frac{x^2}{144} = 1$ (Graph.)
23. $\frac{y^2}{9} - x^2 = 1$
24. $\frac{y^2}{64} - \frac{x^2}{36} = 1$

25. $4x^2 - 9y^2 = 36$ (Graph.)

26. $4y^2 - 16x^2 = 16$ (Graph.)

27. $25x^2 - 100y^2 = 100$

28. $9y^2 - 36x^2 = 36$

29. $4x^2 - 25y^2 = 100$

30. $y^2 - 25x^2 = 25$

31. $\dfrac{x^2}{25} - \dfrac{y^2}{24} = 1$

32. $x^2 - y^2 = 1$

15.5 INTERSECTING FIGURES

The intersections of a line with a conic section or the intersections of two conic sections can be determined by solving a system of equations. For, each such figure is represented by an equation. As in systems of linear equations (Section 10.1), the pair (x_0, y_0) is called a **solution of the system** if (x_0, y_0) is a solution of (or satisfies) *both* equations of the system. *A solution of the system corresponds to a point of intersection of the figures.*

EXAMPLE 1. Determine the intersections of the parabola given by

$$y = x^2$$

with the line given by

$$y = 3x.$$

SOLUTION. Replace y by $3x$ in the equation of the parabola

$$y = x^2.$$

$$3x = x^2$$

$$x^2 - 3x = 0$$

$$x(x - 3) = 0$$

$$x = 0 \quad \text{or} \quad x = 3$$

From the equation of the line, when $x = 0$, $y = 3(0) = 0$; when $x = 3$, $y = 3(3) = 9$. The intersection points are these two solutions: (0, 0) and (3, 9).

CHECK. Because you used the equation of the line to find the y-value corresponding to the x-values 0 and 3, now use the equation of the parabola.

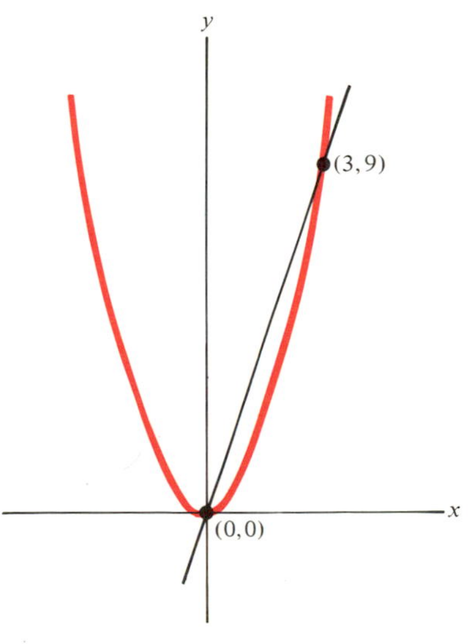

FIGURE 15.36

$(0, 0)$: $\mathbf{0} \stackrel{?}{=} 0^2$ $\quad$ $0 \stackrel{\checkmark}{=} 0$ | $(3, 9)$: $\mathbf{9} \stackrel{?}{=} 3^2$ $\quad$ $9 \stackrel{\checkmark}{=} 9$

EXAMPLE 2. Determine the intersections of the circle given by

$$x^2 + y^2 = 16$$

with the line given by

$$y = 2 - x.$$

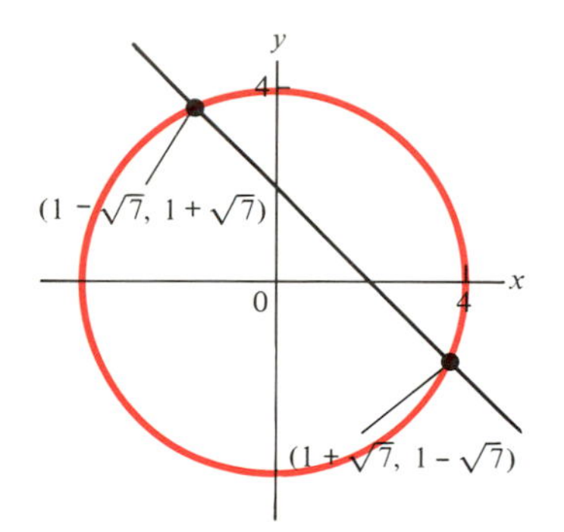

FIGURE 15.37

SOLUTION. From the equation of the line, substitute $2 - x$ for y in the equation of the circle,

$$x^2 + y^2 = 16$$
$$x^2 + (2 - x)^2 = 16$$
$$x^2 + 4 - 4x + x^2 = 16$$
$$2x^2 - 4x - 12 = 0$$
$$x^2 - 2x - 6 = 0$$

By the quadratic formula,

$$x = \frac{2 \pm \sqrt{4 - 4(1)(-6)}}{2}$$
$$= \frac{2 \pm \sqrt{28}}{2}$$
$$= \frac{2 \pm 2\sqrt{7}}{2}$$
$$= 1 \pm \sqrt{7}$$

When $x = 1 + \sqrt{7}$,

$$y = 2 - x = 2 - (1 + \sqrt{7}) = 1 - \sqrt{7}$$

When $x = 1 - \sqrt{7}$,

$$y = 2 - x = 2 - (1 - \sqrt{7}) = 1 + \sqrt{7}$$

The solutions are

$$(1 + \sqrt{7}, 1 - \sqrt{7}) \quad \text{and} \quad (1 - \sqrt{7}, 1 + \sqrt{7}).$$

Because $\sqrt{7} \approx 2.6$, the intersection points are *approximately*

$$(3.6, -1.6) \quad \text{and} \quad (-1.6, 3.6).$$

EXAMPLE 3. Show that the line given by

$$y = 2x$$

does not intersect the circle given by

$$(x - 2)^2 + (y + 4)^2 = 4$$

(a) by drawing the graphs on the same coordinate plane,
(b) by solving a system of equations.

SOLUTION.

(a) See Figure 15.38 .
(b) Replace y by $2x$ in the equation of the circle

$$(x - 2)^2 + (y + 4)^2 = 4.$$

$$(x - 2)^2 + (2x + 4)^2 = 4$$

$$x^2 - 4x + 4 + 4x^2 + 16x + 16 = 4$$

$$5x^2 + 12x + 16 = 0$$

$$x = \frac{-12 \pm \sqrt{144 - 4(5)(16)}}{10}$$

$$= \frac{-12 \pm \sqrt{-176}}{10}$$

The solutions are not real because the discriminant, -176, is negative. There is no intersection point. ■

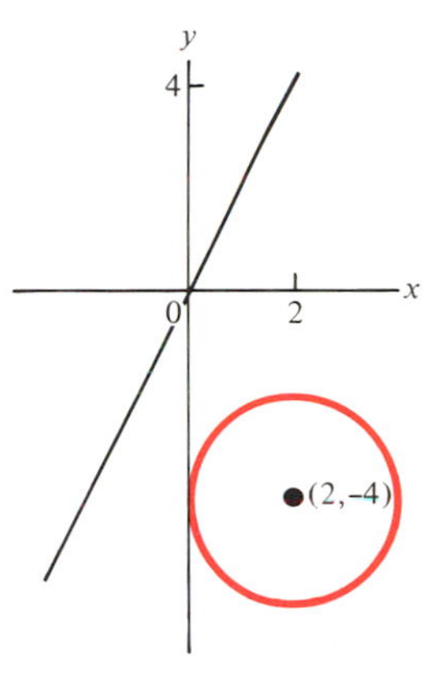

FIGURE 15.38 The line given by $y = 2x$ does not intersect the circle given by $(x - 2)^2 + (y + 4)^2 = 4$.

EXAMPLE 4. Determine the intersections of the ellipse given by

$$\frac{x^2}{4} + \frac{y^2}{9} = 1$$

with the hyperbola given by

$$\frac{x^2}{4} - \frac{y^2}{9} = 1.$$

SOLUTION. Add these equations:

$$\frac{x^2}{4} + \frac{y^2}{9} = 1$$

$$\frac{x^2}{4} - \frac{y^2}{9} = 1$$

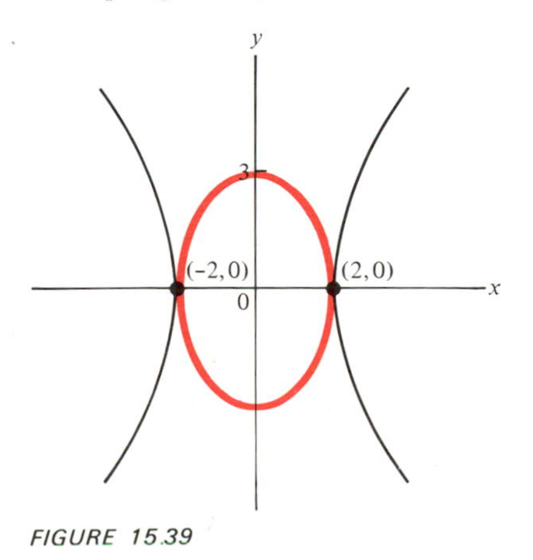

FIGURE 15.39

$$\frac{2x^2}{4} = 2$$

$$\frac{x^2}{4} = 1$$

$$x^2 = 4$$

$$x = \pm 2$$

Replace x by 2, and then by -2 in the equation of the ellipse. Because only the second power of x appears, the corresponding y-values are the same.

$$\frac{\overset{1}{\cancel{4}}}{\underset{1}{\cancel{4}}} + \frac{y^2}{9} = 1$$

$$\frac{y^2}{9} = 0$$

$$y = 0$$

The intersections are (2, 0) and (−2, 0). [See Figure 15.39 on page 537.] ■

EXERCISES

In Exercises 1–18:

(a) Determine all intersections.

(b) Indicate what figure (line, circle, parabola) is represented by each equation.

(c) When called for, graph the figures on the same coordinate system to illustrate the points of intersection.

1. $x^2 + y^2 = 25$ and $x = 3$ (Graph.)
2. $y = 3x^2$ and $y = 5$ (Graph.)
3. $y = -8x^2$ and $y = x$
4. $x^2 + y^2 = 81$ and $y = 3x$
5. $x^2 + y^2 = 49$ and $y = 7 - x$ (Graph.)
6. $y = 4x^2$ and $y = 2x$
7. $x = -2y^2$ and $y = -x$
8. $x^2 + y^2 = 9$ and $y = 3 - x$ (Graph.)
9. $x^2 + y^2 = 100$ and $y = 10 - 3x$
10. $y = -x^2$ and $y = 2$ (Graph.)
11. $y = -x^2$ and $x + y = 0$ (Graph.)
12. $x^2 + y^2 = 2$ and $x = y^2$

13. $x^2 + y^2 = 2$ and $x = -y^2$
14. $x^2 + y^2 = 16$ and $x^2 + y^2 = 25$ (Graph.)
15. $(x + 2)^2 + (y - 3)^2 = 1$ and $y = 2x$
16. $(x + 4)^2 + (y + 1)^2 = 25$ and $y = 4 - x$
17. $x^2 + y^2 = 4$ and $(x - 4)^2 + y^2 = 4$ (Graph.)
18. $x^2 = 5y$ and $x^2 + y^2 = 6$

In Exercises 19–28:

(a) Determine all intersections.

(b) Indicate what figure (line, circle, parabola, ellipse, hyperbola) is represented by each equation.

(c) When called for, graph the figures on the same coordinate system to illustrate the points of intersection.

19. $\dfrac{x^2}{4} + y^2 = 1$ and $x = 1$
20. $x^2 + \dfrac{y^2}{9} = 1$ and $y = 3$
21. $x^2 - y^2 = 1$ and $x = -1$ (Graph.)
22. $x^2 - \dfrac{y^2}{4} = 1$ and $y = x$
23. $\dfrac{x^2}{16} + \dfrac{y^2}{9} = 1$ and $y = -x$
24. $\dfrac{x^2}{9} - \dfrac{y^2}{25} = 1$ and $y = 5x$
25. $x^2 + y^2 = 4$ and $\dfrac{x^2}{9} + \dfrac{y^2}{16} = 1$ (Graph.)
26. $x^2 + y^2 = 16$ and $\dfrac{x^2}{16} + \dfrac{y^2}{9} = 1$ (Graph.)
27. $\dfrac{x^2}{64} - \dfrac{y^2}{36} = 1$ and $\dfrac{x^2}{64} + \dfrac{y^2}{36} = 1$
28. $\dfrac{y^2}{9} - \dfrac{x^2}{16} = 1$ and $\dfrac{x^2}{16} + \dfrac{y^2}{9} = 1$

REVIEW EXERCISES FOR CHAPTER 15

15.1 DISTANCE

In Exercises 1–3 determine $\text{dist}(P_1, P_2)$.

1. $P_1 = (1, 4), P_2 = (-2, 4)$
2. $P_1 = (-3, 1), P_2 = (-3, 6)$
3. $P_1 = (2, 4), P_2 = (1, 7)$
4. Determine $\text{dist}(P, L)$, where $P = (1, 3)$ and L is the line given by

$$x = 4.$$

15.2 CIRCLES

In Exercises 5 and 6 determine (a) the center, (b) the radius length of the circle given by each equation.

5. $x^2 + (y - 2)^2 = 36$

6. $(x - 4)^2 + (y + 1)^2 = 25$

7. Determine the equation of the circle consisting of all points that are 3 units from $(2, -2)$.

8. Complete the square to determine the center and radius length of the circle given by

$$x^2 + y^2 + 8x = 0 .$$

15.3 PARABOLAS

9. (a) Determine the vertex, axis, focus, and directrix of the parabola given by

$$x^2 = 4y.$$

(b) Graph this parabola.

10. Determine the equation of the parabola with focus $(0, 4)$ and directrix given by $y = -4$.

11. Determine the equation of the parabola with vertex at the origin and focus $(8, 0)$.

12. Determine the equation of the parabola with vertex at the origin and directrix given by $y = \frac{3}{2}$.

15.4 ELLIPSES AND HYPERBOLAS

13. (a) Locate the vertices of the ellipse given by

$$\frac{x^2}{100} + \frac{y^2}{36} = 1 .$$

(b) Graph this ellipse.

14. Locate the vertices of the ellipse given by

$$\frac{x^2}{16} + \frac{y^2}{25} = 1 .$$

15. Consider the hyperbola given by

$$\frac{x^2}{25} - \frac{y^2}{9} = 1 .$$

(a) Locate the vertices.

(b) What are the asymptotes?

(c) Graph this hyperbola.

16. Consider the hyperbola given by

$$y^2 - \frac{x^2}{2} = 1 .$$

(a) Locate the vertices.

(b) What are the asymptotes?

15.5 INTERSECTING FIGURES

In Exercises 17–19:

(a) Determine all intersections.

(b) Indicate what figure (line, circle, parabola, ellipse, hyperbola) is represented by each equation.

(c) When called for, graph the figures on the same coordinate system to illustrate the points of intersection.

17. $x^2 + y^2 = 36$ and $y = x$ (Graph.)

18. $y = 4x^2$ and $y = 2x$

19. $\frac{x^2}{9} + \frac{y^2}{4} = 1$ and $x^2 + y^2 = 4$ (Graph.)

TEST YOURSELF ON CHAPTER 15.

1. Determine dist(P_1, P_2):

$$P_1 = (3, -5), \qquad P_2 = (5, -2)$$

2. Determine

(a) the center, (b) the radius length

of the circle given by

$$(x - 3)^2 + (y + 3)^2 = 9 .$$

3. (a) Determine the vertex, axis, focus, and directrix of the parabola given by

$$y^2 = 8x.$$

(b) Graph this parabola.

Choose any 2 of the last 3 problems. Problem 4 is based on Section 15.4.

4. Consider the hyperbola given by

$$\frac{x^2}{9} - \frac{y^2}{25} = 1 .$$

(a) Locate the vertices.

(b) What are the asymptotes?

5. (a) Determine all intersections of the figures given by $x^2 + y^2 = 9$ and $y = x + 3$.

(b) Indicate what figure (line, circle, parabola, ellipse, hyperbola) is represented by each equation.

(c) Graph these figures on the same coordinate system to illustrate the points of intersection.

6. (a) Determine the equation of the circle centered at $(4, -1)$ and with radius length 5.

   (b) Determine the equation of the parabola with the origin as vertex and with directrix $x = -6$.

16
variation

16.1 DIRECT VARIATION

There are many situations in science and everyday life in which two variables, x and y, change, yet the quotient $\frac{y}{x}$ (or product xy) remains constant.

For example, if a rectangular piece of material is cut from a roll of silk of fixed width w, the area A of the piece depends on its length l.
Thus in Figure 16.1 on page 544,

$A_1 = l_1 w$	and $A_2 = l_2 w$
or	or
$\dfrac{A_1}{l_1} = w$	$\dfrac{A_2}{l_2} = w$

Therefore

$$\frac{A_1}{l_1} = \frac{A_2}{l_2} = w$$

No matter where the material is cut,

$$\frac{\text{area}}{\text{length}} \text{ equals (the constant) width.}$$

DEFINITION. Let x and y be variables. Then y **varies directly as** x if

$$y = kx$$

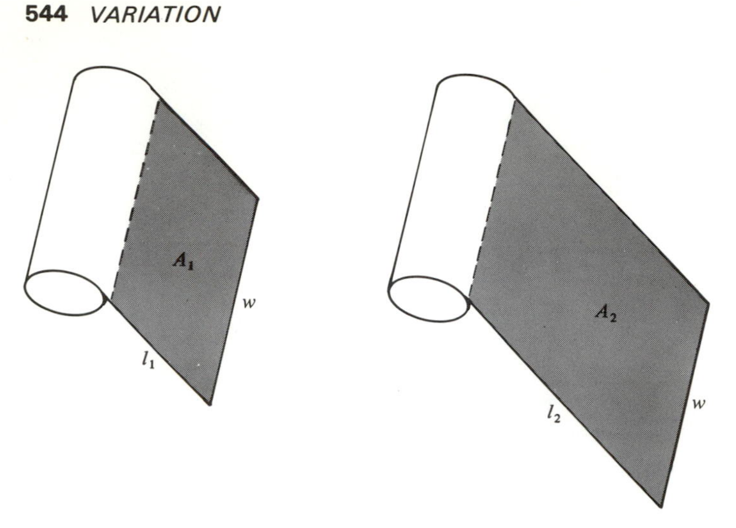

FIGURE 16.1

for some constant k ($\neq 0$). In variational formulas, k is called the **constant of variation.**

Note that if y varies directly as x, then x also varies directly as y. In fact,

$$x = \frac{1}{k}y,$$

and $\frac{1}{k}$ ($\neq 0$) is the constant of variation. Also,

$$\frac{y}{x} = k,$$

when $x \neq 0$. Thus *if y varies directly as x, the quotient $\frac{y}{x}$ remains constant.* Therefore, if x_1 and x_2 are (nonzero) values of x and if y_1 and y_2 are the corresponding values of y, then

$$\frac{y_1}{x_1} = \frac{y_2}{x_2} = k.$$

Corresponding values of y and x are proportional.

EXAMPLE 1. Suppose $y = 2x$ for $x > 0$. Then y varies directly as x. The constant of variation is 2. As x varies over positive numbers, the corresponding values of y and x are proportional. The ratio $\frac{y}{x}$ remains 2. A few of these corresponding values are given in Table 16.1. Fill in the blanks.

x	1	2	$\frac{5}{2}$	4.1	
$y = 2x$	2	4	5		14

TABLE 16.1

SOLUTION.

$$y = 2x$$

When $x = 4.1$,

$$y = 2(4.1) = 8.2$$

When $y = 14$,

$$14 = 2x$$

and

$$7 = x$$

■

When y varies directly as x, a function is defined. The function is given by the equation

$$y = kx.$$

Recall that the graph of this function is the straight line through the origin with slope k. In Figure 16.2, the function described in Example 1 (with $k = 2$) is graphed.

Sometimes y varies directly, not as x, but as a power of x.

DEFINITION. Let x and y be variables and let n be a positive integer. Then y **varies directly as the nth power of x** if

$$y = kx^n$$

for some nonzero constant k.

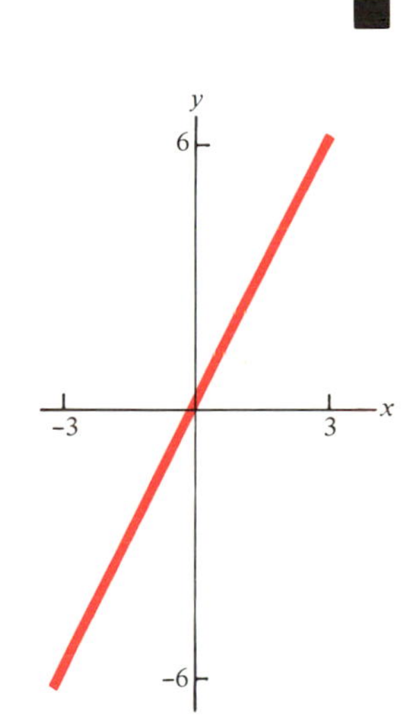

FIGURE 16.2 The graph of the function $y = 2x$.

EXAMPLE 2. Suppose y varies directly as x^2. Suppose, further, that $y = 12$ when $x = 2$. Find y when $x = 3$.

SOLUTION.

$$y = kx^2$$

Substitute **12** for y and 2 for x.

$$\mathbf{12} = k(2^2)$$

$$12 = 4k$$

$$3 = k$$

The constant of variation is 3, and

$$y = 3x^2.$$

Let $x = 3$.

$$y = 3(3^2)$$

$$= 3 \cdot 9$$

$$y = 27$$

■

EXAMPLE 3. The cost of building an expansion bridge varies as the cube of its length. If a 100-foot expansion bridge costs $40 000 to build, how much does a 500-foot bridge cost?

SOLUTION. Let c be the cost (in dollars) and let l be the length of the bridge (in feet). Then

$$c = kl^3$$

When $l = 100$, $\mathbf{c = 40\,000}$.

$$\mathbf{40\,000} = k(100^3)$$

$$40\,000 = 1\,000\,000k$$

$$\frac{4}{100} = k$$

$$\frac{1}{25} = k$$

Now let $l = 500$ in the formula

$$c = \frac{l^3}{25}.$$

Then

$$c = \frac{500^3}{25} = \frac{125\,000\,000}{25}$$

$$c = 5\,000\,000$$

The cost of a 500-foot bridge is $5 000 000.

EXERCISES

In Exercises 1–8 assume y varies directly as x. Determine the constant of variation in each case.

1. $y = 6$ when $x = 2$
2. $y = -6$ when $x = 3$
3. $y = 5$ when $x = 10$
4. $y = -3$ when $x = 3$
5. $y = 100$ when $x = 4$
6. $y = -1$ when $x = -3$
7. $y = 9$ when $x = 4$
8. $y = -5$ when $x = 7$

In Exercises 9–16 assume y varies directly as x.

9. Suppose $y = 8$ when $x = 4$. Find y when $x = 6$.
10. Suppose $y = 9$ when $x = 3$. Find y when $x = 9$.
11. Suppose $y = 5$ when $x = 2$. Find y when $x = 10$.
12. Suppose $y = -4$ when $x = 3$. Find y when $x = -12$.
13. Suppose $y = 10$ when $x = 2$. Find x when $y = 30$.
14. Suppose $y = -2$ when $x = 1$. Find x when $y = 1$.
15. Suppose $y = 18$ when $x = 6$. Find x when $y = 6$.
16. Suppose $y = 5$ when $x = 3$. Find x when $y = 3$.

In Exercises 17–22 assume y varies directly as x^2. Determine the constant of variation in each case.

17. $y = 4$ when $x = 2$
18. $y = 4$ when $x = 1$
19. $y = 4$ when $x = -1$
20. $y = 18$ when $x = 3$
21. $y = 24$ when $x = -2$
22. $y = 12$ when $x = 3$

In Exercises 23–26 assume y varies directly as x^2.

23. Suppose $y = 12$ when $x = 2$. Find y when $x = 4$.
24. Suppose $y = 45$ when $x = 3$. Find y when $x = 1$.
25. Suppose $y = 2$ when $x = 4$. Find y when $x = 1$.
26. Suppose $y = 1$ when $x = -1$. Find y when $x = 2$.
27. Suppose y varies directly as x^3 and suppose $y = 54$ when $x = 3$. What is the constant of variation?
28. Suppose y varies directly as x^4. If $y = 8$ when $x = 2$, find y when $x = 1$.
29. Suppose y varies directly as x, and $y = 6$ when x is 2.
    (a) What function is defined?
    (b) Graph this function.
30. Suppose y varies directly as x, and $y = -5$ when $x = 5$.
    (a) What function is defined?
    (b) Graph this function.

31. Let y vary directly as x. Find y when $x = 0$.
32. Let y vary directly as x^5. Find y when $x = 0$.
33. Fill in:
    (a) The ________________ of a circle varies directly as the radius length.
    (b) The ________________ of a circle varies directly as the radius length squared.
34. The amount of bread consumed in a town varies directly as the population. If 80 000 pounds per week are consumed in a town of 20 000 inhabitants, how many pounds of bread per week are consumed in a town of 15 000 inhabitants?
35. At a constant rate the distance traveled varies directly as time. A car travels 110 miles in 2 hours. At this rate, how far does it travel in $3\frac{1}{2}$ hours?
36. A typist finds that the (approximate) number of errors she makes varies as the square of her typing speed. If she makes 2 errors per page at 40 words per minute, how many errors per page does she make at 80 words per minute?

16.2 INVERSE VARIATION

In addition to *direct* variation, you will also encounter *inverse* variation.

DEFINITION. Let x and y be variables. Then y **varies inversely as** x if

$$y = \frac{k}{x}$$

for some nonzero constant k.

Note that if y varies inversely as x, then x varies inversely as y with the same constant of variation k. In fact, by cross-multiplication,

$$x = \frac{k}{y}$$

Furthermore, *the product of x and y remains constant:*

$$xy = k$$

EXAMPLE 1. Suppose $y = \frac{12}{x}$.

(a) Determine y when $x = 2$ and when $x = 6$.
(b) Determine x when $y = 4$ and when $y = 9$.

SOLUTION.

$$y = \frac{12}{x}$$

Here y varies inversely as x. The constant of variation is 12.

(a) When $x = 2$,

$$y = \frac{12}{2} = 6$$

When $x = 6$,

$$y = \frac{12}{6} = 2$$

(b) When $y = 4$,

$$4 = \frac{12}{x}$$

Cross-multiply to obtain

$$x = \frac{12}{4} = 3.$$

When $y = 9$,

$$9 = \frac{12}{x}$$

$$x = \frac{12}{9} = \frac{4}{3}$$

■

The graph of

$$y = \frac{12}{x}$$

or

$$xy = 12$$

is a hyperbola whose asymptotes are the coordinate axes. [See Figure 16.3.] In general, when y varies inversely as x, a function is defined by

$$y = \frac{k}{x}.$$

The graph of this function is a hyperbola. ■

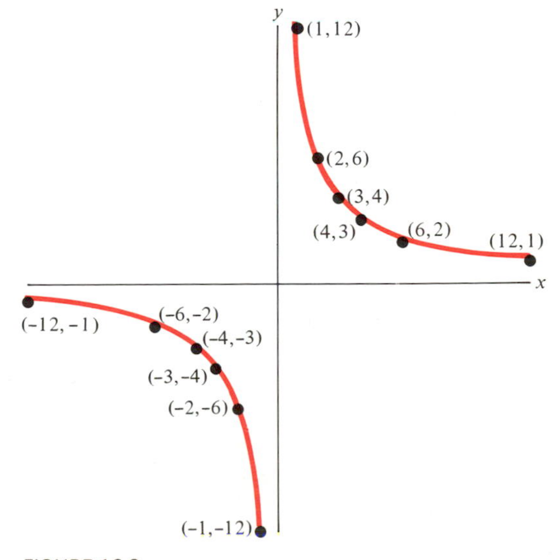

FIGURE 16.3

EXAMPLE 2. Boyle's Law states that the pressure p of a compressed gas varies inversely as the volume v of gas. The pressure is 50 pounds per square inch when the volume is 200 cubic inches. Determine the pressure when the gas is compressed to 125 cubic inches.

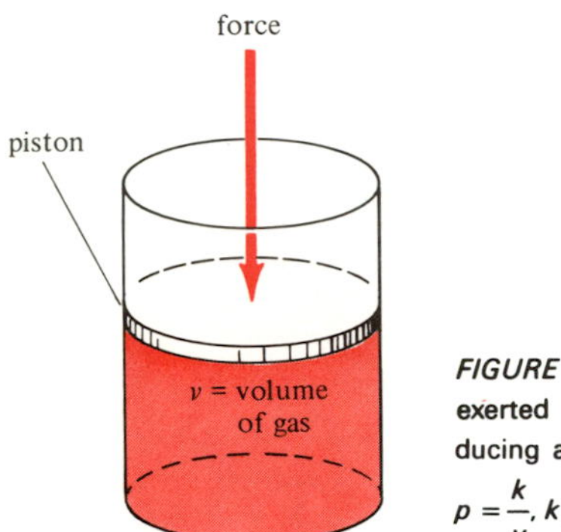

FIGURE 16.4 A force is exerted on a piston, producing a pressure p, where $p = \frac{k}{v}$, $k (\neq 0)$ constant.

SOLUTION. p varies inversely as v.

$$p = \frac{k}{v}$$

When $\mathbf{p} = \mathbf{50}, v = 200$:

$$\mathbf{50} = \frac{k}{200}$$

$$10\,000 = k$$

Thus

$$p = \frac{10\,000}{v}$$

Let $v = \mathbf{125}$.

$$p = \frac{10\,000}{\mathbf{125}} = 80$$

When the volume of gas is 125 cubic inches, the pressure is 80 pounds per square inch. ■

Just as y can vary *directly* as a *power of x*, so too, y can vary *inversely* as a *power of x*.

DEFINITION. Let x and y be variables and let n be a positive integer. Then y varies inversely as x^n if

$$y = \frac{k}{x^n}$$

for some constant $k \neq 0$.

EXAMPLE 3. Suppose $y = \dfrac{-72}{x^3}$, $x \neq 0$. Determine y when:

(a) $x = -1$ (b) $x = 2$ (c) $x = -3$

SOLUTION. Here y varies inversely as x^3. The constant of variation is -72.

(a) Let $x = -1$.

$$y = \frac{-72}{(-1)^3} = \frac{-72}{-1} = 72$$

(b) Let $x = 2$.

$$y = \frac{-72}{2^3} = \frac{-72}{8} = -9$$

(c) Let $x = -3$.

$$y = \frac{-72}{(-3)^3} = \frac{-72}{-27} = \frac{8}{3}$$

■

EXERCISES

In Exercises 1–6 assume that y varies inversely as x. Determine the constant of variation in each case.

1. $y = 4$ when $x = 2$
2. $y = 5$ when $x = 5$
3. $y = -3$ when $x = -2$
4. $y = 9$ when $x = -4$
5. $y = \frac{1}{2}$ when $x = \frac{1}{3}$
6. $y = \frac{1}{20}$ when $x = 10$

In Exercises 7–12 assume that y varies inversely as x.

7. Suppose $y = 3$ when $x = 2$. Find y when $x = 3$.
8. Suppose $y = 12$ when $x = 4$. Find y when $x = 16$.
9. Suppose $y = -20$ when $x = 2$. Find y when $x = -4$.
10. Suppose $y = \frac{1}{2}$ when $x = 4$. Find y when $x = 8$.
11. Suppose $y = 9$ when $x = 4$. Find x when $y = 12$.
12. Suppose $y = 10$ when $x = 7$. Find x when $y = 35$.

In Exercises 13–16 assume that y varies inversely as x^2. Determine the constant of variation in each case.

13. $y = 5$ when $x = 2$

14. $y = 4$ when $x = 3$

15. $y = -2$ when $x = 5$

16. $y = \frac{1}{5}$ when $x = 10$

In Exercises 17–20 assume that y varies inversely as x^2.

17. Suppose $y = 4$ when $x = 4$. Find y when $x = 8$.
18. Suppose $y = 10$ when $x = 2$. Find y when $x = 1$.
19. Suppose $y = 4$ when $x = 2$. Find y when $x = 8$.
20. Suppose $y = 1$ when $x = 1$. Find y when $x = 3$.
21. Suppose y varies inversely as x, and $y = 3$ when $x = 2$.
    (a) What function is defined?
    (b) Graph this function.
    (c) What figure is represented?
22. Suppose y varies inversely as x, and $y = \frac{1}{2}$ when $x = 2$.
    (a) What function is defined?
    (b) Graph this function.
23. Let y vary inversely as x. Can $y = 0$?

In Exercises 24–27 fill in "directly" or "inversely".

24. When length is held constant, the area of a rectangle varies ______________ as the width.
25. When width is held constant, the length of a rectangle varies ______________ as the area.
26. When area is held constant, the length of a rectangle varies ______________ as the width.
27. If all seats are $2.50, the gross receipts of a movie theatre vary ______________ as the number of customers.
28. The beetle population of a field varies inversely as the amount of pesticide used. If 1000 beetles are in the field when 40 pounds of pesticide are used, how many pounds of pesticide are needed to bring the beetle population down to 100?
29. The weight of a body varies inversely as its distance from the center of the earth. Assume the length of the earth's radius is 4000 miles. If a man weighs 150 pounds on earth, how much does he weigh 1000 miles above the surface of the earth?
30. The resistance of an electrical wire of fixed length varies inversely as the square of the radius length of the wire. When the radius length is .005 inch the resistance is 20 ohms. What is the resistance of a wire of radius length .01 inch?

*16.3 JOINT VARIATION

Three or more variables may change so as to preserve a constant quotient.

DEFINITION. Let x, y, z, t be variables. Then t **varies jointly as** x **and** y if

$$t = kxy$$

for some constant k ($\neq 0$). Similarly, t varies jointly as x, y, and z if

$$t = kxyz$$

for some constant k ($\neq 0$).

EXAMPLE 1. Assume t varies jointly as x and y. Suppose $t = 500$ when $x = 5$ and $y = 20$. Find t when $x = 10$ and $y = 15$.

SOLUTION.

$$t = kxy$$

Let $t = 500$, $x = 5$, $y = 20$. Then

$$500 = k \cdot 5 \cdot 20$$

$$500 = 100k$$

$$5 = k$$

Thus

$$t = 5xy$$

When $x = 10$ and $y = 15$,

$$t = 5 \cdot 10 \cdot 15$$

$$t = 750$$

■

t can vary jointly as various combinations of *powers* of x, y, z. For example, t varies jointly as x and the square of y if

$$t = kxy^2$$

for some constant k ($\neq 0$). Similarly, t varies jointly as the square of y and the cube of z if

$$t = ky^2z^3$$

for some $k \neq 0$.

*Optional topic.

EXAMPLE 2. Assume t varies jointly as x, y, and z^2. If $t = 450$ when $x = 2$, $y = 3$, $z = 5$, find t when $x = 4$, $y = 1$, $z = 10$.

SOLUTION.

$$t = kxyz^2$$

Let $t = 450$, $x = 2$, $y = 3$, $z = 5$. Then

$$450 = k \cdot 2 \cdot 3 \cdot 5^2$$

$$450 = 150k$$

$$3 = k$$

Therefore

$$t = 3xyz^2$$

When $x = 4$, $y = 1$, $z = 10$,

$$t = 3 \cdot 4 \cdot 1 \cdot 10^2$$

$$t = 1200$$

■

DEFINITION. Let x, y, t be variables. Then t **varies inversely as** x **and** y if

$$t = \frac{k}{xy}$$

for some constant k ($\neq 0$).

(The word "jointly" is not usually used in the case of inverse variation.)

EXAMPLE 3. Assume t varies inversely as x and y. If $t = 6$ when $x = 2$ and $y = 5$, find x when $t = 15$ and $y = 2$.

SOLUTION.

$$t = \frac{k}{xy}$$

Let $t = 6$, $x = 2$, $y = 5$. Then

$$6 = \frac{k}{2 \cdot 5}$$

$$60 = k$$

$$t = \frac{60}{xy}$$

Let $t = 15$, $y = 2$. Then

$$15 = \frac{60}{2x}$$

$$x = \frac{60}{2 \cdot 15} = 2$$ ■

Other possibilities for inverse variation can occur. For example, t varies inversely as x, y, and the cube of z, if

$$t = \frac{k}{xyz^3}$$

for some constant $k\,(\neq 0)$.

Finally combinations of direct and inverse variations can occur. For example, t varies directly as x and inversely as y if

$$t = \frac{kx}{y}$$

for some constant $k\,(\neq 0)$. Also, t varies jointly as x and the square of y and inversely as z if

$$t = \frac{kxy^2}{z}$$

for some constant $k\,(\neq 0)$.

EXAMPLE 4. The resistance of an electrical wire varies directly as the length of the wire and inversely as the square of its radius length.

The resistance is 40 ohms when the wire is 2000 feet long and the radius length is .1 inch. Determine the resistance when the wire is 5000 feet long and the radius length is .2 inch.

SOLUTION. Let R = resistance, l = length of the wire (in inches), r = radius length (in inches). Then

$$R = \frac{kl}{r^2}$$

Also, $R = 40$ when $l = 2000 \times 12 = 24\,000$ (inches) and $r = .1$. Therefore

$$40 = \frac{k \cdot 24\,000}{(.1)^2}$$

$$40(.01) = 24\,000k$$

$$\frac{.4}{24\,000} = k$$

$$\frac{1}{60\,000} = k$$

Thus

$$R = \frac{l}{60\,000r^2}$$

When $l = 5000 \times 12 = 60\,000$ and $r = .2$,

$$R = \frac{60\,000}{60\,000\ (.2)^2}$$

$$R = \frac{1}{.04} = \frac{100}{4} = 25$$

The resistance is 25 ohms. ■

EXERCISES

In Exercises 1–6 assume t varies jointly as x and y. Determine the constant of variation in each case.

1. $t = 24$ when $x = 2$ and $y = 3$
2. $t = 90$ when $x = 3$ and $y = 6$
3. $t = 132$ when $x = -11$ and $y = -4$
4. $t = -100$ when $x = -2$ and $y = -20$
5. $t = 1$ when $x = 4$ and $y = \frac{1}{2}$
6. $t = 3$ when $x = 2$ and $y = 5$

In Exercises 7–12 assume t varies jointly as x and y.

7. Suppose $t = 18$ when $x = 2$ and $y = 3$. Find t when $x = 3$ and $y = 4$.
8. Suppose $t = 40$ when $x = 5$ and $y = 2$. Find t when $x = 7$ and $y = 9$.
9. Suppose $t = 200$ when $x = 50$ and $y = 8$. Find t when $x = 100$ and $y = 10$.
10. Suppose $t = 30$ when $x = 5$ and $y = 2$. Find y when $t = 45$ and $x = 3$.
11. Suppose $t = 12$ when $x = 8$ and $y = 9$. Find y when $t = 2$ and $x = 3$.
12. Suppose $t = .1$ when $x = .02$ and $y = .5$. Find t when $x = .4$ and $y = .05$.
13. Suppose t varies jointly as x, y, and z. If $t = 16$ when $x = 2$, $y = 1$, $z = 2$, find t when $x = 4$, $y = 3$, $z = 3$.
14. Suppose t varies jointly as u, v, and w. If $t = 20$ when $u = \frac{1}{2}$ $v = 4$, $w = 5$, find u when $t = 40$, $v = 10$, $w = 4$.
15. Suppose t varies jointly as x and y^2. If $t = 96$ when $x = 3$ and $y = 4$, find t when $x = 5$ and $y = -3$.
16. Suppose t varies jointly as x and y^3. If $t = 270$ when $x = 5$ and $y = 3$, find t when $x = 3$ and $y = 2$.
17. Suppose t varies jointly as x, y, and z^2. If $t = 400$ when $x = 5$, $y = 2$, $z = 4$, find t when $x = -1$, $y = -4$, $z = 5$.
18. Suppose t varies jointly as x^2, y^3, and z^4. If $t = 96$ when $x = \frac{1}{2}$, $y = 2$, $z = -2$, find t when $x = \frac{1}{3}$, $y = 3$, $z = -3$.

In Exercises 19–24 assume t varies inversely as x and y. Determine the constant of variation in each case.

19. $t = 5, x = 2, y = 2$

20. $t = 4, x = 1, y = 3$

21. $t = 6, x = -2, y = \frac{1}{2}$

22. $t = 9, x = 2, y = \frac{1}{3}$

23. $t = 40, x = .1, y = .2$

24. $t = \frac{2}{3}, x = \frac{1}{4}, y = 9$

In Exercises 25–30 assume t varies inversely as x and y.

25. If $t = 12$ when $x = 1$ and $y = 3$, find t when $x = 2$ and $y = 6$.

26. If $t = 9$ when $x = 2$ and $y = 5$, find t when $x = 15$ and $y = 3$.

27. If $t = 20$ when $x = 4$ and $y = 7$, find t when $x = 14$ and $y = 4$.

28. If $t = -2$ when $x = -5$ and $y = -3$, find t when $x = 6$ and $y = \frac{1}{2}$.

29. If $t = \frac{1}{2}$ when $x = \frac{1}{2}$ and $y = \frac{1}{2}$, find x when $t = 1$ and $y = \frac{1}{4}$.

30. If $t = 50$ when $x = \frac{1}{2}$ and $y = \frac{1}{5}$, find y when $t = 2$ and $x = 5$.

31. Suppose t varies inversely as x, y, and z. If $t = 4$ when $x = 2$, $y = 6$, $z = 3$, find t when $x = 9, y = 2, z = \frac{4}{3}$.

32. Suppose t varies inversely as x, y, and z^3. If $t = 6$ when $x = 2$, $y = -1$, $z = 1$, find t when $x = 4$, $y = 2$, $z = -2$.

33. Suppose t varies directly as x and inversely as y. If $t = 40$ when $x = 4$ and $y = 3$, find t when $x = 6$ and $y = 2$.

34. Suppose t varies directly as x and inversely as y. If $t = 100$ when $x = 5$ and $y = 2$, find x when $t = 20$ and $y = 4$.

35. Suppose t varies jointly as x and y and inversely as z. If $t = 10$ when $x = 5$, $y = 4$, $z = 8$, find t when $x = 20$, $y = 2$, $z = 5$.

36. Suppose t varies jointly as x and y and inversely as z^2. If $t = 100$ when $x = 5$, $y = 10, z = 3$, find t when $x = 9, y = 4, z = 6$.

37. Suppose t varies jointly as x^2 and y and inversely as z. If $t = 10$ when $x = 2$, $y = 5, z = 8$, find t when $x = 4, y = 2, z = 16$.

38. Suppose t varies directly as x^3 and inversely as y and z^2. If $t = 12$ when $x = 2$, $y = 4, z = 3$, find t when $x = 3, y = 18, z = 3$.

39. A manufacturer constructs rectangular cartons of the same height. The volume of these cartons varies jointly as their length and width. If the volume is 400 cubic inches when the length is 8 inches and the width is 5 inches, what is the volume when the length is 1 foot and the width 9 inches?

40. The volume of a right circular cylinder varies jointly as its height and the square of its base radius length. The volume is 144π cubic inches when the height is 4 inches

and the base radius length is 6 inches. Find the height when the volume is 1200π cubic inches and the base radius length is 10 inches.

41. Several particles rotate about an axis, each with constant velocity. The kinetic energy of each particle varies jointly as its mass and the square of its velocity. One particle has mass 5 kilograms, velocity 12 meters per second, and kinetic energy 30 joules. Find the kinetic energy of a second particle with mass 30 kilograms and velocity 10 meters per second.

42. Let d be the distance between two particles of masses m and M. Gravitational attraction varies jointly as m and M and inversely as d^2. A 10-pound mass and a 5-pound mass that are 100 feet apart attract each other with a force F. Determine the gravitational attraction for a 40-pound mass and a 15-pound mass that are 200 feet apart. (Your answer should be a multiple of F.)

43. The safe-load of a rectangular beam varies jointly as its width and the square of its depth, and inversely as the length between supports. For a beam of width 8 feet, of depth 5 feet, and of length 20 feet between supports, the safe-load is 3000 pounds. What is the safe-load of a beam that is of width 10 feet, of depth 10 feet, and of length 30 feet between supports?

REVIEW EXERCISES FOR CHAPTER 16

16.1 DIRECT VARIATION

1. Suppose y varies directly as x. Determine the constant of variation if $y = 8$ when $x = 2$.

2. Suppose y varies directly as x, and $y = 12$ when $x = -3$. Find y when $x = 6$.

3. Suppose y varies directly as x^2. Determine the constant of variation if $y = 8$ when $x = 2$.

4. Suppose y varies directly as x^2, and $y = 20$ when $x = 2$. Find y when $x = 3$.

5. Suppose y varies directly as x, and $y = 2$ when $x = -4$. What function is defined?

16.2 INVERSE VARIATION

6. Suppose y varies inversely as x. Determine the constant of variation if $y = 9$ when $x = 3$.

7. Suppose y varies inversely as x, and $y = 2$ when $x = 3$. Find y when $x = -6$.

8. Suppose y varies inversely as x^2. Determine the constant of variation if $y = 4$ when $x = 4$.

9. Suppose y varies inversely as x^2, and $y = 2$ when $x = 4$. Find x when $y = 4$.

10. Fill in "directly" or "inversely".

    (a) The distance a car travels at a constant rate varies ____________ as time.

    (b) When area is held constant, the width of a rectangle varies ____________ as the length.

16.3 JOINT VARIATION

11. Assume t varies jointly as x and y. Suppose $t = 48$ when $x = 3$ and $y = 4$. Determine the constant of variation.

12. Assume t varies jointly as x and y. Suppose $t = 30$ when $x = 2$ and $y = 3$. Find t when $x = 10$ and $y = 1$.

13. Assume t varies jointly as x, y, and z. Suppose $t = 100$ when $x = 2$, $y = 25$, $z = \frac{1}{2}$. Find t when $x = 3, y = 4, z = 5$.

14. Assume t varies inversely as x and y. Suppose $t = 1$ when $x = 2$ and $y = 3$. Determine the constant of variation.

15. Suppose t varies inversely as x and y. If $t = 4$ when $x = 2$ and $y = 1$, find t when $x = 2$ and $y = 4$.

16. Suppose t varies jointly as x and y and inversely as z. If $t = 30$ when $x = 10$, $y = 3, z = 2$, find t when $x = 2, y = 5, z = 1$.

17. A manufacturer constructs rectangular boxes of the same height. The volume of these boxes varies jointly as their length and width. If the volume is 120 cubic inches when the length is 8 inches and the width is 3 inches, what is the volume when the length is 1 foot and the width is 10 inches?

TEST YOURSELF ON CHAPTER 16.

Choose any 7 of these. Problems 8 and 9 are based on Section 16.3.

1. Suppose y varies directly as x. Determine the constant of variation if $y = 20$ when $x = 2$.

2. Suppose y varies directly as x, and $y = 100$ when $x = 40$. Find y when $x = 8$.

3. Suppose y varies directly as x^2, and $y = 12$ when $x = 2$. Find y when $x = 6$.

4. Suppose y varies directly as x, and $y = 8$ when $x = 2$. What function is defined?

5. Suppose y varies inversely as x. Determine the constant of variation if $y = 10$ when $x = 6$.

6. Suppose y varies inversely as x, and $y = 9$ when $x = 1$. Find x when $y = 6$.

7. Suppose y varies inversely as x^2, and $y = 8$ when $x = \frac{1}{2}$. Find y when $x = 2$.

8. Assume t varies jointly as x and y. Suppose $t = 60$ when $x = 5$ and $y = 3$. Find t when $x = 8$ and $y = 5$.

9. Assume t varies directly as x and inversely as y. Suppose $t = 3$ when $x = 6$ and $y = 10$. Find t when $x = 3$ and $y = 20$.

17

progressions

17.1 SEQUENCES

DEFINITION. A **sequence** is a function whose domain is the set of positive integers.

The letter "a" is usually used to denote a sequence. Instead of writing the function values of a sequence as $a(1)$, $a(2)$, $a(3)$, and so on, it is customary to write:

$$a_1, a_2, a_3, \ldots, a_i, \ldots$$

In each place the three dots signify "and so on". Thus a sequence consists of infinitely many pairs:

$$(1, a_1), (2, a_2), (3, a_3), \ldots$$

DEFINITION. The function values of a sequence,

$$a_1, a_2, \ldots, a_i, \ldots$$

are called the **terms** of the sequence. a_1 is called the **first term,** a_2 is the **second** term, a_i is the **ith term.**

A sequence is often defined by indicating the ith term a_i, as in Example 1.

EXAMPLE 1. Consider the sequence defined by

$$a_i = \frac{i}{2}.$$

Its terms are given by

$$a_1 = \frac{1}{2},\ a_2 = \frac{2}{2} = 1,\ a_3 = \frac{3}{2},\ a_4 = 2,\ a_5 = \frac{5}{2},\ a_6 = 3,\ \ldots .$$

Thus $a_{47} = \dfrac{47}{2}$ and $a_{48} = 24$ ■

A sequence is a function and can therefore be graphed. The graph consists of the points

$$(1, a_1),\ (2, a_2),\ (3, a_3),\ \ldots$$

corresponding to the pairs of the sequence. Of course, only finitely many points can be shown in a graph. But often, the general picture can be suggested.

EXAMPLE 2. Graph the sequence of Example 1, given by

$$a_i = \frac{i}{2}.$$

SOLUTION. See Figure 17.1. ■

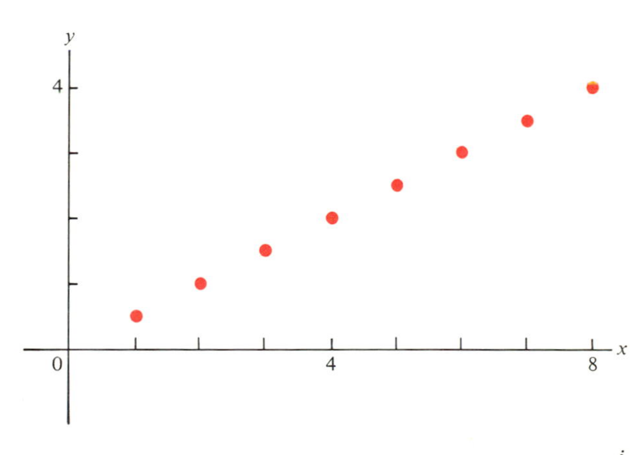

FIGURE 17.1 The graph of the sequence given by $a_i = \dfrac{i}{2}$

EXAMPLE 3.

(a) Determine the first 8 terms of the sequence given by

$$a_i = i - 4.$$

(b) Graph the sequence.

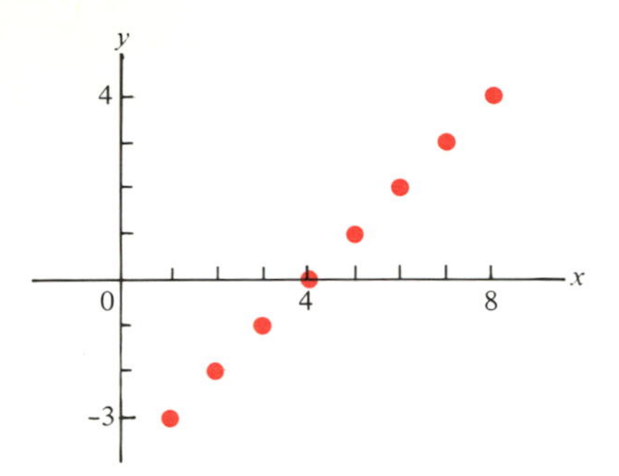

FIGURE 17.2 The graph of the sequence given by $a_i = i - 4$

SOLUTION.

(a) $a_1 = 1 - 4 = -3$ $\quad a_5 = 1$

$a_2 = 2 - 4 = -2$ $\quad a_6 = 2$

$a_3 = 3 - 4 = -1$ $\quad a_7 = 3$

$a_4 = 4 - 4 = 0$ $\quad a_8 = 4$

(b) See Figure 17.2. ■

EXAMPLE 4.

(a) Determine the first 8 terms of the sequence given by

$$a_i = 5 - \frac{i}{2}.$$

(b) Graph this sequence.

SOLUTION.

(a) $a_1 = 5 - \frac{1}{2} = 4.5$ $\quad a_2 = 5 - \frac{2}{2} = 4$ $\quad a_3 = 5 - \frac{3}{2} = 3.5$

$a_4 = 5 - \frac{4}{2} = 3$ $\quad a_5 = 2.5$ $\quad a_6 = 2$

$a_7 = 1.5$ $\quad a_8 = 1$

(b) See Figure 17.3. ■

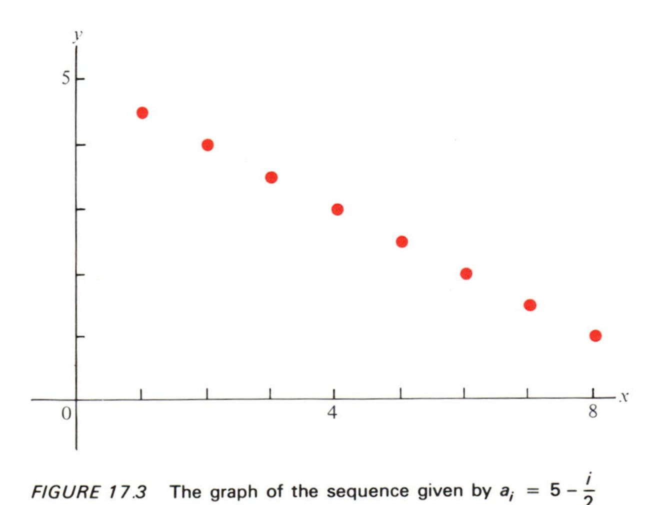

FIGURE 17.3 The graph of the sequence given by $a_i = 5 - \frac{i}{2}$

DEFINITION. A sequence $a_1, a_2, a_3, \ldots, a_i, \ldots$ is said to be **increasing** if

$$a_i < a_{i+1}$$

for *all* $i = 1, 2, \ldots$, and **decreasing** if

$$a_i > a_{i+1}$$

for *all* $i = 1, 2, \ldots$.

Thus for an increasing sequence, the terms get *larger* as the subscripts i get *larger*. For a decreasing sequence, the terms get *smaller* as the subscripts get *larger*.

EXAMPLE 5.

(a) The sequences

$$a_i = \frac{i}{2}$$

of Examples 1 and 2 and

$$a_i = i - 4$$

of Example 3 are increasing.

(b) The sequence

$$a_i = 5 - \frac{i}{2}$$

of Example 4 is decreasing. ■

Here is a sequence that is neither increasing nor decreasing.

EXAMPLE 6. Consider the sequence given by

$$a_i = (-1)^i.$$

$$a_1 = (-1)^1 = -1$$
$$a_2 = (-1)^2 = 1$$
$$a_3 = (-1)^3 = -1$$
$$a_4 = (-1)^4 = 1$$
$$a_5 = (-1)^5 = -1$$
$$a_6 = (-1)^6 = 1$$
$$a_7 = (-1)^7 = -1$$
$$a_8 = (-1)^8 = 1$$

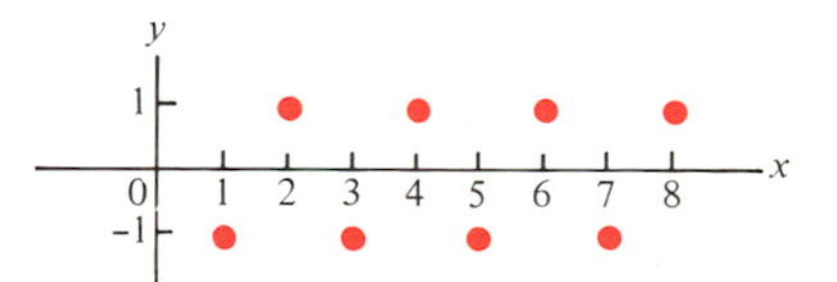

FIGURE 17.4 The graph of the sequence given by $a_i = (-1)^i$

In general,

$$a_i = -1, \text{ if } i \text{ is } \textit{odd};$$

$$a_i = 1, \text{ if } i \text{ is } \textit{even}$$

The sequence is also given by:

$$-1,\ 1,\ -1,\ 1,\ -1,\ 1,\ -1,\ \ldots$$

Observe that

$$a_3 < a_2.$$

Therefore the sequence is *not* increasing. Also,

$$a_2 > a_1$$

Therefore the sequence is *not* decreasing. ■

EXAMPLE 7. Describe the ith term a_i of the suggested sequences.

(a) $5,\ 10,\ 15,\ 20,\ 25,\ 30,\ \ldots$

(b) $4,\ 3,\ 2,\ 1,\ 0,\ -1,\ -2,\ \ldots$

(c) $\frac{1}{2},\ \frac{1}{4},\ \frac{1}{8},\ \frac{1}{16},\ \frac{1}{32},\ \frac{1}{64},\ \ldots$

(d) $-2,\ 4,\ -8,\ 16,\ -32,\ 64,\ \ldots$

SOLUTION.

(a) Let $a_i = 5i$ for $i = 1, 2, \ldots$.

(b) Let $a_i = 5 - i$ for $i = 1, 2, \ldots$. Then $a_1 = 5 - 1 = 4$, $a_2 = 5 - 2 = 3$, $a_3 = 5 - 3 = 2$, and so on.

(c) Let $a_i = \left(\frac{1}{2}\right)^i \left(\text{or } \frac{1}{2^i}\right)$ for $i = 1, 2, \ldots$.

(d) 2^i would yield $2, 4, 8, 16, \ldots$.

To get alternating signs, take

$$a_i = (-2)^i, \text{ for } i = 1, 2, 3, \ldots .$$

■

EXERCISES

In Exercises 1–26:

(a) Determine the first 8 terms of the indicated sequence.

(b) Is the sequence (i) increasing, (ii) decreasing, (iii) neither?

(c) Graph the sequence, if so indicated.

1. $a_i = i$ (Graph.)

2. $a_i = i + 2$ (Graph.)

3. $a_i = i + 5$

4. $a_i = i + 1$

5. $a_i = -i$ (Graph.)

6. $a_i = 10 - i$ (Graph.)

7. $a_i = 2i$ (Graph.)

8. $a_i = -2i$ (Graph.)

9. $a_i = 3i$

10. $a_i = \frac{-i}{3}$

11. $a_i = 5$, for all i

12. $a_i = -1$, for all i (Graph.)

13. $a_i = \begin{cases} 5, & \text{if } i \text{ is odd,} \\ 3, & \text{if } i \text{ is even.} \end{cases}$

14. $a_i = \begin{cases} 1, & \text{if } i \text{ is odd,} \\ 0, & \text{if } i \text{ is even.} \end{cases}$ (Graph.)

15. $a_i = \frac{1}{i}$

16. $a_i = 1 - \frac{1}{i}$

17. $a_i = 2^i$

18. $a_i = (-2)^i$

19. $a_i = \left(\frac{1}{2}\right)^i$

20. $a_i = \left(\frac{-1}{2}\right)^i$

21. $a_i = 6 + 3i$

22. $a_i = 5 - 2i$

23. $a_i = 1 + \frac{i}{2}$

24. $a_i = -1 + \frac{i}{2}$

25. $a_i = 2 + .1i$

26. $a_i = 4 - .2i$

In Exercises 27–38 for the indicated sequences, determine the specified terms.

27. $a_i = 5i$

(a) a_4 (b) a_{10} (c) a_{20} (d) a_{100}

28. $a_i = -10i$

(a) a_1 (b) a_{12} (c) a_{100} (d) a_{200}

29. $a_i = 3 + 4i$

(a) a_1 (b) a_5 (c) a_{10} (d) a_{20}

30. $a_i = 1 - 3i$

(a) a_1 (b) a_4 (c) a_7 (d) a_{10}

31. $a_i = 2 + \frac{5i}{2}$

(a) a_1 (b) a_4 (c) a_8 (d) a_9

32. $a_i = 3 - \frac{i}{3}$

(a) a_1 (b) a_3 (c) a_9 (d) a_{27}

33. $a_i = 1 + (-1)^i$

(a) a_1 (b) a_2 (c) a_{25} (d) a_{26}

34. $a_i = 4 + (-1)^i$

(a) a_1 (b) a_2 (c) a_{99} (d) a_{100}

35. $a_i = \dfrac{12}{i}$

(a) a_2 (b) a_6 (c) a_{12} (d) a_{48}

36. $a_i = \dfrac{-144}{i}$

(a) a_2 (b) a_9 (c) a_{16} (d) a_{72}

37. $a_i = \dfrac{i}{i + 1}$

(a) a_1 (b) a_2 (c) a_9 (d) a_{10}

38. $a_i = \dfrac{3i - 1}{5 + i}$

(a) a_1 (b) a_3 (c) a_5 (d) a_{15}

In each of the Exercises 39–50, describe the ith term, a_i, of the suggested sequence.

39. 2, 3, 4, 5, 6, 7, ...

40. −1, −2, −3, −4, −5, −6, ...

41. 5, 6, 7, 8, 9, 10, ...

42. −3, −2, −1, 0, 1, 2, ...

43. 3, 6, 9, 12, 15, 18, ...

44. −7, −14, −21, −28, −35, −42, ...

45. −4, 8, −12, 16, −20, 24, ...

46. 5, 7, 9, 11, 13, 15, ...

47. 2, 5, 8, 11, 14, 17, ...

48. 3, 9, 27, 81, 243, 729, ...

49. $2, \frac{3}{2}, \frac{4}{3}, \frac{5}{4}, \frac{6}{5}, \frac{7}{6}, \ldots$

50. 1, 4, 9, 16, 25, 36, ...

17.2 SIGMA NOTATION

There is a compact way of indicating addition.

DEFINITION. Let $a_1, a_2, a_3, \ldots, a_n$ be any n numbers. Define

$$\sum_{i=1}^{n} a_i$$

to stand for

$$a_1 + a_2 + a_3 + \cdots + a_n.$$

Read

$$\sum_{i=1}^{n} a_i$$

as

summation a_i, $i = 1$ to n.

In this notation, the symbol $\sum$ (a stylized Greek letter Σ) stands for "summation". Moreover,

$$\sum_{i=1}^{n} a_i$$

indicates that the terms a_i are to be added, beginning with a_1 and ending with a_n. The variable subscript i, which ranges over the integers 1, 2, 3, . . . , n, is called the **index of summation.**

EXAMPLE 1. Let

$$a_1 = 4, a_2 = 3, a_3 = -2, a_4 = 1, a_5 = 5, a_6 = 3\ .$$

Then

$$\begin{aligned}\sum_{i=1}^{6} a_i &= a_1 + a_2 + a_3 + a_4 + a_5 + a_6 \\ &= 4 + 3 + (-2) + 1 + 5 + 3 \\ &= 14\end{aligned}$$

■

Sometimes, the index of summation ranges over the integers 0, 1, 2, 3, . . . , n, instead of 1, 2, 3, . . . , n.

DEFINITION. Let $a_0, a_1, a_2, a_3, \ldots, a_n$ be any $n + 1$ numbers. Define

$$\sum_{i=0}^{n} a_i$$

to stand for

$$a_0 + a_1 + a_2 + a_3 + \cdots + a_n.$$

EXAMPLE 2. Let

$$a_0 = 4, a_1 = -3, a_2 = 5, a_3 = -1,$$
$$a_4 = 2, a_5 = 6, a_6 = 0, a_7 = 12, a_8 = -10\ .$$

Then

$$\begin{aligned}\sum_{i=0}^{8} a_i &= a_0 + a_1 + a_2 + a_3 + a_4 + a_5 + a_6 + a_7 + a_8 \\ &= 4 + (-3) + 5 + (-1) + 2 + 6 + 0 + 12 + (-10) \\ &= 15\end{aligned}$$

■

This "sigma notation" (or Σ-notation) can be generalized.

DEFINITION. Let

$$a_1, a_2, a_3, \ldots, a_i, \ldots$$

be a sequence. For any integers k and n, $1 \leq k \leq n$, define

$$\sum_{i=k}^{n} a_i$$

to stand for

$$a_k + a_{k+1} + a_{k+2} + \cdots + a_n.$$

Here the index of summation i ranges over the integers $k, k + 1, k + 2, \ldots, n$.
Note that if $k = 1$, the original definition applies.

EXAMPLE 3. Consider the sequence given by

$$a_i = 3i.$$

(a) $$\sum_{i=1}^{10} a_i = a_1 + a_2 + a_3 + \cdots + a_{10} = 3 + 6 + 9 + \cdots + 30$$

Here $k = 1$ and $n = 10$. There are 10 terms to be added.

(b) $$\sum_{i=1}^{100} a_i = a_1 + a_2 + a_3 + \cdots + a_{100}$$

$$= 3 + 6 + 9 + \cdots + 300$$

Here $k = 1$ and $n = 100$. There are 100 terms to be added.

(c) $$\sum_{i=2}^{15} a_i = a_2 + a_3 + a_4 + \cdots + a_{15}$$

$$= 6 + 9 + 12 + \cdots + 45$$

Here $k = 2$ and $n = 15$. The index of summation ranges over the integers $2, 3, 4, \ldots, 15$. There are 14 terms to be added.

(d) $$\sum_{i=6}^{33} a_i = a_6 + a_7 + a_8 + \cdots + a_{33}$$

$$= 18 + 21 + 24 + \cdots + 99$$

Here $k = 6$ and $n = 33$. How many terms are to be added?

(e) To express the single term a_k in sigma notation, write:

$$\sum_{i=k}^{k} a_i$$

For example, for the above sequence,

$$\sum_{i=12}^{12} a_i = a_{12} = 36$$ ■

In the next section you will learn a systematic way of determining the sums of Example 3.

For any sequence,

$$a_1, a_2, a_3, \ldots, a_i, \ldots$$

and for any integers k and n, $1 \leq k \leq n$,

$$\sum_{i=1}^{n} a_i = \sum_{i=1}^{k} a_i + \sum_{i=k+1}^{n} a_i$$

(*The sum from a_1 to a_n equals the sum from a_1 to a_k plus the sum from a_{k+1} to a_n.*) In fact,

$$\sum_{i=1}^{n} a_i = a_1 + a_2 + a_3 + \cdots + a_k + a_{k+1} + a_{k+2} + \cdots + a_n$$

$$= (a_1 + a_2 + a_3 + \cdots + a_k) + (a_{k+1} + a_{k+2} + \cdots + a_n)$$

$$= \sum_{i=1}^{k} a_i + \sum_{i=k+1}^{n} a_i$$ ■

Thus also,

$$\sum_{i=k+1}^{n} a_i = \sum_{i=1}^{n} a_i - \sum_{i=1}^{k} a_i$$

When the terms of a sequence are given by a formula, you can use this formula in writing Σ-notation. Thus, if $a_i = 3i$, for $i = 1, 2, \ldots,$

$$\sum_{i=1}^{n} a_i = \sum_{i=1}^{n} 3i$$

and for $1 \leq k \leq n$,

$$\sum_{i=k}^{n} a_i = \sum_{i=k}^{n} 3i$$

EXAMPLE 4. Consider the sequence of Example 3, given by

$$a_i = 3i.$$

As you will see in the next section,

$$\sum_{i=1}^{99} a_i = \sum_{i=1}^{99} 3i = 14\,850,$$

$$\sum_{i=1}^{9} a_i = \sum_{i=1}^{9} 3i = 135.$$

Therefore

$$\sum_{i=10}^{99} a_i = \sum_{i=1}^{99} a_i - \sum_{i=1}^{9} a_i = 14\,850 - 135 = 14\,715$$

In other words,

$$30 + 33 + 36 + \cdots + 297 = 14\,715$$ ■

EXAMPLE 5. Write in Σ-notation:

$$3 + 5 + 7 + 9 + \cdots + 21$$

SOLUTION. Find the (suggested) sequence whose first few terms are

$$3, 5, 7, 9, \ldots, 21.$$

This sequence is given by

$$a_i = 2i + 1.$$

Thus

$$a_1 = 2 \cdot 1 + 1 = 3$$

$$a_2 = 2 \cdot 2 + 1 = 5$$

$$a_3 = 2 \cdot 3 + 1 = 7$$

$$\ldots$$

$$a_{10} = 2 \cdot 10 + 1 = 21$$

and

$$\sum_{i=1}^{10} a_i = \sum_{i=1}^{10} (2i + 1)$$

$$= 3 + 5 + 7 + \cdots + 21$$ ■

You can *add* or *subtract* in Σ-notation. You can also *multiply* the terms *by a constant.*

Let $a_1, a_2, \ldots, a_n, b_1, b_2, \ldots, b_n$

be any numbers. Then

$$\sum_{i=1}^{n} (a_i + b_i) = \sum_{i=1}^{n} a_i + \sum_{i=1}^{n} b_i$$

For, by the Associative and Commutative Laws,

$$\sum_{i=1}^{n} (a_i + b_i) = (a_1 + b_1) + (a_2 + b_2) + (a_3 + b_3) + \cdots + (a_n + b_n)$$

$$= (a_1 + a_2 + \cdots + a_n) + (b_1 + b_2 + \cdots + b_n)$$

$$= \sum_{i=1}^{n} a_i + \sum_{i=1}^{n} b_i \qquad \blacksquare$$

Also, for any number c,

$$\sum_{i=1}^{n} ca_i = c \sum_{i=1}^{n} a_i$$

For, by the Distributive Laws,

$$\sum_{i=1}^{n} ca_i = ca_1 + ca_2 + ca_3 + \cdots + ca_n$$

$$= c(a_1 + a_2 + a_3 + \cdots + a_n)$$

$$= c \cdot \sum_{i=1}^{n} a_i \qquad \blacksquare$$

EXAMPLE 6. Consider the sequences given by

$$a_i = 4i,$$

$$b_i = i + 3.$$

For each i, $4i + (i + 3) = 5i + 3$. Thus

$$\sum_{i=1}^{50} (5i + 3) = \sum_{i=1}^{50} [4i + (i + 3)]$$

$$= \sum_{i=1}^{50} 4i + \sum_{i=1}^{50} (i + 3)$$

$$= 4 \sum_{i=1}^{50} i + \sum_{i=1}^{50} (i + 3) \qquad \blacksquare$$

EXERCISES

In Exercises 1–10 express each in Σ-notation.

1. $a_1 + a_2 + a_3 + a_4$
2. $a_1 + a_2 + a_3 + \cdots + a_{30}$
3. $b_0 + b_1 + b_2 + b_3 + \cdots + b_{11}$
4. $c_1 + c_2 + c_3 + \cdots + c_{2000}$
5. $a_1 + a_2$
6. b_1

7. $a_5 + a_6 + a_7 + \cdots + a_{18}$

8. $b_{19} + b_{20} + b_{21}$

9. $c_{101} + c_{102} + c_{103} + \cdots + c_{200}$

10. a_{93}

In Exercises 11–20 find the numerical value of each sum.

11. $\sum_{i=1}^{6} i$

12. $\sum_{i=1}^{5} 2i$

13. $\sum_{i=0}^{6} (i + 2)$

14. $\sum_{i=1}^{4} 5i$

15. $\sum_{i=1}^{5} (3i + 2)$

16. $\sum_{i=1}^{6} i^2$

17. $\sum_{i=1}^{3} \frac{i}{i + 1}$

18. $\sum_{i=0}^{4} 2^i$

19. $\sum_{i=4}^{10} (i + 3)$

20. $\sum_{i=7}^{10} (5i - 4)$

In Exercises 21–32 write in Σ-notation.

21. $3 + 4 + 5 + \cdots + 12$

22. $-1 + 0 + 1 + 2 + \cdots + 17$

23. $4 + 8 + 12 + 16 + \cdots + 64$

24. $-2 - 4 - 6 - 8 - \cdots - 34$

25. $1 + 4 + 9 + 16 + \cdots + 100$

26. $1^3 + 2^3 + 3^3 + 4^3 + \cdots + 20^3$

27. $\frac{1}{3} + \frac{1}{4} + \frac{1}{5} + \frac{1}{6} + \cdots + \frac{1}{92}$

28. $\frac{1}{2} + \frac{2}{3} + \frac{3}{4} + \frac{4}{5} + \cdots + \frac{29}{30}$

29. $4 + 6 + 8 + 10 + \cdots + 42$

30. $3 + 8 + 13 + 18 + \cdots + 103$

31. $\frac{1}{2} + \frac{1}{4} + \frac{1}{8} + \frac{1}{16} + \cdots + \frac{1}{256}$

32. $\frac{3}{2} + \frac{5}{4} + \frac{7}{8} + \frac{9}{16} + \cdots + \frac{17}{256}$

In Exercises 33–40 determine the indicated number.

33. $7 \sum_{i=1}^{60} a_i = \sum_{i=1}^{60} ca_i$. Find c.

34. $\sum_{i=1}^{20} a_i + \sum_{i=21}^{40} a_i = \sum_{i=1}^{n} a_i$. Find n.

35. $-\sum_{i=1}^{28} a_i = \sum_{i=1}^{28} ca_i$. Find c.

36. Find $\sum_{i=1}^{41} a_i$ if $\sum_{i=1}^{40} a_i = 200$ and $a_{41} = -7$.

37. $\sum_{i=1}^{54} a_i = \sum_{i=1}^{k} a_i + \sum_{i=26}^{54} a_i$. Find k.

38. $\sum_{i=3}^{10} b_i = \sum_{i=1}^{k} b_i - b_1 - b_2$. Find k.

39. $\sum_{i=1}^{64} 5i = \sum_{i=1}^{64} ci + 2 \sum_{i=1}^{64} i$. Find c.

40. $\sum_{i=1}^{17} a_i = \sum_{i=1}^{18} a_i$. Find a_{18}.

17.3 ARITHMETIC PROGRESSIONS

In some of the sequences

$$a_1, a_2, a_3, \ldots, a_i, \ldots$$

that you have seen, each new term (other than the first) is obtained from the preceding one by adding a constant—that is,

$$a_{i+1} - a_i$$

is *the same* for all i. For example, each of the sequences

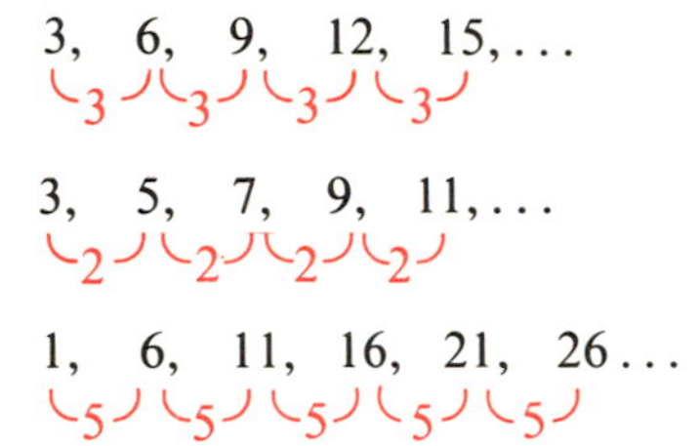

has this property.

DEFINITION. An **arithmetic progression** is a sequence

$$a_1, a_2, a_3, \ldots, a_i, \ldots$$

in which the difference

$$a_{i+1} - a_i$$

is the same for $i = 1, 2, 3, \ldots$. In an arithmetic progression, $a_{i+1} - a_i$ is called the **common difference,** and is usually denoted by d.

$$d = a_{i+1} - a_i$$

or

$$a_{i+1} = a_i + d$$

Thus you add the common difference to a_i to obtain the next term, a_{i+1}. Here are some further examples of *arithmetic progressions*.

EXAMPLE 1.

(a) $6, 12, 18, 24, \ldots, 6i, \ldots$

The first term is 6 and the common difference is 6.

(b) $7, 11, 15, 19, \ldots, 4i + 3, \ldots$

The first term is 7 and the common difference is 4.

(c) $1, -1, -3, -5, \ldots, -2i + 3, \ldots$

The first term is 1 and the common difference is -2.

(d) $5, 5, 5, 5, \ldots, 5, \ldots$

The first term is 5 and the common difference is 0. ■

EXAMPLE 2. A sequence

$$a_1, a_2, a_3, \ldots, a_i, \ldots$$

whose first five terms are

$$1, 3, 4, 8, 9$$

is not an arithmetic progression because

$$a_2 - a_1 = 3 - 1 = 2,$$

whereas

$$a_3 - a_2 = 4 - 3 = 1.$$

Thus $a_{i+1} - a_i$ is not the same for all i. ■

In an arithmetic progression with first term a_1 and common difference d, the terms can be listed as

$$a_1, a_1 + d, a_1 + 2d, a_1 + 3d, \ldots, a_1 + (i - 1)d, \ldots.$$

For, suppose the arithmetic progression is given by

$$a_1, a_2, a_3, \ldots, a_i, \ldots.$$

Then

$$a_2 - a_1 = d. \qquad \text{Thus } a_2 = a_1 + d$$

$$a_3 - a_2 = d. \qquad \text{Thus } a_3 = a_2 + d = a_1 + 2d$$

$$a_4 - a_3 = d. \qquad \text{Thus } a_4 = a_3 + d = a_1 + 3d$$

In general,

$$a_i = a_1 + (i - 1)d$$

EXAMPLE 3. Determine the 17th term of the arithmetic progression with first term 4 and common difference 3.

SOLUTION.

$$a_1 = 4, \qquad d = 3$$

$$a_{17} = a_1 + (17 - 1)d = 4 + 16 \cdot 3 = 52$$

■

An arithmetic progression

$$a_1, a_1 + d, a_1 + 2d, \ldots, a_1 + (i - 1)d$$

is an increasing sequence if $d > 0$; *it is a decreasing sequence if* $d < 0$.

The formula that you next derive will enable you to find the sum of the first n terms of any arithmetic progression—even if n is very large.

Consider the sum of the first four terms of the arithmetic progression

$$5, 9, 13, 17.$$

(Obviously, you could find the sum quickly by adding the four numbers. But the following is the way the formula you will use is obtained.) Denote the sum by S and write it in two ways.

$$\begin{array}{rrcl} & S &=& 5 + 9 + 13 + 17 \\ & S &=& 17 + 13 + 9 + 5 \\ \hline \text{Add:} & 2S &=& 22 + 22 + 22 + 22 \end{array}$$

or

$$2S = 22 \cdot 4 = 88$$

Thus

$$S = \frac{1}{2} \cdot 88 = 44$$

In general, suppose you want the sum of the first n terms of an arithmetic progression $a_1, a_2, a_3, \ldots$. Then you want to find $a_1 + a_2 + a_3 + \cdots + a_{n-2} + a_{n-1} + a_n$. Again denote the sum by S. Because the terms form an arithmetic progression, $a_2 = a_1 + d$ and $a_3 = a_1 + 2d$, where d is the common difference. Moreover, $a_{n-1} = a_n - d$ and $a_{n-2} = a_n - 2d$. Thus write S in two ways:

$$\begin{array}{rrcl} & S &=& a_1 + (a_1 + d) + (a_1 + 2d) + \cdots + (a_n - 2d) + (a_n - d) + a_n \\ & S &=& a_n + (a_n - d) + (a_n - 2d) + \cdots + (a_1 + 2d) + (a_1 + d) + a_1 \\ \hline \text{Add:} & 2S &=& \underbrace{(a_1 + a_n) + (a_1 + a_n) + (a_1 + a_n) + \cdots + (a_1 + a_n) + (a_1 + n)}_{n \text{ times}} \end{array}$$

$$2S = n(a_1 + a_n)$$

$$S = \frac{1}{2} n(a_1 + a_n)$$

Because the sum of the first n terms can also be written as $\sum_{i=1}^{n} a_i$, the preceding formula for S can be written as

$$\text{(A)} \qquad \sum_{i=1}^{n} a_i = \frac{n}{2}(a_1 + a_n).$$

In words, *the sum of the first n terms of an arithmetic progression is half the number of terms times the sum of the first and last terms.*

(If you rewrite (A) as $\sum_{i=1}^{n} a_i = n\left[\frac{1}{2}(a_1 + a_n)\right]$ and note that $\frac{1}{2}(a_1 + a_n)$ is the average of the first and last terms, you could say that *the sum is the number of terms times the average of the first and last terms.*)

EXAMPLE 4. Express the sum of the first 99 terms of the arithmetic progression (of examples 3 and 4 of Section 17.2):

$$3, 6, 9, 12, 15, \ldots, 3i, \ldots$$

SOLUTION. Here $a_1 = 3$, $n = 99$, and $a_{99} = 3 \cdot 99 = 297$. From (A),

$$\begin{aligned}\sum_{i=1}^{99} a_i &= \frac{1}{2} \cdot 99(3 + 297)\\ &= \frac{1}{\cancel{2}} \cdot 99 \cdot \overset{150}{\cancel{300}}\\ &= 99 \cdot 150\\ &= 14\,850\end{aligned}$$

■

EXAMPLE 5.

(a) Find the sum of the first 50 terms of the arithmetic progression with first term -3 and 50th term 340.
(b) Find the common difference for this progression.

SOLUTION.

(a) You are given the first and last terms. Use formula (A) [with $n = 50$]:

$$\sum_{i=1}^{50} a_i = \frac{50}{2}(a_1 + a_{50})$$

Thus

$$\sum_{i=1}^{50} a_i = 25\,(-3 + 340) = 25 \cdot 337 = 8425$$

(b) $$a_n = a_1 + (n - 1)d$$

Therefore

$$\frac{a_n - a_1}{n - 1} = d$$

$$d = \frac{340 - (-3)}{49}$$

$$d = \frac{343}{49}$$

$$d = 7$$ ■

EXAMPLE 6. Consider the arithmetic progression given by

$$a_i = 3 + 5(i - 1).$$

Determine:

$$\sum_{i=101}^{200} a_i$$

SOLUTION. Interpret (A) to mean that the sum is half the number of terms times the sum of the first and last terms. Here the first term, a_{101}, is $3 + 5(101 - 1)$, or = 503. The last term, a_{200}, is $3 + 5(200 - 1)$, or 998. The number of terms is 100. (To see this, note that $\sum_{i=1}^{200} a_i$ has 200 terms and $\sum_{i=1}^{100} a_i$ has 100 terms. When you calculate $\sum_{i=101}^{200} a_i$, you can think of this as taking 200 terms, and then discarding the first 100 terms. Thus you start with the 101st term, and 100 terms remain.) Therefore the sum is $\frac{1}{2} \cdot 100(503 + 998)$, or 75 050. ■

EXERCISES

In each of the Exercises 1–14 a sequence is indicated by means of its first few terms. (Assume a regular pattern, as suggested.) Determine whether the sequence is an arithmetic progression. If so, find the common difference.

1. 4, 8, 12, 16, ...
2. 3, 6, 12, 24, 48, ...
3. 6, 8, 10, 12, 14, ...
4. −3, −1, 1, 3, 5, 7, ...
5. 9, 6, 3, 0, −3, −6, −9, ...
6. 100, 200, 300, 400, ...
7. 1, 5, 11, 15, 21, 25, ...
8. 2, 4, 8, 16, 32, 64, ...
9. 1, 0, 1, 0, 1, 0, ...
10. 13, 17, 21, 25, 29, 33, ...
11. 1, −4, −9, −14, −19, ...
12. 1, 1.5, 2, 2.5, 3, 3.5, ...
13. $\frac{-1}{3}, \frac{-2}{3}, -1, \frac{-4}{3}, \frac{-5}{3}, -2, \frac{-7}{3}, \ldots$

14. $2, 2 + \pi, 2 + 2\pi, 2 + 3\pi, 2 + 4\pi, \ldots$

15. Determine the 10th term of the arithmetic progression:

$$4, 9, 14, 19, 24, \ldots$$

16. Determine the 17th term of the arithmetic progression:

$$-3, 1, 5, 9, 13, \ldots$$

17. Determine the 80th term of the arithmetic progression:

$$2, -1, -4, -7, -10, \ldots$$

18. Determine the 25th and 50th terms of the arithmetic progression:

$$-15, -7, 1, 9, 17, 25, \ldots$$

19. Determine the 30th term of the arithmetic progression with first term 2 and common difference 3.

20. Determine the 40th term of the arithmetic progression with first term 9 and common difference -2.

21. Determine the 20th term of the arithmetic progression with first term 5 and 10th term 95.

22. Determine the 18th term of the arithmetic progression with first term -4 and 20th term 53.

23. Determine the 33rd term of the arithmetic progression with 2nd term 9 and 4th term 13.

24. Determine the 9th term of the arithmetic progression with 16th term 50 and 20th term 38.

In Exercises 25–28 determine the sum of the first 20 terms of the indicated arithmetic progressions.

25. $7, 9, 11, 13, 15, 17, \ldots$

26. $4, 1, -2, -5, -8, -11, \ldots$

27. $8, 8.5, 9, 9.5, 10, 10.5, \ldots$

28. $3, -2, -7, -12, -17, -22, \ldots$

In Exercises 29–32 determine the sum of the first 40 terms of the indicated arithmetic progressions.

29. $5, 14, 23, 32, 41, 50, \ldots$

30. $2 + 6(i - 1),\ i = 1, 2, 3, \ldots$

31. $-3 + 2i,\ i = 1, 2, 3, \ldots$

32. $a_1 = 4,\ d = 5$

In Exercises 33–36 determine the sum of the first 100 terms of the arithmetic progression:

$$a_1, a_2, a_3, \ldots, a_i, \ldots$$

33. $a_1 = 4,\ a_{100} = 301$

34. $a_1 = 9,\ a_{100} = 207$

35. $a_1 = 50,\ a_{100} = -148$

36. $a_1 = 3,\ a_4 = 21$

In Exercises 37–40 evaluate

(a) $\sum_{i=4}^{12} a_i$, (b) $\sum_{i=10}^{20} a_i$

for the arithmetic progression:

$$a_1, a_2, \ldots, a_i, \ldots$$

37. $a_i = 15 + 3(i - 1),\ i = 1, 2, 3, \ldots$

38. $a_1 = 9,\ a_2 = 1$

39. $a_1 = 15,\ a_{10} = 60$

40. $a_1 = -7,\ a_8 = 21$

41. Find the sum of the first 50 even positive integers.

42. Find the sum of all odd integers between 5 and 95 (inclusive).

43. A 20-year-old man earns $9000 a year. Each year he receives an increment of $1200 per year. How much does he earn at age 30?

17.4 GEOMETRIC PROGRESSIONS

You can also determine the sum of the first n terms of a sequence

$$a_1, a_2, a_3, \ldots, a_i, \ldots$$

for which the *ratio*

$$\frac{a_{i+1}}{a_i}$$

is *the same* for all i. For example, each of the suggested sequences

$$2, 4, 8, 16, 32, \ldots \qquad \frac{a_{i+1}}{a_i} = 2$$

$$\frac{1}{2}, \frac{1}{4}, \frac{1}{8}, \frac{1}{16}, \frac{1}{32}, \ldots \qquad \frac{a_{i+1}}{a_i} = \frac{1}{2}$$

$$-3, 1, \frac{-1}{3}, \frac{1}{9}, \frac{-1}{27}, \frac{1}{81}, \ldots \qquad \frac{a_{i+1}}{a_i} = \frac{-1}{3}$$

has this property. In this kind of sequence, each new term (other than the first) is obtained by multiplying the preceding term by a constant.

DEFINITION. A **geometric progression** is a sequence of nonzero terms

$$a_1, a_2, a_3, \ldots, a_i, \ldots$$

in which the ratio

$$\frac{a_{i+1}}{a_i}$$

is the same for $i = 1, 2, 3, \ldots$. In a geometric progression, $\frac{a_{i+1}}{a_i}$ is called the **common ratio,** and is usually denoted by r.

$$r = \frac{a_{i+1}}{a_i} \neq 0$$

EXAMPLE 1. Each of the indicated sequences is a geometric progression:

(a) $1, 5, 25, 125, \ldots, 5^{i-1}, \ldots$

The first term is 1 and the common ratio is 5. To obtain a_{i+1}, multiply a_i by 5.

(b) $2, 6, 18, 54, \ldots, 2(3^{i-1}), \ldots$

The first term is 2 and the common ratio is 3.

(c) $2, -1, \frac{1}{2}, \frac{-1}{4}, \ldots, 2\left(\frac{-1}{2}\right)^{i-1}, \ldots$

The first term is 2 and the common ratio is $\frac{-1}{2}$. ■

EXAMPLE 2. The sequence indicated by

$$2, 4, 6, 8, 10, \ldots$$

is not a geometric progression because

$$\frac{a_2}{a_1} = \frac{4}{2} = 2,$$

whereas

$$\frac{a_3}{a_2} = \frac{6}{4} = \frac{3}{2}.$$

(Note that it is an arithmetic progression with common difference 2.) ■

In a geometric progression with first term a_1 and common ratio r, the terms can be listed as

$$a_1, a_1r, a_1r^2, a_1r^3, \ldots, a_1r^{i-1}, \ldots.$$

For, suppose the geometric progression is given by

$$a_1, a_2, a_3, \ldots, a_i, \ldots.$$

Then

$$\frac{a_2}{a_1} = r. \qquad \text{Thus } a_2 = a_1 r$$

$$\frac{a_3}{a_2} = r. \qquad \text{Thus } a_3 = a_2 r = a_1 r^2$$

$$\frac{a_4}{a_3} = r. \qquad \text{Thus } a_4 = a_3 r = a_1 r^3$$

In general,

$$a_i = a_1 r^{i-1}$$

EXAMPLE 3. Determine the 7th term of the geometric progression with first term 5 and common ratio 2.

SOLUTION.

$$a_1 = 5, \; r = 2$$

$$a_7 = 5(2^{7-1}) = 5 \cdot 64 = 320$$

■

A geometric progression

$$a_1, a_1 r, a_1 r^2, a_1 r^3, \ldots, a_1 r^{i-1}, \ldots$$

is an increasing sequence if $r > 1$; *it is a decreasing sequence if* $0 < r < 1$; *it is neither increasing nor decreasing if* $r < 0$ *or if* $r = 1$.

EXAMPLE 4.

(a) The geometric progression

$$2, 4, 8, 16, 32, \ldots$$

is an increasing sequence. Here $r = 2$

(b) The geometric progression

$$2, 1, \frac{1}{2}, \frac{1}{4}, \frac{1}{8}, \ldots$$

is a decreasing sequence. Here $r = \frac{1}{2}$

(c) The geometric progression

$$2, -4, 8, -16, 32, \ldots$$

is neither increasing nor decreasing. Here $r = -2$

(d) The geometric progression

$$2, 2, 2, 2, 2, \ldots$$

is neither increasing nor decreasing. Here $r = 1$. Notice that this is also an arithmetic progression with common difference 0. ■

Next, you derive the formula for the sum of the first n terms of a geometric progression.

To find the sum of the first four terms of the geometric progression

$$1, 3, 9, 27, \ldots,$$

it would be simplest just to add the four numbers. However, the following approach will show the method used to develop the general formula.

Let S represent the sum. Then

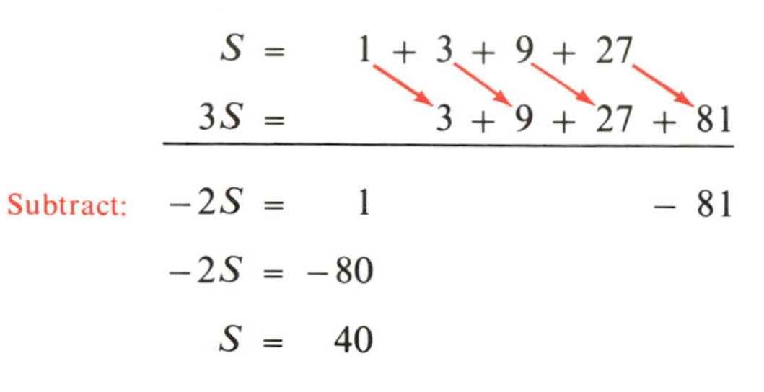

In general, suppose you want to add the first n terms of a geometric progression $a_1, a_1r, a_1r^2, \ldots$ (where $r \neq 1$). The nth term is a_1r^{n-1}. Denote the sum by S:

$$S = a_1 + a_1r + a_1r^2 + \cdots + a_1r^{n-1}$$

$$rS = a_1r + a_1r^2 + \cdots + a_1r^{n-1} + a_1r^n$$

Multiply both sides by r.

Subtract:

$$S - rS = a_1 - a_1r^n$$

$$S(1 - r) = a_1(1 - r^n)$$

$$S = \frac{a_1(1 - r^n)}{(1 - r)}$$

Now isolate the common factor on each side.

Thus

$$S = \sum_{i=1}^{n} a_i = \frac{a_1(1 - r^n)}{1 - r}, r \neq 1$$

If $r = 1$, then

$$a_1 = a_2 = a_3 = \cdots = a_n$$

and

$$\sum_{i=1}^{n} a_i = na_1$$

EXAMPLE 5. Determine the sum of the first 6 terms of the sequence

$$\frac{1}{2}, \frac{1}{4}, \frac{1}{8}, \ldots, \frac{1}{2^i}, \ldots.$$

SOLUTION. This is a geometric sequence with $a_1 = \frac{1}{2}$ and $r = \frac{1}{2}$. Therefore

$$\begin{aligned}\sum_{i=1}^{6} a_i &= \frac{a_1(1 - r^6)}{1 - r} \\ &= \frac{\frac{1}{2}\left[1 - \left(\frac{1}{2}\right)^6\right]}{1 - \frac{1}{2}} \\ &= \frac{\frac{1}{2}\left(1 - \frac{1}{64}\right)}{\frac{1}{2}} \\ &= \frac{63}{64}\end{aligned}$$

■

EXAMPLE 6. Determine the sum of the first 5 terms of the geometric sequence with first term 2 and fourth term 54 .

SOLUTION.

$$\begin{aligned}a_1 &= 2 \\ a_4 &= a_1 r^3 \\ 54 &= 2r^3 \\ 27 &= r^3 \\ 3 &= r\end{aligned}$$

$$\begin{aligned}\sum_{i=1}^{5} a_i &= \frac{a_1(1 - r^5)}{1 - r} \\ &= \frac{2(1 - 3^5)}{1 - 3} \\ &= \frac{2(-242)}{-2} \\ &= 242\end{aligned}$$

■

EXERCISES

In each of the Exercises 1–14 a sequence is indicated by means of its first few terms. (Assume a regular pattern, as suggested.) Determine whether the sequence is a geometric progression. If so, find the common ratio. Also, indicate which sequences are arithmetic progressions.

1. $5, 10, 20, 40, 80, \ldots$
2. $1, 4, 16, 64, 256, \ldots$
3. $1, 4, 7, 10, 13, \ldots$
4. $-2, -4, -6, -8, -10, \ldots$
5. $4, 2, 1, \frac{1}{2}, \frac{1}{4}, \ldots$
6. $9, -3, 1, \frac{-1}{3}, \frac{1}{9}, \ldots$
7. $1, -1, 1, -1, 1, -1, \ldots$
8. $7, -7, 7, -7, 7, -7, \ldots$
9. $15, -30, 60, -120, 240, \ldots$
10. $1, .1, .01, .001, .0001, \ldots$
11. $25, 5, 1, \frac{1}{5}, \frac{1}{25}, \ldots$
12. $\frac{2}{3}, \frac{4}{3}, \frac{8}{3}, \frac{16}{3}, \frac{32}{3}, \ldots$
13. $.2, .4, .8, .16, .32, .64, \ldots$
14. $6, 4, \frac{8}{3}, \frac{16}{9}, \frac{32}{27}, \ldots$
15. Determine the 4th term of the geometric progression with first term 1 and common ratio 5.
16. Determine the 6th term of the geometric progression with first term 5 and common ratio 2.
17. Determine the 7th term of the geometric progression with first term 12 and common ratio $\frac{1}{2}$.
18. Determine the 10th term of the geometric progression with first term $\frac{1}{16}$ and common ratio -2.
19. Determine the 5th term of the geometric progression with 2nd term 15 and common ratio:

    (a) 3 (b) -3 (c) $\frac{1}{3}$ (d) $\frac{-1}{3}$
20. Determine the 7th term of the geometric progression with 3rd term 24 and common ratio 4.
21. Determine the 10th term of the geometric progression with 8th term 60 006 and common ratio 4.
22. Determine the 101st term of the geometric progression with 18th term -1 and common ratio -1.
23. Determine the first term of the geometric progression with 4th and 5th terms 108 and 324, respectively.
24. Determine the 2nd term of the geometric progression with 6th term $\frac{3}{16}$ and common ratio $\frac{1}{2}$.

In Exercises 25–42 evaluate the indicated sum.

25. $\sum_{i=1}^{7} 2^i$

26. $\sum_{i=1}^{5} 5^i$

27. $\sum_{i=1}^{8} (3 \cdot 2^i)$

28. $\sum_{i=1}^{6} 3^i(-2)$

29. $\sum_{i=1}^{6} 4^{i-1}$

30. $\sum_{i=1}^{7} \frac{3}{2^{i-1}}$

31. $\sum_{i=1}^{6} (11 \cdot 2^{i-1})$

32. $\sum_{i=1}^{5} \left(\frac{-5}{2}\right)^{i-1}$

33. $\sum_{i=1}^{9} 2^{i-3}$

34. $\sum_{i=1}^{9} \left(\frac{1}{2}\right)^{i-4}$

35. $\sum_{i=1}^{63} 9(-1)^i$

36. $\sum_{i=1}^{64} 9(-1)^i$

37. $\sum_{i=1}^{5} (-3)^i$

38. $\sum_{i=1}^{5} \left(\frac{-1}{3}\right)^i$

39. $\sum_{i=1}^{6} (.1)^i$

40. $\sum_{i=1}^{6} (.2)^i$

41. $\sum_{i=1}^{8} 2^{i-1}$

42. $\sum_{i=3}^{8} 2^{i-1}$

43. Suppose you are to work for 20 days. You are paid 1¢ for the first day. Each day afterward, your wages are doubled. How much do you receive for the entire job?

44. Suppose you are to work for 10 days. You are paid 1¢ for the first day. Each day afterward, your wages are tripled. How much do you receive for the entire job?

45. The population of tsetse flies in a swamp region is decreasing at the constant annual rate of 50%. If the number of tsetse flies present in 1976 is 64 000 000, how many will there be in 1984?

46. If you deposit $1000 in a bank at an annual interest rate of 5% and do not withdraw, how much money do you have in the bank 3 years later?

*17.5 BINOMIAL EXPANSION

In this section you will learn a systematic way to find the powers of a binomial, $a + b$. First you must consider some preliminary notions.

DEFINITION. Let n be a positive integer. Define:

$$n! = n(n-1)(n-2)\ldots 3 \cdot 2 \cdot 1$$

The symbol $n!$ is read "n factorial".

*Optional topic.

EXAMPLE 1.

(a) $5! = 5 \cdot 4 \cdot 3 \cdot 2 \cdot 1 = 120$

(b) $6! = 6 \cdot 5 \cdot 4 \cdot 3 \cdot 2 \cdot 1 = 720$

Observe that $6! = 6(5!)$.

(c) $100! = 100 \cdot 99 \cdot 98 \ldots 3 \cdot 2 \cdot 1$

(d) $1! = 1$ ■

The following definition is introduced to simplify later notation.

DEFINITION. $0! = 1$

Example 1(b) can be generalized. For $n = 0, 1, 2, \ldots,$

$$(n + 1)! = (n + 1)n!$$

Thus, to obtain $(n + 1)!$ from $n!$, simply multiply $n!$ by $n + 1$.

In general, for integers i and n such that $0 < i < n$,

$$\frac{n!}{i!} = n(n - 1)(n - 2) \ldots (i + 1)$$

In fact,

$$\frac{n!}{i!} = \frac{n(n - 1)(n - 2) \ldots (i + 1)\overset{1}{\cancel{i(i - 1)(i - 2) \ldots 3 \cdot 2 \cdot 1}}}{\underset{1}{\cancel{i(i - 1)(i - 2) \ldots 3 \cdot 2 \cdot 1}}}$$

$$= n(n - 1)(n - 2) \ldots (i + 1)$$

EXAMPLE 2.

(a) Evaluate $\frac{12!}{9!}$.

(b) Evaluate $\frac{101!}{99!}$.

(c) Express $10 \cdot 9 \cdot 8 \cdot 7$ in terms of factorials.

(d) Evaluate $\frac{20!}{18!2!}$.

SOLUTION.

(a) $$\frac{12!}{9!} = \frac{12 \cdot 11 \cdot 10 \cdot \overset{1}{\cancel{9 \cdot 8 \cdot 7 \cdot 6 \cdot 5 \cdot 4 \cdot 3 \cdot 2 \cdot 1}}}{\underset{1}{\cancel{9 \cdot 8 \cdot 7 \cdot 6 \cdot 5 \cdot 4 \cdot 3 \cdot 2 \cdot 1}}} = 1320$$

(b) $$\frac{101!}{99!} = 101 \cdot 100 = 10\,100$$

(c) To express $10 \cdot 9 \cdot 8 \cdot 7$, let $n = 10$ and $i + 1 = 7$.

$$\frac{10!}{6!} = 10 \cdot 9 \cdot 8 \cdot 7$$

(d)
$$\frac{20!}{18!2!} = \frac{20!}{18!} \cdot \frac{1}{2!}$$

$$= \overset{10}{\cancel{20}} \cdot 19 \cdot \frac{1}{\cancel{2} \cdot 1}$$

$$= 190$$

■

As you will see, quotients such as in Example 2(d) are important. A special symbol is given to them. (Note that $20 = 18 + 2$.)

DEFINITION. Let i and n be integers such that

$$0 \leq i \leq n.$$

Define

$$\binom{n}{i} = \frac{n!}{i!(n-i)!}.$$

Observe that

$$i + (n - i) = n.$$

Also,

$$\binom{n}{i} = \binom{n}{n-i}$$

For,

$$\binom{n}{i} = \frac{n!}{i!(n-i)!}$$

and

$$\binom{n}{n-i} = \frac{n!}{(n-i)!\underbrace{[n-(n-i)]}_{i}!} = \frac{n!}{(n-i)!i!}$$

EXAMPLE 3.

(a)
$$\binom{7}{4} = \frac{7!}{4!3!}$$

$$= \frac{7 \cdot 6 \cdot 5}{3 \cdot 2 \cdot 1}$$

$$= 35$$

(b)
$$\binom{7}{3} = \binom{7}{4} = 35$$

Here $3 + 4 = 7$

■

EXAMPLE 4.

(a)
$$\binom{10}{2} = \frac{10!}{2!\,8!} = \frac{10\cdot 9}{2\cdot 1} = 45$$

(b)
$$\binom{8}{4} = \frac{8!}{4!\,4!} = \frac{8\cdot 7\cdot 6\cdot 5}{4\cdot 3\cdot 2\cdot 1} = 70$$

■

The following computations play a key role in expanding powers of $a + b$.

EXAMPLE 5.

(a)
$$\binom{n}{0} = \binom{n}{n} = \frac{n!}{0!\,n!} = \frac{1}{1} = 1$$

(b)
$$\binom{n}{1} = \binom{n}{n-1} = \frac{n!}{1!(n-1)!} = \frac{n\cdot(n-1)!}{(n-1)!} = n$$

Thus

$$\binom{2}{1} = 2; \quad \binom{3}{1} = \binom{3}{2} = 3; \quad \binom{4}{1} = \binom{4}{3} = 4; \quad \binom{5}{1} = \binom{5}{4} = 5$$

(c)
$$\binom{4}{2} = \frac{4!}{2!\,2!} = \frac{4\cdot 3}{2\cdot 1} = 6$$

(d)
$$\binom{5}{2} = \binom{5}{3} = \frac{5!}{2!\,3!} = \frac{5\cdot 4}{2\cdot 1} = 10$$

■

Next, multiply $a + b$ by itself five times. You obtain the following powers of $a + b$, beginning with $(a + b)^0$. In each case, the first result is obtained by ordinary multiplication. Then the coefficients are rewritten in the form $\binom{n}{i}$. Note the pattern.

	Number of Terms
$(a + b)^0 = 1$	
$= \binom{0}{0} a^0 b^0$	1
$(a + b)^1 = a + b$	
$= \binom{1}{0} a^1 b^0 + \binom{1}{1} a^0 b^1$	2

$$(a+b)^2 = a^2 + 2ab + b^2 = \binom{2}{0}a^2b^0 + \binom{2}{1}a^1b^1 + \binom{2}{2}a^0b^2 \quad 3$$

$$(a+b)^3 = a^3 + 3a^2b + 3ab^2 + b^3 = \binom{3}{0}a^3b^0 + \binom{3}{1}a^2b^1 + \binom{3}{2}a^1b^2 + \binom{3}{3}a^0b^3 \quad 4$$

$$(a+b)^4 = a^4 + 4a^3b + 6a^2b^2 + 4ab^3 + b^4 = \binom{4}{0}a^4b^0 + \binom{4}{1}a^3b^1 + \binom{4}{2}a^2b^2 + \binom{4}{3}a^1b^3 + \binom{4}{4}a^0b^4 \quad 5$$

$$(a+b)^5 = a^5 + 5a^4b + 10a^3b^2 + 10a^2b^3 + 5ab^4 + b^5 = \binom{5}{0}a^5b^0 + \binom{5}{1}a^4b^1 + \binom{5}{2}a^3b^2 + \binom{5}{3}a^2b^3 + \binom{5}{4}a^1b^4 + \binom{5}{5}a^0b^5 \quad 6$$

In each expansion of $(a+b)^n$:

1. There are $(n+1)$ terms.
2. The exponents of a decrease by 1 and the exponents of b increase by 1 in each subsequent term.
3. The sum of the exponents of a and b in each term is n.
4. The coefficient of $a^{n-i}b^i$ is $\binom{n}{i}$.

The same pattern continues for larger n. Thus, for $n = 0, 1, 2, \ldots,$

$$(a+b)^n = \binom{n}{0}a^nb^0 + \binom{n}{1}a^{n-1}b^1 + \binom{n}{2}a^{n-2}b^2 + \cdots + \binom{n}{n-2}a^2b^{n-2} + \binom{n}{n-1}a^1b^{n-1} + \binom{n}{n}a^0b^n$$

This formula is called the **binomial expansion.** The coefficients $\binom{n}{i}$ are therefore called the **binomial coefficients.** In Σ-notation this becomes

$$(a+b)^n = \sum_{i=0}^{n}\binom{n}{i}a^{n-i}b^i.$$

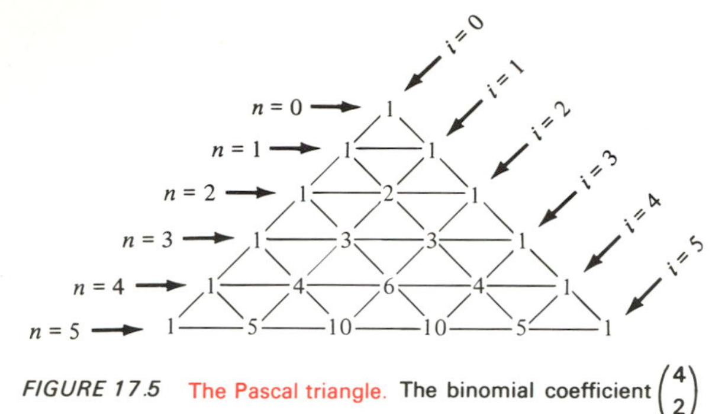

FIGURE 17.5 The Pascal triangle. The binomial coefficient $\binom{4}{2}$ or 6, lies in the row of $n = 4$ and the diagonal of $i = 2$. Note that 6 is the sum of 3 and 3, the two adjacent numbers just above 6.

The binomial coefficients can be determined from the **Pascal triangle** of Figure 17.5.

First write down the 1's along two sides of the triangle. Every other number in the triangle is the sum of the two adjacent numbers just above it. The number in the nth row and ith diagonal is the binomial coefficient $\binom{n}{i}$. This triangle can be enlarged.

EXAMPLE 6. Determine the next row of the Pascal triangle.

SOLUTION. The row of $n = 6$ is obtained by adding each pair of adjacent numbers of the row of $n = 5$. The end numbers (along the sides of the triangle) are again 1.

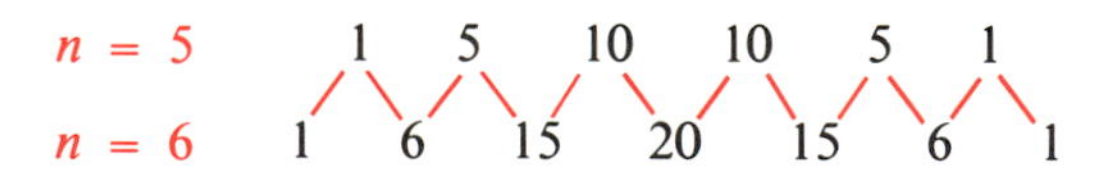

■

EXAMPLE 7. Determine: $(x + 2)^6$

SOLUTION. Let $a = x$ and $b = 2$ in the binomial expansion. Recall that

$$\binom{n}{i} = \binom{n}{n - i}.$$

For example,

$$\binom{6}{4} = \binom{6}{2}$$

Thus

$$(x+2)^6 = \sum_{i=0}^{6} \binom{6}{i} x^{6-i}2^i$$

$$= \underbrace{x^6}_{i=0} + \underbrace{6 \cdot 2x^5}_{i=1} + \underbrace{\frac{6 \cdot 5}{2} 2^2 x^4}_{i=2} + \underbrace{\frac{6 \cdot 5 \cdot 4}{3 \cdot 2} 2^3 x^3}_{i=3}$$

$$+ \underbrace{\frac{6 \cdot 5}{2} 2^4 x^2}_{i=4} + \underbrace{6 \cdot 2^5 x}_{i=5} + \underbrace{2^6}_{i=6}$$

$$= x^6 + 12x^5 + 60x^4 + 160x^3 + 240x^2 + 192x + 64$$ ■

EXAMPLE 8.

(a) Express $(y-1)^{10}$ in Σ-notation.
(b) Determine the sum of the first 4 terms of this expansion.

SOLUTION.

(a) Let $a = y$ and $b = -1$ in the binomial expansion.

$$(y-1)^{10} = \sum_{i=0}^{10} \binom{10}{i} y^{10-i}(-1)^i$$

$$= \sum_{i=0}^{10} (-1)^i \binom{10}{i} y^{10-i}$$

The signs of the terms will alternate:

$+ - + - \cdots - +$

(b) The sum of the first 4 terms is given by:

$$y^{10} - 10y^9 + \frac{10 \cdot 9}{2} y^8 - \frac{10 \cdot 9 \cdot 8}{3 \cdot 2} y^7 = y^{10} - 10y^9 + 45y^8 - 120y^7$$ ■

EXAMPLE 9. Use the binomial expansion to determine $(.99)^5$ to 4 decimal places.

SOLUTION. Let $a = 1, b = -.01$ in the binomial expansion.

$$(.99)^5 = (1-.01)^5 = \sum_{i=0}^{5} \binom{5}{i} \underbrace{1^{5-i}}_{=1}(-.01)^i$$

$$= \sum_{i=0}^{5} \binom{5}{i} (-.01)^i$$

The first few terms are:

$$1 - 5(.01) + 10(.01)^2 - 10(.01)^3 = 1 - .05 + .001 - .000\,01$$
$$= .95\,099$$

To 4 decimal places,

$$(.99)^5 = .9510$$

Note that larger powers of .01 are extremely small and contribute insignificantly to the sum. ■

EXERCISES

In Exercises 1–8 evaluate each expression.

1. $4!$
2. $7!$
3. $8!$
4. $10!$
5. $\dfrac{9!}{5!}$
6. $\dfrac{12!}{5!7!}$
7. $\dfrac{10!}{6!4!}$
8. $\dfrac{(3!)^2}{(2!)^3}$

In Exercises 9–14 express each product as a quotient of the form $\dfrac{n!}{m!}$.

9. $5 \cdot 4$
10. $7 \cdot 6 \cdot 5$
11. $12 \cdot 11 \cdot 10 \cdot 9$
12. $63 \cdot 62 \cdot 61 \cdot 60 \cdot 59 \cdot 58 \cdot 57$
13. $90 \cdot 91 \cdot 92 \cdot 93 \cdot 94 \cdot 95 \cdot 96 \cdot 97 \cdot 98 \cdot 99$
14. $201 \cdot 202 \cdot 203 \ldots 299 \cdot 300$

In Exercises 15–30 evaluate each binomial coefficient.

15. $\binom{3}{2}$
16. $\binom{4}{4}$
17. $\binom{4}{2}$
18. $\binom{5}{1}$
19. $\binom{6}{0}$
20. $\binom{6}{3}$
21. $\binom{7}{7}$
22. $\binom{8}{3}$
23. $\binom{8}{6}$
24. $\binom{10}{3}$
25. $\binom{12}{5}$
26. $\binom{20}{10}$
27. $\binom{396}{0}$
28. $\binom{396}{1}$
29. $\binom{396}{395}$
30. $\binom{396}{396}$

In Exercises 31–34 suppose n is an integer and $n > 2$. Simplify each expression.

31. $\dfrac{n!}{(n-2)!}$
32. $\dfrac{(n+1)!}{(n-1)!}$
33. $\dfrac{(n+3)!}{(n+3)(n+2)}$
34. $\dfrac{n!}{(n-2)!(n-1)}$
35. Determine the row of $n = 7$ of the Pascal triangle.
36. Determine the row of $n = 8$ of the Pascal triangle.
37. Determine the row of $n = 9$ of the Pascal triangle.

In Exercises 38–50 use the binomial expansion to expand each power.

38. $(x + y)^3$
39. $(x - a)^4$
40. $(x - a)^5$
41. $(x + 1)^4$
42. $(x + 1)^6$
43. $(x - 1)^5$
44. $(x - 1)^6$
45. $(x + 3)^4$
46. $(x - 2)^6$
47. $(x + 2y)^4$
48. $\left(x + \frac{1}{2}\right)^4$
49. $(a^2 - b^2)^4$
50. $(xy^2z^3 - a^5b^4)^3$

In Exercises 51–60:

(a) Express each power in Σ-notation.

(b) Determine the sum of the first 4 terms of this expansion.

51. $(x + y)^8$
52. $(x - y)^7$
53. $(x + 1)^{12}$
54. $(a - 1)^9$
55. $(x - 2)^9$
56. $(xy - 1)^{12}$
57. $(x^2y + 1)^8$
58. $(x^3y + az^2)^{10}$
59. $\left(x + \frac{1}{2}\right)^8$
60. $\left(x - \frac{a}{2}\right)^8$

In Exercises 61–68 use the binomial expansion to determine each power to 4 decimal places.

61. $(1.01)^4$
62. $(.99)^4$
63. $(1.1)^6$
64. $(1.1)^{10}$
65. $(1.01)^7$
66. $(.99)^8$
67. $(2.1)^6$
68. $(2.01)^7$

REVIEW EXERCISES FOR CHAPTER 17

17.1 SEQUENCES

In Exercises 1–3:

(a) Determine the first 8 terms of the indicated sequence.

(b) Is the sequence (i) increasing, (ii) decreasing, (iii) neither increasing nor decreasing?

(c) Graph the sequence.

1. $a_i = i + 3$
2. $a_i = 1 - i$
3. $a_i = \frac{2}{i}$

4. Consider the sequences given by: $a_i = 4i$, $i = 1, 2, \ldots$ Determine:

   (a) a_2 (b) a_4

   (c) a_{10} (d) a_{20}

5. Describe the ith term of the suggested sequence whose first six terms are 5, 7, 9, 11, 13, 15.

17.2 SIGMA NOTATION

6. Express $a_1 + a_2 + a_3 + a_4 + a_5$ in Σ-notation.
7. Express $b_7 + b_8 + b_9 + b_{10}$ in Σ-notation.
8. Determine the numerical value of $\sum_{i=1}^{4} 3i$.
9. Write in Σ-notation: $3 + 6 + 9 + 12 + 15 + 18$
10. Determine n such that

$$\sum_{i=1}^{10} a_i + \sum_{i=11}^{20} a_i = \sum_{i=1}^{n} a_i.$$

17.3 ARITHMETIC PROGRESSIONS

11. Determine the common difference of the arithmetic progression indicated by:

    5, 8, 11, 14, 17, 20, ...

12. Determine the 12th term of the arithmetic progression indicated by:

    40, 45, 50, 55, 60, 65, ...

13. Determine the 4th term of the arithmetic progression with first term 10 and 9th term 2.
14. Determine the sum of the first 20 terms of the arithmetic progression indicated by:

    6, 10, 14, 18, 22, 26, ...

15. Find the sum of the first 30 odd positive integers.

17.4 GEOMETRIC PROGRESSIONS

16. Determine the common ratio of the geometric progression indicated by:

    2, 6, 18, 54, 162, ...

17. Determine the 5th term of the geometric progression with first term 3 and common ratio $\frac{1}{3}$.
18. Determine the first term of the geometric progression with 4th term 2000 and 5th term 10 000.
19. Evaluate $\sum_{i=1}^{6} 3^i$.
20. Evaluate $\sum_{i=1}^{7} \frac{5}{2^{i-1}}$.

17.5 BINOMIAL EXPANSION

21. Evaluate:

    (a) $\frac{6!}{4!}$ (b) $\frac{10!}{5!3!}$ (c) $\frac{20!}{12!8!4!}$

22. Express each product as a quotient of the form $\frac{n!}{m!}$.

(a) $8 \cdot 7 \cdot 6 \cdot 5$ (b) $20 \cdot 19 \cdot 18 \cdot 17 \cdot 16$

(c) $100 \cdot 101 \cdot 102 \ldots 399 \cdot 400$

23. Evaluate each binomial coefficient:

(a) $\binom{5}{3}$ (b) $\binom{7}{0}$ (c) $\binom{20}{4}$

In Exercises 24–26 use the binomial expansion to expand each power:

24. $(x + a)^3$ 25. $(x - 1)^4$ 26. $(2xy^2 + 1)^5$

27. (a) Express $(x - 2)^{10}$ in Σ-notation.

(b) Determine the sum of the first 4 terms of this expansion.

28. Use the binomial expansion to determine $(1.1)^5$ to 4 decimal places.

TEST YOURSELF ON CHAPTER 17.

Answer any 8 of these. Problem 9 is based on Section 17.5.

1. Consider the sequence given by

$$a_i = 2i + 3.$$

Determine:

(a) a_3 (b) a_{10} (c) a_{100}

2. Express in Σ-notation:

(a) $a_1 + a_2 + a_3 + a_4 + a_5 + a_6 + a_7 + a_8$

(b) $b_{20} + b_{21} + b_{22} + b_{23} + b_{24}$

3. Evaluate $\sum_{i=1}^{4} (i + 5)$.

4. Which of the following sequences are (i) arithmetic progressions, (ii) geometric progressions, (iii) neither arithmetic nor geometric progressions?

(a) $1, -1, 1, -1, 1, -1, \ldots$

(b) $1, 0, 1, 0, 1, 0, \ldots$

(c) $2, -4, 6, -8, 10, -12, \ldots$

(d) $2, -4, 8, -16, 32, -64, \ldots$

5. Determine the 12th term of the arithmetic progression indicated by:

$$3, 5, 7, 9, 11, 13, \ldots$$

6. Determine the sum of the first 25 terms of the arithmetic progression indicated by:

$$4, 7, 10, 13, 16, 19, \ldots$$

7. Determine the 5th term of the geometric progression with first term $\frac{1}{4}$ and common ratio 4.

8. Evaluate $\sum_{i=1}^{8} \frac{1}{2^i}$.

9. Use the binomial expansion to expand $(a - 1)^5$.

answers

CHAPTER I

SECTION 1.1, p. 7

1. See Figure 1A.

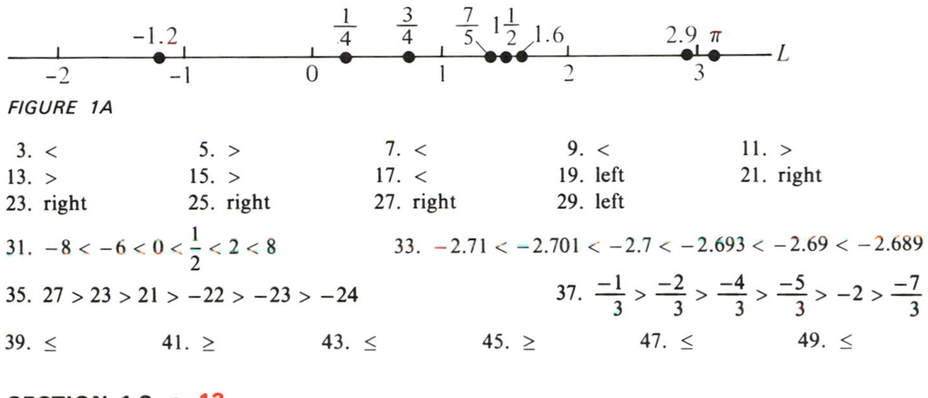

FIGURE 1A

3. $<$	5. $>$	7. $<$	9. $<$	11. $>$
13. $>$	15. $>$	17. $<$	19. left	21. right
23. right	25. right	27. right	29. left	

31. $-8 < -6 < 0 < \frac{1}{2} < 2 < 8$

33. $-2.71 < -2.701 < -2.7 < -2.693 < -2.69 < -2.689$

35. $27 > 23 > 21 > -22 > -23 > -24$

37. $\frac{-1}{3} > \frac{-2}{3} > \frac{-4}{3} > \frac{-5}{3} > -2 > \frac{-7}{3}$

39. $\leq$	41. $\geq$	43. $\leq$	45. $\geq$	47. $\leq$	49. $\leq$

SECTION 1.2, p. 12

1. -4	3. $\frac{-2}{3}$	5. -7	7. 0
9. $\frac{-3}{5}$	11. 19	13. 0	15. $\frac{2}{5}$
17. $\frac{1}{10}$	19. 2 000 000	21. $=$	23. $<$
25. $>$	27. $<$	29. $>$	31. -7

SECTION 1.3, p. 18

1. 495	3. 11 818	5. -95	7. -265
9. 1478	11. 512	13. -2156	15. -147
17. 136	19. 14	21. 5570	23. -4034
25. -207	27. 3	29. -7	31. 109
33. -3981	35. -2963	37. 206	39. -408
41. -14	43. 20 miles west of Denver		

SECTION 1.4, p. 22

1. 171	3. 544	5. -97	7. -597
9. -1575	11. 175 767	13. 2 066 068	15. 511
17. 4	19. -3	21. -2	23. 3
25. 9	27. 9	29. 82	31. -24
33. -15	35. 54	37. 15	39. -53
41. \$1			

SECTION 1.5, p. 30

1. 182 3. -182 5. -437 7. 0
9. 60 11. 8 13. 48 15. -48
17. 2788 19. $-10\,692$ 21. 5 013 057 23. -2
25. 2 27. 2 29. 0 31. undefined
33. 3 35. 15 37. -39 39. 2
41. 9 43. 1 45. $163 47. 2

SECTION 1.6, p. 33

1. 18 3. 26 5. -9
7. -9 9. 4 11. 11
13. -5 15. -23 17. 2
19. 7 21. 3 23. 42
25. 1 27. -84 29. -27
31. -10 33. 70 35. 6

SECTION 1.7, p. 36

1. 64 3. 8 5. 0 7. 10 000
9. -32 11. 256 13. 36 15. -216
17. 900 19. 49 21. 9 23. 1
25. 100 27. 43 29. 54 31. 243
33. -12 35. 1 37. -232 39. -65
41. $\frac{29}{3}$ 43. $\frac{-3}{2}$ 45. 17 47. 32

REVIEW EXERCISES, p. 37

1. (a) $<$ (b) $<$ (c) $<$ (d) $>$
3. $\frac{1}{6} < \frac{1}{5} < \frac{1}{3} < \frac{2}{5} < \frac{1}{2} < \frac{2}{3}$
5. (a) -5 (b) 5 (c) 0 (d) -5
7. (a) $=$ (b) $=$ (c) $<$ (d) $>$
9. 119 11. 4586 13. 35 miles south of Tulsa 15. 6764
17. 35 19. -74 21. $-827\,415$ 23. -90
25. -30 27. -24 29. -35 31. -49
33. $\frac{7}{3000}$

TEST YOURSELF, p. 40

1. $-3.88 < -3.87 < -3.8 < -3.78 < -3.77 < -3.7$
2. (a) 1.4 (b) 0 (c) $\frac{1}{2}$
3. 8, -8 4. 2453 5. 15 6. 361 422
7. -4 8. 1 9. -18 10. -48
11. -4 12. 400

CHAPTER 2

SECTION 2.1, p. 46

1. term 3. term 5. term 7. not a term

9. not a term 11. term 13. 15 15. $\frac{1}{3}$

17. -7 19. 0 21. 40 23. $\frac{1}{3}$

25. $10xy$ 27. $-xy$

29. polynomial 31. polynomial 33. polynomial

35. not a polynomial 37. polynomial

SECTION 2.2, p. 48

1. 17 3. 7 5. 499

7. -1 9. 15 11. -6

13. (a) 10 (b) 6 (c) 100

15. (a) 8 (b) 40 (c) 9805

17. (a) 1991 (b) 11 (c) 1 099 901

19. -3 21. 64 23. 54

25. 1 27. 89 29. -8

31. -133 33. (a) -1 (b) 0 35. (a) -20 (b) -8

37. 121 square feet 39. 670 square inches 41. 20 000 feet

43. 1200π cubic inches

SECTION 2.3, p. 54

1. similar 3. similar 5. not similar

9. $14, \frac{1}{2}$ | $x^2, -x^2$ | $x, \frac{1}{2}x$

11. $-r^2st, 2r^2ts$ | $rs^2t, 3s^2rt$ | $4rst^2$

13. $9a$ 15. $-9x$ 17. $14b$ 19. $-7xy$

21. $2b + 1$ 23. $8c - 5d$ 25. $5r - 5s$

27. $-6xy$ 29. $6a + 4b$ 31. $5yz$

33. $6x^2 - 6xy - y^2 - z$ 35. $x + 2y$ 37. $x + 2z$

39. $m - 4n + r - s - t$ 41. $-2a$ 43. $-2a - 5b$

45. $abc - ab + ac - 4$ 47. $3b$ 49. $2m - r + 4$

51. $7a + 3c$ 53. $4a - 4b + 3$ 55. $5x + 5y - z$

57. x^3 59. $-3x - 3z$ 61. 40

SECTION 2.4, p. 60

1. x^3 3. z^{11} 5. b^{12}

7. z^9 9. x^2y^2 11. y^4z^3

13. $2x^3y^8$ 15. $42a^3b^4$ 17. $-24t^3x^4y^3z^4$

19. $5x + 5y$ 21. $3ax + 15y$ 23. $x^2 + 3x$

25. $2x^3 + 2x^4$ 27. $3a^3bc + 3ab^3c$ 29. $3r^3s^2t^2 - 18r^2s^2t^3 + 6r^5s^2t^{12}$

31. $m^2 + 9m + 8$ 33. $-2a^2 + 8a + 90$ 35. $x^2 + 12x + 36$

37. $x^2 - 81$ 39. $a^2 - 121$ 41. $u^8 - 1$

43. $x^3 + 6x^2 + 4x - 5$
45. $x^2 - xy - 2y^2$
47. $-10a^2 + 36ab - 18b^2$
49. $x^2 + 2xy + xz + y^2 + yz$
51. $2a^2 + 7ab + 5ac + 3b^2 + 15bc$
53. $x^4 + 4x^3y + 4x^2y^2 + 4xy^3 + 3y^4$
55. $2a - 3b - 6c$
57. $3x^2 - 6xy + 12x$
59. $x^2 - x + y^2 + y - 4$
61. $-2a^2 + b^2 + 4ab$
63. $2x^2 + 7x + 5$
65. 42
67. -676

SECTION 2.5, p. 63

1. x
3. z^3
5. x^4
7. y^5
9. a^5
11. x^{11}
13. $2x^2$
15. $-2x^2$
17. $3a^3$
19. $2y^3$
21. 1
23. 1
25. 3
27. -1
29. $\frac{a}{3}$

SECTION 2.6, p. 65

1. 5
3. 4
5. 0
7. 4
9. 26
11. 22
13. $-5x + 1$
15. $-3x + 18$
17. $y^3 - y^2 - y + 3$
19. $6m^6 - 4m^4 + 2m^2$
21. $-t^8 - t^6 + t^5 + t^3 - t^2 + 14$
23. $x^8 - 5x^4 - \frac{x^2}{2} + 3x + 6$
25. 4
27. 7
29. 3
31. 7
33. 4
35. 12

SECTION 2.7, p. 71. (The checks are given below.)

1. $x + 1$
3. $2x - 3$
5. $-5a^2 - a + 1$
7. $x + 3$
9. $a - 3$
11. $z + 10$
13. $a^2 + a + 2$
15. $x^2 - 3x + 5$
17. $2x + 4$
19. $t^2 - 2t - 9$
21. $a + 5$
23. $m + 1$
25. $x^2 - x + 2$
27. $x + 2 + \frac{1}{x + 2}$
29. $x - 9 + \frac{7}{x}$
31. $x + 5 + \frac{-4}{x + 4}$
33. $c + 2$
35. $2x^2 - 2x + 3 + \frac{-2x - 1}{x^2 + x + 1}$
37. $y^2 - 2 + \frac{2y - 8}{y^2 - 2}$
39. $1 + \frac{1}{z + 1}$
41. $1 + \frac{-1}{10t + 1}$
43. $t - 3$
45. $z^2 + 2z - 1 + \frac{-10z + 12}{2z^2 - z + 4}$
47. $a^3 + a^2 + 5a + 4 + \frac{a^2 - 5a - 2}{a^3 - a^2 + 1}$
49. $x^2 + 5x + 6$

CHECKS.

1. $\underbrace{(x + 1)}_{\text{quotient}} \; \underbrace{x}_{\text{divisor}} = \underbrace{x^2 + x}_{\text{dividend}}$

3. $\underbrace{(2x - 3)}_{\text{quotient}} \; \underbrace{2x}_{\text{divisor}} = \underbrace{4x^2 - 6x}_{\text{dividend}}$

13.
$$\begin{array}{rl}
a^2 + a + 2 & \leftarrow \text{quotient} \\
\underline{a \ + 4} & \leftarrow \text{divisor} \\
a^3 + \ a^2 + 2a & \\
\underline{4a^2 + 4a + 8} & \\
a^3 + 5a^2 + 6a + 8 & \leftarrow \text{dividend}
\end{array}$$

23.
$$\begin{array}{rl}
m \ + 1 & \leftarrow \text{quotient} \\
\underline{2m^2 - 5} & \leftarrow \text{divisor} \\
2m^3 + 2m^2 & \\
\underline{- 5m - 5} & \\
2m^3 + 2m^2 - 5m - 5 & \leftarrow \text{dividend}
\end{array}$$

27.
$$\begin{array}{rl}
x + 2 & \leftarrow \text{quotient} \\
\underline{x + 2} & \leftarrow \text{divisor} \\
x^2 + 4x + 4 & \\
+ \quad \underline{1} & \leftarrow \text{remainder} \\
x^2 + 4x + 5 & \leftarrow \text{dividend}
\end{array}$$

35.
$$\begin{array}{rl}
2x^2 - 2x + 3 & \leftarrow \text{quotient} \\
\underline{x^2 + \ x + 1} & \leftarrow \text{divisor} \\
2x^4 - 2x^3 + 3x^2 & \\
2x^3 - 2x^2 + 3x & \\
\underline{2x^2 - 2x + 3} & \\
2x^4 \qquad + 3x^2 + \ x + 3 & \\
+ \quad \underline{-2x - 1} & \leftarrow \text{remainder} \\
2x^4 \qquad + 3x^2 - \ x + 2 & \leftarrow \text{dividend}
\end{array}$$

37.
$$\begin{array}{rl}
y^2 - 2 & \leftarrow \text{quotient} \\
\underline{y^2 - 2} & \leftarrow \text{divisor} \\
y^4 - 4y^2 \qquad + 4 & \\
+ \quad \underline{2y - 8} & \leftarrow \text{remainder} \\
y^4 - 4y^2 + 2y - 4 & \leftarrow \text{dividend}
\end{array}$$

SECTION 2.8, p. 76

1. $\underline{-1}\rfloor\ 1 \ 2 \ 4$
3. $\underline{2}\rfloor\ 1 \ -2 \ 2 \ 4$
5. $\underline{0}\rfloor\ 1 \ 5 \ 1 \ 1 \ -1$
7. $\underline{-4}\rfloor\ 1 \ -3 \ 0 \ 4$
9. $x + 2$ (no remainder)
11. $2x + 2$, remainder 1
13. $x^3 + x$ (no remainder)
15. $x^5 + x^2 + 2$, remainder 3
17. $x + 3$
19. $y - 3$
21. $x^2 + x + 1$
23. $t^2 + 1$
25. $x^3 - x^2 + 2x - 3$
27. $a^2 + 2a + 3$
29. $y^7 + y^6 + y^5 + y^4 + y^3 + y^2 + y + 1$
31. $m^3 - m^2 + 2m - 2 + \dfrac{3}{m + 1}$
33. $a^2 + 2a + 1$
35. $z^3 + 3z^2 + 3 + \dfrac{9}{z - 3}$

REVIEW EXERCISES, p. 78

1. (a) term (b) not a term (c) term (d) term
3. (a) $6x^2y$ (b) $24xyz$ (c) $\dfrac{1}{4}xy$ (d) $-4xyz$

5. 5
7. (a) 4 (b) -2 (c) 13 (d) -11
9. 400
11. (a) $2m$ (b) $2x$ (c) $3m^2 + 4m$ (d) $3x + z$
13. $-3x^2y + 2x - 4y$
15. $9x + 4y - 5$
17. y^7
19. $x^3y^2 + x^3y - 3x^4$
21. $2m^2 + mn - n^2$
23. $x^4 - 2x^3 + x^2$
25. $4x^5$
27. 2
29. (a) $-t^2 + 1$ (b) $4x^4 + 3x^3 + 2x^2 - 2x - 1$ (c) $\frac{-x^9}{3} + x^6 + x^4 - 8$ (d) $-m^5 + 2m^4$
31. $x + 3$
33. $m + 5 + \frac{-10}{m^2 + 2}$
35. $x + 7$
37. $x^2 + 4$

TEST YOURSELF, p. 80

1. (a) -2 (b) -6 (c) $\frac{3}{4}$
2. (a) 31 (b) 0 (c) 1083
3. (a) 8 (b) 50
4. (a) $-x$ (b) $6x$ (c) $2x - 3y - 2z$
5. $4x + 4y + 3z$
6. $x^2 + 7x - 17$
7. $-5x^6$
8. $4x^2 - 3xy - y^2$
9. -2
10. (a) 4 (b) 3 (c) 4
11. $x^2 + 3x + 2 + \frac{2}{x - 4}$

CHAPTER 3

SECTION 3.1, p. 85

1. 2^3
3. $3 \cdot 5$
5. $2 \cdot 13$
7. $-2 \cdot 3 \cdot 5$
9. $-5 \cdot 7$
11. $2^3 \cdot 5$
13. $2^2 \cdot 3 \cdot 5$
15. $-2^3 \cdot 3^2$
17. $2^5 \cdot 3$
19. $2^2 \cdot 3^3$
21. $2^3 \cdot 3 \cdot 5$
23. 5^3
25. $2^2 \cdot 3 \cdot 11$
27. $2^5 \cdot 5$
29. $-2^4 \cdot 5^2$
31. 8
33. 2
35. 6
37. 4
39. 12
41. 1
43. 5
45. 25
47. 2
49. 2
51. 5
53. 1
55. 2
57. 30
59. 2, 3, 5, 7, 11, 13, 17, 19, 23, 29, 31, 37, 41, 43, 47
61. (a) All nonzero integers (b) 1, -1 (c) 1, -1

SECTION 3.2, p. 88

1. $2(2x + 1)$
3. $10(2b^2 + 3)$
5. $x(3x + 5)$
7. $a^2(a - 1)$
9. $2x(x + 4)$
11. $10a^2(a - 1)$
13. $8c^3(8c^2 - 9)$
15. $11t^6(7t + 11)$

17. $xy(x + y)$
19. $st(r^2s^2 - t)$
21. $3xy(2x - 3)$
23. $10mn^4(5m^4 - 3n)$
25. $ab^3(ab + a^2 - 3b)$
27. $x^7y^4(x^6 + x^3y^3 + y^6)$
29. $rs^2t^2(r^3 + 1 - s^2t^2)$
31. $4(2x^2 + x + 3)$
33. $7(2z^3 - 3z^2 + 5)$
35. $15(5a^2 + a - 6)$
37. $2x(2x^2 + 5x + 6)$
39. $3a(a^2 - 2a + 1)$
41. $5xy(x + 2 - 3y)$
43. $4mn(4m^2n^2 + mn + 3)$
45. $5xy^2(y^2 - 2z + 4xz^2)$
47. $ad(a^2bc^3d + 4c - 3bd)$
49. $4(4a^3c^2b^4 + 5a^5b - 11a^3c^4 + 10)$

SECTION 3.4, p. 93

1. $(x + 1)(x - 1)$
3. $(y + 7)(y - 7)$
5. $(x + y)(x - y)$
7. $(s + 6)(s - 6)$
9. $(2x + y)(2x - y)$
11. $(5x + 2a)(5x - 2a)$
13. $4(5a + 4b)(5a - 4b)$
15. $(a^2 + b)(a^2 - b)$
17. $(3 + y^3)(3 - y^3)$
19. $(a^2 + b^4)(a + b^2)(a - b^2)$
21. $(s^{12} + t^{11})(s^{12} - t^{11})$
23. $(x^2 + 2y^3)(x^2 - 2y^3)$
25. $(10 + 7a^3)(10 - 7a^3)$
27. $(3x + 8y)(3x - 8y)$
29. $(11a^4 + 10b^3)(11a^4 - 10b^3)$
31. $2(x + y)(x - y)$
33. $3(a + 2x)(a - 2x)$
35. $11(2x^2 + 3y^3)(2x^2 - 3y^3)$
37. $x(x + 1)(x - 1)$
39. $3a(a + 2)(a - 2)$
41. $3a(x + y)(x - y)$
43. $(a^2 + b^2)(a + b)(a - b)$
45. $(m^2 + 4)(m + 2)(m - 2)$
47. $(4y^6 + 1)(2y^3 + 1)(2y^3 - 1)$
49. $(2x + 3 + y)(2x + 3 - y)$
51. $4ab$

SECTION 3.5, p. 97

1. $(x + 2)(x + 1)$
3. $(a + 6)(a + 1)$
5. $(x + 2)^2$
7. $(m + 2)(m - 1)$
9. $(s + 4)(s - 3)$
11. $(a - 5)^2$
13. $(z + 7)(z - 2)$
15. $2(x + 3)(x + 1)$
17. $2(t - 3)^2$
19. $x(x + 3)^2$
21. $2z^3(z + 4)(z - 7)$
23. $5y^2z^2(x - 6)(x - 2)$
25. $(7 + a)(3 + a)$
27. $-2(y + 7)(y + 5)$
29. $3(t + 9)(t + 2)$
31. $9x^2(x + 2)(x + 1)$
33. DOES NOT FACTOR
35. $(z + 1)(z - 3)$
37. $(m + 2)^2$
39. DOES NOT FACTOR
41. DOES NOT FACTOR

SECTION 3.6, p. 102

1. $(2x + 1)(x + 1)$
3. $(2a + 3)(a + 3)$
5. $(2m - 1)(m + 3)$
7. $(2a - 1)(a + 4)$
9. $(3x - 1)^2$
11. $(3m - 7)(m + 1)$
13. $(3a - 2)(2a + 3)$
15. $(5x + 4)(2x + 1)$
17. $(5a + 6)(a + 2)$
19. $2(3b + 2)(b + 1)$
21. $5(3x + 1)^2$
23. $4y^2(2y - 5)(y + 5)$
25. $a^4(4a - 3)(3a - 4)$
27. $(x + y)^2$
29. $(2u + v)^2$
31. $(y + 2z)(y + z)$
33. $(a + 4b)(a - b)$
35. $(r + 5s)(r - 2s)$
37. $(3a + 2b)(3a - 10b)$
39. $(ab + cd)^2$
41. $5(x + 3y)^2$
43. $3a(x + 2y)^2$
45. DOES NOT FACTOR
47. $(3a - 1)(a - 1)$
49. $(3x - 2)^2$
51. DOES NOT FACTOR

SECTION 3.7, p. 104

1. $(c + d)(u + v)$
3. $(a - b)(x - y)$
5. $(y + b)(y - a)$
7. $(x + 3y)(c - d)$

9. $2(2a - 3b)(x + y)$
11. $2(9a - 2b)(x - 3y)$
13. $(a + b)(2x + 3y)$
15. $(x + 2y)(u - 3v)$
17. $(a + b)(a + 5)$
19. $(u + v)(v + 1)$
21. $(a^2 + b)(a + 1)$
23. $(a + b)(u + v)(u - v)$
25. $(a + 1)(x + 2y)(x - 2y)$
27. $(2a + b)(2x + 3y)(2x - 3y)$
29. $(s + t)(s - t)(x + y)(x - y)$
31. $(x^2 + y^2)(r + s)(r - s)$
33. $(a + b)(x^2 + xy + y^2)$
35. $(a + b)(x + 2)(x + 1)$
37. $(m + 2n)(s + 3t)^2$
39. $(a + b)(a - b)(x + 1)^2$
41. $(x + 2)(x - 2)(y + 5)(y + 2)$
43. $z^2(a + b)(x + y)$
45. $u^3(x^2 + y^2)(a + b)(a - b)$

SECTION 3.8, p. 108

1. $(a^2 + a + 1)(a - 1)$
3. $(y^2 - 2y + 4)(y + 2)$
5. $(c^2 - 3c + 9)(c + 3)$
7. $(x^2 - 10x + 100)(x + 10)$
9. $(x^2 + 2xz + 4z^2)(x - 2z)$
11. $(9a^2 + 12ab + 16b^2)(3a - 4b)$
13. $(y^2 + yz + z^2)(y^2 - yz + z^2)(y + z)(y - z)$
15. $(c^6 - c^3 + 1)(c^2 - c + 1)(c + 1)$
17. $(x^2 + 2x + 4)(x^2 - 2x + 4)(x + 2)(x - 2)$
19. $[(3u)^{10} + (3u)^5 + 1][(3u)^5 - 1]$
21. $(x^6y^6 - x^3y^3 + 1)(x^6y^6 + x^3y^3 + 1)(x^2y^2 - xy + 1)(x^2y^2 + xy + 1)(xy + 1)(xy - 1)$
23. $7(z^2 + 2zy + 4y^2)(z - 2y)$
25. $x^2(a^2 + 10a + 100)(a - 10)$
27. $20(x^4y^4 - 2x^2y^2 + 4)(x^2y^2 + 2)$
29. $(a + 5)(x - y)(x^2 + xy + y^2)$
31. $(a + 4)(m + n)(m^2 - mn + n^2)$
33. $(2a + b)(2x + y)(4x^2 - 2xy + y^2)$
35. $(y - 4z)(10a - 3)(100a^2 + 30a + 9)$
37. $(a + b)(a - b)(x + y)(x^2 - xy + y^2)$
39. $5a(x + 2y)(m - n)(m^2 + mn + n^2)$
41. $(a - b - 1)[a^2 + a(b + 1) + (b + 1)^2]$

REVIEW EXERCISES, p. 108

SET I

1. (a) prime (b) not a prime (c) prime (d) prime
3. (a) 5 (b) 8 (c) 7 (d) 25
5. $3x(3x^2 - 3x + 2)$
7. $5x^2(3y^2 - 2y + 4)$
9. $(a + 3)(a - 3)$
11. $(5m + 4n)(5m - 4n)$
13. $10(x^3 + y)(x^3 - y)$
15. $(x + 3)(x + 1)$
17. $4(m - 4)(m - 3)$
19. $2(3x + 1)(x + 2)$
21. $(a + b)(x + y)$
23. $(x + y)(y + 1)$
25. $4(a + b)(m^2 + n^2)$
27. $(x + y)(x^2 - xy + y^2)$
29. $(a + 10b)(a^2 - 10ab + 100b^2)$
31. $(a + b)(a^2 - ab + b^2)(a - b)(a^2 + ab + b^2)$

SET II

1. $5x(x - 5)$
3. DOES NOT FACTOR
5. $(y + 4)(y - 5)$
7. $s^2(t + 4)(t - 4)$

9. DOES NOT FACTOR
11. $x^2y(y^2 + y + x)$
13. $(2x + 3)(2x + 1)$
15. $(x + 1)(y + 1)$
17. $(2m + 7)(m + 3)$
19. $(y - a)(y - 6a)$

TEST YOURSELF, **p. 110**

1. (a) $12 = 2^2 \cdot 3$ (b) $80 = 2^4 \cdot 5$ (c) $-132 = -2^2 \cdot 3 \cdot 11$
2. (a) 5 (b) 3
3. (a) $9(2x^2 - 3)$ (b) $5x(5x^2 + 8x + 20)$
4. $(3x + 5y)(3x - 5y)$
5. $(x + 3)(x - 2)$
6. $5(a + 2b)(x + y)$
7. $(a - 2b)(a^2 + 2ab + 4b^2)$
8. $(a^2 + 4b^2)(a + 2b)(a - 2b)$
9. $4(m - 5n)(m - 2n)$
10. $19a(x + y)(x - y)$
11. $(x - a)(y - b)$
12. $(5a + 1)(3a - 2)$

CHAPTER 4

SECTION 4.1, **p. 118**

1. $\frac{-1}{2}$
3. $\frac{\sqrt{3}}{5}$
5. $\frac{-5}{8}$
7. $\frac{5}{8}$
9. $\frac{-2}{3}$
11. $\frac{1}{2}$
13. $\frac{1}{2}$
15. $\frac{1}{2}$
17. $\frac{1}{3}$
19. $\frac{5}{4}$
21. $\frac{-1}{4}$
23. $\frac{-5}{8}$
25. $\frac{7}{16}$
27. $\frac{3}{8}$
29. $\frac{16}{25}$
31. 8
33. $\frac{1}{25}$
35. $\frac{1}{3}$
37. $\frac{1}{27}$
39. 36
41. $\frac{4}{3}$
43. $\frac{16}{15}$
45. $\frac{10}{21}$

SECTION 4.2, **p. 121**

1. 3
3. 2
5. $\frac{-3}{2}$
7. $\frac{14}{5}$
9. $\frac{3}{7}$
11. (a) -1 (b) $\frac{-1}{3}$ (c) $\frac{-1}{3}$
13. (a) $\frac{1}{5}$ (b) $\frac{-1}{15}$ (c) $\frac{1}{20}$
15. (a) $\frac{1}{5}$ (b) 2 (c) $\frac{-6}{5}$
17. (a) 0, (b) 4 (c) -2
19. (a) $\frac{10\,099}{20}$ (b) $\frac{-10\,099}{20}$ (c) $\frac{19}{4}$
21. $\frac{1}{2}$
23. $\frac{5}{8}$
25. $\frac{-1}{3}$
27. $\frac{-5}{121}$
29. $\frac{-7}{5}$
31. (a) $\frac{-1}{2}$ (b) $\frac{1}{10}$

33. (a) 0 (b) $\frac{-2}{7}$

35. (a) 1 (b) $\frac{-1}{2}$

37. (a) 0 (b) $\frac{8}{3}$

39. 9 feet per second, per second. (Every second the velocity increases by 9 feet per second.)

41. (a) 2000 (b) 1000

SECTION 4.3, p. 127

1. $2x^2$
3. $5z^3$
5. $-2a^3b$
7. $\frac{-5ad}{3}$
9. y
11. $-abc$
13. a^3c
15. $\frac{rt}{s}$
17. $\frac{3mn^2}{2}$
19. $\frac{c^2}{2a}$
21. $\frac{-5q}{p}$
23. $\frac{-xz}{2y}$
25. $\frac{-2}{3bc}$
27. $\frac{xz}{y^2}$
29. $\frac{ab^2}{4c}$
31. $\frac{5xaz}{9y}$
33. $\frac{1}{2a}$
35. $a + b$
37. $\frac{1}{(a-b)^3}$
39. $4(a + c)$
41. $\frac{a-b}{(x+y)^2}$
43. $\frac{y(x-y)}{2x}$
45. $\frac{p+q}{a+b}$
47. $(a + b)^2(m - n)^2(x + y)$

SECTION 4.4, p. 129

1. $ab(b + 1)$
3. $a(b + 1)$
5. $3x + 5y$
7. $\frac{8x^2 + 3y}{y}$
9. $3b(3 + b^2)$
11. $\frac{y^2(x + y)}{x}$
13. $2(2x + 1)$
15. $-(3a^3 + 2b)$
17. $a(b - 1)$
19. $a(-b + 1)$
21. $\frac{1}{y(x^2 + 2y)}$
23. $3xy + 5$
25. $a(ab^2 - 2)$
27. $\frac{x^2y^2 - 2}{2}$
29. $10x^2 - 8x + 5$
31. $\frac{3r^3 + 5r - 1}{r}$
33. $\frac{1}{5n + p - 1}$
35. $\frac{5(2s^4 + 3st^3 + 4t^4)}{s}$
37. $\frac{abc(a^8b^6c^4 - 2a^4b^3c^2 + 1)}{2}$
39. $4x^2y + 2xyz + 3z^2$
41. $\frac{-9x^2 + 7x - 11}{2}$
43. $\frac{5x^3 - 6x^2 + 4x - 10}{4x}$
45. $\frac{2}{-5t^2u^2 + 3tu - 4u + 5}$
47. $\frac{3xy(3x + 2y^3)}{8}$

SECTION 4.5, p. 132

1. $\frac{a-b}{a+b}$
3. a^2bc
5. $a + 2$

7. $\dfrac{2(m-n)}{m+n}$ 9. $\dfrac{x-y}{(x+y)^2}$ 11. $\dfrac{a+5}{a+2}$

13. $\dfrac{1}{x-a}$ 15. $\dfrac{c-2}{c+1}$ 17. $\dfrac{5-a}{2(a+1)}$

19. $\dfrac{ab}{1-b}$ 21. -1 23. $\dfrac{2+3c}{(2-3c)^2}$

25. $\dfrac{a^2(x^2+3x+9)}{4(x+3)}$ 27. $\dfrac{1}{x^2+2y^2}$ 29. $\dfrac{z^2-az+a^2}{2(a-z)}$

31. $\dfrac{-1}{x^2+y^2}$ 33. $\dfrac{a+1}{b-c}$ 35. $\dfrac{r+s}{y-x}$

37. $\dfrac{(x^2+y^2)(x+y)}{(x-y)^2}$ 39. $\dfrac{y-z}{a}$

REVIEW EXERCISES, p. 133

1. (a) . (c)

3. (a) $\frac{1}{2}$ (b) $\frac{-3}{4}$ (c) $\frac{-2}{5}$ (d) $\frac{3}{4}$

5. (a) 1 (b) 6 (c) $\frac{121}{50}$

7. -12 9. $4x^3$ 11. $4xz$ 13. $\dfrac{a+b}{a-b}$

15. $x+2$ 17. $\dfrac{abc+bc-a}{c}$ 19. $2(a-b)$ 21. $\dfrac{-(x+a)}{a^2+ax+x^2}$

TEST YOURSELF, p. 134

1. (a) $\frac{3}{5}$ (b) $\frac{2}{3}$ (c) $\frac{75}{77}$

2. (a) -1 (b) $\frac{7}{8}$ 3. $\dfrac{-23}{5}$

4. $5y^2$ 5. $x-y$ 6. $\dfrac{3a(a+b)(a-c)^3}{2b(a+c)}$

7. $\dfrac{4xy+6y-3x}{2xy}$ 8. $\dfrac{x+4}{x+3}$

CHAPTER 5

SECTION 5.1, p. 142

1. $\dfrac{3}{10}$ 3. $\dfrac{-2}{21}$ 5. $\dfrac{-7}{50}$ 7. $\dfrac{56}{3}$

9. $\dfrac{160}{117}$ 11. $\dfrac{xa}{yb}$ 13. $\dfrac{2}{xy}$ 15. $\dfrac{xa^2}{2}$

17. $\dfrac{-35ax^2}{y}$

19. $\dfrac{a^2(x-1)}{3}$

21. $3ab^2(x-a)$

23. $a^2(a-x)(x+1)$

25. $\dfrac{(x-4)(a+1)}{5(1-a)(x+1)}$

27. $\dfrac{(2x-3)(x-1)}{7a^2x}$

29. $\dfrac{-(a-b)^2}{3ab(a+b)}$

31. $\dfrac{15}{2}$

33. $\dfrac{-2}{3}$

35. $\dfrac{32}{27}$

37. $\dfrac{-5}{6}$

39. $\dfrac{14}{5}$

41. $\dfrac{5a^2c}{3b}$

43. $\dfrac{10ab^2}{c^2}$

45. $\dfrac{a^2cy}{27bx}$

47. $\dfrac{x-a}{2}$

49. $\dfrac{(x+1)^3}{(x-3)^2}$

51. $\dfrac{(a-2)b}{a^2(a+2)}$

53. $\dfrac{9(y-3)}{4}$

55. $\dfrac{4(1-y)(a-1)}{(a^2+a+1)a}$

57. $\left(\dfrac{x+1}{x}\right)^2$

59. $\dfrac{(x+1)^2(x+3)}{x(x-1)}$

61. $x^3+5x^2-25x-125$ [or $(x+5)^2(x-5)$]

63. $\left(a+\dfrac{1}{4}\right)\left(a-\dfrac{1}{4}\right)$ $\left[\text{or } \dfrac{(4a+1)(4a-1)}{16}\right]$

65. $\left(\dfrac{2y}{3}+\dfrac{1}{5}\right)\left(\dfrac{2y}{3}-\dfrac{1}{5}\right)$ $\left[\text{or } \dfrac{(10y+3)(10y-3)}{225}\right]$

SECTION 5.2, p. 149

1. 20
3. 10
5. 60
7. 288
9. 20
11. 60
13. 2000
15. 720
17. xy^3
19. x^2
21. $x^2y^2z^2$
23. $(x-a)^2$
25. $a(x+3)(x-3)$
27. $(x+4)(x+1)^2$
29. $50a^4(x+2)(x-2)(x-3)$
31. $(y+2)^2(y-2)(x+1)$
33. (a) 6 (b) $\dfrac{3}{6}, \dfrac{2}{6}$
35. (a) 8 (b) $\dfrac{3}{8}, \dfrac{-6}{8}$
37. (a) 196 (b) $\dfrac{63}{196}, \dfrac{6}{196}$
39. (a) 12 (b) $\dfrac{6}{12}, \dfrac{8}{12}, \dfrac{-9}{12}$
41. (a) 336 (b) $\dfrac{105}{336}, \dfrac{-14}{336}, \dfrac{88}{336}$
43. (a) 180 (b) $\dfrac{50}{180}, \dfrac{68}{180}, \dfrac{21}{180}$
45. (a) $2^2 \cdot 3 \cdot 5^2$ (b) $\dfrac{2^5}{2^2 \cdot 3 \cdot 5^2}, \dfrac{-2 \cdot 5 \cdot 7}{2^2 \cdot 3 \cdot 5^2}, \dfrac{-3^2}{2^2 \cdot 3 \cdot 5^2}$
47. (a) ax (b) $\dfrac{x}{ax}, \dfrac{a}{ax}$
49. (a) x^2 (b) $\dfrac{a}{x^2}, \dfrac{bx}{x^2}$
51. (a) x^2y^2 (b) $\dfrac{ay}{x^2y^2}, \dfrac{bx}{x^2y^2}$
53. (a) $x^2(x-a)^2$ (b) $\dfrac{x-a}{x^2(x-a)^2}, \dfrac{-x}{x^2(x-a)^2}$
55. (a) $(x-a)(x+a)$ (b) $\dfrac{(x+a)^2}{(x-a)(x+a)}, \dfrac{(x-a)^2}{(x-a)(x+a)}$
57. (a) $(x-3)^2(x-2)$ (b) $\dfrac{4(x-3)}{(x-3)^2(x-2)}, \dfrac{2-x}{(x-3)^2(x-2)}$

59. (a) $u^2(u+1)(u-1)$ (b) $\dfrac{2u}{u^2(u+1)(u-1)}, \dfrac{-3u^2}{u^2(u+1)(u-1)}, \dfrac{-2(u+1)(u-1)}{u^2(u+1)(u-1)}$

61. (a) $y(y-7)$ (b) $\dfrac{y+2}{y(y-7)}, \dfrac{-y}{y(y-7)}, \dfrac{-2y}{y(y-7)}$

SECTION 5.3, p. 154

1. 1
3. $\frac{11}{10}$
5. $\frac{13}{24}$
7. $\frac{7}{12}$
9. $\frac{4}{3}$
11. $\frac{17}{48}$
13. $\frac{11}{24}$
15. 1
17. $\frac{49}{40}$
19. $\frac{37}{28}$
21. $\frac{9}{200}$
23. $\dfrac{-31}{2^5 \cdot 3^2 \cdot 5^4}$
25. $\dfrac{3}{m}$
27. $\dfrac{n+m}{mn}$
29. $\dfrac{5+x}{x^2}$
31. $\dfrac{2x+y}{x^2y^2}$
33. $\dfrac{bc+ac-ab}{abc}$
35. $\dfrac{bv+u^2-t^2}{tuv}$
37. $\dfrac{3}{x-a}$
39. 0
41. $\dfrac{2x+5}{(x+2)(x-2)}$
43. $\dfrac{x+3}{(x+1)(x+4)}$
45. $\dfrac{-(x+4)}{(x+2)(x+3)}$
47. $\dfrac{x^2+ux-x+2u}{(x+u)(x-u)}$
49. $\dfrac{1-2y}{(x+y)(x-y)}$
51. $\dfrac{a^2+a-18}{a^3(a-9)}$
53. $\dfrac{3x-(a+b+c)}{(x-a)(x-b)(x-c)}$
55. $\dfrac{2x^2+4x+3}{x(x+3)^2}$
57. $\dfrac{x(2x+1)}{(x+1)^2(x-1)}$
59. $\dfrac{-t^4}{2}$
61. $\left(x+\frac{1}{2}\right)^2$
63. $\left(x-\frac{1}{3}\right)^2$

SECTION 5.4, p. 159

1. $\frac{5}{16}$
3. 6
5. 2
7. $\frac{2}{5}$
9. $\frac{16}{63}$
11. $\frac{3}{4}$
13. $\frac{15}{8}$
15. $\frac{13}{34}$
17. $\frac{-4}{33}$
19. $\dfrac{ac}{b}$
21. $\dfrac{ad}{bc}$
23. $\dfrac{x}{az}$
25. $\dfrac{abcxy}{z}$
27. $\dfrac{3c}{2xya}$
29. $\dfrac{y+x}{y(x-1)}$
31. $\dfrac{1}{a^2-a+1}$
33. -2
35. -6
37. $5y^2$
39. $\dfrac{2x-y}{2x}$
41. $\dfrac{2-x}{x+2}$
43. $\dfrac{n-m}{n+m}$

SECTION 5.5, p. 165

1. .6
3. .8
5. 16.57
7. 2.55
9. 1.158abc
11. .36
13. 1.103
15. .0828ab

17. 8.76 19. 450 21. $.424x^2$ 23. .03
25. $.000\,21xyz$ 27. .000 01 29. .008 31. .1616
33. .21 35. .2901 37. .03 39. 30
41. -170 43. 1500 45. -39 47. $-.155$
49. 20.239

REVIEW EXERCISES, p. 166

1. $\frac{14}{15}$ 3. $\frac{(a+b)(1-b)}{3(a-b)}$ 5. $\frac{(x^2+y^2)(x+y)}{3}$

7. (a) 8 (b) 96 (c) 30

9. (a) LCD is 8; $\frac{3}{4}=\frac{6}{8}, \frac{-5}{8}=\frac{-5}{8}$

(b) LCD is 144; $\frac{1}{48}=\frac{3}{144}, \frac{-1}{36}=\frac{-4}{144}$

(c) LCD is 120; $\frac{1}{20}=\frac{6}{120}, \frac{1}{30}=\frac{4}{120}, \frac{3}{40}=\frac{9}{120}$

11. $\frac{2}{3}$ 13. $\frac{11}{150}$ 15. $\frac{1+2x}{x^2}$ 17. $\frac{2a+x}{(x+a)(x-a)}$

19. $\frac{1}{(x+1)^2(x-1)}$ 21. $\frac{3}{2}$ 23. $\frac{4xb}{3az}$ 25. 2.489

27. 1.136

TEST YOURSELF, p. 168

1. $\frac{1}{6}$ 2. $\frac{-a}{x+1}$ 3. $\frac{-3a}{20b(a+2)}$ 4. (a) 24 (b) 90

5. (a) LCD is y^3; $\frac{a}{y^2}=\frac{ay}{y^3}, \frac{-1}{y^3}=\frac{-1}{y^3}$

(b) LCD is $a^2(x+5)(x-5)$; $\frac{1}{ax-5a}=\frac{a(x+5)}{a^2(x+5)(x-5)}, \frac{2}{a^2}=\frac{2(x+5)(x-5)}{a^2(x+5)(x-5)}, \frac{-1}{x^2-25}=\frac{-a^2}{a^2(x+5)(x-5)}$

6. $\frac{3}{25}$ 7. $\frac{x+4}{(x+1)(x-1)}$ 8. $\frac{3x^3-4y^2}{12}$

9. $\frac{25ay}{27cx^2}$ 10. $\frac{1}{a+1}$

CHAPTER 6

SECTION 6.1, p. 172

1. no; ii. yes 3. yes; i. true 5. yes; i. true
7. no; ii. yes 9. no; ii. no 11. no; ii. no

13. root
15. root
17. not a root
19. not a root
21. root
23. root
25. not a root
27. not a root
29. root
31. root
33. not a root

SECTION 6.2, p. 178. (The checks are given below.)

1. 10
3. -5
5. -10
7. 2
9. -8
11. -12
13. $\frac{2}{5}$
15. 3
17. 3
19. 5
21. 1
23. -7
25. 5
27. $\frac{13}{40}$
29. -4
31. 4
33. 22
35. -2
37. 0
39. -5
41. -2
43. No. 0 is a root of the first equation, but not of the second. Restrictions must be placed on the Addition Principle, as stated on p. 174.

CHECKS.

1. $10 + 1 \stackrel{?}{=} 11$

$11 \stackrel{\checkmark}{=} 11$

5. $-10 + 5 \stackrel{?}{=} -5$

$-5 \stackrel{\checkmark}{=} -5$

11. $\frac{1}{3}(-12) \stackrel{?}{=} -4$

$-4 \stackrel{\checkmark}{=} -4$

25. $5 - [2 - 3(5)] \stackrel{?}{=} 18$

$5 - [2 - 15] \stackrel{?}{=} 18$

$5 + 13 \stackrel{?}{=} 18$

$18 \stackrel{\checkmark}{=} 18$

27. $10\left(\frac{13}{40} - \frac{1}{4}\right) \stackrel{?}{=} \frac{3}{4}$

$10\frac{(13 - 10)}{40} \stackrel{?}{=} \frac{3}{4}$

$\frac{3}{4} \stackrel{\checkmark}{=} \frac{3}{4}$

31. $4 - [2 - (4 - 3)] \stackrel{?}{=} 7 - 4$

$4 - [2 - 1] \stackrel{?}{=} 3$

$3 \stackrel{\checkmark}{=} 3$

35. $(-2)^2 + 5(-2) \stackrel{?}{=} (-2)^2 - 10$

$4 - 10 \stackrel{\checkmark}{=} 4 - 10$

SECTION 6.3, p. 182. (The checks are given below.)

1. -18
3. 4
5. $\frac{-4}{3}$
7. 20
9. $\frac{19}{12}$
11. 8
13. 20
15. 12
17. $\frac{3}{2}$
19. 7
21. 8
23. -6
25. $\frac{-1}{2}$
27. 3
29. 11
31. 5
33. -3
35. 16
37. 4
39. $\frac{1}{4}$

41. $\frac{45}{11}$ 43. 3 45. 4 47. 6

49. $\frac{8}{7}$ 51. $\frac{10}{7}$

CHECKS.

1. $\frac{-18}{9} \overset{?}{=} -2$

$-2 \overset{\checkmark}{=} -2$

15. $\frac{5}{6}(12) + \frac{3(12)}{4} \overset{?}{=} \frac{5(12)}{3} - 1$

$10 + 9 \overset{?}{=} 20 - 1$

$19 \overset{\checkmark}{=} 19$

19. $\frac{35}{7} \overset{?}{=} 5$

$5 \overset{\checkmark}{=} 5$

25. $\frac{\frac{-1}{2}}{\frac{-1}{2} + 1} \overset{?}{=} -1$

$\frac{\frac{-1}{2}}{\frac{1}{2}} \overset{?}{=} -1$

$\frac{-1}{2} \cdot \frac{2}{1} \overset{?}{=} -1$

$-1 \overset{\checkmark}{=} -1$

33. $\frac{-3+7}{-3+2} \overset{?}{=} -4$

$\frac{4}{-1} \overset{?}{=} -4$

$-4 \overset{\checkmark}{=} -4$

37. $\frac{2}{4} + \frac{6}{4} \overset{?}{=} 2$

$\frac{1}{2} + \frac{3}{2} \overset{?}{=} 2$

$\frac{4}{2} \overset{?}{=} 2$

$2 \overset{\checkmark}{=} 2$

SECTION 6.4, p. 188. (The checks are given below.)

1. $\frac{c}{5}$ 3. $b - c - 13$ 5. $\frac{9}{3 - a}$

7. $\frac{3 + b}{1 - 2a}$ 9. $\frac{9b + a}{3}$ 11. $\frac{4b + 6c}{33}$

13. $\frac{4}{a + b}$ 15. $\frac{-a(bc + 1)}{bc}$ 17. $\frac{9}{4a^2b}$

19. $4 - \frac{a}{b(7 - a)}$ 21. $\frac{a - 4 - b}{2}$ 23. $\frac{3}{a + b - 9}$

25. (a) $1 + t - 2x$ (b) $\frac{1 + t - y}{2}$ 27. (a) $\frac{cz - by}{a}$ (b) $\frac{cz - ax}{b}$

29. (a) $\frac{1 + 3xz}{5a}$ (b) $\frac{5ay - 1}{3x}$ 31. (a) $\frac{9 - y^2 - yz}{y + z}$ (b) $\frac{9 - xy - y^2}{x + y}$

33. (a) $\frac{-bt + b^2 + 1 + a^2}{a}$ (b) $\frac{a^2 - ax + 1 + b^2}{b}$ 35. (a) $\frac{Kz - 3b(x - 1)}{5z}$ (b) $\frac{3b(x - 1)}{K - 5t}$

37. $\dfrac{C}{2\pi}$

39. $\dfrac{v}{\pi r^2}$

41. (a) $\dfrac{c(2gh - v^2)}{2g}$ (b) $\dfrac{35}{16}$

CHECKS.

1. $5\left(\dfrac{c}{5}\right) \stackrel{?}{=} c$

$c \stackrel{\checkmark}{=} c$

5. $a\left(\dfrac{9}{3-a}\right) + 7 \stackrel{?}{=} 3\left(\dfrac{9}{3-a}\right) - 2$

$a\left(\dfrac{9}{3-a}\right) + 9 \stackrel{?}{=} 3\left(\dfrac{9}{3-a}\right)$

$\dfrac{9a + 9(3-a)}{3-a} \stackrel{?}{=} \dfrac{27}{3-a}$

$9a + 27 - 9a \stackrel{?}{=} 27$

$27 \stackrel{\checkmark}{=} 27$

15. $\dfrac{\dfrac{-a(bc+1)}{bc}}{\dfrac{-a(bc+1)}{bc} + a} \stackrel{?}{=} bc + 1$

$\dfrac{\dfrac{-a(bc+1)}{bc}}{\dfrac{-\cancel{abc} - a + \cancel{abc}}{bc}} \stackrel{?}{=} bc + 1$

$\dfrac{-\cancel{a}(bc+1)}{\cancel{bc}} \cdot \dfrac{\cancel{bc}}{-\cancel{a}} \stackrel{?}{=} bc + 1$

$bc + 1 \stackrel{\checkmark}{=} bc + 1$

SECTION 6.5, p. 193

1. 1
3. 2
5. 1
7. 0
9. 2
11. $(x-1)(x-2)(x-3)(x-4)(x-5)(x-6)(x-7)(x-8) = 0$
13. $x + 10 = x - 3$

$10 \stackrel{x}{=} -3$

15. $3(u-3) + 1 - u = 2(u+5)$

$-8 \stackrel{x}{=} 10$

17. $(x+1)(x-1) = (x+1)^2 - 2x$

$x^2 - 1 = x^2 + 2x + 1 - 2x$

$-1 \stackrel{x}{=} 1$

19. $x(x-5) = (x-2)^2 + 3 - x$

$x^2 - 5x = x^2 - 4x + 4 + 3 - x$

$0 \stackrel{x}{=} 7$

21. $5x - 6 = 3(x-2) + 2x$

$-6 \stackrel{\checkmark}{=} -6$

23. $y - (2 - 3y) = 4(y-1) + 2$

$-2 \stackrel{\checkmark}{=} -2$

25. $(t-4)^2 + 16t = (t+4)^2$

$t^2 - 8t + 16 + 16t = t^2 + 8t + 16$

$16 \stackrel{\checkmark}{=} 16$

27. $\dfrac{y+4}{2} + 1 - y = \dfrac{6-y}{2}$

$\dfrac{y + 4 + 2 - 2y}{2} = \dfrac{6-y}{2}$

$6 \stackrel{\checkmark}{=} 6$

29. (a)
31. (c)
33. (a)
35. (a)
37. (b)
39. (b)

REVIEW EXERCISES, p. 194. **(The checks in Exercises 9 and 17 are given below.)**

1. yes; i. true
3. no; ii. no
5. root
7. root
9. 20
11. -15
13. $\frac{12}{5}$
15. 4
17. 5
19. 6
21. $\dfrac{1+2c}{3}$
23. $\dfrac{5(2y-1)}{3}$
25. (a) $\dfrac{2t+3-3y+3z}{5}$ (b) $\dfrac{2t+3-5x+3z}{3}$
27. 2
29. 3
31. (a)

CHECKS.

9. $20 - 10 \stackrel{?}{=} 10$

$10 \stackrel{\checkmark}{=} 10$

17. $\dfrac{5+3}{8} \stackrel{?}{=} \dfrac{5-2}{3}$

$1 \stackrel{\checkmark}{=} 1$

TEST YOURSELF, p. 195. **(The checks in Exercises 2 and 5 are given below.)**

1. (a) no (b) yes
2. $\dfrac{-1}{7}$
3. -12
4. $\frac{5}{4}$
5. 16
6. 8
7. $\frac{5}{3}$
8. $\dfrac{c+4-a+b}{6}$
9. (a) $\dfrac{1-4y}{5y}$ (b) $\dfrac{1}{4+5x}$
10. Equations (b) and (d) have a single root.

CHECKS.

2. $4\left(\dfrac{-1}{7}\right) + 2 \stackrel{?}{=} 1 - 3\left(\dfrac{-1}{7}\right)$

$\dfrac{-4+14}{7} \stackrel{?}{=} \dfrac{7+3}{7}$

$\dfrac{10}{7} \stackrel{\checkmark}{=} \dfrac{10}{7}$

5. $\dfrac{4}{16} \stackrel{?}{=} \dfrac{3}{16-4}$

$\dfrac{1}{4} \stackrel{?}{=} \dfrac{3}{12}$

$\dfrac{1}{4} \stackrel{\checkmark}{=} \dfrac{1}{4}$

CHAPTER 7

SECTION 7.1, p. 201

1. (a) $x+5$ (b) $x-8$ (c) $3x$ (d) $\dfrac{x}{4}$ (e) $x+3$ (f) $x-10$
3. (a) $x+1$ (b) $x+(x+1)+(x+2)$(or $3x+3$) (c) $x+2$, $x+4$
5. 18, 19
7. 25, 27
9. 17
11. 8, 9
13. $x+5$
15. $x-3$

17. The daughter is 5 years old and her father is 25 years old.
19. The man is 23 years old.
21. Harry is 18 years old.

SECTION 7.2, p. 205

1. \$2.27
3. 10 nickels and 15 quarters
5. 18 quarters and 16 dimes
7. 18 quarters
9. 14 pennies
11. 62 ice cream bars
13. 30 pounds of coffee at 75 cents per pound and 20 pounds of coffee at \$1.25 per pound
15. \$1.80 per pound

SECTION 7.3, p. 210

1. $\frac{1}{2}$ hour
3. $2\frac{1}{4}$ hours
5. 50 miles
7. 75 miles per hour
9. 70 miles per hour
11. 750 feet per minute

SECTION 7.4, p. 213

1. \$33.60
3. \$3570
5. \$687.66
7. \$1600
9. \$7000 at 7%; \$9500 at 6%
11. 7%

SECTION 7.5, p. 216

1. 54 gallons
3. 1 gallon
5. $8\frac{3}{4}$ ounces
7. $\frac{5}{3}$ gallons
9. 120 tons of the first alloy (2 parts zinc to 1 part tin), 80 tons of the second alloy

SECTION 7.6, p. 219

1. $\frac{60}{11}$ hours
3. 6 hours
5. 6 days
7. $\frac{36}{11}$ hours
9. 30 minutes
11. 90 minutes

REVIEW EXERCISES, p. 220

1. 25, 27, 29
3. 12 nickels, 3 dimes, 6 quarters
5. 70 miles per hour
7. \$3600
9. 36 gallons
11. $\frac{24}{7}$ hours

TEST YOURSELF, p. 221

1. 41, 49
2. 30
3. 70 miles per hour
4. 24 gallons

CHAPTER 8.

SECTION 8.1, p. 225

1. {Bob, Carol, Ted, Alice}
3. {Maine, New Hampshire, Vermont, Massachusetts, Connecticut, Rhode Island}
5. {left field, center field, right field}
7. {1, 2, 3, 4, 5, 6, 7, 8, 9, 10}
9. {−9, −7, −5, −3, −1}
11. function
13. function
15. function
17. function
19. function
21. (a) {1, 2, 3, 4} (b) {7, 8, 9, 10}
23. (a) {100, 200, 300, 400} (b) {1, 2, 3, 4}
25. (a) {−1, −2, −3, −4, −5, −6} (b) {0, 1}
27. $1 \to 1$ $\quad 1 \to 2$
    $2 \to 2$ $\quad 2 \to 1$
29. 6

SECTION 8.2, p. 230

1. (a) 1, 2, 3, 4 (b) 0 (c) 0
3. (a) 2 (b) 4 (c) 6 (d) 8
5. (a) −10 (b) −5 (c) 0 (d) −20
7. (a) −1 (b) 24 (c) 0 (d) 1
9. (a) 1 (b) 2 (c) 5 (d) 10
11. (a) 54 (b) 104 (c) 1004 (d) −16
13. (a) 0 (b) −5 (c) $\frac{5}{8}$ (d) $\frac{25}{64}$
15. (a) 1 (b) $\frac{1}{4}$ (c) 4 (d) 10
17. (a) $\frac{-1}{5}$ (b) $\frac{-1}{6}$ (c) $\frac{-1}{2}$ (d) $\frac{1}{10}$
19. (a) 0 (b) $\frac{-1}{3}$ (c) $\frac{2}{3}$ (d) $\frac{40}{9}$
21. (a) $\frac{-1}{2}$ (b) $\frac{-1}{2}$ (c) $\frac{-1}{2}$ (d) $\frac{-1}{2}$
23. (a) $f(s) = s^2$ (b) 81 (c) $\frac{1}{81}$
25. (a) $f(h) = \frac{h^2}{3}$ (b) $\frac{3b^2}{4}$ (c) 108 (d) 243
27. (a) $f(h) = h + 5$ (b) $f(H) = H - 5$
29. (a) $f(t) = 2 + 4t$ (b) 210
31. The set of all real numbers other than 1
33. The set of all real numbers other than 0

SECTION 8.3, p. 237

1. $P = (1, 1)$, $Q = (-1, -1)$, $R = (3, -2)$, $S = \left(5, \frac{1}{2}\right)$, $T = (-4, -3)$, $U = \left(0, \frac{-9}{2}\right)$, $V = \left(\frac{5}{2}, -5\right)$

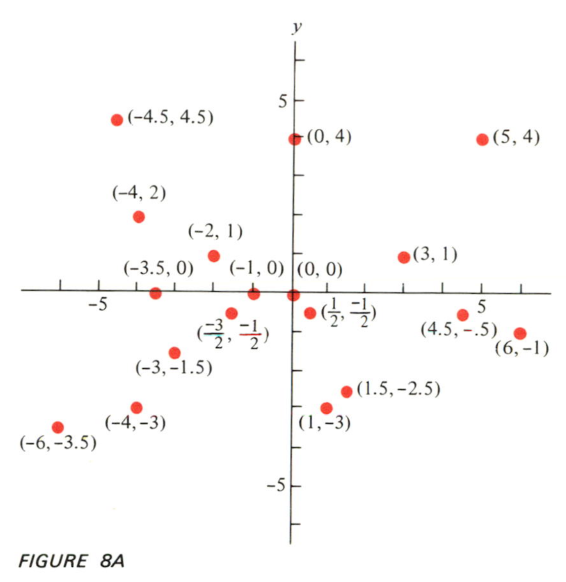

FIGURE 8A

3.–20. See Figure 8A.

21. IV	23. *y*-axis	25. III	27. *x*-axis
29. IV	31. I	33. I	

SECTION 8.4, p. 244

1. See Figure 8B.

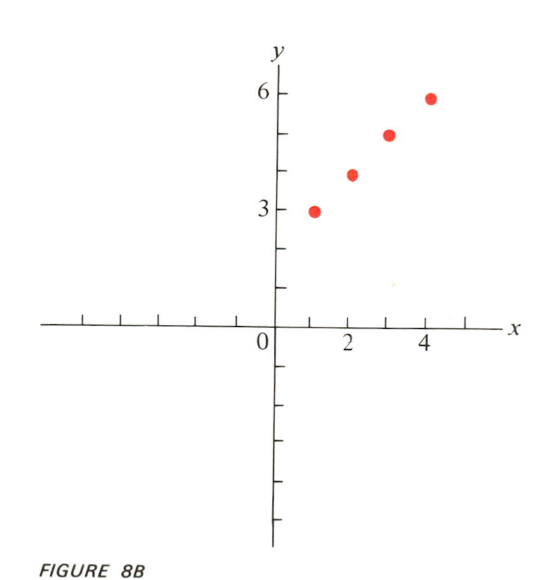

FIGURE 8B

3. See Figure 8C on page A22.
5. See Figure 8D on page A22.
7. See Figure 8E on page A22.

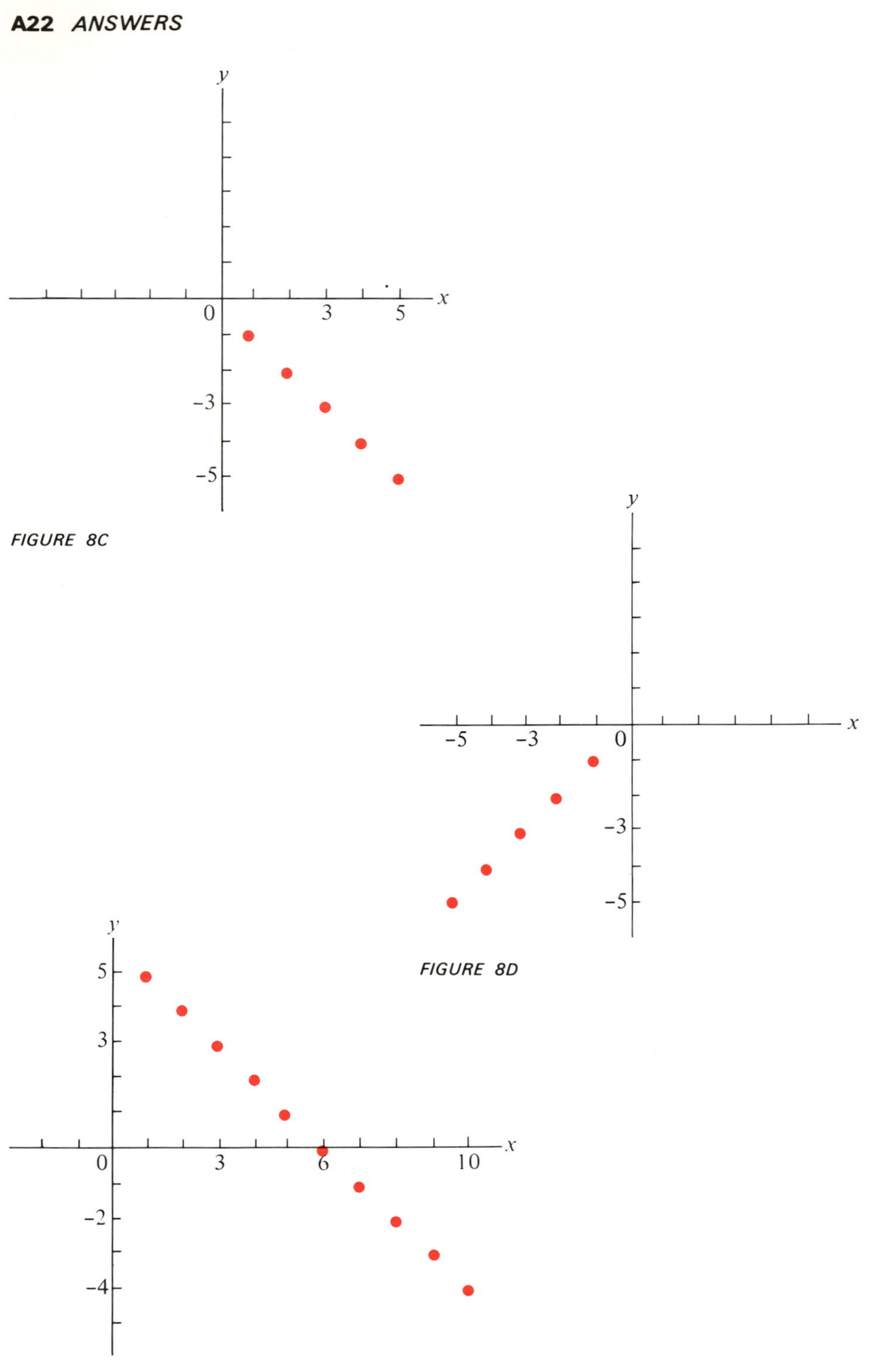

FIGURE 8C

FIGURE 8D

FIGURE 8E

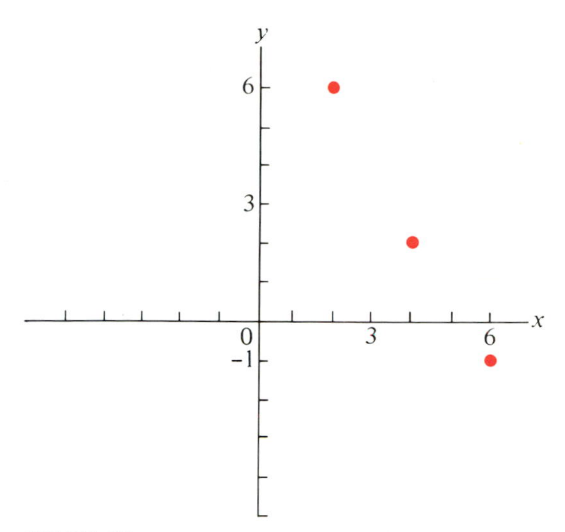

FIGURE 8F

9. See Figure 8F.
11. function
13. function
15. function
17. not a function
19. See Figure 8G.

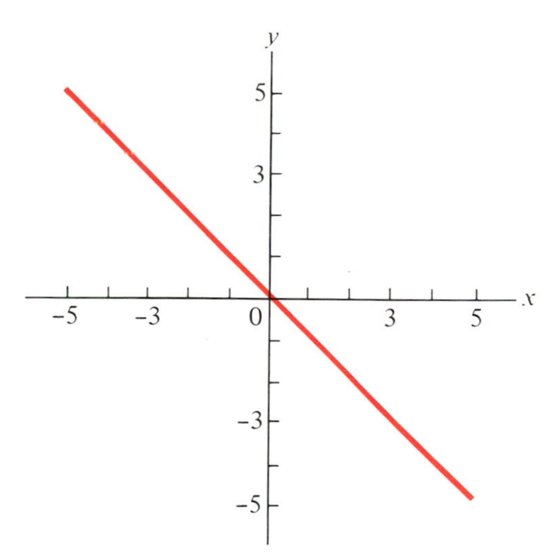

FIGURE 8G

21. See Figure 8H on page A24.
23. See Figure 8I on page A24.
25. See Figure 8J on page A24.
27. See Figure 8K on page A25.
29. See Figure 8L on page A25.
31. See Figure 8M on page A25.
33. See Figure 8N on page A25.
35. See Figure 8O on page A26.

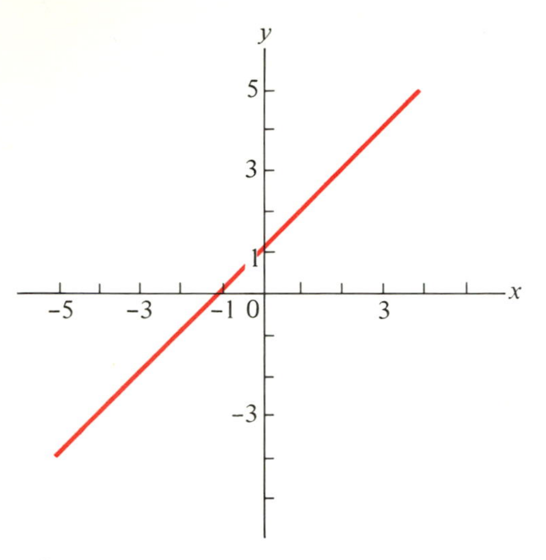

FIGURE 8H

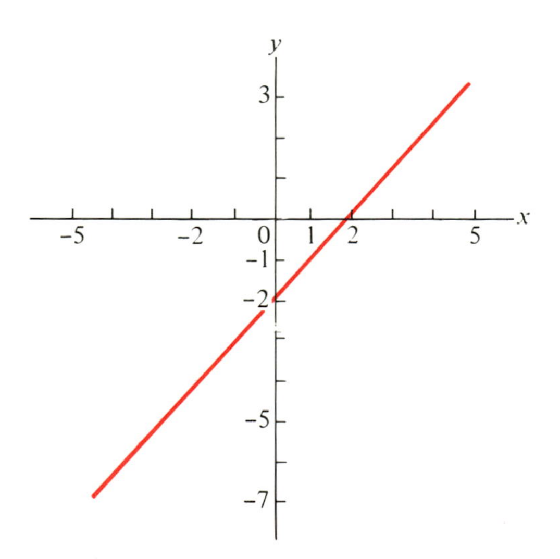

FIGURE 8I

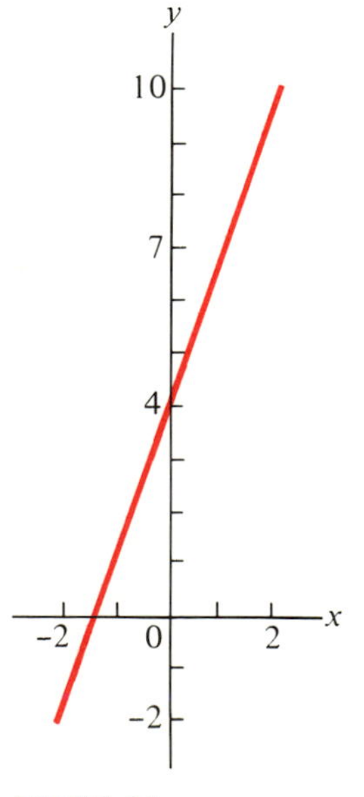

FIGURE 8J

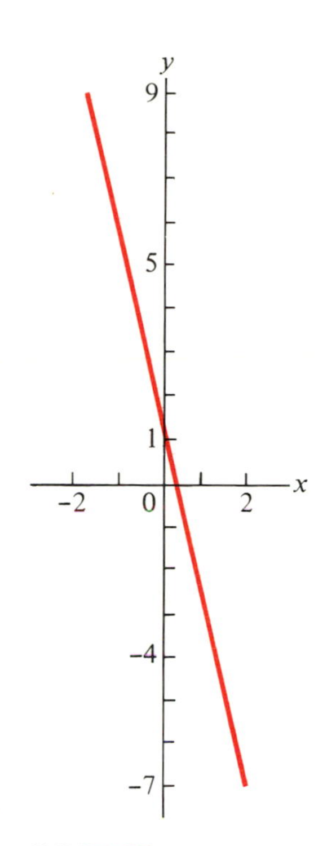

FIGURE 8K

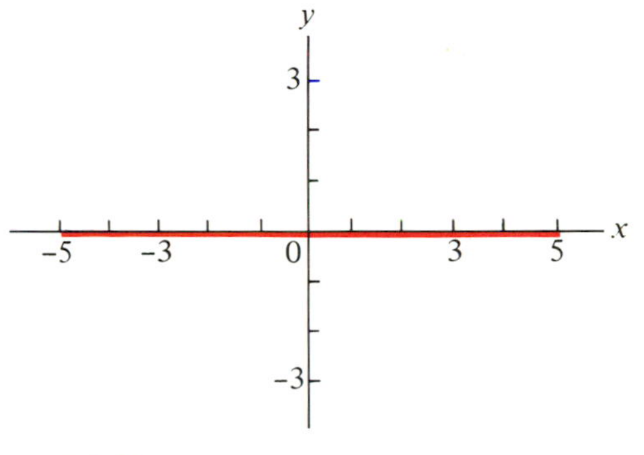

FIGURE 8L

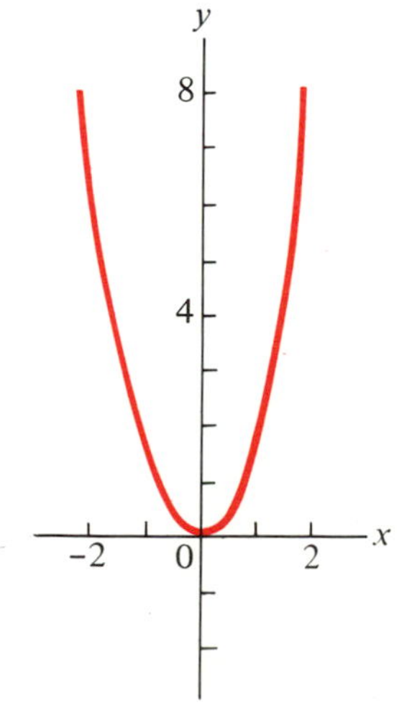

FIGURE 8M

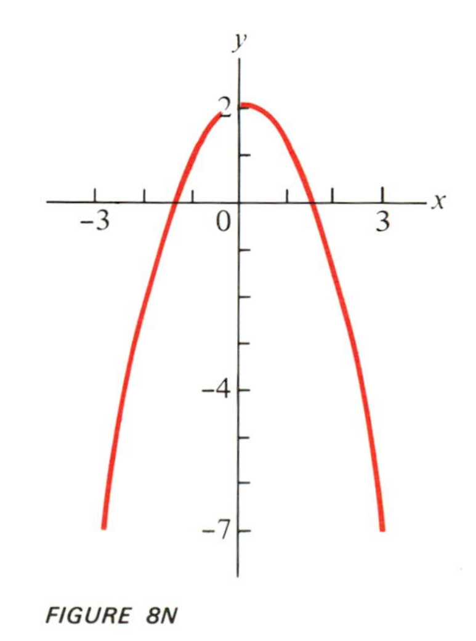

FIGURE 8N

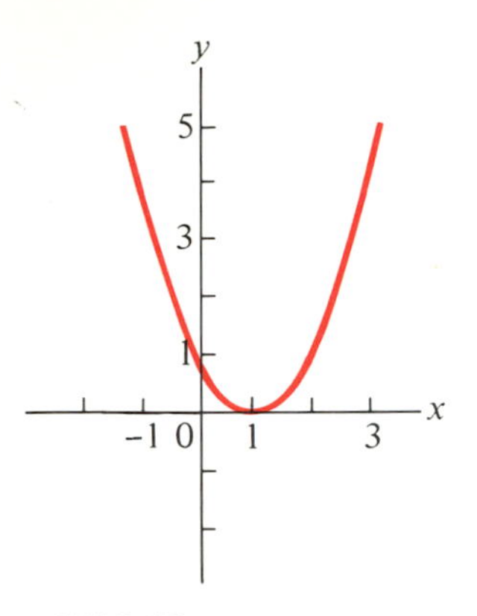

FIGURE 80

37. function
39. function
41. not a function
43. function
45. function
47. not a function

SECTION 8.5, p. 252

1. one–one
3. not one–one
5. one–one
7. one–one
9. one–one
11. one–one
13. not one–one
15. $f(1) = f(2) = 4$
17. $f(1) = f(-1) = -2$
19. (a) and (e) are inverses, (b) and (f) are inverses, (c) and (d) are inverses.
21. $\{(2, 1), (4, 2), (8, 3)\}$
23. $\{(1, 1), (2, 2), (3, 3), (4, 4), (5, 5)\}$
25. $\{(4, 2)\}$
27. $y = \frac{x}{2}$
29. $y = x - 8$
31. $y = 9 - x$
33. $y = \frac{x - 7}{3}$
35. $y = \frac{7 - x}{2}$
37. $y = 4\left(x - \frac{1}{2}\right)$

REVIEW EXERCISES, p. 254

1. (a) {red, white, blue} (b) $\{1, 2, 3, 4, 5, 6\}$ (c) $\{2, 4, 6, 8, 10, 12, 14\}$
3. The domain is $\{5, 10, 15, 20, 25\}$; the range is $\{1, 2, 3\}$.
5. (a) 2, 4, 6, 8 (b) 2 (c) 1
7. (a) 9 (b) 25 (c) 6 (d) 4
9. (a) $f(s) = s^3$ (b) 125 cubic inches (c) .008 cubic inch
11. $P = (3, 2)$, $Q = (2, 3)$, $R = (-1, -3)$, $S = (4.5, -3)$, $T = (4.5, 0)$, $U = \left(0, \frac{-1}{2}\right)$, $V = (-2, 4)$
13. (a) I (b) II (c) III (d) y-axis (e) IV (f) I
15. See Figure 8P on page A27.
17. (a) function (b) not a function (c) function (d) not a function
19. (a) not one–one (b) one–one (c) one–one
21. $y = \frac{x + 4}{3}$

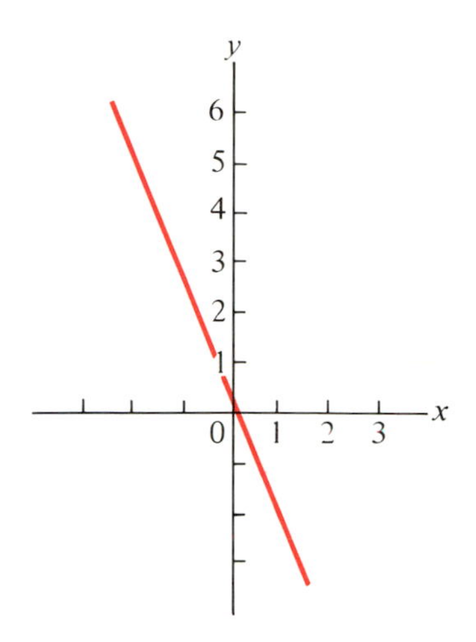

FIGURE 8P

TEST YOURSELF, p. 258

1. (a) function (b) not a function
2. The domain is {2, 4, 6, 8, 10}; the range is {3, 5, 7, 9}.
3. (a) -3 (b) 5 (c) 0 (d) 41
4. (a) -4 (b) -1 (c) -7 (d) 8
5. (a) $f(r) = 2\pi r$ (b) 10π
6. See Figure 8Q.
7. See Figure 8R.

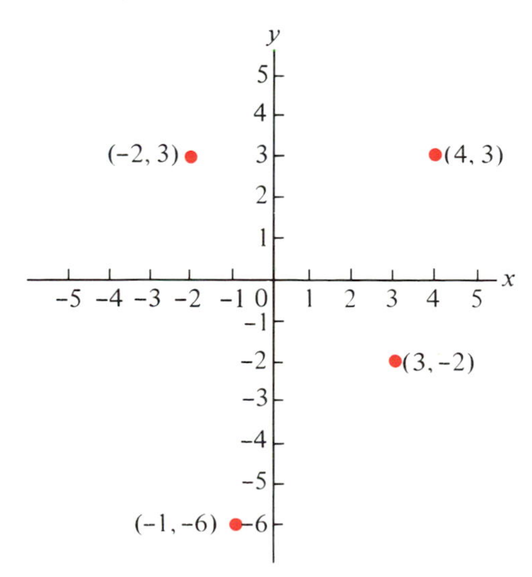

FIGURE 8Q

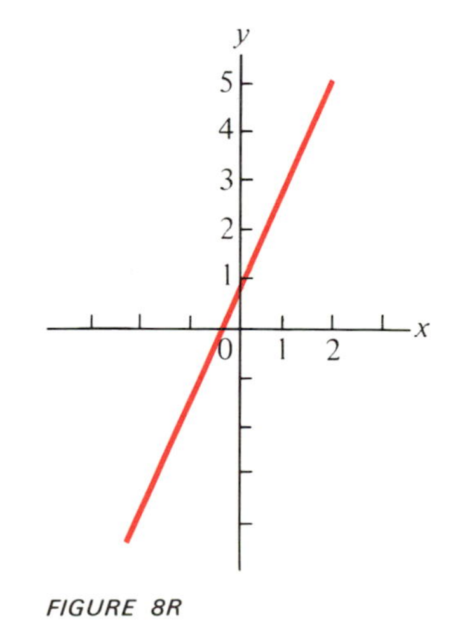

FIGURE 8R

8. (a) function (b) not a function (c) function

CHAPTER 9

SECTION 9.1, p. 264

1. 6
3. $\frac{7}{5}$
5. 10
7. 6
9. 2
11. 8
13. −7
15. 1
17. $\frac{42}{5}$
19. $\frac{-3}{7}$
21. 4, −4
23. .2
25. 24¢
27. $18
29. $1.25
31. $31.40
33. 45
35. 21 feet, 6 feet
37. 24
39. $a = 3$, $c = 16.5$
41. $a = 9$, $c = 15$

SECTION 9.2, p. 273

1. 1
3. $\frac{1}{6}$
5. $\frac{-1}{6}$
7. −7
9. $\frac{1}{2}$
11. $\frac{3}{5}$
13. $\frac{-1}{12}$
15. $\frac{1}{2}$
17. 13
19. −1
21. −15
23. $y - 4 = 3x$
25. $y - 3 = 2(x + 1)$
27. $y - 2 = \frac{1}{2}(x - 1)$
29. $y - 1 = 10(x - 1)$
31. $y + 1 = -3(x + 1)$
33. $y = 3$
35. $y + 1 = \frac{2}{3}(x - 1)$
37. (a) $y - 3 = 2(x - 3)$ (b) See Figure 9A.

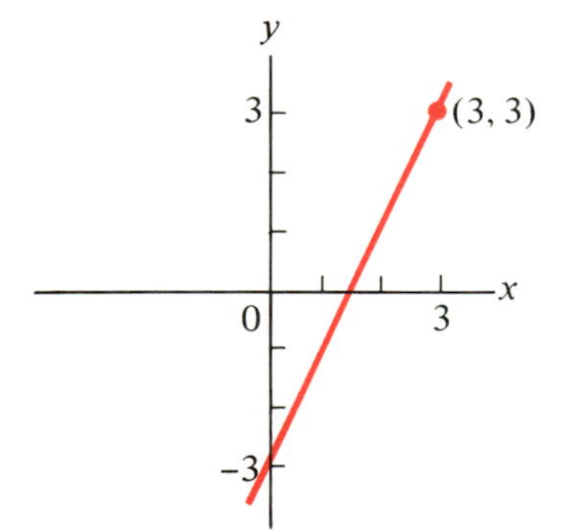

FIGURE 9A

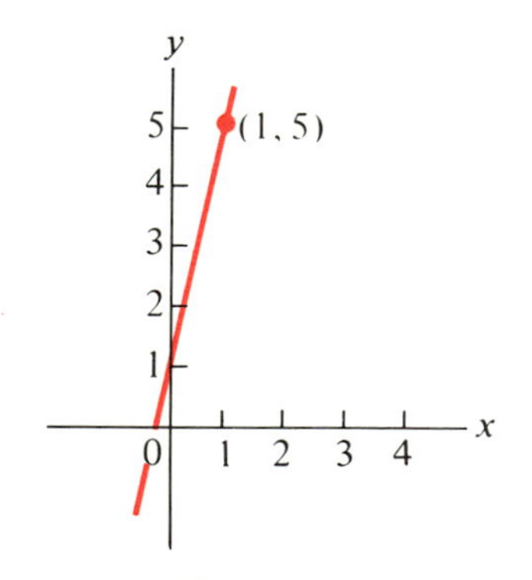

FIGURE 9B

39. (a) $y - 5 = 4(x - 1)$ (b) See Figure 9B.
41. (a) $y - 2 = \frac{-1}{2}(x - 4)$ (b) See Figure 9C on page A29.
43. (a) neither horizontal nor vertical
    (b) neither horizontal nor vertical
    (c) horizontal (d) vertical
    (e) horizontal (f) vertical
45. $y = 5$
47. $x = 1$

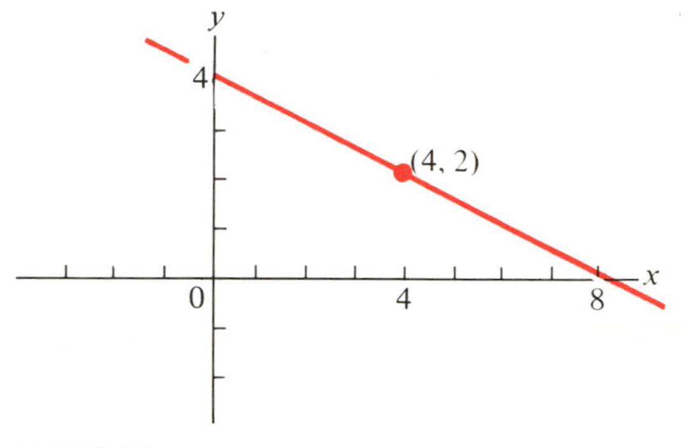

FIGURE 9C

SECTION 9.3, p. 281

1. (a) -2 (b) 2
3. (a) -7 (b) 14
5. (a) $\frac{5}{2}$ (b) -5
7. (a) $\frac{-5}{2}$ (b) $\frac{-5}{3}$
9. $y = 2x + 3$
11. $y = 2x + 18$
13. $y = 3x + 1$
15. $y = \frac{-3}{5}x + \frac{29}{5}$
17. $y - 4 = 2(x - 1)$
19. $y = -5(x - 1)$
21. $y = 2(x + 3)$
23. $x - y + 10 = 0$
25. $3x + y - 8 = 0$
27. $8x - y = 0$
29. $y - 5 = 0$
31. (a) 5 (b) See Figure 9D.

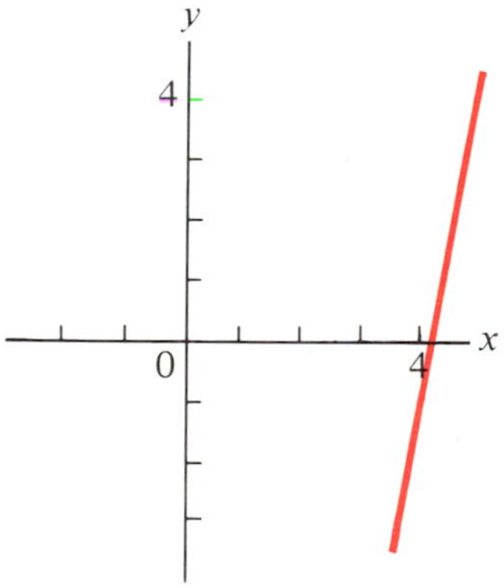

FIGURE 9D

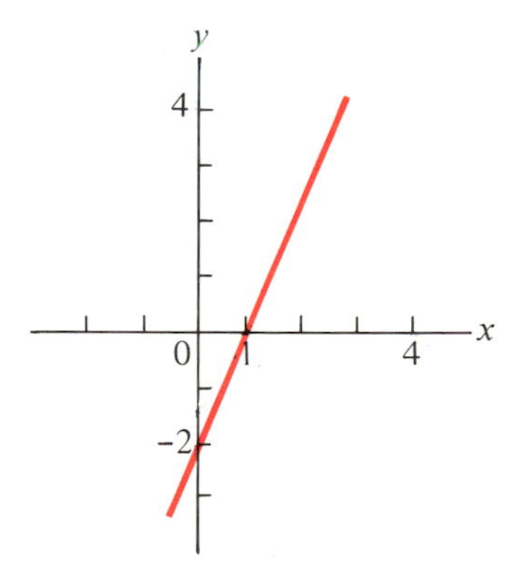

FIGURE 9E

33. (a) 2 (b) See Figure 9E.
35. (a) $\frac{8}{3}$ (b) See Figure 9F on page A30.
37. (a) $\frac{-1}{5}$ (b) See Figure 9G on page A30.
39. $y + 1 = \frac{3}{5}(x + 1)$
41. $y = -4x + 5$
43. $y = -2$
45. $y = 5$

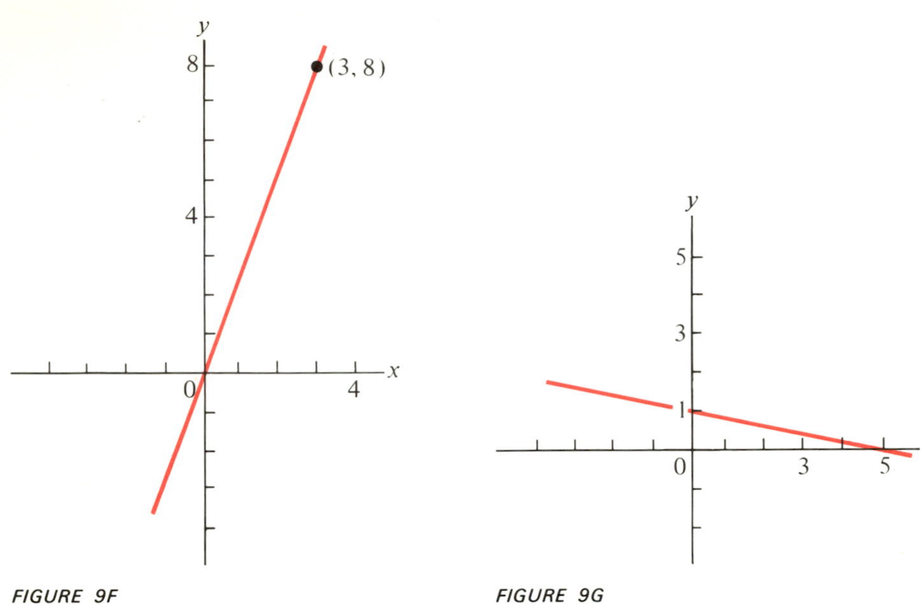

FIGURE 9F

FIGURE 9G

47. (a) $y - 9 = -4(x - 1)$ (b) $y - 9 = -4x + 4$ Add 4 to both sides.
$y - 5 = -4x + 8$
$y - 5 = -4(x - 2)$

SECTION 9.4, p. 291

1. parallel
3. perpendicular
5. parallel
7. neither parallel nor perpendicular
9. parallel
11. parallel
13. neither parallel nor perpendicular
15. neither parallel nor perpendicular
17. neither parallel nor perpendicular
19. perpendicular
21. perpendicular
23. neither parallel nor perpendicular
25. neither parallel nor perpendicular
27. parallel
29. (a) $(y - 1) = 2(x - 1)$ (b) $(y - 1) = \frac{-1}{2}(x - 1)$
31. (a) $y + 3 = -2x$ (b) $y + 3 = \frac{x}{2}$
33. (a) $y - 5 = -(x - 2)$ (b) $y - 5 = x - 2$
35. (a) $y = 8$ (b) $x = 2$
37. (a) $x = 0$, (b) $y = 0$
39. $L_1 \parallel L_4$; $L_1 \perp L_5$, $L_4 \perp L_5$, $L_2 \perp L_3$
41. $L_1 \parallel L_3$, $L_1 \parallel L_5$, $L_3 \parallel L_5$, $L_2 \parallel L_4$; $L_1 \perp L_2$, $L_1 \perp L_4$, $L_3 \perp L_2$, $L_3 \perp L_4$, $L_5 \perp L_2$, $L_5 \perp L_4$
43. No. For, L_1 can be the same line as L_3. However, if these lines are distinct, then they are parallel.
45. $L_1 \perp L_3$

REVIEW EXERCISES, p. 293

1. $\frac{-9}{5}$ 3. 25 cents   5. $\frac{1}{2}$ 7. -1

9. (a) $y - 2 = \frac{-1}{2}(x - 1)$ (b) See Figure 9H.

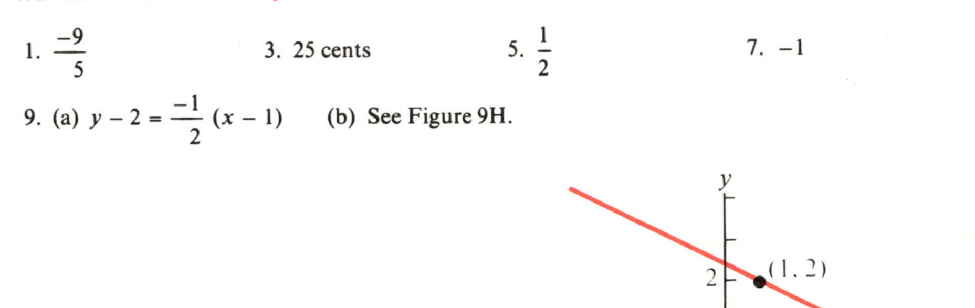

FIGURE 9H

11. $x = 1$ 13. $y = -x + 7$ 15. $3y - 5x - 8 = 0$
17. (a)   19. (b)

21. (a) $y - 5 = 2(x - 2)$ (b) $y - 5 = \frac{-1}{2}(x - 2)$

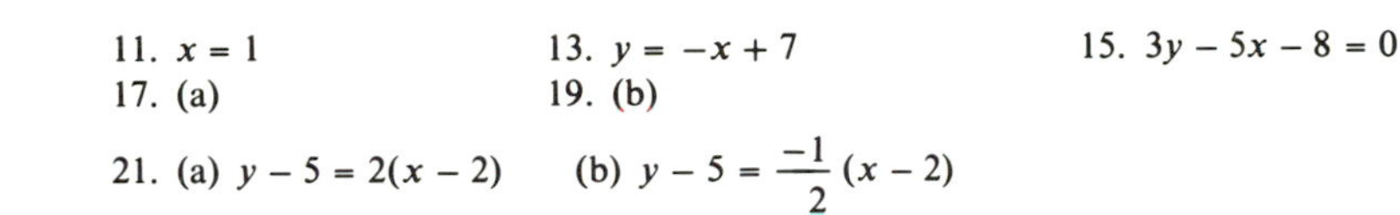

TEST YOURSELF, p. 294

1. (a) 6   (b) 6   2. $\frac{4}{3}$ 3. $y + 2 = 5(x - 2)$
4. $y = -2$ 5. (a) -3 (b) -3 6. $5x + 3y - 7 = 0$
7. $y - 6 = 5(x - 2)$ 8. (b)   9. $y + 3 = 4(x - 2)$
10. $L_2 \parallel L_3$; $L_1 \perp L_2$, $L_1 \perp L_3$, $L_4 \perp L_5$

CHAPTER 10

SECTION 10.1, p. 304

1. See Figure 10A on page A32.
3. See Figure 10B on page A32.
5. See Figure 10C on page A32.
7. See Figure 10D on page A32.
9. See Figure 10E on page A33.

11. (1, 2)   13. (5, 2)   15. (6, 0)   17. $\left(\frac{1}{2}, \frac{1}{4}\right)$
19. (2, 5)   21. (1, 1)   23. (9, −3)   25. (4, 3)
27. (0, 0)   29. (6, −2)   31. (6, −3)   33. (3, −4)
35. (5, −5)   37. (11, 12)   39. 17, 13

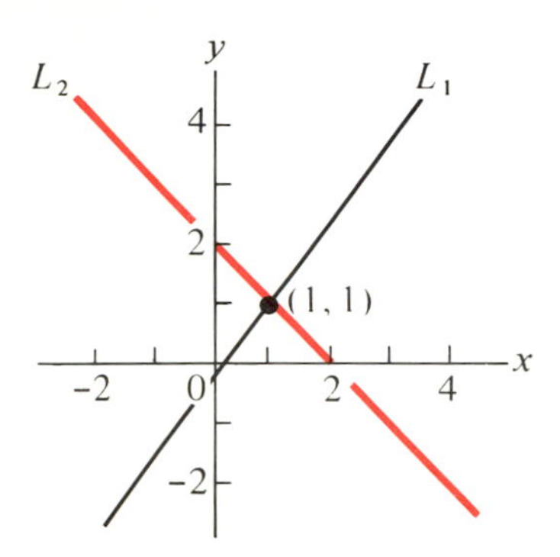

FIGURE 10A

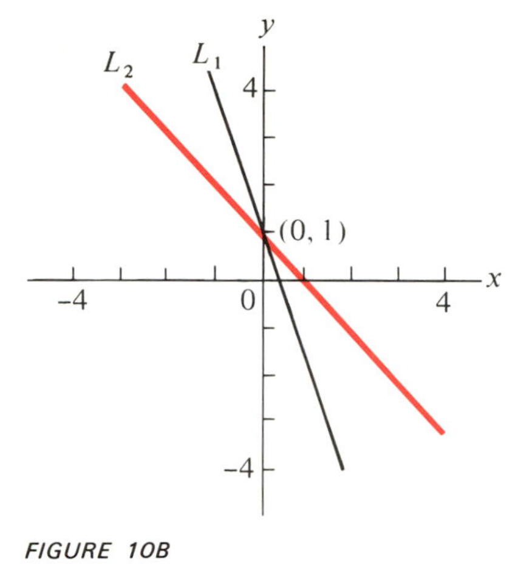

FIGURE 10B

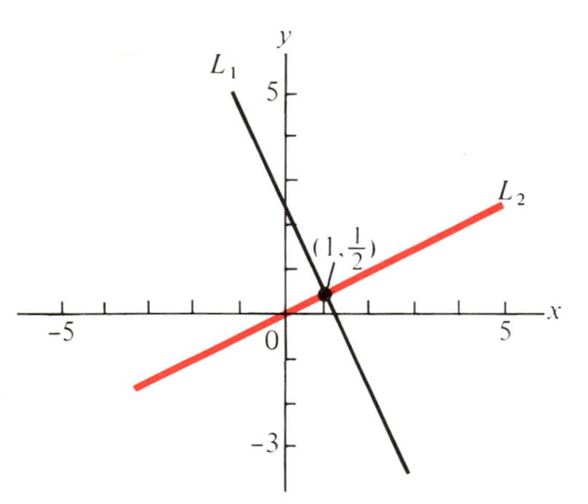

FIGURE 10C

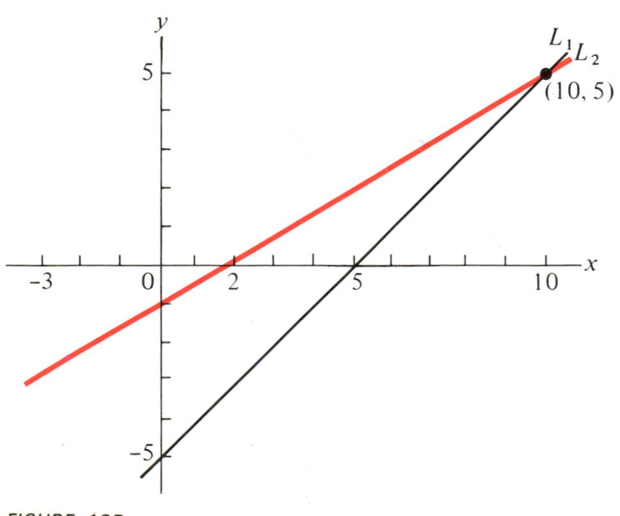

FIGURE 10D

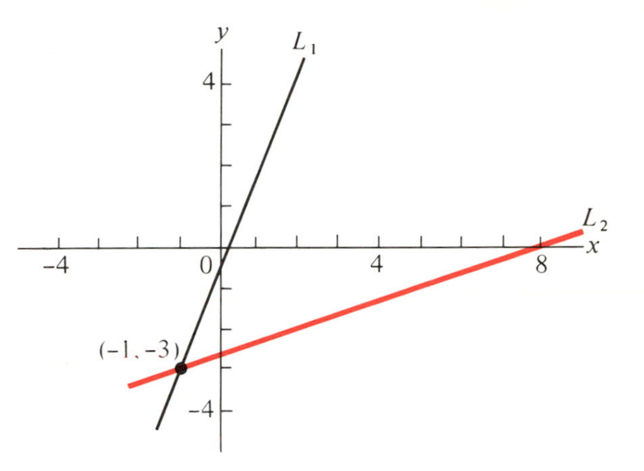

FIGURE 10E

41. Ted is 12; his brother is 6.

43. The dictionary is $1\frac{1}{2}$ inches thick. The encyclopedia is 2 inches thick.

45. $a = 4,\ b = -3$ 47. $a = 1,\ b = 1$ 49. $a = 1,\ b = 1$

SECTION 10.2, p. 311

1. (1, 1, 4) 3. (1, −1, 3) 5. (6, −1, −2)
7. (8, 2, 3) 9. (1, 4, 8) 11. (4, 7, 2)
13. (20, 1, 2) 15. (1, 9, −2) 17. (−2, −1, −4)
19. $\left(2, \frac{1}{2}, -1\right)$ 21. 2, 4, and 6 23. 6 nickels, 3 dimes, and 8 quarters

SECTION 10.3, p. 317

1. −1 3. 6 5. 0 7. 5
9. −2 11. 80 13. $\frac{-2}{5}$ 15. 10
17. −1 19. 4 21. 9 23. 0

25. $\begin{vmatrix} a & c \\ b & d \end{vmatrix} = -\begin{vmatrix} c & a \\ d & b \end{vmatrix}$

27. (a) Cramer's Rule applies. (b) (3, 2)

29. (a) Cramer's Rule applies. (b) $\left(\frac{1}{3}, \frac{1}{3}\right)$

31. (a) Cramer's Rule does not apply.

33. (a) Cramer's Rule applies. (b) $\left(-7, \frac{4}{3}\right)$

35. (a) Cramer's Rule applies. (b) (0, 1)

37. (a) Cramer's Rule applies. (b) $\left(\frac{10}{3}, \frac{-5}{2}\right)$

39. (a) Cramer's Rule applies. (b) (−1, 2)

SECTION 10.4, p. 325

1. −1 3. 2 5. 0 7. 0
9. −60 11. −.08 13. 2 15. 5
17. 0

19. $\begin{vmatrix} ka_1 & b_1 & c_1 \\ ka_2 & b_2 & c_2 \\ ka_3 & b_3 & c_3 \end{vmatrix} = k \begin{vmatrix} a_1 & b_1 & c_1 \\ a_2 & b_2 & c_2 \\ a_3 & b_3 & c_3 \end{vmatrix}$

21. (a) Cramer's Rule applies. (b) (1, 1, 1)
23. (a) Cramer's Rule applies. (b) (−1, 2, 0)
25. (a) Cramer's Rule applies. (b) (6, 0, −6)
27. (a) Cramer's Rule does not apply.
29. (a) Cramer's Rule does not apply.
31. (a) Cramer's Rule applies. (b) (−1, −1, −1)

REVIEW EXERCISES, p. 326

1. See Figure 10F.

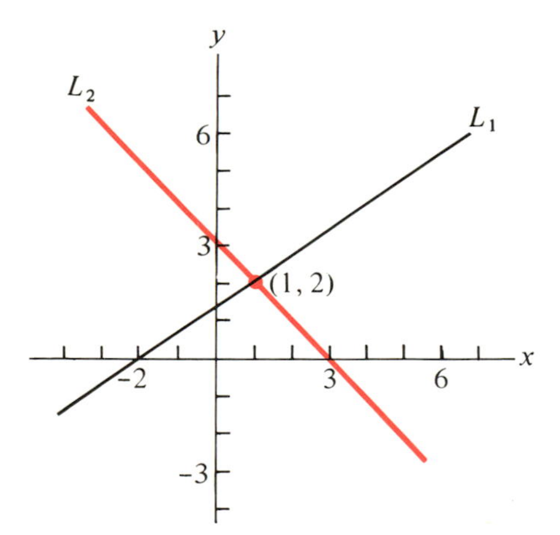

FIGURE 10F

3. (1, 3) 5. (−2, 2) 7. $a = 2$, $b = -3$
9. (1, 5, 2) 11. 7, 11, 13 13. 14
15. 2 17. (a) Cramer's Rules does not apply.
19. 2 21. 5
23. Cramer's Rule does not apply.

TEST YOURSELF, p. 328

1. (2, 3) 2. (5, 4) 3. (6, −1) 4. (4, −2)
5. (1, 2) 6. (a) −8 (b)10
7. (1, 0, 5) 8. (−1, 2, 3) 9. (1, 2, 5)

CHAPTER 11

SECTION 11.1, p. 334

1. $\frac{1}{10}$ 3. 1 5. $\frac{1}{32}$ 7. 1 9. $\frac{1}{144}$ 11. $\frac{1}{10\,000}$ 13. $\frac{-1}{5}$ 15. $\frac{-1}{8}$ 17. 2 19. $\frac{5}{2}$ 21. 1 23. $\frac{1}{12}$ 25. 64 27. -1 29. -1 31. 0 33. -2 35. -1 37. -5 41. y^{-6} 43. x^2y^{-2} 45. $a^{-2}b^{-1}$ 47. $(x+y)^{-1}$

SECTION 11.2, p. 338

1. a^3 3. a^9 5. $\frac{1}{a}$ 7. $\frac{1}{a^{10}}$ 9. a^{10} 11. a^4x^2 13. $\frac{a^3}{100}$ 15. b^2 17. $4x^2y^2$ 19. $2ab^2$ 21. $\frac{2a^7}{b^6}$ 23. a^{18} 25. $\frac{1}{x+y}$ 27. $\frac{10}{3}$ 29. 0 31. $\frac{5}{3x^2y^2}$ 33. $b-a$ 35. (a) 64 (b) 64

SECTION 11.3, p. 343

1. 1.49×10 3. 5.084×10^3 5. 6.95×10^{-1} 7. 9×10^{-1} 9. 1×10 11. 2.48×10^5 13. $8.200\,01 \times 10^{-4}$ 15. 8 17. 2.79×10^{-1} 19. 6×10^{-1} 21. 4×10^8 23. 83.4 25. 444 000 000 27. 5.15 29. .8181 31. 474 000 33. .062 35. 11.1 37. .005 87 39. 389.624 41. 10^{10} 43. 6×10^{12} 45. 1 47. 5×10^2 49. (b), (c), (d) 51. 1.5×10^{15} miles

SECTION 11.4, p. 348

1. 4 3. 1 5. 9 7. 11 9. 2 11. 5 13. 2 15. 16 17. 1 19. 4 21. $\frac{1}{4}$ 23. $\frac{1}{10}$ 25. $\frac{1}{2}$ 27. $\frac{1}{2}$ 29. $\frac{1}{3}$ 31. $\frac{1}{27}$ 33. $\frac{1}{1000}$ 35. $\frac{1}{6}$ 37. 20 39. 25 41. 60 inches (or 5 feet) 43. 12π 45. 1 47. 1

SECTION 11.5, p. 351

1. x 3. $x^{9/4}$ 5. $y^{1/2}$ 7. a^4
9. a 11. $x^{3/4}$ 13. $y^{1/4}$ 15. $x^{5/6}$
17. $z^{13/12}$ 19. $x^{1/2}$ 21. $x^{5/12}$ 23. $b^{1/2}$
25. $a^{3/2}$ 27. $y^{5/12}$ 29. y 31. $\dfrac{c^{1/2}}{a^{1/3}b^{2/3}}$
33. $x^{10/3}y^{4/3}$ 35. $x^5z^{1/6}$ 37. $3\cdot 3^{1/2}$ 39. $4\cdot 2^{1/2}\cdot a^2$
41. $\dfrac{1}{5\cdot 3^{1/2}ab^{5/2}}$ 43. $4x^{1/2}$ 45. $\dfrac{8}{27}$ 47. 25
49. 3 51. 25

SECTION 11.6, p. 355

1. 5 3. 3 5. 2 7. 2
9. 4 11. 1 13. $\sqrt[5]{a}$ 15. $\sqrt[3]{c^4}$
17. $\sqrt[4]{c^{-3}}\left(\text{or } \dfrac{1}{\sqrt[4]{c^3}}\right)$ 19. $3\sqrt[5]{(ab)^2}$ 21. $\sqrt{a+b}$
23. $\sqrt[4]{(a+5b)^{-1}}$ or $\left(\dfrac{1}{\sqrt[4]{a+5b}}\right)$
25. $\sqrt[3]{a^2}\sqrt[4]{b^3}$ 27. $x^{1/2}$ 29. $(xy)^{1/3}$ 31. $x^{1/3}z^{1/4}$
33. $(5x^2y)^{2/3}$ 35. $(2+x)^{1/2}$ 37. $x^{1/2}+y^{1/2}$ 39. $\dfrac{3}{x^{1/2}}$
41. $5a$ 43. abc^2 45. $\dfrac{2a^3b}{c^4}$ 47. $\dfrac{a^2b}{2c}$
49. $a+b$ 51. .1 53. $20a^6$ 55. $2a^2c^3\sqrt{3c}$
57. $-a^2\sqrt{ab}$ 59. $\dfrac{1}{125a^6}$ 61. $2ab^2\sqrt[4]{2a^3}$ 63. $\dfrac{3ab^3c^4}{2}$
65. $3a^2$ 67. $\sqrt[3]{a^3+b^9}$

SECTION 11.7, p. 359

1. $5\sqrt{3}$ 3. $\sqrt{7}$ 5. $5\sqrt{x}$
7. $5\sqrt{a}$ 9. $-3ab\sqrt{b}$ 11. $\dfrac{\sqrt{7}}{2}$
13. $\dfrac{5\sqrt[4]{5}}{2}$ 15. $\dfrac{5\sqrt{11}}{6}$ 17. $\dfrac{-9\sqrt{3}}{20}$
19. $3(5^{1/3})$ 21. $3\sqrt{2}+3\sqrt{7}$ 23. $3\sqrt{3}+3-\sqrt{6}$
25. $\sqrt{2}x+\sqrt{2ax}+\sqrt{ax}+a$ 27. $x+4\sqrt{xy}+4y$ 29. $x+x^{1/2}$
31. $a-a^{5/4}$ 33. $x-1$ 35. $x^{1/2}-2(xy)^{1/4}+y^{1/2}$
37. $9\sqrt{2}$ 39. $\sqrt{6}(1+\sqrt{2}-\sqrt{10})$ 41. $\sqrt{b}(a+3)$
43. $xy^2\sqrt{z}(3+x)$ 45. $\sqrt{6xy}(2xy-3+5x)$ 47. $4+5\sqrt{3}$
49. $3+\sqrt{2}$ 51. $1-\sqrt{y}$ 53. $\dfrac{\sqrt{b}(\sqrt{2}-a)}{a^3}$
55. $\dfrac{4b-a}{2}$ 57. $3\sqrt{2}$ 59. $6\sqrt{7}$

61. $22\sqrt{6}$

63. 0

65. $xy\sqrt[3]{2}(3 + xy^2)$

67. $\frac{ab\sqrt{c}}{2}(5b - 2)$

SECTION 11.8, p. 363

1. $\frac{2\sqrt{3}}{3}$

3. $\frac{7(5^{1/2})}{5}$

5. $\frac{3\sqrt{2}}{4}$

7. $\frac{\sqrt{10}}{6}$

9. $\frac{\sqrt{a}}{a}$

11. $\frac{x\sqrt{ab}}{ab}$

13. $\frac{-5\sqrt{x}}{2x}$

15. $\frac{x^{1/2}(1 + 2x)}{x}$

17. $\frac{\sqrt{y} - y\sqrt{2}}{2y}$

19. $\frac{9^{1/3}}{3}$

21. $\frac{3\sqrt[3]{x}}{x}$

23. $\frac{x^{1/2}(x - a)^{2/3}}{x - a}$

25. $-1 + \sqrt{2}$

27. $\frac{5(3 + \sqrt{2})}{7}$

29. $\frac{4(1 - \sqrt{a})}{1 - a}$

31. $\frac{\sqrt{x}(1 - \sqrt{x})}{1 - x}$

33. $\frac{-3(\sqrt{x} - 1)}{x - 1}$

35. $\frac{-5\sqrt{x - y}}{x - y}$

37. $\frac{3(a + \sqrt{b + 1})}{a^2 - b - 1}$

39. $\frac{\sqrt{x}(\sqrt{x} - \sqrt{y})}{x - y}$

41. $\frac{6(3 - 2\sqrt{x})}{9 - 4x}$

43. $\frac{2(5 - 7\sqrt{2})}{73}$

45. $\frac{2(7\sqrt{2} - 5\sqrt{3})}{23}$

47. $\frac{(a + b)(a\sqrt{x} + b\sqrt{y})}{a^2x - b^2y}$

SECTION 11.9, p. 371

1. $3i$

3. $2\sqrt{2}i$

5. $\frac{1}{2}i$

7. $\frac{2}{3}i$

9. (a) 1 (b) 1 (c) $1 - i$

11. (a) 3 (b) 3 (c) $3 - 3i$

13. (a) -2 (b) 4 (c) $-2 - 4i$

15. (a) 0 (b) 2 (c) $-2i$

17. (a) $\frac{1}{2}$ (b) $\sqrt{2}$ (c) $\frac{1}{2} - \sqrt{2}i$

19. (a) $-\sqrt{2}$ (b) $\sqrt{3}$ (c) $-\sqrt{2} - \sqrt{3}i$

21. $3 + 6i$

23. 3

25. $9 + 7i$

27. $\frac{3}{4} + \frac{3}{2}i$

29. $1 + 3i$

31. $-1 + 6i$

33. $\frac{1}{4} - \frac{1}{4}i$

35. -6

37. $-8 - 20i$

39. 2

41. $5 + 10i$

43. $-5 - 15i$

45. -9

47. $-3 + 4i$

49. $8 - 6i$

51. i

53. $-5 - 3i$

55. $\frac{-8}{5} + \frac{9}{5}i$

57. $\frac{5}{26} - \frac{1}{26}i$

59. $-i$

61. 1

63. $1 + i$

65. -1

67. 1

REVIEW EXERCISES, p. 372

1. $\frac{1}{16}$
3. $\frac{7}{16}$
5. -2
7. $\frac{1}{a^6}$
9. $\frac{64}{b^2}$
11. 1.018×10^{-1}
13. 719
15. .000 121
17. 8
19. $a^{5/4}$
21. $8a^{1/8}b^{1/2}c$
23. (a) 5 (b) 4
25. (a) $(xyz)^{3/5}$ (b) $\frac{(x+y)^{1/3}}{x^{1/2}}$
27. $3a^2bc^4$
29. (a) $3\sqrt{5}$ (b) $\frac{5\sqrt{3}}{3}$
31. (a) $4(1-\sqrt{2})$ (b) $x\sqrt{yz}(yz^2+3x)$
33. (a) $8\sqrt{2}$ (b) $10z^2+9\sqrt{y}$
35. $\frac{1-\sqrt{3}}{2}$
37. (a) $9i$ (b) $\frac{1}{3}i$
39. $4+7i$
41. $\frac{5}{2}-\frac{3}{2}i$

TEST YOURSELF, p. 374

1. (a) a^3 (b) $\frac{4}{125}$
2. (a) 9.49×10^{-1} (b) 31 400
3. $\frac{a}{b}$
4. (a) 8 (b) $ab^2(a+b)$
5. $\frac{3(\sqrt{2}+2)}{2}$
6. $3a^{2/3}b^{4/3}+b$
7. $\frac{\sqrt{b}(2\sqrt{2}a-1)}{2}$
8. $\frac{3(1+\sqrt{a})}{1-a}$
9. (a) $\sqrt{5}$ (b) -2 (c) $\sqrt{5}+2i$
10. $10+10i$

CHAPTER 12

SECTION 12.1, p. 379

1. $\log_2 16 = 4$
3. $\log_{10} 1\,000\,000 = 6$
5. $\log_{10} \frac{1}{10} = -1$
7. $\log_4 2 = \frac{1}{2}$
9. $\log_{16} 8 = \frac{3}{4}$
11. $\log_9 \frac{1}{3} = \frac{-1}{2}$
13. $\log_{1/3} 9 = -2$
15. $\log_{4/7} \frac{7}{4} = -1$
17. $\log_{.1} .0001 = 4$
19. $\log_{.01} .1 = \frac{1}{2}$
21. $2^4 = 16$
23. $8^{1/3} = 2$

25. $7^{-1} = \frac{1}{7}$

27. $49^{1/2} = 7$

29. $10^{-2} = .01$

31. $81^{3/4} = 27$

33. 2
35. 4
37. 3
39. $\frac{3}{2}$

41. -6
43. 0
45. 3
47. -3

49. 25
51. 1
53. $\sqrt{3}$
55. 0

57. 2
59. .8
61. .2
63. .1

SECTION 12.2, p. 384

3. (a) $6 = 2 + 4$ (b) $\log_b n_1 n_2 = \log_b n_1 + \log_b n_2$. Here $b = 2$, $n_1 = 4$, $n_2 = 16$

5. (a) $3 = 5 - 2$ (b) $\log_b \frac{n_1}{n_2} = \log_b n_1 - \log_b n_2$. Here $b = 10$, $n_1 = 100\,000$, $n_2 = 100$

7. (a) $6 = 2 \cdot 3$ (b) $\log_b n^r = r \log_b n$. Here $b = 2$, $n = 8$, $r = 2$

9. (a) $4 = 2 \cdot 2$ (b) $\log_b n^r = r \log_b n$. Here $b = 3$, $n = 9$, $r = 2$

13. $\log_b x - \log_b y$

15. $\log_{10} a + \log_{10} b - \log_{10} c$

17. $3 \log_{10} x$

19. $\log_5 x + 4 \log_5 y$

21. $2 \log_3 x + \frac{1}{2} \log_3 y - \log_3 z$

23. $\frac{1}{2}(\log_b 5 + \log_b x - \log_b 3)$

25. $4 \log_{10} a + 2 \log_{10} b - 3 \log_{10} c$

27. $\frac{1}{2} \log_b x - \frac{1}{3} \log_b y + \frac{3}{4} \log_b z$

29. $-5\left(\frac{1}{2} \log_b x + 2 \log_b y - 3 \log_b z\right)$

33. $2 \log_b 5$

35. $\log_b 2 - \log_b 3$
37. $2 \log_b 2 + \log_b 3$
39. $4 \log_b 2 + \log_b 3 - 2 \log_b 5$

41. $-\log_b 3$
43. $\log_b st$
45. $\log_b rst$

47. $\log_b \frac{10}{ax}$
49. $\log_b x^2$
51. $\log_b \frac{x^{1/4} z}{y^{2/3}}$

53. $\log_b \left(\frac{10}{7}\right)^{1/3}$
55. $\log_3 \frac{(7 \cdot 5^{1/3})^2}{17}$

SECTION 12.3, p. 393

1. .0128
3. .6345
5. .4771
7. .9731

9. .7803
11. .8621
13. .0374
15. 1

17. 4
19. -2
21. -4
23. 6

25. 8
27. 6
29. 3
31. 2.9405

33. 1.2279
35. 8.4082
37. $.9708 - 1$
39. $.1335 - 2$

41. .8055
43. $.9201 - 4$
45. 1.7924
47. $.6920 - 3$

49. $.9222 - 3$

SECTION 12.4, p. 397

1. 1.97
3. 7.44
5. 2.72
7. 3.65

9. 36 500
11. .003 65
13. 6.09
15. 6 090 000

17. .0609
19. 10.9
21. 956
23. .007 06

25. .0956
27. 1.07
29. 3.65
31. 446 000

33. 644 000 000

SECTION 12.5, p. 403

1. .6133
3. .9098
5. 1.6167
7. .8473 − 1
9. 5.0153
11. .8864 − 3
13. .8452
15. .8452 − 1
17. .5550
19. .5550 − 1
21. 3.9338
23. .9300 − 3
25. 7.5963
27. .9225 − 5
29. 9.9913
31. 4.001
33. 9.004
35. 95.42
37. 97 180
39. .099 63
41. 6.653
43. 66 530
45. 6.668
47. .066 68
49. 1 064 000
51. .2472
53. 1.549
55. .9998
57. 4 641 000
59. .083 64

SECTION 12.6, p. 411. (Interpolation was used to obtain these answers.)

1. 55.01
3. 911.2
5. 486.8
7. 45 390
9. 47 480 000
11. 4.748
13. .005 398
15. .000 969 4
17. 20 440 000
19. 166.2
21. 2.801×10^{-11}
23. 1.951
25. .7488
27. .8294
29. 18.71
31. .013 40
33. $2655
35. 1.325×10^{12}

SECTION 12.8, p. 420

1. 2.21
3. 1.16
5. .53
7. .51
9. 1.16
11. −.67
13. 10 000
15. −5
17. .025
19. 2000
21. $\frac{10}{3}$
23. 3.86 hours
25. 8.31 years
27. 1.59
29. 1.83
31. 2.47
33. 1.72
37. $\frac{1}{2}\log_2 29$
39. $\frac{\log_2 93}{\log_2 10}$
41. $\frac{\log_{12} 9.02}{\log_{12} 7}$
43. x

REVIEW EXERCISES, p. 421

1. $\log_{10} 1000 = 3$
3. $\log_{16} 4 = \frac{1}{2}$
5. 4
7. 2
9. $\frac{1}{2}$
11. $\log_b x + \log_b y - \log_b 2$
13. $2\log_b 2 + \log_b 3 + \log_b 5$
15. .9253
17. 2.7882
19. .8000 − 2
21. 4590
23. .9141
25. 5.005
27. 58.36
29. .0141
31. 200 000
33. $\frac{\log_3 7}{\log_3 2}$

TEST YOURSELF, p. 423

1. −4
2. 7
3. $3\log_3 2 + \log_3 5$
4. 2.9284
5. 90.2
6. .4630
7. 1701

CHAPTER 13

SECTION 13.1, p. 429

1. 0, 2
3. 0, 7
5. 1, 2
7. $\frac{-1}{2}, \frac{1}{2}$
9. $\frac{2}{3}$, 2
11. A, B
13. $\frac{B}{A}, \frac{-B}{A}$
15. 2, 3, 4
17. 0, −3, $\frac{-1}{2}, \frac{4}{3}$
19. 0, 3
21. 0, $\frac{9}{4}$
23. 1, 7
25. 2, 3
27. 7, −3
29. 4
31. 0, $\frac{36}{25}$
33. $\frac{1}{2}$, −3
35. 1, $\frac{-1}{5}$
37. $\frac{1}{2}, \frac{-1}{4}$
39. 1
41. 1, 2
43. 4, −1
45. 0, −3
47. 0, 3, −5
49. $x^2 - 3x + 2 = 0$
51. $x^2 + x - 12 = 0$
53. $3x^2 - 5x + 2 = 0$
55. $x^2 - 6x + 9 = 0$

SECTION 13.2, p. 433

1. ± 1
3. $\pm\sqrt{11}$
5. $\pm 2\sqrt{2}$
7. ± 2
9. 0
11. $\frac{\pm 5}{3}$
13. $\pm 2\sqrt{7}$
15. $\frac{\pm 7\sqrt{2}}{4}$
17. $\pm 4B$
19. $\pm 9AB^2C^3$
21. 2, −8
23. 2, 3
25. 2, −6
27. $\pm A$
29. $\frac{\pm B}{A}$
31. $\frac{\pm 7AB}{2}$
33. $C \pm \frac{4}{AB}$
35. $\pm 4i$
37. $\pm 2\sqrt{6}\,i$
39. $2 \pm 2i$
41. ± 11
43. 0, −2
45. $1 \pm 2\sqrt{3}$
47. $B \pm AC$
49. $\pm 8i$
51. ± 4

SECTION 13.3, p. 438

1. $1 \pm \sqrt{2}$
3. $-2 \pm \sqrt{6}$
5. $2 \pm \sqrt{5}$
7. $4 \pm \sqrt{3}$
9. $1 \pm \sqrt{5}$
11. $\frac{-1 \pm \sqrt{5}}{2}$
13. $\frac{-5 \pm \sqrt{17}}{2}$
15. $\frac{5 \pm \sqrt{5}}{2}$
17. $\frac{-1 \pm \sqrt{17}}{4}$
19. $\frac{-1 \pm \sqrt{19}}{3}$
21. $-2 \pm \sqrt{2}$
23. $-1 \pm i$
25. $\frac{3 \pm \sqrt{3}\,i}{2}$
27. $-1 \pm \frac{\sqrt{2}}{2}$
29. $\frac{3 \pm \sqrt{7}}{2}$
31. $-1 \pm \frac{\sqrt{5}}{2}$
33. $\frac{-1 \pm \sqrt{2}\,i}{2}$
35. $-1 \pm \sqrt{2}$

SECTION 13.4, p. 442

1. (a) 53 (b) 2 distinct real roots
3. (a) 5 (b) 2 distinct real roots
5. (a) 0 (b) exactly 1 real root
7. (a) 17 (b) 2 distinct real roots
9. (a) 0 (b) exactly 1 real root
11. (a) 112 (b) 2 distinct real roots
13. $\frac{-1 \pm \sqrt{7}i}{2}$
15. $1 \pm \sqrt{3}$
17. $\frac{5 \pm \sqrt{65}}{2}$
19. $\frac{\pm\sqrt{2}i}{2}$
21. $\frac{-9 \pm \sqrt{97}}{4}$
23. $\frac{-5 \pm \sqrt{10}}{5}$
25. -7
27. $\frac{1}{2}$
29. -6
31. $-1, -2$
33. $\frac{-B \pm \sqrt{B^2 + 8B}}{4}$
35. ± 8
37. ± 2
39. ± 4

SECTION 13.5, p. 447

1. 2, 10
3. 5, -5
5. -1
7. $2 \pm \sqrt{3}$
9. $-8, -7$
11. 10 inches by 5 inches
13. 50 feet by 20 feet, 40 feet by 25 feet
15. 20 feet by 8 feet
17. 60 miles per hour
19. 5 seconds
21. 6 hours
23. 15 seconds

SECTION 13.6, p. 453

1. 2

CHECK: $\sqrt{2+2} \stackrel{?}{=} 2$

$\sqrt{4} \stackrel{?}{=} 2$

$2 \stackrel{\checkmark}{=} 2$

3. 7

CHECK: $\sqrt{2(7)+2} \stackrel{?}{=} 4$

$\sqrt{16} \stackrel{?}{=} 4$

$4 \stackrel{\checkmark}{=} 4$

5. 17

CHECK: $\sqrt{17+8} - 5 \stackrel{?}{=} 0$

$\sqrt{25} - 5 \stackrel{?}{=} 0$

$5 - 5 \stackrel{\checkmark}{=} 0$

7. 3

CHECK: $[13(3) - 3]^{1/2} - 6 \stackrel{?}{=} 0$

$36^{1/2} = 6$

$6 \stackrel{\checkmark}{=} 6$

9. 5

CHECK: $\sqrt{2(5)+2} \stackrel{?}{=} 2\sqrt{3}$

$\sqrt{12} \stackrel{?}{=} 2\sqrt{3}$

$2\sqrt{3} \stackrel{\checkmark}{=} 2\sqrt{3}$

11. 5

CHECK: $5 + \sqrt{5-1} \stackrel{?}{=} 7$

$\sqrt{4} \stackrel{?}{=} 2$

$2 \stackrel{\checkmark}{=} 2$

(10 is not a root.)

$10 + \sqrt{10-1} \stackrel{?}{=} 7$

$\sqrt{9} \stackrel{?}{=} -3$

$3 \stackrel{x}{=} -3$

13. 5

CHECK: $\sqrt{2(5)-1}-8 \stackrel{?}{=} -5$

$\sqrt{9} \stackrel{?}{=} 3$

$3 \stackrel{\checkmark}{=} 3$

(13 is not a root.)

$\sqrt{2(13)-1}-8 \stackrel{?}{=} -13$

$\sqrt{25} \stackrel{?}{=} -5$

$5 \stackrel{x}{=} -5$

15. -4

CHECK: $\sqrt{1-2(-4)}+(-4)+1 \stackrel{?}{=} 0$

$\sqrt{9} \stackrel{?}{=} 3$

$3 \stackrel{\checkmark}{=} 3$

(0 is not a root.)

$\sqrt{1-2(0)}+0+1 \stackrel{?}{=} 0$

$\sqrt{1} \stackrel{?}{=} -1$

$1 \stackrel{x}{=} -1$

17. 2

CHECK: $\sqrt{2(2)}-\sqrt{2-1} \stackrel{?}{=} 1$

$2-1 \stackrel{\checkmark}{=} 1$

19. 0

CHECK: $\sqrt{4(0)+1}+3 \stackrel{?}{=} \sqrt{16+0}$

$1+3 \stackrel{\checkmark}{=} 4$

(20 is not a root.)

$\sqrt{4(20)+1}+3 \stackrel{?}{=} \sqrt{16+20}$

$\sqrt{81}+3 \stackrel{?}{=} \sqrt{36}$

$9+3 \stackrel{x}{=} 6$

21. 16

CHECK: $\sqrt{1+3(16)}-3 \stackrel{?}{=} \sqrt{16}$

$\sqrt{49}-3 \stackrel{?}{=} 4$

$7-3 \stackrel{\checkmark}{=} 4$

(1 is not a root.)

$\sqrt{1+3(1)}-3 \stackrel{?}{=} \sqrt{1}$

$2-3 \stackrel{x}{=} 1$

23. 4

CHECK: $\sqrt{4}\sqrt{4-3} \stackrel{?}{=} 2$

$2 \cdot 1 \stackrel{\checkmark}{=} 2$

(-1 is not a root.)

$\sqrt{-1}\sqrt{-1-3} \stackrel{?}{=} 2$

$i \cdot 2i \stackrel{?}{=} 2$

$-2 \stackrel{x}{=} 2$

25. 13

CHECK: $\sqrt{13+3}\sqrt{13-9} \stackrel{?}{=} 8$

$4 \cdot 2 \stackrel{\checkmark}{=} 8$

(-7 is not a root.)

$\sqrt{-7+3}\sqrt{-7-9} \stackrel{?}{=} 8$

$2i \cdot 4i \stackrel{?}{=} 8$

$-8 \stackrel{x}{=} 8$

27. $\frac{1}{3}$

CHECK: $$\sqrt{3\left(\frac{1}{3}\right)}\sqrt{9\left(\frac{1}{3}\right)+1} \stackrel{?}{=} 2$$

$$\sqrt{1}\sqrt{4} \stackrel{?}{=} 2$$

$$1 \cdot 2 \stackrel{\checkmark}{=} 2$$

$\left(\frac{-4}{9} \text{ is not a root.}\right)$

$$\sqrt{3\left(\frac{-4}{9}\right)}\sqrt{9\left(\frac{-4}{9}\right)+1} \stackrel{?}{=} 2$$

$$\sqrt{\frac{-4}{3}}\sqrt{-3} \stackrel{?}{=} 2$$

$$\frac{2\sqrt{3}i}{3} \cdot \sqrt{3}i \stackrel{?}{=} 2$$

$$-2 \stackrel{x}{=} 2$$

29. 10

CHECK: $$5\sqrt{3(10)+6} \stackrel{?}{=} 3\sqrt{9(10)+10}$$

$$5\sqrt{36} \stackrel{?}{=} 3\sqrt{100}$$

$$5 \cdot 6 \stackrel{\checkmark}{=} 3 \cdot 10$$

REVIEW EXERCISES, p. 453

1. $2, -3$
3. $-1, -4$
5. $x^2 + x - 20 = 0$
7. $\pm 5A$
9. $\pm 4\sqrt{3}i$
11. $-3 \pm 2\sqrt{5}$
13. $\frac{-4 \pm \sqrt{10}}{2}$
15. (a) 5 (b) 2 distinct real roots
17. (a) 0 (b) exactly 1 real root
19. $\frac{9 \pm 3\sqrt{5}}{2}$
21. $\frac{5 \pm \sqrt{15}i}{2}$
23. $-13, -11$
25. 6

CHECK: $$\sqrt{6+3} - 3 \stackrel{?}{=} 0$$

$$3 - 3 \stackrel{?}{=} 0$$

$$0 \stackrel{\checkmark}{=} 0$$

27. No root

CHECK: (0 is not a root.)

$$\sqrt{0+1} - \sqrt{0} + 1 \stackrel{?}{=} 0$$

$$1 - 0 + 1 \stackrel{?}{=} 0$$

$$2 \stackrel{x}{=} 0$$

TEST YOURSELF, p. 455

1. $3, -6$
2. $\pm 3\sqrt{5}A$
3. $\frac{-3 \pm \sqrt{7}i}{2}$
4. $\frac{-3 \pm \sqrt{5}}{2}$
5. $\frac{-3 \pm \sqrt{7}}{2}$
6. $5, 9$
7. 1

CHECK: $$\sqrt{1+3} - \sqrt{1} \stackrel{?}{=} 1$$

$$2 - 1 \stackrel{?}{=} 1$$

$$1 \stackrel{\checkmark}{=} 1$$

CHAPTER 14

SECTION 14.1, p. 460

1. (a)	3. (a)	5. (f)	7. (e)
9. (b)	11. (e)	13. (c)	15. (e)
17. true	19. true	21. true	23. true
25. true	27. false	29. false	31. [2, 9]
33. [−1, 0]	35. (−∞, −6)	37. [8, ∞)	39. (−7, −5)

41. (a) See Figure 14A. (b) [−2, 2]

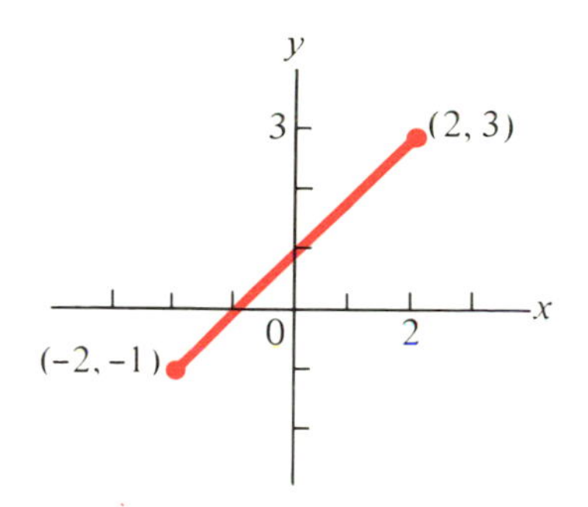

FIGURE 14A

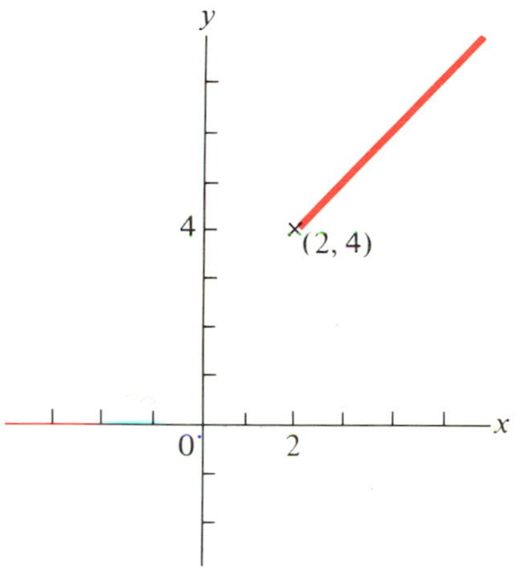

FIGURE 14B

43. (a) See Figure 14B. (b) (2, ∞)
45. (a) See Figure 14C. (b) (−∞, 1)

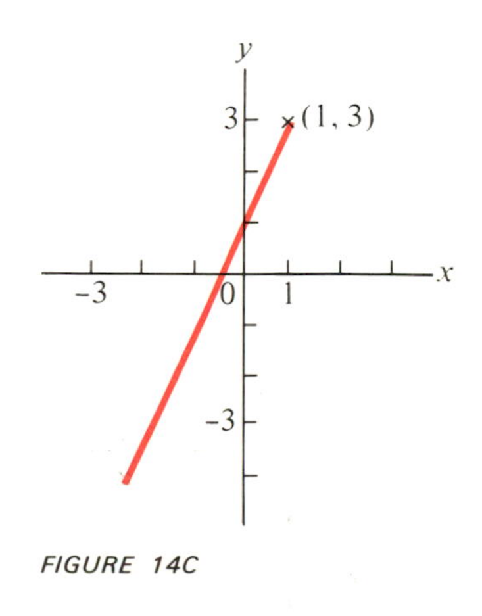

FIGURE 14C

SECTION 14.2, p. 466

1. <	3. <	5. >	7. >

9. $<$ 11. $>$ 13. $>$ 15. $>$
17. $>$ 19. $<$ 21. $<$
23. $x + 2 < 4 + 2$
25. $2a \leq 2(4)$
27. $-2x \geq -2(5)$ because -2 is *negative* and multiplication by a negative number reverses the sense of the inequality.
29. $2x < 3$. Therefore, $x < \frac{3}{2} < 2$. Thus $x < 2$

SECTION 14.3, p. 471

1. $(-\infty, 4)$ 3. $(-\infty, 13)$ 5. $(-\infty, 4]$ 7. $(1, \infty)$
9. $[0, \infty)$ 11. $(-\infty, 3)$ 13. $(8, \infty)$ 15. $(-\infty, 0)$
17. $(-1, \infty)$ 19. $(-\infty, -1)$ 21. $\left[\frac{-1}{2}, \infty\right)$ 23. $\left(\frac{3}{4}, \infty\right)$
25. $(-\infty, -3]$ 27. $\left[\frac{-9}{4}, \infty\right)$ 29. $\left(-\infty, \frac{-11}{4}\right)$ 31. $\left(-\infty, \frac{-1}{2}\right]$
33. $\left[\frac{-14}{27}, \infty\right)$ 35. $\left(-\infty, \frac{1}{2}\right)$ 37. $\left[\frac{1}{4}, \infty\right)$ 39. $\left[\frac{-2}{5}, \infty\right)$
41. $(0, 6)$ 43. $[-9, -3]$ 45. $(3, 4)$ 47. $(-3, 3)$
49. $\left(-1, \frac{7}{2}\right)$ 51. $\left(\frac{-8}{5}, 2\right)$ 53. $[-2, 1]$ 55. $(-4, 0)$
57. $\left(\frac{-1}{5}, 1\right)$ 59. $\left(\frac{-1}{2}, \frac{1}{2}\right)$ 61. $[120, 180]$ 63. $[32, 41]$
65. $[2, 8]$ 67. $(-\infty, 0)$

SECTION 14.4, p. 475

1. (a)
3. (a), (d)
5. (a), (d)
7. (a), (b), (c)
9. (a), (b), (c), (d)
11. (a), (c)
13. (a), (d)
15. (a), (b), (c), (d)
17. (a), (b), (d)
19. (b), (d)
21. See Figure 14D.
23. See Figure 14E.
25. See Figure 14F on page A47.
27. See Figure 14G on page A47.
29. See Figure 14H on page A47.
31. See Figure 14I on page A47.
33. See Figure 14J on page A47.
35. See Figure 14K on page A47.

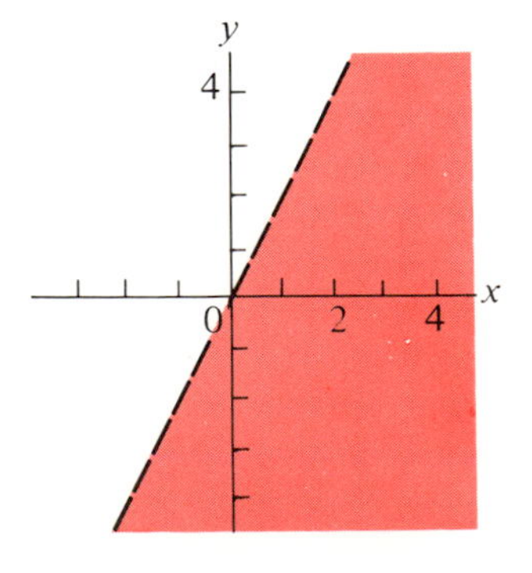

FIGURE 14D

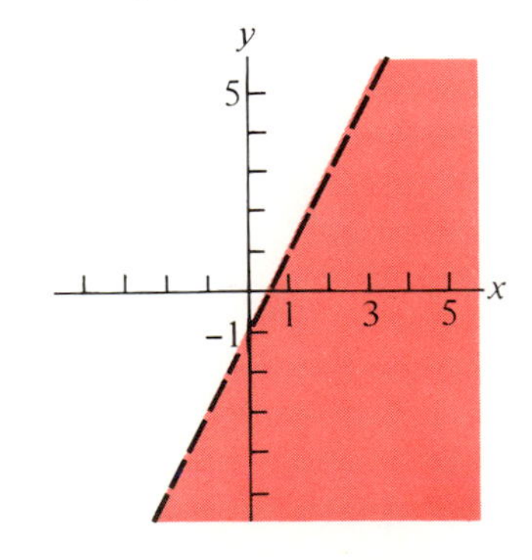

FIGURE 14E

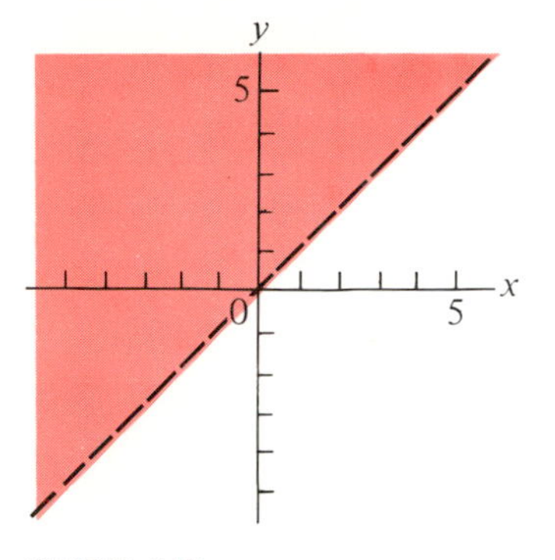

FIGURE 14F

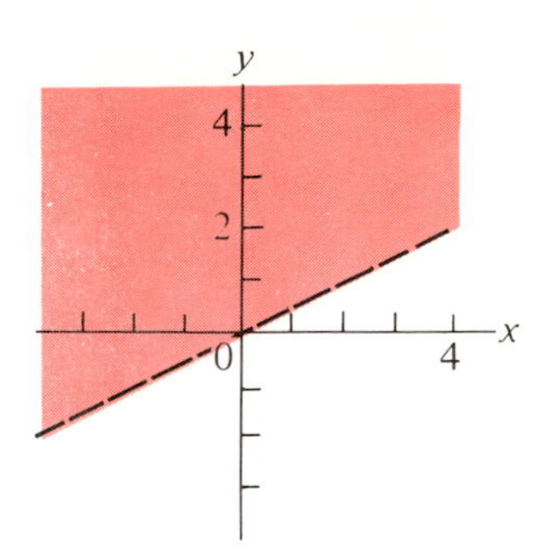

FIGURE 14G

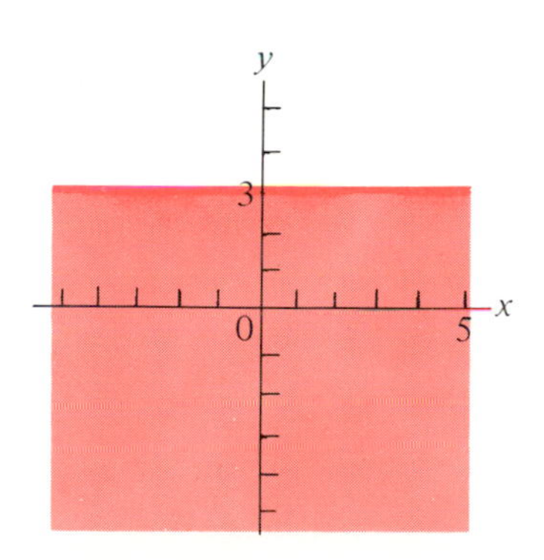

FIGURE 14H

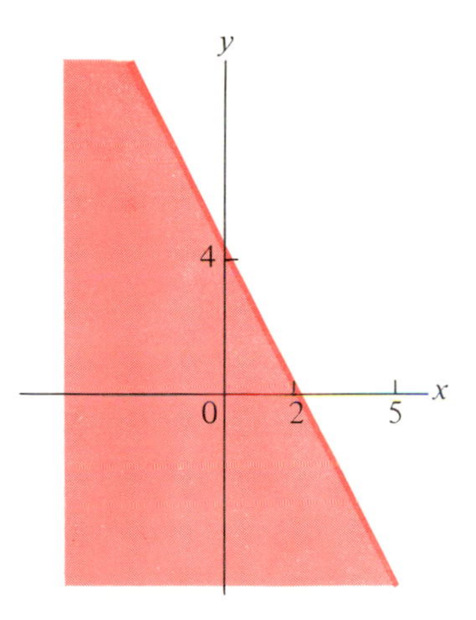

FIGURE 14I

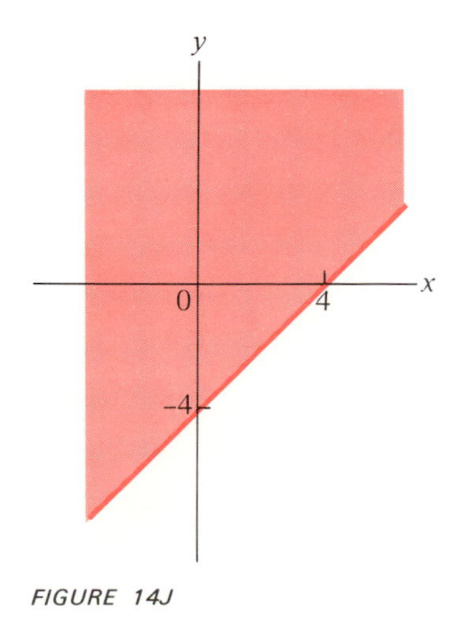

FIGURE 14J

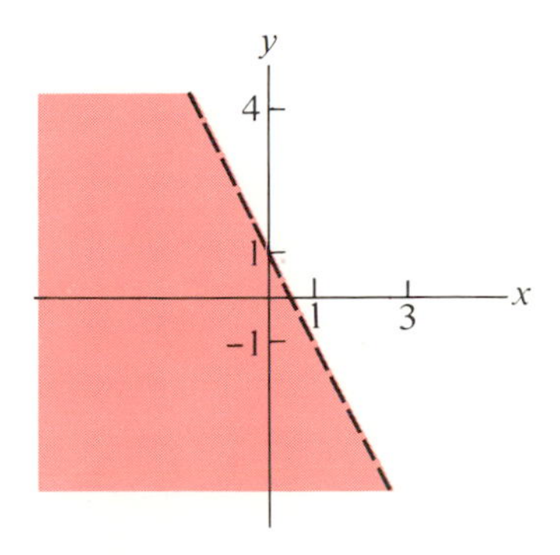

FIGURE 14K

SECTION 14.5, p. 481

1. $(-7, 3)$
3. $(0, 5)$
5. $[0, 4]$
7. no solution
9. all x
11. $(-\infty, -3) \cup (-1, \infty)$
13. $(2, 5)$
15. $[-2, 6]$
17. $\left(-1, \frac{-1}{2}\right)$
19. $\left(-\infty, \frac{1}{2}\right) \cup \left(\frac{3}{2}, \infty\right)$
21. $(-\infty, 0) \cup (3, \infty)$
23. $(-\infty, -4) \cup (0, \infty)$
25. $(-\infty, -1) \cup (0, \infty)$
27. $(-\infty, -4) \cup [-3, \infty)$
29. $(-2, \infty)$
31. $(-\infty, -10) \cup (-5, \infty)$
33. $(-\infty, -2)$
35. $(-\infty, -2) \cup (1, \infty)$
37. $(-\infty, -5) \cup (-4, \infty)$
39. all x
41. $(0, 2)$

SECTION 14.6, p. 487. (The checks are given below.)

1. 3, 7
3. 0, 6
5. -5, 3
7. $\frac{3}{2}, \frac{5}{2}$
9. $\frac{-1}{4}, \frac{5}{4}$
11. 7, -7
13. ± 5.3
15. no root
17. 2, -2
19. 7, -5
21. 1, 7
23. 2, -6
25. 5, -4
27. 4, 16
29. $\frac{-1}{5}$
31. $\frac{1}{2}$ $\left(\frac{-1}{6}\right.$ is not a root.$\left.\right)$
33. $\frac{4}{3}$ (6 is not a root.)
35. $\frac{4}{3}$ $\left(\frac{-6}{5}\right.$ is not a root.$\left.\right)$
37. no root $\left(\frac{-1}{2}\right.$ is not a root.$\left.\right)$
39. The roots are the nonnegative numbers.
41. 1, $\frac{-1}{3}$
43. $\frac{9}{7}$ and $\frac{-9}{5}$
45. $\frac{-1}{2}$

CHECKS.

19. $|7-1| \stackrel{?}{=} 6$, $6 \stackrel{\checkmark}{=} 6$ | $|-5-1| \stackrel{?}{=} 6$, $|-6| \stackrel{?}{=} 6$, $6 \stackrel{\checkmark}{=} 6$

25. $|2(5)-1| \stackrel{?}{=} 9$, $9 \stackrel{\checkmark}{=} 9$ | $|2(-4)-1| \stackrel{?}{=} 9$, $|-9| \stackrel{?}{=} 9$, $9 \stackrel{\checkmark}{=} 9$

31. $\left|2\left(\frac{1}{2}\right)+1\right| \stackrel{?}{=} 4\left(\frac{1}{2}\right)$, $2 \stackrel{\checkmark}{=} 2$ | $\left|2\left(\frac{-1}{6}\right)+1\right| \stackrel{?}{=} 4\left(\frac{-1}{6}\right)$, $\left|\frac{2}{3}\right| \stackrel{?}{=} \frac{-2}{3}$, $\frac{2}{3} \stackrel{x}{=} \frac{-2}{3}$

33. $\left|\frac{4}{3}+1\right| \stackrel{?}{=} 5-2\left(\frac{4}{3}\right)$ | $|6+1| \stackrel{?}{=} 5-2(6)$

$\frac{7}{3} \stackrel{\checkmark}{=} \frac{7}{3}$ | $7 \stackrel{x}{=} -7$

35. $\left|\frac{4}{3}+5\right| \stackrel{?}{=} 4\left(\frac{4}{3}\right)+1$ | $\left|\left(\frac{-6}{5}\right)+5\right| \stackrel{?}{=} 4\left(\frac{-6}{5}\right)+1$

$\frac{19}{3} \stackrel{\checkmark}{=} \frac{19}{3}$ | $\frac{19}{5} \stackrel{x}{=} \frac{-19}{5}$

37. $\left|\frac{-1}{2}+1\right| \stackrel{?}{=} \frac{-1}{2}$

$\frac{1}{2} \stackrel{x}{=} \frac{-1}{2}$

SECTION 14.7, p. 493

1. $(-10, 10)$
3. $[-4, 4]$
5. $(-6, 8)$
7. $(-11, 5)$
9. $\left(\frac{-7}{2}, \frac{9}{2}\right)$
11. $(-2, 3)$
13. $\left(\frac{-7}{5}, 1\right)$
15. no solution
17. $\left(\frac{-7}{3}, 3\right)$
19. $(-\infty, -5) \cup (5, \infty)$
21. $(-\infty, -40] \cup [40, \infty)$
23. $(-\infty, -1] \cup [1, \infty)$
25. $(-\infty, -5) \cup (7, \infty)$
27. $\left(-\infty, \frac{-5}{3}\right) \cup (1, \infty)$
29. $(-\infty, -3) \cup \left(\frac{9}{2}, \infty\right)$
31. $(-\infty, 2) \cup \left(\frac{10}{3}, \infty\right)$
33. all real numbers
35. all real numbers
37. 5
39. 9
41. 5
43. 4
45. -6
47. 5
49. $\frac{1}{2}$
51. 36

SECTION 14.8, p. 497

1. 18
3. 0
5. (a) 5 (b) 10 (c) 3 (d) 0 (e) 1
7. (a) 0 (b) 4 (c) 10 (d) 4 (e) 10
9. (a) 1 (b) 4 (c) $\frac{2}{3}$ (d) 1 (e) $\frac{5}{3}$
11. See Figure 14L on page A50.
13. See Figure 14M on page A50.
15. See Figure 14N on page A50.
17. See Figure 14O on page A50.
19. See Figure 14P on page A50.
21. See Figure 14Q on page A50.
23. true
25. false
27. true
29. true
31. true
33. 2 and $\frac{-2}{3}$

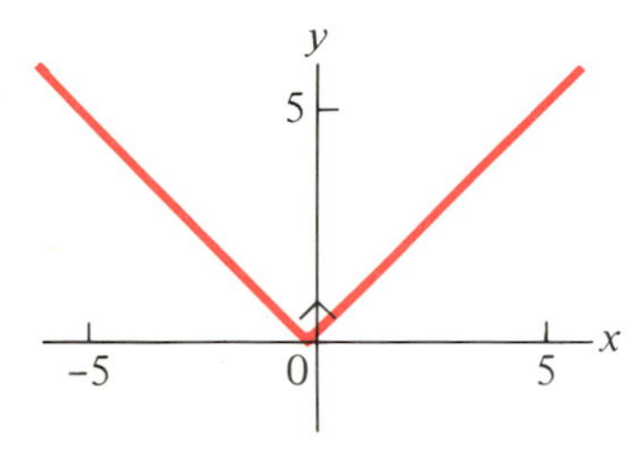

FIGURE 14L

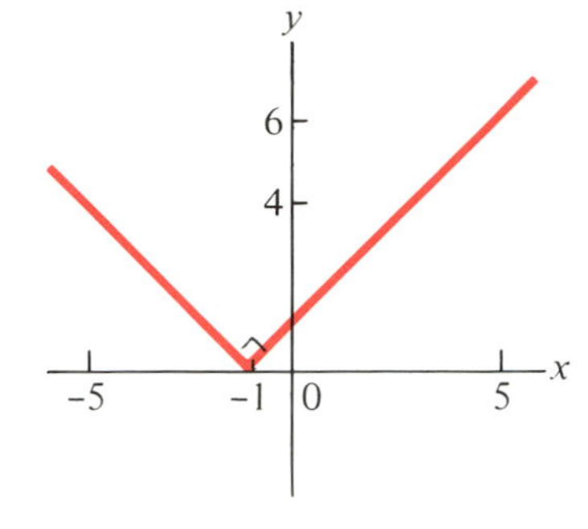

FIGURE 14M

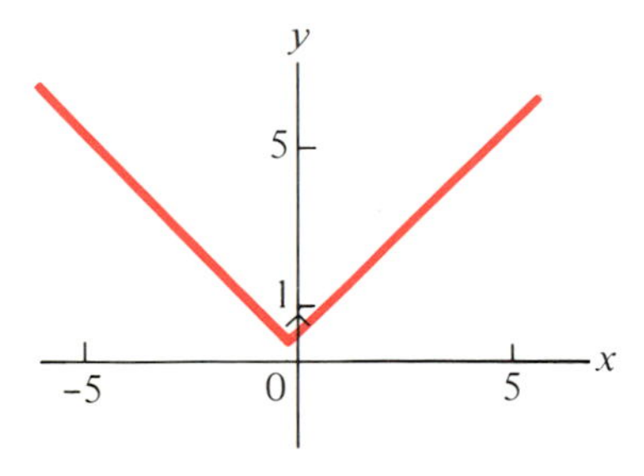

FIGURE 14N

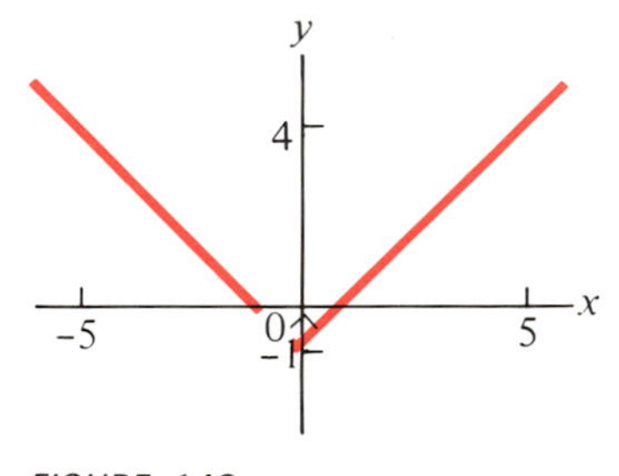

FIGURE 14O

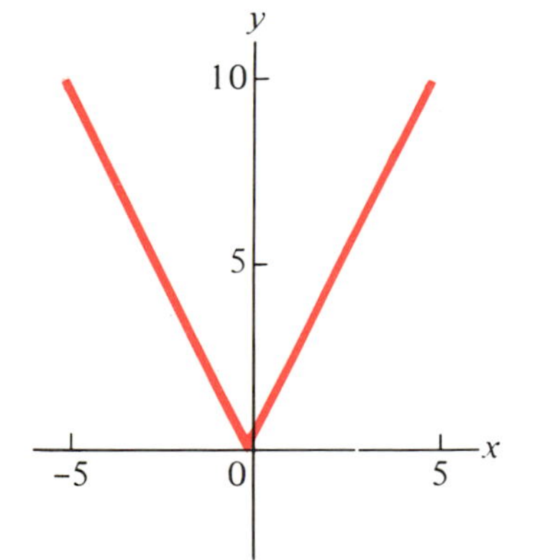

FIGURE 14P

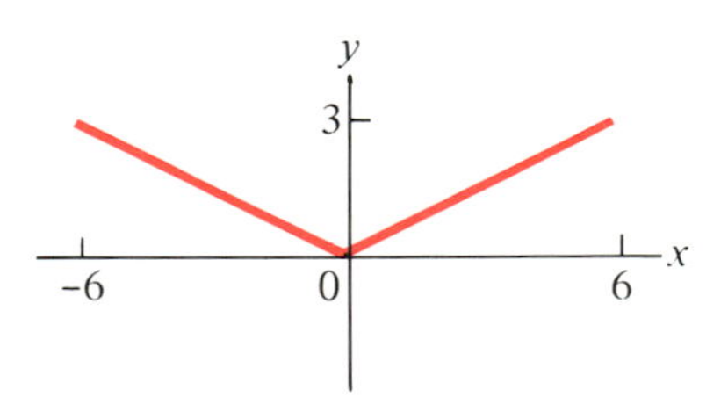

FIGURE 14Q

35. -2 and $\frac{-10}{7}$

37. $\frac{9}{2}$ and $\frac{7}{6}$

39. $\frac{5}{11}$ and $\frac{1}{3}$

41. $\frac{34}{9}$ and $\frac{22}{15}$

43. $\frac{3}{4}$

REVIEW EXERCISES, p. 499

1. (b)

3. (a)

5. (a) false (b) true (c) false (d) true

7. >
9. >
11. $(-\infty, 3]$
13. $[1, \infty)$
15. $[-1, 3]$
17. (c)
19. See Figure 14R.
21. $(-\infty, -3) \cup (-2, \infty)$
23. $(-\infty, 1) \cup (2, \infty)$
25. 10, −2
27. 0, 14
29. 1

CHECK: $|1 + 1| \stackrel{?}{=} |1 - 3|$

$$|2| \stackrel{?}{=} |-2|$$

$$2 \stackrel{\checkmark}{=} 2$$

31. $(-\infty, -6] \cup [0, \infty)$
33. 3
35. (a) 0 (b) 1 (c) 3 (d) 4
37. See Figure 14S.
39. $\frac{-1}{2}$

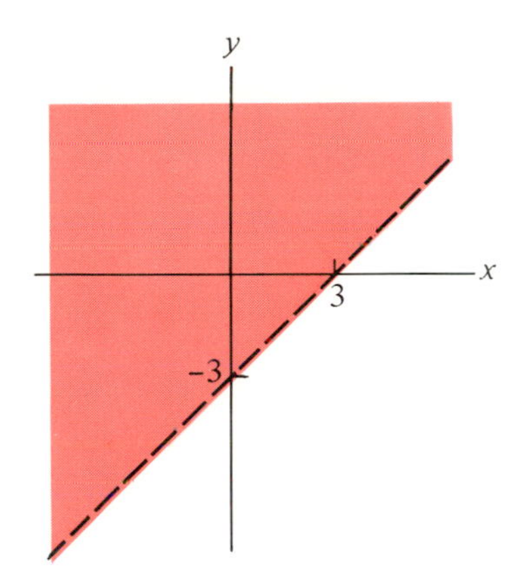

FIGURE 14R

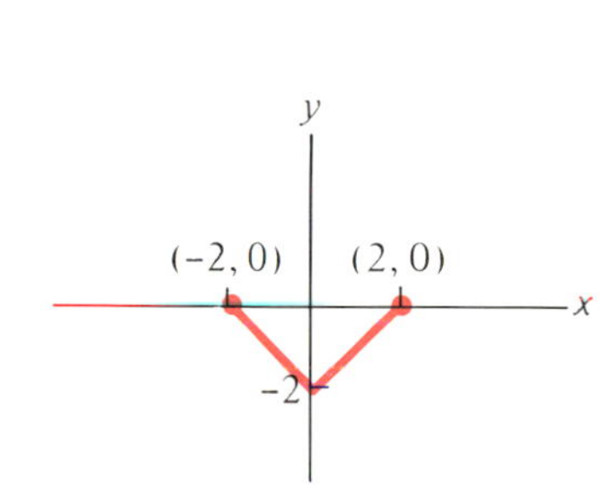

FIGURE 14S

TEST YOURSELF, p. 501

1. (a) true (b) false (c) true
2. (a) < (b) <
3. $[4, \infty)$
4. $[4, 10]$
5. $(-2, \infty)$
6. 0, 4
7. $(-\infty, -6) \cup (4, \infty)$
8. 1
9. (a) 0 (b) 6
10. $\frac{\pm 3}{2}$

CHAPTER 15

SECTION 15.1, p. 509

1. 2
3. 5
5. 3
7. 12
9. $\frac{1}{4}$
11. 5
13. $3\sqrt{2}$
15. $2\sqrt{2}$
17. $2\sqrt{5}$
19. 5
21. $2\sqrt{13}$
23. $(.2)\sqrt{2}$
25. 3
27. 8
29. 2
31. 3

33. 2

35. $\frac{3}{2}$

37. These are the points on the y-axis.

SECTION 15.2, p. 516

1. (a) (0, 0) (b) 3

3. (a) (0, 0) (b) $2\sqrt{2}$

5. (a) (1, 0) (b) 4

7. (a) (2, 4) (b) 7

9. (a) $(-8, 7)$ (b) 12

11. (a) $\left(\frac{-5}{3}, \frac{1}{3}\right)$ (b) $\sqrt{2}$

13. (a) $\left(\frac{1}{4}, \frac{1}{2}\right)$ (b) $\frac{1}{2}$

15. (a) $(-2.7, 1.7)$ (b) .1

17. $x^2 + y^2 = 16$

19. $(x-2)^2 + (y-3)^2 = 25$

21. $\left(x - \frac{1}{4}\right)^2 + \left(y - \frac{1}{2}\right)^2 = \frac{1}{4}$

23. See Figure 15A.

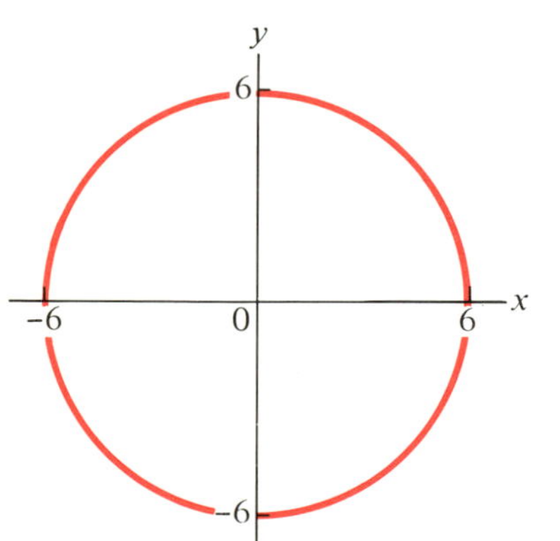

FIGURE 15A

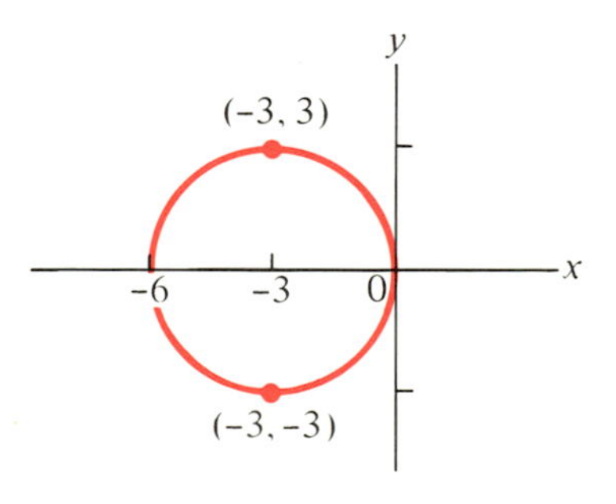

FIGURE 15B

25. See Figure 15B.

27. See Figure 15C.

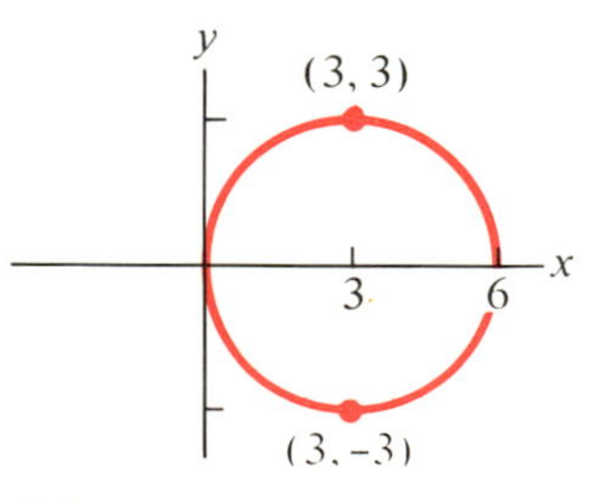

FIGURE 15C

29. $(x+5)^2 + y^2 = 25$; center $(-5, 0)$, radius 5

31. $(x+1)^2 + (y+3)^2 = 10$; center $(-1, -3)$, radius $\sqrt{10}$

33. $(x+3)^2 + (y+9)^2 = 90$; center $(-3, -9)$, radius $3\sqrt{10}$

35. $(x+3)^2 + \left(y + \frac{5}{2}\right)^2 = \frac{61}{4}$; center $\left(-3, \frac{-5}{2}\right)$, radius $\frac{\sqrt{61}}{2}$

37. $(x+2)^2 + \left(y - \frac{3}{2}\right)^2 = \frac{29}{4}$; center $\left(-2, \frac{3}{2}\right)$, radius $\frac{\sqrt{29}}{2}$

39. $(x+2)^2 + (y+10)^2 = 104$; center $(-2, -10)$, radius $2\sqrt{26}$

SECTION 15.3, p. 524

1. vertex (0, 0); axis: y-axis; focus (0, 2); directrix: $y = -2$ [See Figure 15D.]

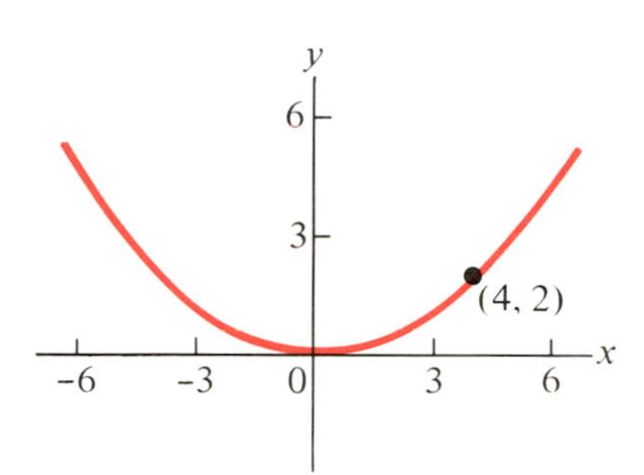

FIGURE 15D

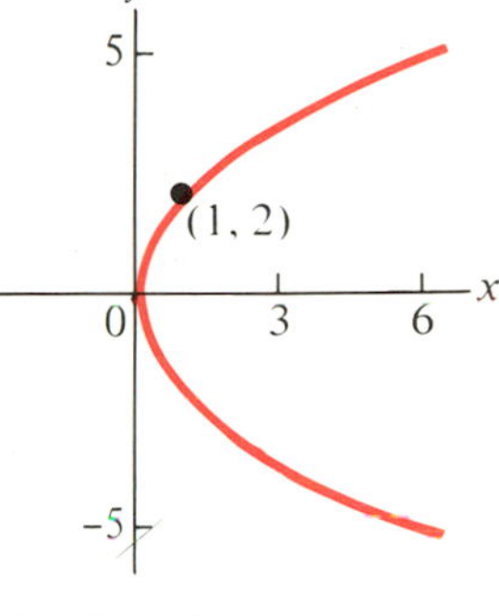

FIGURE 15E

3. vertex (0, 0); axis: y-axis; focus (0, 3); directrix: $y = -3$
5. vertex (0, 0); axis: x-axis: focus (1, 0); directrix: $x = -1$ [See Figure 15E.]
7. vertex (0, 0); axis: x-axis; focus $\left(\frac{1}{4}, 0\right)$; directrix: $x = \frac{-1}{4}$
9. vertex (0, 0); axis: y-axis; focus $\left(0, \frac{3}{4}\right)$; directrix: $y = \frac{-3}{4}$
11. vertex (0, 0); axis: x-axis; focus $\left(\frac{1}{2}, 0\right)$; directrix: $x = \frac{-1}{2}$

13. $y^2 = 12x$
15. $x^2 = 12y$
17. $y^2 = 20x$
19. $x^2 = 40y$
21. $y^2 = 32x$
23. $y^2 = -16x$
25. $x^2 = 20y$
27. $x^2 = 4y$
29. $y^2 = 24x$

SECTION 15.4, p. 534

1. (5, 0), (−5, 0), (0, 4), (0, −4). [See Figure 15F on page A54.]
3. (3, 0), (−3, 0), (0, 2), (0, −2)
5. (13, 0), (−13, 0), (0, 12), (0, −12). [See Figure 15G on page A54.]
7. (10, 0), (−10, 0), (0, 6), (0, −6). [See Figure 15H on page A54.]
9. (1, 0), (−1, 0), (0, 2), (0, −2)
11. $(\sqrt{13}, 0)$, $(-\sqrt{13}, 0)$, (0, 2), (0, −2)
13. (2, 0), (−2, 0), (0, 3), (0, −3)
15. (2, 0), (−2, 0), (0, 1), (0, −1)
17. (a) (3, 0), (−3, 0) (b) $y = \frac{4}{3}x$, $y = \frac{-4}{3}x$ (c) [See Figure 15I on page A54.]
19. (a) (1, 0), (−1, 0) (b) $y = 2x$, $y = -2x$
21. (0, 4), (0, −4) (b) $y = \frac{4}{3}x$, $y = \frac{-4}{3}x$, (c) [See Figure 15J on page A55.]
23. (a) (0, 3), (0, −3) (b) $y = 3x$, $y = -3x$

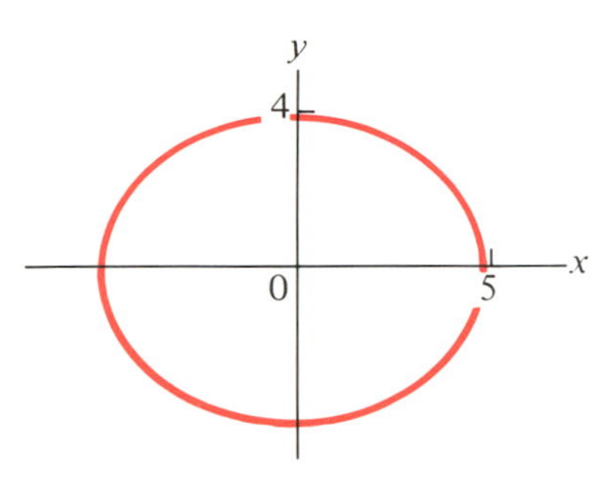

FIGURE 15F

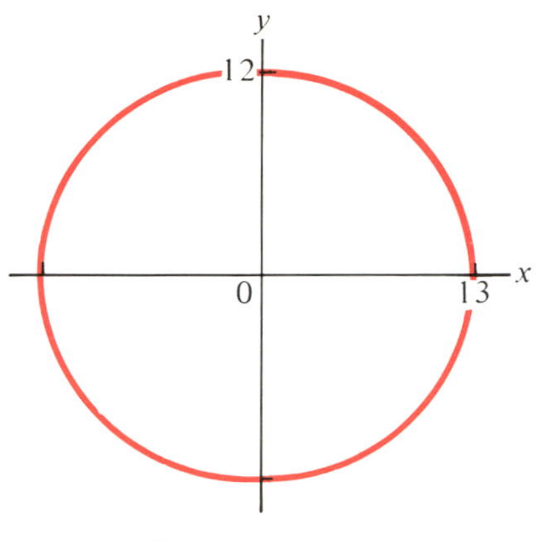

FIGURE 15G

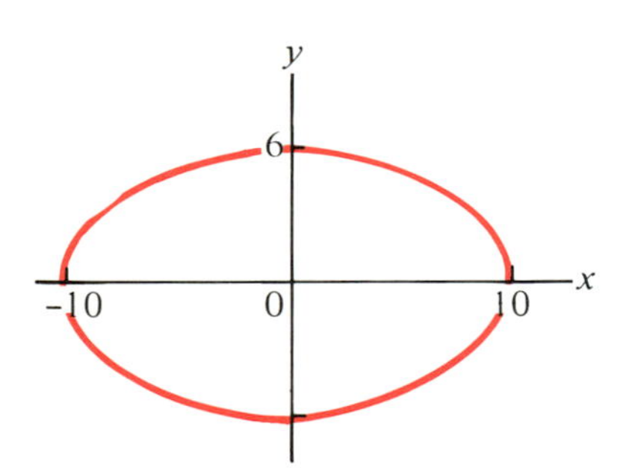

FIGURE 15H

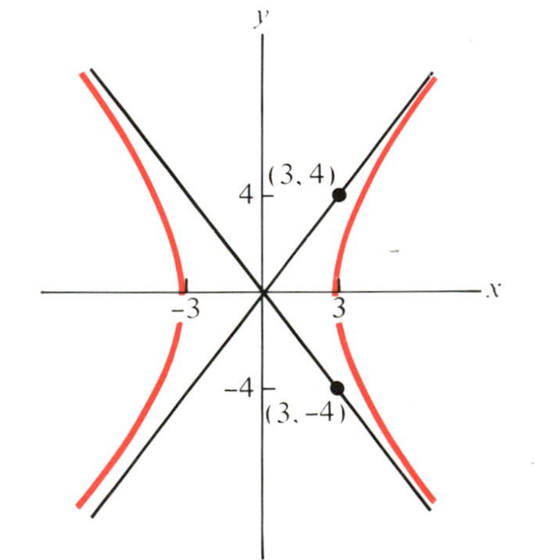

FIGURE 15I

25. (a) (3, 0), (−3, 0) (b) $y = \frac{2}{3}x$, $y = \frac{-2}{3}x$ (c) [See Figure 15K on page A55.]

27. (a) (2, 0), (−2, 0) (b) $y = \frac{1}{2}x$, $y = \frac{-1}{2}x$

29. (a) (5, 0), (−5, 0) (b) $y = \frac{2}{5}x$, $y = \frac{-2}{5}x$

31. (a) (5, 0), (−5, 0) (b) $y = \frac{2\sqrt{6}}{5}x$, $y = \frac{-2\sqrt{6}}{5}x$

SECTION 15.5, p. 538

1. (a) (3, 4), (3, −4) (b) $x^2 + y^2 = 25$ (circle), $x = 3$ (line) (c) See Figure 15L.
3. (a) (0, 0), $\left(\frac{-1}{8}, \frac{-1}{8}\right)$ (b) $y = -8x^2$ (parabola), $y = x$ (line)

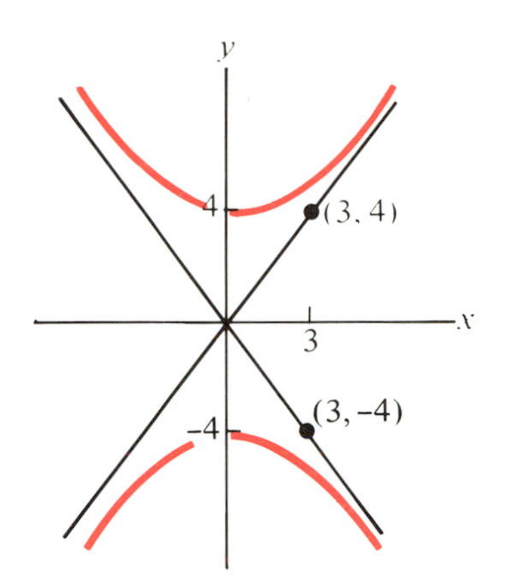

FIGURE 15J

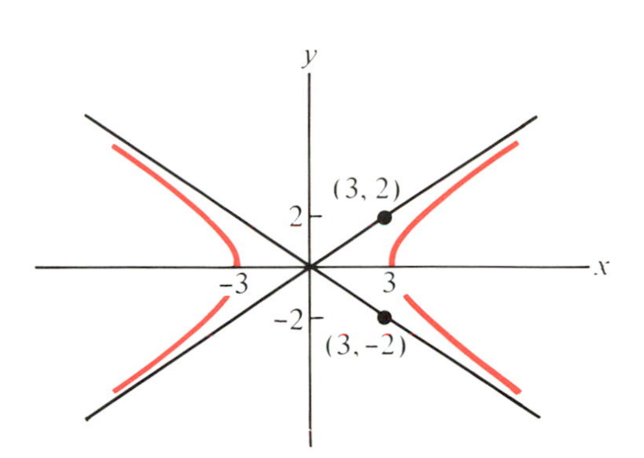

FIGURE 15K

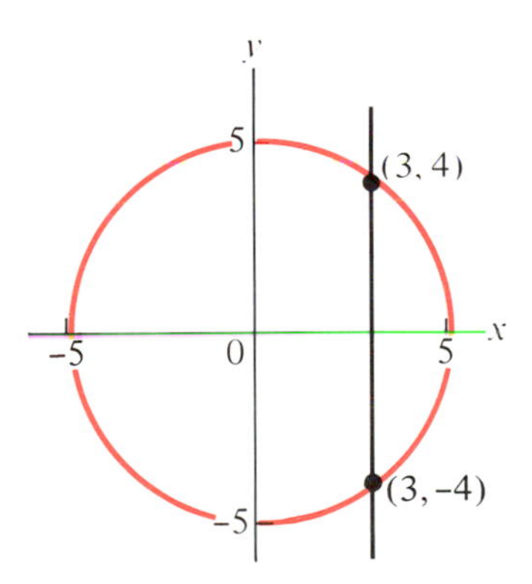

FIGURE 15L

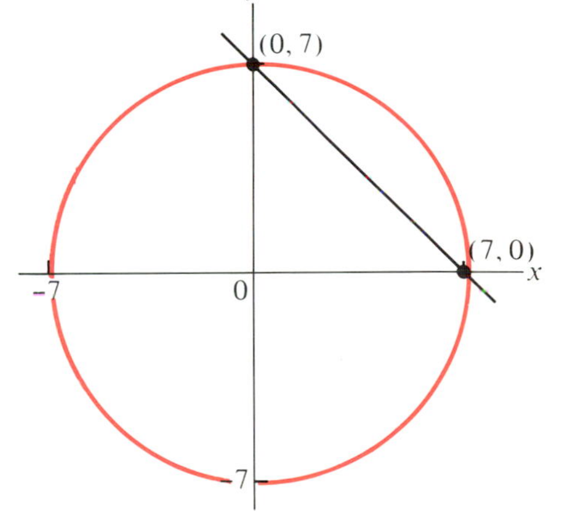

FIGURE 15M

5. (a) (0, 7), (7, 0) (b) $x^2 + y^2 = 49$ (circle), $y = 7 - x$ (line) (c) See Figure 15M.

7. (a) (0, 0), $\left(\frac{-1}{2}, \frac{1}{2}\right)$ (b) $x = -2y^2$ (parabola), $y = -x$ (line)

9. (a) (0, 10), (6, −8) (b) $x^2 + y^2 = 100$ (circle), $y = 10 - 3x$ (line)

11. (a) (0, 0), (1, −1) (b) $y = -x^2$ (parabola), $x + y = 0$ (line) (c) See Figure 15N on page A56.

13. (a) (−1, 1), (−1, −1) (b) $x^2 + y^2 = 2$ (circle), $x = -y^2$ (parabola)

15. (a) No intersection. Apply the quadratic formula to the equation $5x^2 - 8x + 12 = 0$, obtained by substituting $2x$ for y in the equation $(x + 2)^2 + (y - 3)^2 = 1$. The roots are not real.
(b) $(x + 2)^2 + (y - 3)^2 = 1$ (circle), $y = 2x$ (line)

17. (a) (2, 0) (b) $x^2 + y^2 = 4$ (circle), $(x - 4)^2 + y^2 = 4$ (circle)
(c) See Figure 15O on page A56.

19. (a) $\left(1, \frac{\sqrt{3}}{2}\right)$, $\left(1, \frac{-\sqrt{3}}{2}\right)$ (b) $\frac{x^2}{4} + y^2 = 1$ (ellipse), $x = 1$ (line)

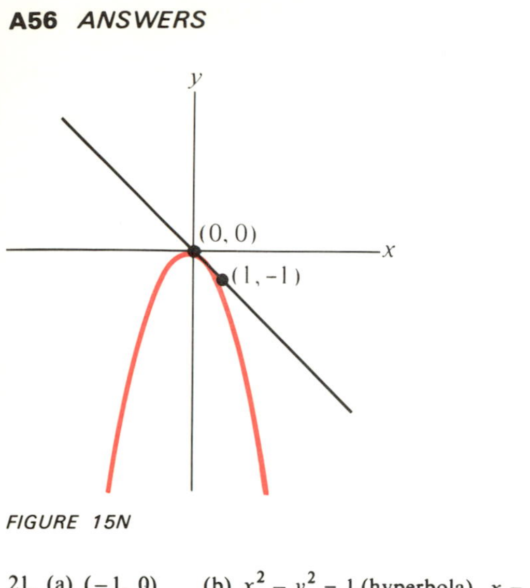

FIGURE 15N

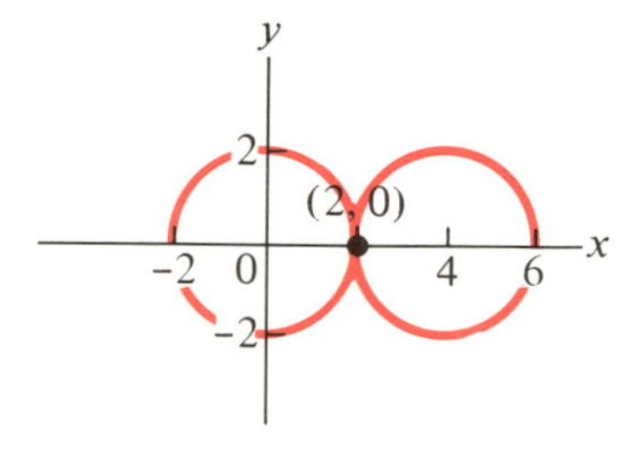

FIGURE 15O

21. (a) $(-1, 0)$ (b) $x^2 - y^2 = 1$ (hyperbola), $x = -1$ (line) (c) See Figure 15P.
23. (a) $\left(\frac{12}{5}, \frac{-12}{5}\right)$, $\left(\frac{-12}{5}, \frac{12}{5}\right)$ (b) $\frac{x^2}{16} + \frac{y^2}{9} = 1$ (ellipse), $y = -x$ (line)
25. (a) no intersections (b) $x^2 + y^2 = 4$ (circle), $\frac{x^2}{9} + \frac{y^2}{16} = 1$ (ellipse) (c) See Figure 15Q.
27. (a) $(8, 0)$, $(-8, 0)$ (b) $\frac{x^2}{64} - \frac{y^2}{36} = 1$ (hyperbola), $\frac{x^2}{64} + \frac{y^2}{36} = 1$ (ellipse)

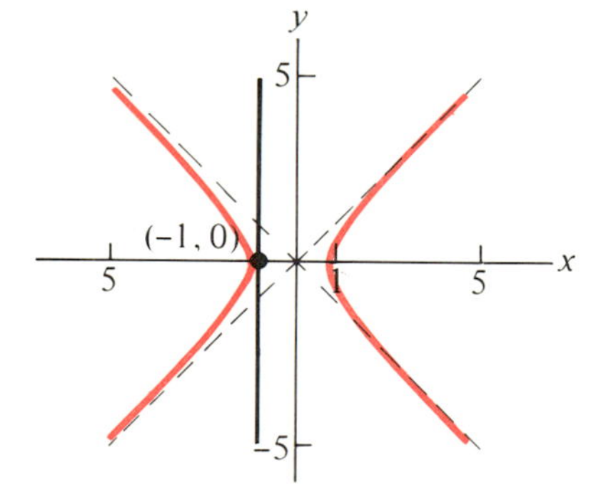

FIGURE 15P

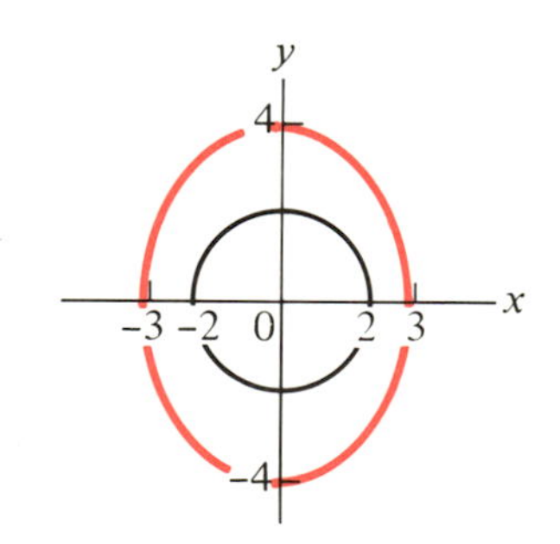

FIGURE 15Q

REVIEW EXERCISES, **p. 539**

1. 3
3. $\sqrt{10}$
5. (a) $(0, 2)$ (b) 6
7. $(x - 2)^2 + (y + 2)^2 = 9$
9. (a) vertex: origin; axis: y-axis; focus: $(0, 1)$; directrix: $y = -1$
   (b) [See Figure 15R on page A57.]
11. $y^2 = 32x$
13. (a) $(10, 0)$, $(-10, 0)$, $(0, 6)$, $(0, -6)$
    (b) See Figure 15S on page A57.

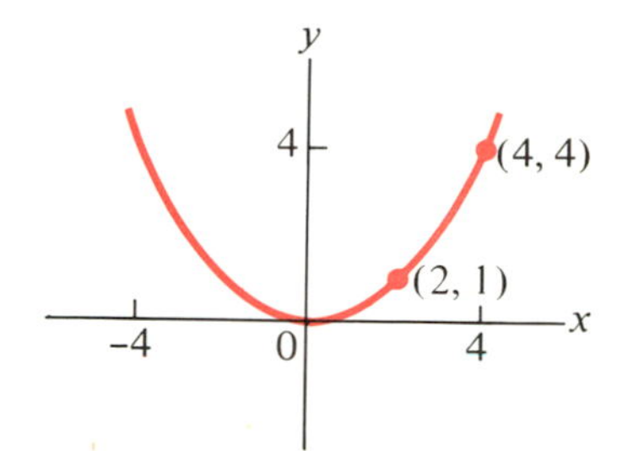

FIGURE 15R

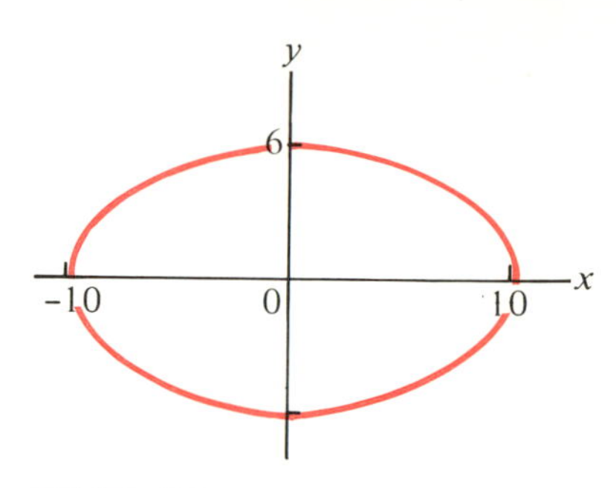

FIGURE 15S

15. (a) $(-5, 0)$, $(5, 0)$ (b) $y = \frac{3}{5}x$, $y = \frac{-3}{5}x$
 (c) See Figure 15T.

17. (a) $(3\sqrt{2}, 3\sqrt{2})$, $(-3\sqrt{2}, -3\sqrt{2})$ (b) $x^2 + y^2 = 36$ (circle), $y = x$ (line)
 (c) See Figure 15U.

19. (a) $(0, 2)$, $(0, -2)$ (b) $\frac{x^2}{9} + \frac{y^2}{4} = 1$ (ellipse), $x^2 + y^2 = 4$ (circle) (c) See Figure 15V on page A58.

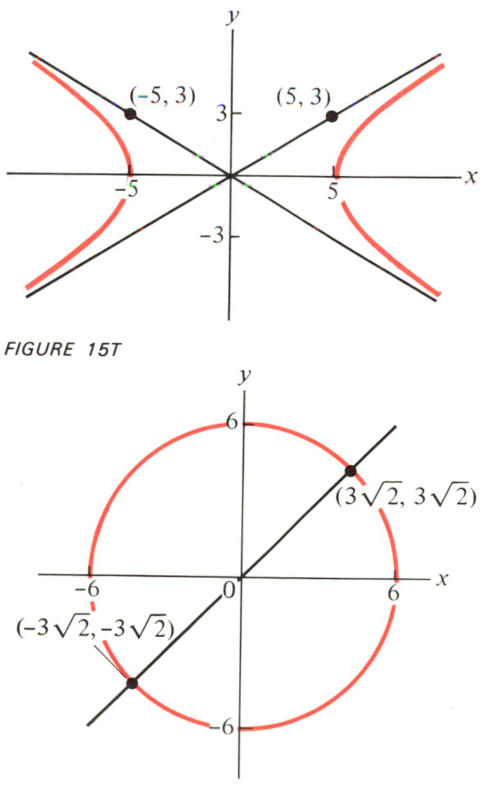

FIGURE 15T

FIGURE 15U

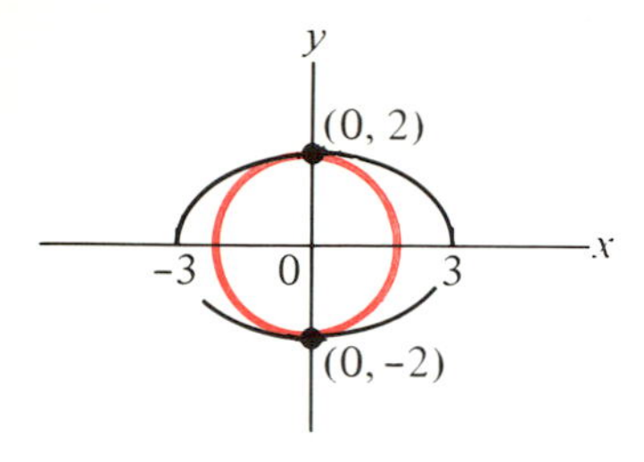

FIGURE 15V

TEST YOURSELF, p. 541

1. $\sqrt{13}$
2. (a) $(3, -3)$ (b) 3
3. (a) vertex: origin; axis: x-axis; focus $(2, 0)$; directrix: $x = -2$ (b) See Figure 15W.
4. (a) $(3, 0)$, $(-3, 0)$ (b) $y = \frac{5}{3}x$, $y = \frac{-5}{3}x$
5. (a) $(0, 3)$, $(-3, 0)$ (b) $x^2 + y^2 = 9$ (circle), $y = x + 3$ (line) (c) See Figure 15X.
6. (a) $(x - 4)^2 + (y + 1)^2 = 25$ (b) $y^2 = 24x$

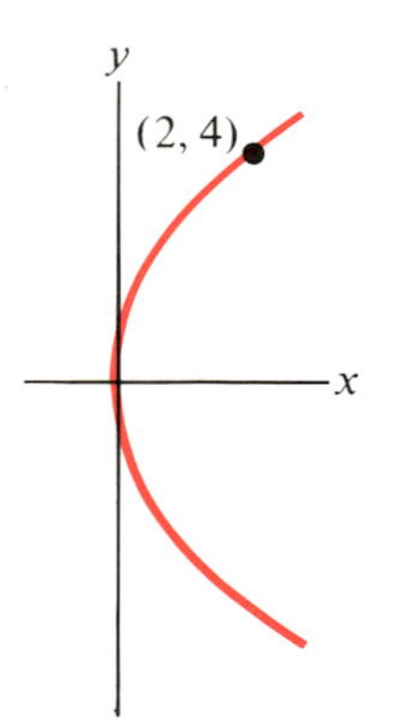

FIGURE 15W

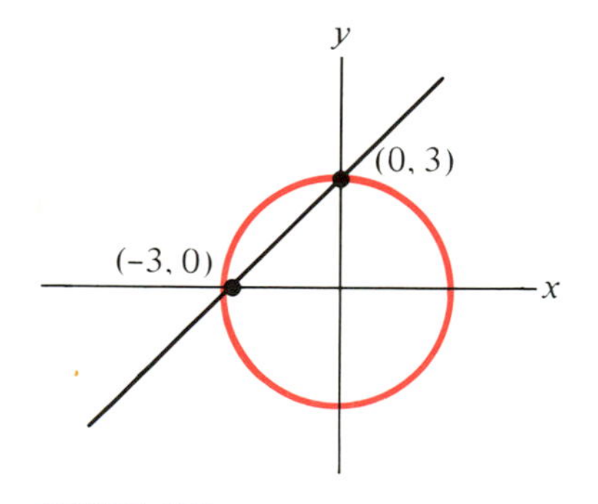

FIGURE 15X

CHAPTER 16

SECTION 16.1, p. 547

1. 3
3. $\frac{1}{2}$
5. 25
7. $\frac{9}{4}$
9. 12
11. 25
13. 6
15. 2
17. 1
19. 4
21. 6
23. 48
25. $\frac{1}{8}$
27. 2
29. (a) $y = 3x$ (b) See Figure 16A on page A59.
31. 0
33. (a) circumference (or diameter) (b) area
35. $192\frac{1}{2}$ miles

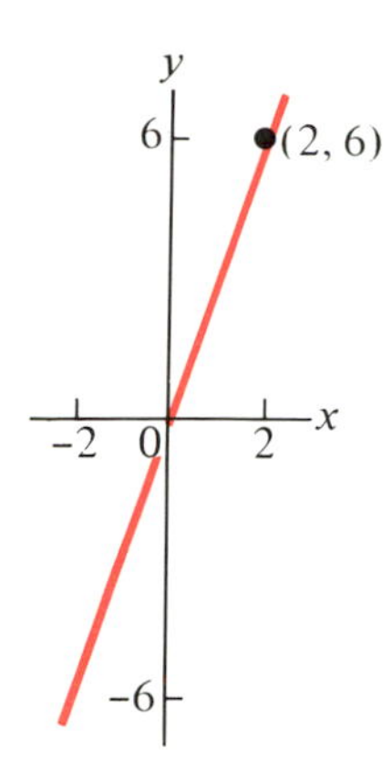

FIGURE 16A

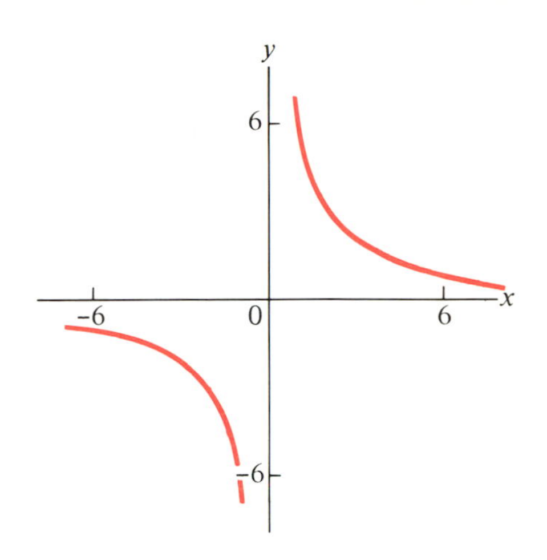

FIGURE 16B

SECTION 16.2, p. 551

1. 8 3. 6 5. $\frac{1}{6}$ 7. 2
9. 10 11. 3 13. 20 15. -50
17. 1 19. $\frac{1}{4}$

21. (a) $y = \frac{6}{x}$ (b) See Figure 16B. (c) hyperbola

23. no 25. directly 27. directly
29. 120 pounds

SECTION 16.3, p. 556

1. 4 3. 3 5. $\frac{1}{2}$ 7. 36
9. 500 11. 4 13. 144 15. 90
17. 250 19. 20 21. -6 23. .8
25. 3 27. 10 29. $\frac{1}{2}$ 31. 6
33. 90 35. 32 37. 8 39. 1080 cubic inches
41. 125 joules 43. 10000 pounds

REVIEW EXERCISES, p. 558

1. 4 3. 2 5. $y = \frac{-x}{2}$ 7. -1
9. $2\sqrt{2}$ 11. 4 13. 240 15. 1
17. 600 cubic inches

TEST YOURSELF, p. 559

1. 10 2. 20 3. 108 4. $y = 4x$

5. 60 6. $\frac{3}{2}$ 7. $\frac{1}{2}$ 8. 160

9. $\frac{3}{4}$

CHAPTER 17

SECTION 17.1, p. 564

1. (a) 1, 2, 3, 4, 5, 6, 7, 8 (b) i (c) See Figure 17A.
3. (a) 6, 7, 8, 9, 10, 11, 12, 13 (b) i
5. (a) $-1, -2, -3, -4, -5, -6, -7, -8$ (b) ii (c) See Figure 17B.

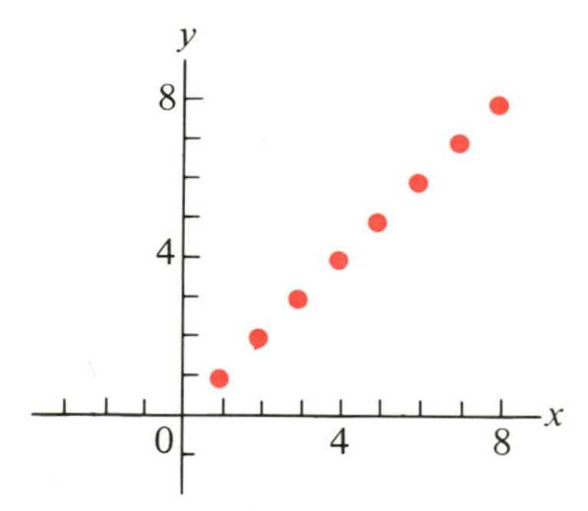

FIGURE 17A

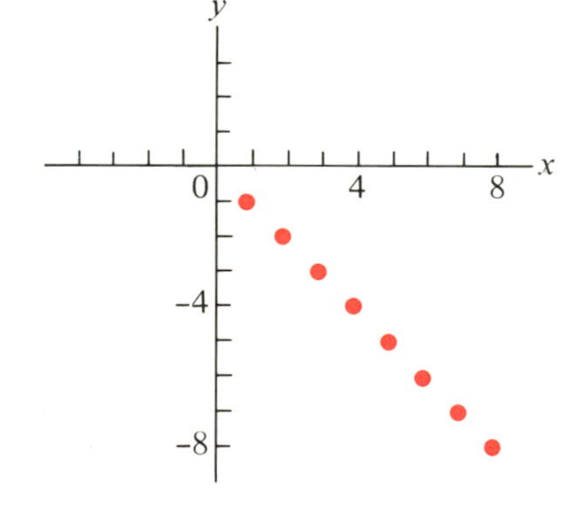

FIGURE 17B

7. (a) 2, 4, 6, 8, 10, 12, 14, 16 (b) i (c) See Figure 17C on page A61.
9. (a) 3, 6, 9, 12, 15, 18, 21, 24 (b) i
11. (a) 5, 5, 5, 5, 5, 5, 5, 5 (b) iii
13. (a) 5, 3, 5, 3, 5, 3, 5, 3 (b) iii
15. (a) $1, \frac{1}{2}, \frac{1}{3}, \frac{1}{4}, \frac{1}{5}, \frac{1}{6}, \frac{1}{7}, \frac{1}{8}$ (b) ii
17. (a) 2, 4, 8, 16, 32, 64, 128, 256 (b) i
19. (a) $\frac{1}{2}, \frac{1}{4}, \frac{1}{8}, \frac{1}{16}, \frac{1}{32}, \frac{1}{64}, \frac{1}{128}, \frac{1}{256}$ (b) ii
21. (a) 9, 12, 15, 18, 21, 24, 27, 30 (b) i
23. (a) $\frac{3}{2}, 2, \frac{5}{2}, 3, \frac{7}{2}, 4, \frac{9}{2}, 5$ (b) i
25. (a) 2.1, 2.2, 2.3, 2.4, 2.5, 2.6, 2.7, 2.8 (b) i
27. (a) 20 (b) 50 (c) 100 (d) 500

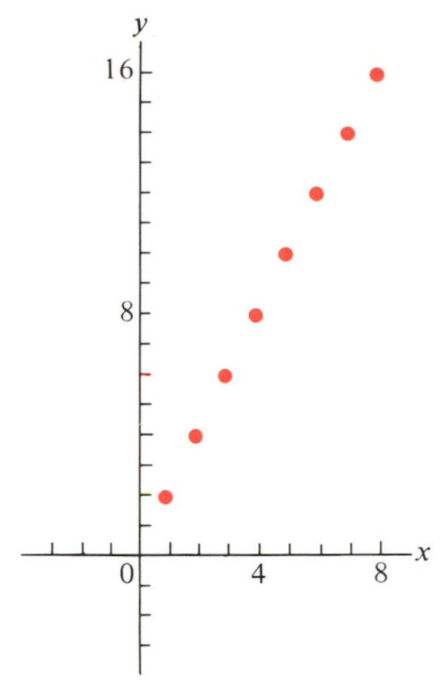

FIGURE 17C

29. (a) 7 (b) 23 (c) 43 (d) 83

31. (a) $\frac{9}{2}$ (b) 12 (c) 22 (d) $\frac{49}{2}$

33. (a) 0 (b) 2 (c) 0 (d) 2

35. (a) 6 (b) 2 (c) 1 (d) $\frac{1}{4}$

37. (a) $\frac{1}{2}$ (b) $\frac{2}{3}$ (c) $\frac{9}{10}$ (d) $\frac{10}{11}$

39. $a_i = i + 1$

41. $a_i = i + 4$

43. $a_i = 3i$

45. $a_i = (-1)^i \cdot 4i$

47. $a_i = -1 + 3i$

49. $a_i = \frac{i+1}{i}$

SECTION 17.2, p. 571

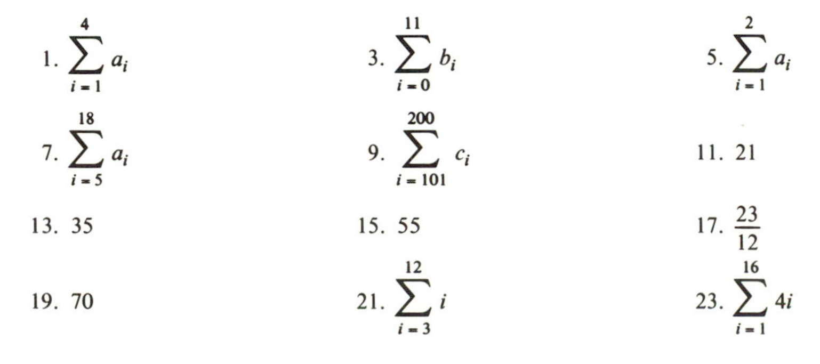

1. $\sum_{i=1}^{4} a_i$

3. $\sum_{i=0}^{11} b_i$

5. $\sum_{i=1}^{2} a_i$

7. $\sum_{i=5}^{18} a_i$

9. $\sum_{i=101}^{200} c_i$

11. 21

13. 35

15. 55

17. $\frac{23}{12}$

19. 70

21. $\sum_{i=3}^{12} i$

23. $\sum_{i=1}^{16} 4i$

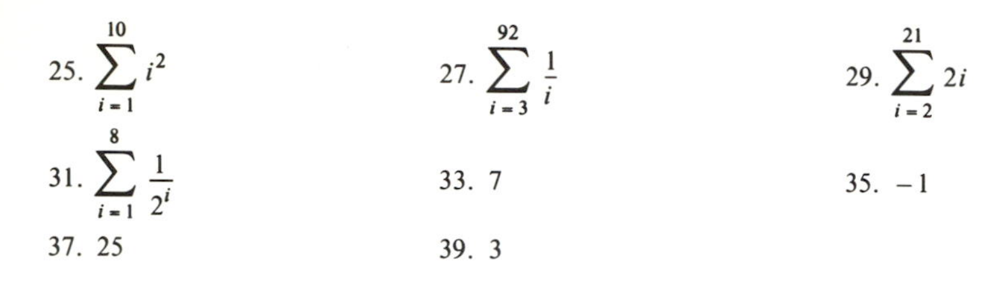

25. $\sum_{i=1}^{10} i^2$ 27. $\sum_{i=3}^{92} \frac{1}{i}$ 29. $\sum_{i=2}^{21} 2i$

31. $\sum_{i=1}^{8} \frac{1}{2^i}$ 33. 7 35. -1

37. 25 39. 3

SECTION 17.3, p. 577

1. arithmetic progression with common difference 4
3. arithmetic progression with common difference 2
5. arithmetic progression with common difference -3
7. not an arithmetic progression
9. not an arithmetic progression
11. arithmetic progression with common difference -5
13. arithmetic progression with common difference $\frac{-1}{3}$

15. 49 17. -235 19. 89 21. 195
23. 71 25. 520 27. 255 29. 7220
31. 1520 33. 15 250 35. -4900 37. (a) 324 (b) 627
39. (a) 450 (b) 935 41. 2550 43. $21 000 per year

SECTION 17.4, p. 584

1. geometric progression with common ratio 2
3. not a geometric progression; arithmetic progression
5. geometric progression with common ratio $\frac{1}{2}$
7. geometric progression with common ratio -1
9. geometric progression with common ratio -2
11. geometric progression with common ratio $\frac{1}{5}$
13. not a geometric progression

15. 125 17. $\frac{3}{16}$

19. (a) 405 (b) -405 (c) $\frac{5}{9}$ (d) $\frac{-5}{9}$

21. 960 096 23. 4 25. 254 27. 1530
29. 1365 31. 693 33. $\frac{511}{4}$ 35. -9
37. -183 39. .111 111 41. 255 43. $10 485.75
45. 250 000

SECTION 17.5, p. 592

1. 24 3. 40 320 5. 3024 7. 210
9. $\frac{5!}{3!}$ 11. $\frac{12!}{8!}$ 13. $\frac{99!}{89!}$ 15. 3

17. 6
19. 1
21. 1
23. 28
25. 792
27. 1
29. 396
31. $n(n-1)$
33. $(n+1)!$
35. 1 7 21 35 35 21 7 1
37. 1 9 36 84 126 126 84 36 9 1
39. $x^4 - 4x^3a + 6x^2a^2 - 4xa^3 + a^4$
41. $x^4 + 4x^3 + 6x^2 + 4x + 1$
43. $x^5 - 5x^4 + 10x^3 - 10x^2 + 5x - 1$
45. $x^4 + 12x^3 + 54x^2 + 108x + 81$
47. $x^4 + 8x^3y + 24x^2y^2 + 32xy^3 + 16y^4$
49. $a^8 - 4a^6b^2 + 6a^4b^4 - 4a^2b^6 + b^8$
51. (a) $\sum_{i=0}^{8} \binom{8}{i} x^{8-i}y^i$ (b) $x^8 + 8x^7y + 28x^6y^2 + 56x^5y^3$
53. (a) $\sum_{i=0}^{12} \binom{12}{i} x^{12-i}$ (b) $x^{12} + 12x^{11} + 66x^{10} + 220x^9$
55. (a) $\sum_{i=0}^{9} (-1)^i \binom{9}{i} x^{9-i}2^i$ (b) $x^9 - 18x^8 + 144x^7 - 672x^6$
57. (a) $\sum_{i=0}^{8} \binom{8}{i} x^{16-2i}y^{8-i}$ (b) $x^{16}y^8 + 8x^{14}y^7 + 28x^{12}y^6 + 56x^{10}y^5$
59. (a) $\sum_{i=0}^{8} \binom{8}{i} \frac{x^{8-i}}{2^i}$ (b) $x^8 + 4x^7 + 7x^6 + 7x^5$
61. 1.0406
63. 1.7716
65. 1.0721
67. 85.7661

REVIEW EXERCISES, p. 593

1. (a) 4, 5, 6, 7, 8, 9, 10, 11 (b) increasing (c) See Figure 17D.
3. (a) 2, 1, $\frac{2}{3}$, $\frac{1}{2}$, $\frac{2}{5}$, $\frac{1}{3}$, $\frac{2}{7}$, $\frac{1}{4}$ (b) decreasing (c) See Figure 17E.

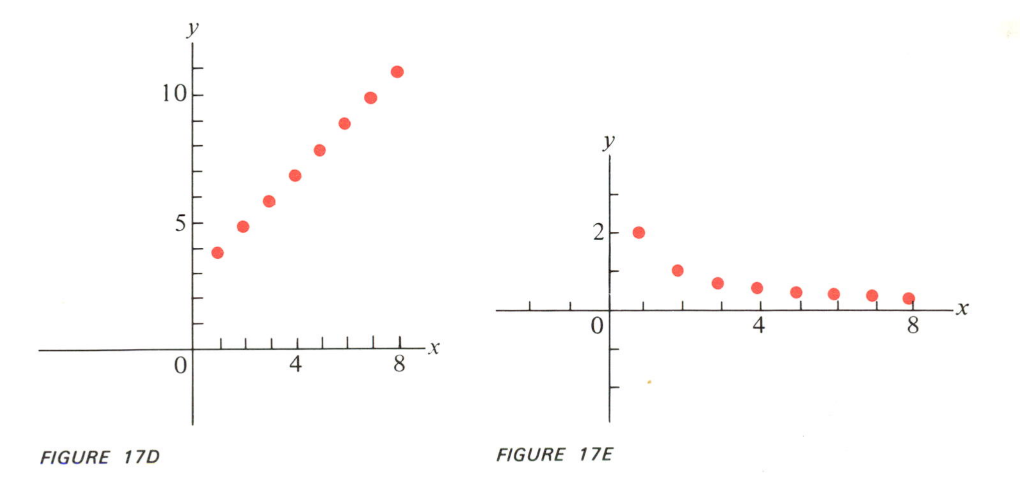

FIGURE 17D

FIGURE 17E

5. $a_i = 3 + 2i$

7. $\sum_{i=7}^{10} b_i$ or $\sum_{i=1}^{4} b_{i+6}$

9. $\sum_{i=1}^{6} 3i$

11. 3

13. 7

15. 900

17. $\frac{1}{27}$

19. 1092

21. (a) 30 (b) 5040 (c) $\frac{20\,995}{4}$

23. (a) 10 (b) 1 (c) 4845

25. $x^4 - 4x^3 + 6x^2 - 4x + 1$

27. (a) $\sum_{i=0}^{10} (-1)^i \binom{10}{i} x^{10-i} 2^i$ (b) $x^{10} - 20x^9 + 180x^8 - 960x^7$

TEST YOURSELF, p. 595

1. (a) 9 (b) 23 (c) 203
2. (a) $\sum_{i=1}^{8} a_i$ (b) $\sum_{i=20}^{24} b_i$ or $\sum_{i=1}^{5} b_{i+19}$
3. 30
4. (a) ii (b) iii (c) iii (d) ii
5. 25
6. 1000
7. 64
8. $\frac{255}{256}$
9. $a^5 - 5a^4 + 10a^3 - 10a^2 + 5a - 1$

index

Common Logarithms

n	0	1	2	3	4	5	6	7	8	9
1.0	.0000	.0043	.0086	.0128	.0170	.0212	.0253	.0294	.0334	.0374
1.1	.0414	.0453	.0492	.0531	.0569	.0607	.0645	.0682	.0719	.0755
1.2	.0792	.0828	.0864	.0899	.0934	.0969	.1004	.1038	.1072	.1106
1.3	.1139	.1173	.1206	.1239	.1271	.1303	.1335	.1367	.1399	.1430
1.4	.1461	.1492	.1523	.1553	.1584	.1614	.1644	.1673	.1703	.1732
1.5	.1761	.1790	.1818	.1847	.1875	.1903	.1931	.1959	.1987	.2014
1.6	.2041	.2068	.2095	.2122	.2148	.2175	.2201	.2227	.2253	.2279
1.7	.2304	.2330	.2355	.2380	.2405	.2430	.2455	.2480	.2504	.2529
1.8	.2553	.2577	.2601	.2625	.2648	.2672	.2695	.2718	.2742	.2765
1.9	.2788	.2810	.2833	.2856	.2878	.2900	.2923	.2945	.2967	.2989
2.0	.3010	.3032	.3054	.3075	.3096	.3118	.3139	.3160	.3181	.3201
2.1	.3222	.3243	.3263	.3284	.3304	.3324	.3345	.3365	.3385	.3404
2.2	.3424	.3444	.3464	.3483	.3502	.3522	.3541	.3560	.3579	.3598
2.3	.3617	.3636	.3655	.3674	.3692	.3711	.3729	.3747	.3766	.3784
2.4	.3802	.3820	.3838	.3856	.3874	.3892	.3909	.3927	.3945	.3962
2.5	.3979	.3997	.4014	.4031	.4048	.4065	.4082	.4099	.4116	.4133
2.6	.4150	.4166	.4183	.4200	.4216	.4232	.4249	.4265	.4281	.4298
2.7	.4314	.4330	.4346	.4362	.4378	.4393	.4409	.4425	.4440	.4456
2.8	.4472	.4487	.4502	.4518	.4533	.4548	.4564	.4579	.4594	.4609
2.9	.4624	.4639	.4654	.4669	.4683	.4698	.4713	.4728	.4742	.4757
3.0	.4771	.4786	.4800	.4814	.4829	.4843	.4857	.4871	.4886	.4900
3.1	.4914	.4928	.4942	.4955	.4969	.4983	.4997	.5011	.5024	.5038
3.2	.5051	.5065	.5079	.5092	.5105	.5119	.5132	.5145	.5159	.5172
3.3	.5185	.5198	.5211	.5224	.5237	.5250	.5263	.5276	.5289	.5302
3.4	.5315	.5328	.5340	.5353	.5366	.5378	.5391	.5403	.5416	.5428
3.5	.5441	.5453	.5465	.5478	.5490	.5502	.5514	.5527	.5539	.5551
3.6	.5563	.5575	.5587	.5599	.5611	.5623	.5635	.5647	.5658	.5670
3.7	.5682	.5694	.5705	.5717	.5729	.5740	.5752	.5763	.5775	.5786
3.8	.5798	.5809	.5821	.5832	.5843	.5855	.5866	.5877	.5888	.5899
3.9	.5911	.5922	.5933	.5944	.5955	.5966	.5977	.5988	.5999	.6010
4.0	.6021	.6031	.6042	.6053	.6064	.6075	.6085	.6096	.6107	.6117
4.1	.6128	.6138	.6149	.6160	.6170	.6180	.6191	.6201	.6212	.6222
4.2	.6232	.6243	.6253	.6263	.6274	.6284	.6294	.6304	.6314	.6325
4.3	.6335	.6345	.6355	.6365	.6375	.6385	.6395	.6405	.6415	.6425
4.4	.6435	.6444	.6454	.6464	.6474	.6484	.6493	.6503	.6513	.6522
4.5	.6532	.6542	.6551	.6561	.6571	.6580	.6590	.6599	.6609	.6618
4.6	.6628	.6637	.6646	.6656	.6665	.6675	.6684	.6693	.6702	.6712
4.7	.6721	.6730	.6739	.6749	.6758	.6767	.6776	.6785	.6794	.6803
4.8	.6812	.6821	.6830	.6839	.6848	.6857	.6866	.6875	.6884	.6893
4.9	.6902	.6911	.6920	.6928	.6937	.6946	.6955	.6964	.6972	.6981
5.0	.6990	.6998	.7007	.7016	.7024	.7033	.7042	.7050	.7059	.7067
5.1	.7076	.7084	.7093	.7101	.7110	.7118	.7126	.7135	.7143	.7152
5.2	.7160	.7168	.7177	.7185	.7193	.7202	.7210	.7218	.7226	.7235
5.3	.7243	.7251	.7259	.7267	.7275	.7284	.7292	.7300	.7308	.7316
5.4	.7324	.7332	.7340	.7348	.7356	.7364	.7372	.7380	.7388	.7396
n	0	1	2	3	4	5	6	7	8	9